Lecture Notes in Physics

Springer
Berlin
Heidelberg
New York
Barcelona
Hong Kong
London
Milan
Paris
Tokyo

Physics and Astronomy ONLINE LIBRARY

http://www.springer.de/phys/

The Editorial Policy for Edited Volumes

The series *Lecture Notes in Physics* (LNP), founded in 1969, reports new developments in physics research and teaching - quickly, informally but with a high degree of quality. Manuscripts to be considered for publication are topical volumes consisting of a limited number of contributions, carefully edited and closely related to each other. Each contribution should contain at least partly original and previously unpublished material, be written in a clear, pedagogical style and aimed at a broader readership, especially graduate students and nonspecialist researchers wishing to familiarize themselves with the topic concerned. For this reason, traditional proceedings cannot be considered for this series though volumes to appear in this series are often based on material presented at conferences, workshops and schools.

Acceptance

A project can only be accepted tentatively for publication, by both the editorial board and the publisher, following thorough examination of the material submitted. The book proposal sent to the publisher should consist at least of a preliminary table of contents outlining the structure of the book together with abstracts of all contributions to be included. Final acceptance is issued by the series editor in charge, in consultation with the publisher, only after receiving the complete manuscript. Final acceptance, possibly requiring minor corrections, usually follows the tentative acceptance unless the final manuscript differs significantly from expectations (project outline). In particular, the series editors are entitled to reject individual contributions if they do not meet the high quality standards of this series. The final manuscript must be ready to print, and should include both an informative introduction and a sufficiently detailed subject index.

Contractual Aspects

Publication in LNP is free of charge. There is no formal contract, no royalties are paid, and no bulk orders are required, although special discounts are offered in this case. The volume editors receive jointly 30 free copies for their personal use and are entitled, as are the contributing authors, to purchase Springer books at a reduced rate. The publisher secures the copyright for each volume. As a rule, no reprints of individual contributions can be supplied.

Manuscript Submission

The manuscript in its final and approved version must be submitted in ready to print form. The corresponding electronic source files are also required for the production process, in particular the online version. Technical assistance in compiling the final manuscript can be provided by the publisher's production editor(s), especially with regard to the publisher's own LaTeX macro package which has been specially designed for this series.

LNP Homepage (http://www.springerlink.com/series/lnp/)

On the LNP homepage you will find:
–The LNP online archive. It contains the full texts (PDF) of all volumes published since 2000. Abstracts, table of contents and prefaces are accessible free of charge to everyone. Information about the availability of printed volumes can be obtained.
–The subscription information. The online archive is free of charge to all subscribers of the printed volumes.
–The editorial contacts, with respect to both scientific and technical matters.
–The author's / editor's instructions.

G. Contopoulos N. Voglis (Eds.)

Galaxies and Chaos

Editors

George Contopoulos
and Nikos Voglis
Academy of Athens
Research Centre for Astronomy
Anagnostopoulou Street 14
106 73 Athens, Greece

A catalog record for this book is available from the Library of Congress.

Bibliographic information published by Die Deutsche Bibliothek
Die Deutsche Bibliothek lists this publication in the Deutsche Nationalbibliografie; detailed bibliographic data is available in the Internet at http://dnb.ddb.de

ISSN 0075-8450
ISBN 3-540-40470-8 Springer-Verlag Berlin Heidelberg New York

Springer-Verlag Berlin Heidelberg New York
a member of BertelsmannSpringer Science+Business Media GmbH

http://www.springer.de

Printed in Germany

Typesetting: Camera-ready by the authors/editor
Camera-data conversion by Steingraeber Satztechnik GmbH Heidelberg
Cover design: *design & production*, Heidelberg

Printed on acid-free paper
54/3141/du - 5 4 3 2 1 0

Preface

During the last two decades the science of nonlinear and chaotic dynamics has had a spectacular development. Many important ideas and tools appeared in the literature, helping to give a deeper understanding of the role of order and chaos in dynamical systems. One of the most fruitful applications of these ideas and tools has been in the field of dynamical astronomy, namely in galactic dynamics and in the dynamics of the solar system. On the other hand recent observational studies of galaxies and of exosolar systems have come to the point of detecting order and chaos in these systems. For this reason the members of the Research Center of Astronomy of the Academy of Athens decided to organize an international workshop on this subject. This workshop "Galaxies and Chaos. Theory and Observations" was held in Athens in September 16-19, 2002, (see http://www.cc.uoa.gr/gc2002/). A total number of 77 participants from 21 countries from all over the World attended the workshop, namely from Europe, U.S.A, Australia, Japan and Chile. There were 45 talks (23 of them invited talks) and 10 posters. The workshop brought together the experience of people working on galactic dynamics and galaxy formation (theory and observations) with the experience of people working on nonlinear dynamical systems. The talks summarized the most recent developments in both theoretical and observational aspects of galactic dynamics with emphasis on the role of chaos in galaxies. Studies of chaos in galaxies use methods similar to those frequently used in celestial mechanics, or other branches of physics and astronomy. For this reason we invited some speakers from related fields of research. A few interesting papers on some of the most up-to-date problems of celestial mechanics are included in this volume.

The Scientific Organizing Committee was composed of: G. Contopoulos (chairman, Academy of Athens), E. Athanassoula (Observatoire de Marseille, France), A. Bosma (Observatoire de Marseille, France), H. Dejonghe (University of Ghent, Belgium), A. Fridman (Russian Academy of Sciences), P. Grosbøl (ESO, Germany), P.O. Lindblad (Stockholm Observatory, Sweden), D. Lynden-Bell (University of Cambridge, UK), D. Merritt (Rutgers University, USA), and N. Voglis (Academy of Athens). The Local Organizing Committee was composed of: N. Voglis (chairman), H. Dara, Ch. Efthymiopoulos, P. Patsis, V. Tritakis and M. Zoulias.

The Academy of Athens covered a considerable part of the expenses of the workshop. But we are grateful also to several other institutions and persons, namely: The University of Athens, in particular the vice-rector Dr. G. Dermitzakis, that provided both financial and substructure support, the Hellenic Ministry of Culture, the A.G. Leventis Foundation, the City of Athens, Siemens S.A. in Athens, the European Physical Society and private donors. With their help we could in particular organize an archeological tour of Athens, a closing dinner at the terrace of a hotel in the center of the city, and provide free hotel rooms and free lunches to many participants. We thank heartily all of them.

The Editors
G. Contopoulos, N. Voglis

Contents

Part I Order and Chaos

Order and Chaos in Astronomy
George Contopoulos 3

Critical Ergos Curves and Chaos at Corotation
Donald Lynden-Bell, Jaideep M. Barot 30

Stellar Dynamics and Molecular Dynamics: Possible Analogies
Andrea Carati, Luigi Galgani 44

Waves Derived from Galactic Orbits. Solitons and Breathers
Nikos Voglis 56

Discrete Breathers and Homoclinic Dynamics
Tassos Bountis, Jeroen M. Bergamin 75

Chaos or Order in Double Barred Galaxies?
Witold Maciejewski 91

Jeans Solutions for Triaxial Galaxies
Glenn van de Ven, Chris Hunter, Ellen Verolme, Tim de Zeeuw 101

Nonlinear Response of the Interstellar Gas Flow to Galactic Spiral Density Waves
Edward Liverts, Michael Mond 109

The Level of Chaos in N-Body Models of Elliptical Galaxies
Nikos Voglis, Costas Kalapotharakos, Iioannis Stavropoulos, Christos Efthymiopoulos 117

Low Frequency Power Spectra and Classification of Hamiltonian Trajectories
George Voyatzis 126

Part II Orbit Theory

Disk-Crossing Orbits
Chris Hunter ... 137

Chaos and Chaotic Phase Mixing in Galaxy Evolution and Charged Particle Beams
Henry E. Kandrup ... 154

Black Hole Motions
Richard H. Miller ... 169

Weak Homology of Bright Elliptical Galaxies
Giuseppe Bertin ... 185

Part III Observations

Observing Chaos in Disk Galaxies
Preben Grosbøl ... 201

Observational Determination of the Gravitational Potential and Pattern Speed in Strongly Barred Galaxies
Per A.B. Lindblad, Per Olof Lindblad ... 213

Observational Manifestation of Chaos in Grand Design Spiral Galaxies
Alexei M. Fridman, Roald Z. Sagdeev, Oleg V. Khoruzii, Evgenii V. Polyachenko ... 223

Quarter-Turn Spirals in Galaxies
Evgenii Polyachenko ... 235

Dynamics of Galaxies: From Observations to Distribution Functions
Herwig Dejonghe, Veronique De Bruyne ... 243

Arp 158: A Study of the HI
Mansie G. Iyer, Caroline E. Simpson, Stephen T. Gottesman, Benjamin K. Malphrus ... 263

Dark Matter in Spiral Galaxies
Albert Bosma ... 271

A SAURON View of Galaxies
Ellen K. Verolme, Michele Cappellari, Glenn van de Ven, P. Tim de Zeeuw, Roland Bacon, Martin Bureau, Yanick Copin, Roger L. Davies, Eric Emsellem, Harald Kuntschner, Richard McDermid, Bryan W. Miller, Reynier F. Peletier 279

Photometric Properties of Karachentsev's Mixed Pairs of Galaxies
Alfredo Franco-Balderas, Deborah Dultzin-Hacyan, Héctor M. Hernández-Toledo 286

Spline Histogram Method for Reconstruction of Probability Density Functions of Clusters of Galaxies
Dmitrijs Docenko, Kārlis Bērziņš 294

Stars Close to the Massive Black Hole at the Center of the Milky Way
Nelly Mouawad, Andreas Eckart, Susanne Pfalzner, Christian Straubmeier, Rainer Spurzem, Reinhard Genzel, Thomas Ott, Rainer Schödel 302

Part IV Formation and Evolution of Galaxies

Angular Momentum Redistribution and the Evolution and Morphology of Bars
Lia Athanassoula 313

Major Mergers and the Origin of Elliptical Galaxies
Andreas Burkert, Thorsten Naab 327

Dynamical Evolution of Galaxies: Supercomputer N-Body Simulations
Edward Liverts, Evgeny Griv, Michael Gedalin, David Eichler 340

Formation of the Halo Stellar Population in Spiral and Elliptical Galaxies
Tetyana Nykytyuk 348

Model of Ejection of Matter from Dense Stellar Cluster and Chaotic Motion of Gravitating Shells
Maxim V. Barkov, Vladimir A. Belinski, Genadii S. Bisnovatyi-Kogan, Anatoly I. Neishtadt 357

Direct vs Merger Mechanism Forming Counterrotating Galaxies
Maria Harsoula, Nikos Voglis 365

Pitch Angle of Spiral Galaxies as Viewed from Global Instabilities of Flat Stellar Disks
Shunsuke Hozumi 380

Collisionless Evaporation from Cluster Elliptical Galaxies
Veruska Muccione, Luca Ciotti 387

Part V Solar System Dynamics

Chaos in Solar System Dynamics
Rudolf Dvorak 395

Dynamics of Extrasolar Planetary Systems: 2/1 Resonant Motion
John D. Hadjidemetriou, Dionyssia Psychoyos 412

The "Third" Integral in the Restricted Three-Body Problem Revisited
Harry Varvoglis, Kleomenis Tsiganis, John D. Hadjidemetriou 433

List of Contributors

Athanassoula Lia
Observatoire de Marseille,
France
lia@obmara.cnrs-mrs.fr

Bacon Roland
Centre de Recherche Astronomique de Lyon,
France
bacon@obs.univ-lyon1.fr

Barkov Maxim
Space Research Institute,
Russia
barmv@sai.msu.ru

Barot J.M.
Westminster School,
U.K.
jaideep.barot@westminster.org.uk

Belinski Vladimir A.
National Institute of Nuclearphysics (INFN) and International Center of Relativistic
Astrophysics (ICRA),
Italy
volodia@vxrmg9.icra.it

Bergamin Jeroen
University of Patras,
Greece

Bērziņš Kārlis
Ventspils International Radio Astronomy Center,
Latvia
kberzins@latnet.lv

Bisnovatyi-Kogan Genadii
Space Research Institute,
Russia
gkogan@mx.iki.rssi.ru

Bertin Giuseppe
Universit'a degli Studi di Milano,
Italy
bertin@sns.it

Bosma Albert
Observatoire de Marseille,
France
bosma@batis.cnrs-mrs.fr

Bountis Tassos
University of Patras,
Greece
bountis@math.upatras.gr

Bureau Martin
University of Columbia,
U.S.A.
bureau@astro.columbia.edu

Burkert Andreas
MPI Heidelberg
Germany,
burkert@mpia-hd.mpg.de

Cappellari Michele
Leiden University,
Netherlands
cappellari@strw.leidenuniv.nl

Carati Andrea
Universita di Milano,
Italy
carati@mat.unimi.it

Ciotti Luca
Universit'a di Bologna,
Italy
ciotti@bo.astro.it

Contopoulos George
Academy of Athens,
Research Center for Astronomy,
Greece
gcontop@cc.uoa.gr

Copin Yannick
Institut de Physique Nucl'eaire de Lyon,
France
y.copin@ipnl.in2p3.fr

Davies Roger L.
University of Oxford,
U.K.
rld@astro.ox.ac.uk

De Bruyne Veronique
University of Ghent,
Astronomical Observatory,
Belgium
Veronique.DeBruyne@rug.ac.be

De Zeeuw Tim
Leiden University,
Netherlands
tim@strw.leidenuniv.nl

Dejonghe Herwig
University of Ghent,
Astronomical Observatory,
Belgium
Herwig.Dejonghe@rug.ac.be

Docenko Dmitrijs
University of Latvia,
Latvia
dima@latnet.lv

Dultzin-Hacyan Deborah
Instituto de Astronomia de la Univ. Nacional Autonoma di Mexico,
Mexico
deborah@astroscu.unam.mx

Dvorak Rudolf
University of Vienna,
Austria
dvorak@astro.univie.ac.at

Eckart Andreas
University of Cologne,
Germany
eckart@ph1.uni-koeln.de

Efthymiopoulos Christos
Academy of Athens,
Research Center for Astronomy,
Greece
cefthim@cc.uoa.gr

Eichler David
Ben-Gurion University,
Israel
eichler@bgumail.bgu.ac.il

Emsellem Eric
Centre de Recherche Astronomique de Lyon,
France
emsellem@obs.univ-lyon1.fr

Franco-Balderas Alfredo
Instituto de Astronomia de la Univ. Nacional Autonoma di Mexico,
Mexico
alfred@astroscu.unam.mx

Fridman Alexei
Russian Academy of Sciences,
Russia
fxela@online.ru

Galgani Luigi
Universita di Milano,
Italy
galgani@berlioz.mat.unimi.it

Gedalin Michael
Ben-Gurion University,
Israel
gedalin@bgumail.bgu.ac.il

Genzel Reinhard
Max-Planck-Institut für extraterrestrische Physik,
Germany
genzel@imprs-astro.mpg.de

Gottesman Stephen
University of Florida,
U.S.A.
gott@astro.ufl.edu

Griv Evgeny
Ben-Gurion University,
Israel
griv@bgumail.bgu.ac.il

Grosbøl Preben
ESO,
Garching,
Germany
pgrosbol@eso.org

Hadjidemetriou John
University of Thessaloniki,
Greece
hadjidem@physics.auth.gr

Harsoula Maria
Academy of Athens,
Research Center for Astronomy,
Greece
mharsoul@phys.uoa.gr

Hernández-Toledo Héctor
Instituto de Astronomia de la Univ. Nacional Autonoma di Mexico,
Mexico

Hozumi Shunsuke
Max-Planck-Institut fuer Astronomie,
Heidelberg,
Germany
hozumi@mpia-hd.mpg.de

Hunter Chris
Florida State University,
U.S.A.
hunter@math.fsu.edu

Iyer Mansie
Florida International University,
U.S.A.
miyer01@fiu.edu

Kalapotharakos Costas
University of Athens,
Greece
ckalapot@cc.uoa.gr

Kandrup Henry Emil
University of Florida,
U.S.A.
kandrup@astro.ufl.edu

Khoruzhii Oleg
Troitsk Institute for Innovation and Thermonuclear Researches,
Russia
okhor@inasan.rssi.ru

Kuntschner Harald
European Southern Observatory,
Germany

Lindblad Per A.B.
Stockholm University Observatory,
Sweden

Lindblad Per Olof
Stockholm University Observatory,
Sweden
po@astro.su.se

Liverts Edward
Ben-Gurion University,
Israel
eliverts@bgumail.bgu.ac.il

Lynden-Bell Donald
University of Cambridge,
U.K.
dlb@ast.cam.ac.uk

Maciejewski Witold
Osservatorio Astrofisico di Arcetri,
Italy
witold@arcetri.astro.it

Malphrus Benjamin
Morehead State University,
U.S.A.
b.malphr@morehead-st.edu

McDermid Richard
Leiden University,
Netherlands
mcdermid@strw.leidenuniv.nl

Miller Richard H.
University of Chicago,
U.S.A.
rhm@oddjob.uchicago.edu

Miller Bryan W.
Gemini Observatory,
Chile
bmiller@gemini.edu

Mond Michael
Ben-Gurion University,
Israel
mond@menix.bgu.ac.il

Mouawad Nelly
University of Cologne,
Germany
nelly@ph1.uni-koeln.de

Muccione Veruska
Observatoire de Geneve,
Switzerland
veruska.muccione@obs.unige.ch

Naab Thorsten
Institute of Astronomy,
Cambridge,
UK
naab@ast.cam.ac.uk

Neishtadt Anatoly
Space Research Institute,
Russia
aneishta@vm1.iki.rssi.ru

Nykytyuk Tetyana
National Academy of Sciences of
Ukraine, Ukraine,
nikita@mao.kiev.ua

Ott Thomas
Max-Planck-Institut für extraterrestrische Physik,
Germany
ott@mpe.mpg.de

Peletier Reynier
University of Nottingham,
U.K.
reynier.peletier@nottingham.ac.uk

Pfalzner Susanne
University of Cologne,
Germany
pfalzner@ph1.uni-koeln.de

Polyachenko Evgenii
Institute of Astronomy,
RAS,
Russia
epolyach@inasan.rssi.ru

Psyhoyos Dionysia
University of Thessaloniki,
Greece
dpsyc@skiathos.physics.auth.gr

Sagdeev Roald
University of Maryland,
U.S.A.
rzs@umd.edu

Schödel Rainer
Max-Planck-Institut für extraterrestrische Physik,
Germany
rainer@mpe.mpg.de

Simpson Caroline
Florida International University,
U.S.A.
simpsonc@fiu.edu

Spurzem Rainer
Astronomisches Recheninstitut,
Germany
spurzem@ari.uni-heidelberg.de

Stavropoulos Ioannis
University of Athens,
Greece
istavrop@cc.uoa.gr

Straubmeier Christian
University of Cologne,
Germany
cstraubm@ph1.uni-koeln.de

Kleomenis Tsiganis
Aristotle University of Thessaloniki,
Greece,
tsiganis@astro.auth.gr

Van de Ven Glenn
Leiden University,
Netherlands
vdven@strw.leidenuniv.nl

Varvoglis Harry
University of Thessaloniki,
Greece
varvogli@astro.auth.gr

Verolme Ellen
Leiden University,
Netherlands
verolme@strw.leidenuniv.nl

Voglis Nikos
Academy of Athens,
Research Center for Astronomy,
Greece
nvogl@cc.uoa.gr

Voyatzis George
University of Thessaloniki,
Greece
voyatzis@auth.gr

Part I

Order and Chaos

Order and Chaos in Astronomy

George Contopoulos

Research Center of Astronomy, Academy of Athens, 14, Anagnostopoulou St.,
Athens, GR-10673, Greece

Abstract. We review the applications of order and chaos in various branches of astronomy. Order and chaos appear in generic dynamical systems, including the sun and other stars, the solar system and galaxies, up to the whole Universe. We discuss in particular the various types of orbits in galaxies, emphasizing the role of diffusion of chaotic orbits and the escapes to infinity. Then we consider chaos in dissipative systems, like gas in a galaxy, chaos in relativity and cosmology, and chaos in stellar pulsations and in the solar activity.

1 Introduction

The study of order and chaos had an explosive development in recent years. Thousands of papers were published on this subject. Particular problems of interest for Astronomy have been studied in various fields. Such fields are:

1. Celestial Mechanics
2. Galactic Dynamics
3. Relativity
4. Cosmology
5. Stellar Pulsations
6. Solar Activity

In the present review we will discuss several problems from these fields.

A special book by G. Contopoulos on "Order and Chaos in Dynamical Astronomy" (Springer Verlag, 2002) [20] has just appeared. This book of 624 pages and 305 figures has about 1200 selected references for further reading on this subject.

2 Celestial Mechanics

There are two very different traditions in Dynamical Astronomy. One deals mainly with regular phenomena (periodic and quasi-periodic motions) and the other with irregular phenomena (chaotic motions).

The basic example of order was provided by celestial mechanics. Strictly speaking only in integrable systems all motions are regular. But the systems considered in celestial mechanics are assumed to be close to integrable, and most motions are close to quasi-periodic. Thus the solar system was considered for a long time a paradigm of order.

Only in recent years chaos has been found in the solar system. The first chaotic phenomena referred to the irregular rotation of some satellites and to the distribution of the asteroids ([81], [48], [82], [46], [40], [65]). On the other hand the motions of the planets seem to be stable over several billions of years. But chaos is present in the motions of the planets also. The time scale for chaos (the so-called Lyapunov time) is relatively short in the case of Pluto (some 10^7 years). Nevertheless Pluto does not escape from the solar system, because of some resonance effects. Thus Pluto is an example of "stable chaos".

A similar phenomenon of "stable chaos" appears also in the case of certain asteroids that fill some resonances (and do not form gaps there), despite their short Lyapunov time [63]. A new explanation of this phenomenon will be given by H. Varvoglis during this workshop.

Another planet with Lyapunov time shorter than a Hubble time is Mercury. According to Laskar [55] Mercury's orbit should approach eccentricity 1 after about 3.5 billion years, therefore Mercury should either plunge into the sun, or collide with Venus. However, R. Dvorak, will present numerical evidence in this workshop that Mercury will be stable for several Hubble times.

There are two more phenomena that affect our Earth, and deal with order and chaos. One is the origin of the meteorites that hit the Earth. These meteorites and the dust of the zodiacal light come from very different parts of the solar system, and their orbits are in general chaotic [34].

The other phenomenon deals with the obliquity of the Earth's axis, which is stabilized by the action of the Moon [54]. Without the Moon the Earth's orientation would change completely, in a chaotic way, and this should affect considerably the evolution of life. However, with the Moon in its present position the chaotic variation of the Earth's obliquity does not exceed certain limits, and the corresponding climatic changes are also limited.

3 Classification of Chaotic Systems

Chaotic behaviour is a general characteristic of nonintegrable dynamical systems. There are various degrees of chaos that are classified in classical books of statistical mechanics on regular and chaotic motion (e.g. [58]). Namely nonintegrable systems are considered to be ergodic, mixing, Kolmogorov, or Anosov (each class is a subclass of the previous class). They are ergodic if most motions go everywhere in the available phase space, mixing if two nearby particles deviate considerably, Kolmogorov if this deviation is exponential in time, and Anosov if the average exponential factor (the so-called Lyapunov characteristic number) is always larger than a nonzero number. In random systems we have the limit of an infinite Lyapunov characteristic number.

However, this classification is now obsolete. In fact ergodic systems (and mixing, Kolmogorov and Anosov systems) appear only rarely, and generic dynamical systems contain both order and chaos. Namely the set of ordered motions is not zero in general, and this is the opposite of what happens in ergodic systems. This conclusion is based on the famous KAM theorem ([53], [1], [67]), that proves

OLD CLASSIFICATION

Integrable systems
Ergodic systems
Mixing systems
Kolmogorov systems
Anosov systems

NEW CLASSIFICATION

	ORDERED	CHAOTIC		
COMPACT	Integrable	(General Case) Systems with Divided Phase Space	(Limiting Cases) Ergodic Mixing Kolmogorov Anosov	RANDOM
NONCOMPACT	Integrable with escapes	Nonintegrable with escapes (Chaotic scattering)	—	

Fig. 1. Old and New Classification of Dynamical Systems.

the existence of a finite set of quasi-periodic motions in generic dynamical systems. Thus the presently accepted classification of dynamical systems is shown in Fig. 1.

We call chaotic systems those having chaotic domains, but possibly also ordered domains. In limiting cases we have only chaotic (ergodic) motions.

A system is close to integrable if its chaotic domains are small and it is close to ergodic if its ordered domains are small. The random systems are extremely chaotic. Their Lyapunov time is zero. As regards the noncompact systems (i.e. systems with escapes) they may be integrable or nonintegrable. In the latter case we have the phenomenon of "*chaotic scattering*".

A general theorem that has been proved recently in the case of some simple maps, namely the logistic map [51] and the standard map [36] is the following. If the perturbation parameter K is large there are many values of K for which there are no stable periodic orbits and the system is completely chaotic (i.e. the ordered orbits have a measure zero). Nevertheless there is no interval ΔK of values of K without any islands of stability. This theorem seems to be applicable in generic dynamical systems.

Examples of Hamiltonian systems with such a behaviour have been found in recent years [25] and [44]. These systems have a central periodic orbit which is alternatively stable and unstable, up to arbitrarily large values of the perturbation.

A classical case where we have both order and chaos is the standard map

$$
\begin{aligned}
x' &= x + y' \\
& \qquad (mod 1) \\
y' &= y + \frac{K}{2\pi} sin(2x)
\end{aligned}
\tag{1}
$$

For small K (Fig. 2a) this system is mostly ordered, with only a little chaos near the unstable periodic orbit (0,0) and the asymptotic curves emanating from it.

As K increases chaos increases also (Figs. 2b,c,d,e) and for K=8 it seems that chaos is complete.

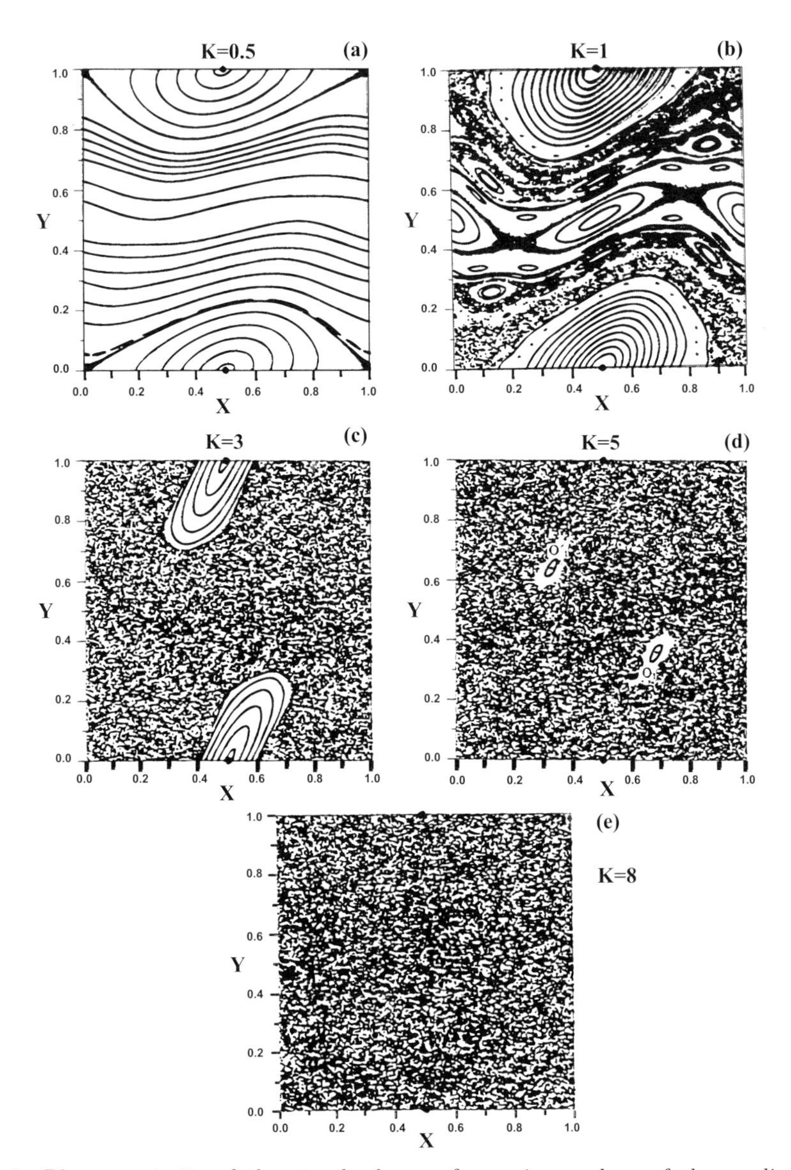

Fig. 2. Phase portraits of the standard map for various values of the nonlinearity parameter K: (a)K=0.5, (b)K=l, (c)K=3, (d)K=5, (e)K=8.

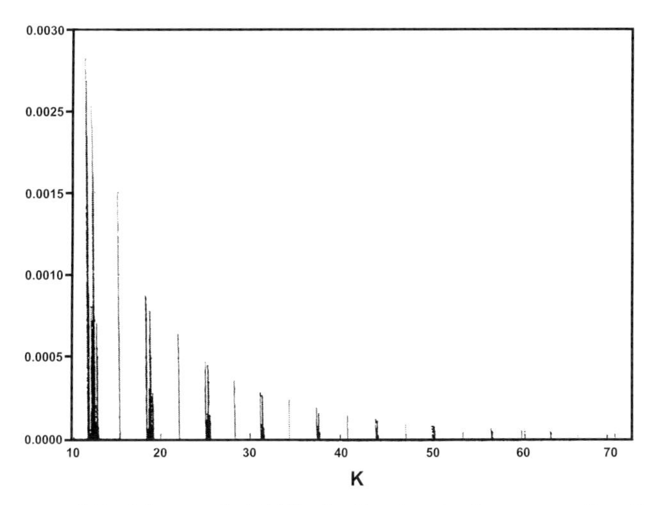

Fig. 3. The area of the islands of stability in the standard map as a function of K [with unit the square (0,l)x(0,l)].

Nevertheless small islands of order have been found for larger values of K in a recurrent way (Fig. 3). Namely islands appear in particular regions of the phase space whenever K increases by 2π [37].

Therefore chaos is never complete, despite a widespread opinion that beyond a limiting perturbation the system should remain completely chaotic.

4 Galactic Dynamics

Order and chaos play an important role in galactic dynamics. The appearance of order in galactic dynamics is based on the "third integral" of motion [14].

One of the first applications of the third integral was in the velocity ellipsoid of stars near the sun. If there are only two integrals of motion, the energy

$$E = \frac{1}{2}(R^2 + Z^2 + \Theta^2) + V \tag{2}$$

and the angular momentum

$$J = r\Theta \tag{3}$$

one should have a distribution function of the form

$$f = f(E, J) \tag{4}$$

In particular an ellipsoidal distribution should be of the form

$$f = f(R^2 + Z^2 + k(\Theta - \Theta_0)^2 \tag{5}$$

with the two equal axes along R and Z. However, the observations had shown definitely that the Z axis is much shorter than R. This indicated the existence of a third integral of motion

$$I = Z^2 + \dots \tag{6}$$

that should be included in the distribution function f for the velocity [5], so that the distribution function should become

$$f = f(R^2 + (1 + k')Z^2 + k(\Theta - \Theta_0)^2) \tag{7}$$

My involvement in the subject of the third integral in galaxies started during my first visit to Stockholm in 1956. I worked with Professor Bertil Lindblad on a generalization of the epicylic theory of planar galactic orbits. But when I tried to extend the theory to three dimensions I could not do very much analytically. At that time Per Olof Lindblad was calculating planar orbits in galaxies to explain the spiral arms. His first calculations were presented at an IAU Symposium in 1956 [59] At my request he calculated for me two orbits in three dimensions. The surprising thing was that these orbits (Fig. 4) were not ergodic, as I expected, but indicated the existence of a third integral of motion in generic dynamical systems ([13],[14]). I realized later that Whittaker [80] had already found a third integral (that he called adelphic integral) in particular cases by a different method.

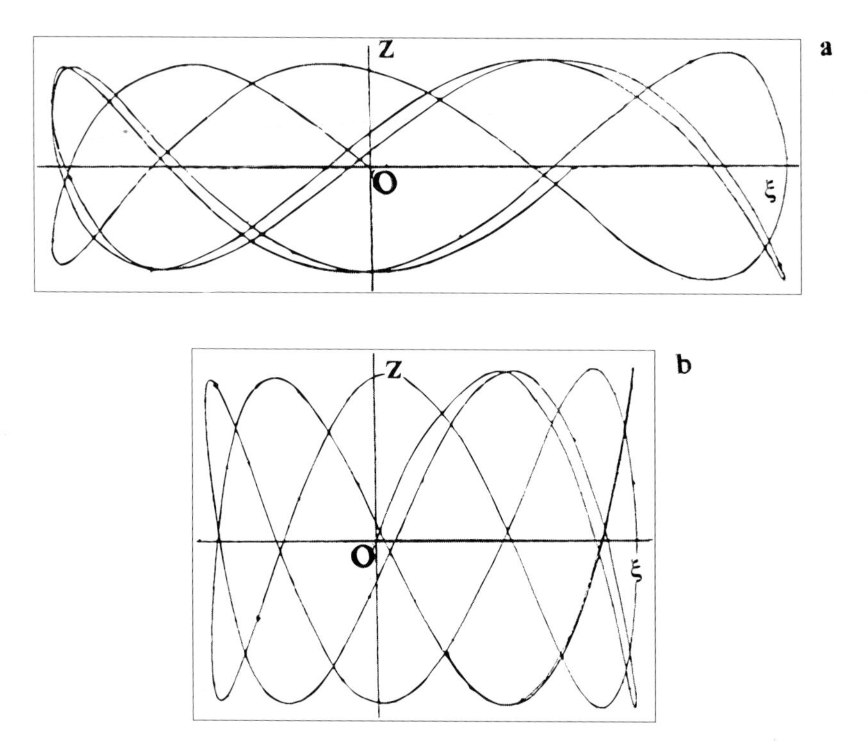

Fig. 4. The first calculated orbits in the meridian plane of an axisymmetric galaxy (1956) are like deformed Lissajous figures.

It is remarkable that Birkhoff [7] had also found a similar integral by another different method, but he never believed in its usefulness. In fact the third integral is in general not exact, but only a formal series. According to Birkhoff this indicated that such an integral would be applicable only over limited times. In particular he believed that a linearly stable periodic orbit in a system of two degrees of freedom would be stable for all times only in integrable systems.

It was only through the KAM theorem that the complete stability of linear stable orbits was established for generic two-dimensional dynamical systems. But even after the KAM theorem was established, it was not clear what was the usefulness of such a formal integral.

I remember that when I first spoke to Moser about the third integral in galactic dynamics he was doubtful of its usefulness, as it is only an asymptotic series that does not converge. But at that time I could calculate by computer algebra ([21], [16]) the higher order terms of the third integral (this was done by simple Fortran, without the present packages of computer algebra, like mathematica etc). The results were remarkable, indicating an excellent agreement between theoretical and numerical orbits when higher order terms of the third integral were calculated. This impressed Moser and he remarked in one of his papers ([68]) that because of this work in galactic dynamics the subject of the third integral "received renewed interest".

At the same time a student of Arnold, Nekhoroshev, studied the usefulness of truncated third integrals. He found the best truncation, and he showed the applicability of these integrals over exponentially long times, a result that is the central element of the Nekhoroshev [69] theory.

In some cases one has to reach very high orders of truncation of the third integral in order to find the Nekhoroshev limit and deviations beyond it.

Another independent development started with the Fermi-Pasta-Ulam paper [39] on coupled oscillators that revealed the existence of ordered orbits in systems of many degrees of freedom. This result was also explained, later, by means of formal integrals of motion, of the third integral type.

5 Chaotic Orbits in Galaxies

A recent example refers to an application of the third integral to self-consistent models of galaxies, generated by N-body simulations [28]

Up to now in most applications the potential of the galaxy was assumed to be given by a simple analytic formula. Then a third integral could be constructed in order to find the structure of the regular orbits. But in our present studies we start with models produced by the collapse of a protogalaxy. This gives not only the density and the potential of the final model, after the collapse, but also the distribution of the velocities. This method is very different from other methods, like the Schwarzschild (1979) method, that try to construct self-consistent models by populating various initial conditions of orbits in an appropriate way.

The success of the original Schwarzschild method, indicated that most orbits are regular, i.e. his self-consistent models were very close to integrable. At the same time several integrable models, the so-called Stäckel models, were introduced in galactic dynamics ([35], [8], [78], [50], etc). In particular, several self-consistent models were constructed. This led to the conjecture that chaos is unimportant in galactic dynamics. It was stated loosely that somehow, Nature avoids chaos in galaxies and forms only integrable models.

However, our studies of models generated by N-body simulations indicate that in many cases both ordered and chaotic orbits are present. There are three main types of orbits in a galaxy, box orbits (Fig. 5a), tube orbits (Fig. 5b) and chaotic orbits (Fig. 5c). Box orbits appear in nonrotating systems (e.g. elliptical galaxies). Tube orbits appear near various resonances (see also Figs. 6a,b,c). Chaotic orbits are due to an interaction of resonances. E.g. in Fig. 7 a chaotic orbit joins the bar region of a galaxy with the outer near circular orbits. Chaos is small in some cases, but not negligible. On the other hand no model was found that is completely chaotic.

The general conclusion is that Nature does not form integrable models like those derived from Stäckel potentials, nor ergodic systems, like those used in statistical mechanics, except in an approximate way.

There are three types of chaotic orbits in galaxies:

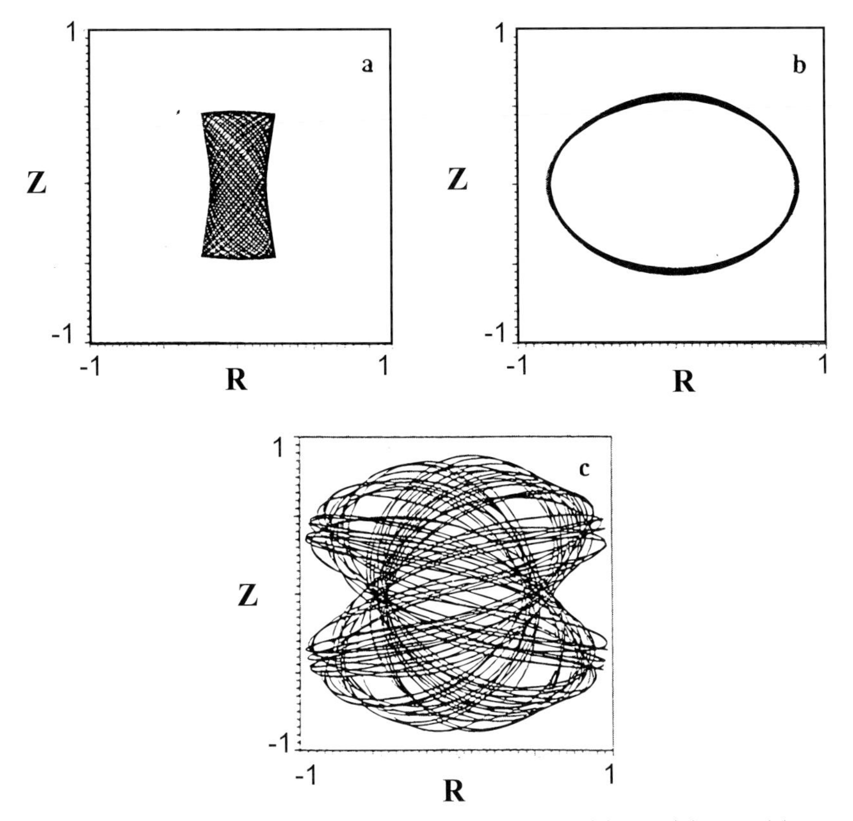

Fig. 5. Typical orbits in a self-consistent galactic model: (a)box, (b)tube, (c)chaotic.

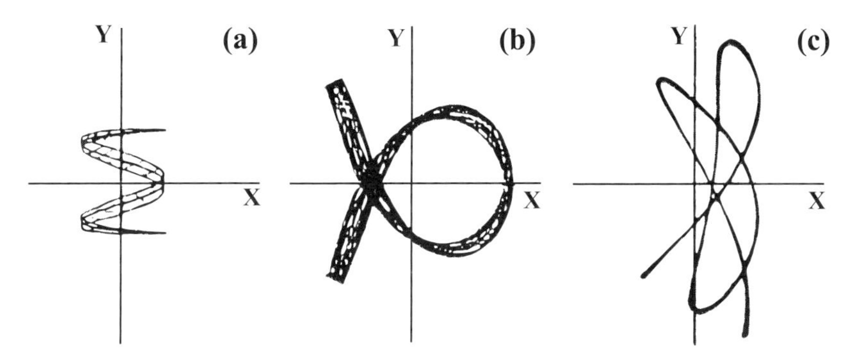

Fig. 6. Three further types of tube orbits.

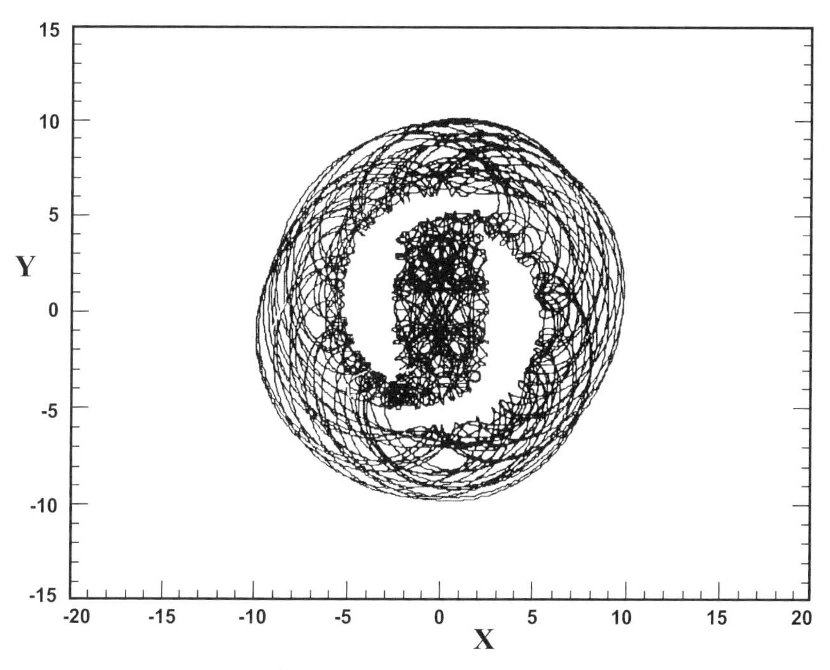

Fig. 7. A chaotic orbit in a barred galaxy.

1. Orbits near corotation in rotating spiral, or barred galaxies [17]. Such orbits are important in terminating the bars close to corotation.
2. Elongated orbits passing near the center of a galaxy ([61], [43], [62]). Such orbits are particularly important when there is a strong mass concentration near the center, e.g. a black hole. A large black hole in a nonspherical galaxy tends to change the character of the orbits, from boxes filling curvilinear parallelograms far the center, to almost Keplerian orbits near the center. The mixing of two types of orbital behaviour along the same orbit produces chaos. One can even estimate the size of a black hole by the degree of chaos produced around it.

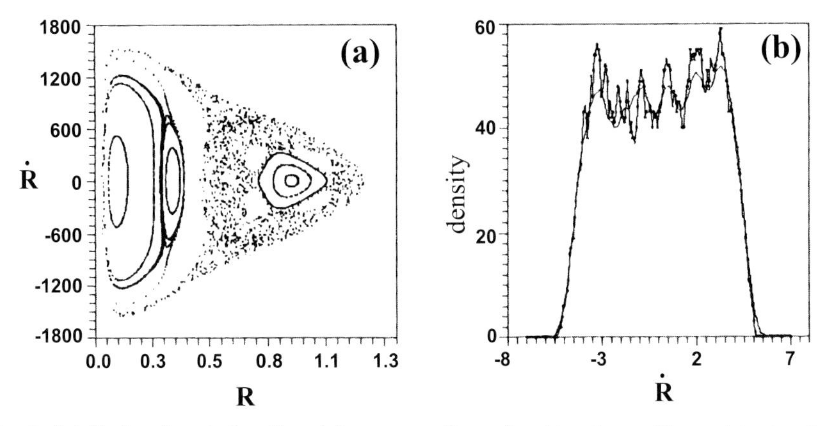

Fig. 8. (a) Ordered and chaotic orbits on a surface of section in a self-consistent galactic model. (b) The density in the chaotic domain (R=0.6) is roughly constant.

3. A third chaotic region is the region between the box orbits, that cover a large inner part of a galaxy (except the very central region) and the tube orbits that refer mainly to orbits circulating around the main galaxy in the outer parts (Fig. 8). We notice that chaotic orbit produce a constant density in the chaotic domain.

6 Diffusion of Galactic Orbits

When an integrable system is slightly perturbed it develops a small degree of chaos. This is introduced as follows. In an integrable case the Poincaré surface of section is filled with invariant curves. We may consider a general case in which the invariant curves close around a central invariant point O. The rotation number along successive invariant curves varies smoothly. For every rational rotation number n/m all the orbits starting on the corresponding invariant curve are periodic, of period m (Figs. 9a,b). When a generic perturbation of order ϵ is introduced, only two periodic orbits of period m are left, one stable and one unstable (Fig. 9c). (In some cases there are more pairs of stable-unstable orbits). The stable orbits are surrounded by islands, while near the unstable points some chaos appears.

As the perturbation increases the islands increase in size. The chaotic regions also increase in size forming zones of instability. However between the main zones of instability there are still invariant curves, around the center,and the various chaotic regions are separated.

When the perturbation ϵ goes beyond a critical value ϵ_{crit} the various chaotic regions communicate (Fig. 9c) and chaos becomes important abruptly. This is the phenomenon of "resonance overlap" ([16], [74], [84], [11]). This leads to a diffusion of the orbits that is called "resonance overlap diffusion". This is the main mechanism that introduces a large degree of chaos.

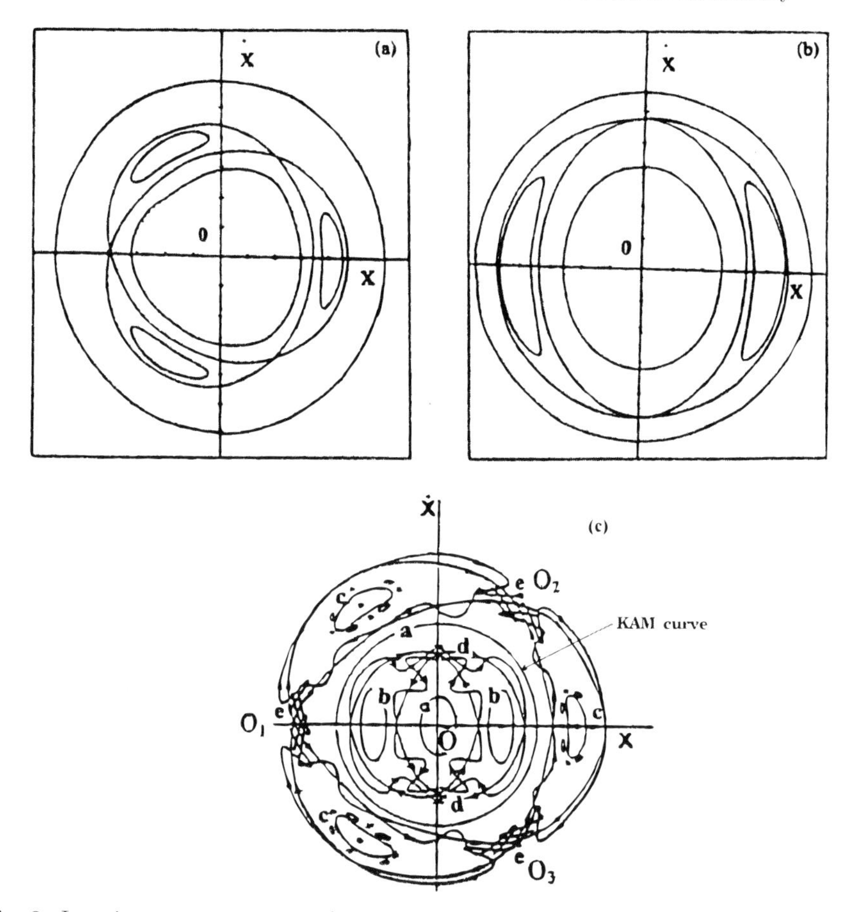

Fig. 9. Invariant curves on a surface of section of two resonant integrable cases and one nonintegrable case. (a) Case with three islands, (b) Case with two islands. (c) A case with both double and triple islands. For small perturbations the chaotic domains near the unstable double and triple orbits are separated by KAM curves. But for larger perturbations the last KAM curve is destroyed and large chaos, due to resonance overlap, is produced.

In three or more degrees of freedom there is one more mechanism that introduces chaos, namely Arnold diffusion [3]. In fact, while in systems of two degrees of freedom the KAM surfaces are two-dimensional and separate the interior from the exterior in the 3-D phase space, in three degrees of freedom the KAM surfaces are three-dimensional and do not separate the phase space, which is now five-dimensional. Thus the various chaotic regions always overlap, even for arbitrarily small perturbation ϵ.

However the time scale of Arnold diffusion is very large, of order

$$T \propto exp(\frac{1}{\epsilon}) \tag{8}$$

i.e. it is exponentially long ([3], [69], [11]). In the close neighbourhood of invariant tori the time scale is even superexponential, of the form

$$T \propto exp(exp(\frac{1}{\epsilon})) \tag{9}$$

or even longer [66]. In galactic problems this time is much longer than the age of the Universe, therefore Arnold diffusion is not important. Only in plasma physics this diffusion may have observable consequences.

Although Arnold diffusion is also due to resonance overlap, nevertheless we distinguish clearly the resonance overlap diffusion from Arnold diffusion ([54], [22]) even in systems of three or more degrees of freedom. In fact, Arnold diffusion appears only very close to the resonance lines of the Arnold web (Fig. 10) while resonance overlap extends over large regions of the phase space.

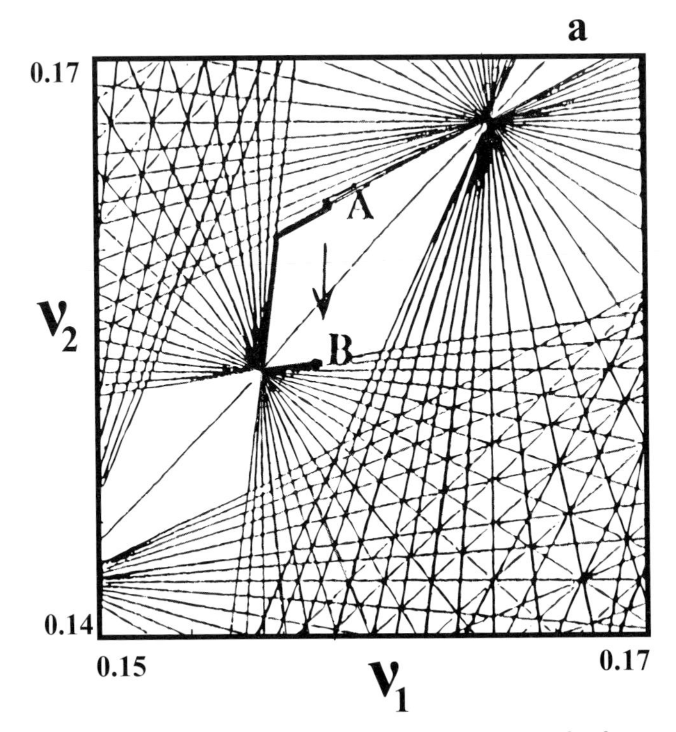

Fig. 10. The Arnold web contains the resonant lines between the frequencies ν_1 and ν_2. If the perturbation is small there is a slow Arnold diffusion from A to B along the thick lines. But if the perturbation is large the resonant lines become very thick and allow resonance overlap diffusion, directly from A to B.

A numerical example was given in the case of two coupled standard maps [22]

$$x_1{}' = x_1 + y_1{}' \quad , \quad y_1{}' = y_1 + \frac{K}{2\pi} sin 2\pi x_1 - \frac{\beta}{\pi} sin 2\pi (x_2 - x_1)$$

$$(\,mod 1) \quad (10)$$

$$x_2{}' = x_2 + y_2{}' \quad , \quad y_2{}' = y_2 + \frac{K}{2\pi} sin 2\pi x_2 - \frac{\beta}{\pi} sin 2\pi (x_1 - x_2)$$

where K is the nonlinearity and β the coupling parameter.

The diffusion time T is given empirically by the formula

$$T = exp[a - b(\beta - \beta_{crit})] \quad (11)$$

where K=3, a=4.14, b=4160, and $\beta_{crit} = 0.305124$ is a critical value of the coupling parameter (Fig. 10).

For $\beta > \beta_{crit}$ the diffusion time decreases with β according to the exponential law (11) (Fig. 11) and becomes very small for relatively large β. However, for ($\beta < \beta_{crit}$ the time T increases superexponentially as β decreases and for $\beta < 0.305$ it becomes larger than T=10^{10} iterations, i.e. it is very difficult to calculate numerically.

In systems of two degrees of freedom there is also a slow diffusion, like Arnold diffusion, when the orbits have to cross the holes of cantori, surrounding islands of stability.

Cantori appear near the outermost invariant curve surrounding a stable invariant point, when the perturbation K increases beyond a critical value K_{crit}. Then the outermost invariant curve is destroyed and it becomes a cantorus with

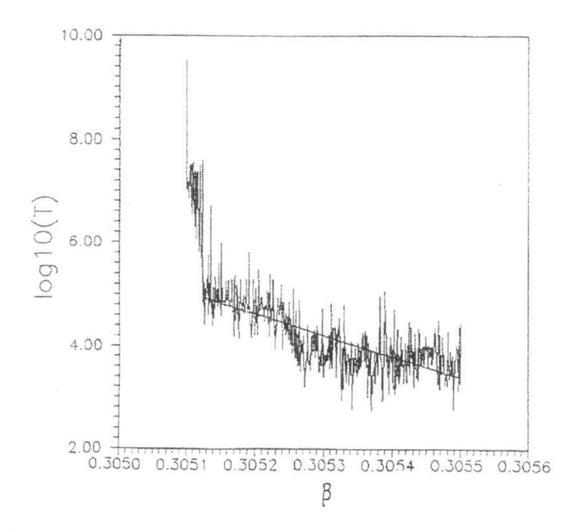

Fig. 11. The diffusion time T as a function of the coupling parameter β increases exponentially for decreasing β, if $\beta > 0.305124$, and superexponentially for smaller β.

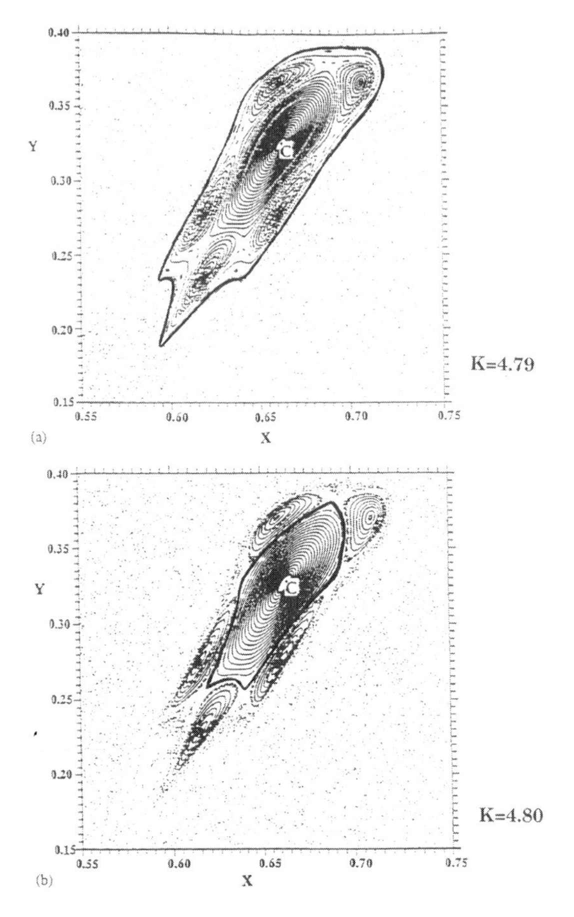

Fig. 12. An island of stability in the standard map inside the large chaotic sea is limited by the last KAM curve, (a) For K=4.79 the last KAM curve surrounds 5 islands of stability, (b) for K=4.80 the last KAM curve is inside these islands.

infinite holes. Chaotic orbits close to this cantorus can cross it from inside outwards, or from outside inwards. Thus we see two different chaotic domains outside the main island of stability. A "sticky zone" between the island and the cantorus and a "large chaotic sea" outside the cantorus (Fig. 12). When K goes beyond K_{crit} the size of the island decreases abruptly and a sticky zone is formed between the cantorus and the new outermost limit of the island. The diffusion time increases exponentially as we deviate from the cantorus inwards and approach the new boundary of the island.

On the other hand if the perturbation increases, the holes of the cantorus become larger and the diffusion time becomes abruptly much shorter.

As an example we consider the cantori around an important island in the standard map (see (1)) for K=5 and K=4.998. In the first case we found orbits that escape to the chaotic sea through the cantorus in a time T=2, while in the second case this time increases to T=2x10^5 [28].

Despite these variations the stickiness effect plays a role in practical applications. For example in galactic dynamics some orbits stay close to the boundary of a resonant island for a long time, and can be considered as regular during the life-time of the galaxy, although in the long run they may become very chaotic and may even escape from the system.

7 Escapes

Another interesting subject related to chaos, deals with escapes. If the energy h of a star in a galaxy is larger than the escape energy h_{esc}, the star may escape to infinity. However, there are many stars, with energy larger than h_{esc}, which do not escape, because they are trapped around a stable periodic orbit.

One way to study the problem of escapes is by using a simple model. An example is given by the Hamiltonian

$$H = \frac{1}{2}(\dot{x}^2 + \dot{y}^2 + Ax^2 + By^2) - \epsilon xy^2 = h \tag{12}$$

(A=1.6, B=0.9, ϵ = 0.08 and various values of the energy h). The escape energy in this case is h_{esc}=25.31.

If h is larger than h_{esc} the CZV has two openings (Fig. 13). At the openings , there are two unstable periodic orbits, that are called Lyapunov orbits. Any particle crossing a Lyapunov orbit outwards escapes from the system and never returns to it.

For an energy slightly smaller than the escape energy most initial conditions generate chaotic orbits. Most of these orbits escape when the energy increases slightly above the escape energy. However, the escape may take a long time. There are sets of orbits that escape rather fast, after one or two iterations, but there are also orbits that escape only after many iterations. The domains of escape are limited by orbits that are asymptotic to one or the other Lyapunov orbits.

In Fig. 15 we show the escape domains for orbits escaping after one or two iterations in the forward time direction from the upper opening only.

By counting the escaping particles out of a large number of bodies we found empirically the escape rate and the number of remaining particles N out of an initial number N_0 [24]

$$N = N_0 e^{-pt} \tag{13}$$

If the energy is larger than a critical value h_{crit}, much larger than escape energy h_{esc}, practically all particles escape according to the law (13) and the escape rate p is proportional to a power of (h-h_{crit}),

$$p \propto (h - h_{crit})^2 \tag{14}$$

where a=0.5, practically the same for several potentials. Thus the exponent a may be a universal number [52].

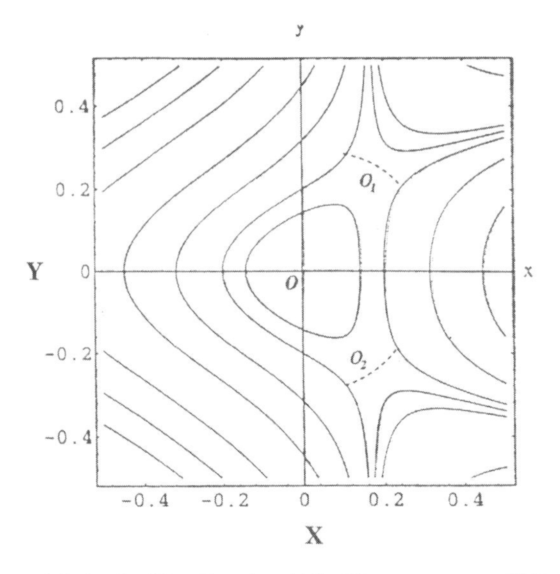

Fig. 13. Equipotentials in the Hamiltonian (12). These are open if the energy is larger than the energy of escape. At the openings, there are two unstable periodic orbits, O_1 and O_2, called Lyapunov orbits.

If the energy is larger than h_{esc}, but smaller than h_{crit}, there is a set of particles N_{non}, that never escape. Then the escape law is of the form

$$N - N_{non} = (N_0 - N_{non})e^{-pt} \tag{15}$$

The transition at h_{crit} is like a phase transition and it is connected with the size of the islands of stability. For $h < h_{crit}$ there are important islands of stability, while for $h > h_{crit}$ the size of the islands becomes very small, and most islands disappear altogether.

The fact that some particles escape from the system has as a consequence the nonconservation of areas on a surface of section (Fig. 14). Namely, while for $h < h_{esc}$ the plane y=0 is a Poincaré surface of section and the areas are conserved, for $h > h_{esc}$ this surface is no more a Poincaré surface of section and the areas are reduced.

Thus for $h > h_{esc}$ the 2-D map on the surface y=0 looks like a dissipative system although in the full 3-D space the volumes are preserved. Furthermore, although conservative systems do not have attractors, nevertheless in systems with escapes the infinity acts as an attractor. And if there are more than one openings of the curves of zero velocity we may speak of a corresponding number of co-existing attractors.

8 Dissipative Systems

The main difference between dissipative systems and conservative systems is the existence of attractors at a finite distance.

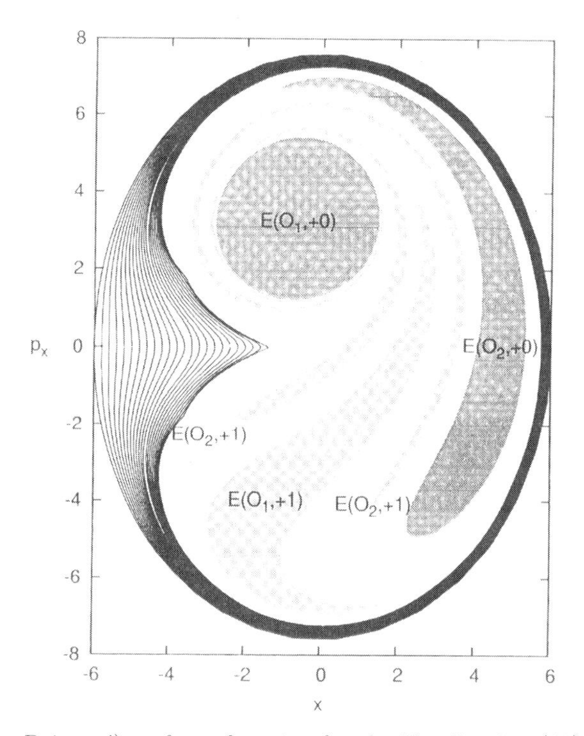

Fig. 14. A (non Poincaré) surface of section for the Hamiltonian (12) and an energy larger than the escape energy. The orbits starting in the domains $E(O_1,n)$ escape through the Lyapunov orbit O_1 after n further crossings with the surface of section. Near the boundary there are invariant curves of nonescaping orbits.

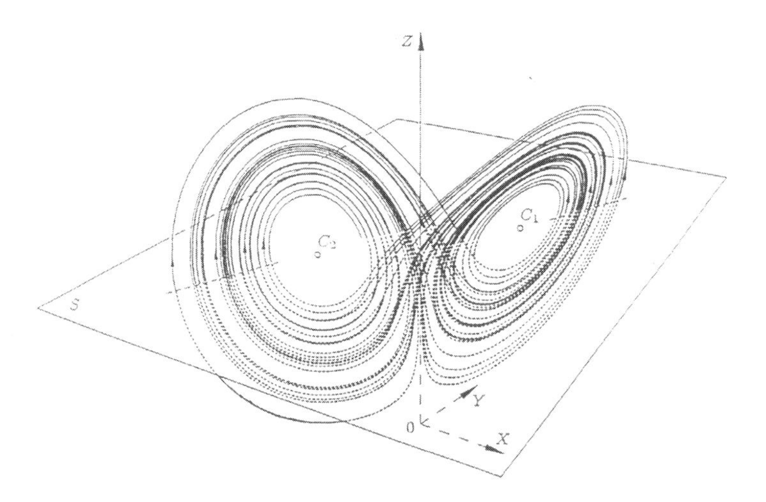

Fig. 15. A Lorentz strange attractor.

In general, dissipative systems have three types of attractors (1) Point attractors, (2) Limit cycles (in two dimensional flows) and (3) strange attractors. In more dimensions we may have limiting structures of higher than one dimension.

We have two types of dissipative systems (1) those given by nonlinear differential equation, and (2) those given by nonlinear maps.

A simple dissipative system of differential equations is the Lorentz system [60]

$$\begin{aligned} dx/dt &= \sigma(y-x) \\ dy/dt &= \rho x - y - xz \\ dz/dt &= -\beta z + xy \end{aligned} \tag{16}$$

(where σ, ρ, $\beta > 0$).

A most simple dissipative map is the Hénon map [47]

$$x' = y - Kx^2 + 1 \quad , \quad y' = bx \tag{17}$$

with $0 < b < l$.

These systems were the first to provide strange attractors (Fig. 15).

In dissipative cases the volume, or surface, in phase space is shrinking. The Lyapunov characteristic number is negative in the cases of point attractors or limit cycles, but it is positive in the case of strange 'attractor. Namely, the moving points approach the attractor, but nearby points on the attractor deviate exponentially. The Hénon map (17) has a limiting conservative case for b=l. In this case no attractor appears.

There are some classical books on chaos in dissipative systems, like the books of Lichtenberg and Lieberman [58] and Guckenheimer and Holmes [45].

In galaxies the gas is a dissipative system. Thus, we have a secular evolution of the gas that is different from the behaviour of the stars. In particular we may have an attractor at the center of the galaxy, or a limit cycle in the case of a stationary flow of gas. A special case is the appearance of vortices near corotation (Fig. 16) in spiral and barred galaxies ([38], [41], [42]), a subject that will be discussed by Fridman during this workshop.

9 Chaos in Relativity and Cosmology

Much work has been done on this subject in recent year. An example of this activity is the volume on "Deterministic Chaos in General Relativity" edited by Hobill et al. [49]. A more recent review was provided by Contopoulos et al. [27].

One of the first cases where chaos was found in General Relativity was the case of two fixed black holes ([18], [19]). It is remarkable that the relativistic problem is chaotic, while the corresponding classical problem of two fixed centers is completely integrable. In particular the relativistic motions of photons are completely chaotic, but the phase space extends to infinity. On the other hand the motions of particles with nonzero rest mass are in part ordered and in part chaotic.

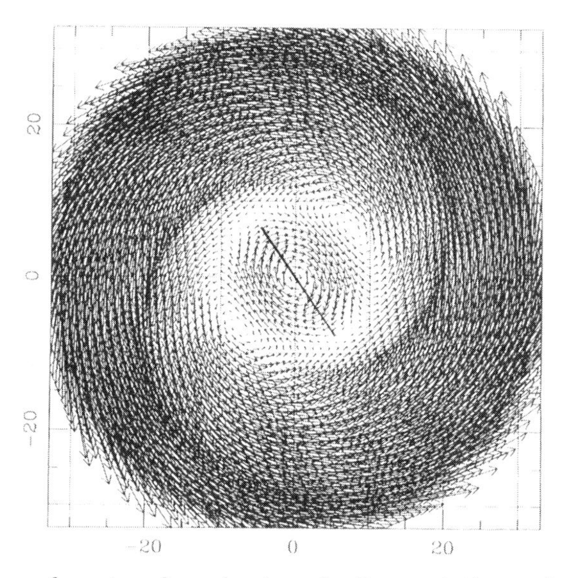

Fig. 16. The flow of gas in a barred galaxy. In the corotation region there are two cyclones along the bar and two anticyclones perpendicularly to the bar.

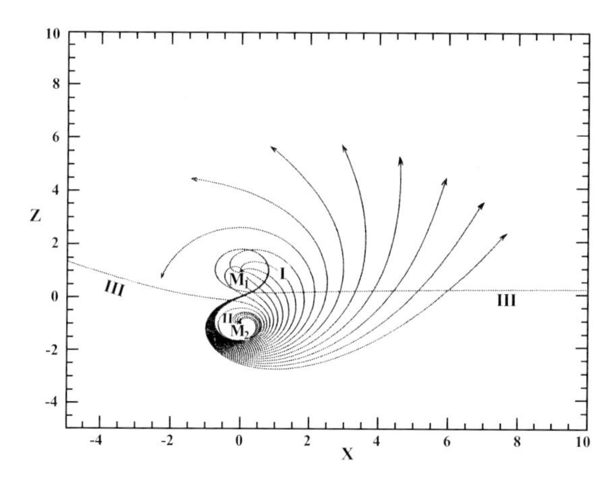

Fig. 17. Chaos in the case of two fixed black holes. Orbits of photons coming from infinity fall into the black hole M_1 (orbits of type I), or M_2 (orbits of type II), or escape to infinity (orbits of type III).

In the first case the photons may escape to one of the two black holes, or to infinity. A beam of photons (Fig. 17) coming from infinity is separated into three sets, one (I) leading to the black hole M_1 the second (II) leading to the black hole M_2, and the third (III) leading to infinity.

The three sets are fractal and their basic property is that between two orbits of different sets there is one orbit of the third set. For example in Fig. 17 between the orbits 1 leading to M_1 and the orbit 2 leading to infinity, there is an orbit

3 leading to M_2 (in fact there is a set of orbits (III) leading to M_2). Similarly between the orbit 1 (M_1) and the orbit 3(M_2) there is the orbit 4 leading to infinity, etc.

The three sets of orbits I, II and III are intertwined in a fractal way.

In the case of particles of nonzero rest mass, with energy smaller than the escape energy, the orbits can escape to the black holes M_1 and M_2 only. In such a case there are some stable periodic orbits and orbits close to them do not escape to any of the black holes.

In recent years much more work has been done in this problem and in other relativistic problems where chaos is present.

An important case with application to cosmology is the Mixmaster Universe model ([6], [64]). This represents a particular solution of Einstein's equations that has three different scale factors α, β, γ along the axes x, y, z. These equations are

$$\begin{aligned} 2\alpha &= (e^{2\beta} - e^{2\gamma})^2 - e^{4\alpha} \\ 2\beta &= (e^{2\gamma} - e^{2\alpha})^2 - e^{4\beta} \\ 2\gamma &= (e^{2\alpha} - e^{2\beta})^2 - e^{4\gamma} \end{aligned} \tag{18}$$

The basic property of this model is that the two scale factors are positive and one negative (or two negative and one positive). Therefore, the model expands along certain directions and contracts along others. But the directions of expansion and contraction change in a chaotic way.

There has been much analytical and numerical work on this model. A strange feature of this model is that the Lyapunov characteristic number is zero, yet the system seems to be chaotic. For some time it was expected that the Mixmaster model may be integrable, but more recently it was established that it is nonintegrable [57], [26], [31]. On the other hand the Mixmaster model is not ergodic [33], and has no periodic orbits.

These somewhat conflicting properties can be understood if we notice that the Mixmaster model is not compact, i.e. its orbits escape in general to infinity. In particular this property explains why the Lyapunov characteristic number is zero. In fact, if a particle tends to infinity, a nearby particle deviates from it linearly in time, i.e.

$$\xi = \xi_0 t \tag{19}$$

as in the case of two nearby orbits escaping from a galaxy. Therefore the Lyapunov characteristic number is given by

$$LCN = \lim_{t \to \infty} \frac{ln|\frac{\xi}{\xi_0}|}{t} = \lim_{t \to \infty} \frac{lnt}{t} = 0 \tag{20}$$

and this is equal to zero. But the finite time LCN is positive, and this indicates that the Mixmaster model is chaotic for all finite times (Fig. 18). This behaviour is similar to a chaotic scattering case [26].

There are more general cosmological models where chaos is important. In fact the role of chaos in Cosmology seems to be more and more appreciated in recent years.

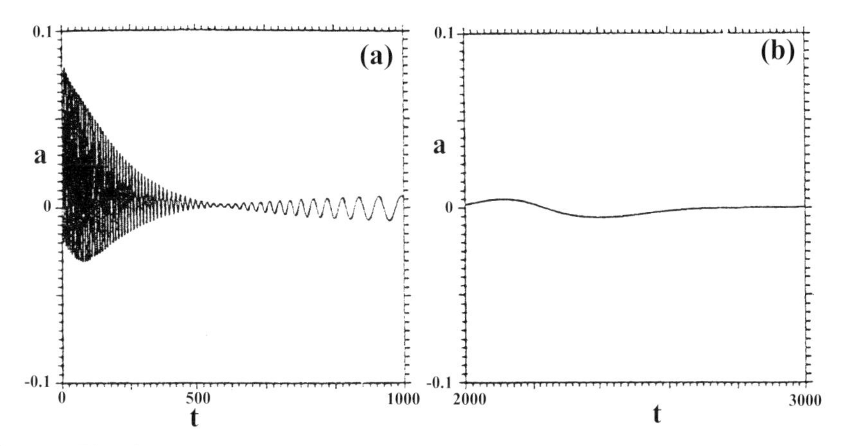

Fig. 18. The short time Lyapunov characteristic number in the mixmaster model as a function of t for (a) 0¡t¡1000, and (b) 2000¡t¡3000. The average value is positive and tends to zero as t.

In a similar way in the modern unified theories of the Universe (superstrings, quantum gravity, etc) chaos seems to play a significant role that only recently has started to be explored.

10 Chaos in Stellar Pulsations

The variable stars have only rarely completely regular pulsations. In most cases they have irregularities in the period and in the amplitude of the pulsations.

In many cases it is even impossible to define an average period of the variable stars. Various types of irregular variables appear in specific regions of the H-R diagram.

A review of the various studies of irregular variable stars was given some years ago by Perdang [70].

A classical method to study the variations in stellar models is numerical hydrodynamics. In the case of radial oscillations one separates the variable star in a number of concentric shells and follows their evolution in time [12] by solving the hydrodynamic and heat flow equations (Fig. 19). These solutions give good agreement with observations.

An alternative method is to consider the equations for the amplitudes of the most important excited modes. By solving numerically the so-called "amplitude equations" one finds results that are consistent with the hydrodynamic results [32], [9], [10].

The amplitude equations are of the form

$$\frac{d\alpha_1}{dt} = \kappa_i \alpha_i + N_i \tag{21}$$

where N_i are functions containing the non-linear terms. What is remarkable is that accurate results are found if one considers only two or three modes (i=2, or 3), and N_i contains only quadratic and cubic nonlinearities.

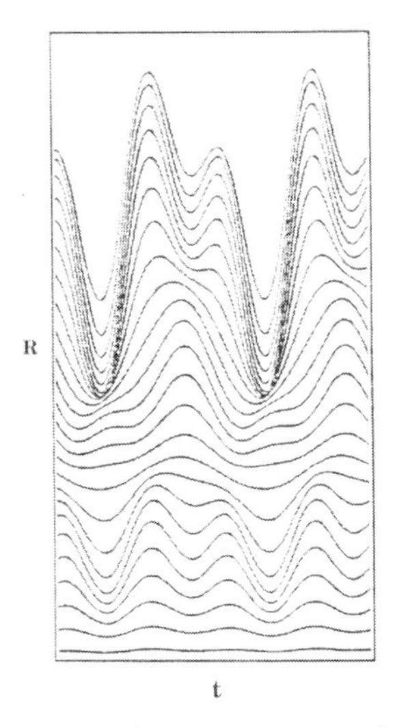

Fig. 19. The variations of the radii of various layers of a pulsating variable star as a function of t.

Thus, instead of the partial differential equations of hydrodynamics, we reach a problem of coupled ordinary differential equation, which is very similar to the problems of dissipative systems of particle dynamics, that we mentioned above.

A special case of particular interest is the case of radial perturbations around a state of hydrodynamical equilibrium. As it was pointed out by Woltjer [83] long ago, this problem can be formulated as a Hamiltonian system of a finite number N of degrees of freedom.

The Hamiltonian takes the form

$$H = \frac{1}{2}\sum_{i=1}^{N} {p_i}^2 + V \qquad (22)$$

where V contains terms of order 2,3, and possibly higher order terms

$$V = \frac{1}{2}\sum_{i=1}^{N} {\omega_i}^2 {q_i}^2 + \frac{1}{3}\sum_{i,j,k=1}^{N} {V_{ijk}}^{(3)} q_i q_j q_k + ... \qquad (23)$$

This Hamiltonian is of the same form, as those used in galactic dynamics. An extensive use of such a Hamiltonian in studying stellar pulsations was made by Perdang and Blacher [71], [72].

We will not discuss this subject further, but only refer to a related review article by Perdang [70]. Perdang and his associates found chaotic oscillations that can be described by such a Hamiltonian formalism. The analogy with our dynamical problem of galactic astronomy is astonishing.

The reduction of the partial differential equations of hydrodynamics (like those encountered in stellar pulsations) to a set of ordinary differential equations can be extended to more general problems. In fact there is an analogy between partial and ordinary differential equations. There are integrable and nonintegrable systems of partial differential equations, as there are integrable and nonintegrable systems of ordinary differential equations. In the integrable cases there are infinite conserved quantities, like the integrals of motion of the ordinary differential equations. These systems have as particular solutions the solitons, i.e. solitary waves that move unchanged, even after interacting nonlinearly with each other.

Of even greater importance is the fact that many nonintegrable systems of partial differential equations are close to integrable and they have soliton solutions, which decay only slowly in time.

This subject has a close analogy with the nonintegrable systems of ordinary differential equations that are close to integrable, like the systems encountered in galactic dynamics.

11 Chaos in Solar Activity

It is well known that the solar cycle repeats itself every 11 years, but not exactly in the same way (Fig. 20). This phenomenon can be explained by assuming that the solar cycle is due to a strange attractor ([76], [77]). This can be shown by reconstructing the sunspot data M_i for the successive days i in a 3-dimensional space (M_i, M_{i+d}, M_{i+rd}), where d is a fixed delay time. A projection of the reconstructed manifold on a plane is shown in Fig. 21a. Spiegel and Wolf [77] selected the delay interval d=1200 days, after some experimentation. In Fig. 21a

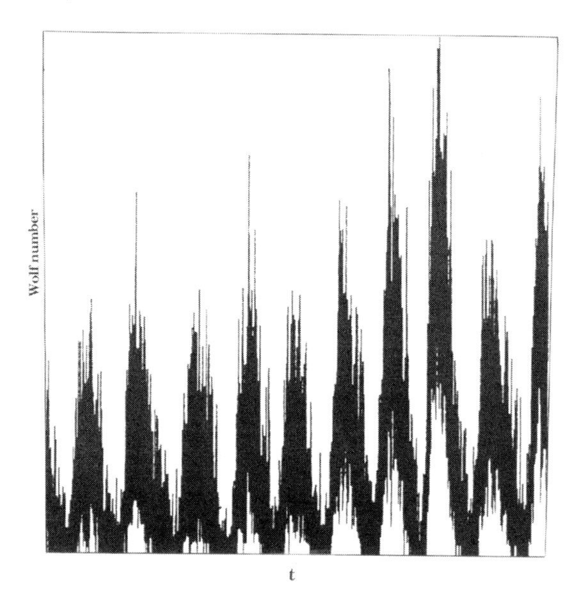

Fig. 20. The solar activity (Wolf numbers) follows the 11-year cycle, but different cycles are not the same.

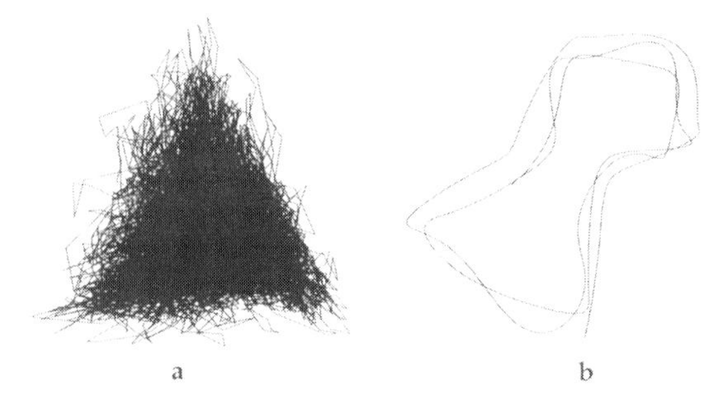

Fig. 21. Reconstruction of the sunspot data (Wolf numbers M_i, M_{i+d}, M_{i+2d} for all days i and a delay d=1200 days), (a) Projection of the lines joining successive points on a plane, (b) After filtering the Fourier components by a cutoff at 1 year the figure tends to a strange attractor.

we see a chaotic behaviour, but no detailed information can be drawn from this figure. However, after using a filter that eliminates all Fourier components with period one year or less, one finds the curve of Fig. 21b, which is reminiscent of the Lorentz strange attractor of Fig. 15.

This strange attractor explains the variations of the sunspot cycles, and even more important variations like the Maunder minimum. This refers to the fact that the solar activity was practically nonexistent during the reign of Louis IV, the king that chose the sun as his emblem. During that period the sun was not covered by sunspots. This phenomenon can now be explained by means of a strange attractor with a 5-dimensional fractal dimension [77].

Therefore, the theory of chaos has important applications in explaining the solar activity. However, the work done in this field is still rather limited.

Another application refers to the solar flares. This phenomenon is a manifestation of an important chaotic process that is called "self-organized criticality"([58],[4]). Namely the energy of the magnetic field is continuously accumulated in certain regions and it is released at irregular time intervals in the form of flares. This phenomenon is very similar to earthquakes, that also accumulate energy from the motion of the plates, and release it at irregular time intervals. This subject will be discussed during our workshop by Dr. Papazachos.

Therefore, the theory of chaos has useful applications in the most important solar phenomena. Similar phenomena appear in other types of stars, and in active galaxies, like active galactic nuclei and quasars.

However, a much more detailed study of these phenomena is required before we understand them completely. I only stress here the similarity of these problems with the corresponding phenomena of galactic dynamics from a mathematical point of view.

A conclusion from my review is that Order and Chaos play a very important role in Astronomy. Much work has been done already but the prospects of this new field are practically unlimited. I do not doubt that the study of Order and

Chaos will help to unify Physics and Astronomy from their most elementary constituents to the whole Universe.

References

1. Arnold, V.I.: Sov. Math. Dokl. **2**, 245 (1961)
2. Arnold, V.I.: Sov. Math. Dokl. **5**, 581 (1964)
3. Arnold, V.I. and Avez, A.: *Ergodic Problems of Classical Mechanics* (Benjamin, New York, 1968)
4. Bak, P.:*How Nature Works* (Springer Verlag, New York, 1996)
5. Barbanis, B.: Z. Astrophys. **56**, 56 (1962)
6. Belinskii, V.A. and Khalatnikov, I.M.: Sov. Phys. JETP **29**, 911 (1969)
7. Birkhoff, G.D.: *Dynamical Systems* (Amer.Math.Soc., Providence, R.I., 1927)
8. Bishop, J.L.: Astrophys. J. **305**, 14 (1986)
9. Buchler, J.R.: in *Chaos in Astrophysics*, ed. by Buchler, J.R., Perdang, J. and Spiegel, E.A. (Reidel, Dordrecht, 1985) p. 11
10. Buchler, J.R.: in *Chaotic Phenomena in Astrophysics*, ed. by Buchler,J.R and Eichhorn, H., N.Y. Acad.Sci. Annals **497**, 37 (1987)
11. Chirikov, B.V.: Phys.Rep. **52**, 263 (1979)
12. Christy, F.R. : Quart. R.Astr. Soc. **9**, 13 (1968)
13. Contopoulos, G.: Stockholm Ann. **20**, No 5 (1958)
14. Contopoulos, G.: Z. Astrophys. **49**, 273 (1960)
15. Contopoulos, G.: Astrophys. J. Suppl. **13**, 503 (1966)
16. Contopoulos, G.: in *Les Nouvelles Methodes de la Dynamique Stellaire*, ed. by Hénon M. and Nahon, F. Bull.Astron. (3) **2**, 223 (Besancon, CNRS, 1967)
17. Contopoulos, G.: Astron.Astrophys. **117**, 89 (1983)
18. Contopoulos, G.: Proc. Roy. Soc. London A**431**, 183 (1990)
19. Contopoulos, G.: Proc. Roy. Soc. London A**435**, 551 (1991)
20. Contopoulos, G.: *Order and Chaos in Dynamical Astronomy* (Springer Verlag, New York, 2002)
21. Contopoulos, G. and Moutsoulas, M.: Astron.J. **70**, 817 (1965)
22. Contopoulos, G. and Voglis, N.: Cel. Mech.Dyn.Astron. **64**, 1 (1996)
23. Contopoulos, G., Galgani, L. and Giorgilli, A.: Phys.Rev. A **18**, 1183 (1978)
24. Contopoulos, G., Kandrup, H.E. and Kaufmann, D.: Physica D **64**, 310 (1993)
25. Contopoulos, G., Papadaki, H. and Polymilis, C. : Cel.Mech.Dyn.Astron. **60**, 249 (1994)
26. Contopoulos, G. Grammaticos, B. and Ramani, A.: J. Phys. A **28**, 5313 (1995)
27. Contopoulos, G., Voglis, N. and Efthymiopoulos, C.: Cel.Mech.Dyn.Astr. **73**, 1 (1999)
28. Contopoulos, G., Efthymiopoulos, C. and Voglis, N.: Cel.Mech.Dyn.Astron. **78**, 243 (2000)
29. Contopoulos, G., Harsoula, M. and Voglis, N.:Cel.Mech.Dyn.Astron. **78**,197 (2000)
30. Contopoulos, G., Voglis, N. and Kalapotharakos, C.: Cel.Mech.Dyn.Astr. **83**, 191 (2002)
31. Cornish, N.J. and Levin, J.J.: Phys.Rev. D **55**, 7489 (1998)
32. Coullet, F. and Spiegel, E.A.: SIAM J.Appl.Math. **43**, 776 (1983)
33. Cushman, R. and Sniatycki, J.: Rep.Math.Phys. **36**,75 (1995)
34. Dermott, S.F. and Nicholson, P.D.: Highlights of Astronomy **8**, 259 (1989)
35. de Zeeuw, T.: Mon.Not.R.Astr.Soc. **216**, 273 (1985)

36. Duarte, P.: Ann.Inst.Henri Poincaré **11**, 359 (1994)
37. Dvorak, R., Freistetter, F., Funk, B. and Contopoulos G., in *Proceedings of the 3rd Austrian-Hungarian Workshop on Trojans and related topics*, ed. by Freistetter, F., Dvorak, R. and Erdi, B. (2003)
38. England, M.N., Hunter,J.H. Jr. and Contopoulos, G.: Astrophys. J. **540**, 154 (2000)
39. Fermi, E., Pasta, J. and Ulam, S.: Los Alamos Lab. Rep. LA 1949 (1955)
40. Ferraz-Mello, S.: in *Asteroids, Comets, Meteors 1993* ed. by Milani, A., di Martino, M. and Cellino, A. (Kluwer, Dordrecht, 1994) p. 175
41. Fridman, A.M., Khoruzhii, O.V., Polyachenko, V.L., Zasov, A.V., Silchenko, O.K., Moiseev, A.V., Burlak, A.N., Afanasiev, V.L., Dodonov, S.N. and Knapen, J.H.: Mon. Not R. Astr. Soc. **323**, 651 (2001)
42. Fridman, A.M., Khoruzhii, O.V., Lyakhovich, V.V., V.L., Silchenko, O.K.,Zasov, A.V., Afanasiev, V.L., Dodonov, S.N. and Boulesteix, J.: Astron. Astrophys. **771**, 538 (2001)
43. Gerhard, O.E. and Binney, J.: Mon.Not.R.Astr.Soc. **216**, 467 (1985)
44. Grousousakou, E. and Contopoulos, G.: in *The Dynamics of Small Bodies in the Solar System* ed. by Steves, B.A. and Roy, A.E. (Kluwer, Dordrecht, 1999) p. 535
45. Guckenheimer, J. and Holmes, P.: *Nonlinear Oscillations, Dynamical Systems and Bifurcations of Vector Fields* (Springer Verlag, N.York, 1983)
46. Hadjidemetriou, J.: Cel.Mech.Dyn.Astron. **56**, 563 (1993)
47. Hénon, M.: Comm.Math.Phys. **50**, 69 (1976)
48. Henrard, J.: in *Long Term Behaviour of Natural and Artificial N-Body Systems*, ed. by Roy A.E. (Kluwer, Dordrecht, 1983) p. 405
49. Hobill, D., Burd,A. and Coley, A. (eds): *Deterministic Chaos in General Relativity* (Plenum Press, New York, 1994)
50. Hunter, C. and de Zeeuw, P.T.: Astrophys. J. **389**, 79 (1992)
51. Jacobson, M.V.: Commun.Math.Phys. **81**, 39 (1981)
52. Kandrup, H.E., Siopis, C., Contopoulos, G. and Dvorak, R.: Chaos **9**, 381 (1999)
53. Kolmogorov, A.N.: Dokl. Akad. Nauk. SSSR **98**, 527 (1954)
54. Laskar, J. : Physica D **67**, 257 (1993)
55. Laskar, J. : Astron.Astrophys. **287**, L9 (1994)
56. Laskar, J., Joutel, F. and Robutel, P.: Nature **361**, 615 (1993)
57. Latifi, A., Musette, M. and Conte, R.: Phys. Lett. A **194**, 83 (1994)
58. Lichtenberg, A.J and Lieberman, M.A. : *Regular and Chaotic Dynamics*, 2nd Ed. (Springer Verlag, New York, 1992)
59. Lindblad, B. and Lindblad, P.O.: in *IAU Symposium 5*, 8 (1958)
60. Lorentz, E.N.: J.Atmos. Sci. **20**, 130 (1963)
61. Mayer, F. and Martinet, L.: Astron.Astrophys. **27**, 199 (1973)
62. Merritt, D. and Fridman, T.: Astrophys.J. **460**, 136 (1996)
63. Milani, A. and Nobili, A.M.: Nature **357**, 569 (1992)
64. Misner, C.M.: Phys. Rev. Lett. **22**, 1071 (1969)
65. Moons, M. and Morbidelli, A.: Icarus **114**, 33 (1995)
66. Morbidelli, A. and Giorgilli, A.: J. Stat. Phys. **78**, 1607 (1995)
67. Moser, J. : Nachr. Acad. Wiss. Gottingen II. Math. Phys. Kl. 1 (1962)
68. Moser, J. : Mem. Amer. Math. Soc. **81**, 1 (1968)
69. Nekhoroshev, N.N. : Russ. Math. Surv. **32**(6), 1 (1977)
70. Perdang, J.: in "Chaos in Astrophysics" ed. by Buchler,J.R, Perdang, J. and Spiegel, E.A., (Reidel, Dordrecht, 1985) p. 11
71. Perdang, J. and Blacher, S.: Astron.Astrophys. **112**, 35 (1982)

72. Perdang, J. and Blacher, S.: Astron.Astrophys. **136**, 263 (1984)
73. Pettini, M. and Vulpiani, A.: Phys.Lett.A **106**, 207 (1984)
74. Rosenbluth, M.N., Sagdeev, R.A., Taylor, J.B. and Zaslavsky, G.M.: Nucl. Fusion **6**, 217 (1966)
75. Schwarzschild, M.: Astrophys. J. **232**, 236 (1979)
76. Spiegel, E.A.: in Spiegel, E.A. and Zahn, J-P. (eds), "Lecture Notes in Physics", **71**, 3 Springer Verlag, New York (1977)
77. Spiegel, E.A. and Wolf, A.: in *Chaotic Phenomena in Astrophysics* ed. by Buchler,J.R and Eichhorn, H., N.Y. Acad.Sci. Annals **497**, 55, (1987)
78. Statler, T.S.: Astrophys. J. **321**, 113 (1987)
79. Sussman, G.J. and Wisdom, J.: Science **241**, 433 (1988)
80. Whittaker, E.T.: *A Treatise on the Analytical Dynamics of Particles and Rigid Bodies*, 4th Ed., (Cambridge Univ. Press, Cambridge, 1937)
81. Wisdom, J.: Astron. J. **87**,577 (1982)
82. Wisdom, J., Peale, S.J. and Mignard, F.: Icarus **58**,137 (1984)
83. Woltjer, J.: Mon.Not.R.Astr.Soc. **95**,260 (1935)
84. Zaslavsky, G.M. and Chirikov, B.V.: Sov.Phys.Uspekhi **14**, 549 (1972)

Critical Ergos Curves and Chaos at Corotation

Donald Lynden-Bell[1,2,3] and Jaideep M. Barot[1,4]

[1] Institute of Astronomy, The Observatories, Cambridge, CB3 0HA.
[2] Clare College, dlb@ast.cam.ac.uk.
[3] Physics Dept, The Queen's University, Belfast, BT7 1NN.
[4] Westminster School, jaideep.barot@westminster.org.uk.

Abstract. The theory of adiabatic invariants is developed to cover the gyration of a star about a nearly equipotential orbit in a galaxy with a strong bar. The guiding centres for such orbits follow curves of constant Ergos. The energy and the gyration adiabatic invariant give two constants of the motion. Critical Ergos curves have a pair of X-type gravitational neutral points which provide switches between trajectories that have the star circulating forward or backward relative to the corotating frame of the bar and those that liberate back to remain on one side of the galaxy's centre.

An attempt to discover the dynamical basis of the apparently random switching, that has been observed in computations of orbits with finite amplitudes of gyration, FAILS to find any such chaos at small gyration amplitudes, where Ergos curves give a good description of guiding centre motion.

1 Introduction

Eddington [15] looked for solutions of the collisionless Boltzmann equation that lacked axial symmetry but were steady in non-rotating axes. He introduced the idea of principal velocity surfaces to which the principal axes of the velocity ellipsoid were orthogonal. He then proved that, if the velocity ellipsoids were triaxial corresponding to three independent integrals quadratic in the velocities, the principal velocity surfaces had to be confocal quadrics. Also the potential had to be of a special form corresponding to Stackle's separable systems. Chandrasekhar [6] vehemently criticised Eddington's assumption that principal velocity surfaces existed but the analysis without that assumption produced no new solutions of interest. Meanwhile Clarke [7] derived the algebraic integrals corresponding to Eddington's system which were exploited to great effect by Kuzmin [20] and others. Lynden-Bell [21] gave a new analysis without assuming that the integrals were quadratic, but while he derived all six integrals and showed that the turning points lay on the confocal quadrics, he again found no new systems. It was de Zeeuw [12] & [13] who's careful categorisations of the orbital structure in these separable systems that revived interest in them. For an elementary derivation in axial symmetry, see Lynden-Bell [25]. Rather less is known about the analytic form of the integrals of the motion in systems that are only steady when viewed in rotating axes. Freeman [16] gave a fine analysis of the special systems in which the forces are linear functions of position which form a natural development of Riemann's homogeneous ellipsoids. Vandervoort [28] discovered a Stackle system in rotating axes which was further developed by Contopoulos and Vandervoort [9] but the density corresponding to this special potential is not positive everywhere. de Zeeuw and Merritt [11] developed a theory suitable for the cores of rotating systems, while Berman and Mark [3] analysed nearly circular orbits trapped in weakly non-linear spiral waves and gave analytical approximations to the slightly non-circular motion of the guiding centres. Binney & Tremaine [4] gave a general discussion of computed orbits.

For individual orbits a significant advance was made by L.S. Hall [17] who asked for invariant relations for one energy rather than an integral of the motion for all energies. This gave him a far wider class of potentials than those for which exact integrals exist Whittaker [29] Marshall & Wojciechowski [26].

Here we develop the adiabatically invariant gyration of a star about a guiding centre to give us an approximate integral independent of the energy, which is especially useful in the complicated region of barred galaxies close to corotation. The analysis of orbits into a gyration about a guiding centre's motion shows a bifurcation at the gravitational neutral points at the ends of the bar. Could it be that it is the phase of the gyration motion as the star enters the bifurcation region that determines which way the orbit goes? If so, we have a natural origin for the chaos that has been observed in orbits near corotation Contopoulos et al., [10]. In this paper, Sect. 2 is devoted to exact special cases in which the two dimensional motion in the galactic plane is integrable, Antonov & Shanshiev [2]. Section 3 develops the theory of guiding centre motion; when the gyration is of zero amplitude this motion is along Ergos curves which are not far from equipotential Lynden-Bell [22]. Section 4 considers the finite gyration about slightly modified Ergos curves while Sect. 5 analyses the motion near saddle points and the behaviour of the switch that directs the orbit into libration or circulation.

In the related problem in which a charged particle moves in an electromagnetic field some progress has been made in classifying the separable systems' scalar and vector potentials but even for axial symmetry Lynden-Bell [23] such classification is far from complete, although the charged Kerr Metric with $G = 0$ provides a very interesting special case Lynden-Bell [24].

2 Exact Special Cases

In rotating axes the equations of motion of a star in a galactic plane may be written

$$\ddot{\mathbf{R}} = \boldsymbol{\nabla}\Phi - 2\boldsymbol{\Omega} \times \dot{\mathbf{R}} \tag{1}$$

where $\Phi = \psi + \frac{1}{2}\Omega^2 R^2$ is the gravitational plus centrifugal potential measured in the sense that Φ is large in those regions to which particles are attracted by gravitational or by centrifugal forces. Two special cases give the clue as to what to do next

1. When $\boldsymbol{\nabla}\Phi = \mathbf{g}$ is a constant then we may orient the y axis upwards, i.e., along $-\mathbf{g}$. We then have a case analogues to the $\mathbf{E} \times \mathbf{B}/B^2$ drift of plasma

physics, writing $g_y = -g = d\Phi/dy$

$$\ddot{x} = 2\Omega\dot{y} , \tag{2}$$

$$\ddot{y} = -g - 2\Omega\dot{x} . \tag{3}$$

We integrate the first and insert it into the second to find with c a constant

$$\dot{x} = 2\Omega(y - c) , \tag{4}$$

$$\ddot{y} + 4\Omega^2(y - c + \tfrac{1}{4}g\Omega^{-2}) = 0 , \tag{5}$$

so y oscillates harmonically about the value $c - \frac{1}{4}g\Omega^{-2} = y_0$.
In plasma physics (2) and (3) are commonly combined by writing $\zeta = x + iy$. Then

$$\ddot{\zeta} + 2i\Omega\dot{\zeta} = -ig ,$$

so

$$\frac{d}{dt}\left(e^{2i\Omega t}\dot{\zeta}\right) = -ige^{2i\Omega t} ,$$

which may readily be integrated twice to give

$$\zeta = -\tfrac{1}{2}g\Omega^{-1}t + ae^{-2i\Omega t} + \zeta_0 , \tag{6}$$

where a and ζ_0 are complex integration constants. Thus the motion consists of a circular gyration of amplitude $|a|$ and frequency 2Ω about a guiding centre that moves with velocity $\mathbf{v}_d = -\frac{1}{2}g\Omega^{-1}\hat{\mathbf{x}}$ starting from point $\zeta = \zeta_0$. Notice that we may write this drift velocity in the form $\mathbf{g} \times (2\boldsymbol{\Omega})/4\Omega^2$ in analogy to $\mathbf{E}\times\mathbf{B}/B^2$. The fact that $\mathbf{g} = \nabla\Phi$ means that the guiding centre's motion is along an equipotential but that is only true when the equipotentials are of constant curvature as we show presently. When $\mathbf{g} = \nabla\Phi$ is not constant but Φ is a non-linear function of y, (2) and (4) are still valid and (3) may be replaced by

$$\ddot{y} = d\Phi/dy - 4\Omega^2(y - c) = \frac{d}{dy}\left[\Phi - 2\Omega^2y^2 + 4c\Omega^2y\right] . \tag{7}$$

In general we now have a non-linear oscillator with an energy–like integral

$$\tfrac{1}{2}\dot{y}^2 - \Phi(y) + 2\Omega^2y^2 - 4c\Omega^2y = I = \text{constant} , \tag{8}$$

but let us start with the simplest case in which g is expanded to first order about $y = y_0$ the trajectory of the guiding centre. Then

$$\Phi = \Phi_0 - g_0(y - y_0) + \tfrac{1}{2}\Phi_0''(y - y_0)^2 .$$

Equation (5) then takes the form

$$\ddot{y} + \kappa^2(y - y_0) = 0$$

where $\kappa^2 = 4\Omega^2 - \Phi_0''$ evidently

$$y - y_0 = \mathcal{I}m(ae^{i\kappa t}) \; .$$

and by (4)

$$\dot{x} = 2\Omega(y - y_0) + 2\Omega(y_0 - c) \; ,$$

$$x = \frac{2\Omega}{\kappa}\mathcal{R}e\left(ae^{i\kappa t}\right) + 2\Omega(y_0 - c)t + x_0 \; . \qquad (9)$$

If we write

$$\zeta = x + i\frac{2\Omega}{\kappa}(y - y_0) \; ,$$

then

$$\zeta = (2\Omega/\kappa)ae^{i\kappa t} + v_d t + \zeta_0 \; ,$$

where the first term represents an elliptical gyration at angular frequency κ and the remainder is the drift motion of the guiding centre at velocity

$$v_d = -g_0 2\Omega/\kappa^2$$

along $y = y_0$. In the non–linear case (8) y has some mean value which we may again call y_0 and $\langle\dot{x}\rangle = 2\Omega(y_0 - c)$ where $\langle\dot{x}\rangle$ indicates the temporal mean.
Evidently $\dot{x} - \langle\dot{x}\rangle = 2\Omega(y - y_0)$ so x executes an oscillation out of phase with $y - y_0$, making a closed curve which moves with the guiding centre. More generally again Φ_0'' might depend on x. Then (2) and (3) would be replaced by

$$\ddot{x} = 2\Omega\dot{y} + \frac{\partial\Phi_0''}{\partial x}\tfrac{1}{2}(y - y_0)^2 \; ,$$

$$\ddot{y} = \frac{\partial\Phi_0}{\partial y} - 2\Omega\dot{x} \; .$$

if we again write $\dot{x} = 2\Omega(y - c)$ then c must vary, albeit slowly, since $(y - y_0)^2$ is small. We again get

$$\ddot{y} + \kappa^2(y - y_0) = 0$$

but now y_0 may depend weakly on time.
We form the adiabatic invariant

$$J = \frac{1}{2\pi}\oint \dot{y}dy \; , \qquad (10)$$

which will depend on y_0 through the value of κ^2 . We then use the invariance of J and the exact conservation of the energy $\frac{1}{2}\dot{\mathbf{R}}^2 - \Phi = E_R$ to determine the small changes in c and y_0.

2. In the above, the equipotentials were lines of constant y (or almost so). More generally suppose that the equipotentials Φ = constant are curved with radius of curvature r at the position considered. If $\Phi = \Phi(r)$ we may solve (1) in cylindrical polar coordinates centred at the centre of curvature. With ϕ the azimuthal angle we have

$$r^{-1} d/dt(r^2 \dot{\phi}) = -2\Omega \dot{r}$$

so

$$r^2(\dot{\phi} + \Omega) = h = \text{constant} \; , \tag{11}$$

and

$$\ddot{r} - r\dot{\phi}^2 = \Phi'(r) + 2\Omega r \dot{\phi} \; ,$$

hence

$$\ddot{r} = d/dr \left[\Phi - \tfrac{1}{2} \left(hr^{-1} - \Omega r \right)^2 \right] \; , \tag{12}$$

so

$$\tfrac{1}{2}\dot{r}^2 - \left[\Phi - \tfrac{1}{2} \left(hr^{-1} - \Omega r \right)^2 \right] = E_R \; .$$

Notice that if the centre of curvature were the galaxy's centre then $r = R$. This energy is $\frac{1}{2}(\dot{r}^2 + r^2\dot{\phi}^2) - \Phi$, precisely the energy in the rotating axes. We have written motion in an axially symmetrical potential in this complicated way (in rotating axes) not merely to see the analogy with problem (1) but also because we now wish to consider problems lacking any global axial symmetry which are nevertheless steady when viewed from rotating axes. Our results are in a suitable form for applications to barred spiral galaxies and to galaxies with strong non-radial gravity fields.

We shall now generalise the above results to cases where the equipotentials are not of constant curvature but have their curvatures varying continuously along the orbits. Provided that the epicyclic motion is rapid compared with the drift motion of the guiding centre along, or almost along, the equipotential we expect an adiabatic invariant for the oscillation across the equipotentials of the form $J = (2\pi)^{-1} \oint \dot{y} dy$. This together with the exact conservation of the energy relative to the rotating axes, gives two integrals of the motion and allows the calculation of the orbits generally and of the drift trajectories of the guiding centres in particular. In the next section we shall concentrate on finding the drift trajectories of the guiding centres. Among all possible orbits will be some for which the adiabatic invariant governing the gyration about the guiding centre is and remains zero. Thus there will be a one parameter family of non-oscillating trajectories.

3 Drift Trajectories – Ergos Curves

Near corotation, drift velocities are slow and guiding centre accelerations negligible, so (1) can be rewritten in the galactostrophic approximation in which Coriolis force balances the gradient of the potential

$$2\boldsymbol{\Omega} \times \dot{\mathbf{R}} = \boldsymbol{\nabla} \boldsymbol{\Phi} \; , \tag{13}$$

so

$$\dot{\mathbf{R}} = \nabla\Phi \times \boldsymbol{\Omega}/(2\Omega^2) \; . \tag{14}$$

The drift velocity $\dot{\mathbf{R}}$ is thus along an equipotential (of constant Φ) - this is just the $\mathbf{E}\times\mathbf{B}/B^2$ drift of plasma physics. However, here this approximation is unsatisfactory since it actually conflicts with the exact conservation of energy whenever $|\nabla\Phi|$ varies along an equipotential. $\dot{\mathbf{R}}^2$ as given by (14) clearly varies along an equipotential so $E_R = \frac{1}{2}\dot{\mathbf{R}}^2 - \Phi$ clearly varies along an equipotential. But E_R is strictly conserved along any trajectory so the approximation that gave the drift trajectories along equipotentials conflicts with exact conservation of energy. We now give a treatment free of such conflict.

We start again but now suppose that the drift trajectories lie at small angles to the equipotentials rather than along them. Let $\hat{\mathbf{n}}(x,y)$ be the unit normal to the drift trajectories with the sense that $\hat{\mathbf{n}} \times \hat{\boldsymbol{\Omega}} \equiv \hat{\mathbf{t}}$, gives the direction of the drift velocity. $\hat{\boldsymbol{\Omega}}$ is the vector $\boldsymbol{\Omega}/\Omega$. Then $\hat{\mathbf{n}}$ lies at a small angle to $\nabla\Phi$ c.f. (14). Further we shall define the curvature vector of the drift trajectories $\mathbf{K}(x,y)$. $\mathbf{K}$ is perpendicular to the trajectory and points towards its centre of curvature from (x,y) . The magnitude of K is the reciprocal of the radius of curvature of the drift trajectory at (x,y). A star travelling along a drift trajectory at velocity v will have a transverse acceleration $\mathbf{K}v^2$ towards that centre of curvature. Taking components of (1) along the trajectory's normal $\hat{\mathbf{n}}$ we thus find

$$\mathbf{K}\cdot\hat{\mathbf{n}}v^2 = \hat{\mathbf{n}}\cdot\nabla\Phi - 2\Omega v \; . \tag{15}$$

To simplify this notation we put $\mathbf{K}\cdot\hat{\mathbf{n}} = K$ noting that $\mathbf{K}$ and $\hat{\mathbf{n}}$ are both perpendicular to the trajectory; K is either $|\mathbf{K}|$ or $-|\mathbf{K}|$ depending on the sense of the trajectory's curvature. Solving for v we find

$$v = \frac{1}{K}\left[\sqrt{\mathbf{K}\cdot\nabla\Phi + \Omega^2} - \Omega\right] = \frac{\hat{\mathbf{n}}\cdot\nabla\Phi}{\sqrt{\mathbf{K}\cdot\nabla\Phi + \Omega^2} + \Omega} \; . \tag{16}$$

Notice the close correspondence between this expression and (14) which gives $v = |\nabla\Phi|/(2\Omega)$. Evidently if the angle between $\hat{\mathbf{n}}$ and $\nabla\Phi$ is small enough to have its square neglected, and if $|\mathbf{K}\nabla\Phi| \ll \Omega^2$ the two expressions become equal. However (16) is exact while (14) was approximate. We now use exact energy conservation relative to the rotating axes

$$E_R = \tfrac{1}{2}v^2 - \Phi = \tfrac{1}{2}\frac{(\hat{\mathbf{n}}\cdot\nabla\Phi)^2}{\left[\sqrt{\mathbf{K}\cdot\nabla\Phi + \Omega^2} + \Omega\right]^2} - \Phi \equiv \mathcal{E}(x,y) \; . \tag{17}$$

The function $\mathcal{E}(x,y)$ is called the Ergos (Lynden-Bell, [22]). The definition is implicit since $\hat{\mathbf{n}}$ is the normal to the trajectories along which $\mathcal{E}$ is constant and whose curvatures are given by $\mathbf{K}(x,y)$. So far all is exact; the Ergos curves along which $\mathcal{E}$ is constant give the drift trajectories of the (zero amplitude) guiding centres. Now for any function F that is $-\Phi$ or any better approximation to the Ergos, writing suffixes to denote differentiation, and

$$s = (F_x^2 + F_y^2)^{\frac{1}{2}} \; , \tag{18}$$

$$\hat{\mathbf{n}} = -s^{-1}(F_x,\ F_y);\ \hat{\mathbf{n}} \times \hat{\boldsymbol{\Omega}} = \hat{\mathbf{t}}\ , \tag{19}$$

we have

$$\mathbf{K} = \left(\hat{\mathbf{t}} \cdot \boldsymbol{\nabla}\right) \hat{\mathbf{t}} = s^{-3} \left(F_x^2 F_{yy} - 2F_x F_y F_{xy} + F_y^2 F_{xx}\right) \hat{\mathbf{n}}\ . \tag{20}$$

Wherever $|\mathbf{K} \cdot \boldsymbol{\nabla}\Phi|$ is not as large as Ω^2 it is easy to find approximations to the Ergos. At zero order we use $-\Phi$ for F and calculate first approximates to $\hat{\mathbf{n}}$ and $\mathbf{K}$ from the above formulae. Substituting them into (17) we find a first approximation to the Ergos $\mathcal{E}_1(x, y)$. Using $\mathcal{E}_1$ for F in the above formulae we calculate 2nd approximations to $\hat{\mathbf{n}}$ and $\mathbf{K}$ and putting them into (17) we get $\mathcal{E}_2(x, y)$. Near corotation this will converge quite quickly to give the Ergos and the level surfaces of it give the Ergos curves along which the guiding centre trajectories lie. For related work on such systems see Antonov & Shanshiev [2]. Very close to gravitational neutral points where $\nabla\Phi = 0$ it is easiest to calculate the Ergos curves as trajectories with zero gyration directly. Figs. 1 and 2 give the equipotentials and the Ergos Curves.

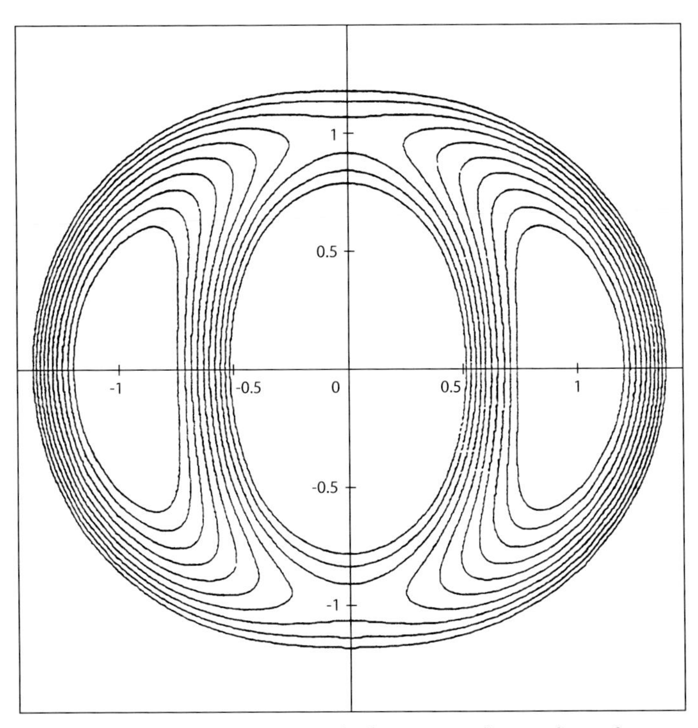

Fig. 1. Equipotentials of $\Phi = \psi + \frac{1}{2}\Omega^2 R^2$ where $R^2 = x^2 + y^2$; $\psi = \mathcal{G}M(b + s)^{-1}\left[1 - 0.02b^2(x^2 - y^2)s^{-4}\right]$; and $s^2 = R^2 + b^2$. The angular velocity of the bar, Ω, is chosen so that $\Omega^2 s = \mathcal{G}M/(b + s)^2$, that is $\Omega^2 = \mathcal{G}Mb^{-3}/(4 + 3\sqrt{2})$. In the diagram $\mathcal{G}M = 1$ and $b = 1$.

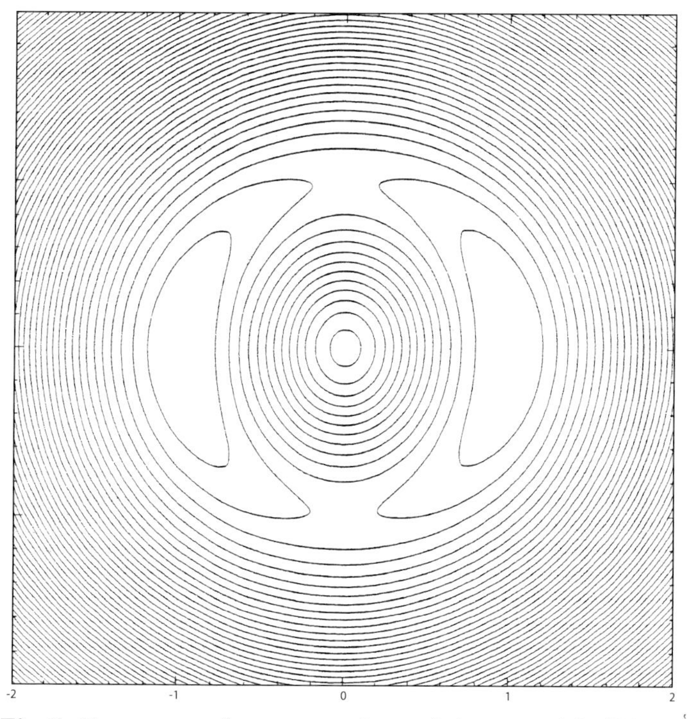

Fig. 2. Ergos curves for zero gyration and the potential of Fig. 1.

4 The Gyration Adiabatic Invariant

At any point $\mathbf{R}_0$ the equipotentials have some curvature $K_\circ$ and near there Φ can be approximated as being a function of r the distance to the centre of curvature. Thus, in a region near $\mathbf{R}_0$ the angular momentum about that centre of curvature will be approximately conserved. Taking the cross product of (1) by $\mathbf{r}$ the vectorial distance from that centre of curvature and using $\dot{R} = \dot{r}$ we find

$$d/dt(\mathbf{r} \times \dot{\mathbf{r}}) = -2\mathbf{r} \times (\boldsymbol{\Omega} \times \dot{\mathbf{r}}) + O(\epsilon^2) \ ,$$

so

$$\mathbf{r} \times (\dot{\mathbf{r}} + \boldsymbol{\Omega} \times \mathbf{r}) = \mathbf{h} + O(\epsilon^2) \ , \tag{21}$$

as in (11), but now h is only approximately constant locally. (21) will be just as true of the motion of the guiding centre as it is of the motion of the star that gyrates about that centre. Let the guiding centre be at r_0 and the star at $r_0 + \eta$ then working to first order in η writing $r = r_0 + \eta$ in (12)

$$\ddot{\eta} + \kappa^2 \eta = -(\bar{h} r_0^{-2} - \Omega)\delta h \ ,$$

where $\kappa^2 = -d^2\Phi/dr^2 - h^2 r_0^{-3} + \Omega^2 r_0$. We could have chosen to compare the motion of our star with that of a guiding centre with the same h and put

$\delta h = 0$ but as neither h nor δh are quite constant we have chosen not to do that. Evidently η vibrates harmonically about $\kappa^{-2}(\Omega - hr_0^{-2})\delta h = \eta_0$. We shall now assume that this η vibration is sufficiently fast that the corresponding action is adiabatically invariant. So the invariant is

$$J = \frac{1}{2\pi} \oint \dot{\eta} d\eta = \oint \sqrt{2[E_R + \Phi] - (hr^{-1} - \Omega r)^2} dr$$
$$= \Delta E_R / \kappa$$

where ΔE_R is the excess energy above that of the guiding centre. The integral is evaluated with E_R and h fixed and with $\Phi = \Phi(r, \phi, \mathbf{R}_0)$ only weakly dependent on ϕ and expanded about the point R_0 and ϕ . In the integration R_0 and ϕ are held fixed and only r varies. Henceforth any dependence on ϕ may be incorporated into the R_0 dependence. Thus we find

$$J = J(E_R, h, \mathbf{R}_0) \ .$$

ΔE_R and J are second order in the displacement from the guiding centre. We are now able to give a correction of this order to the guiding centre's motion which we earlier determined in the limit when J was zero. When J is non-zero the vibration about the guiding centre has extra energy $\Delta E_R = \kappa J$. While J is fixed; κ still varies from point to point. Thus the effective potential for the guiding centres motion is

$$\widetilde{\Phi} = \Phi - \kappa J \ ,$$

so that the energy of the total motion is

$$E_R = \tfrac{1}{2}\dot{\mathbf{R}}^2 - \Phi = \tfrac{1}{2}\dot{\mathbf{R}}_0^2 - \widetilde{\Phi} \ ,$$

where $\dot{\mathbf{R}}_0$ is the motion of the guiding centre. Thus the Ergos curves for guiding centres of given J should be calculated with $\widetilde{\Phi}$ replacing Φ . Fig. 3 shows a banana orbit in which one can see the gyration especially near the ends of the banana. Fig. 4 shows an orbit that starts librating in a banana close to the critical ergos curve but then switches to circulation outside corotation.

5 Is the Saddle-Point Switch Chaotic?

When in the 1960's Michel Hénon [19], [18] and George Contopoulos [8] discovered the fascination of the onset of chaos in stellar dynamical orbits I saw that a new branch of mathematics would develop, (Drazin [14]), but by that time I was more interested in the astrophysical problems cast up by astronomy than in the purely mathematical ones. I have never regretted that decision, though I have watched with admiration the developments pioneered by my more mathematical colleagues. One of the early examples of chaos was in Doug Allen's thesis [1] on the behaviour of coupled disk dynamos. The problem was suggested by Bullard and its solutions gave some indication of why the Earth's magnetic field suffers chaotic reversals. Later I learned of the pioneering studies of Mary

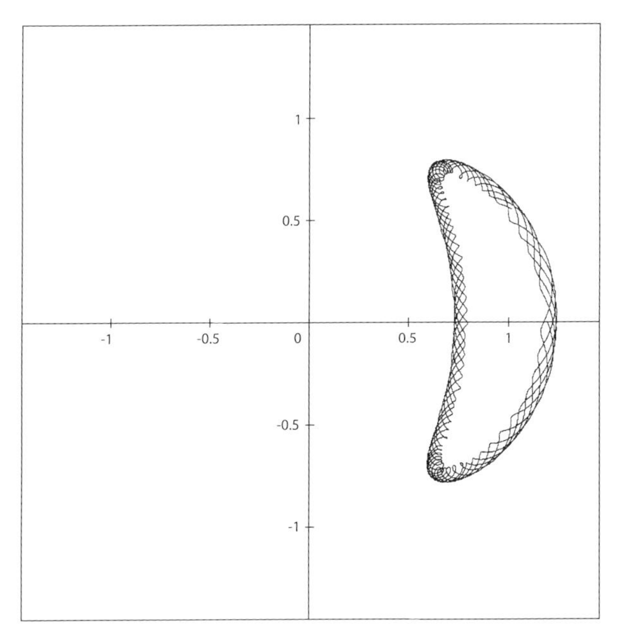

Fig. 3. A banana orbit showing the effects of gyration about the moving guiding centre especially near the ends of the banana.

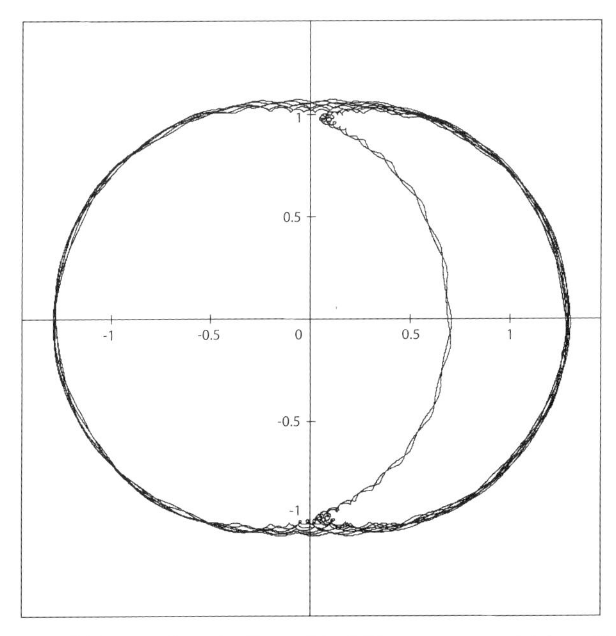

Fig. 4. A banana orbit very close to the critical Ergos curve, which switched from a librating orbit to one circulating outside corotation. Although integration was continued much longer it did not switch back. Either the orbit must hit a very small hole to cross back or the inaccuracy of the integrator allowed a small change in the guiding–centre motion so that it no longer came close to the critical switch.

Cartwright [5] who came upon the chaotic behaviour in the more mathematical context of differential equations. A deep mathematical study of the conditions that generate chaos in such systems was made by Colin Sparrow [27] under the aegis of Peter Swinnerton-Dyer and DLB gained a taste for their mathematical rigour by attending some of their lectures on chaos. What little he remembers involved orbits that continually came back into a critical region from which they could emerge in one of several different directions. It was the critical switching between these that led to chaos in the solutions. Over the years he has heard George Contopoulos talk about chaos many times, and chaos near corotation in orbits that get close to the ends of the bar in barred spiral galaxies has often been found. When George spoke on the subject at the Saltsjobaden Meeting in December 1995 [10], DLB had the belief that the saddle points in the gravitational potential provided just that critical switch with two very different outcomes that Sparrow needed. I thought the gyrations of the stellar orbit about its guiding centre would provide just that wobble between one side of the separatrix and the other needed to give chaos. The pressure of preparing this talk provided the stimulus needed to work this out properly. We start by analysing the switch at one of the gravitational points shown in Fig. 1. Centering our coordinates x, y on the upper saddle point Φ may be expanded for x, y small in the form

$$\Phi = \Phi_0 + \tfrac{1}{2}\alpha^2 x^2 + \tfrac{1}{2}\beta^2 y^2 \ ,$$

so the equations of motion (1) take the form

$$\ddot{x} = -\alpha^2 x + 2\Omega \dot{y} \ ,$$

$$\ddot{y} = \beta^2 y - 2\Omega \dot{x}$$

writing D for d/dt we see that

$$(D^4 + \omega_0^2 D^2 - \alpha^2\beta^2)x = 0 \ ,$$

where $\omega_0^2 = \alpha^2 + 4\Omega^2 - \beta^2$, and y obeys the same equation. In practice $\alpha^2 + 4\Omega^2 - \beta^2 > 0$. Writing $D = iw$ we see that, for w^2 there is one positive root

$$w^2 = w_1^2 = \tfrac{1}{2} w_0^2 \left(1 + \sqrt{1 + 4\alpha^2\beta^2\omega_0^{-4}}\right)$$

and a negative one with

$$-\omega^2 = \gamma^2 = 2\omega_0^{-2}\alpha^2\beta^2 / \left(1 + \sqrt{1 + 4\alpha^2\beta^2\omega_0^{-4}}\right) \ .$$

The dying solution, $\gamma > 0$, $e^{-\gamma t}$ corresponds to a contraction of the points along the separatrix line from upper left or bottom right while the growing $e^{\gamma t}$ solution corresponds to expansion along the separatrix line from the saddle both to lower left and upper right. Together these motions give $x = \gamma(Ae^{-\gamma t} + Be^{\gamma t})$, $2\Omega y = (\gamma^2 - \alpha^2)Ae^{-\gamma t} + (\gamma^2 + \alpha^2)Be^{\gamma t}$ where A & B are arbitrary constants with the separatrices given by $B = 0$ and $A = 0$ respectively. This flow is drawn in Fig. 6. At the saddle the flow switches to left or to right depending on the sign of B which decides on which side of the separatrix the guiding centre approaches. However, superposed on these motions is an elliptical gyration due to the real roots $\omega^2 = \omega_1^2$, these give $x = \omega_1 \mathcal{C} e^{i\omega_1 t}$ and $2\Omega y = (\omega_1^2 - \alpha^2) i\mathcal{C} e^{i\omega_1 t}$ where $\mathcal{C}$ is an arbitrary complex constant and the real x and real y are the real parts of the

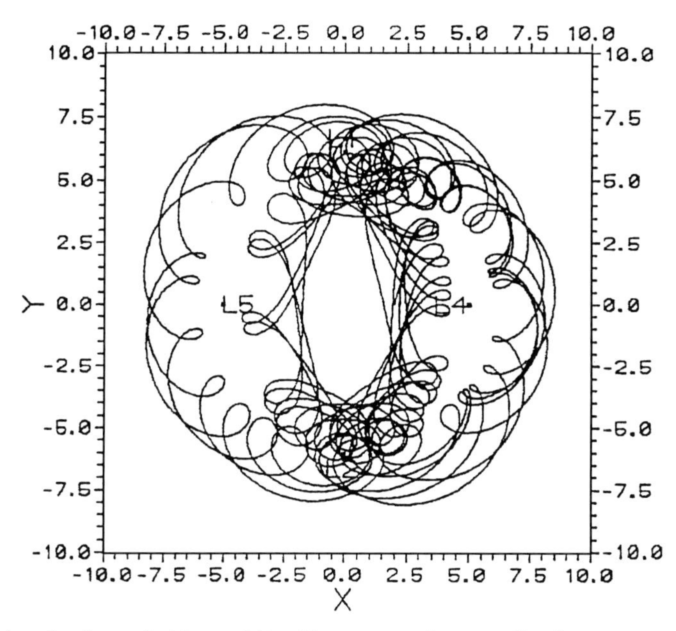

Fig. 5. A chaotically switching orbit of large gyration amplitude computed by Contopoulos et al 1996.

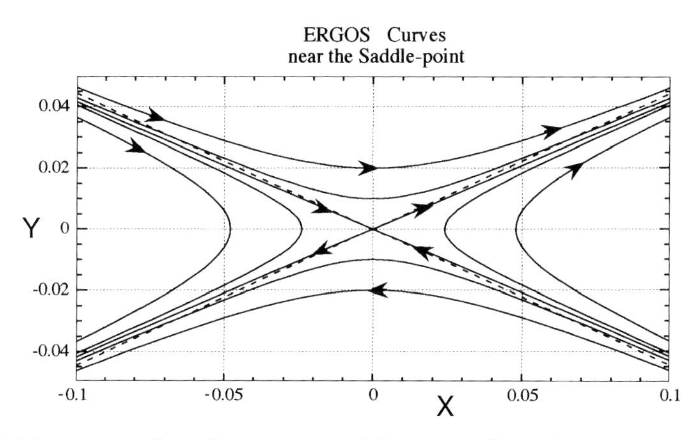

Fig. 6. Guiding centre flow close to the saddle point. Critical equipotentials (*shown dashed*) are close to the critical Ergos curves.

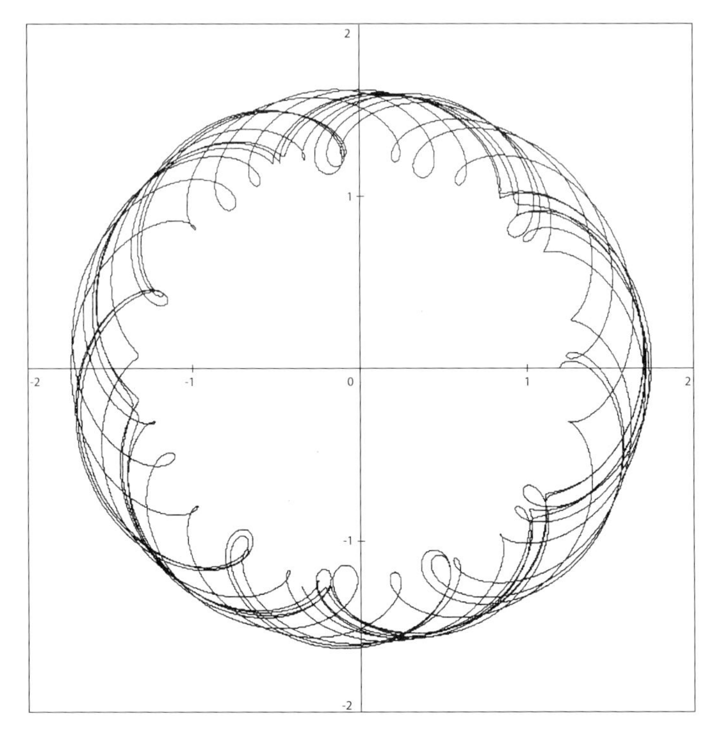

Fig. 7. An orbit outside corotation shows gyrations of slightly variable amplitude.

expressions given. Interestingly this elliptical[1] gyration continues unaffected by the saddle point.

Thus the switch to left or right is determined **not** by the position of the star but by the position of its guiding centre. DL-B's concept at the start of this investigation was that the switch would act on the star's position, so that the phase of the gyration as the star approached the saddle point would be crucial. Now this concept is seen to be false there is no random switching because the guiding centres follow the ergos curves. What then is the origin of the apparent switching of orbits seen in Figs. 4 & 5?

Three possibilities are

1. At finite gyration amplitudes there are resonances between the gyration and the motions of the guiding centres which lead to oscillations in the value of J and of the energy of the guiding centre's orbit which allow it to cross the separatrix before approaching the saddle–point switch.
2. The zero gyration motion of the guiding centre along an ergos curve is itself unstable.
3. For bars with significant non-radial forces the motions along the ergos curves are too rapid for the good conservation of the adiabatic invariant. Accurate separation between a guiding centre motion and a gyration is not possible

[1] For Fig. 6 the ellipse is almost round being only 1% flattened in y.

except close to the saddle–points. Orbits close to the separatrix will return on different sides of it on different approaches to the saddle–points.

Thus we have been unable to **isolate** the origin of the apparent randomness in the switching. However, we hope we have added some understanding of the orbits and of their integrals of motion.

Figure 7 shows an orbit that circulates "backward" outside corotation. Notice that even at the same azimuth there are small differences in the gyration amplitude; this may be due to inexact conservation of the adiabatic invariant.

References

1. D.W. Allen: Proc. Camb. Phil. Soc. 68, 671 (1962)
2. V. A. Antonov, & F.T. Shanshiev: 1993 Celestial Mechanics & Dynamical Astronomy, **59**, 209 (1994)
3. R.B. Berman & J.W. Mark: ApJ, **216**, 257 (1977)
4. J.J. Binney & S.D. Tremaine: *Galactic Dynamics* (Princeton University Press, Princeton 1987)
5. M. Cartwright & D.E. Littlewood: J.London Math.Soc., 20, 180, (1946)
6. S. Chandrasekhar: Principles of Stellar Dynamics p132, Dover NY 196 (1942)
7. G.L. Clarke: MNRAS **97**, 182 (1937)
8. G. Contopoulos: Astron. J., **76**, 147 (1971)
9. G. Contopoulos & P.O. Vandervoort: ApJ, **389**, 119 (1992)
10. G. Contopoulos, N. Voglis & C. Efthymiopoulos: *B*arred Galaxies & Circumnuclear Activity eds. Aa Sandqvist & P.O. Lindblad, Proceedings of Nobel Symposium 98, Springer, Berlin (1996)
11. P.T. de Zeeuw & D. Merritt: ApJ, **267**, 571 (1983)
12. P.T. de Zeeuw: MNRAS **216**, 273 (1985)
13. P.T. de Zeeuw: MNRAS **216**, 599 (1985)
14. P.G. Drazin: *Non Linear Systems* (Cambridge University Press, Cambridge 1992)
15. A.S. Eddington: MNRAS **76**, 37 (1915)
16. K.C. Freeman: MNRAS, **134**, 1 (1966)
17. L.S. Hall: Physica 8D, 90 (1983)
18. M. Hénon: Quart. App. Math, **27**, 291 (1969)
19. M. Hénon & C Heiles: Astron J, **69**, 73 (1964)
20. G.G. Kuzmin: Astron. Zh **33**, 27 (1956)
21. D. Lynden-Bell: MNRAS **124**, 95 (1962)
22. D. Lynden-Bell: *Barred Galaxies & Circumnuclear Activity* eds. Aa Sandqvist & P.O. Lindblad, Proceedings of Nobel Symposium 98, (Springer, Berlin, 1996)
23. D. Lynden-Bell: MNRAS, **312**, 301 (1999)
24. D. Lynden-Bell: *New Developments in Astrophysical Fluid Dynamics*, (eds. Dalsgaard & Thompson), C.U.P., Cambridge, in press, (2002), (astro-ph/0207064)
25. D. Lynden-Bell: MNRAS, accepted, (2003), (astro-ph/0210417)
26. I. Marshall & S. Wojciechowski: J.Math.Phys., **29**, 1338 (1988)
27. C. Sparrow: *The Lorentz Equations Bifurcations Chaos and Strange Attractions*, (Springer Verlag, New York App Math, 41, 1982)
28. P.O. Vandervoort: ApJ, **232**, 91 (1979)
29. E.T. Whittaker: *Analytical Dynamics*, (Cambridge University Press, Cambridge, 1959)

Stellar Dynamics and Molecular Dynamics: Possible Analogies

Andrea Carati and Luigi Galgani

Università di Milano, Dipartimento di Matematica, Via Saldini 50,
I–20133 Milano, Italy

Abstract. In stellar dynamics one is accostumed to deal with the Lynden–Bell distribution, which presents two peculiar characteristics: a) it resembles the quantum Fermi–Dirac distribution, and b) describes a state of metaequilibrium, that is expected to evolve on much longer time scales to a standard Maxwell–Boltzmann equilibrium. Here it is illustrated how an analogous situation seems to occur in molecular dynamics, described within the context of classical mechanics. The problem concerns the contribution of the internal degrees of freedom, typically the vibrations, to the specific heat; correspondingly, the metaequilibrium state leads to a Bose–Einstein–like rather than to a Fermi–Dirac–like distribution. We also point out that in molecular dynamics the "nonclassical" features seem to be related to the fact that the evolution of the energy is dominated by the presence of rare but conspicuous jumps, as in processes of Lévy type; this too has some analogies with stellar dynamics.

1 Introduction

About thirty years ago it occurred to one of the present authors, in collaboration with A. Scotti and C. Cercignani [1][2], to observe a quantum–like feature in the problem of the specific heats studied in the context of classical mechanics. The model considered was that of Fermi–Pasta–Ulam [3], describing a one–dimensional crystal with nonlinear interactions between adjacent atoms. The observation was that, if the specific energy was small enough and the energy was given initially to the lowest–frequency mode, the distribution of energy among the modes, estimated by numerical solutions of the equations of motion, turned out to have a Planck–like form. Moreover, to a great astonishment, even the action entering such a distribution turned out to be of the order of magnitude of Planck's constant. It took some months to become convinced that the latter fact was neither a mistake nor an accident, as we briefly recall now. The relevant point is that, in the model, realistic molecular parameters had been introduced; for example, for a model of crystal Argon, one introduces the mass m of Argon and a realistic Lennard–Jones interatomic potential $V(r) = 4\epsilon[(\sigma/r)^{12} - (\sigma/r)^6]$, with certain parameters ϵ and σ given in the literature. Now, one immediately checks that the action A naturally built up from the given parameters m, ϵ, σ is just $A = \sqrt{m\epsilon}\,\sigma$; on the other hand it turns out that for realistic parameters there exists the relation $A = 2Z\hbar$ where $\hbar$ is the (rationalized) Planck's constant and Z the atomic number. This remark explains how in classical models of molecular dynamics Planck's constant is introduced, so to say by hands, through the values of the molecular parameters.

In such a way an adventure was started, which consisted in trying to prove that, at least in the problem of the specific heats (see however [4], where the electrodynamics of point particles is also discussed), classical mechanics might not be inconsistent with quantum mechanics, or at least in understanding how something like this might make sense. Having explained in which way Planck's constant had been introduced, somehow by hands, through the molecular parameters, the main problem was then to understand how the dynamics itself, described by Newton's equations with given potentials, might make the job. This is the reason why intense contacts were started with scientists working in the field of dynamical systems, from mathematics to celestial mechanics. Perhaps the first among them was G. Contopoulos, with whom an everlasting friendship and scientific collaboration was initiated. The adventure had since then several phases, with several incursions into purely mathematical aspects of the theory of dynamical systems; a long collaboration with G. Benettin and A. Giorgilli took place, and finally the first of the present authors joined the party.

The present phase seems to be characterized by the realization that there should exist two (or several) well distinct relaxation times in the problem, as is now familiar in the physics of glasses or spin glasses. The point we want to stress here is that the existence of two relaxation times is a familiar feature in stellar dynamics too. Indeed, one there refers to a first "violent" relaxation to a metaequilibrium state of the type of Lynden–Bell [5] (during which the collisions can be neglected), that should then be followed by an extremely slow relaxation to a standard equilibrium state governed by the collisions. Moreover, such a phenomenon of the existence of a metaequilibrium state turns out to occur, in stellar dynamics, just in conjunction with the appearence of a quantum–like feature, namely the Fermi–like distribution of Lynden–Bell. These two features, namely the existence of a metaequilibrium state and its actual quantum–like aspect, constitute the analogies of stellar dynamics with the problem of the specific heats in classical mechanics which motivated the present talk, mainly addressed to people working in the field of stellar dynamics.

2 Planck's Law, Its Interpretation by Einstein, and the Points of View of Boltzmann and of Nernst

Planck's law is concerned with the mean energy U of a system of N harmonic oscillators of angular frequency ω at absolute temperature T. In terms of inverse temperature $\beta = 1/k_B T$ and of the quantum of energy $\hbar\omega$, where k_B is the Boltzmann's constant, it asserts that the mean energy U has the form

$$U = N \frac{\hbar\omega}{e^{\beta\hbar\omega} - 1} \ . \tag{1}$$

The relevant feature of this law is that in the limit of high temperatures or low frequencies, i.e. for $\beta\hbar\omega \ll 1$, it leads to the "classical" value $U = Nk_B T$, independent of frequency (this realizes the so–called equipartition of energy), while it gives degeneration, i.e. the vanishing of U (in a frequency–dependent way), for $\beta\hbar\omega \gg 1$.

Everyone knows how Planck's law is usually deduced, by the standard equilibrium argument using the Maxwell–Boltzmann principle, if energy is assumed to be quantized, i.e. if one admits that the allowed values for the energy E are $E_n = n\hbar\omega$, $n = 1, 2, \cdots$. This standard argument in fact constitutes the second deduction given by Planck himself; a variant of it, which takes into account the zero–point energy $N\hbar\omega/2$, comes about if the energy levels are assumed to be given by $E_n = (n + 1/2)\hbar\omega$.

So, Planck's law turns out to be a consequence of quantization. Conversely, it was shown by Poincaré [6] (in one of his last papers, that he wrote under the stimulus of the discussions at the first Solvay Conference [7]) that quantization is necessary if one has to recover Planck's law at all. Actually, an accurate analysis of the paper of Poincaré shows that the situation is not so clear, and for example Einstein never was convinced of this necessity of quantization. This is witnessed by some remarks he made in his scientific autobiography [8]. Indeed, after having recalled how he himself had "*showed in a definitive and direct way that it is necessary to attribute a certain immediate concreteness to Planck's quanta and that, under the energetic aspect, radiation possesses a sort of molecular structure*", after a few lines he adds: "*This interpretation, that almost all contemporary physicists consider as essentially definitive, to me appears instead as a simple provisional way out*".

What did Einstein have in mind in saying these words? In our opinion the answer is found in his contribution to the Solvay Conference, where he showed how Planck's law can be obtained by arguments which make no reference to quantization at all. The key point is a physical interpretation of the procedure that had been followed by Planck in the first deduction of his law, on October 19, 1900. Let us recall that in that paper Planck had obtained his formula as a solution of the ordinary differential equation (we are using here a contamination of the notations of Planck and of Einstein)

$$\frac{\mathrm{d}U}{\mathrm{d}\beta} = -\big(\hbar\omega\, U + \frac{U^2}{N}\big)\,, \tag{2}$$

to which he had arrived, with no real physical interpretation, by a purely formal interpolation between two limit equations, well adapted to the cases of high frequencies and low frequencies respectively. What Einstein did, was to split such an equation into a system of two equations, namely

$$\frac{\mathrm{d}U}{\mathrm{d}\beta} = -\sigma_E^2 \tag{3a}$$

$$\sigma_E^2 = \hbar\omega\, U + U^2/N\,, \tag{3b}$$

where there appears a further quantity σ_E^2 having a well definite physical meaning, namely the variance of energy. Indeed the former equation (26a), relating specific heat to variance of energy, had been discovered in the year 1903 by Einstein himself in one of his first papers, as an identity in the canonical ensemble, and was conceived by him as a kind of a general thermodynamic relation that should have some more general validity. In his mind, the second relation (77) should instead have a dynamical character, and might in principle be deducible from a microscopic dynamics. In his very words [7]: these two relations "*exhaust the thermodynamic content of Planck's*" formula; and "*a mechanics compatible with the energy fluctuation* $\sigma_E^2 = \hbar\omega U + U^2/N$ *must then necessarily lead to Planck's*" formula. So the main idea is that the energy exchanged with a reservoir, i.e. the specific heat $\frac{\mathrm{d}U}{\mathrm{d}\beta}$, should be related to the energy fluctuations. In turn, the functional dependence of the energy fluctuations on the mean energy should fix the functional form of the mean energy in terms of temperature and of frequency. Clearly, no reference to quantization is made here.

Another key point enters now, which goes back to Boltzmann; we refer to the role of nonequilibrium. Indeed, being confronted with the phenomenological lack of equipartition of energy in crystals and in polyatomic molecules, Boltzmann conceived the idea that in such cases one was actually dealing with situations in which equilibium had not yet been reached. In his very words: [9] "*The constituents of the molecule are by no means connected together as absolutely undeformable bodies, but rather this connection is so intimate that during the time of observation these constituents do not move noticeably with respect to each other, and later on their thermal equilibrium with the progressive motion is established so slowly that this process is not accessible to observation*". In modern terms, the situations of nonequipartition of energy should be understood as analogous to the metaequilibrium situations occurring in glasses and spin glasses.

Let us recall that the idea of Boltzmann, according to which the situations of nonequipartition would correspond to states of metaequilibrium, was pursued for several years by Jeans [10], with the explicit aim of avoiding a recourse to quantization. But the work of Poincaré on the necessity of quantization made so strong an impression on him that he found himself forced to make a public retractation [11] and to abandon any further attempt in that direction. We will recall below how the problem was reopened much later, in the year 1954, by the work of Fermi, Pasta and Ulam. We would also like to mention that the idea of Boltzmann that there should exist a "time–dependent specific heat" is today accepted as a trivial fact, with no mention to Boltzmann at all (see for example [12]).

We add now here some further comments. The first remark is that there seems to be a strong relation between the point of view of Boltzmann and that of Einstein: the key point is the role of the dynamics in the problem of the specific heats. According to Boltzmann, what is relevant for the specific heat of a body is not the energy it possesses, but rather the energy it can exchange through the dynamical interaction with a heat reservoir within a given observation time. In turn, the latter energy, the exchangeable one, is related to the dynamical fluctuations of the energy of the body. This is in fact essentially the statement of the well known Fluctuation Dissipation Relation (see for example [13], of which the relation (26a) of Einstein seems to be a precursor. In fact, the Fluctuation Dissipation Relation has a form very similar to (26a), the main difference being that it involves quantities having a dynamical character. Consider a system at inverse temperature β, and denote by $E(t)$ its energy at time t, and by $U(t)$ the mean energy exchanged up to time t with a reservoir at inverse temperature $\beta + \mathrm{d}\beta$. Then, the Fluctuation Dissipation Relation reads

$$\frac{\mathrm{d}U}{\mathrm{d}\beta} = -\frac{1}{2}\langle\big(E(t) - E(0)\big)^2\rangle \ , \tag{4}$$

the mean $\langle\cdot\rangle$ being taken with respect to initial data with a Maxwell–Boltzmann distribution at inverse temperature β. From this, the static relation (26a) of Einstein is recovered at times so large that the autocorrelation of the energy vanishes, so that $E(t)$ and $E(0)$ become independent and thus the quantity $1/2\langle\big(E(t) - E(0)\big)^2\rangle$ reduces to the static canonical variance σ_E^2.

A final comment concerns the role of the Maxwell–Boltzmann distribution as a statistical measure for the initial data, irrespective of the dynamics. The fact is that, for a system of independent harmonic oscillators distributed according to Maxwell–Boltzmann, one has for the mean energy U the value Nk_BT, i.e. formally equipartition of energy. But such an equipartition, as Boltzmann would say, concerns the mechanical energy possessed by the system just in virtue of the choice of the initial data, and has a priori nothing to do with the thermodynamic energy, which should be defined as the exchangeable one within the observation time. The latter, i.e. the thermodynamic or exchangeable energy, is instead measured by the dynamical fluctuations of energy. The fact that the initial distribution of energy presents a nonvanishing variance σ_E^2 is of no relevance for the specific heat, which depends on the exchangeable energy, i.e. on the dynamics.

This remark explains how one can have a situation in which there is both equipartition of energy in relation to the initial data, and Planck's law in relation to the exchangeable energy, as was first conceived by Nernst [14] in an extremely deep, almost unknown, work (see also [15]). In particular, Nernst also introduced a deep conception of the energy which, in the sense of Boltzmann, does not contribute to the specific heat: on the one hand it should be characterized as being, from the dynamical point of view, of ordered type (*geordnete*); on the other hand it would constitute a classical analog of the quantum zero-point energy. It is worth recalling the argument by Nernst. He assumes that the quantum of energy $\hbar\omega$ plays the dynamical role of a stochasticity threshold for the harmonic oscillators; the motions would be of ordered type below threshold and of disordered type (*ungeordnete*) above it. Furthermore, he assumes that the oscillators are distributed according to Maxwell–Boltzmann, so that one has equipartition for their mechanical energy. Now, one immediately computes the mean disordered energy (per oscillator) E_1, namely the mean energy conditioned by $E > \hbar\omega$ and the mean ordered energy (per oscillator) E_0, namely the mean energy conditioned by $E < \hbar\omega$, and one finds

$$E_1 = k_BT + \hbar\omega \quad , \quad E_0 = k_BT - \frac{\hbar\omega}{e^{\hbar\omega/k_BT} - 1} \quad . \tag{5}$$

Similarly one finds that the fraction n_1 of oscillators above threshold is $n_1 = \exp(-\hbar\omega/k_B T)$. Then the exchangeable energy U can be assumed to be defined by $U = N n_1 (E_1 - E_0)$, which coincides with Planck's law.

3 The Fermi–Pasta–Ulam Problem

The problem of a dynamical foundation for the principle of equipartition of energy in classical mechanics was reopened in the year 1954 by Fermi, Pasta and Ulam [3]. The interest of Fermi for this problem was indeed a rather old one, since it goes back to his work [16] of the year 1923 (which is sometimes misunderstood in the literature; see however [17]), where he improved the theorem of Poincaré on the integrals of motion of a Hamiltonian system. So Fermi came back to the problem when for the first time he had the facilities of a computer for the numerical integration of the equations of motion of a rather large system of particles.

As mentioned above, Fermi, Pasta and Ulam considered a system of N points (atoms) on a line, with a nonlinear interaction between adjacent atoms and certain boundary conditions; typically, the positions of the extreme atoms were kept fixed and the number N of moving atoms was 64. The interaction potential energy had the form $V(r) = r^2/2 + \alpha r^3/3 + \beta r^4/4$, with given constants α, β. For $\alpha = \beta = 0$ one has a linear system which, by a familiar argument, is equivalent to a system of N uncoupled harmonic oscillators (normal modes) having certain frequencies. The problem is then how many normal modes take part in the energy sharing, which should occur in virtue of the nonlinear interaction. The authors considered initial conditions in which the energy was given just to the lowest normal mode (i.e. to the mode of lowest frequency), and the aim was to observe, by numerical solutions of the equations of motion, the rate of the flow of energy towards the modes of higher frequency, which was expected to occur in order to establish the equipartition of energy among all the modes. They found the unexpected result that, up to the times considered, the energy appeared to be distributed just among a packet of normal modes of low frequency without flowing to the high frequency modes. They also gave a figure reporting the mean (in time average) energy versus frequency, which exhibited an exponential decay.

After this original work two more works had, in our opinion, a particularly relevant role, namely that of Izrailev and Chirikov [18] of the year 1966, and that of Bocchieri, Scotti and Loinger [19] of the year 1970. F.M. Izrailev and B. Chirikov understood that there existed the problem of an energy threshold. By analogy with the situations occurring in perturbation theory, in connection with the existence of ordered motions in the sense of Kolmogorov, Arnold and Moser, they conjectured that equipartition would be obtained if the initial energy E was larger than a certain threshold energy E^c. Thus the result of Fermi, Pasta and Ulam was explained as being due to the fact that only small energies, with $E < E^c$, had been considered. The crucial point is then to understand how does the critical energy E^c depend on the number N of degrees of freedom, because in situations of physical interest N should be of the order of the Avogadro number. So one has to look at the specific critical energy $\epsilon^c = E^c/N$ (we are using here the symbol ϵ for the specific energy, with no relation to the parameter ϵ of the Lennard–Jones potential mentioned in the Introduction). The authors had clearly in mind the idea that one might prove that $\epsilon^c \to 0$ as $N \to \infty$ (at least for initial excitations of the high frequency modes). Indeed, if this were true, then in any physically meaningful system one would always have equipartition of energy. Then everybody would be happy, because this would prove that classical mechanics predicts a wrong result, as everyone has learned at school.

P. Bocchieri, A. Scotti and A. Loinger (working with a Lennard–Jones interaction potential) gave a strong indication in the sense that, on the contrary, the specific energy threshold ϵ^c should tend to a finite nonvanishing value in the limit $N \to \infty$. Shortly after such a work, in the paper [1] it was shown that, just in situations in which according to Bocchieri, Scotti and Loinger one does not have equipartition, the distribution of energy among the normal modes has a Planck-like form, and even with an action of the order of magnitude of Planck's constant. Thus the adventure was started of looking for a deeper understanding of the relations between classical and quantum mechanics.

After such "old" works, many other works followed (see for example [20] and [21]), mostly with the intent of establishing whether $\epsilon^c \to 0$ as $N \to \infty$, or not. The theoretical framework also changed a lot. Indeed, initially reference was made to KAM theory, while later the point of view of N.N. Nekhoroshev entered the game [22]. The attention was thus shifted towards the idea that one might always have equipartition as $t \to \infty$, but with the possibility that the relaxation time might increase exponentially fast as the specific energy decreases.

From this point of view, a recent relevant result was given, in our opinion, in the paper [23]. Here it is confirmed that the results depend on the specific energy ϵ, and actually in the following way. There exists a specific energy threshold ϵ^c such that, if the energy is initially given to a small packet of modes of very low frequency, the relaxation time to equilibrium (i.e. to equipartition) increases as a power of $1/\epsilon$ if $\epsilon > \epsilon^c$. Instead, if $\epsilon < \epsilon^c$, one has a first rapid (violent) relaxation to a natural packet extending up to a maximal frequency $\overline{\omega}(\epsilon) \simeq \epsilon^{1/4}$, while only on much larger time scales one would get equipartition. Moreover one starts now having analytical results confirming such a scenario, exactly in terms of the specific energy ϵ in the limit $N \to \infty$; such analytical results also give a precise analytical form for the spectrum corresponding to the natural packet mentioned above.

In conclusion, it seems that below a certain critical specific energy ϵ^c one deals with a metaequilibrium state, which is only later followed, on much longer time scales, by a real Maxwell–Boltzmann equilibrium. Such a scenario has many similarities with the one that is familiar for glasses and spin glasses, as was first pointed out in the paper [24].

4 The Landau–Teller Model of Molecular Collisions

We finally give a short review on the contribution of the internal degrees of freedom to the specific heat of polyatomic molecules.

It was recalled above that we owe to Boltzmann the fundamental idea that the contribution of the internal degrees of freedom to the specific heat would manifest itself only on long time scales, in contrast to the degrees of freedom of the center of mass (also called external degrees of freedom), which are known to relax to equilibrium after a few collisions. We also recalled how Jeans gave support to the point of view of Boltzmann, but later made a public retractation after the work of Poincaré on the necessity of quantization. It is also of interest to mention that the problem of the existence of long relaxation times was discussed from the experimental point of view at the first Solvay Conference, by Nernst and others. The opinion was there expressed that there was no evidence at all for such longer relaxation times.

Actually such longer relaxation times were observed experimentally in the year 1925, in studies of the dispersion (and anomalous diffusion) of sound in diatomic gases [25], and it turned out that the times were even 6 orders of magnitude larger than the mean collision time. In looking for an explanation of such longer relaxation times, in a period in which the discussions of Boltzmann and Jeans had been completely forgotten, a most relevant contribution was given by Landau and Teller [26]. They considered an extremely simplified model capturing the essence of the problem, namely the exchange of energy in a collision between an atom and a linear spring, with an exponential interatomic potential. For the energy exchange δE they actually found an expression of a form already indicated by Jeans, namely $\delta E \simeq \exp(-\omega a/v)$, where ω is the frequency of the spring, a the range of the potential and v the velocity of the impinging atom. A relaxation time was then extracted from such a formula of the exchanged energy, and the common opinion was formed that the theory fits rather well the experimental data [27]. Serious doubts were however raised concerning the goodness of the agreement, as is witnessed for example by the following quotation [28]: "*It is impossible to determine whether the choice of the potential parameters is physically significant, because all errors in the theory are compensated by adjustable potential parameters*".

In any case, the common opinion is that the approach to equilibrium should be controlled by a single relaxation time, say τ_L. Correspondingly, the law of temporal approach should be exponential [29], of the type $\exp(-t/\tau_L)$, as is familiar in the Onsager theory. We are of the opinion that the situation is here, however much more delicate. Indeed the Onsager theory is well suited for the approach to equilibrium of systems presenting a completely chaotic dynamics, while we are here confronted with the opposite situation, namely with systems that are nearly integrable, for which no one was able up to now to produce a statistical mechanics compatible with the dynamics (see however the papers [30], where indications are given that the statistics might be given according to the ideas of Einstein recalled above).

Further results were recently given in the paper [31]. There it is shown that the statistics induced by the dynamics in the Landau–Teller model of molecular collisions is a rather complex one, because over very long times the processes of the energy exchanges have many similarities to the well known Lévy processes, which are dominated by the presence of rare, but highly conspicuous, jumps. For an analogous situation in stellar dynamics see [32]. We are still working on the Landau–Teller model, and we hope to be able to show that here too one meets with two relaxation times, the shorter one leading to a state of metaequilibrium characterized by a "nonclassical" statistics, and the second one leading to the final Mazwell–Boltzmann equilibrium. This fact might be of interest for the phenomenology of sound dispersion.

5 Conclusions

We have recalled above how strong was the impact of the theorem of Poincaré concerning the necessity of quantization if the phenomenological law of Planck is to be recovered. In fact, after that work essentially all attempts at providing a classical understanding of Planck's law along the lines indicated by Boltzmann were abandoned. Peculiar exceptions were Einstein, who never proved convinced, and Nernst, who introduced the dynamical interpretation for the zero–point energy illustrated above.

Now, if one looks at the proof of Poincaré's theorem, it seems evident that the fundamental hypothesisis there made is that one should be dealing with a real equilibrium (i.e., almost by definition, with the Maxwell–Boltzmann distribution). And in fact on several occasions Poincaré had stressed that, if one recedes from equilibrium, one cannot have any thermodynamics at all: everything would become fuzzy. It seems difficult to disagree. Let us quote Poincaré himself [33]: "*Jeans tried to reconcile things, by supposing that what we observe is not a statistical equilibrium, but a kind of provisional equilibrium. It is difficult to take this point of view; his theory, being unable to foresee anything, is not contradicted by experience, but leaves without explanation all known laws*".

But in fact it seems to us that there is a possibility of getting a thermodynamics without a full equilibrium. The possibility is that one would be actually dealing, not just with a nonequilibrium, but rather with a situation of metaequilibrium, as for example in the case of glasses. In such a case one can have situations which apparently are indistinguishable from situations of a true equilibrium, so that the critique of Poincaré could be overcome. This is exactly what we are proposing, and is the reason why in this review we insisted in a particular strong way on the relevance of being able to findi at least two time scales in the problem of diatomic molecules (the existence ot two time scales in the Fermi–Pasta–Ulam problem being, by now, almost granted): only on an extremely long time scale would one get a true equilibrium, while on a first, short, time scale one would reach a metastable state (as in stellar dynamics). In turn, the termodynamics of the metastable state could not be described by the usual procedure of the type of Onsager, beacuse the latter makes reference to a chaotic dynamics, as stressed by Bowen, Ruelle, Sinai and by Gallavotti. In the state of metastability one is instead dealing, from a dynamical point of view, with the other extreme situation, namely with nearly integrable systems, and the suitable thermodynamics could perhaps be recovered along the lines suggested by Einstein and by Nernst. So, in conclusion, our suggestion is that, according to classical mechanics, Planck's law would describe a metaequilibrium state, at variance with quantum mechanics which interprets it as referring to a true equilibrium. It would be of a certain interest to ascertain whether the phenomenology might prove to be consistent with the scenario of metaequilibrium.

Finally we add a comment about Poincaré. It would appear that, if we are right in suggesting that the metaequilibrium scenario is a priori theoretically consistent, then Poincaré would be wrong. Thus we were very glad in discovering that Poincaré himself had some doubts about his attitude, essentially because he had to admit that, after all, there are more things in heaven and earth than his philosophy could imagine. Indeed, just a few lines after the destructive sentence concernig Jeans quoted above, in connection with the quantization of energy he added [33]: "*Will discontinuity reign over the physical universe and will its triumph be definitive? Or rather will it be recognised that such a discontinuity is only an appearence and that it dissimulates a series of continuous processes? The first person that saw a collision believed to be observing a discontinuous phenomenon, although we know today that the person was actually seeing the effect of very rapid changes of velocity, yet continuous ones*". It is true that Poincaré also adds the skeptical conclusion: "*To try to express today an opinion about these problems would mean to be wasting one's ink*". But at least we are comforted in learning that he admitted that other scenarios (such as that of Boltzmann, Einstein and Nernst, we would say) might be consistent.

References

1. L. Galgani, A.Scotti: Phys. Rev. Lett. **28**, 1173 (1972)
2. C. Cercignani, L. Galgani, A.Scotti: Phys. Lett. A **38**, 403 (1972); L. Galgani and A. Scotti: *Recent progress in classical nonlinear dynamics*, Rivista Nuovo Cim. **2**, 189 (1972)
3. E. Fermi, J. Pasta and S. Ulam, in E. Fermi: Collected Papers (University of Chicago Press, Chicago, 1965); Lect. Appl. Math. **15**, 143 (1974)
4. A. Carati, L. Galgani: Found. Phys. **31**, 69 (2001)
5. D. Lynden Bell: Mon. Not. R. Astr. Soc. **136**, 101 (1967)
6. H. Poincaré, J. Phys. Th. Appl. **2**, 5 (1912), in *Oeuvres* IX, 626–653
7. *La théorie du rayonnement et les quanta*, ed. by P. Langevin and M. de Broglie (Gauthier–Villars, Paris 1912)
8. A. Einstein. In: *Albert Einstein: philosopher–scientist*, ed. by P.A. Schilpp (Tudor P.C., New York, 1949)
9. L. Boltzmann: *Lectures on Gas Theory*, transl. by S.G. Brush (University of California Press, Berkeley 1964); Nature **51**, 413 (1895)
10. J.H. Jeans: Phil. Mag. **35**, 279 (1903); Phil. Mag. **10**, 91 (1905)
11. Physics at the British Association: Nature **92**, 304 (1913); P.P. Ewald: *Bericht uber die Tagung der British Association in Birmingham (10-17 September)* Phys. Z. **14**, 1297 (1913), page 1298

12. N.O. Birge, S.R. Nagel: Phys. Rev. Lett. **54**, 3674 (1985); N.O. Birge: Phys. Rev. B **34**, 1631 (1986)
13. M. Doi, S.F. Edwards: *The Theory of Polymer Dynamics* (Clarendon Press, Oxford 1896)
14. W. Nernst: Verh. Dtsch. Phys. Ges, **18**, 83 (1916)
15. L. Galgani, G. Benettin: Lettere Nuovo Cim. **35**, 93 (1982); L. Galgani: Nuovo Cim. B **62**, 306 (1981); Lettere Nuovo Cim. **31**, 65 (1981); L.Galgani: in *Stochastic processes in classical and quantum systems*, ed. by S. Albeverio, G. Casati, D. Merlini, Lecture Notes in Physics N. 262 (Springer, Berlin 1986)
16. E. Fermi: Nuovo Cim. **25**, 267 (1923); Phys. Z. **24**, 261 (1923)
17. G. Benettin, G. Ferrari, L. Galgani, A. Giorgilli: Nuovo Cim. B **72** 137 (1982); G. Benettin, L. Galgani, A. Giorgilli: Poincaré's Non-Existence Theorem and Classical Perturbation Theory in Nearly-Integrable Hamiltonian Systems, in *Advances in Nonlinear Dynamics and Stochastic Processes*, ed. by R. Livi and A. Politi (World Scientific, Singapore 1985)
18. F.M. Izrailev and B.V. Chirikov: Sov. Phys. Dokl. **11**, 30 (1966)
19. P.Bocchieri, A.Scotti, B.Bearzi. A.Loinger: Phys. Rev. A **2**, 2013 (1970)
20. R. Livi, M. Pettini, S. Ruffo and A. Vulpiani: J. Stat. Phys. **48**, 539 (1987); D. Escande, H. Kantz, R. Livi, S. Ruffo: J. Stat. Phys. **76**, 605 (1994); D. Poggi, S. Ruffo, H. Kantz: Phys. Rev. E **52**, 307 (1995); J. De Luca, A.J. Lichtenberg, S. Ruffo: Phys. Rev. E **60**, 3781 (1999); L. Casetti, M. Cerruti–Sola, M. Modugno, G. Pettini, M. Pettini, R. Gatto: Rivista Nuovo Cim. **22**, 1 (1999); G. Parisi: Europhys. Lett. **40**, 357 (1997); A. Perronace, A. Tenenbaum: Phys. Rev. E **57** (1998) D.L. Shepelyansky: Nonlinearity **10**, 1331 (1997); P.R. Kramers, J.A. Biello, Y. Lvov: *Proceed. Fourth Int. Conf. on Dyn. Syst. and Diff. Eq., May 24–27, 2002, Wilmington, N.C.*, Discrete Cont. Dyn. Systems (in print)
21. L. Galgani, A. Giorgilli, A. Martinoli, S. Vanzini: Physica D **59**, 334 (1992)
22. G. Benettin, L. Galgani, A. Giorgilli: Nature **311**, 444 (1984); L. Galgani: in *Non-Linear Evolution and Chaotic Phenomena*, ed. by G. Gallavotti and P.F. Zweifel NATO ASI Series B: Vol. 176 (Plenum Press, New York 1988); G. Benettin, L. Galgani, A. Giorgilli: Comm. Math. Phys. **121**, 557 (1989); G. Benettin, L. Galgani, A. Giorgilli: Phys. Lett. A **120**, 23 (1987)
23. L. Berchialla, L. Galgani, A. Giorgilli: *Proceed. Fourth Int. Conf. on Dyn. Syst. and Diff. Eq., May 24–27, 2002, Wilmington, N.C.*, Discrete Cont. Dyn. Systems (in print)
24. A. Carati, L. Galgani: J. Stat, Phys. **94**, 859 (1999)
25. Pierce: Proc. Amer. Acad. **60**, 271 (1925); P.S.H. Henry, Nature **129**, 200 (1932)
26. L.D. Landau, E. Teller: Phy. Z. Sowjet. **10**, 34 (1936), in *Collected Papers of L.D. Landau*, ed. by ter Haar (Pergamon Press, Oxford 1965), page 147
27. K.F. Herzfeld, T.A. Litovitz: *Absorption and dispersion of ultrasonic waves* (Academic Press, New York and London, 1959); H.O. Kneser: in *Rendiconti della Scuola Internazionale di Fisica "Enrico Fermi": XXVII, Dispersion and absorption of sound by molecular processes* (Academic Press, New York and London, 1963); D. Rapp, T. Kassal: Chem. Rev. **64**, 61 (1969); A.B. Bhatia: *Ultrasonic Absorption* (Clarendon Press, Oxford, 1967); J.D. Lambert: *Vibrational and rotational relaxation in gases* (Clarendon Press, Oxford 1977); V.A. Krasilnikov, *Sound and ultrasound waves* (Moscow 1960, and Israel Program for Scientific Translations, Jerusalem 1963); H.O. Kneser: Schallabsorption und Dispersion in Gases, in *Handbuch der Physik XI–I* (Springer–Verlag, Berlin 1961)
28. D. Rapp, T. Kassal: Chem. Rev. **64**, 61 (1969)

29. O. Baldan, G. Benettin: J. Stat. Phys. **62**, 201 (1991); G. Benettin, A. Carati, P. Sempio: J. Stat. Phys. **73**, 175 (1993); G. Benettin, A. Carati, G. Gallavotti: Nonlinearity **10**, 479 (1997); G. Benettin, P. Hjorth, P. Sempio: J. Stat. Phys. **94**, 871 (1999)
30. A. Carati, L. Galgani: Phys. Rev. E **61**, 4791 (2000); A.Carati, L. Galgani: Physica A **280**, 105 (2000); A. Carati, L. Galgani: in *Chance in Physics*, ed. by J. Bricmont et al., Lecture Notes in Physics (Springer–Verlag, Berlin 2001)
31. A. Carati, L. Galgani, B. Pozzi: Phys. Rev. Lett. (2003, in print)
32. Dynamics and Thermodynamics of Systems with Long–Range Interactions, ed. by T.Dauxois, S Ruffo, E. Arimondo. M. Wilkens, Lecture Notes in Physics (Springer–Verlag, Berlin 2002)
33. H. Poincaré: Revue Scientifique **17**, 225 (1912), in *Oeuvres IX*, pp. 654–668

Waves Derived from Galactic Orbits. Solitons and Breathers

Nikos Voglis

Academy of Athens, Research Center for Astronomy, 14 Anagnostopoulou Str., Athens, GR-10673, Greece

Abstract. We show how it is possible to define collections of non-interacting particles moving in the same potential so that solitons or breathers are formed. The displacements of particles in this case obey a partial differential equation (PDE). Thus it is possible to derive PDEs from a given Hamiltonian. We demonstrate the above methodology using, as an example, a cubic potential and we derive a Korteweg-de Vries equation.

We apply this methodology in galactic models and show that near resonances, where "Third Integrals" of motion can be defined, the motion of stars can be described in terms of solutions of a Sine-Gordon PDE. This equation admits kink or anti-kink solitons solutions.

In particular in the case of the Inner Lindblad Resonance (ILR), applying the Third Integral on a string of stars having as initial conditions the successive consequents of one orbit on a Poincaré surface of section a Frenkel-Kontorova Hamiltonian is constructed. The corresponding equations of motion are an infinite set of discrete Sine-Gordon equations. This set of equations admits solutions that represent localized oscillations on a grid that are known as Discrete Breathers. An analytic breather solution is derived and compared with the corresponding numerical solution in the case of a perturbed isochrone model.

The advantage of this methodology is that it takes into account the distribution of phases of stars moving under the same value of the third integral. Because of their nature, soliton solutions resist to dispersion and they can be a natural building block to construct more stable non-linear density waves in galaxies.

1 Introduction

Bars or spiral arms are of the main features in the internal structure of galaxies. Their stability is one of the most interesting problem in Galactic Dynamics. Efforts to explain this stability, particularly in the case of the spiral arms, give less stable structures than it is expected from the frequency of their observed occurrence. In reasonable galactic models spiral arms can survive for no more than several rotational periods (5 to 6).(See, for example, [19], [8], [17]).

As it is known, bars and spiral arms are maxima of density composed of different stars at different times, that is, they are maxima of density waves travelling along the azimuthal direction with pattern velocity which is different in general than the mean angular velocity of stars. This mechanism implies some phase correlations in the motion of stars in the galaxy. The main reason for which bars or spiral arms can be destroyed is because of the dispersion of velocities among stars that does not allow particular phase correlations to live for long.

A well known mechanism able to suppress the consequences of dispersion in dynamical systems composed of many particles is by the competition of the dispersion with the nonlinearity of the dynamical system. Such a competition is responsible, for example, for the robustness of solitons. If the spiral arms are formed by density waves of solitary nature one expects to have increased robustness compared with the linear waves. A special class of localized soliton-like solutions are the breathers (see, for example, [15]).

Solitons are solutions of nonlinear integrable Partial Differential Equations (PDEs). These solutions represent localized (not very extensive) waves travelling in a medium and preserving their identity (their basic parameters) even after interactions with other similar waves.

Nonlinearity appears in nature almost everywhere, while integrability is not very common. In most cases, nonlinearity leads to non-integrable dynamical systems. Non-integrability is characterized by chaos, i.e. a sensitive dependence on the initial conditions expressed by an exponential divergence of initially neighboring orbits. Thus, the question arises: Is non-integrability compatible with solitons?

Non-integrable systems contain both stable and unstable periodic orbits. Chaos appears only around the unstable periodic orbits. Around the stable periodic orbits the equations of motion are nearly integrable. It is worth looking for the possibility that solutions of integrable PDEs can describe the basic features of the behavior of dynamical system around stable periodic orbits.

The behavior of a dynamical system around stable periodic orbits is well known. Due to the KAM theorem, a stable periodic orbit is surrounded by invariant tori. Motion on these tori is regular, i.e. it resembles the motion in an integrable system. The respective integral, that keeps the motion on a torus, is the so called 'third integral' [5].

If the theory of solitons can describe the behavior of a dynamical system in the areas of regular motion, then this theory must be directly related to the theory of the third integral. Can we produce integrable PDEs and soliton solutions from the third integral?

In this paper we show how this can be obtained. We first give, in Sect. 2, an example to show how a sequence of non-interacting particles, moving in a potential $V(y)$, can be defined so that they form a soliton travelling along the direction x on the $x-y$ plane.

We apply the same methodology in the epicyclic theory of the motion of stars in galactic potentials. Such potentials correspond to non-integrable Dynamical Systems. The motion of stars in galaxies can be either regular or chaotic, depending on the particular location in their phase space. The level of chaos, however, is low. We have found, for example, [20] that the Lyapunov Characteristic Numbers (LCN) of the orbits in galactic N-Body simulations is less than 0.1. This corresponds to a Lyapunov time (the time necessary for chaos to be effective) of the order more than 10 dynamical times. A black hole at the center of the galaxies can enhance the level of chaos, but even in the case of a huge black hole (e.g. of mass 1% of the total mass of the galaxy), only a very small number of stars describe orbits with LCN somehow larger than 0.1 [21].

In such an environment where order or weakly chaotic motion dominates it is possible to define nonlinear waves as we see below. In Sect. 3 we summarize the linear theory of the epicyclic motion and we show how a Klein-Gordon chain can be defined in terms of the Poincaré Surface Of Section (PSOS). In Sect. 4 we discuss in brief the notion of galactic resonances and the physical meaning of the 'slow angle'. Sections 3 and 4 serve mainly in giving the definitions and in understanding the role of various quantities in the problem. In Sect. 5 the non-linear theory near the ILR in galactic models is discussed reviewing in brief the paper [6]. In this paper Contopoulos, using post-epicyclic approximation terms, derives a third integral Φ describing the motion of stars near the ILR. In Sect. 6 we show how a Sine-Gordon PDE and a corresponding Hamiltonian density can be written, based on Φ. When Φ is applied for an infinite sequence of stars on the Poincaré S.O.S, a Frenkel-Kontorova Hamiltonian [10] can be constructed. The corresponding equations of motion are an infinite set of discrete Sine-Gordon equations which admit discrete soliton or discrete breather solutions. Analytic soliton or breather solutions are given in Sect. 7 and a comparison with a numerical application is given in Sect. 8. A summary and discussion is given in Sect. 9.

2 Solitons of Non-interacting Particles Moving in a Given Potential

In this section we show that it is possible to construct solitons of non-interacting particles that move in an 1-dimensional potential $V(y)$ on the x-y plane.

Consider, for example, the Hamiltonian

$$H = \frac{\dot{y}^2}{2c^2} - \frac{cy^2}{2} + \frac{y^3}{6} \tag{1}$$

where c is a constant. For the value $H = 0$ of the Hamiltonian we get

$$\frac{dy}{y\sqrt{(1 - y/3c)}} = \pm c^{3/2} dt. \tag{2}$$

If we set

$$y = \frac{3c}{\cosh^2 a\phi} \tag{3}$$

we get

$$\phi = \phi_0 \pm \frac{c^{3/2}}{2a} t \tag{4}$$

where ϕ_0 is the value of ϕ at $t = 0$.

On the x-y plane we consider a continuous sequence of non-interacting particles with coordinates $x = \phi_0$ and $y_0 = 3c/\cosh^2 ax$ and velocities $\dot{x} = 0$ and $\dot{y}_0 = \pm y_0\sqrt{1 - y_0/3c)}$. The sign (+) or (-) is chosen to be the sign of x. This sequence of particles forms a soliton travelling along the x direction with velocity $v = c^{3/2}/2a = c$, for $2a = c^{1/2}$. This soliton is described by the equation

$$y = \frac{3c}{\cosh^2 \frac{\sqrt{c}}{2}(x - ct)} \tag{5}$$

and it can be derived as a particular solution

$$u = y(x - ct) = y(\phi) \tag{6}$$

of the well known Korteweg-de Vries PDE

$$\frac{\partial u}{\partial t} + u\frac{\partial u}{\partial x} + \frac{\partial^3 u}{\partial x^3} = 0 \tag{7}$$

Notice that, under the transformation (6) the above PDE can be integrated twice to give

$$\frac{y_\phi^2}{2} - c\frac{y^2}{2} + \frac{y^3}{6} = Ay + B \tag{8}$$

where A and B are the constants of integrations. In the particular case of $A = 0$ at a fixed value of x we can replace y_ϕ by $y_\phi = -\dot{y}/c$ to find the Hamiltonian (1).

3 Linear Epicyclic Waves in Galactic Models

The idea of forming waves with collections of non-interacting particles moving in the same potential can be applied in galactic models. In this section we show how this can be obtained in the simple case of the linear epicyclic motion.

In a rotating galactic model (either bared or spiral) with a plane of symmetry (r, θ) in polar coordinates the motion of stars on this plane can be given by the Hamiltonian

$$H = \frac{1}{2}(\dot{r}^2 + \frac{J_0^2}{r^2}) - \Omega_s J_0 + V_0(r) + V_1(r, \theta) = H_0 + V_1(r, \theta) \tag{9}$$

where J_0 is the angular momentum of a star with respect to an inertial frame, $V_0(r)$ is the axisymmetric component of the potential, $V_1(r, \theta)$ is the perturbation of the potential due either to the bar of the galaxy or to the spiral arms that form a pattern rotating with angular velocity Ω_s. H_0 is the unperturbed (axisymmetric) part of the Hamiltonian.

We define a cartesian coordinate system XY rotating with the pattern, i.e. with angular velocity Ω_s. The polar angle θ is measured in the frame XY from the X-axis.

In the unperturbed potential $V_0(r)$ a star of $H_0 = h$ (the Jacobi integral) and angular momentum $J_0 = J_c$ describes a circular orbit with radius r_c such that

$$h = \frac{J_c^2}{2r_c^2} - \Omega_s J_c + V_0(r_c) \tag{10}$$

where r_c is the root of the equation

$$J_c^2 = r_c^3 \frac{dV_0(r_c)}{dr_c} \tag{11}$$

We define the radial action J_1 as

$$J_1 = \frac{1}{2\pi} \oint \dot{r} dr \tag{12}$$

and the azimuthal action J_2 as $J_2 = J_c$. If the unperturbed part H_0 of the Hamiltonian (9) is expanded up to the linear term in the radial action in Taylor series around the radius r_c, it becomes

$$H_0 = \frac{J_2^2}{2r_c^2} - \Omega_s J_2 + V_0(r_c) + \omega_1 J_1 \tag{13}$$

where ω_1 is the epicyclic frequency given by

$$\omega_1^2 = V_0''(r_c) + \frac{3J_2^2}{r_c^4} \tag{14}$$

(a prime denotes the derivative with respect to r).

In terms of the radial epicyclic coordinate $x = r - r_c$ and its conjugate momentum p the radial action is

$$J_1 = \frac{1}{2}(\omega_1 x^2 + \frac{p^2}{\omega_1}) \tag{15}$$

The guiding center (i.e. the center of the epicycle) describes the circular orbit r_c with respect to an inertial frame with angular velocity

$$\Omega_c = \frac{J_2}{r_c^2} \tag{16}$$

while with respect to the rotating XY frame with angular velocity

$$\omega_2 = \Omega_c - \Omega_s \tag{17}$$

The epicyclic angle θ_1 and the azimuthal angle θ_2 conjugate to J_1 and J_2 respectively, are determined by the equations

$$\dot{\theta}_1 = \frac{\partial H_0}{\partial J_1} = \omega_1 \ , \qquad \dot{\theta}_2 = \frac{\partial H_0}{\partial J_2} = \omega_2 \tag{18}$$

from which we get

$$\theta_1 = \omega_1(t - t_a) \ , \qquad \theta_2 = \omega_2 t \tag{19}$$

where t_a is a constant.

The azimuthal angle θ_2 gives the position of the guiding center of the epicycle. We can define the origin of θ_2 so that the guiding center crosses the X-axis at $t = 0$.

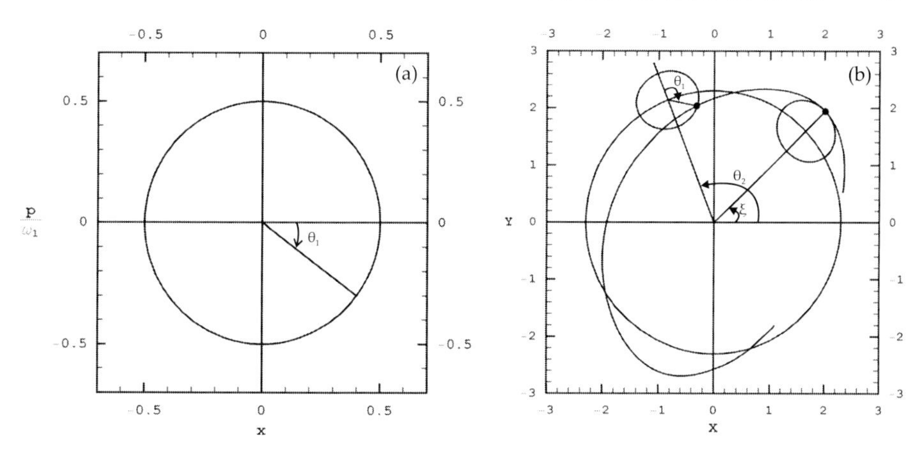

Fig. 1. (a) The phase space $(x, p/\omega_1)$ of the epicyclic motion and the definition of the epicyclic angle θ_1. (b) The azimuthal angle θ_2 gives the position of the guiding center. The angle ξ is the value of θ_2 when the star is a its apocenter.

The angle θ_1 is the phase angle on the phase space $(x, p/\omega_1)$, (Fig.1a). This angle measures the phase of the star on its orbit. For example, if we define θ_1 so that $\theta_1 = 0$ when the star is at its apocenter, then, when $\theta_1 = \pi$, the star is at its pericenter. All other values of θ_1 cover all the intermediate phases. Under these definitions the constant t_a in (19) is the time when the star reaches its apocenter.

If ξ is the value of θ_2 at the time when the star is at a particular phase of its orbit, e.g. at its apocenter, then $\xi = \omega_2 t_a$(Fig.1b). The epicyclic angle θ_1 is written as

$$\theta_1 = \omega_1 t - k\xi \tag{20}$$

where $k = \frac{\omega_1}{\omega_2}$. The epicyclic coordinate x and its conjugate momentum p are expressed in terms of the action J_1 and the angle θ_1 as

$$x = \sqrt{\frac{2J_1}{\omega_1}} \cos\theta_1 = \sqrt{\frac{2J_1}{\omega_1}} \cos(\omega_1 t - k\xi) , \tag{21a}$$

$$\frac{p}{\omega_1} = -\sqrt{\frac{2J_1}{\omega_1}} \sin\theta_1 = -\sqrt{\frac{2J_1}{\omega_1}} \sin(\omega_1 t - k\xi) . \tag{21b}$$

The angle ξ can be used as an independent variable. Let us consider a continuous sequence of stars with the same actions J_1 and J_2, uniformly distributed along the values of ξ at a given time. All these stars follow the same orbit but with different initial phases. According to (21a), their motions (in and out the circular orbit r_c) form a continuous sinusoidal wave, with wavenumber k, travelling along the ξ-axis.

We can easily show that (21a) is a solution of the Klein-Gordon wave equation

$$x_{tt} + \omega_2^2 x_{\xi\xi} + 2\omega_1^2 x = 0 \tag{22}$$

The term $\omega_2^2 x_{\xi\xi}$ in this equation expresses a kind of phase interaction between neighboring stars in the sequence.

Consider the Poincaré surface of section (PSOS) of the orbit of a single star, with actions J_1 and J_2, when for example $\theta_2 = 0$, that is, when the star crosses the positive X-axis with a positive velocity component along the Y-axis ($\dot{y} > 0$). If the star starts from its apocenter that is initially on the above PSOS then at $t = 0$, $\theta_1 = 0$ and $\theta_2 = 0$. This means that the value of ξ for this star is $\xi = 0$.

For irrational values of $k = \omega_1/\omega_2$, the successive consequents of this orbit on the PSOS, that occur in times $t_n = 2\pi n/\omega_2$ with $n = 0, \pm 1, \pm 2, ...$, belong to an invariant curve of action J_1 on the plane $(x, p/\omega_1)$. The rotation angle of these consequents on the invariant curve is θ_1 and takes the values

$$\theta_{1n} = \omega_1 t_n = 2\pi n k \tag{23}$$

Consider now an infinite sequence of stars numbered as $n = 0, \pm 1, \pm 2, ...$ with the same actions J_1, J_2, having as initial conditions (at $t = 0$) the above consequents. From (20) and (23) we get

$$\xi_n = -2\pi n \tag{24}$$

Therefore the epicyclic angles of these stars as functions of time are

$$\theta_{1n}(t) = \omega_1 t + 2\pi n k \tag{25}$$

and their equations (21a and 21b) become

$$x_n = \sqrt{\frac{2J_1}{\omega_1}} \cos\left(\omega_1 t + k 2\pi n\right) , \tag{26a}$$

$$\frac{p_n}{\omega_1} = -\sqrt{\frac{2J_1}{\omega_1}} \sin\left(\omega_1 t + k 2\pi n\right) . \tag{26b}$$

Equation (26a) describes a discrete sinusoidal wave travelling along an 1-dimensional grid of particles. This equation is a solution of the discrete set of Klein-Gordon ordinary differential equations

$$\ddot{x}_n + \frac{\omega_2^2}{4\pi^2}(x_{n+1} + x_{n-1} - 2x_n) + \omega^2 x_n = 0 \tag{27}$$

where ω is given by the equation

$$\omega^2 = \omega_1^2 \left(1 + \frac{\sin^2 \frac{\omega}{\omega_2}\pi}{(\frac{\omega}{\omega_2}\pi)^2}\right) \tag{28}$$

that tends to the value of $\omega = \sqrt{2}\omega_1$ provided that the wavenumber $k = \frac{\omega}{\omega_2}$ satisfies the condition $k \ll 1$.

It is remarkable that the second term in (27) represents a phase interaction between neighboring stars in the sequence, as if this interaction were due to repulsing elastic forces.

Thus, up to now, we have shown that using the invariance of the epicyclic action J_1 and the azimuthal action J_2, we can define continuous epicyclic waves satisfying a Klein-Gordon equation (22). Furthermore, in terms of the Poincaré S.O.S., we can construct a Klein-Gordon chain of stars satisfying a corresponding set of ordinary differential equations (27).

The above theory can be generalized in the non-linear case as it will be described in Sect. 5.

4 Resonances in the Unperturbed Potential. The Inner Lindblad Resonance and the Slow Angle ψ

In a central potential $V_0(r)$ the apocenters of an orbit in general precess. The radial frequency Ω_r and the mean azimuthal frequency Ω_a of an orbit are defined as

$$\Omega_r = \frac{2\pi}{T_r}, \qquad \Omega_a = \frac{\phi}{T_r} \tag{29}$$

where T_r is the time from one apocenter to the next, called radial period, and ϕ is the azimuthal angle between two successive apocenters.

In general this orbits is not closed, i.e. is not periodic. However, this orbit closes, if it is observed in a frame rotating with angular velocity

$$\Omega_s = \frac{\phi - 2\pi}{T_r} = \Omega_a - \Omega_r \tag{30}$$

In other words it becomes periodic resembling in this case a Keplerian ellipse. Such a closed orbit is called 1:1 resonant periodic orbit.

If the frame rotates with angular velocity

$$\Omega_s = \frac{2\phi - 2\pi}{2T_r} = \Omega_a - \frac{\Omega_r}{2} \tag{31}$$

the orbit closes again but in this case it resembles an ellipse symmetric with respect to the center. This is a 2:1 resonant periodic orbit. Equation (31) in terms of the frequencies $\omega_1 = \Omega_r$ and $\omega_2 = \Omega_a - \Omega_s$ used in the previous section can be written as

$$\omega_1 = 2\omega_2 \tag{32}$$

(Notice that the mean azimuthal frequency Ω_a is identical to the angular velocity Ω_c of the guiding center). Equation (32) expresses the well known Inner Lindblad Resonance (ILR) in galaxies.

Consider a 2:1 resonant orbit, i.e. an orbit exactly at the ILR of a galaxy in the XY plane in the unperturbed potential $V_0(r)$. The angle between the major axis of the orbit and the X-axis is exactly the angle ξ, defined in the previous section that determines the orientation of the apocenter. This orbit can be analyzed in circular and epicyclic motion as in the previous section.

The slow angle ψ (introduced by Lynden-Bell [13]) is defined as

$$\psi = \theta_1 - 2\theta_2 \tag{33}$$

Using (19) and (20) we get

$$\psi = (\omega_1 - 2\omega_2)t - 2\xi \tag{34}$$

Since the orbit is exactly at the resonance ($\omega_1 - 2\omega_2 = 0$), we have

$$\psi = -2\xi \tag{35}$$

This means that the angle ψ for this orbit is constant and determines the orientation of the major axis of the orbit. For other orbits, however, near to the resonance, the angle ψ varies slowly with $\dot{\psi}$ measuring the slow precession of the major axis of the orbit. For this reason the angle ψ is called slow angle or precession angle [13], [6], [14], [16].

On the PSOS as defined in the previous section, i.e. when $\theta_2 = 0$ the values of the angle ψ coincide with the corresponding angle of θ_{1n} of (23). Thus the angle ψ on the PSOS $(x, p/\omega_1)$ plays the role of the rotation angle on an invariant curve.

5 Non-linear Epicyclic Motion near the Inner Lindblad Resonance

Any perturbation $V_1(r, \theta)$ imposed on the axisymmetric potential $V_0(r)$ can be analyzed in Fourier modes [11] as

$$V_1(r, \theta_1, \theta_2) = \epsilon \sum_{lm} V_{lm}(J_1, J_2) \cos(l\theta_1 - m\theta_2) \tag{36}$$

where ϵ is a small positive quantity and l, m are integers. For the orbits near the ILR the most important term of the Fourier modes in the expansion (36) (i.e. the mode that can pump more energy on such orbits than all other terms) is the term with $l = 1$ and $m = 2$, namely, the term

$$\epsilon V_{12}(J_1, J_2) \cos(\theta_1 - 2\theta_2) = \epsilon V_{12}(J_1, J_2) \cos \psi \tag{37}$$

This is the resonant term in the ILR.

Using post epicyclic approximations Contopoulos [6] has shown that the Hamiltonian (9) can be written in the form

$$H = h + \omega_1 I_1 + \omega_2 I_2 + a I_1^2 + 2b I_1 I_2 + c I_2^2 + ... + V_1(r, \theta) \tag{38}$$

where

$$I_1 = \frac{1}{2\pi} \oint \dot{r} dr = J_1 \tag{39}$$

is the radial action exactly the action J_1 defined in Sect. 2.

The azimuthal action I_2 in the Hamiltonian (38) is the excess from the angular momentum of the circular orbit

$$I_2 = J_0 - J_c \tag{40}$$

which is of the same order as J_1. The frequencies ω_1 and ω_2 have the same meaning as in Sect. 2.

Contopoulos [6] shows that near the ILR under proper canonical transformations the Hamiltonian (38) leads to an equivalent Hamiltonian $\mathcal{H}$ up to terms of $O(\epsilon)$

$$\mathcal{H} = \omega_2 J_2 + c J_2^2 + \Phi(J_1, \psi) = 0 \tag{41}$$

where $\Phi(J_1, \psi)$ is a third integral given by

$$\Phi(J_1, \psi) = \gamma J_1 + \alpha J_1^2 + \epsilon_1 (\frac{2J_1}{\omega_1})^{1/2} (J_{20} - 2J_1) \cos \psi \tag{42}$$

describing the epicyclic motion near the inner Lindblad resonance in post epicyclic approximation terms. In these expressions ϵ_1 is a small positive constant proportional to ϵ in (37), γ is defined as

$$\gamma = \omega_1 - 2\omega_2 \tag{43}$$

and α, c, J_{20} are constants.

The azimuthal action is $J_2 = I_2 + 2J_1$, the so called fast action. Its conjugate fast angle ψ_2 (equal to the azimuthal angle θ_2)is ignorable in the Hamiltonian (41). Therefore the azimuthal action J_2 is a constant and the problem up to terms of $O(\epsilon)$ is integrable.

The angle ψ is the slow angle given by (33). Its conjugate action is J_1 that can be written as

$$J_1 = \frac{1}{2} (\frac{\overline{p}^2}{\omega_1} + \omega_1 \overline{x}^2) \ . \tag{44}$$

in terms of the epicyclic coordinate

$$\overline{x} = (\frac{2J_1}{\omega_1})^{1/2} \cos \psi \tag{45}$$

and its conjugate momentum

$$\frac{\overline{p}}{\omega_1} = -(\frac{2J_1}{\omega_1})^{1/2} \sin \psi \ . \tag{46}$$

In terms of $\overline{x}$ and $\overline{p}$ the third integral (42) is a fourth order polynomial. A periodic orbit of the system is located at an extreme value of $\Phi(J_1, \psi)$. In other words periodic orbits are found as the roots of algebraic system

$$\Phi_x = 0, \qquad \Phi_p = 0 \tag{47}$$

where the index denotes the corresponding partial derivative. A root of this system gives a stable or an unstable periodic orbit depending on the sign of the quantity $S = \Phi_{xx}\Phi_{pp} - \Phi_{xp}^2$ (stable for $S > 0$).

For very small values of the perturbation parameter ϵ_1 the system (47) has only one real root giving a stable periodic orbit called x_1, while for larger values of ϵ_1 three roots appear, by a saddle point bifurcation, giving two stable orbits called x_1 and x_2 and one unstable orbit called x_3 (Figs.1a,b in [6]).

6 A Sine-Gordon Equation

The equations of motion derived from $\mathcal{H}$ in (41) give

$$\dot{\psi}_2 = \dot{\theta}_2 = w = \frac{\partial \mathcal{H}}{\partial J_2} = \omega_2 + 2cJ_2 = constant \tag{48}$$

$$\dot{\psi} = \frac{\partial \mathcal{H}}{\partial J_1} = \frac{\partial \Phi}{\partial J_1} = \gamma + 2\alpha J_1 + \epsilon_1 A' \cos\psi \tag{49}$$

$$\dot{J}_1 = -\frac{\partial \mathcal{H}}{\partial \psi} = -\frac{\partial \Phi}{\partial \psi} = \epsilon_1 A(J_1) \sin\psi \tag{50}$$

where

$$A = (\frac{2J_1}{\omega_1})^{1/2}(J_{20} - 2J_1) \tag{51}$$

and A' is the derivative of A with respect to J_1.

The second time derivative of ψ obeys the pendulum equation up to $O(\epsilon_1)$ terms, i.e.

$$\ddot{\psi} - \omega_0^2 \sin\psi = 0 \tag{52}$$

where

$$\omega_0^2 = \epsilon_1[2\alpha A(J_1) - (\gamma + 2\alpha J_1)A'] \tag{53}$$

A pendulum equation similar to (52) is known in Galactic Dynamics [16] derived by alternative arguments. It expresses the possibility that an orbit precesses so that its major axis either librates around a fixed direction at $\psi = \pi$, or rotates around the center of the galaxy.

In the case when the three periodic orbits x_1,x_2,x_3 appear, one can find both librating or rotating orbits with a separatrix between them passing through the unstable orbits x_3.

If (52) is integrated once we get

$$\frac{\dot{\psi}^2}{2} = -\omega_0^2 cos\psi + C \tag{54}$$

where C is the integration constant. This constant is a measure of the value of the third integral $\Phi(J_1, \psi)$ at a given orbit. It can be easily shown that, up to terms of $O(\epsilon)$,

$$\gamma^2 + 4\alpha\Phi = 2C \tag{55}$$

In the general case (for any value of C) (54) can be further integrated in term of the Jacobian elliptic functions (e.g. see [1]). On the separatrix, ψ reaches the value $\psi = 0$ with $\dot{\psi} = 0$, thus we get $C = \omega_0^2$. For this value of C the solution of (54) is

$$\tan\frac{\psi}{4} = e^{\pm[\omega_0(t - t_0)]} \tag{56}$$

where t_0 is the time when $\psi = \pi$. On the PSOS, according to (45), this value of ψ corresponds to the pericenter of the orbit, since $\overline{x}$ becomes minimum. The value $\psi = 0$, on the other hand, corresponds to the apocenter of the orbit. Such an orientation, of an orbit, on the separatrix, is approached as time goes to infinity.

Introducing the angle ξ as in the previous sections

$$\xi = wt_0, \qquad k_0 = \frac{\omega_0}{w} \tag{57}$$

$$\tan\frac{\psi}{4} = e^{\pm(\omega_0 t - k_0\xi)} \tag{58}$$

We consider an continuous sequence of stars with the same constant $C = \omega_0^2$ (on the separatrix) with characteristic phase angles $\phi = \omega_0 t - k_0 \xi$. Equation (58) represents a kink or antikink soliton travelling along ξ and is a solution of the Sine-Gordon wave equation

$$\ddot{\psi} + w^2 \psi_{\xi\xi} - 2\omega_0^2 \sin\psi = 0 \tag{59}$$

The Lagrangian density function from which this equation can be derived is

$$\mathcal{L} = \frac{\dot{\psi}^2}{2} + \frac{w^2 \psi_\xi^2}{2} - 2\omega_0^2 \cos\psi \tag{60}$$

and the corresponding Hamiltonian density

$$\Phi_* = \dot{\psi}\frac{\partial L}{\partial \dot{\psi}} - \mathcal{L} = \frac{\dot{\psi}^2}{2} - \frac{w^2 \psi_\xi^2}{2} + 2\omega_0^2 \cos\psi \tag{61}$$

It is easy to show that this Hamiltonian density can be expressed up to $O(\epsilon)$ in terms of the constant C in (54) or the third integral Φ in (42) as

$$\Phi_* = 2C - \frac{1}{2}(\dot{\psi}^2 + w^2 \psi_\xi^2) = \gamma^2 + 4\alpha\Phi - \frac{1}{2}(\dot{\psi}^2 + w^2 \psi_\xi^2) \tag{62}$$

If we consider an infinite number of stars starting at $t = 0$ with $\xi = -2\pi n$, $n = 0, \pm 1, \pm 2, ...$, i.e. on the separatrix of the Poincaré S.O.S. as described in the previous sections, (58) becomes

$$\tan\frac{\psi_n}{4} = e^{\pm(\omega_0 t + k_0 2\pi n)} \tag{63}$$

Since $k_0 \ll 1$ this is a solution of the discrete set of the ordinary Sine-Gordon equations

$$\ddot{\psi}_n + \frac{w^2}{4\pi^2}(\psi_{n+1} + \psi_{n-1} - 2\psi_n) - 2\omega_0^2 \sin\psi_n = 0 \tag{64}$$

These equations can also be derived directly from a Fenkel-Kontorova Hamiltonian constructed in terms of the Hamiltonian density (61) by replacing the partial derivative ψ_ξ by its discrete equivalent and sum over all the n stars with initial conditions on a single invariant curve of the Poincaré S.O.S. (same value of Φ_*). Thus we get

$$\Phi_{FK} = \sum_n \Phi_{*n} = \sum_n [\frac{\dot{\psi}_n^2}{2} - \frac{w^2}{8\pi^2}(\psi_{n+1} - \psi_n)^2 + 2\omega_0^2 \cos\psi_n] \tag{65}$$

If ψ_n is a periodic function with frequency ω and amplitude X_n, i.e. $\psi_n = X_n \exp \omega t$, the dynamics of ψ_n in (64) can be studied through the following map produced by direct substitution in (64)

$$\begin{aligned} Y_{n+1} &= Y_n + K \sin X_n + \Lambda X_n \ , \\ X_{n+1} &= X_n + Y_{n+1} \ , \end{aligned} \tag{66}$$

where $K = 2(\omega_0 T)^2$ and $\Lambda = (\omega T)^2$, with $T = 2\pi/w$.

7 Analytic Solutions

If we make the transformation

$$\phi = \omega_0 t - k_0 \xi \tag{67}$$

the Sine-Gordon equation (59) after a first integration becomes

$$\psi_\phi^2 = 2[c + 1 - 2\sin^2 \frac{\psi + \pi}{2}] \tag{68}$$

where c is the integration constant and $c \geq -1$. The solution of this equation is given by the Jacobian sn-oidal Elliptic Functions. Namely, if

$$m = (c+1)/2 \tag{69}$$

the solution is given by

$$\cos \frac{\psi}{2} = m^{1/2} sn(\phi \mid m) \tag{70}$$

This solution for ψ is a soliton travelling along ξ with velocity $d\xi/dt = k_0/\omega_0 = w$. Using (33) and (48) the epicyclic coordinate $x = x_m \cos\theta_1$ can be expressed as

$$x = x_m \cos(\psi + 2wt) = x_m(\cos\psi \cos 2wt - \sin\psi \sin 2wt) \tag{71}$$

where $x_m = \sqrt{2J_1/\omega_1}$ is the amplitude of the epicyclic coordinate x. This amplitude varies since the action J_1 is a function of time according to (50). Integrating (50) we finally get

$$x_m = x_0 \pm \beta m^{1/2} \sqrt{1 - sn^2(\phi \mid m)} \tag{72}$$

where x_0 is the epicyclic coordinate for $\psi = 0$ and β measures the amplitude of variations of x_m around the value of x_0. The proper choice of sign in (72) depends on the sign of $\dot{\psi}$ (minus for $\dot{\psi} < 0$).

From (70) we have

$$\cos\psi = 2m sn^2(\phi \mid m) - 1 \tag{73}$$

$$\sin\psi = 2m^{1/2} sn(\phi \mid m)\sqrt{1 - m sn^2(\phi \mid m)} \tag{74}$$

In the case of $m \ll 1$ we get the harmonic oscillation limit

$$\cos\psi = 2m^{1/2} \sin^2\phi - 1 \tag{75}$$

$$\sin\psi = 2m^{1/2} \sin\phi (1 - \frac{m}{2} \sin^2\phi) \tag{76}$$

In the particular case of $m = 1$ the motion occurs on one of the two separatrices. In this case the expressions (73) and (74) become

$$\cos\psi = 1 - \frac{2}{\cosh^2\phi}, \qquad \sin\psi = \frac{2\tanh\phi}{\cosh\phi}\,. \tag{77}$$

Let R_g be the radius of the circle described by the guiding center. The radius R_3 to a star describing the unstable periodic orbit x_3 can be approximated by

$$R_3 = R_g + x_0 \cos 2wt \tag{78}$$

The radius to any star describing an epicyclic motion with the same guiding center can be written as

$$R = R_g + x_m \cos \theta_1 \tag{79}$$

For the stars moving on the separatrix surrounding the periodic orbits x_2 (where $\dot{\psi} < 0$) the correct sign in (72) is (-). In this case the difference $\Delta R = R - R_3 = x_m \cos\theta_1 - x_0 \cos 2wt$ can be written as

$$\Delta R = [-2(x_0 - \frac{\beta}{\cosh\phi})\frac{1}{\cosh^2\phi} - \frac{\beta}{\cosh\phi}] \cos 2wt$$

$$+2(x_0 - \frac{\beta}{\cosh\phi})\frac{\tanh\phi}{\cosh\phi} \sin 2wt \tag{80}$$

The r.h.s in (80) represents a breather (composed of two simpler superposed breathers). This is a breather of a continuous sequence of stars along ξ. For the discrete sequence $\{\mathbf{n}\}$ of stars the angle ϕ is

$$\phi = \omega_0 t + k_0 2\pi n \tag{81}$$

In this case we have a discrete breather on an one-dimensional grid of stars, similar to breathers found in other branches of Physics. Studies on discrete breathers is currently a very active field of research (see e.g. [18], [22], [2], [15], [9], [3], [4], [33], [23], see also Bountis and Bergamin in this volume). One-dimensional discrete breathers are defined as common frequency oscillations of a not very large number of particles localized in a region of an infinite grid of interacting particles. The localization is due to the fact that the amplitudes of the oscillations decrease exponentially with the distance from a given point of the grid due to the nonlinearity of the potential. Discrete breathers can be either stationary, remaining at the same region of the grid or they can travel along the grid. In the present case the breather travels along $-n$ with a phase velocity $d\xi/dt = -2\pi\Delta n/\Delta t = \omega_0/k_0 = w$.

8 A Numerical Application and Comparison with the Analytic Results

We present below a comparison of the above solution (80) with the numerical calculations in a particular galactic model. In this model the axisymmetric component of the potential is taken to be the well known isochrone model

$$V_0(r) = -\frac{1}{1+\sqrt{1+r^2}}\ , \tag{82}$$

on which a bar-like perturbation

$$V_1(r,\theta) = \epsilon r^{1/2}(16 - r)\cos 2\theta \tag{83}$$

is superposed. This model has been used in [7] in studying galactic orbits in week and strong bars. The adopted values of ϵ and Ω_s are $\epsilon = 0.00001$ and $\Omega_s = 0.05$. For a value of the Jacobi constant $h = -0.28$, motion occurs near the ILR. The three periodic orbits x_1, x_2 (stable) and x_3 (unstable) on the XY plane are shown in Fig. 2, while the corresponding Poincaré S.O.S. $(X, \dot{X})$ is shown in Fig.3. In this figure the axes are $X = R_g + x$ and $\dot{X} = \dot{x}$.

This system is non-integrable. For the above choice of parameters, chaos in the region of the unstable orbit x_3 at ($X_3 \approx 2.33$, $\dot{X} = 0$) is very small, so that it can be neglected. The stable periodic orbits x_1 and x_2 are at the centers of the two islands at ($X \approx 1.03$, $\dot{X} = 0$) and ($X \approx 1.82$, $\dot{X} = 0$), respectively. These islands are limited by two separatrices. (In fact, what we call a separatrix here, is a very thin homoclinic tangle formed by the unstable and the stable asymptotic curves emanating from the unstable periodic orbit x_3). The dots in Fig.3 correspond to the successive consequents of a single orbit on each separatrix.

Let us focus on the points of one separatrix only, for example, the separatrix surrounding the periodic orbit x_2. We number these consequents as $n = 0, \pm 1, \pm 2, \ldots$ composing the sequence $\{\mathbf{n}\}$ of stars. The consequent $n = 0$ is at $\dot{X} \approx 0$ and $X_{min} = 1.27$. Positive or negative n means a consequent in the future or in the past, respectively, relative to the consequent at $n = 0$.

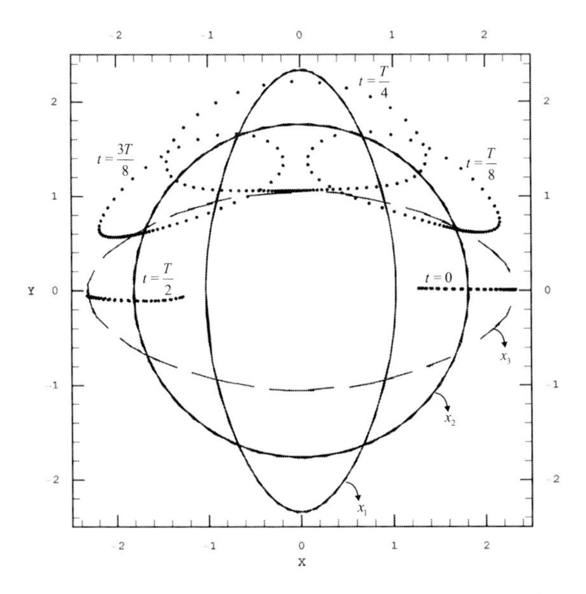

Fig. 2. The three periodic orbits x_1,x_2 (stable) and x_3 (unstable) on the XY plane. The chain of stars forming the breather of Fig.4 is shown at several snapshots.

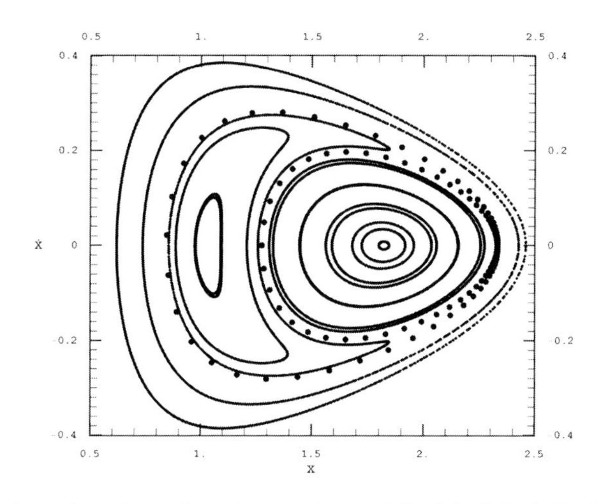

Fig. 3. The Poincaré surface of section in the model of (82),(83) for $h = -0.28$. The three periodic orbits x_1,x_2, and x_3 are at $X = 1.03$, $X = 1.82$, $X = 2.33$, respectively, with $\dot{X} = 0$. Dots represent the successive consequents of an orbit starting on each separatrix. The dots on the separatrix surrounding x_2 are the initial conditions of the chain of stars forming the breather shown in Fig.4.

We run simultaneously the orbits of a chain of $2n_{max} + 1 = 145$ stars with initial conditions the above consequents and we calculated

$$\Delta R(n,t) = R_n(t) - R_{x_3}(t) \tag{84}$$

where $R_n(t)$ is the radial distance from the center at time t of the star n and $R_{x_3}(t)$ is the same as $R_n(t)$, but of a star moving exactly on the unstable periodic orbit x_3.

In Fig. 4 the line with dots gives $\Delta R(n,t)$ as a function of n at four snapshots at times $t = 0,\ T/8,\ 2T/8,\ 3T/8$, where T is the azimuthal period corresponding to the frequency w. The thin solid line gives at the same times ΔR as a function of ξ evaluated from (80). The values of x_0 and β are taken from the data of Fig. 3. They are $x_0 = 0.65$ and $\beta = 0.24$. The agreement between the numerical $\Delta R(n,t)$ and the analytic ΔR is quite good. The small differences are mainly due to the approximate representation of the orbit x_3 by the equation (78) but also to the fact that the analytic solution contains only $O(\epsilon)$ terms. Similar results are found if we consider the other separatrix of Fig.3.

We see therefore that proper chains of stars can form breathers travelling along the chain with a speed w equal to the angular velocity of the guiding center. The chain of stars used in the numerical breather of Fig.4 is shown in real space in Fig.2 at $t = 0,\ T/8,\ 2T/8,\ 3T/8, T/2$. The stars of this chain lie initially ($t = 0$) on a straight line along the X axis. The chain evolves to a loop at $T/4$ and forms again an almost straight line along X at $T/2$ and so on, being successively inside or outside the unstable periodic orbit x_3.

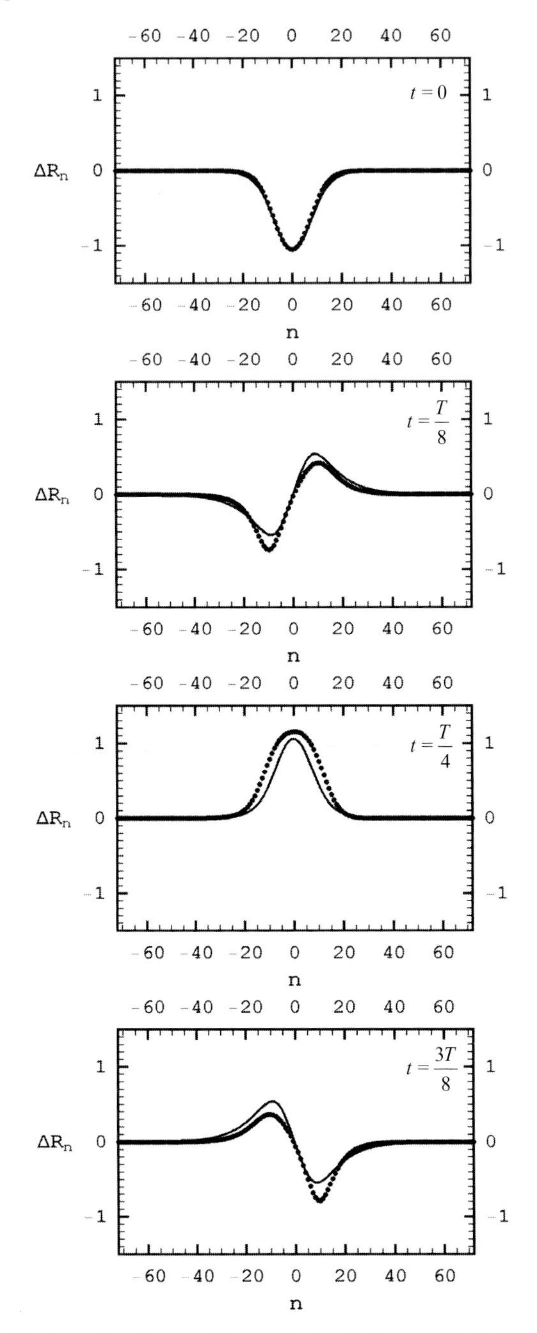

Fig. 4. Dots give the breather formed by a chain of stars with initial conditions on the separatrix surrounding the periodic orbit x_2. The thin solid line represents the analytic solution in (80).

If the sequence $\{\mathbf{n}\}$ of stars is defined with initial conditions on any invariant curve of Fig.3, then sn-oidal solitons are formed according to the solution (70) travelling along the chain of stars with velocity w. In other words stars pass through a constant phase ϕ of the soliton with speed w. This means that a constant phase ϕ is stationary with respect to the X-Y frame. This is a very convenient property to define nonlinear density waves or solitary density waves in galaxies.

9 Summary and Discussion

We have shown how it is possible to write PDEs governing the motion of collections of non-interacting particles moving in a given potential. In other words, we have shown how one can pass from the theory of orbits to the theory of nonlinear waves.

In particular, we have shown that, on the basis of the third integral, it is possible to write PDEs, governing the motion of stars in galactic models. The fact that the Third Integral can be directly related to the theory of solitons joins different points of view and bridges two different fields of experience.

In the case of the Inner Lindblad resonances a Sine-Gordon PDE is derived for the slow angle of precession of the orbits. The corresponding Lagrangian or Hamiltonian densities are given.

Using the successive consequents of an orbit on the PSOS, we can define a Klein-Gordon chain of stars, that is, an one-dimensional grid obeying a Frenkel-Kontorova Hamiltonian.

Breathers can be constructed in galactic models for sets of stars on the separatrices of unstable periodic orbits. An analytic breather solution is obtained near the ILR which is in good agreement with the corresponding numerical results.

This analysis opens the possibility that collections of orbits forming solitons and breathers can be superimposed, instead of single orbits, to construct bars or spiral arms in galaxies. This approach takes into account the fact that many stars in a galaxy can move in phase correlated orbits, consistent to the structure of phase space. Such collections of stars resist to dispersion as long as the Third Integral is conserved. This feature is important for nonlinear density waves, which can form longer living patterns. The surface of section is only used as a tool to realize the phase correlations of orbits. Notice, that the distribution of stars along an invariant curve can be far from uniform, seriously deformed by the nonlinear effects. This is taken into account as well. In this aspect, the solutions of solitons and breathers, we have found, are more natural building blocks of nonlinear density waves.

Acknowledgments

I wish to thank Prof. G. Contopoulos and Dr. Ch. Efthymiopoulos for their comments and helpful discussions. I also wish to thank C. Kalapotharakos for his help in preparing the figures of this paper.

References

1. Abramowitz M. and Stegun I.A.: "Handbook of Mathematical Functions" (Dover Publications, New York, 1972)
2. Aubry S.: Physica D, **71**, 196 (1994)
3. Bountis T., Capel H.W., Kollemann J.C., Bergamin, J.M. and van der Weele J.P.: Phys. Let. A, **268**, 50 (2000)
4. Bountis T., Bergamin J.M. and Basios V.: Phys. Let. A, **295**, 115 (2002)
5. Contopoulos G.: Zs. f. Ap., **49**, 273 (1960)
6. Contopoulos G.: ApJ, **201**, 566 (1975)
7. Contopoulos G. and Papayannopoulos Th.: A&A, **92**, 33 (1980)
8. Elmegreen B.G. and Thomason M.: A&A, **272**, 37 (1993)
9. Flash S. and Willis C.R.: Phys. Rep., **295**, 181 (1998)
10. Frenkel Ya. and Kontorova T.: Fiz. Zh., **1**, 137 (1939)
11. Kalnais A.: ApJ., **166**, 275 (1971)
12. Kevrekidis P.G., Saxena A. and Bishop A.R.: Phys. Rev. E, **64**, 026611 (2001)
13. Lynden-Bell D. in "Dynamical Structrure and Evolution of Stellar Systems", Saas Fee (Geneva Observatory, 1973)
14. Lynden-Bell D., in "Galactic Dynamics and N-body Simulations", G. Contopoulos, N.K. Spyrou, L. Vlahos (Eds), EADN Astroph. School VI, (Springer-Verlag, 1994)
15. MacKay R.S. and Aubry S.: Nonlinearity, **7**, 1623 (1994)
16. Palmer, P.L.: "Stability of Collisionless Stellar Systems" (Kluwer Academic Publishers, 1994)
17. Patsis P.A., Kaufmann D.E.: A&A, **352**, 469 (1999)
18. Takeno S., Kisoda K. and Sievers A. J., 1988, Prog. Theor. Phys. Suppl. **94**, 242 (1999)
19. Thomason M., Elmegreen, B.G., Donner K.J. and Sundelius B.: ApJ, **356**, L9 (1990)
20. Voglis N., Kalapotharakos C. and Stavropoulos I., 2002, MNRAS, **337**, 619 (2002)
21. Volgis N., Kalapotharakos C., Stavropoulos I. and Efthymiopoulos Ch.: in "Galaxies and Chaos" G. Contopoulos, N. Voglis (eds) Lecture Notes in Physics (Springer-Verlag, 2002)
22. Sievers A.J. and Takeno S.: Phys. Rev. Lett., **61**, 970 (1988)
23. Tsironis G.P.: J. Phys. A: Math. Gen. **35**, 951 (2002)

Discrete Breathers in Nonlinear Lattices: A Review and Recent Results

Tassos Bountis and Jeroen M. Bergamin

Department of Mathematics and Center for Research and Applications of Nonlinear Systems, University of Patras, 26500 Patras, Greece

Abstract. Localization phenomena in systems of many (often infinite) degrees of freedom have attracted attention in solid state physics, nonlinear optics, superconductivity and quantum mechanics. The type of localization we are concerned with here is *dynamic* and refers to oscillations occurring not because of the presence of some defect, but due to the interaction between nonlinearity and resonances. In particular, we shall describe an entity called *discrete breathers*, which represent localized periodic oscillations in nonlinear lattices. As suggested by other authors in this volume, this type of behavior may be observed in density fluctuations of stars rotating in a galaxy in the discrete or continuum approximation. Since the reader may not be too familiar with these concepts, we have chosen first to review the history of discrete breathers in the second half of last century and then present an account of our recent results on the efficient computation of breathers in multi-dimensional lattices using homoclinic orbits. This allows us to make a much more detailed study and classification of discrete breathers than had previously been possible, as well as accurately follow their existence and stability properties as certain physical parameters of the problem are varied.

1 The History of Energy Localization

In 1955, Fermi, Pasta and Ulam (FPU) [1] presented the first systematic study of the energy properties of a chain of 32 identical particles, interacting through linear and nonlinear forces and attached to fixed boundaries, according to the equations of motion:

$$\begin{aligned} \ddot{u}_n &= (u_{n+1} - 2u_n + u_{n-1}) + \alpha\left((u_{n+1} - u_n)^2 - (u_n - u_{n-1})^2\right) , \quad n = 1, ..., 31 \\ \dot{u}_0 &= \dot{u}_{32} = 0 \\ u_0 &= u_{32} = 0 \, , \end{aligned} \tag{1}$$

where $u_n = u_n(t)$ represents each particle's displacement from equilibrium and dots denote differentiation with respect to time t. Using the newly developed computers of the Los Alamos Laboratories in the USA, they integrated (1), starting with the initial condition $u_n = \sin\left(\frac{\pi n}{32}\right)$, and observed a remarkable near-recurrence of the solutions to the initial condition after relatively short time periods (see Fig. 1 below).

All expectations of the theory of statistical mechanics before these experiments predicted that higher order modes of oscillation (i.e. states with $u_n = \sin\left((2k+1)\frac{\pi n}{32}\right)$, $k \in \mathbb{N}$ larger than one) would equally share the energy of

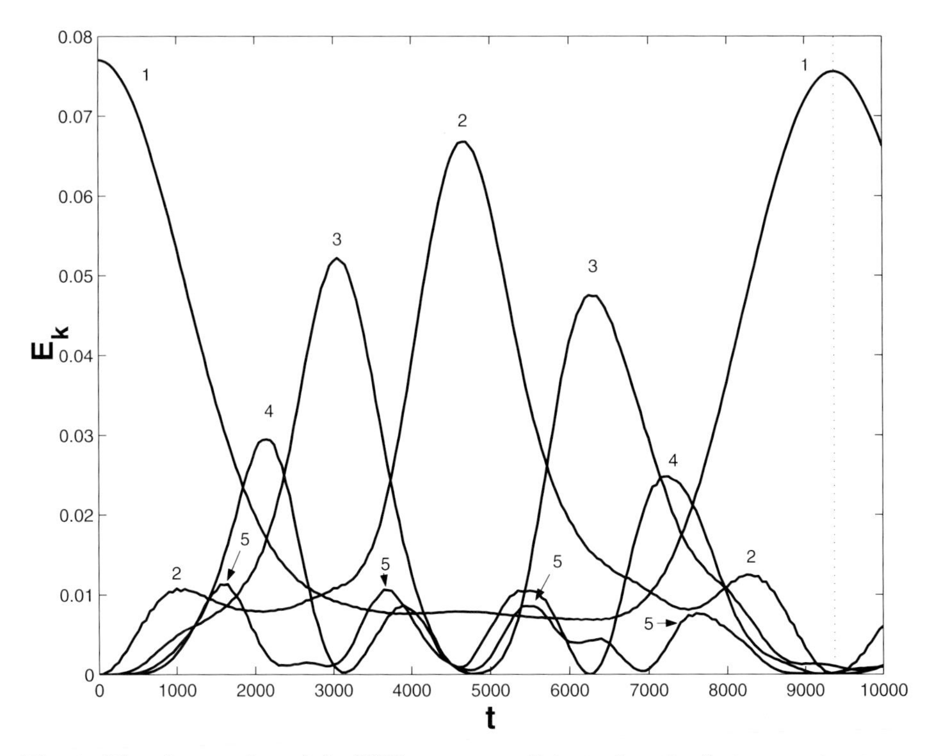

Fig. 1. Time-integration of the FPU system until just after the first approximate recurrence of the initial state. Shown here is how the energy $E_k = \frac{1}{2}\left(\dot{a}_k^2 + 2a_k^2 \sin^2\left(\frac{k\pi}{2N}\right)\right)$ is divided over the first five modes $a_k = \sum_{i=1}^{N} x_i \sin\left(\frac{ik\pi}{N}\right)$. The numbers in the figure indicate the wave-number k

the system, thus achieving finally a situation of thermodynamic equilibrium. Of course, the Poincaré recurrence theorem dictated that the initial state would again emerge, but after much longer times than the recurrences observed by Fermi, Pasta and Ulam. It was therefore understandable that this discovery created a great excitement within the scientific community of the period.

In 1965, attempting to explain the results of the FPU experiment, Zabusky and Kruskal [2] derived the Korteweg–De Vries (KdV) equation of shallow water waves in the long wavelength and small amplitude approximation,

$$q_\tau + qq_x + \delta^2 q_{xxx} = 0 \quad \delta^2 \ll 1 \, , \tag{2}$$

as a continuum limit of the FPU system (1). They observed that solitary traveling wave solutions, now known as *solitons*, exist, whose interaction properties result in analogous recurrences of initial states, see Fig. 2. In taking their continuum limit, however, Zabusky and Kruskal overlooked an important aspect of the FPU chain: the *discrete nature* of the system.

In 1969, Ovchinnikov [3] showed – in a study of coupled nonlinear oscillators modeling finite-sized molecules – how discreteness in combination with an

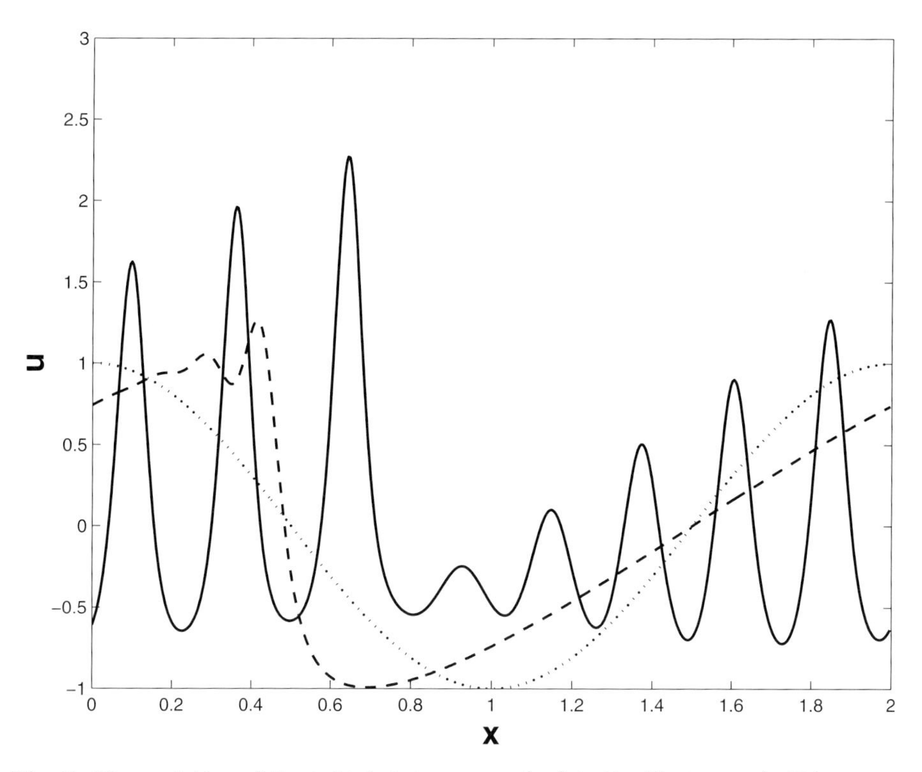

Fig. 2. The evolution of the initial state $u = \cos(\pi x)$ in the Korteweg–de Vries equation, at the times $t = 0$ (*dotted line*) $t = 1/\pi$ (*dashed line*). At an intermediate stage ($t = 3.6/\pi$, *solid line*), the solitons are maximally separated, while at $t \approx 30.4/\pi$ the initial state is nearly recovered

intrinsic nonlinearity of the system can cause energy localization. Due to this combination, resonances between neighboring oscillators are avoided when certain frequency bands (the so-called *phonon bands*) are outside the spectrum of vibrations of the system (see Fig. 3). Thus, the recurrence phenomena in the FPU experiments can be explained as the result of limited energy transport between Fourier modes, caused by discreteness and non-resonance effects.

1.1 The Discovery of Discrete Breathers

Twenty years later (1988), the subject of localized oscillations in nonlinear lattices was revived in a paper by Sievers and Takeno [4], in which they used analytical arguments to show that energy localization occurs generically in FPU systems of *infinitely many* particles in one dimension, obeying the equations:

$$\ddot{u}_n = (u_{n+1} - 2u_n + u_{n-1}) + \alpha \left((u_{n+1} - u_n)^2 - (u_n - u_{n-1})^2 \right) , \qquad (3)$$

cf. (1), for $-\infty < n < \infty$.

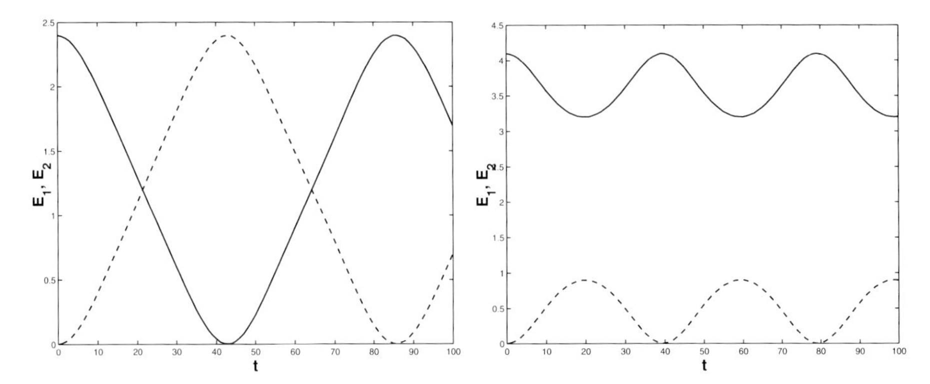

Fig. 3. Local energies $E_1(t)$ (*solid line*) and $E_2(t)$ (*dashed line*) of the two coupled nonlinear oscillators considered by Ovchinnikov [3], showing how energy transfer is impeded by discreteness and nonlinearity. **Left:** Complete energy transfer when the first oscillator has initial amplitude $a = 2.0$ ($E_1(0) = 2.4$). **Right:** Incomplete energy transfer when the first oscillator has initial amplitude $a = 2.5$ ($E_1(0) \approx 4.1$)

Combining their perturbative analysis with numerical experiments, they demonstrated that a new type of solution, the so-called *discrete breathers* exist as oscillations which are both time-periodic and spatially localized. In their simplest form, such solutions can exhibit significant oscillations only of the middle ($n = 0$) and nearby ($n = -1, +1$) particles. However, a great many patterns are possible (the so-called *multibreathers*) in which several particles around the middle one oscillate with large amplitudes, as shown here in Fig. 4. How can one determine all the possible shapes? Which of them are stable under small perturbations? These are the kind of questions that we set out to answer in our research.

Besides the FPU system, the existence of these localized oscillations was soon verified numerically by other research groups on a variety of lattices, including the Klein-Gordon (KG) system

$$\ddot{u}_n = -V'(u_n) + \alpha(u_{n+1} - 2u_n + u_{n-1}) \ , \tag{4}$$

where V is an on-site potential and α is a parameter indicating the coupling strength.

It was not, however, until 1994, that a mathematical proof of the existence of discrete breathers was published by MacKay and Aubry [5] in the case of one-dimensional lattices of the type (4). Under the general assumptions of nonlinearity and non-resonance, such chains of interacting oscillators were rigorously shown to possess discrete breather solutions for small enough values of the coupling parameter $\alpha > 0$, as a continuation of their obvious existence at $\alpha = 0$.

Section 2 below contains the main part of our contribution to the field of discrete breathers. We have found that a very convenient way to construct them and study their stability properties, away from the $\alpha = 0$ limit, is through the "*geometry*" of the *homoclinic solutions* of *nonlinear recurrence relations* [6].

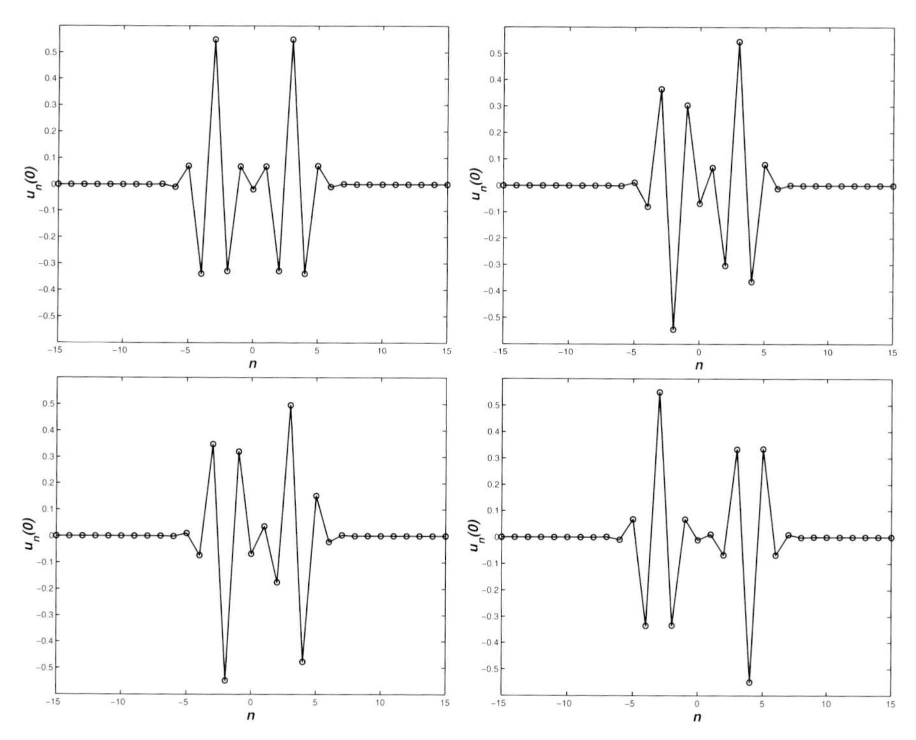

Fig. 4. Several breathers of an FPU lattice of the form (3) with cubic (rather than quadratic) interactions, obtained by starting a Newton-Raphson search using homoclinic orbits of a map of the form (6) as an initial guess

2 The Connection with Homoclinic Dynamics

Focusing on the property of spatial localization, Flach was the first to show that discrete breathers in simple one-dimensional chains can be actually represented by homoclinic orbits in the Fourier amplitude space of time-periodic functions [7]. Indeed, inserting a Fourier series

$$u_n(t) = \sum_{k=\infty}^{\infty} A_n(k) \exp(\mathrm{i}k\omega t) \tag{5}$$

into the equations of motion of either the FPU (3) or KG (4) lattice and setting the amplitudes of terms with the same frequency equal to zero, leads to the system of equations

$$-k^2\omega^2 A_n(k) = \langle -V'(u_n) + W'(u_{n+1} - u_n) - W'(u_n - u_{n-1}), \exp(\mathrm{i}k\omega t)\rangle \quad \forall k, n \in \mathbb{Z}\,.$$

This is an infinite-dimensional mapping of the Fourier coefficients $A_n(k)$ with the brackets $\langle .,. \rangle$ indicating a properly normalized inner product. Time-periodicity is ensured by the Fourier basis functions $\exp(ik\omega t)$. Spatial localization requires that $A_n(k) \to 0$ exponentially as $n \to \pm\infty$. Hence a discrete breather is a homoclinic orbit in the space of Fourier coefficients, i.e. a doubly infinite sequence of points beginning at 0 for $n \to -\infty$ and ending at 0 for $n \to +\infty$. In fact, keeping only the Fourier term ($k = 1$) with the largest amplitude reduces, in some cases, the above system to a simple 2-dimensional mapping

$$a_{n+1} = g(a_n, a_{n-1}) , \tag{6}$$

whose invariant manifolds of the saddle fixed point at the origin can be easily plotted, as in Fig. 5.

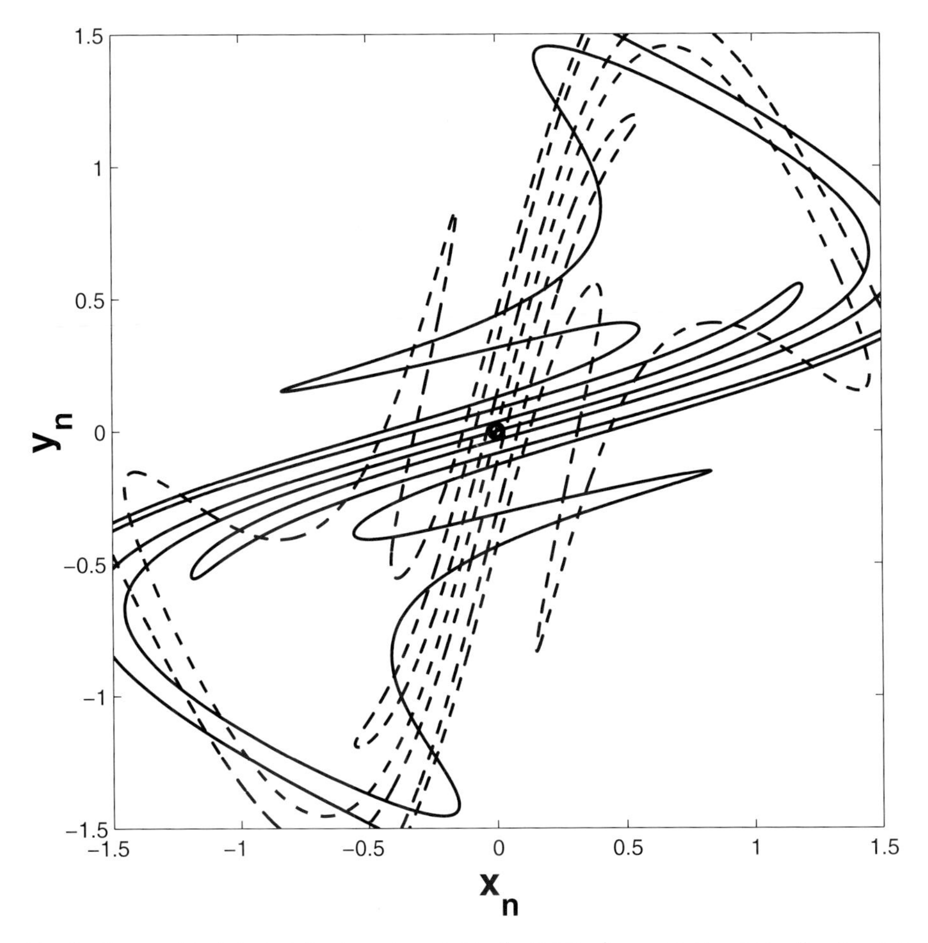

Fig. 5. The stable (*dashed line*) and unstable (*solid line*) manifolds of the fixed point at (0, 0) of a 2-dimensional map of the form (6), with $x_n = a_n$ and $y_n = a_{n+1}$. The manifolds are clearly seen to intersect at infinitely many points, hence a wealth of homoclinic orbits exists

2.1 How to Construct Homoclinic Orbits

Realizing the importance of homoclinic orbits to the subject of discrete breathers, we have been able to develop in [8] efficient numerical methods for locating homoclinic orbits of invertible maps of arbitrary but finite dimension. To this end, we found it particularly useful to exploit symmetry properties of the maps and understand the geometry of the invariant manifolds near the origin, which is always a fixed point of the mappings of the saddle type.

To see how this is done, let us consider a general first order map (or recurrence relation)

$$\boldsymbol{x}_n = f(\boldsymbol{x}_n) , \quad \boldsymbol{x}_n \in \mathbb{R}^d , \tag{7}$$

with d a positive integer. Recall that homoclinic orbits are solutions for which $\boldsymbol{x}_n \to 0$ as $n \to \pm\infty$ and concentrate on all orbits satisfying a symmetry condition of the form

$$\boldsymbol{x}_n = \mathrm{M}\boldsymbol{x}_{-n} , \tag{8}$$

where M is a $d \times d$-matrix with constant entries and $\det(\mathrm{M}) \neq 0$. Observe that if an orbit obeys such a symmetry and $\boldsymbol{x}_n \to 0$ as $n \to -\infty$, then it is also true that $\boldsymbol{x}_n \to 0$ as $n \to \infty$. Hence, any orbit which obeys this symmetry and satisfies $\boldsymbol{x}_n \to 0$ as $n \to -\infty$ is a homoclinic orbit.

To obtain such an orbit, we first need to specify numerically its asymptotic behavior as $n \to -\infty$. In other words, it has to be on the *unstable manifold* of $\boldsymbol{x} = 0$. Furthermore, a necessary requirement for a homoclinic orbit to exist is that the origin be a saddle fixed point of the map. Then it is known that the unstable manifold of the origin is well approximated in its vicinity by the unstable Euclidean eigenspace of the linearized equations, which is tangent to the nonlinear manifold and has the same number of dimensions.

Now, the dimension of the linear unstable eigenspace equals the number of coordinates necessary to determine a point $\boldsymbol{x}_{-N}$, $N \gg 1$ uniquely. Thus, by choosing this point to be on the linear unstable manifold very close to the origin, it follows that it will also be approximately on the corresponding nonlinear manifold. Thus, when mapped forward $N+1$ times, we can test whether it satisfies the above symmetry relation (8). By this approach, locating symmetric homoclinic orbits becomes a search for solutions of the system

$$\begin{cases} \boldsymbol{x}_1 - \mathrm{M}\boldsymbol{x}_{-1} = 0 \\ \boldsymbol{x}_0 - \mathrm{M}\boldsymbol{x}_0 = 0 \end{cases} , \tag{9}$$

since, given $\boldsymbol{x}_{-N}$, the values of $\boldsymbol{x}_1$, $\boldsymbol{x}_0$ and $\boldsymbol{x}_{-1}$ are uniquely obtained by direct iteration of the map (7).

In the case of an invertible map we can use this method to find also all *asymmetric* homoclinic orbits, i.e. those which do not obey the symmetry condition (9). This can be done by introducing the new "sum" and "difference" variables

$$\begin{cases} \boldsymbol{v}_n = \boldsymbol{x}_n + \boldsymbol{x}_{-n} \\ \boldsymbol{w}_n = \boldsymbol{x}_n - \boldsymbol{x}_{-n} \end{cases} ,$$

which always possess the symmetry

$$\begin{cases} \boldsymbol{v}_n = \boldsymbol{v}_{-n} \\ \boldsymbol{w}_n = -\boldsymbol{w}_{-n} \end{cases} . \tag{10}$$

In this way, we can apply again the above strategy and look for symmetric homoclinic orbits of a new map (of *double the dimension* of the original f) described by the equations

$$F : \begin{cases} \boldsymbol{v}_{n+1} = f\left(\frac{\boldsymbol{v}_n+\boldsymbol{w}_n}{2}\right) + f^{-1}\left(\frac{\boldsymbol{v}_n-\boldsymbol{w}_n}{2}\right) \\ \boldsymbol{w}_{n+1} = f\left(\frac{\boldsymbol{v}_n+\boldsymbol{w}_n}{2}\right) - f^{-1}\left(\frac{\boldsymbol{v}_n-\boldsymbol{w}_n}{2}\right) \end{cases} , \tag{11}$$

yielding homoclinic orbits $\boldsymbol{x}_n$ of the original map f, (7), that are not themselves necessarily symmetric. On the other hand, each homoclinic orbit of f is a symmetric homoclinic orbit of the new map F. Therefore, by determining all *symmetric* homoclinic orbits of F, we find *all* homoclinic orbits of f. Following this approach, we have also been able to *classify* all possible homoclinic orbits by assigning to them *symbolic sequences* in a systematic way, according to their complexity (see [8,9] for more details and Fig. 6 as an example of the results of these papers).

Thus, we now come to our second major contribution on this topic, described in Sect. 3 below. This concerns a new approach to the computation of discrete breathers, which can be efficiently applied to lattices of more than one spatial dimension and systems with vector valued variables assigned to each lattice site. The main idea is to write a breather solution as a product of a space-dependent and a time-dependent part and reduce the problem to finding the homoclinic orbits of a 2 dimensional map, under the constraint that the given ODEs possess simple periodic oscillations of a well-defined type and of specified period.

3 A New Approach Is Introduced

As was mentioned above, it is possible to write down a map in Fourier amplitude space linking discrete breathers with homoclinic orbits. This map can be reduced to a finite-dimensional recurrence relation, by neglecting Fourier components with a wave number k larger than some cutoff value k_{max}. Then, one can use the methods of our papers [8,9] to approximate discrete breather solutions by finding all the homoclinic solutions of these recurrence relations.

Recently, however, this problem has been re-examined from a different perspective: In 2002, inspired by the work of other authors like Flach [10] and Kivshar [11], Tsironis [12] suggested a new way to approximately *separate* amplitude from time-dependence, yielding in some cases ODEs with known solutions (for example elliptic functions) while keeping the dimension of the recurrence relation as low as possible. This led to an improved accuracy of the calculations and provided analytical expressions of discrete breathers for a special class of FPU and KG systems.

In a very recent paper [13], Bergamin extended Tsironis' work by developing a numerical procedure for which the time-dependent functions need not be

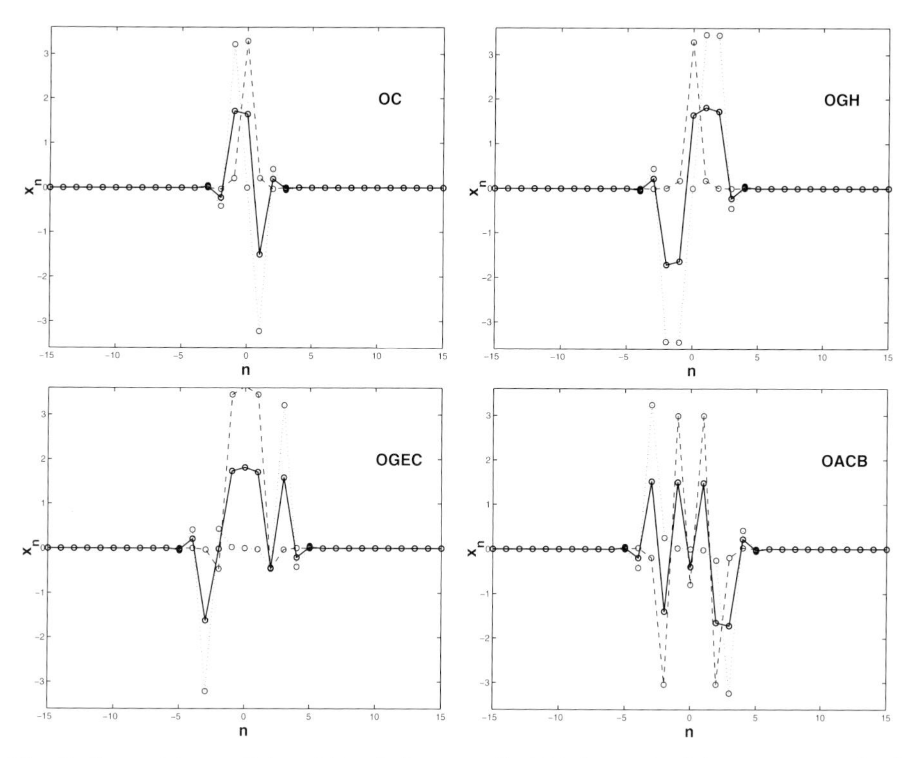

Fig. 6. Several homoclinic orbits of the map F, (11) determined by a zero-search of the system of equations $v_1 - v_{-1} = 0$ and $w_0 = 0$, related by $v_n = x_n + x_{-n}$ and $w_n = x_n - x_{-n}$. Shown here are v_n (*dashed*), w_n (*dotted*) and x_n (*solid*). Also indicated is the symbolic name of the orbit, assigned by the procedure given in [8]

known analytically. In this way, a much wider class of nonlinear lattices can now be treated involving scalar or vector valued variables in one or more spatial dimensions.

In particular, the approximation proposed by Tsironis can be more precisely formulated as follows

$$\begin{cases} u_{n+1}(t) - u_n(t) \approx (a_{n+1} - a_n) T_n(t) \\ u_{n-1}(t) - u_n(t) \approx (a_{n-1} - a_n) T_n(t) \end{cases}, \tag{12}$$

where a_n denotes the time-independent amplitude of $u_n(t)$ and $T_n(t)$ is its time-dependence, defined by $T_n(0) = 1$ and $\dot{T}_n(0) = 0$.

Note now, that all FPU and KG systems are derived from a potential function of the form

$$U = \sum_{n=-\infty}^{\infty} V(u_n) + W(u_{n+1} - u_n) .$$

Using the approximation (12), the equations of motion

$$\ddot{u}_n = -V'(u_n) + W'(u_{n+1} - u_n) - W'(u_n - u_{n-1})$$

are transformed into

$$a_n \ddot{T}_n = -V'(a_n T_n) + W'((a_{n+1} - a_n) T_n) - W'((a_n - a_{n-1}) T_n) \ . \quad (13)$$

This is an ordinary differential equation (ODE) for $T_n(t)$ which can, in principle, be solved since the initial conditions are known.

Of course, an analytical solution of ODE (13) is in general very difficult to obtain. However, since we are primarily interested in the amplitudes a_n, what we ultimately need to do is develop a numerical procedure to find a recurrence relation linking a_n, a_{n+1} and a_{n-1} *without having to solve the ODE beforehand.*

Let us observe first, that the knowledge of a_n, a_{n+1} and a_{n-1} permits us to solve the above ODE numerically. Under mild conditions, solutions $T_n(t)$ can thus be obtained, which are time-periodic, while for a discrete breather all functions $T_n(t)$ have the same period. Choosing a specific value for this period, allows us to *invert the process* and determine a_{n+1} as a function of a_n and a_{n-1}, similar to (6). In the same way, we also determine a_{n-1} as a function of a_n and a_{n+1}. Thus, a *two-dimensional invertible map* for the a_n has been constructed, ensuring that all oscillators have the same frequency.

As is explicitly shown in [13,14], on a variety of examples, the homoclinic orbits of this map provide highly accurate approximations to the discrete breather solutions with the given period and the initial state $u_n(0) = a_n$, $\dot{u}_n(0) = 0$. In fact, we can now apply this approach to more complicated potentials and higher dimensional lattices, as we demostrate in Fig. 7, where we compute a discrete breather solution of a 2-dimensional lattice, with indices n, m in the x, y directions and dependent variable $u_{n,m}(t)$.

Having thus discovered new and efficient ways of calculating discrete breathers in a wide class of nonlinear lattices, we now turn to the study of their stability, control and continuation properties in parameter space. More specifically, we shall show that it is possible to use our methods to extend the domain of existence of breathers to parameter ranges that cannot easily reached by other more standard continuation techniques.

3.1 Stability and Existence of Discrete Breathers Using Control

So far, we have seen that transforming nonlinear lattice equations to low-dimensional maps and using numerical methods to compute their homoclinic orbits provides an efficient tool for approximating discrete breathers in any (finite) dimension and classifying them in a systematic way. This clears the path for an investigation of important properties of large numbers of discrete breathers of increasing complexity. One such property, which is relevant to many applications and requires that a solution be known to great accuracy, is stability in time.

In our recent work [15], the accurate knowledge of a discrete breather solution was used in a rather uncommon way to study stability properties: As is well known, in Physics, Electronics and Engineering, a familiar task is to try to influence the behavior of a system, by applying control methods. By control we mean here the addition of an *external force* to the system which allows us

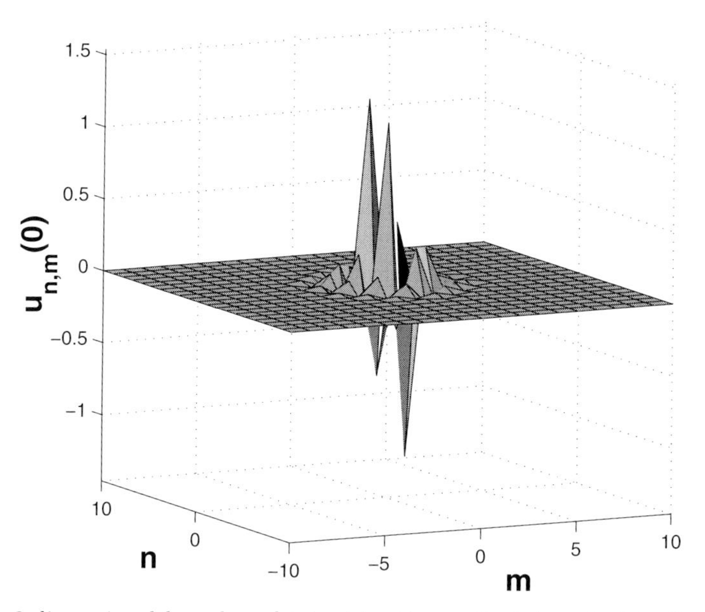

Fig. 7. A 2-dimensional breather obtained by the method described in the text, for equations of motion of the KG type where the particles at each lattice site n, m experience harmonic interactions with its 4 nearest neighbors and a quartic on-site potential.

to influence its dynamics. In particular, the objective of our control will be to change the *stability type* of a discrete breather solution of the controlled system, compared with the uncontrolled one.

The system is altered in such a way that the solution itself does not change. In other words, in the controlled system the solution is exactly the same as in the uncontrolled and for this reason the control is of the feedback type. If a solution is unstable in the uncontrolled system, the extra terms added to the equations in the controlled case can cause the solution to become stable and vice versa. The system we have studied [15] is based on the KG equations of motion written as

$$\ddot{u}_n = -V'(u_n) + \alpha(u_{n+1} - 2u_n + u_{n-1}) + L\frac{d}{dt}(\hat{u}_n - u_n) , \qquad (14)$$

where $\hat{u}_n$ is the known discrete breather solution of the equations when $L = 0$. The parameter L indicates how strongly the control term influences the KG system. Thus, for any value of L, $u_n = \hat{u}_n$ is clearly seen to be a solution of both the controlled as well as the uncontrolled lattice equations. Clearly, for $L > 0$, the $L\frac{d}{dt}u_n$ term introduces dissipation, while the $L\frac{d}{dt}\hat{u}_n$ represents periodic forcing. It is therefore reasonable to investigate whether, by increasing L, the dissipative part of the process will force the system to converge in time to a stable solution. If this solution is the original $\hat{u}_n$, the latter is stable. If this does not happen, the original solution is unstable. In [14,17] the following proposition is proved, though in a slightly more general formulation:

Proposition 1. *Let $u_n = \hat{u}_n$ be a periodic solution of the lattice equations*

$$\ddot{u}_n = -V'(u_n) + \alpha (u_{n+1} - 2u_n + u_{n-1}) \ .$$

Then there exists an $L > 0$ such that $u_n = \hat{u}_n$ is an asymptotically stable solution of the modified (controlled) system

$$\ddot{u}_n = -V'(u_n) + \alpha (u_{n+1} - 2u_n + u_{n-1}) + L\frac{d}{dt}(\hat{u}_n - u_n) \ .$$

This result clearly implies that, by increasing L, it is possible to stabilize the original solution, independent of its stability in the uncontrolled situation. Let us demonstrate this by taking the breather of Fig. 8, which is unstable at $L = 0$, substitute its (known) form $\hat{u}_n(t)$ in the above equations and increase the value of L. As we see in Fig. 5, it is quite easy to stabilize it at $L = 1.17$, since increasing the value of L gradually brings all eigenvalues of the monodromy matrix of the solution inside the unit circle.

The above Proposition has an additional significant advantage: It gives us the opportunity to address the question of the *existence* of discrete breathers in ranges of the coupling parameter α where other techniques do not apply. In order to do this, it is important to recall first how this question was originally answered (though partially) in the existence proof of MacKay and Aubry, based on the notion of the so - called anti-continuum limit $\alpha = 0$ [5].

Let us observe that the KG equations of motion

$$\ddot{x}_i = -\ V'(x_i) + \alpha (x_{i+1} - 2x_i + x_{i-1})$$

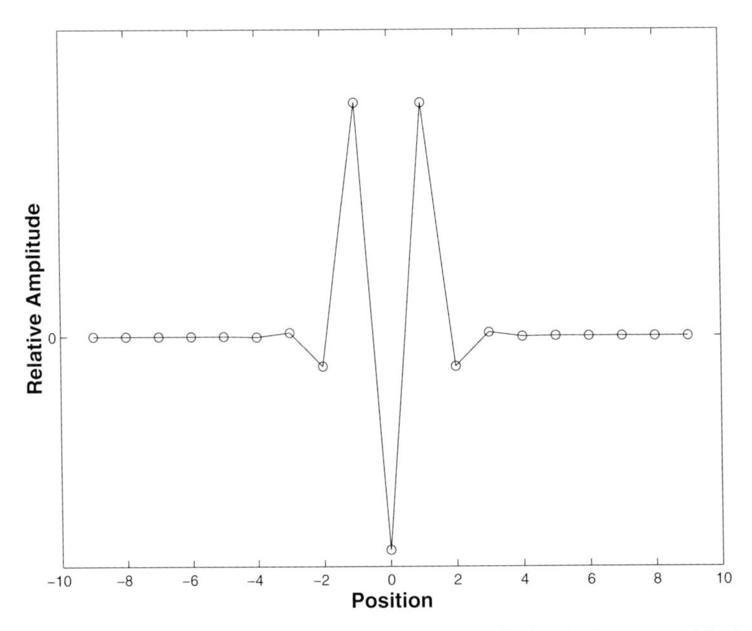

Fig. 8. A breather shape $u_n(t) = \hat{u}_n(t)$ of the system (14) which is unstable for $L = 0$

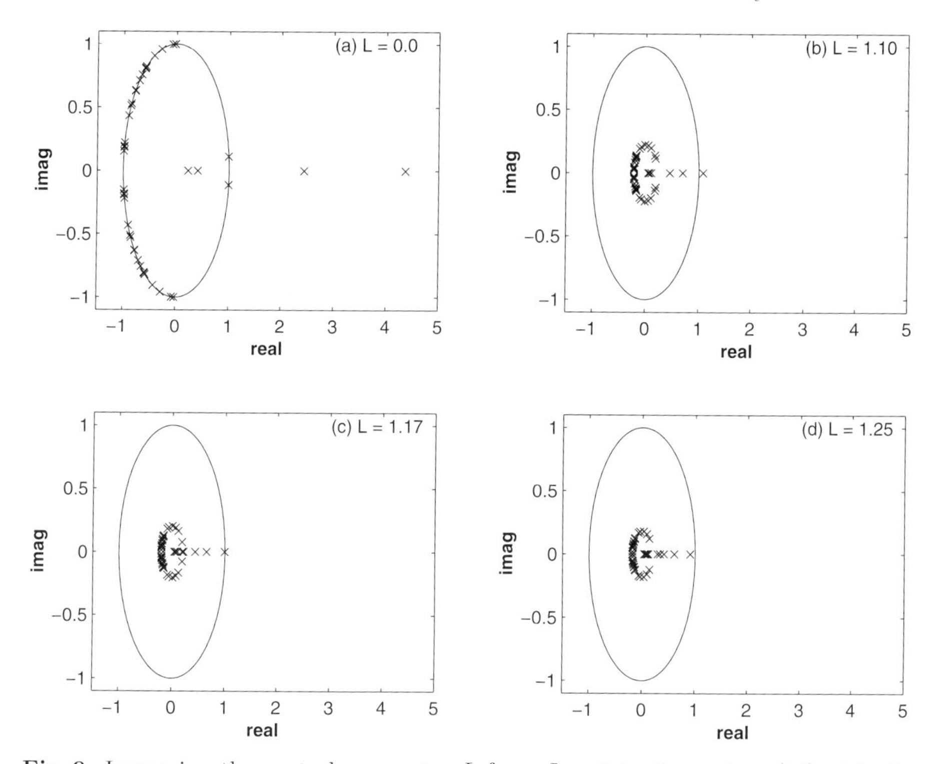

Fig. 9. Increasing the control parameter L from $L = 0$ in the system (14) with the initial shape given in Fig. 8 moves the eigenvalues of the orbit's monodromy matrix inside the unit circle. Initially, some eigenvalues are outside the unit circle, but eventually, for $L > 1.17$, all eigenvalues attain magnitudes less than one, thus achieving stability and control

describe a system of uncoupled oscillators for $\alpha = 0$. Obviously, in that case, any initial condition, where only a finite number of oscillators have a non-zero amplitude, is a discrete breather solution. In their celebrated paper of 1994, MacKay and Aubry prove that, under the conditions of nonresonance with the phonon band and nonlinearity of the function $V'(x)$, this solution can be continued to the regime where $\alpha > 0$.

According to their approach, however, continuation for $\alpha > 0$ is possible, only as long as the eigenvalues of the Floquet matrix of the solution do not cross the value $+1$. This means that the typical occurrence of a bifurcation, through which the breather becomes unstable, prevents MacKay and Aubry's continuation method from following the breather beyond that value of α. This is where the control method we have proposed comes to the rescue: As we can see in Fig. 7, by following a path in $\alpha > 0$ and $L > 0$ space, a discrete breather solution can be continued to a higher value of α, by choosing L in the controlled system such that the solution remains stable!

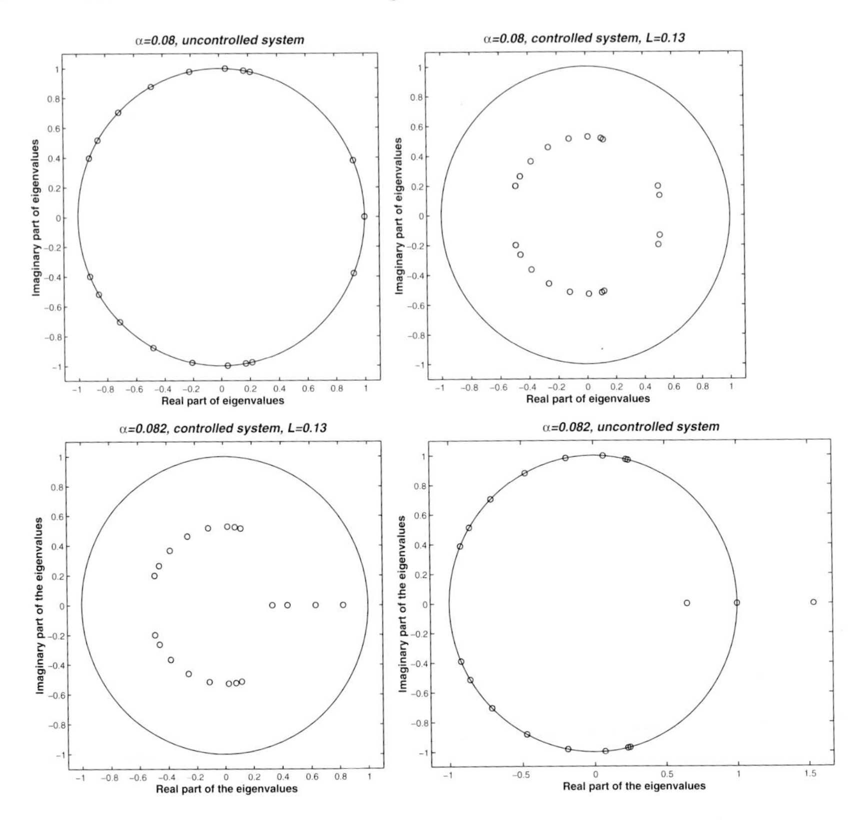

Fig. 10. When the uncontrolled system approaches a bifurcation point for $\alpha = 0.08$ and $L = 0$ (**upper left figure**), the bifurcation can be avoided by increasing the control parameter, for example to $L = 0.13$, such that the breather solution is asymptotically stable (**upper right**). When the coupling strength is increased to $\alpha = 0.082$ (**lower left**), control can be switched off to return to the breather of the uncontrolled system, which is now unstable (**lower right**)

Therefore, since by the above Proposition one can always find L such that stability is possible for any coupling α, this allows the continuation of any solution of the $\alpha = 0$ case to $\alpha > 0$ values beyond bifurcation, demonstrating the existence of breather solutions in the corresponding parameter regime. The fact that $\hat{u}_n$, which is a (stable) solution of the controlled system is by definition also a (unstable) solution of the uncontrolled system, implies that we have succeeded in continuing a discrete breather solution to higher values of the coupling parameter.

In Fig. 6, we show in L, α space the regions of stability of this particular breather. As is well-known by the work of Segur and Kruskal [16], breathers are not expected to exist in these systems in the continuum limit of α going to infinity. At what coupling value though and how do they disappear? Can we use our control aided continuation methods to follow them at arbitrarily high α to

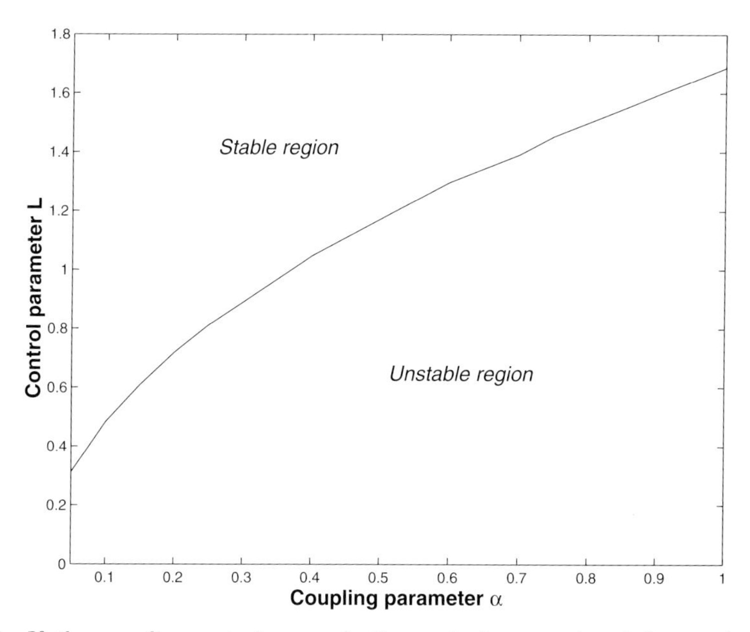

Fig. 11. If the coupling α is increased, the control parameter L has to be larger to achieve stability and successful control. Shown here are the regions for which the breather of Fig. 8 is a stable or unstable solution $u_n(t) = \hat{u}_n(t)$ of (14)

be able to answer such questions? Currently, we are working on this problem and results are expected to appear in a future publication [17].

Acknowledgements

One of the authors (T.B.) wishes to thank G. Contopoulos and N. Voglis for their interest in this work and their invitation to present it in a conference on Galaxies and Chaos. The other author (J.M.B.) acknowledges partial support by a "Karatheodory" graduate research fellowship of the University of Patras.

References

1. E. Fermi, J. Pasta and S. Ulam: Tech. Rep. Los Alamos Nat. Lab. **LA 1940**, (1955)
2. N.J. Zabusky and M.D. Kruskal: Phys. Rev. Lett. **15**, 240 (1965)
3. A.A. Ovchinnikov: Zh. Eksp. Teor. Fiz. **57**, 263 (1969) Sov. Phys. JETP **30**, 147 (1970)
4. A.J. Sievers and S. Takeno: Phys. Rev. Lett. **61**, 970 (1988)
5. R.S. MacKay and S. Aubry: Nonlinearity **7**, 1623 (1994)
6. T. Bountis, H.W. Capel, M. Kollmann, J. Ross, J.M. Bergamin and J.P. van der Weele: Phys. Lett. A **268**, 50 (2000)
7. S. Flach: Phys. Rev. E **51**, 3579 (1995)
8. J.M. Bergamin, T. Bountis and M.N. Vrahatis: Nonlinearity **15**, 1603 (2002)

9. J.M. Bergamin, T. Bountis and C. Jung: J. Phys. A: Math. Gen. **33**, 8059 (2000)
10. S. Flach: Phys. Rev. E **50**, 3134 (1994)
11. Y.S. Kivshar: Phys. Rev. E **48**, R43 (1993)
12. G.P. Tsironis: Journal of Physics A **35**, 951 (2002)
13. J.M. Bergamin: Numerical approximation of breathers in lattices with nearest-neighbor interactions, to appear in Phys. Rev. E (2003)
14. J.M. Bergamin: Localization in nonlinear lattices and homoclinic dynamics. PhD thesis, University of Patras, Greece (2003)
15. T. Bountis, J.M. Bergamin and V. Basios: Phys. Lett. A **295**, 115 (2002)
16. H. Segur and M.D. Kruskal: Phys. Rev. Lett. **58**, 747 (1987)
17. J.M. Bergamin, T. Bountis and H.W. Capel: Continuation of discrete breathers using control methods, in preparation (2003).

Chaos or Order in Double Barred Galaxies?

Witold Maciejewski

[1] INAF – Osservatorio Astrofisico di Arcetri, Largo E. Fermi 5, 50125 Firenze, Italy,
[2] Obserwatorium Astronomiczne Uniwersytetu Jagiellońskiego, Kraków, Poland

Abstract. Bars in galaxies are mainly supported by particles trapped around closed periodic orbits. These orbits respond to the bar's forcing frequency only and lack free oscillations. We show that a similar situation takes place in double bars: particles get trapped around orbits which only respond to the forcing from the two bars and lack free oscillations. We find that writing the successive positions of a particle on such an orbit every time the bars align generates a closed curve, which we call a loop. Loops allow us to verify consistency of the potential. As maps of doubly periodic orbits, loops can be used to search the phase-space in double bars in order to determine the fraction occupied by ordered motions.

1 Introduction

Bars within bars appear to be a common phenomenon in galaxies. Recent surveys show that up to 30% of early-type barred galaxies contain nested bars [4]. The relative orientation of the two bars is random, therefore it is likely that the bars rotate with different pattern speeds. Inner bars, like large bars, are made of relatively old stellar populations, since they remain distinct in near infrared [5]. Galaxies with two independently rotating bars do not conserve the Jacobi integral, and it is a complex dynamical task to explain how such systems are sustained. To account for their longevity, one has to find sets of particles that support the shape of the potential in which they move. Particle motion in a potential of double bars belongs to the general problem of motion in a pulsating potential [6] [9], of which the restricted elliptical 3-body problem is the best known example. Families of closed periodic orbits have been found in this last problem, where the test particle moves in the potential of a binary star with components on elliptical orbits [2]. However, such families are parameterized by values that also characterize the potential (i.e. ellipticity of the stellar orbit and the mass ratio of the stars), and their orbital periods are commensurate with the pulsation period of the potential. For a given potential, these families are reduced to single orbits separated in phase-space. The solution for double bars is formally identical, and there an orbit can close only when the orbital period is commensurate with the relative period of the bars. Such orbits are separated in phase-space, and therefore families of closed periodic orbits are unlikely to provide orbital support for nested bars. Another difficulty in supporting nested bars is caused by the piling up of resonances created by each bar, which leads to considerable chaotic zones. In order to minimize the number of chaotic zones, resonant coupling between the bars has been proposed [10], so that the resonance generated by one bar overlaps with that caused by the other bar.

Finding support for nested bars has been hampered by the fact that closed periodic orbits are scarce there. However, it is particles, not orbits, which create density distributions that support the potential. The concept of closed periodic orbit is too limiting in investigation of nested bars, and another description of particle motion, which does not have its limitations, is needed. Naturally, in systems with two forcing frequencies, double-periodic orbits play a fundamental role. Thus in double bars a large fraction of particle trajectories gets trapped around a class of double-periodic orbits. Although such orbits do not close in any reference frame, they can be conveniently mapped onto the loops [8], which are an efficient descriptor of orbital structure in a pulsating potential. The loop is a closed curve that is made of particles moving in the potential of a doubly barred galaxy, and which pulsates with the relative period of the bars. Orbital support for nested bars can be provided by placing particles on the loops.

Here I give a systematic description of the loop approach, which recovers families of stable double-periodic orbits, and which can be applied to any pulsating potential. In Sect. 2 I use the epicyclic approximation to introduce the basic concepts, and in Sect. 3 I outline the general method.

2 The Epicyclic Solution for Any Number of Bars

If a galaxy has a bar that rotates with a constant pattern speed, it is convenient to study particle orbits in the reference frame rotating with the bar. If two or more bars are present, and each rotates with its own pattern speed, there is no reference frame in which the potential remains unchanged. In order to point out formal similarities in solutions for one and many bars, I solve the linearized equations in the inertial frame. This is equivalent to the solution in any rotating frame, and the transformation is particularly simple: in the rotating frame the centrifugal and Coriolis terms are equivalent to the Doppler shift of the angular velocity. It is convenient to show it in cylindrical coordinates (R, φ, z): if $\mathbf{e}_z$ is the rotation axis, then the R and φ components of the equation of motion for the rotating frame, $\ddot{\mathbf{r}} = -\nabla\Phi - 2(\mathbf{\Omega_B} \times \dot{\mathbf{r}}) - \mathbf{\Omega_B} \times (\mathbf{\Omega_B} \times \mathbf{r})$, can be written as

$$\ddot{R} - R(\dot{\varphi} + \Omega_B)^2 = -\frac{\partial \Phi}{\partial R},$$
$$R\ddot{\varphi} + 2\dot{R}(\dot{\varphi} + \Omega_B) = -\frac{1}{R}\frac{\partial \Phi}{\partial \varphi}.$$

These equations are identical with the components of the equation of motion in the inertial frame,

$$\ddot{\mathbf{r}} = -\nabla\Phi, \tag{1}$$

where clearly the angular velocity $\dot{\varphi}$ in the rotating frame corresponds to $\dot{\varphi} + \Omega_B$ in the inertial frame. For the rest of this section I assume the inertial frame, in which the equation of motion (1) has the following R and φ components in cylindrical coordinates

$$\ddot{R} - R\dot{\varphi}^2 = -\frac{\partial \Phi}{\partial R}, \tag{2}$$

$$R\ddot{\varphi} + 2\dot{R}\dot{\varphi} = -\frac{1}{R}\frac{\partial \Phi}{\partial \varphi}. \tag{3}$$

The z component in any frame is $\ddot{z} = -\partial\Phi/\partial z$, but I consider here motions in the plane of the disc only, hence I neglect the dependence on z.

To linearize equations (2) and (3), one needs expansions of R, φ and Φ to first order terms. The epicyclic approximation is valid for particles whose trajectories oscillate around circular orbits. For such particles one can write

$$R(t) = R_0 + R_I(t), \tag{4}$$

$$\varphi(t) = \varphi_{00} + \Omega_0 t + \varphi_I(t), \tag{5}$$

$$\Phi(R,\varphi,t) = \Phi_0(R) + \Phi_I(R,\varphi,t), \tag{6}$$

where terms with index I are small to the first order, and second- and higher-order terms were neglected. The parameter φ_{00} allows the particle to start from any position angle at time $t=0$, so that $\varphi_0 = \varphi_{00} + \Omega_0 t$. Asymmetry Φ_I in the potential is small and may be time-dependent. The angular velocity Ω_0 on the circular orbit of radius R_0 relates to the potential Φ_0 through the zeroth order of (2): $\Omega_0^2 = (1/R_0)(\partial\Phi_0/\partial R)\mid_{R_0}$. The zeroth order of (3) is identically equal to zero, and the first order corrections to (2) and (3) take respectively forms

$$\ddot{R}_I - 4A\Omega_0 R_I - 2R_0\Omega_0\dot{\varphi}_I = -\frac{\partial \Phi_I}{\partial R}\mid_{R_0,\varphi_0}, \tag{7}$$

$$R_0\ddot{\varphi}_I + 2\Omega_0\dot{R}_I = -\frac{1}{R_0}\frac{\partial \Phi_I}{\partial \varphi}\mid_{R_0,\varphi_0}, \tag{8}$$

where A is the Oort constant defined by $4A\Omega_0 = \Omega_0^2 - \frac{\partial^2\Phi_0}{\partial R^2}\mid_{R_0}$.

We assume that the bars are point-symmetric with respect to the galaxy centre. Thus to first order the departure of the barred potential from axial symmetry can be described by a term $\cos(2\varphi)$. If multiple bars, indexed by i, rotate independently as solid bodies with angular velocities Ω_i, the time-dependent first-order correction Φ_I to the potential can be written as

$$\Phi_I(R,\varphi,t) = \sum_i \Psi_i(R)\cos[2(\varphi - \Omega_i t)], \tag{9}$$

where the radial dependence $\Psi_i(R)$ has been separated from the angle dependence. No phase in the trigonometric functions above means that we define $t=0$ when all the bars are aligned. Derivatives of (9) enter right-hand sides of (7) and (8), which after introducing $\omega_i = 2(\Omega_0 - \Omega_i)$ take the form

$$\ddot{R}_I - 4A\Omega_0 R_I - 2R_0\Omega_0\dot{\varphi}_I = -\sum_i \frac{\partial \Psi_i}{\partial R}\mid_{R_0}\cos(\omega_i t + 2\varphi_{00}), \tag{10}$$

$$R_0\ddot{\varphi}_I + 2\Omega_0\dot{R}_I = \frac{2}{R_0}\sum_i \Psi_i(R_0)\sin(\omega_i t + 2\varphi_{00}). \tag{11}$$

In order to solve the set of equations (10,11), one can integrate (11) and get an expression for $R_0\dot{\varphi}_I$, which furthermore can be substituted to (10). This substitution eliminates φ_I, and one gets a single second order equation for R_I, which can be written schematically as

$$\ddot{R}_I + \kappa_0^2 R_I = \sum_i A_i \cos(\omega_i t + 2\varphi_{00}) + C_\varphi, \tag{12}$$

where $A_i = -\frac{4\Omega_0 \Psi_i}{\omega_i R_0} - \frac{\partial \Psi_i}{\partial R}_{|R_0}$, $\kappa_0^2 = 4\Omega_0(\Omega_0 - A)$, and $C_\varphi/2\Omega_0$ is the integration constant that appears after integrating (11). This is the equation of a harmonic oscillator with multiple forcing terms, whose solution is well known. It can be written as

$$R_I(t) = C_1 \cos(\kappa_0 t + \delta) + \sum_i M_i \cos(\omega_i t + 2\varphi_{00}) + C_\varphi/\kappa_0^2. \tag{13}$$

The first term of this solution corresponds to a free oscillation at the local epicyclic frequency κ_0, and C_1 is unconstrained. The terms under the sum describe oscillations resulting from the forcing terms in (9), and M_i are functions of A_i. Hereafter I focus on solutions without free oscillations, thus I assume that $C_1 = 0$. These solutions will lead to closed periodic orbits and to loops. The formula for $\varphi_I(t)$ can be obtained by substituting (1) into the time-integrated (11). As a result, one gets

$$\dot{\varphi}_I = \sum_i N_i \cos(\omega_i t + 2\varphi_{00}) - \frac{2AC_\varphi}{\kappa_0^2 R_0}, \tag{14}$$

where again N_i are determined by the coefficients of the equations above. Note that to the first order $\Omega_0[R_0 + C_\varphi/\kappa_0^2] = \Omega_0[R_0] - 2AC_\varphi/\kappa_0^2 R_0$, thus the integration constants entering (1) and (14) correspond to a change in the guiding radius R_0, and to the appropriate change in the angular velocity Ω_0. They all can be incorporated into R_0, and in effect the unique solutions for R_I and φ_I are

$$R_I(t) = \sum_i M_i \cos(\omega_i t + 2\varphi_{00}), \tag{15}$$

$$\varphi_I(t) = \sum_i N_i' \cos(\omega_i t + 2\varphi_{00}) + const, \tag{16}$$

where free oscillations have been neglected. The integration constant in (16) is an unconstrained parameter of the order of φ_I.

2.1 Closed Periodic Orbits in a Single Bar

In a potential with a single bar there is only one term in the sums (15) and (16), hereafter indexed with B. Consider the change in values of R_I and φ_I for a given particle after half of its period in the frame corotating with the bar. This interval is taken because the bar is bisymmetric, so its forcing is periodic with the period π in angle. After replacing t by $t + \pi/(\Omega_0 - \Omega_B)$ one gets

$$\begin{aligned} R_I &= M_B \cos[\omega_B (t + \frac{\pi}{\Omega_0 - \Omega_B}) + 2\varphi_{00}] \\ &= M_B \cos(\omega_B t + 2\pi + 2\varphi_{00}). \end{aligned}$$

Thus the solution for R_I after time $\pi/(\Omega_0 - \Omega_B)$ returns its starting value, and the same holds true for φ_I. After twice that time, i.e. in a full period of this particle in the bar frame, the epicycle centre returns to its starting point and the orbit closes. Thus (15) and (16) describe closed periodic orbits in the linearized problem of a particle motion in a single bar.

2.2 Loops in Double Bars

When two independently rotating bars coexist in a galaxy (hereafter indexed by B and S), there is no reference frame in which the potential is constant. Thus when a term from one bar in (15) and (16) returns to its starting value, the term from the other bar does not (unless the frequencies of the bars are commensurate). Therefore the particle's trajectory does not close in any reference frame. However, consider the change in value of R_I and φ_I after time $\pi/(\Omega_S - \Omega_B)$, which is the relative period of the bars. One gets

$$\begin{aligned} R_I &= M_B \cos[\omega_B (t + \frac{\pi}{\Omega_S - \Omega_B}) + 2\varphi_{00}] + M_S \cos[\omega_S (t + \frac{\pi}{\Omega_S - \Omega_B}) + 2\varphi_{00}] \\ &= M_B \cos(\omega_B t + 2\pi \frac{\Omega_0 - \Omega_B}{\Omega_S - \Omega_B} + 2\varphi_{00}) + M_S \cos(\omega_S t + 2\pi \frac{\Omega_0 - \Omega_S}{\Omega_S - \Omega_B} + 2\varphi_{00}) \\ &= M_B \cos(\omega_B t + 2\pi + 2\varphi_{01}) + M_S \cos(\omega_S t + 2\varphi_{01}), \end{aligned}$$

where $\varphi_{01} = \varphi_{00} + \pi \frac{\Omega_0 - \Omega_S}{\Omega_S - \Omega_B}$. The same result can be obtained for φ_I. This means that the time transformation $t \rightarrow t + \pi/(\Omega_S - \Omega_B)$ is equivalent to the change in the starting position angle of a particle from φ_{00} to φ_{01}. Consider motion of a set of particles that have the same guiding radius R_0, but start at various position angles φ_{00}. This is a one-parameter set, therefore in the disc plane it is represented by a curve, and because of continuity of (15) and (16) this curve is closed. After time $\pi/(\Omega_S - \Omega_B)$, a particle starting at angle φ_{00} will take the place of the particle which started at φ_{01}, a particle starting at φ_{01} will take the place of another particle from this curve and so on. The whole curve will regain its shape and position every $\pi/(\Omega_S - \Omega_B)$ time interval, although positions of particles on the curve will shift. This curve is the epicyclic approximation to the *loop*: a curve made of particles moving in a given potential, such that the curve returns to its original shape and position periodically. In the case of two bars, the period is the relative period of the bars, and the loop is made out of particles having the same guiding radius R_0. Particles on the loop respond to the forcing from the two bars, but they lack any free oscillation. An example of a set of loops in a doubly barred galaxy in the epicyclic approximation can be seen in [7]. Since they occupy a significant part of the disc, one should anticipate large zones of ordered motions also in the general, non-linear solution for double bars.

3 Full Nonlinear Solution for Loops in Nested Bars

Tools and concepts useful in the search for ordered motions in double bars are best introduced through the inspection of particle trajectories in such systems. For this inspection I chose the potential of Model 1 defined in [8], where the small bar is 60% in size of the big bar, and pattern speeds of the bars are not commensurate. Consider a particle moving in this potential inside the corotation of the small bar. Simple experiments with various initial velocities show that if the initial velocity is small enough, the particle usually remains bound. A typical trajectory is shown in the left panels of Fig.1 – since it depends on the reference frame, it is written twice, for reference frame of each bar. Further experimenting with initial velocities shows that particle trajectories are often even tidier: they look like those in the right panels of Fig.1, as if the trajectories were trapped around some regular orbit.

Fine adjustments of the initial velocity lead to a highly harmonious trajectory (Fig.2), which looks like a loop orbit in a potential of a single bar (see e.g. Fig.3.7a in [19]). This is only a formal similarity, but understanding it will let us find out what kind of orbit we see in Fig.2. The loop orbit in a single bar forms when a particle oscillates around a closed periodic orbit. Therefore two frequencies are involved: the frequency of the free oscillation, and the forcing frequency of the bar. On the other hand, the Fourier transform of the trajectory from Fig.2 shows two sharp peaks at frequencies equal to the forcing frequencies of the two bars (Fig.3). Thus the trajectory from Fig.2 also has two frequencies: this time these are the forcing frequencies from the two bars, while the free oscillation is absent. This is how the solution in the linear approximation (Sect. 2.2) was constructed. We conclude that in both the linear (epicyclic approximation) case

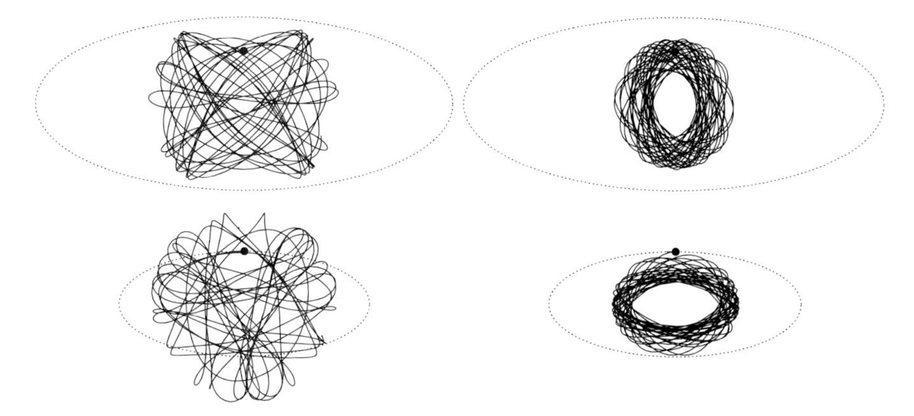

Fig. 1. Two example trajectories (one in the two left panels, one in the two right ones) of a particle that moves in the potential of two independently rotating bars. The particle is followed for 10 relative periods of the bars, and its trajectory is displayed in the frame corotating with the big bar (top panels), and the small bar (bottom panels). Each bar is outlined in its own reference frame by the dotted line. Large dot marks the starting point of the particle.

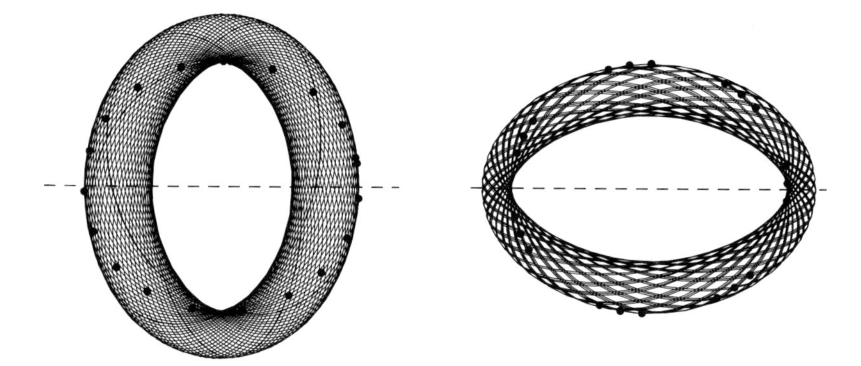

Fig. 2. A doubly periodic orbit in the doubly barred potential, followed for 20 relative periods of the bars, and written in the frame corotating with the big bar (left), and the small bar (right). The long axis of each bar is marked by the dashed line. Dots mark positions of the particle at every alignment of the bars.

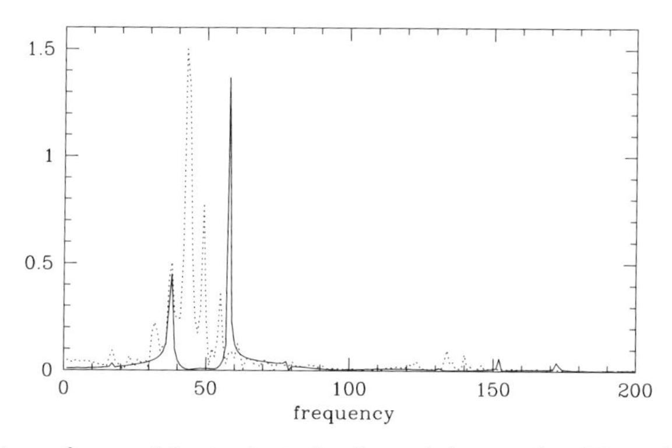

Fig. 3. Fourier transforms of the trajectories from right panels of Fig.1 (dotted line) and from Fig.2 (solid line). The peaks in the solid line are related to the forcing frequencies of the bars, and the peaks in the dotted line are not.

and the general case we are dealing with doubly periodic orbits in an oscillating potential of a double bar, with frequencies equal to the forcing frequencies of the bars. In the epicyclic approximation, these orbits have a nice feature that particles following them populate loops: closed curves that return to their original shape and position at every alignment of the bars. One may therefore expect that also in the general case these particles gather on loops.

If in the general case particles on doubly periodic orbits form a loop, one can construct it by writing positions of a particle on such an orbit every time the bars align. These positions are the initial conditions for particles forming the loop, because after every alignment, the nth particle generated in this way takes the position of particle $n+1$. The first 20 positions of a particle on a doubly periodic orbit are overplotted in Fig.2. They indeed seem to be arranged on a closed ellipse-like curve; the shape of this curve varies in time, but it returns to where it started at every alignment of the bars (Fig.4). This construction shows

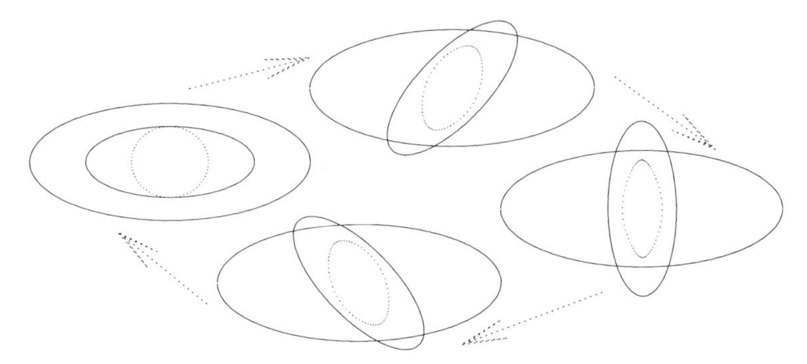

Fig. 4. Evolution of the loop from Fig.2 during one relative period of the bars. The bars, outlined with solid lines, rotate counterclockwise. The loop is made out of points that represent separate particles on doubly periodic orbits.

that in the general case particles on doubly periodic orbits also form loops. Note that positions of particles on other orbits, which involve free oscillations, when written at every alignment of the bars, densely populate some two-dimensional section of the plane, and do not gather on any curve. It is extremely useful for the investigation of the orbital structure in double bars that the appearance of the loop is frame-independent. Loops provide an efficient way to classify doubly periodic orbits, which has been hampered so far by the dependence of the last ones on the reference frame.

It turns out that doubly periodic orbits play crucial role in providing orbital support for the pulsating potential of double bars. No closed periodic orbits have been proposed as candidates for the backbone of such a potential. If in a given potential of two bars there are loops that follow the inner bar, and other loops that follow the outer bar, then one may expect that such a potential is dynamically possible. An example of such a potential has been constructed in [8]. The loop from Fig.4 does not follow either bar in its motion, and therefore it is unlikely that it supports the assumed potential. It can be shown that in that potential, there are no loops which could support the two bars. Thus that potential is not self-consistent. This example shows how efficient is the loop approach in rejecting hypothetical doubly barred systems that have no orbital support.

Doubly periodic orbits in double bars are surrounded by regular orbits in the same way as are the closed periodic orbits in a single bar. In both cases, the trapped regular orbits oscillate around the parent orbit. The trajectory from the right panels of Fig.1 is an example of a regular orbit that is trapped around the doubly periodic orbit from Fig.2. How much of the phase space in double bars is occupied by orbits trapped around doubly periodic orbits? It can be examined by launching a particle from e.g. the minor axis of the bar, in the direction perpendicular to this axis, when the bars are aligned. If the particle is trapped, its positions at every alignment of the two bars lie within a ring containing the loop. The width of this ring depends on the particle's position along the minor axis, and on its velocity. It is displayed in Fig.5 for the potential

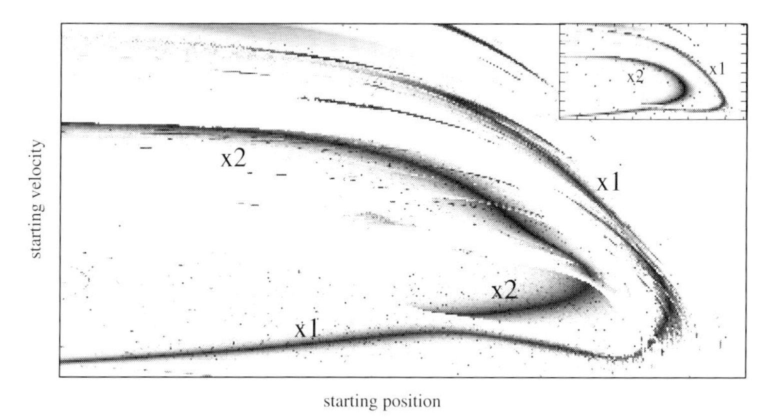

Fig. 5. The width of the ring formed by particles trapped around loops in Model 2 from [8] as a function of the particle's position along the minor axis of the aligned bars, and of its velocity (perpendicular to this axis). Darker color means smaller width. In the insert, the same is shown for rings around closed periodic orbits in a single bar (same model, but inner bar axisymmetric). Regions related to the x_1 and x_2 orbits, and to the loops originating from them, are marked.

of Model 2 defined in [8]. Two stripes of low width appear on the diagram, which correspond to the x_1 and x_2 orbits in a single bar [3] (displayed in the insert). Thus in double bars there are doubly periodic orbits that correspond to closed periodic orbits in single bars. There are possible regions of chaos in double bars (white stripes in Fig.5), but overall loops in double bars and periodic orbits in single bars trap similar volumes of phase-space around them.

4 Conclusions

In a potential of two independently rotating bars, a large fraction of phase space can be occupied by trajectories trapped around parent regular orbits. These orbits are doubly periodic, with the two periods corresponding to the forcing frequencies of the two bars, but they do not close in any reference frame. Like particle trajectories oscillating around closed periodic orbits in a single bar, particle trajectories in double bars oscillate around the doubly periodic parent orbits. The structure of the parent regular orbits can be mapped using the loop approach, which allows us to single out dynamically possible double bars.

Acknowledgments

The concept of the loop as the organized form of motion in double bars benefits from the insight of Linda Sparke. I thank Lia Athanassoula for our collaboration that lead to this paper, and Peter Erwin for comments on the manuscript.

References

1. Binney, J. & Tremaine, S. 1987, Galactic Dynamics (Princeton: Princeton Univ. Press)
2. Broucke, R. A. 1969, Periodic Orbits in the Elliptic Restricted Three-Body Problem, Jet Propulsion Laboratory Report 32-1360
3. Contopoulos, G. & Papayannopoulos, Th. 1980, A&A, 92, 33
4. Erwin, P. & Sparke, L. S. 2002, AJ, 124, 65
5. Friedli, D., Wozniak, H., Rieke, M., Martinet, L., & Bratschi, P. 1996, A&AS, 118, 461
6. Louis, P. D. & Gerhard, O. E. 1988, MNRAS, 233, 337
7. Maciejewski, W. & Sparke, L. S. 1997, ApJL, 484, L117
8. Maciejewski, W. & Sparke, L. S. 2000, MNRAS, 313, 745
9. Sridhar, S. 1989, MNRAS, 238, 1159
10. Sygnet, J. F., Tagger, M., Athanassoula, E., & Pellat, R. 1988, MNRAS, 232, 733

Jeans Solutions for Triaxial Galaxies

Glenn van de Ven[1], Chris Hunter[2], Ellen Verolme[1], and Tim de Zeeuw[1]

[1] Sterrewacht Leiden, Postbus 9513, 2300 RA Leiden, The Netherlands
[2] Department of Mathematics, Florida State University, Tallahassee, FL 32306-4510, USA

Abstract. The Jeans equations relate the second-order velocity moments to the density and potential of a stellar system. For general three-dimensional stellar systems, there are three equations, but these are not very helpful, as they contain six independent moments. By assuming that the potential is triaxial and of separable Stäckel form, the mixed moments vanish in confocal ellipsoidal coordinates. The three Jeans equations and three remaining non-vanishing moments form a closed system of three highly-symmetric coupled first-order partial differential equations in three variables. They were first derived by Lynden–Bell in 1960, but have resisted solution by standard methods. Here we present the general solution by superposition of singular solutions.

1 Introduction

Much has been learned about the mass distribution and internal dynamics of galaxies by modeling their observed kinematics with solutions of the Jeans equations (e.g. [4]). The Jeans equations connect the second-order velocity moments (or the velocity dispersions, if the mean streaming motion is known) directly to the density and the gravitational potential of the galaxy, without the need to know the phase-space distribution function f. In nearly all cases there are fewer Jeans equations than velocity moments, so that additional assumptions have to be made about the degree of anisotropy. Furthermore, the resulting second moments may not correspond to a physical distribution function $f \geq 0$. These significant drawbacks have not prevented wide application of the Jeans approach to the kinematics of spherical and axisymmetric galaxies. Many (components of) galaxies have triaxial shapes ([2], [3]), including early-type bulges, bars, and giant elliptical galaxies. In this geometry, there are three Jeans equations, but little use has been made of them, as they contain six independent second moments, three of which have to be chosen ad-hoc (see e.g. [11]).

An exception is provided by the special set of triaxial mass models that have a gravitational potential of Stäckel form. In these systems, the Hamilton–Jacobi equation separates in confocal ellipsoidal coordinates ([19]), so that all orbits have three exact integrals of motion, which are quadratic in the velocities. The three mixed second-order velocity moments vanish, so that the three Jeans equations for the three remaining second moments form a closed system. Lynden–Bell ([13]) was the first to derive the explicit form of these Jeans equations. He showed that they constitute a highly symmetric set of three first-order partial differential equations for three unknowns, each of which is a function of the ellipsoidal coordinates, but he did not derive solutions.

When it was realized that the orbital structure in the triaxial Stäckel models is very similar to that in numerical models for triaxial galaxies with cores ([6], [16]), interest in the second moments increased, and the Jeans equations were solved for a number of special cases. These include the axisymmetric limits and elliptic discs ([8], [10]), triaxial galaxies with only thin tube orbits ([12]), and the scale-free limit ([11]). In all these cases the equations simplify to a two-dimensional problem, which can be solved with standard techniques after transforming two first-order equations into a single second-order equation in one dependent variable. However, these techniques do not carry over to a single third-order equation in one dependent variable, which is the best that one could expect to have in the general case. As a result, the latter has remained unsolved.

We have solved the two-dimensional case with an alternative solution method, which does not use the standard approach, but instead uses superposition of singular solutions. This approach can be extended to three dimensions, and provides the general solution for the triaxial case in closed form. We present the detailed solution method elsewhere ([23]), and here we summarise the main results. In ongoing work we will apply our solutions, and will use them together with the mean streaming motions ([20]) to study the properties of the observed velocity and dispersion fields of triaxial galaxies.

2 The Jeans Equations for Separable Models

We define confocal ellipsoidal coordinates (λ, μ, ν) as the three roots for τ of

$$\frac{x^2}{\tau+\alpha} + \frac{y^2}{\tau+\beta} + \frac{z^2}{\tau+\gamma} = 1 \, , \tag{1}$$

with (x, y, z) the usual Cartesian coordinates, and with constants α, β and γ such that $-\gamma \leq \nu \leq -\beta \leq \mu \leq -\alpha \leq \lambda$. Surfaces of constant λ are ellipsoids, and surfaces of constant μ and ν are hyperboloids of one and two sheets, respectively. The confocal ellipsoidal coordinates are approximately Cartesian near the origin and become conical at large radii, i.e., equivalent to spherical coordinates.

We consider models with a gravitational potential of Stäckel form

$$V_S(\lambda, \mu, \nu) = -\frac{F(\lambda)}{(\lambda-\mu)(\lambda-\nu)} - \frac{F(\mu)}{(\mu-\nu)(\mu-\lambda)} - \frac{F(\nu)}{(\nu-\lambda)(\nu-\mu)} \, , \tag{2}$$

where $F(\tau)$ is an arbitrary smooth function. This potential is the most general form for which the Hamilton–Jacobi equation separates ([15], [18]) All orbits have three exact isolating integrals of motion, which are quadratic in the velocities (e.g. [6]). There are no irregular orbits, so that Jeans' theorem is strictly valid ([14]), and the distribution function f is a function of the three integrals. Therefore, out of the six symmetric second-order velocity moments, defined as

$$\langle v_i v_j \rangle(\boldsymbol{x}) = \frac{1}{\varrho} \iiint v_i v_j f(\boldsymbol{x}, \boldsymbol{v}) \, \mathrm{d}^3 v \, , \quad (i, j = 1, 2, 3), \tag{3}$$

with density ϱ, the three mixed moments vanish, and we are left with $\langle v_\lambda^2 \rangle$, $\langle v_\mu^2 \rangle$ and $\langle v_\nu^2 \rangle$, related by three Jeans equations. These were first derived by Lynden-Bell ([13]), and can be written in the following form ([23])

$$\frac{\partial S_{\lambda\lambda}}{\partial \lambda} - \frac{S_{\mu\mu}}{2(\lambda-\mu)} - \frac{S_{\nu\nu}}{2(\lambda-\nu)} = g_1(\lambda, \mu, \nu) \,, \tag{4a}$$

$$\frac{\partial S_{\mu\mu}}{\partial \mu} - \frac{S_{\nu\nu}}{2(\mu-\nu)} - \frac{S_{\lambda\lambda}}{2(\mu-\lambda)} = g_2(\lambda, \mu, \nu) \,, \tag{4b}$$

$$\frac{\partial S_{\nu\nu}}{\partial \nu} - \frac{S_{\lambda\lambda}}{2(\nu-\lambda)} - \frac{S_{\mu\mu}}{2(\nu-\mu)} = g_3(\lambda, \mu, \nu) \,, \tag{4c}$$

where we have defined the diagonal components of the stress tensor

$$S_{\tau\tau}(\lambda, \mu, \nu) = \sqrt{(\lambda-\mu)(\lambda-\nu)(\mu-\nu)}\, \varrho \langle v_\tau^2 \rangle \,, \qquad \tau = \lambda, \mu, \nu, \tag{5}$$

and the functions g_1, g_2 and g_3 depend on the density and potential (2) as

$$g_1(\lambda, \mu, \nu) = -\sqrt{(\lambda-\mu)(\lambda-\nu)(\mu-\nu)}\, \varrho \frac{\partial V_S}{\partial \lambda} \,, \tag{6}$$

where g_2 and g_3 follow from g_1 by cyclic permutation $\lambda \to \mu \to \nu \to \lambda$. Similarly, the three Jeans equations follow from each other by cyclic permutation. The stress components have to satisfy the following continuity conditions

$$S_{\lambda\lambda}(-\alpha, -\alpha, \nu) = S_{\mu\mu}(-\alpha, -\alpha, \nu) \,, \quad S_{\mu\mu}(\lambda, -\beta, -\beta) = S_{\nu\nu}(\lambda, -\beta, -\beta) \,, \tag{7}$$

at the focal ellipse ($\lambda = \mu = -\alpha$) and focal hyperbola ($\mu = \nu = -\beta$), respectively.

We prefer the form (5) for the stresses instead of the more common definition without the square root, since it results in more convenient and compact expressions. In self-consistent models, the density ϱ equals ϱ_S, with ϱ_S related to V_S by Poisson's equation. The Jeans equations, however, do not require self-consistency, so that we make no assumptions on the form of ϱ other than that it is triaxial, i.e., a function of (λ, μ, ν), and that it tends to zero at infinity.

3 The Two-Dimensional Case

When two or all three of the constants α, β or γ in (1) are equal, the triaxial Stäckel models reduce to limiting cases with more symmetry and thus with fewer degrees of freedom. Solving the Jeans equations for oblate, prolate, elliptic disc and scale-free models reduces to the same two-dimensional problem ([10], [11], [23]), of which the simplest form is the pair of Jeans equations for Stäckel discs

$$\frac{\partial S_{\lambda\lambda}}{\partial \lambda} - \frac{S_{\mu\mu}}{2(\lambda-\mu)} = g_1(\lambda, \mu) \,, \tag{8a}$$

$$\frac{\partial S_{\mu\mu}}{\partial \mu} - \frac{S_{\lambda\lambda}}{2(\mu-\lambda)} = g_2(\lambda, \mu) \,, \tag{8b}$$

with at the foci ($\lambda = \mu = -\alpha$) the continuity condition

$$S_{\lambda\lambda}(-\alpha, -\alpha) = S_{\mu\mu}(-\alpha, -\alpha) \,. \tag{9}$$

In this case the stress components and the functions g_1 and g_2 are

$$S_{\tau\tau}(\lambda, \mu) = \sqrt{(\lambda-\mu)}\, \varrho \langle v_\tau^2 \rangle \quad (\tau = \lambda, \mu), \quad g_1(\lambda, \mu) = -\sqrt{(\lambda-\mu)}\, \varrho \frac{\partial V_S}{\partial \lambda} \,, \tag{10}$$

where g_2 follows from g_1 by interchanging $\lambda \leftrightarrow \mu$, and ϱ denotes a surface density.

The two Jeans equations (8) can be recast into a single second-order partial differential equation in either $S_{\lambda\lambda}$ or $S_{\mu\mu}$, which can be solved by employing standard techniques like Riemann's method ([5], [23]). However, these standard techniques do not carry over to the triaxial case, and we therefore introduce an alternative method, based on the superposition of singular solutions.

We consider a simpler form of (8) by substituting for g_1 and g_2, respectively $\tilde{g}_1 = 0$ and $\tilde{g}_2 = \delta(\lambda_0 - \lambda)\delta(\mu_0 - \mu)$. We refer to solutions of these simplified Jeans equations as *singular solutions*. Singular solutions can be interpreted as contributions to the stresses at a fixed field point (λ, μ) due to a source point in (λ_0, μ_0) (Fig. 1). The full stress at the field point can be obtained by adding all source point contributions, each with a weight that depends on the local density and potential. Once we know the singular solutions, we can use the superposition principle to construct the the solution of the full Jeans equations (8).

Since the derivative of a step-function $\mathcal{H}$ is equal to a delta-function, it follows that the singular solutions must have the form

$$\begin{aligned} S_{\lambda\lambda} &= A(\lambda, \mu)\mathcal{H}(\lambda_0 - \lambda)\mathcal{H}(\mu_0 - \mu) \,, \\ S_{\mu\mu} &= B(\lambda, \mu)\mathcal{H}(\lambda_0 - \lambda)\mathcal{H}(\mu_0 - \mu) - \delta(\lambda_0 - \lambda)\mathcal{H}(\mu_0 - \mu) \,. \end{aligned} \tag{11}$$

where the functions A and B must solve the homogeneous Jeans equations, i.e., (8) with zero right-hand side, and satisfy the following boundary conditions

$$A(\lambda_0, \mu) = \frac{1}{2(\lambda_0 - \mu)} \,, \qquad B(\lambda, \mu_0) = 0 \,. \tag{12}$$

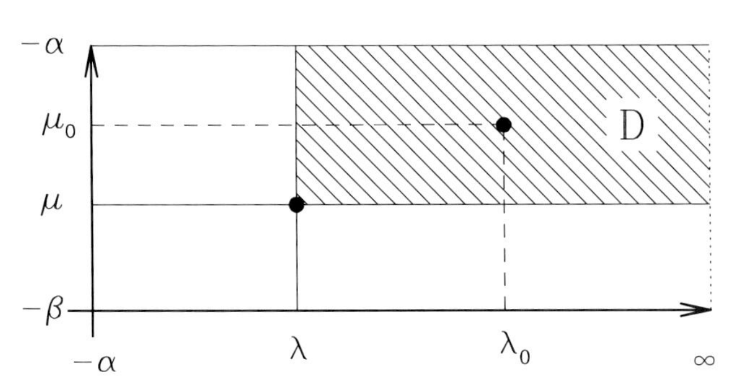

Fig. 1. *The (λ_0, μ_0)-plane. The total stress at a field point (λ, μ) consists of the weighted contributions from source points at (λ_0, μ_0) in the domain D.*

We solve this two-dimensional homogeneous boundary problem by superposition of particular solutions. We first derive a particular solution of the homogeneous Jeans equations with a free parameter z, which we assume to be complex. We then construct a linear combination of these particular solutions by integrating over z. We choose the integration contours in the complex z-plane, such that the boundary conditions (12) are satisfied simultaneously. The resulting homogeneous solutions are complex contour integrals, which can be evaluated in terms of the complete elliptic integral of the second kind, $E(w) \equiv \int_0^{\frac{\pi}{2}} \mathrm{d}\theta \sqrt{1 - w \sin^2 \theta}$, and its derivative $E'(w)$, as

$$A = \frac{E(w)}{\pi(\lambda_0 - \mu)}, \quad B = -\frac{2wE'(w)}{\pi(\lambda_0 - \lambda)}, \quad \text{with} \quad w = \frac{(\lambda_0 - \lambda)(\mu_0 - \mu)}{(\lambda_0 - \mu_0)(\lambda - \mu)}. \tag{13}$$

We obtain a similar system of simplified Jeans equations by interchanging the expressions for $\tilde{g}_1$ and $\tilde{g}_2$. The singular solutions of this simplified system follow from (11) by interchanging $\lambda \leftrightarrow \mu$ and $\lambda_0 \leftrightarrow \mu_0$ at the same time.

To find the solution to the full Jeans equations (8) at (λ, μ), we multiply the latter singular solutions and (11) by $g_1(\lambda_0, \mu_0)$ and $g_2(\lambda_0, \mu_0)$ respectively, and integrate over $D = \{(\lambda_0, \mu_0)\colon \lambda \leq \lambda_0 \leq \infty, \mu \leq \mu_0 \leq -\alpha\}$ (Fig. 1). This gives the first two integrals of the two equations (14a) and (14b) below. The remaining terms are due to the non-vanishing stress at the boundary $\mu = -\alpha$, and are found by multiplying the singular solutions (11), evaluated at $\mu_0 = -\alpha$, by $-S_{\mu\mu}(\lambda_0, -\alpha)$ and integrating over λ_0 in D. The final result for the solution of the Jeans equations (8) for Stäckel discs, after using the evaluations (13), is

$$\begin{aligned} S_{\lambda\lambda}(\lambda, \mu) = &\int_\lambda^\infty \mathrm{d}\lambda_0 \int_\mu^{-\alpha} \mathrm{d}\mu_0 \left[-g_1(\lambda_0, \mu_0) \frac{2wE'(w)}{\pi(\mu_0 - \mu)} + g_2(\lambda_0, \mu_0) \frac{E(w)}{\pi(\lambda_0 - \mu)} \right] \\ &- \int_\lambda^\infty \mathrm{d}\lambda_0 \, g_1(\lambda_0, \mu) - \int_\lambda^\infty \mathrm{d}\lambda_0 \, S_{\mu\mu}(\lambda_0, -\alpha) \left[\frac{E(w)}{\pi(\lambda_0 - \mu)} \right]_{\mu_0 = -\alpha}, \end{aligned} \tag{14a}$$

$$\begin{aligned} S_{\mu\mu}(\lambda, \mu) = &\int_\lambda^\infty \mathrm{d}\lambda_0 \int_\mu^{-\alpha} \mathrm{d}\mu_0 \left[-g_1(\lambda_0, \mu_0) \frac{E(w)}{\pi(\lambda - \mu_0)} - g_2(\lambda_0, \mu_0) \frac{2wE'(w)}{\pi(\lambda_0 - \lambda)} \right] \\ &- \int_\mu^{-\alpha} \mathrm{d}\mu_0 \, g_2(\lambda, \mu_0) + S_{\mu\mu}(\lambda, -\alpha) - \int_\lambda^\infty \mathrm{d}\lambda_0 \, S_{\mu\mu}(\lambda_0, -\alpha) \left[-\frac{2wE'(w)}{\pi(\lambda_0 - \lambda)} \right]_{\mu_0 = -\alpha}. \end{aligned} \tag{14b}$$

The solution depends on ϱ and V_S through g_1 and g_2. This means that, for given ϱ and V_S, the solution is uniquely determined once we have prescribed $S_{\mu\mu}$ at the boundary $\mu = -\alpha$. At this boundary, $S_{\lambda\lambda}$ is related to $S_{\mu\mu}$ by the first Jeans equation (8a), evaluated at $\mu = -\alpha$, up to an integration constant, which is fixed by the continuity condition (9). We are thus free to specify either of the two stress components at $\mu = -\alpha$.

4 The General Case

The singular solution method introduced in the previous section can be extended to three dimensions to solve the Jeans equations (4) for triaxial Stäckel models. Although the calculations are more complex for a triaxial model, the stepwise solution method is similar to that in two dimensions.

We simplify the Jeans equations (4) by setting two of the three functions g_1, g_2 and g_3 to zero and the remaining equal to $\delta(\lambda_0-\lambda)\delta(\mu_0-\mu)\delta(\nu_0-\nu)$. In this way, we obtain three similar simplified systems ($i=1,2,3$), each with three singular solutions $S_i^{\tau\tau}(\lambda,\mu,\nu;\lambda_0,\mu_0,\nu_0)$ ($\tau=\lambda,\mu,\nu$), that describe the stress components at a fixed field point (λ,μ,ν) due to a source point in (λ_0,μ_0,ν_0).

The singular solutions have a form that is similar to that in the two-dimensional case (11). They consist of combinations of step-functions and delta-functions multiplied by functions that are the solutions of homogeneous boundary problems. The functions that must solve a two-dimensional homogeneous boundary problem can be found as in Sect. 3, and can be expressed in terms of complete elliptic integrals, cf. (13). The singular solutions in the general case also contain three functions A, B and C that must solve the triaxial homogeneous Jeans equations, i.e., (4) with zero right-hand side, and satisfy three boundary conditions. This three-dimensional homogeneous boundary problem can be solved by integrating a *two*-parameter particular solution over both its complex parameters, and choosing the combination of contours such that the three boundary conditions are satisfied simultaneously. The resulting homogeneous solutions A, B and C are products of complex contour integrals, and can be evaluated as sums of products of complete *hyper*elliptic integrals.

To find the solution of the full Jeans equations (4) , we multiply each singular solution $S_i^{\tau\tau}$ by $g_i(\lambda_0,\mu_0,\nu_0)$, so that the contribution from the source point naturally depends on the local density and potential. Then, for each coordinate $\tau=\lambda,\mu,\nu$, we add the three weighted singular solutions, and integrate over a finite volume within the physical region $-\gamma\leq\nu\leq-\beta\leq\mu\leq-\alpha\leq\lambda$. This results in the following general solution of the Jeans equations (4) for triaxial Stäckel models

$$\begin{aligned}
S_{\tau\tau}(\lambda,\mu,\nu) = & \int_\lambda^{\lambda_e}\mathrm{d}\lambda_0\int_\mu^{\mu_e}\mathrm{d}\mu_0\int_\nu^{\nu_e}\mathrm{d}\nu_0\sum_{i=1}^{3} g_i(\lambda_0,\mu_0,\nu_0)\,S_i^{\tau\tau}(\lambda,\mu,\nu;\lambda_0,\mu_0,\nu_0)\\
& -\int_\mu^{\mu_e}\mathrm{d}\mu_0\int_\nu^{\nu_e}\mathrm{d}\nu_0\,S_{\lambda\lambda}(\lambda_e,\mu_0,\nu_0)\,S_1^{\tau\tau}(\lambda,\mu,\nu;\lambda_e,\mu_0,\nu_0)\\
& -\int_\nu^{\nu_e}\mathrm{d}\nu_0\int_\lambda^{\lambda_e}\mathrm{d}\lambda_0\,S_{\mu\mu}(\lambda_0,\mu_e,\nu_0)\,S_2^{\tau\tau}(\lambda,\mu,\nu;\lambda_0,\mu_e,\nu_0)\\
& -\int_\lambda^{\lambda_e}\mathrm{d}\lambda_0\int_\mu^{\mu_e}\mathrm{d}\mu_0\,S_{\nu\nu}(\lambda_0,\mu_0,\nu_e)\,S_3^{\tau\tau}(\lambda,\mu,\nu;\lambda_0,\mu_0,\nu_e)\;,
\end{aligned} \tag{15}$$

with $\tau = \lambda, \mu, \nu$. Whereas the integration limits λ, μ and ν are fixed due to the position of the field point, the limits λ_e, μ_e and ν_e are not, and may be any value in the corresponding physical ranges, i.e., $\lambda_e \in [-\alpha, \infty]$, $\mu_e \in [-\beta, -\alpha]$ and $\nu_e \in [-\gamma, -\beta]$, but $\lambda_e \neq -\alpha$. The latter choice would lead to solutions which generally have the incorrect radial fall-off, and hence are non-physical. If we choose $\lambda_e = \infty$, there is no contribution from the second line in (15) due to vanishing stress at large distance. If we furthermore take $\mu_e = -\alpha$ and $\nu_e = -\beta$, the integration volume becomes the three-dimensional extension of D (Fig. 1).

Whereas the volume integral in (15) already solves the inhomogeneous Jeans equations (4) , the three area integrals are needed to obtain the correct values at the boundary surfaces $\lambda = \lambda_e$, $\mu = \mu_e$ and $\nu = \nu_e$. On each of these surfaces the three stress components are related by two of the three Jeans equations (4) and the continuity conditions (7). Since the (weight) functions g_i are known for given ϱ and V_S, this means that the solution (15) yields all three stresses everywhere in the triaxial model, once one of the stress components is prescribed on the three boundary surfaces. If we take $\lambda_e = \infty$ and $\mu_e = \nu_e = -\beta$, the contributing boundary surfaces reduce to the single (x, z)-plane, containing the long and the short axis of the galaxy. This compares well with Schwarzschild ([17]), who used the same plane to start his numerically calculated orbits from.

5 Discussion and Conclusions

Eddington ([9]) showed that the velocity ellipsoid in a triaxial galaxy with a separable potential of Stäckel form is everywhere aligned with the confocal ellipsoidal coordinate system in which the equations of motion separate. Lynden-Bell ([13]) derived the three Jeans equations which relate the three principal stresses to the potential and the density. Solutions were found for the various two-dimensional limiting cases, but with methods that do not carry over to the general case, which remained unsolved. We have presented an alternative solution method, based on the superposition of singular solutions (see [23] for details). This approach, unlike the standard techniques, can be generalised to solve the three-dimensional system. The resulting solutions contain complete (hyper)elliptic integrals, which can be evaluated in a straightforward way.

The general Jeans solution is not unique, but requires specification of principal stresses at certain boundary surfaces, given a separable triaxial potential and a triaxial density distribution (not necessarily the one that generates the potential). These boundary surfaces can be taken to be the plane containing the long and the short axis of the galaxy, and, more specifically, the part that is crossed by all three families of tube orbits and the box orbits.

The set of all Jeans solutions (15) contains all the stresses that are associated with the physical distribution functions $f \geq 0$, but, as in the case of spherical and axisymmetric models, also contains solutions which are unphysical, e.g., those associated with distribution functions that are negative in some parts of phase space. The many examples of the use of spherical and axisymmetric Jeans models in the literature suggest nevertheless that the Jeans solutions can be of significant use.

While triaxial models with a separable potential do not provide an adequate description of the nuclei of galaxies with cusped luminosity profiles and a massive central black hole ([7]), they do catch much of the orbital structure at larger radii, and in some cases even provide a good approximation of the galaxy potential. The solutions for the mean streaming motions, i.e., the first velocity moments of the distribution function, are helpful in understanding the variety of observed velocity fields in giant elliptical galaxies and constraining their intrinsic shapes (e.g. [1], [21], [22]). We expect that the projected velocity dispersion fields that can be derived from our Jeans solutions will be similarly useful, and, in particular, that they can be used to establish which combinations of viewing directions and intrinsic axis ratios are firmly ruled out by the observations.

It is remarkable that the entire Jeans solution can be written down by means of classical methods. This suggests that similar solutions can be found for the higher dimensional analogues of (4) , most likely involving hyperelliptic integrals of higher order. It is also likely that the higher-order velocity moments for the separable triaxial models can be found by similar analytic means, but the effort required may become prohibitive.

References

1. Arnold R., de Zeeuw P. T., Hunter C., 1994, MNRAS, 271, 924
2. Binney J., 1976, MNRAS, 177, 19
3. Binney J., 1978, MNRAS, 183, 501
4. Binney J., Tremaine S., 1987, Galactic Dynamics. Princeton, NJ, Princeton University Press
5. Copson E. T., 1975, Partial Differential Equations. Cambridge, Cambridge University Press
6. de Zeeuw P. T., 1985, MNRAS, 216, 273
7. de Zeeuw P. T., Peletier R., Franx M., 1986, MNRAS, 221, 1001
8. Dejonghe H., de Zeeuw P. T., 1988, ApJ, 333, 90
9. Eddington A. S., 1915, MNRAS, 76, 37
10. Evans N. W., Lynden-Bell D., 1989, MNRAS, 236, 801
11. Evans N. W., Carollo C. M., de Zeeuw P. T., 2000, MNRAS, 318, 1131
12. Hunter C., de Zeeuw P. T., 1992, ApJ, 389, 79
13. Lynden-Bell D., 1960, PhD thesis, Cambridge University
14. Lynden-Bell D., 1962, MNRAS, 124, 1
15. Lynden-Bell D., 1962, MNRAS, 124, 95
16. Schwarzschild M., 1979, ApJ, 232, 236
17. Schwarzschild M., 1993, ApJ, 409, 563
18. Stäckel P., 1890, Math. Ann., 35, 91
19. Stäckel P., 1891, Über die Integration der Hamilton-Jacobischen Differential gleichung mittelst Separation der Variabeln. Habilitationsschrift, Halle
20. Statler T. S., 1994, ApJ, 425, 458
21. Statler T. S., 2001, AJ, 121, 244
22. Statler, T. S., Dejonghe, H., Smecker-Hane, 1999, AJ, 117, 126
23. van de Ven G., Hunter C., Verolme E. K., de Zeeuw P. T., 2003, MNRAS, submitted

Nonlinear Response of the Interstellar Gas Flow to Galactic Spiral Density Waves

Edward Liverts[1,2] and Michael Mond[1]

[1] Ben-Gurion University, of the Negev, P.O. Box 653, Beer-Sheva 84105, Israel
[2] Yaroslav-the-Wise Novgorod State University, Novgorod 173021, Russia

Abstract. Supersonic nonlinear gas flow is studied in order to describe galactic spiral density waves. It is shown analytically that ultra harmonic periodic solutions may exist if nonlinear effects are taken into account. The relevance of those solutions to observed data is discussed.

1 Introduction

An extensive literature has been devoted to simulations of the large-scale flow of interstellar gas in a stellar spiral density wave in galaxies, e.g. [1–4]. This study has played an important role in the understanding of many processes occurring in galaxies such as the structure of dust lanes observed along the inner edges of spiral arms in many galaxies [5,6]; the enhanced synchrotron radiation from spiral arms [7,8]; the radio emission of HI at the wavelength 21 cm [9]; a trigger mechanism for star formation, and the creation the narrow bands of young highly luminous stars [2,10,11]. A review of the problem is presented in [12–14]. It has been shown that spiral density waves, propagating in the galactic disk and interpreted as the galactic spiral arms, may induce large nonlinear perturbations in the gas flow. It was suggested that such nonlinear phenomena take place in the gas even though the amplitude of the spiral stellar field is relatively small. This is so since the response to the gravitational potential induced by the stellar density wave is roughly proportional to a^{-2}, where a is the velocity dispersion for stars, and the sound speed for the gas [4]. For the gas $a \sim 8$ km/s, while for the stars $a \sim 40$ km/s. Hence, if the amplitude of the perturbation field is small, one can use a linear theory to calculate a disturbance in the stellar disk. In contrast, the perturbation in the gaseous component is much stronger and one has to use a nonlinear theory. The steady flow in the nonlinear regime was considered numerically and it was shown, in particular, that a secondary shock wave is possible (this was borne out by numerical calculations carried out for the range of galactocentric radius r from $10.0kpc$ to $12.5kpc$ in the adopted model) see [11]. The nonlinear effect may well account for the origin of the Sagitta-Carina feature and relatively short spurs or feathers (see [11], and references therein). Moreover, the nonlinear effect may provide an answer to the old puzzle of how a two-arm potential drives multiple arms. In optical images we can see primarily a brightness distribution, which, generally speaking, does not reflect the over-all mass distribution. The spiral arms owe their high luminosity to the fact that the brightest objects in the galaxy are concentrated in them: giant stars of the early OB spectral classes and ionized hydrogen HII regions which have high luminosity in their emission line. Certain galaxies show a distinct spiral structure outlined by stars of later spectral class that belong to the old population of galaxy's spherical and disk subsystems. This effect in galactic morphology was first noticed by Zwicky [15] who detected smooth red arms in the disk of the grand design galaxy M51 and showed that the morphology of the evolved disk population need not follow the Hubble classification assigned from the young population tracers. A similar point was also made by Vorontsov-Vel'yaminov [16] and the same picture was later observed in certain other galaxies. Two images of the giant grand design Hubble type Sc galaxy NGC 309 seen almost face-on, one in blue light and the other in near infrared were published [17].

The aim of this paper is to analytically obtain conditions under which spiral density wave may give rise to a nonlinear response of parametrically excited and forced gas flow. That question is considered in Sect. 3 In addition it is aimed to consider gas-star structures presented in Sect. 4 that are typical for real galaxies in contrast to the results obtained from the purely gaseous response calculations without taking into account the nonlinear effects.

2 Basic Equations

Consider the flow of a galactic interstellar gas under the influence the gravitational potential due to the galactic stars. It is assumed that the gas rotates with a given angular velocity $\Omega(r)$ at a distance r from the galactic center, while the stars are assumed to be arranged along two spiral arms that result from a density wave with an angular phase velocity Ω_p. Hence, in a frame that rotates with angular velocity Ω_p, the arms appear stationary. As a result, it is convenient to write the hydrodynamic equations that describe the steady gas flow in a frame that rotates with the angular velocity Ω_p. They are:

$$\nabla \cdot (\rho \mathbf{v}) = 0, \tag{1}$$

$$(\mathbf{v} \cdot \nabla)\mathbf{v} + 2\boldsymbol{\Omega}_p \times \mathbf{v} = -\frac{1}{\rho}\nabla P - \nabla(\Phi - \frac{1}{2}{\Omega_p}^2 r^2). \tag{2}$$

In (1)–(2), ρ, $\mathbf{v}$, and $P = a^2\rho$ are the density, velocity, and pressure of the interstellar gas, respectively, and Φ is the gravitational potential of the stellar subsystem.

It is assumed that the potential Φ is given by the sum of an axisymmetric unperturbed potential $\Phi_0(r)$ and a non axisymmetric perturbation $\Phi_1(\mathbf{r}, t)$. The non axisymmetric part of gravitational potential is due to the stellar spiral density wave and is of the form [2]

$$\Phi_1(\mathbf{r}, t) = \Re\bar{\phi}_1 e^{i(2(\Omega_p t - \varphi) + \int k(r)dr)} \tag{3}$$

where k is the radial wave number of the two-armed trailing spiral wave and is related to the angle of inclination of a spiral arm to the circumferential direction i by the relation $k = -2/r/\tan i$. Here and below r, φ are the galactocentric cylindrical coordinates and the axis of the galactic rotation is along the z-axis. Obviously, the characteristic length in this problem is $1/k \sim \lambda$. According to observations and following [2] a tightly wound spiral perturbation is considered, in which the angle of inclination i is small, hence $\lambda/r \ll 1$ or $|k|r \gg 1$. It is useful to introduce the spiral coordinates (η, ξ). The spiral coordinates are fixed in the rotating frame (with angular velocity Ω_p) such that η is constant along the spiral arms while ξ is constant along lines that are orthogonal to the spiral arms. Using the above definition of the spiral coordinates, the perturbed spiral stellar gravitational potential can be written in the form [2]

$$\Phi_1(\eta) = F \frac{\Omega^2 r \cos^2 i}{k} \cos\eta, \tag{4}$$

where F is the amplitude of the perturbation due to the stellar spiral density wave as a fraction of the unperturbed axisymmetric gravitational potential. The gaseous response depends on the strength of the density wave gravitational field [11,20] and have to be considered as nonlinear perturbation even though the moderate spiral forcing being 10% of the axisymmetric force field [19,20]

We turn now to deriving the equations that describe the gas flow under the stellar gravitational potential in the spiral coordinate system defined above. In order to do that we assume that the gas flow is given by an axisymmetric basic flow that describes the response of the gas to the axisymmetric unperturbed potential Φ_0 plus a perturbation which results from the spiral perturbation in the stellar gravitational potential Φ_1, as given in (4).

The solutions of (1) and (2) for rapidly rotating thin gaseous disk that describes the basic axisymmetric flow is readily obtained in the spiral coordinate system as

$$v_{||0} = (\Omega - \Omega_p) r \cos i, \qquad v_{\perp 0} = (\Omega - \Omega_p) r \sin i = \frac{2}{k}(\Omega - \Omega_p)\cos i, \tag{5}$$

where $v_{||}, v_{\perp}$ are the gas velocity components parallel and perpendicular to the spiral arms, respectively. It is easy to see that $|v_{\perp 0}/v_{||0}| \sim 1/|k|r \ll 1$.

In order to describe the perturbed variables the following angular velocity scale is introduced:

$$\chi = 2\Omega[1 + \frac{r}{2\Omega}\frac{d\Omega}{dr}\cos^2 i]^{1/2}, \tag{6}$$

with the aid of which the following normalization is introduced for the velocity components:

$$u = \frac{v_{\perp 1}}{\bar{v}_{\perp}}, \quad \bar{v}_{\perp} = \chi\frac{\cos i}{k}, \quad v = \frac{v_{||1}}{\bar{v}_{||}}, \quad \bar{v}_{||} = \frac{\chi^2}{\Omega}\frac{\cos i}{k} \tag{7}$$

Example for values for the various parameters can be estimated from the galactic equilibrium parameters at the solar position as given in [11]:

$$\Omega = 24.7\frac{km}{sec\ kpc}, \quad \Omega_p = 13.5\frac{km}{sec\ kpc}, \quad \chi = 31\frac{km}{sec\ kpc}. \tag{8}$$

Inserting (4-7) into (1) and (2), eliminating the perturbed density ρ_1, and assuming that the derivatives across the spiral arms are much bigger than the derivatives along the arms [18] result in the following two equations for the perturbed velocity components:

$$\frac{(-\nu+u)^2-c^2}{-\nu+u}\frac{du}{d\eta}=v-f\sin\,\eta,\quad u+(-\nu+u)\frac{dv}{d\eta}=0, \tag{9}$$

where

$$f=F(\Omega^2/\chi^2)kr,\quad c^2=\frac{a^2k^2}{\chi^2\cos^2 i},\quad \nu=-\frac{v_{\perp 0}}{\bar{v}_\perp}, \tag{10}$$

are the dimensionless amplitude of the stellar density spiral wave, the squared dimensionless sound speed, and the dimensionless basic perpendicular velocity, respectively. The values for those parameters as estimated from (8) are $0.1-0.2$, 0.195, and 0.72, respectively.

Finally, by eliminating v and defining a new variable $y=u/(f\nu)$, a single second order ordinary differential equation is obtained:

$$\begin{aligned}y''&+\omega_0^2(1+2f\cos\eta-f^2y\cos\eta)y=\omega_0^2\cos\eta+f((2\nu^2\omega_0^2+1)yy''+\\&+(2\nu^2\omega_0^2-1)y'^2+\omega_0^2y^2)+f^2\nu^2\omega_0^2(-3y^2y''-2yy'^2)+\\&+f^3\nu^2\omega_0^2(y^3y''+y^2y'^2).\end{aligned} \tag{11}$$

where $\omega_0^2=(\nu^2-c^2)^{-1}$ is the natural frequency of linear oscillations near steady state as a result of a small imbalance between the Coriolis force and the gaseous pressure.

3 Ultra Harmonic Resonances

The solution of(11) for the cases in which $\omega_0\not\approx 2,3,\ldots,n$ was presented in [20]. However, the response of the $n-th$ harmonic (ultra harmonic resonances response) can be expected to be sufficiently large if $\omega_0\approx 2,3,\ldots,n$ as a result of the combined ultra harmonic and the parametric resonances. The existence of this ultra harmonic resonances has been recently demonstrated by the numerical calculations [11]. Equation (11) will be analyzed analytically in this section by the method of multiple scales which is often used in the analysis of weakly nonlinear dynamic systems [21].

3.1 The Resonant Case of $\omega_0\approx 2$

In this the combined ultra harmonic and the parametric resonances response when $\omega_0\approx 2$ will be studied. The solution of (11) to first order in the small parameter f is straight forward and is given by

$$y_0=a_0\cos(\omega\eta)+2\Lambda\cos\eta,\quad \Lambda=\frac{1}{2}\frac{\omega_0^2}{\omega_0^2-1} \tag{12}$$

Substituting (12) into (11), equating coefficient of like powers of f, one obtains an equation for determining the next-order terms of the expansion of the solution $y(\eta)$. However, the fact that ω_0 is close to 2 gives rise to secular terms and small-divisors in the higher order equations. Consequently, in order to proceed further, a detuning parameter σ is introduced

$$\omega_0 = 2 + \sigma, \tag{13}$$

as well as higher order deviations of the frequency from ω_0

$$\omega = \omega_0 + \omega^{(1)} + \omega^{(2)} + \dots . \tag{14}$$

Using the above definitions, the equation for y_1 may be rewritten as:

$$\begin{aligned} y_1'' + \omega_0^2 y_1 &= f(-8\nu^2\omega_0^2\Lambda^2 + 2\omega_0^2\Lambda(\Lambda - 1))\cos 2\eta \\ +2\omega_0 a_0' \sin(\omega_0 + \omega^{(1)})\eta &+ 2\omega_0 a_0 (\omega^{(1)}\eta)' \cos(\omega_0 + \omega^{(1)})\eta + NST \end{aligned} \tag{15}$$

where NST stands for non secular terms. Eliminating the secular terms from last equation yields

$$\begin{aligned} f(-8\nu^2\omega_0^2\Lambda^2 + 2\omega_0^2\Lambda(\Lambda - 1))\sin(\sigma + \omega^{(1)})\eta + 2\omega_0 a_0' &= 0 \\ f(-8\nu^2\omega_0^2\Lambda^2 + 2\omega_0^2\Lambda(\Lambda - 1))\cos(\sigma + \omega^{(1)})\eta + 2\omega_0 a_0 (\omega^{(1)}\eta)' &= 0 \end{aligned} \tag{16}$$

As a_0 as well as $\omega^{(1)}$ are constants, it follows from (16) that

$$\omega^{(1)} = -\sigma. \tag{17}$$

It can be seen that unlike the non resonant response for which the amplitude of 2-nd harmonic is proportional to f which is a small parameter, here this amplitude is of order f/σ which is of order one. Thus, small perturbations in the stellar gravitational potential give rise to finite response in the gas flow.

3.2 The Resonant Case of $\omega_0 \approx 3$

In the case of the combined ultra harmonic and parametric resonances response when $\omega_0 \approx 3$ the small-divisor terms that result from the non linear terms occur at $O(f^2)$ and the amplitudes of the responses have been ordered so that affects of the resonances first occur at $O(f^2)$. The solution of (11) to order $O(f)$ is given by

$$y_1 = a_{10} + 2\Lambda \cos \eta + a_{12} \cos 2\eta + a_{13} \cos \omega\eta, \tag{18}$$

where

$$a_{10} = -2f\Lambda^2, \quad a_{12} = -4f\frac{\Lambda^2}{\omega_0^2 - 4}(2\nu^2\omega_0^2 + \frac{\omega_0^2}{2} - 1) \tag{19}$$

Substituting the solution (18) into (11) yields

$$y_2'' + \omega_0^2 y_2 = A\cos 3\eta + B\cos\omega\eta + 2\omega_0 a_{13}'\sin\omega\eta + \\ +2\omega_0 a_{13}(\omega^{(1)}\eta)'\cos\omega\eta + NST \tag{20}$$

in which

$$A = f\omega_0^2 a_{12}(2\Lambda - 2\nu^2 - 2 - 8\nu^2\Lambda) + f^2\Lambda^2(1 + 10\nu^2\omega_0^2\Lambda)$$
$$B = f(-(2\nu^2\omega_0^2 + 1)\omega^2 a_{10}a_{13} + f\nu^2\omega_0^2(8\Lambda^2 a_{13} + \frac{7}{4}a_{13}{}^3\omega^2))$$

To proceed further, once again a detuning parameter σ and the deviation of the frequency from ω_0 due to non-linearity are introduced according to

$$\omega_0 = 3 + \sigma \\ \omega = \omega_0 + \omega^{(1)} + \omega^{(2)} + \ldots \tag{21}$$

Thus, eliminating the secular terms from (20) yields

$$A\sin(\sigma + \omega^{(1)}) + 2\omega_0 a_{13}' = 0 \\ A\cos(\sigma + \omega^{(1)}) + B + 2\omega_0 a_{13}(\omega^{(1)}\eta)' = 0 \tag{22}$$

whose solution is

$$\omega^{(1)} = -\sigma \tag{23}$$

In this case steady state solutions correspond to the solution of the cubic equation. For such detuning parameter that scales as f^2 the solution for a_{13} is of order one. Thus, once again, small perturbations in the stellar gravitational potential give rise to finite response of the gas flow.

Figure 1 presents the resonant solutions in the velocity plane and gas density profiles compared with non-resonant solutions.

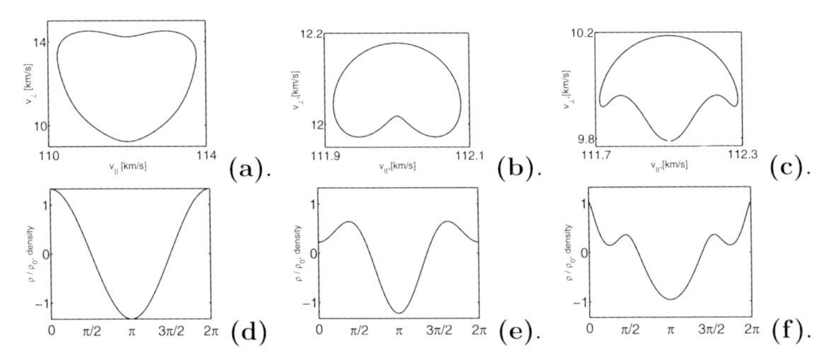

Fig. 1. Flow in velocity plane and normalized gas density profiles. Three cases are presented: 1. **(a)** and **(d)** - without resonance for distance from galactic center, r=10kpc, 2. **(b)** and **(e)** - resonance one-two for r=12.5 kpc, 3. **(c)** and **(f)** - resonance one-three for r=13.5 kpc. The variation of the normalized density ρ/ρ_0 presented as a function of the phase angle η.

4 Interpretation of the Results

The importance of corotation and Lindblad resonances has been recognized long ago see [22–25]. Particularly, it was found by calculation of the orbits that the main families of orbits in realistic model of spiral galaxy enhance the spiral arms up to the resonance $\chi/(\Omega - \Omega_p) = 4/1$. In this paper we have employed the hydrodynamic approach following [11] and have seen that the ultra harmonic resonances exist if the natural frequency ω_0 is a rational multiple of the forcing frequency. In our case the natural frequency is given by

$$\omega_0 \approx \frac{\chi}{2(\Omega - \Omega_p)}(1 + \frac{a^2k^2}{8(\Omega - \Omega_p)^2 \cos^2 i})$$

and forcing frequency is equal to unity. So considered above ultra harmonic resonances $\omega_0 \approx 2$ and $\omega_0 \approx 3$ correspond to 4/1 and 6/1 respectively.

The substantially different spatial arrangement of spirals of young and old objects would appear to be the most remarkable features [15]. This circumstance should be an important factor in the theory of the origin of the spiral structure, in attempts to explain the observational data. Therefore we suppose that non-linear gas response effects on spiral density wave that is created by old objects could be responsible for the appearance of structures similar to the observed ones. To show this we have plotted a chart (Fig. 2) as the variation of the gas density obtained in our calculations. We see from Fig. 2 that the secondary compression associated with the resonance one-two obtained in our calculations for distance from galactic center 10.5 kpc up to 11.5 kpc that produces the arm bifurcation and may well account for major spiral features. Notice that a bifurcation of gaseous spiral arms was modeled numerically in [27] and was associated also with the presence of nonlinear effects at the 4/1 ultraharmonic

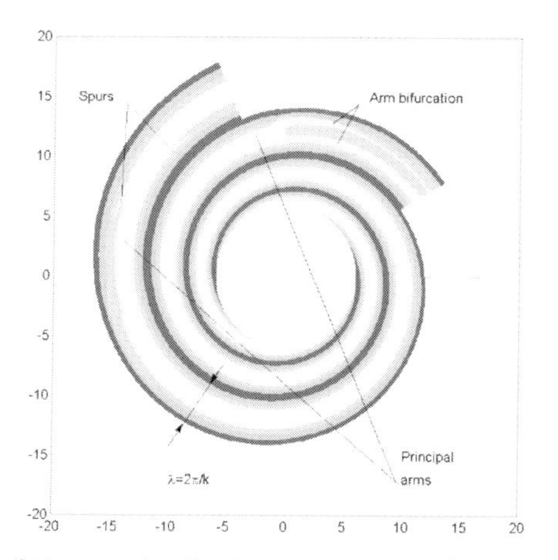

Fig. 2. Distribution of the gas density in response to galactic spiral density wave for two-arms mode and $\Omega_p = 13.5$ km/sec/kpc. The darkness of the chart is proportional to the value of the density (The black line shows the minimum)

resonance. The resonance one-three is found in our calculations to extend in distance from galactic center 13.5kpc up 14 kpc produce relatively short spurs and not major spiral features.

The authors thanks Yuriy N. Efremov and Arthur D. Chernin from the Sternberg Astronomical Institute, Moscow State University for their interest in this work. The authors have benefited from helpful discussions with Michael Gedalin and Evgeny Griv from the Ben-Gurion University of the Negev in Israel. Also, the authors are grateful to an anonymous referee of the paper for his/her useful comments on the original manuscript.

References

1. C.C. Lin, C. Yuan, and F.H. Shu: Astrophys. J. **155**, 721 (1969)
2. W.W. Roberts: Astrophys. J. **158**, 123 (1969)
3. S.B. Pikel'ner: Astron. Zh. **47**, 752 (1970)
4. P.R. Woodward: Astrophys. J. **195**, 61 (1975)
5. M. Fujimoto: 'Modeling of gas flow through a spiral sleeve'. In: *Non-stable Phenomena in Galaxies. IAU Symp. No. 29, Byurakan, May 4-12, 1966.* (The Pubblishing House of the Academy of Sciences of Armenian SSR Yerevan 1968) pp.453-463
6. B.T. Lynds: 'The Distribution of Dark Nebulae in Late-Type Spirals'. In: *The Spiral Structure of our Galaxy, IAU Symp. No.* 38, ed. W.Becker and G.Contopoulos, (Reidel, Dordrecht 1970) pp.26-34
7. D.S. Mathewson, P.S. van der Kruit, and W.N. Brown: Astron. Astrophys., **17**, 468 (1972)
8. Y. Sofue: Publ. Astron. Soc. Japan, **37**, 507 (1985)
9. V.G. Berman, and Yu.N. Mishurov: Astrofisics, **16**, 52 (1980)
10. F.H. Shu, V. Milione, W. Gedel, C. Yuan, D.W. Goldsmith, and W.W. Roberts: Astrophys. J. **173**, 557 (1972)
11. F.H. Shu, V. Milione, and W.W. Roberts: Astrophys. J. **183**, 819 (1973)
12. S.A. Kaplan, S.B. Pikel'ner: Annual review of astronomy and astrophysics **12**, 113 (1974)
13. L.S. Marochnik, A.A. Suchkov: Soviet Physics - Uspekhi **17**, 85 (1974)
14. K. Rohlfs: *Lectures on Density-Wave Theory.* Lect. Notes. Phys. No. 69, (Springer, Berlin 1977)
15. F.Zwicky: *Morphological Astronomy* (Springer-Verlag, Berlin 1957)
16. B.A. Vorontsov-Vel'iaminov: *Extragalactic astronomy* (Harwood Academic, Chur, Switzerland 1987)
17. D.L. Block, R.J. Wainscoat: Nature **353**, Sept. 5, 48 (1991)
18. A.N. Nelson, and T. Matsuda: Month. Not. RAS, **179**, 663 (1977)
19. W.W. Roberts, Jr.: PASP, **105**, 670 (1993)
20. E. Liverts: Astrophys. & Space Sci. **274**, 513 (2000)
21. L.D. Landau, and E.M. Lifshitz: *Mecanics* (Pergamon Press, Oxford 1960)
22. B. Lindblad:, Handbuch der Physik, **53**, 21 (1959)
23. D. Lynden-Bell, and A.J. Kalnajs: Month. Not. RAS, **157**, 1 (1972)
24. G. Contopoulos, and P.Grosbøl: Astron. Astrophys., **155**, 11 (1986)
25. R.J. Allen, B. Canzian, and S.H. Lubow: PASP, **105**, 664 (1993)
26. P. Artymowicz, and S.H. Lubow: Astrophys. J., **389**, 129 (1992)
27. P.A. Patsis, P.Grosbøl, and N. Hiotelis: Astron. Astrophys., **323**, 762, (1997)

The Level of Chaos in N-Body Models of Elliptical Galaxies

Nikos Voglis[1], Costas Kalapotharakos[1,2],
Iioannis Stavropoulos[1,2], and Christos Efthymiopoulos[1]

[1] Academy of Athens, Research Center for Astronomy
[2] University of Athens, Department of Physics, Section of Astrophysics

Abstract. This paper presents a combination of methods that estimate reliably the level of chaos (proportion of particles in chaotic orbits and the corresponding values of the Lyapunov Characteristic Numbers) in self-consistent N-body models of elliptical galaxies. A careful simultaneous use of several numerical tools can induce the proportion of particles in chaotic orbits dynamically important within one Hubble time. In models with smooth centers the mass component in chaotic motion is less than about 30% of the total mass. In models with central black holes this percentage increases up to 70%. Typical Lyapunov characteristic numbers are below 0.1 in units of the inverse crossing time. A remarkable property of the chaotic mass component is that it has a different surface density profile than that of the ordered component. The superposition of the two profiles causes observable humps in the overall profile, which are suggested as a possible observational 'signature' of chaos in elliptical galaxies.

1 Introduction

An important open problem of stellar dynamics is the level of chaos in realistic self-consistent stellar systems. In this paper we study the level of chaos in four different self-consistent N-Body models of elliptical galaxies in equilibrium.

Two of the N-Body systems (Q and C models) are produced from quiet and clumpy cosmological initial conditions respectively [6], [4]. These models are non-rotating and they have a smooth density profile at the center. The other two models (QB1 and QB2) are produced from the Q model by adding a point mass (black hole) at the center, with a mass equal to 0.1% and 1% of the total galactic mass respectively. All the models are triaxial, but the Q and QB models are more elliptical than the C-models.

The self-consistent potential at equilibrium is realized by the N-Body code [1] as a smooth series of a radial plus spherical harmonic expansion. Near equilibrium, the expansion coefficients have almost constant values. Then the system is represented approximately by a 3D autonomous Hamiltonian

$$H = \frac{1}{2}(\dot{x}^2 + \dot{y}^2 + \dot{z}^2) + V(x, y, z) \tag{1}$$

The equilibrium configurations are triaxial in all models, taken as x the direction of the shortest axis and z the direction of the longest axis.

2 Method to Distinguish Regular and Chaotic Orbits

The most chaotic orbits are identified by defining a particular threshold in the values of their Lyapunov Characteristic Number (LCN). Orbits below this threshold, although chaotic, behave macroscopically like regular orbits within a Hubble time.

The method, used to distinguish between regular and chaotic orbits, faces the problem of the very different (by two orders of magnitude) orbital periods of particular orbits within the same system. When one calculates Lyapunov times, must decide whether these times should be expressed in terms of the particular periods of the orbits, or of the half mass crossing time (average dynamical period) in the system. The present method provides a compromise for this problem and gives a not very much biased estimator of the chaoticity of the orbits, as the latter reflects to the macroscopic properties of the system within a Hubble time. The method uses three different indicators.

2.1 Specific Finite Time Lyapunov Characteristic Number

The Specific Finite Time Lyapunov Characteristic Number L_j for the orbit of the particle j is given by the formula

$$L_j(T_{rj}, t_j) = \frac{T_{rj}}{t_j} \sum_{i=1}^{N_j} a_{ij} \tag{2}$$

where T_{rj} is the average radial period of the orbit, t_j is the integration time, N_j is the number of time steps $\Delta t = t_j/N_j$ and a_{ij} is the stretching number [7] at the time step i, $(i = 1, ...N_j)$. The stretching number a_{ij} in (2) is defined in terms of the length of the deviation vector $\xi_j(t_i)$ from the orbit j at the time t_i in the six-dimensional phase space, by the equation

$$a_{ij} = \ln \frac{\xi_j(t_i + \Delta t)}{\xi_j(t_i)} \tag{3}$$

The deviation vectors $\xi_j(t_i)$ are calculated by numerical integration of the variational equations of motion.

The Hubble time is taken equal to 100 half mass crossing times (T_{hmct}) of the system. The values of L_j of all the orbits are calculated for the same number of radial periods $t_j/T_{rj} = 1200$. Thus, the particles of even the shortest radial periods ($T_{rj} \geq T_{hmct}/300$) are integrated for more than 4 Hubble times. This integration time detects chaotic orbits with L_j stabilized at values no smaller than 10^{-3}. This threshold is rather arbitrary but it is satisfactory for all practical purposes.

2.2 Common Unit Finite Time LCN

The L_js are converted in common units by defining the Common Unit Finite Time Lyapunov Characteristic Number L_{cu}, as $L_{cu} = L_j T_{hmct}/T_{rj}$, in units of $1/T_{hmct}$. This number compares the chaotic orbits as regards their efficiency within a Hubble time.

2.3 Alignment Index

The method of Alignment Index [5] is based on some known properties of the time evolution of deviation vectors [8], [9]. Consider the evolution of two arbitrary different initial deviation vectors ξ_{j1} and ξ_{j2} of the same orbit. If an orbit is chaotic, the two deviation vectors tend to become parallel or anti-parallel to each other (depending on the initial values of the deviation vectors). If the orbit is regular, the two vectors tend to become tangent to the surface of an invariant torus and oscillate with respect to each other. This difference is measured by taking the minimum of the quantities $d_{j-}(t) = |\boldsymbol{\xi}_{j1}(t) - \boldsymbol{\xi}_{j2}(t)|$ (parallel deviation vectors) and $d_{j+}(t) = |\boldsymbol{\xi}_{j1}(t) + \boldsymbol{\xi}_{j2}(t)|$ (anti-parallel deviation vectors). This is called the Smaller ALignment Index, (SALI), or simply Alignment Index, (AI). For chaotic orbits, it reaches the limit of the computer accuracy ($\approx 10^{-16}$) at the end of the integration time. For regular orbits it is improbable to be less than 10^{-3} all along the integration time. Thus, the distinction made by the AI method is clear and fast, even for very weakly chaotic orbits.

3 Results

Due to computing time limitations, it is only possible to calculate the orbits of a representative sample of particles of the whole system, namely, one in every four particles uniformly distributed along the whole set of particles. Using as initial conditions the three coordinates and velocities of each particle of the sample, each orbit is integrated in the Hamiltonian (1) for a maximum time $t_j = 1200 T_{rj}$.

In Fig. 1a the values of the L_j of the particles in the sample of the Q-model are plotted against their Alignment Indices at $t_j/T_{rj} = 20$, in log-log scale. Most points appear concentrated in a single group of triangular shape around a mean value of $Log(L_j) \simeq -1.4$ with $Log(AI) > -3$. This is called a **regular group**. A number of points form a lane emanating from the upper end of the regular group towards smaller values of AI. The points on the lane correspond to orbits that have just started indicating their chaotic character.

At $t_j/T_{rj} = 100$ (Fig. 1b) the main part of the regular group is displaced towards lower values of L_j following a t^{-1} law. However, a good number of points have followed a streaming motion along the lane towards smaller values of AI and larger values of L_j and tend to form a **chaotic group**.

As time increases the number of orbits in the chaotic group increases, but more and more slowly. At $t_j/T_{rj} = 1200$ (Fig. 1c) the chaotic group is well separated from the regular group. The streaming of points along the lane becomes slower, but non-negligible. The points on this lane correspond to weakly chaotic orbits.

By introducing a threshold $L_j = 10^{-2.8}$ as the value separating the chaotic from the regular group, and a threshold $AI = 10^{-3}$, we find that the fraction of the chaotic orbits (chaotic component) corresponds to about 32% of the total mass. The rest is called regular component. If the points along the transport lane

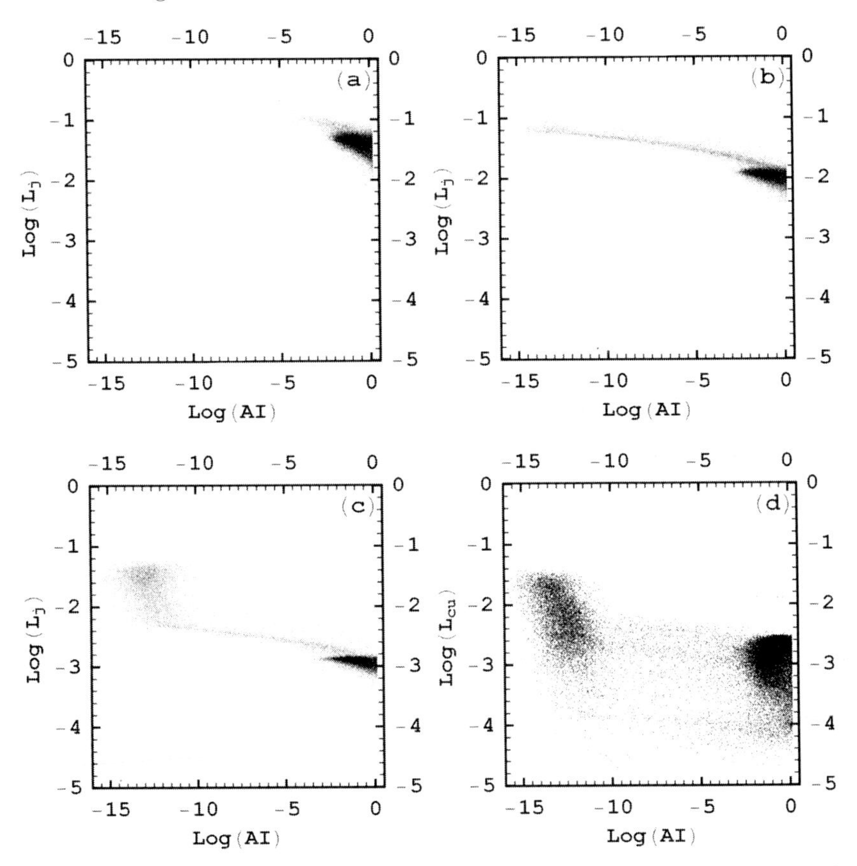

Fig. 1. The evolution of the orbits of the particles in the sample on the plane of the Log(AI)-Log(L) at **(a)** 20, **(b)** 100, **(c)** 1200 radial periods. In **(d)** we plot the Finite Time LCN in common units of $(1/T_{hmct})$, as derived from the data of (c)

are not counted with the chaotic orbits, then we find that the 'strictly chaotic' orbits ($AI \leq 10^{-10}$) are about 26%.

If L_j is converted in units of the inverse half mass crossing time T_{hmct} of the system, we find $L_{cu} = L_j T_{hmct}/T_{rj}$. Then, the results of Fig. 1c are converted to those shown in Fig. 1d. The L_{cu} of the detected chaotic orbits range between $10^{-4.6}$ and $10^{-1.4}$, with a preference above the value of 10^{-3}. It is obvious that a number of orbits that have been characterized as chaotic (with $L_j \geq 10^{-2.8}$) have values of L_{cu}s much smaller than the minimum value of L_j, because of their long radial periods. These small values of the L_{cu}, describe the very small diffusion in a Hubble time.

Figure 2 shows the same planes L_j vs. AI and L_{cu} vs. AI as in the Q model (Fig. 1c,d) for the experiments C (Fig. 2a,b), QB1 (Fig. 2c,d) and QB2 (Fig. 2e,f). The C model contains a smaller number of chaotic orbits and a somewhat smaller appearing maximum value of L_j and L_{cu}. On the other hand, chaos is much more abundant if we add a black hole (Fig. 2c,d model QB1 with a

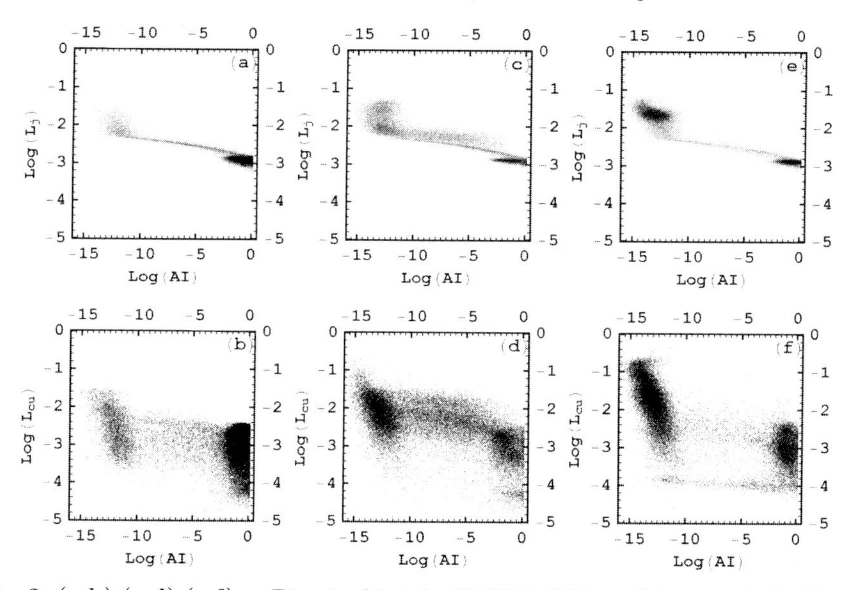

Fig. 2. (a,b),(c,d),(e,f) as Figs. 1c,d but for C, QB1, QB2 model respectively. Notice that the QB models (especially QB2) have orbits with L_{cu}s above 10^{-1}

black hole of 0.1% and Fig. 2e,f. model QB2 with a black hole of 1% of the total galactic mass). As the mass of black hole increases, there are more particles in the region of higher Lyapunov numbers (L_j and L_{cu}). However, even in the QB2 model L_js remain well below the value of 10^{-1}. A small fraction of particles have L_{cu} larger than 10^{-1} due to their very short radial periods. There is a pronounced transport of particles from regions of low L_j (or L_{cu}) to regions of larger L_j in the QB experiments.

Figure 3 shows the **time evolution of the proportion of chaotic orbits** for all the systems. Fig. 3a shows the time evolution of the percentage of the particles in orbits characterized as chaotic in the Q model. The solid line corresponds to a strict criterion of chaoticity, i.e. $AI < 10^{-10}$. The dashed line corresponds to a more flexible criterion, including the particles on the transport lane. About 30% of the particles in the Q model are characterized as chaotic after 1200 half mass crossing times. A smaller percentage (about 25%) is found in the case of the C experiment (Fig. 3b). On the other hand, the percentage of chaotic orbits is much higher if we add a black hole (QB1 and QB2 models, Figs. 3c,d), being as high as 70% in the case of the model QB2.

An important effect of this large increase of the level of chaos in the QB experiments is in **the distribution of the energies** (number density function) of the various systems. The distribution of all the particles along the energy axis for the Q experiment is shown in Fig. 4a by a solid line. The dashed line gives the distribution of the particles of the detected chaotic component. In the region of small energies (i.e. below the energy level of about -60) no chaotic orbits were detected by this threshold of chaoticity. The majority of particles

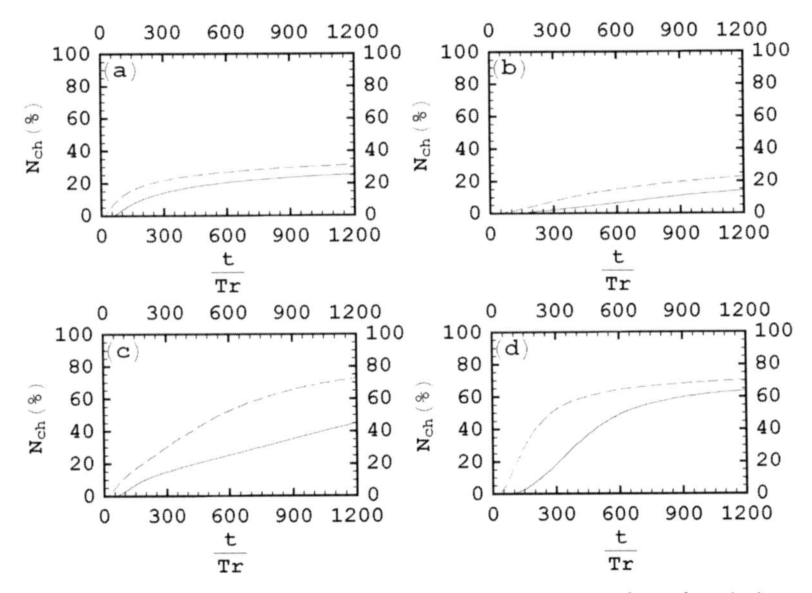

Fig. 3. The evolution of the percentage of orbits characterazed as chaotic in various models (**(a)** Q, **(b)** C, **(c)** QB1, **(d)** QB2). The solid line corresponds to the "strictly" chaotic part ($AI < 10^{-10}$) while the dashed line includes the orbits on the lane. The different rates of growth reflects the difference in the level of chaos

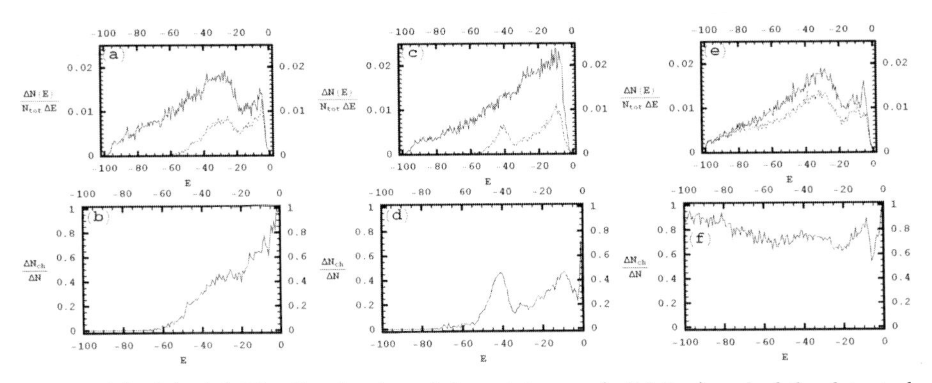

Fig. 4. **(a)**, **(c)**, **(e)** The distribution of the total mass (solid line) and of the detected chaotic part (dashed line) along the energy axis in the Q, C, QB1 model respectively. **(b)**, **(d)**, **(f)** The ratio of the number of chaotic orbits to the total number of orbits at every bin of energy along the energy axis in the Q, C, QB1 model respectively

in chaotic orbits with large binding energies spend most time at large radii and they mainly contribute in forming the halo of the galaxy. This effect is clearly seen in Fig. 4b, which gives the **relative ratio of particles in chaotic orbits** along the energy axis for the Q experiment.

The same distribution for the C model is seen in Figs. 4c,d. There is a remarkable difference between the Q and C experiments. In the Q model (Fig. 4b) the ratio of the detected chaotic orbits to the total number of orbits at every energy bin increases almost monotonically. In contrast, in the C model (Fig. 4d) this ratio has a pronounced maximum at the energy level of ≈ -40 followed by a minimum around the energy level of ≈ -30. This difference can be explained by noticing that in the C model there is a good number of ordered 1:1 tube orbits, at the energy levels around -30. These orbits, together with the chaotic orbits that have an almost spherical distribution, balance the effect of the box or box-like orbits, so that the asphericity of the system is reduced to the value required by the self-consistent equilibrium. On the other hand, in the Q model, at this energy levels, there is an extensive stable area in phase space corresponding to the 1:1 tube orbits, but it is almost empty, i.e. it is occupied by only a small number of orbits [2], [3]. At the same energy levels the Q model possesses a good number of chaotic orbits. These orbits are flexible to follow boxy or circular geometries but they are almost spherically distributed. Thus they prevent large departures of the system from sphericity.

If we add a black hole (Figs. 4e,f, model QB1), the chaotic orbits are no longer limited to small absolute binding energies but their distribution extends all the way to energies corresponding to the central value of the potential. This is because the black hole destroys the regular character of most box orbits close to the center, by causing large deflections of the orbits.

We finally compare the **surface densities** of the projections of the particles on various planes for the representative models Q and QB2. Fig. 5a,c shows the projections of the particles in ordered orbits for the Q experiment on the plane x-z and y-z respectively, while Fig. 5b,d shows the same projections but for the particles in chaotic motion. The main conclusion is that the large ellipticity of this galaxy is due to the regular orbits mainly, while the chaotic component tends to make the galaxy more spherical. On the other hand, the addition of a black hole (model QB2, Figs. 6a,b,c,d) has the effect of increasing the number of chaotic orbits. Thus the galaxy becomes more spherical in the presence of a black hole. Notice that the projection of regular particles on the plane y-z (Fig. 6c) is almost spherical. This is due to the fact that regular orbits move mostly at 1:1 tube orbits. For that reason the models with black hole tends to be more oblate.

The combination of the two surface density profiles (regular and chaotic) has the effect that the logarithmic slope ($s(r) = \frac{d \ln \sigma(r)}{d \ln r}$) of the overall profile $\sigma(r)$ forms an observable hump (Fig. 5e,f,g,h), [10], especially if the surface density profiles are taken along the shortest axis of the projection. Such a hump may be an observable signature of chaos in non-rotating elliptical galaxies.

4 Conclusions

We propose a methodology to obtain reliable estimates on the level of chaos in a self-consistent galactic system. This methodology combines three different numerical methods known in the literature. The combined use of the three methods provides a solution to the problem of estimation of Lyapunov times despite the very different periods of particular orbits within a galactic system.

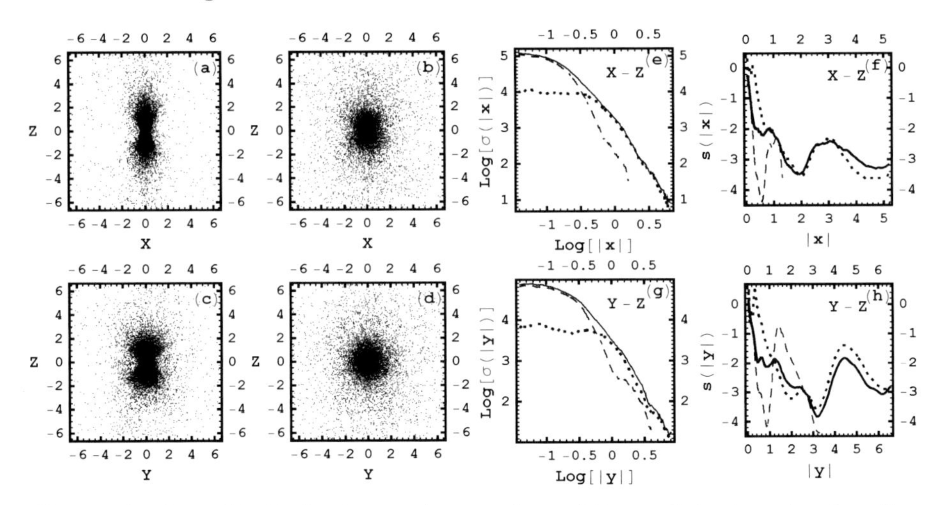

Fig. 5. The Q model. Projection on the x-z plane of the particles in ordered motion **(a)** and of the particles in chaotic motion **(b)**. In **(c)** and **(d)** as in (a) and (b), but on the y-z plane. In **(e)**, **(g)** is shown the surface density profiles along a slit on the short axis of each projection in Log-Log scale. In **(f)**, **(h)** is shown the Logarithmic slope of the surface density as a function of the distance along the same slit. Solid lines refer to the total mass, the dashed lines to the mass in ordered motion and the dotted lines to the mass in chaotic motion, The different slopes of the two components create a hump at about the half mass radius of the system

The models Q and C (with smooth central density profiles) have chaotic orbits only at relatively low absolute energies, i.e. at energy levels exceeding the deepest 30% of the potential well. Below this level most orbits are regular boxes or box-like. In the Q model, the detected chaotic part is about 30% of the total mass. This part has a nearly spherical distribution. It imposes limitations on the maximum ellipticity of the system, despite the fact that only a part less than about 8% of the total mass moves in chaotic orbits able to develop chaotic diffusion within a Hubble time. In the C model, the detected chaotic part is about 25% of the total mass, but only less than 2% can develop chaotic diffusion within a Hubble time.

Chaos is much more pronounced in the QB models with central black holes, and it extends to energies reaching the minimum of the potential well. This has implications on the number of particles in box or 1:1 tube orbits, and it affects the ellipticity of the systems. The overall proportion of particles in chaotic orbits reaches as much as 70% in the QB models. The more massive central black hole model (QB2) produces chaotic orbits with higher values of $L_{cu}s$, but the limit of 10^{-1} is hardly exceeded.

In all the systems, the chaotic components produce different surface density profiles than these of the rest of the mass. The combination of the two profiles produces observable signatures of chaos in non-rotating elliptical galaxies.

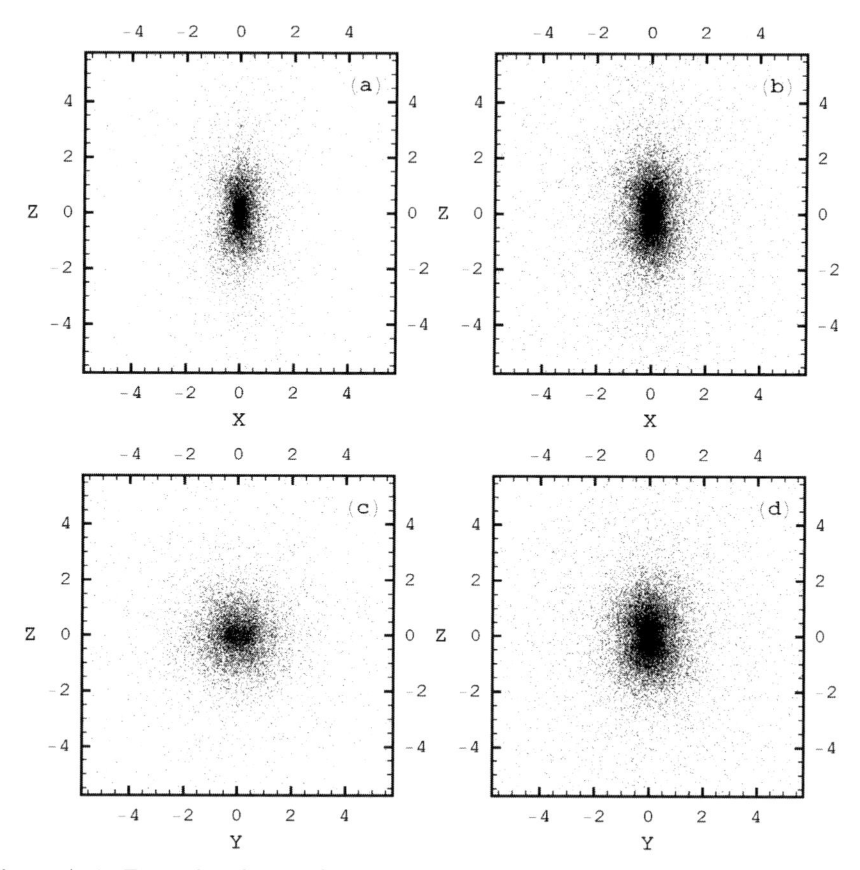

Fig. 6. As in Fig. 5, but for the QB2 model. We see that the projection of the particles in ordered motion forms a nearly spherical distribution because in that case the ordered motions are mostly 1:1 tube

References

1. A. J. Allen, P. L. Palmer, J. Papaloizou: MNRAS **242**, 576 (1990)
2. G. Contopoulos, C. Efthymiopoulos, N. Voglis: CeMDA **78**, 243 (2000)
3. G. Contopoulos, N. Voglis, C. Kalapotharakos: CeMDA **83**, 191 (2002)
4. C. Efthymiopoulos, N. Voglis: A&A **378**, 679 (2002)
5. Ch. Skokos: JPhA **34**, 10029 (2001)
6. N. Voglis: MNRAS **267**, 379 (1994)
7. N. Voglis, G. Contopoulos: JPhA **27**, 4899 (1994)
8. N. Voglis, G. Contopoulos, C. Efthymiopoulos: PhRvE **57**, 372 (1998)
9. N. Voglis, G. Contopoulos, C. Efthymiopoulos: CeMDA **73**, 211 (1999)
10. N. Voglis, C. Kalapotharakos, I. Stavropoulos: MNRAS **337**, 619 (2002)

Low Frequency Power Spectra and Classification of Hamiltonian Trajectories

George Voyatzis

Department of Physics, University of Thessaloniki, 54124 Thessaloniki, Greece

Abstract. We consider the problem of trajectory classification (as regular or chaotic) in Hamiltonian systems through power spectrum analysis. We focus our attention on the low frequency domain and we study the asymptotic behavior of the power spectrum when the frequencies tend to zero. A low frequency power estimator γ is derived that indicates the significance of the relative power included by the low frequencies and we show that it is related to the underlying dynamics of the trajectories. The asymptotic behavior of γ along a trajectory is qualitatively similar to that of the finite time Liapunov characteristic number. The standard map is used as a test model, because it is a typical model for describing Hamiltonian dynamics.

1 Introduction

Considering a variable $x(t)$ along a trajectory of a dynamical system, its (right side) power spectrum, defined as

$$p(f)=\lim_{T\to\infty}\frac{1}{T}\left|\int_0^T x(t)e^{-i2\pi ft}dt\right|^2, f>0, \tag{1}$$

yields valuable information about the underlying local dynamics of the system. Particularly, Hamiltonian systems, having a $2n$-dimensional compact phase space, exhibit mainly two different types of dynamics, regular and chaotic, which are associated with qualitatively different power spectra [11,10,8]. Regular trajectories are wound on invariant tori, they are quasiperiodic and their power spectra are discrete (in the sense that they are described by few and well separated spectral peaks). Chaotic trajectories have spectra with more or less "grassy" background that indicates the existence of continuous spectral components. The above property is of significant importance in semiclassical dynamics [7]. Another important property of regular spectra is their invariant character, i.e. spectral peaks should be located at constant positions for different trajectory time segments and the fundamental frequencies $\omega_i, i=1,..,n$ of a particular torus can be identified. But instead, chaotic spectra may show substantial changes through consequential segments indicating the non-existence of a torus. This characteristic has been proven a very useful tool in the study of long term trajectories in celestial mechanics [23].

From a theoretical point of view, quasiperiodicity implies spectra that are composed of lines at frequencies $f=\sum_{i=1}^{n} m_i\omega_i$, $m_i\in\mathbf{Z}$, i.e. the frequencies, where spectral peaks can appear, form a dense set on the real frequency axis of the spectrum. In general and under the absence of analytical solutions, trajectories are presented by sampled data for finite time intervals. Therefore, on the basis of numerical evidence, the distinction between regular and chaotic spectra becomes quite difficult when it refers to strongly deformed tori or weak chaotic motion. In [2] is shown that the above distinction, via finite time spectra, is impossible in principle and the true dynamics is revealed when $t \to \infty$.

Concerning the above intrinsic limitation, it is natural to think about asymptotic properties of the spectra for long time trajectory evolution. By increasing time, lower frequency modes are revealed in the spectrum. Then, the low frequency spectrum domain should be expected to provide a serious indication for characterizing the underlying dynamics. This paper attempts to handle the low frequency properties of power spectra obtained for bounded trajectories of Hamiltonian systems. Instead of calculating the whole power spectrum, the time evolution of a dynamical quantity $\gamma(t)$ along a trajectory is examined that reveals efficiently the requested information.

2 The Low Frequency Domain and Underlying Dynamics

For dynamical variables $x(t)$ that are well-behaving functions (i.e. they evolve smoothly in time, are bounded and have no singularities) their power spectra converge exponentially to zero as $f \to \infty$. Thus, we may define a high-frequency cut-off f_H such that $p(f) \approx 0$ for $f > f_H$ [10]. This is the case independently on whether the trajectories are regular or chaotic. In a non strict way, the low frequency domain is defined as the frequency interval $\mathcal{L} = (0, f_0)$, where $f_0 \ll f_H$. When a low frequency cut-off $f_L \in \mathcal{L}$ can be defined, such that $p(f)$ converges to zero when $f \to 0$, then the power spectrum is called *convergent*, otherwise, is called *divergent*. There are indications that these two types of spectra are related to the type of the underlying dynamics. In [6] it is shown, through reordered spectra, that the quasiperiodic regime of circle maps corresponds to convergent power spectra ($p(f) \sim f$) while in [4] an exponential convergence is indicated. In [12] the amplitude of peaks, located at low frequencies, are associated with the effect of small denominators in the convergence of the classical perturbation series of near integrable systems. The convergence of these series, which implies that a torus persists the perturbation, implies also convergent spectra. In [1] it is mentioned that chaotic trajectories contains a "central peak" in $\mathcal{L}(f_L)$, which is a peak with considerable amplitude. Generally, by considering more precise computations, the "central peak" is proven to be an erratic continuous portion in the low frequency domain. It's presence has been ascribed to the existence of a nearby separatrix trajectory [10,8]. Additionally, chaotic trajectories may show $1/f^a$ power spectra ($a \approx 1$) even for Hamiltonian systems of few degrees of freedom. Such behavior suggest the existence of a slow diffusion process [3].

In Fig. 1 the rich dynamics of the standard map

$$x_{x+1} = x_n + k\sin(x_n + y_n)\ ,\ \ y_{n+1} = x_n + y_n \ \ (mod\, 2\pi), \tag{2}$$

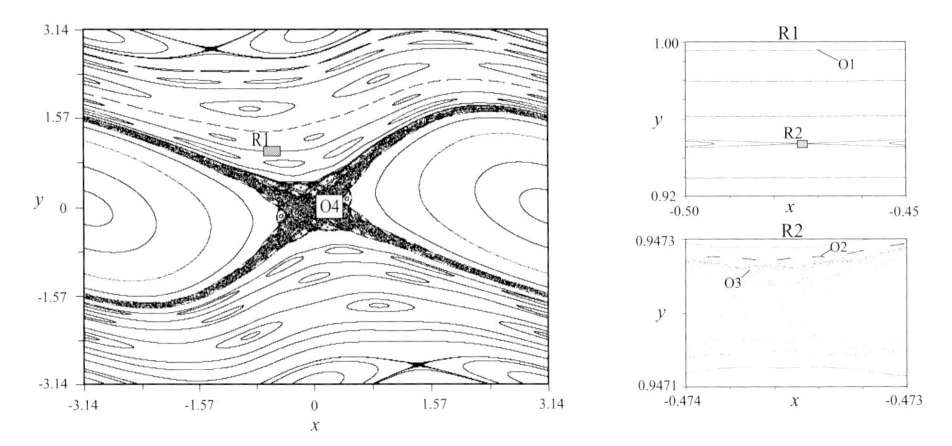

Fig. 1. The standard map dynamics for k =0.7. Magnifications of the regions R1 and R2 are shown and the orbits, referred to spectra of Fig.2, are indicated by O1,O2,O3 and O4.

is presented on the plane $x - y$ for $k = 0.7$. The trajectory O1 is a typical quasiperiodic trajectory that evolves relatively far from a significant resonance. The trajectory O2 forms small islands and is located close to the narrow chaotic zone of the weak resonance 2:9 (region R2 of Fig. 1). The main chaotic region (O4) is obtained around the unstable fixed point at (0,0). The power spectra of these trajectories are shown in Fig. 2. Log-Log scales are used in order to emphasize the low frequency domain. For quasiperiodic trajectories, located far from significant resonances, discrete peaks constitutes the spectrum (Fig. 2a). Furthermore, we obtain that the peaks show a rapid decay in amplitude as $f \to 0$. When a quasiperiodic trajectory evolves on a torus close to a separatrix manifold, much more peaks are present both at high and low frequency domains (Fig. 2b). A typical characteristic in this case is the appearance of a family of peaks in some low frequency domain (e.g. the peaks surrounded by the dotted-line in Fig. 2b). However, the spectrum is convergent and this is always the case as long as the underlying dynamics is regular. Figure 2c refers to a trajectory that evolves inside the homoclinic web of the separatrix at the resonance 2:9 of width $\Delta y \approx 10^{-4}$. The low frequency domain shows a continuous distribution which approximates a Lorentzian spectrum shape [10]. We should note that this characteristic can not be noticed by using normal scales. Finally, Fig. 2d corresponds to the trajectory, evolving in the chaotic region around the unstable fixed point at (0,0). The spectrum seems to follow an $1/f^a (a \approx 0.8)$ divergence, but for this and all other trajectories of the standard map examined, the $1/f^a$-divergence is limited and the spectrum saturates, i.e. it tends to a constant nonzero value as $f \to 0$.

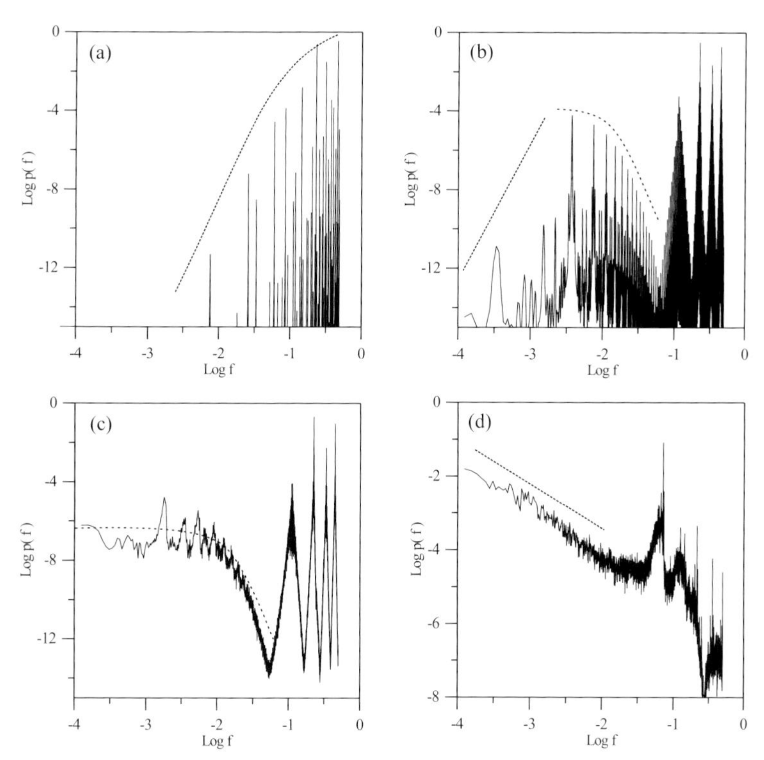

Fig. 2. Typical power spectra (a-d) for the orbits O1,O2,O3 and O4 respectively, shown in Fig.1.

3 Estimation of the Low Frequency Power and Asymptotic Characteristics

The asymptotic behavior of the power spectrum as $f \to 0$ can be estimated by examining the relative power included in $\mathcal{L}$. Considering a power spectrum $p = p(f)$, its normalized power spectral density at a frequency $f = f_0$ is defined as

$$\hat{p}(f_0) = \lim_{\Delta f \to 0} \frac{1}{\Delta f} \frac{I_p(f_0 - \Delta f/2, f_0 + \Delta f/2)}{I_p(0, \infty)}, \tag{3}$$

where we have used the notation

$$I_p(a, b) = \int_a^b p(f) df.$$

Since we are supposed to study the spectrum at low frequencies $f_0 \ll 1$, we may take $f_0 \in \mathcal{L}$ and $\Delta f = f_0$. Then, by taking into account the high frequency cut-off f_H, we approximate (3) by the quantity

$$\gamma^*(f_0) = \frac{1}{f_0} \frac{I(f_0/2, 3f_0/2)}{I(f_0, f_H)}. \tag{4}$$

We may consider $\gamma^*(f_0)$ to be a function of time along the particular trajectory by introducing the "period" $T = 1/f_0$,

$$\gamma(T) = \gamma^*(f_0) \ , \ f_0 \equiv 1/T. \tag{5}$$

In average, the power spectra of Hamiltonian trajectories follows profiles (smooth shapes), which can be described by the form

$$\tilde{p}(f) = \frac{p_0}{(\tau f)^a + \tau^2 f^2}, \ f > 0, \tau > 0, a \in (-a_0, a_0). \tag{6}$$

For high frequencies ($f \to \infty$), $p(f)$ decays rapidly to zero for any value of the parameter a, while τ is a frequency scaling factor [10]. For the low frequency domain we obtain convergent spectra for $a < 0$, the Lorentzian spectrum form for $a = 0$ and $1/f^a$-divergence for $a > 0$. It can be shown, that the corresponding γ estimator, obeys the asymptotic behavior

$$\gamma(T) \propto T^a \ \text{for} \ T \to \infty. \tag{7}$$

Thus, as $T \to \infty$, $\gamma(T)$ approximates the asymptotic behavior of the power spectrum when $f \to 0$.

Under numerical analysis, the variation of quantities along trajectories is given as time-series x_k. Namely, they are restricted in finite time intervals $[0, T]$ and are sampled at equidistant points $t_k = k\Delta t, k = 0, 1, ..., N - 1$, where $N = [T/\Delta t]$. Then, by applying a discrete Fourier transform (DFT) [9], the right side power spectrum is approximated by the relation

$$p(f_n) = \frac{|H_n|^2}{N^2} \ , \ f_n = \frac{n}{N\Delta t}, \ n = 1, ..., N/2, \tag{8}$$

where H_n denote the DFT coefficients. By setting $Dt = 1$, the calculated spectrum is restricted in the frequency domain $[f_1, f_{N/2}]$, where $f_1 = 1/N$ is the lowest available frequency and $f_{N/2} = 1/2$ is the high cutoff (Nyquist frequency). Then, we may let in (4) $f_0 = 1/N$, $f_H = 1/2$, $I_p(f_0/2, 3f_0/2) = p(f_1)$ and $I_p(f_0, f_H) = \sum_{n=1}^{n=N/2} p(f_k)$. Also, by taken into account the discrete form of the Parseval's theorem and writing H_1 in its trigonometric form, we obtain

$$\gamma(N) = \frac{\left(\sum\limits_{k=0}^{N-1} x_k \cos(2k\pi/N) \right)^2 + \left(\sum\limits_{k=0}^{N-1} x_k \sin(2k\pi/N) \right)^2}{\sum\limits_{k=0}^{N-1} x_k^2}, \tag{9}$$

where N is the length of the trajectory sample that corresponds to integration time $T = N\Delta t$. The formula (9) does not allow the simultaneous calculation for both the trajectory and the estimator γ. However, γ should be assumed as a dynamical variable along the trajectory.

4 Numerical Results and Conclusions

In Fig. 3 the evolution of $\gamma(N)$, compared to the finite Liapunov characteristic number $\lambda(N)$ is shown for the same trajectories as that of Fig. 2. For the calculation of $\gamma(N)$ we used the time series $z_n = \sin(x_n)$, $n = 0, ..., N-1$ in order to avoid the discontinuity caused by the modulo operation in (2). Cases (a) and (b) clearly indicate regular dynamics. Both, $\gamma(N)$ and $\lambda(N)$ converge rapidly to zero, almost as $1/N$. In case (c), which corresponds to a weakly chaotic trajectory, we obtain convergence, for both $\gamma(N)$ and $\lambda(N)$, which holds during that time the orbit is sticky. Afterwards, $\lambda(N)$ tends slowly to saturation, while $\gamma(N)$ shows an abrupt increment indicating that the trajectory entered the chaotic channel and reveals its chaotic nature efficiently. In case (d) $\lambda(N)$ seems to tend at a relatively large value indicating the chaotic character of the trajectory. The estimator $\gamma(N)$ shows a remarkable divergence indicating, additionally, a slow diffusion that is apparent at least up to the integration interval. Independently of the parameter k, cases (a) and (b) are typical for all regular trajectories, and, in average, it holds $\gamma \sim N^a$ with $a \approx 1$. It worths to note, that the convergent evolution of $\gamma(N)$ is followed by dense sharp peaks, towards to lower values. Such peaks indicate that the corresponding spectra should be

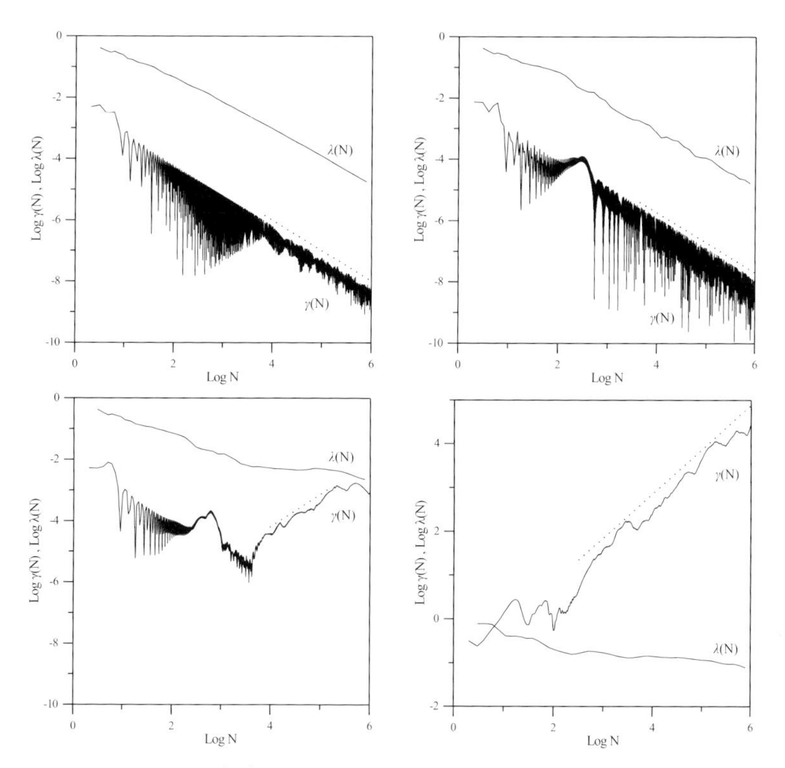

Fig. 3. The evolution of $\gamma(N)$ and the finite time Liapunov characteristic number for the trajectories referred in Fig. 2.

discrete as, indeed, they are. For chaotic trajectories, the evolution of $\gamma(N)$ is quite smooth because of the continuity of the associated spectra.

From Fig. 3 we observe that $\gamma(N)$ and $\lambda(N)$ show the same asymptotic characteristics that can be used to classify the trajectories as ordered or chaotic. Especially for ordered trajectories, $\gamma(N)$ and $\lambda(N)$ show a similar decay with respect to N. For chaotic orbits, $\gamma(N)$ and $\lambda(N)$ seem to reach saturation values in different ways, but nevertheless, we can conclude their non-convergent character. In Fig. 4 we plot the pairs $(\gamma(N), \lambda(N))$ obtained from 2000 randomly selected initial conditions for $k = 0.9$ (crosses) and $k = 1$ (circles) and after $N = 3 \cdot 10^6$ iterations. The 97% of the points belongs either in the domain A (regular trajectories) or in the domain B (chaotic trajectories). The rest of them, which mainly correspond to $k = 0.9$, may be classified as weakly chaotic trajectories. By increasing the number of iterations, the points in region A, and, generally, the points that correspond to ordered motion, move to lower values, since both $\gamma(N)$ and $\lambda(N)$ tend to zero. Therefore, the reliability of the trajectory classification increases as the length of the trajectory increases.

In many cases, γ shows in average N^a-divergence, with $a \approx 1$. However, after a long time evolution, γ shows small oscillations around a constant value $\bar{\gamma}$ (saturation). The value of $\bar{\gamma}$ is proportional to how long the N^a-divergence takes place and it is associated to the $1/f^a$-divergence of the spectrum caused by the slow diffusion through small islands and cantori [8]. Thus, large values of $\bar{\gamma}$ indicate the diffusive character of the trajectory rather than its strong chaotic evolution. In Fig. 5 we can see that the chaotic orbits, which start near the unstable fixed point (0,0) for $k \leq 1$, correspond to large values for $\bar{\gamma}$ (about 10^4), but, as k increases, we observe that $\bar{\gamma} \to 1$. According to the definition of γ, this value is obtained when the power spectrum has the form $p(f) \approx$const (white noise).

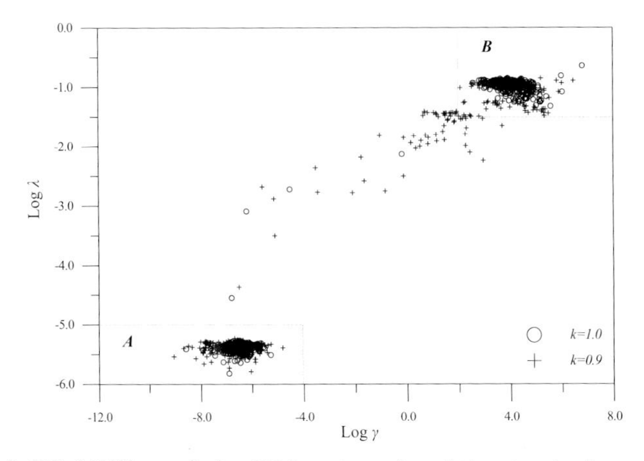

Fig. 4. The $(\gamma(N),\lambda(N))$ graph for 2000 trajectories of the standard map. Points that belong to regions A or B indicates regular and chaotic dynamics, respectively, with great certainty.

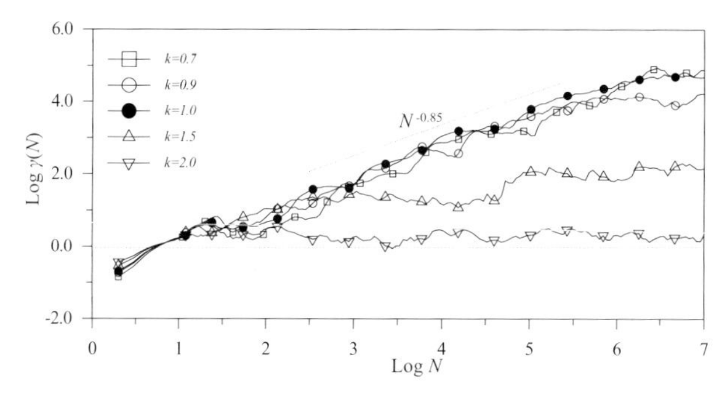

Fig. 5. The $\gamma(N)$ evolution along a chaotic trajectory close to the unstable fixed point (0,0) and for different values of the mapping parameter k.

Conclusively, we may claim that regular dynamics is associated with convergent spectra, in the sense that $p(f) \to 0$ as $f \to 0$ or, equivalently, $\gamma(T) \to 0$ as $T \to \infty$. Chaos is associated with divergent spectra where $p(f) \to$ const.> 0 as $f \to 0$ or, equivalently, $\gamma(T) \to$ const.> 0 as $T \to \infty$. In other words, for Hamiltonian bounded trajectories, discrete spectra are convergent, while, spectra with continuous local domains are divergent. Although there is not a rigorous proof, the numerical results support strongly the above relation.

The low frequency power estimator γ can be used as an indicator for the qualitative character of a trajectory, in a similar manner to that of the Liapunov characteristic number λ. These two quantities are expressed in the same units (time^{-1}) but they have different physical meaning. Namely, γ is associated with the long term regular or irregular evolution rather than the linear or exponential divergence of nearby trajectories and is calculated along a single trajectory.

Acknowledgements

The author would like to thank Dr. S.Ichtiaroglou for the fruitful discussions about this topic.

References

1. V.V.Beloshapkin, G.M.Zaslavsky: Phys. Lett. A **97**, 121 (1983)
2. R.S. Dumont, P. Brumer: J. Chem. Phys. **88**, 1481 (1988)
3. T. Geisel, A.Zacherl, G.Rabons: Phys. Rev. Lett. **59**, 2503 (1987)
4. S. Kim, S. Ostlund, G.Yu: Physica D **31**, 117 (1988)
5. J. Laskar: Physica D **67**, 257 (1993)
6. S.Ostlund, D.Rand, J. Sethna, E. Siggia: Physica D **8**, 303 (1983)
7. R.Roy, B.G.Sumpter, G.a.Pfeffer, S.K. Gray, D.W. Noid: Physics Rep. **205**, 111 (1991)
8. G.E. Powell: Regular and irregular frequency spectra. Ph.D.Thesis , Queen Mary College, London (1980)
9. T.R. Press, S.A. Teukolsky, W.T. Vetterling, B.P.Flannery: *Numerical Recipes in C*, (Cambridge University Press, Cambridge, 1992)

10. R.Z. Sagdeev, D.A. Usikov, G.M. Zaslavsky: *Nonlinear Physics* (Harwood Academic 1988)
11. M. Tabor: *Chaos and integrability in nonlinear dynamics*, (Wiley, New York 1989)
12. G. Voyatzis, S. Ichtiaroglou: J. Phys. A **25**, 5931 (1992)

Part II

Orbit Theory

Disk-Crossing Orbits

Chris Hunter

Department of Mathematics, Florida State University, Tallahassee FL 32306-4510, USA

Abstract. We study orbits in simplified models of galaxies which consist of two components, a disk and a halo. The disk is idealized as razor-thin, though we present evidence that this simplifying assumption is not critical. We find that the presence of the disk causes many more resonances than have been found in similar smooth potentials. Those resonances grow at relatively modest values of energy, overlap, and give rise to many stochastic orbits. A significant range of regular orbits remain and show smooth KAM curves, even though the discontinuous potential due to the razor-thin disk means that current versions of the KAM theorem do not apply.

1 Introduction

Many galaxies have a disk component and one or more other components which are much less flattened. Normal spirals and S0s are examples. Bender et al [1] have found that many ellipticals can be classified as disky, and that their diskiness is consistent with a disk plus bulge model [2]. For simplicity, we use halo as an all-embracing term to describe the whole non-disk component. Orbits of halo stars in such galaxies will necessarily cross back and forth through the disk. As they do, they will experience a fairly abrupt change in the gravitational force field. This paper examines how these changes in the force field affect the dynamics of the orbit. It is a topic which seems to have attracted very little attention so far, with two notable exceptions. Ostriker, Spitzer, and Chevalier [3] have discussed the effect that the compressive gravitational shocks, caused by passage through the disk, have on the internal structure of globular clusters. Our interest is on how that passage affects the orbits of individual stars of the halo population, on the resonances which they can cause, and on the extent to which they can induce chaos. This issue was also studied in the 1970s by L. Martinet and co-authors [4]–[7] who integrated orbits in Schmidt's [8], [9] two models of the Galaxy. These models have highly flattened disk components.

We shall use simple models which have both disk and halo components. These are the Kuzmin-like potentials which were introduced recently by Tohline and Voyages [10]. Their disk components are razor-thin. We describe their potentials and densities in Sect. 2, and their dynamics in Sect. 3. Numerical integrations are needed to study the detailed properties of orbits. We describe the results of those integrations in Sect. 4. In Sect. 5 we show, again by means of orbit integrations, that our findings do not depend critically on the simplifying approximation that the disk component is razor-thin. Sect. 6 sums up, and relates our findings to other work.

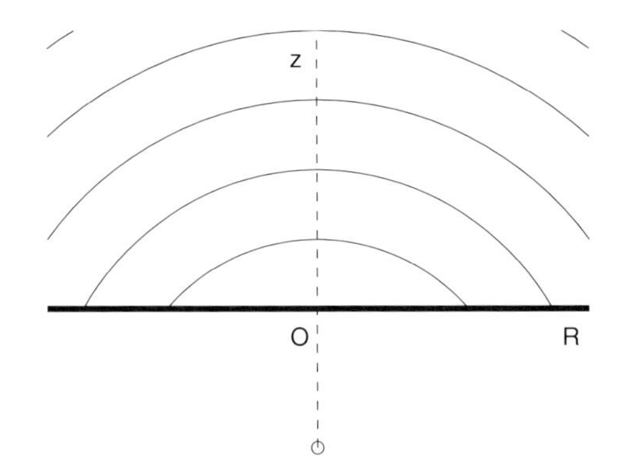

Fig. 1. The thin solid lines in $z > 0$ are equipotential surfaces above Kuzmin's disk. They are circles in (R, z)-space centered on an image mass at $z = -a$, which is shown as a small open circle. The equipotentials in $z < 0$ are their reflections in the disk $z = 0$

2 Models

2.1 Kuzmin's Disk

Kuzmin's disk [11], [3] is a flat density distribution with the remarkable property that the gravitational potential above it is that due to a point mass at a distance a below it, while the potential below it is that due to a point mass at a distance a above it. Its gravitational potential is

$$\Phi = \frac{-GM}{\sqrt{R^2 + (a + |z|)^2}}, \tag{1}$$

when $z = 0$ is the plane of the disk, and R measures radial distance from the z-axis of symmetry. Its equipotential surfaces are spherical, those in $z > 0$ centered on the point mass below the disk, and those in $z < 0$ on the point mass above the disk.

2.2 Kuzmin-Like Potentials

Kuzmin's disk is a particular case of the more general Kuzmin-like potentials of Tohline and Voyages [10], for which

$$\Phi = \Phi(\xi), \qquad \xi = \sqrt{R^2 + (a + |z|)^2}. \tag{2}$$

They have the same spherical equipotential surfaces as Kuzmin's disk, with centers of force on the opposite side of the $z = 0$ plane, but a force law which is more general than the inverse-square of Kuzmin's disk. The variable ξ measures distance from the centers of force. Poisson's equation shows that the potential (2) is generated by a density

$$\varrho = \frac{1}{4\pi G}\Delta\Phi = \frac{1}{4\pi G}\left[\Phi''(\xi) + \frac{2[1 + a\delta(z)]}{\xi}\Phi'(\xi)\right]. \tag{3}$$

The Dirac delta function $\delta(z)$ arises from derivatives of $|z|$ because $\partial|z|/\partial z = \mathrm{sgn}(z)$, and $\partial\,\mathrm{sgn}(z)/\partial z = 2\delta(z)$. Kuzmin-like potentials therefore arise from a volume density which is stratified on the equipotentials, together with a surface density $\sigma = a\Phi'(\xi)/2\pi G\xi$ on the plane $z = 0$. It is this surface density which causes the z-component of force to be discontinuous. The volume density vanishes for the special case of Kuzmin's disk, and there is then only a surface density $\varrho = aM\delta(z)/2\pi(a^2 + R^2)^{3/2}$.

The choice of a specific spherical potential $\Phi(\xi)$ fixes both the volume and the surface densities. One way of judging their relative significance is to compute the mass of each component interior to some equipotential $\xi = \xi_0$. Using the geometric formula of $4\pi\xi(\xi - a)$ for the surface area of the two part-spheres which form the equipotential of constant ξ, we obtain

$$M_{\mathrm{disk}}(\xi_0) = \frac{a}{G}\left[\Phi(\xi_0) - \Phi(a)\right], \qquad M_{\mathrm{total}}(\xi_0) = \frac{\xi_0}{G}(\xi_0 - a)\Phi'(\xi_0), \tag{4}$$

for the disk and total mass interior to the equipotential $\xi = \xi_0$.

2.3 The Logarithmic Kuzmin-Like Potentials

As an example, we shall investigate the logarithmic potential which is widely used as a model in galactic dynamics [3]. It gives a Kuzmin-like model with potential and density

$$\Phi(\xi) = V_0^2 \ln \xi, \qquad \varrho = \frac{V_0^2}{4\pi G\xi^2} + \frac{aV_0^2\delta(z)}{2\pi G(R^2 + a^2)}. \tag{5}$$

As (4) shows, the mass in the disk grows logarithmically with increasing distance whereas the total mass grows linearly. The ratio of these masses is

$$\frac{M_{\mathrm{disk}}(\xi_0)}{M_{\mathrm{total}}(\xi_0)} = \frac{a}{(\xi_0 - a)} \ln\left(\frac{\xi_0}{a}\right), \tag{6}$$

and the relative significance of the disk diminishes with increasing ξ_0. Orbits at higher energies and greater distances experience forces for which the halo is increasingly dominant. Figure 2-a shows how isophotes would appear when viewed at a small angle from edge-on, assuming the same mass-to-light ratio for all mass.

3 Dynamics

Motion in a spherical potential conserves the energy E and the angular momentum vector about the center. The motion is integrable and is confined to the plane through the center of force that is perpendicular to the angular momentum vector. Unless confined always to the plane $z = 0$, an orbit in a Kuzmin-like potential passes continually back and forth between the regions $z > 0$ and $z < 0$,

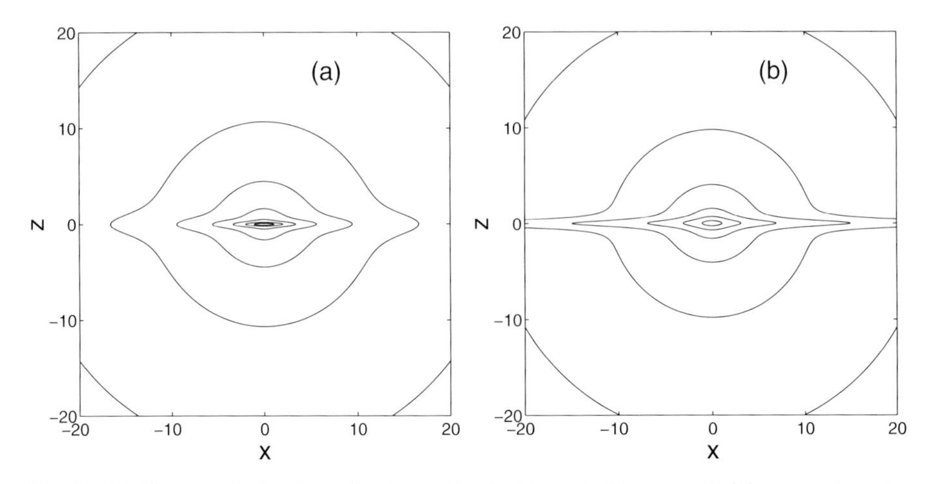

Fig. 2. (a) Projected density of a logarithmic Kuzmin-like model (5), assuming the same mass-to-light ratio for all mass, when viewed at 5° from edge-on (left). Contour levels decrease by factors of $10^{-1/3}$. The projected disk density is infinite when viewed edge-on, and the isophotes become increasingly less pointy as the viewing angle increases. (b) Projected density of the same model after thickening with $b = 0.5$ as described by (12) in Sect. 5 when viewed edge-on. The centermost contours and successive outer ones represent the same levels as in (a). Thickening spreads density out from the disk, giving higher volume densities; the level of the outermost contour of (b) matches the next to outermost of (a)

and hence from one spherical potential field to the other. Different angular momenta are conserved in the two regions. Both are described by the vector

$$\mathbf{J} = [\mathbf{r} + a\,\mathrm{sgn}(z)\mathbf{k}] \times \mathbf{v}, \tag{7}$$

where $\mathbf{r}$ is the position relative to the origin, $\mathbf{v}$ is velocity, and $\mathbf{k}$ is the unit vector in the z-direction. An equation of motion for $\mathbf{J}$ can be derived from the equations of motion in the Kuzmin-like potential. They are

$$\frac{\mathrm{d}\mathbf{r}}{\mathrm{d}t} = \mathbf{v}, \qquad \frac{\mathrm{d}\mathbf{v}}{\mathrm{d}t} = -\nabla\Phi = -\frac{[\mathbf{r} + a\,\mathrm{sgn}(z)\mathbf{k}]}{\xi}\Phi'(\xi). \tag{8}$$

It follows that

$$\frac{\mathrm{d}\mathbf{J}}{\mathrm{d}t} = [\mathbf{r} + a\,\mathrm{sgn}(z)\mathbf{k}] \times \frac{\mathrm{d}\mathbf{v}}{\mathrm{d}t} + \mathbf{v} \times \mathbf{v} + 2a\delta(z)(\mathbf{k} \times \mathbf{v})\frac{\mathrm{d}z}{\mathrm{d}t} = 2a\delta(z)(\mathbf{k} \times \mathbf{v})\frac{\mathrm{d}z}{\mathrm{d}t}. \tag{9}$$

This equation confirms the constancy of the vector $\mathbf{J}$ except when the orbit crosses the disk. Then $\mathbf{J}$ changes discontinuously by an amount $\pm 2a(\mathbf{k} \times \mathbf{v})$ unless $\mathbf{k} \times \mathbf{v} = 0$, that is unless the orbit is then travelling perpendicularly to the disk. The change in $\mathbf{J}$ is always perpendicular to the z-direction because the z-component J_z of $\mathbf{J}$ is conserved; J_z is the angular momentum about the axis of symmetry, and is a constant of the motion, as in any axisymmetric potential.

Since it is known that energy and angular momenta are the only constants of motion in general spherical potentials and, as we have seen, neither J_x nor J_y are conserved for general Kuzmin-like potentials, it follows that, contrary to what is stated in [10], Kuzmin-like potentials are not generally integrable. Kuzmin's disk is exceptional, its potential is Stäckel, and so all orbits in it are integrable. Their integrability relies on the Keplerian nature of its spherical potentials; motions then have an extra integral of motion, the Laplace-Runge-Lenz vector [13], in addition to energy and angular momentum. The Laplace-Runge-Lenz vector changes discontinuously across the disk too. However, there is a linear combination of its z-component and $\mathbf{J}^2$ for which the two discontinuities cancel [14], and this combination provides Kuzmin's third integral [11]. (Another exceptional case is that of a harmonic potential $\Phi \propto \xi^2$. It is not generally relevant for galaxies, except as an approximation for orbits which remain close to the center of the disk.)

Because position and velocity of an orbit are continuous as the disk is crossed, the tangent to an orbit changes smoothly. The hallmark of the discontinuity in the gravitational force is a discontinuity in the curvature of an orbit. Fig. 3-a gives the clearest example of this feature. It shows a particular $J_z = 0$ periodic polar orbit in the potential due to Kuzmin's disk. The orbit in $z < 0$ is part of an ellipse with focus at the center of force in $z > 0$. This ellipse is oriented such that the orbit is travelling directly away from the center of force in $z < 0$ as it crosses to $z > 0$. Hence its trajectory in $z > 0$ remains a straight line away from the lower center of force. The orbit travels along that line until it reaches the maximum extent which its energy allows, after which it returns and retraces its path.

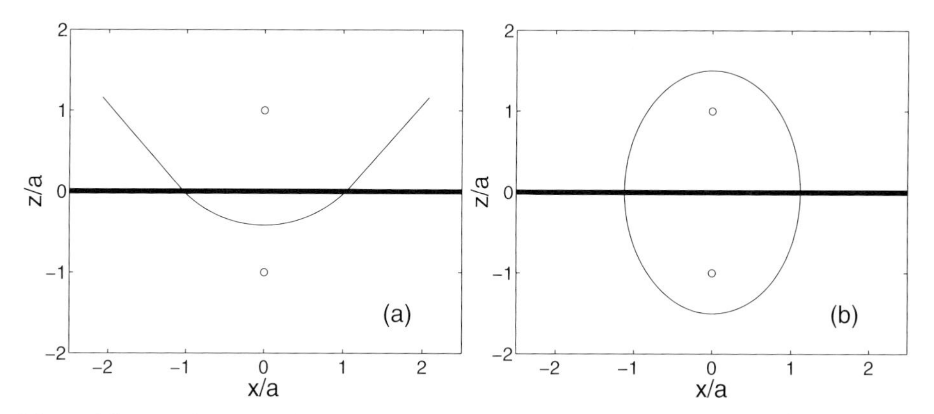

Fig. 3. The simplest periodic disk-crossing polar orbits for Kuzmin's disk: (a) a box orbit, which is a banana in the terminology of Binney [15] and a boxlet in the terminology of Miralda-Escudé and Schwarzschild [16], and (b) an ellipse, for energy $E = -GM/3a$. The open circles again denote the image masses as in Fig. 1. Similarly shaped periodic orbits occur for other Kuzmin-like potentials. The banana boxlet has a twin, which is its mirror image in $z = 0$.

4 Orbits

4.1 Orbits in Kuzmin's Disk

Orbits with $J_z \neq 0$ are short-axis tubes. Polar orbits with $J_z = 0$, which lie in planes through the axis of symmetry, may be either boxes or loops. At low energies $-1 < E < -GM/2a$ for which the orbit along the z-axis is stable, only box orbits occur. Loop orbits become possible at higher energies in the range $-GM/2a < E < 0$ for which the z-axis orbit has become unstable [14] The parent of the loop orbit family is a closed ellipse, as shown in Fig. 3-b. It is not quite a Keplerian ellipse because the orbit is always under the influence of a point mass at the more distant focus. Rather it is the union of two slower halves of a Keplerian ellipse. Although there are box orbits for all energies, the type of banana boxlet shown in Fig. 3-a occurs only for the energy range $-GM/2\sqrt{2}a \leq E < 0$. The curved part of the orbit is a quarter circle for the lower limit of energy, and so is nearly circular for the case illustrated. Its eccentricity increases as E increases, and the linear parts, which extend to a distance $-GM/E$ from the lower center of force, lengthen.

4.2 Orbits in Kuzmin-Like Potentials

In the following subsections, we present numerical results for the logarithmic potential (5) using $z = 0$ surfaces of section. Surfaces of section for Kuzmin's disk can be found analytically as contours on which the third integral is constant, but numerical integration is needed for Kuzmin-like potentials. We computed orbits by a variant of the method described in Appendix A.1 of Hunter et al [17] which integrates accurately from one crossing of $z = 0$ to the next using polar coordinates in (z, v_z)-space. No interpolation is used to find when crossings occur, and there is no uncertainty as to which force formula to use. The surfaces of section for the logarithmic potential (5) contain resonant island chains and stochastic as well as regular regions. They stand in marked contrast to the regular surfaces of section for Kuzmin's disk, for which the only significant feature is the bifurcation when loops first appear with polar orbits. Surfaces of section are plotted for specific values of energy and angular momentum J_z about the axis of symmetry. We express J_z as a fraction k of the angular momentum of the circular orbit in the z-plane., i.e. the maximum possible J_z for the given energy. We use a as unit of length, and for ease of understanding, identify energies by R_c, the radius of a circular orbit at that energy. In terms of R_c, the energy is

$$E = \frac{1}{2}\left[\ln(R_c^2 + 1) + \frac{R_c^2}{R_c^2 + 1}\right]. \tag{10}$$

For the unbounded potential (5), $E \geq 0$, and R_c increases monotonically with the energy, first as $R_c \approx \sqrt{E}$ for small E, but later as $R_c \approx e^E$ for large E. We label periodic orbits by the ratio $m : n$ where m and n are the numbers of the full cycles in z and R (or x in the case of polar orbits) respectively, in a complete cycle.

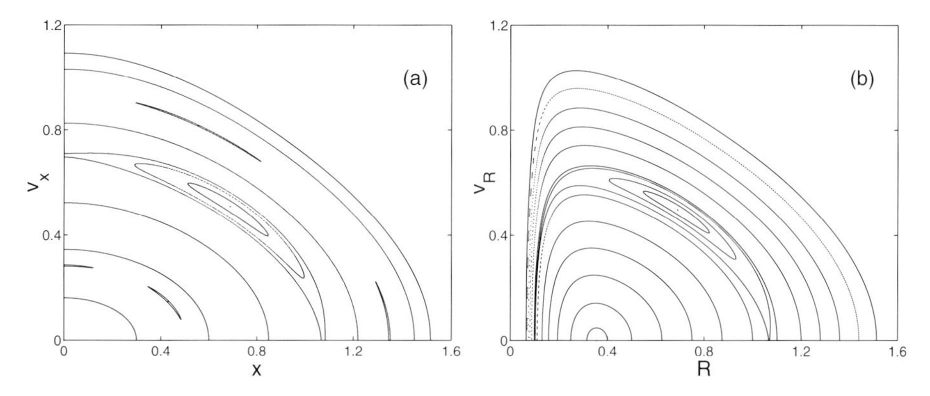

Fig. 4. (a) Surface of section for polar orbits in the logarithmic potential (5) for $R_c = 1$, (b) Surface of section for axisymmetric orbits in the same potential and at the same energy, and with $k = 0.1$, i.e. angular momentum 0.1 of that for a circular orbit at this energy

Polar Orbits

Only box orbits occur at low energies, and (x, v_x) surfaces of section mostly show smooth rings centered on the z-axis orbit at $(0, 0)$. However bifurcations do occur before the z-axis orbit becomes unstable at $R_c = 1.37$. Figure 4-a for $R_c = 1$ shows three rings of island chains. The prominent middle ring is due to the $2 : 1$ banana boxlets, and their unstable figure-of-eight companion. These orbits bifurcate from the z-axis orbit when $R_c = 0.62$. Initially their paths lie close to the z-axis, but their initial island chain cuts the z-axis at $z = 0.35$, and is clearly removed from $(0, 0)$. The smooth curves surrounding $(0, 0)$ remain, indicating that the stability of the z-axis orbit is unaffected by this bifurcation. The outer narrow ring corresponds to $3 : 1$ orbits. They bifurcate from the z-axis orbit at a lower energy when $R_c < 0.45$. The inner narrow ring corresponds to $3 : 2$ orbits of a well-known type; an unstable fish and a stable antifish [16]. A separatrix between box and loop orbits is formed once the z-axis orbit becomes unstable at $R_c = 1.37$, and a visible stochastic region develops around the separatrix [18]. Figure 5 shows that resonant islands take up a significant part of the box orbit region at $R_c = 3$. The $2 : 1$ banana boxlet at $x = 1.84$, $v_x = 1.15$ is still prominent, while the $3 : 2$ family with periodic points at $x = 2.54$, $v_x = 0.79$ and $x = 0$, $v_x = 1.14$ has grown in significance. Figure 6 shows that most of the box orbit region has become stochastic when energy has increased to $R_c = 10$, as a result of substantial overlapping of the resonant regions which are so prominent in Fig. 5. There are many resonant islands in Fig. 6, and even islands within islands, but the stochastic sea dominates the outer part of the surface of section. The island for the banana boxlets, which is here centered on $x = 3.2$, $v_x = 1.7$, has moved further outwards and become thinner. The outward migration is due to the fact that these orbits become flatter with increasing energy; the outer boundary of the surfaces of section is the orbit which is always in the plane

$z = 0$. A few box orbits remain in the outer parts of the surface of section. They too are flat and never depart far from the $z = 0$ plane.

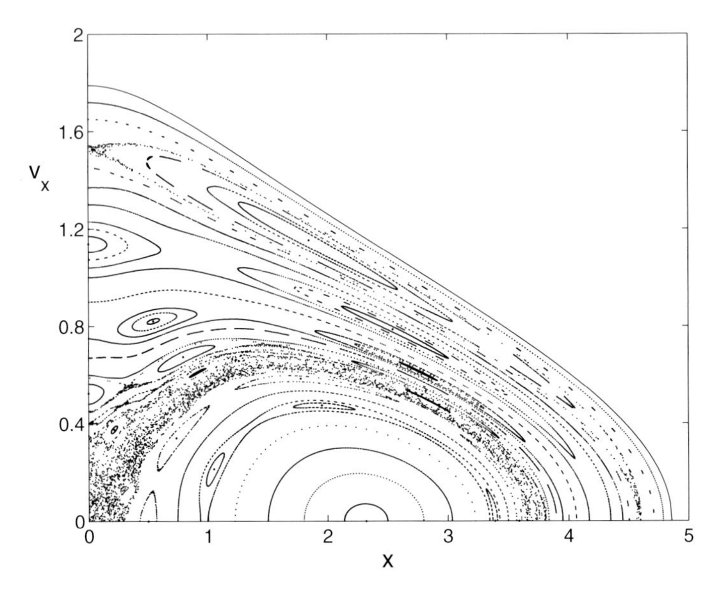

Fig. 5. Surface of section for polar orbits in the logarithmic potential (5) for $R_c = 3$

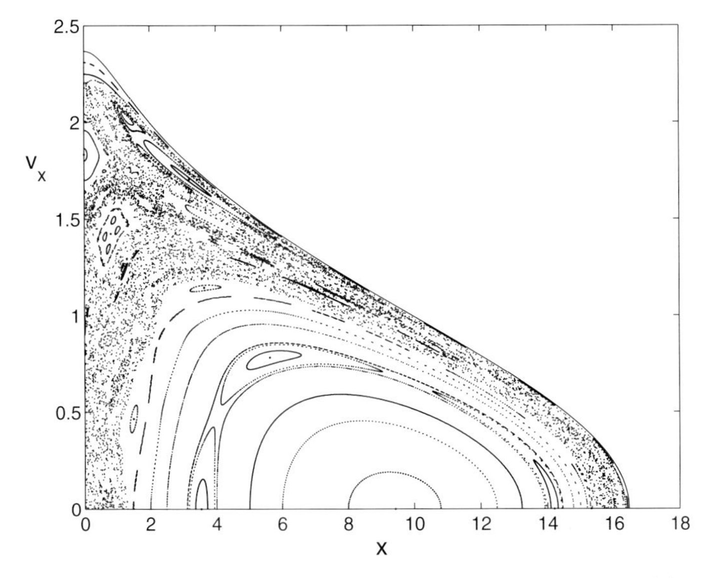

Fig. 6. Surface of section for polar orbits in the logarithmic potential (5) for $R_c = 10$

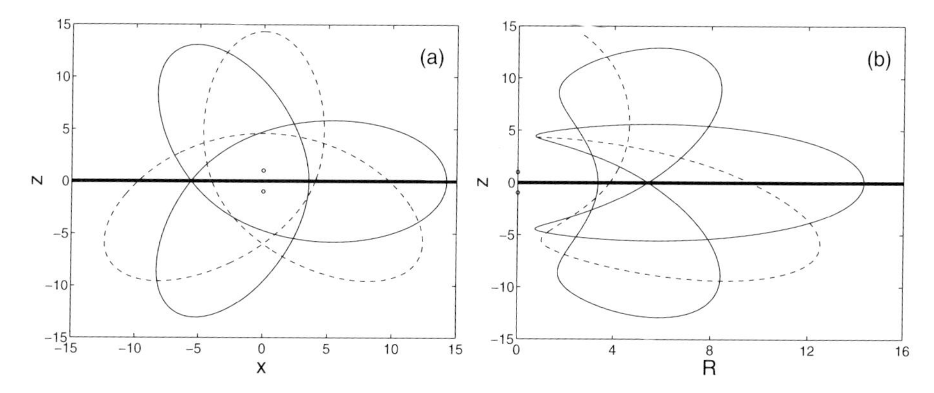

Fig. 7. (a) Stable and unstable 2 : 2 orbits, full and broken lines respectively, which occupy the centermost chain of islands in Fig. 6. The reflection of the unstable orbit in $z = 0$ gives another unstable 2 : 2 orbit. (b) The stable and unstable 2 : 4 orbits, full and broken lines respectively, which occupy the centermost chain of islands in Fig. 11

While the box orbit region is becoming stochastic, the loop orbit region, centered around the 1 : 1 periodic closed loop, remains largely regular, though with some resonant islands. The first out from the center and most prominent in Fig. 6 are due to the 2 : 2 periodic orbits whose three-petalled rosette form is shown in Fig. 7-a.

Delay Plots of Angular Momentum provide a third way of analyzing orbits in Kuzmin-like potentials, which supplement plots in position space and surfaces of section. While the orbit is above the disk, J_y is its angular momentum about the lower center of force, and remains constant. The next time the orbit returns to $z > 0$, it has a new value of J_y, except for exceptional cases such as the elliptical orbit of Fig. 3-b. The sequence of successive values of J_y characterizes an orbit, and generates a delay plot, similar to those used for analyzing time series generated by other dynamical systems [19] [20], in which two successive values are plotted as a point, the earlier value giving the abscissa and its successor the ordinate. Figure 8 displays J_y delay plots for four orbits which are included in the surface of section of Fig. 6. The physical forms of the first three are shown in Fig. 9. The orbits are arranged by decreasing stochasticity and increasing regularity. Orbit (a) inhabits the stochastic region just outside the regular central core of loop orbits. The points in its delay plot are scattered fairly uniformly. Figure 9-a shows that it almost fills the area within the zero-velocity curve, apart from narrow slivers near the left and right edges. Orbit (b) from well inside the stochastic region fills less than half the region within the zero-velocity curve. Its delay plot shows much randomness, but also a double ring structure. Orbit (c) is something of a surprise because it seems to start in a stochastic region, yet its delay plot is highly ordered. Its pretzel shape as seen in Fig. 9-c shows it to belong to the family of 4 : 3 periodic orbits parented by the periodic orbit which crosses the surface of section of Fig. 6 at $(0.81, 1.42)$ and $(6.38, 1.21)$. Orbit (c)

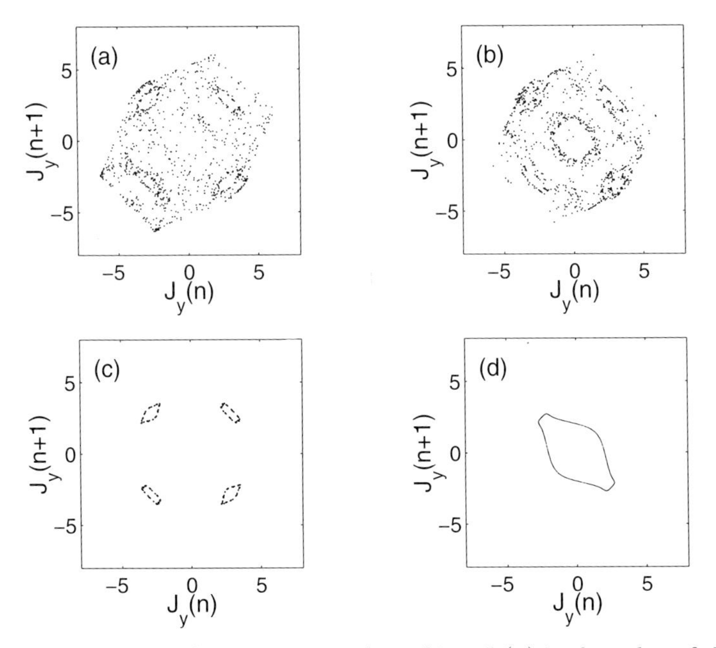

Fig. 8. J_y delay plots for four $R_c = 10$ polar orbits. $J_y(n)$ is the value of J_y during the n-th passage through $z > 0$. The orbits start in $z = 0$ from (a) $x = 2$, $v_x = 1$, (b) $x = 1$, $v_x = 1.7$, (c) $x = 4$, $v_x = 1.4$, (d) $x = 0$, $v_x = 2.25$

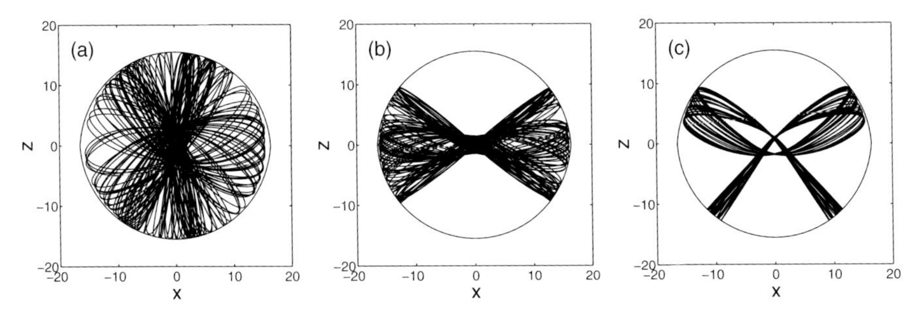

Fig. 9. Orbits for first three delay plots of Fig. 8, after 200 crossings of $z = 0$. The outer boundary is the zero-velocity curve

generates the set of broken lines which surrounds the four island chain around $(0.81, 1.42)$, and broken lines appear in its delay plot too. The starting point of this orbit at $(4, 1.4)$ lies at the left end of a long and narrow island surrounding the other periodic point at $(6.38, 1.21)$. That island is lost in the complexity of Fig. 6. Finally, orbit (d) is a regular box orbit which gives a smooth curve around the outer part of the surface of section. Its delay plot is also smooth and regular. Being close to the outer boundary, this orbit is a flat box which stays close to the disk at all times.

Axisymmetric Orbits

The surfaces of section Fig. 4-b, Fig. 10, and Fig. 11 for the same three energies of $R_c = 1$, 3, and 10, respectively, and for J_z angular momenta which are one-

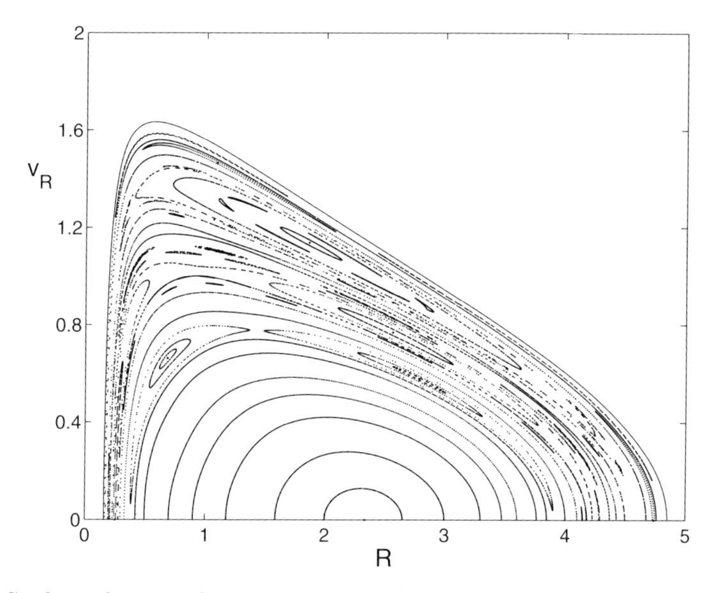

Fig. 10. Surface of section for axisymmetric orbits in the logarithmic potential (5) for $R_c = 3$ and $k = 0.1$

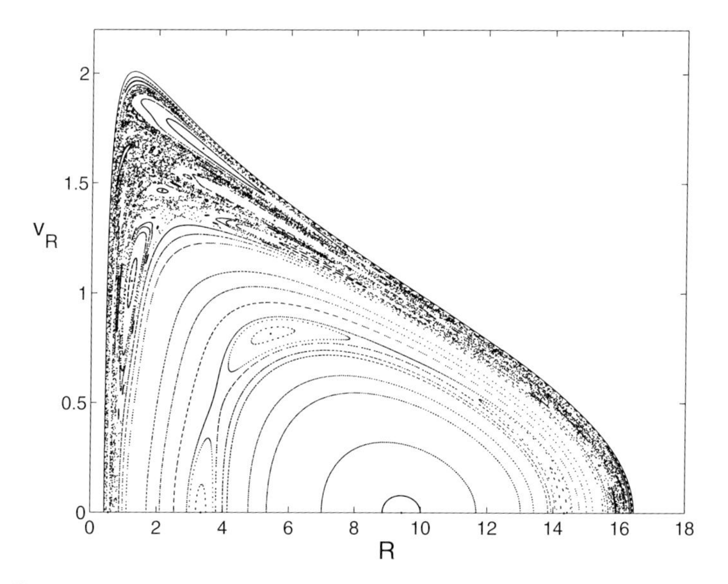

Fig. 11. Surface of section for axisymmetric orbits in the logarithmic potential (5) for $R_c = 10$ and $k = 0.1$

tenth of the maximum possible ($k = 0.1$) show similar trends with increasing energy to the corresponding figures for polar orbits. The predominant class of regular orbits are short axis tubes [14]. They give the smooth rings surrounding the center which represents a thin tube or shell. Figure 4-b shows a prominent island at the lowest energy. This is due to the 1 : 1 saucer orbit [21] which is obtained when the angular momentum barrier stops a banana orbit before it gets to the z-axis and reflects it back along its path after its first crossing of the disk. The $R_c = 3$ surface of section in Fig. 10 shows its outer part to be occupied by resonant island chains. Those resonances have overlapped when the energy has increased to $R_c = 10$, and most of the outer part is stochastic. However there is a large regular core with a prominent island chain. That chain corresponds to the axisymmetric counterparts of the 2 : 2 rosettes of the polar case. The angular momentum barrier truncates and reflects these orbits to give the 2 : 4 orbits shown in Fig. 7-b, a stable spaceship and an unstable pair of twisted fish. Note that this island chain is present already as the innermost island chain in Fig. 10 for the lower energy of $R_c = 3$. There is a difference in that there the unsymmetric twisted fish are stable, and the spaceship is unstable.

Delay plots can be constructed for this case too, using the quantity $J_x^2 + J_y^2$ which is now constant during each passage through $z > 0$. Figure 12 shows delay plots for two orbits from the outer stochastic part of Fig. 11. The left delay plot shows wide scattering, with some concentration at the boundaries, and an empty hole at the top right, while the right plot shows far less scattering. Figure 13 shows that neither orbit comes close to filling the region within the zero-velocity curve. Orbit (b) with the more organized delay plot, ranges more widely, but less randomly, than orbit (a).

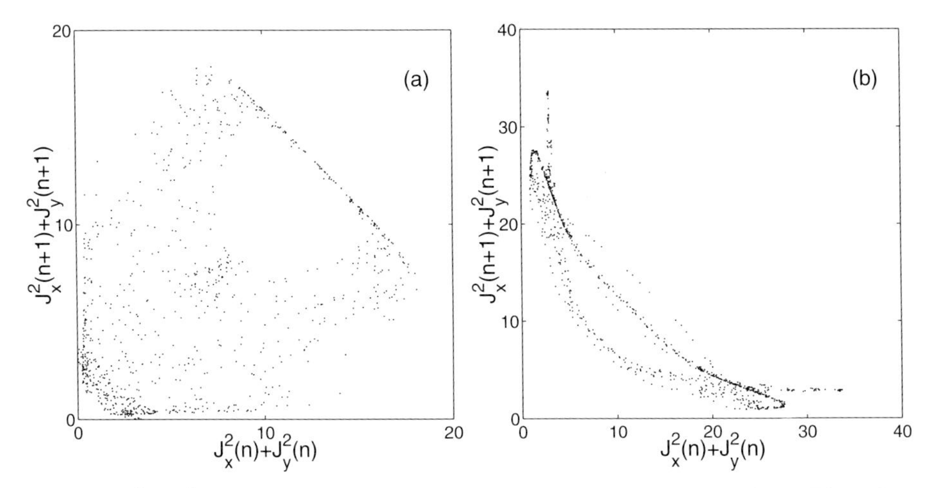

Fig. 12. $J_x^2 + J_y^2$ delay plots for two $R_c = 10$, $k = 0.1$ axisymmetric orbits. The orbits start in $z = 0$ from (a) $R = 1$, $v_R = 1.7$, (b) $R = 1.6$, $v_R = 1$

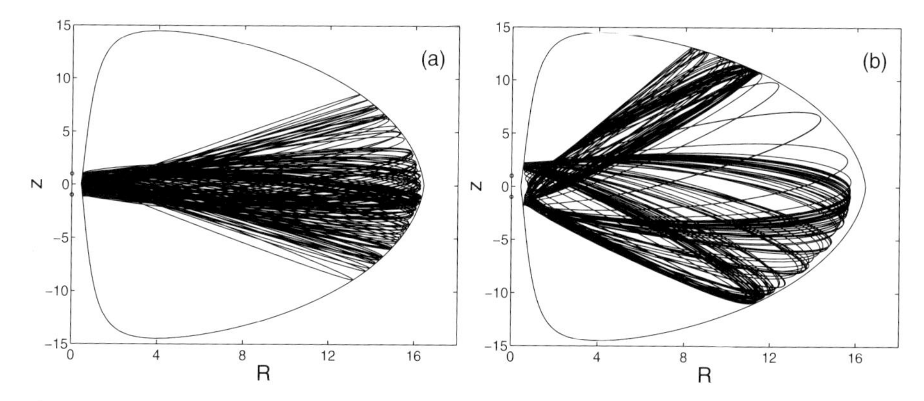

Fig. 13. Orbits for the two delay plots of Fig. 12 after 200 crossings of $z = 0$. The outer boundary is the zero-velocity curve

5 Thickened Disks

We have so far been investigating force fields which have sharp discontinuities. That is an idealization of a galaxy with a disk-like component, and so one must wonder how critically do the results which we have found depend on the assumption of an abrupt discontinuity. We investigate this issue by thickening the disk, borrowing a device introduced by Miyamoto and Nagai [22]. We add a length b, which is zero in the limit of a razor-thin disk, to the definition of the variable ξ on which the potential depends, so that now

$$\Phi = \Phi(\xi), \qquad \xi = \sqrt{R^2 + (a + \sqrt{z^2 + b^2})^2}. \tag{11}$$

This potential is due to the density field

$$\varrho = \frac{1}{4\pi G}\left[\Phi'' + \frac{2\Phi'}{\xi} + \frac{ab^2\Phi'}{\xi(z^2+b^2)^{3/2}} - \frac{b^2}{\xi}\left(\frac{\Phi''}{\xi} - \frac{\Phi'}{\xi^2}\right)\left(1 + \frac{a}{\sqrt{z^2+b^2}}\right)^2\right] \tag{12}$$

The first three components give the density (3) in the limit $b \to 0$ when $b^2/(z^2 + b^2)^{3/2} \to 2\delta(z)$. The right half of Fig. 2 shows contours of the projected density (11) for the logarithmic potential of (5) when viewed edge-on. Comparison with the unthickened case in the left half shows that there has been a considerable lowering of density contrasts. Even though ξ can no longer be interpreted as a specific distance, it does tend to a radial distance at large distances where most of the equipotentials become quite spherical. Despite the changes in the potential for the displayed case of $b = 0.5$, which is not much smaller than the displacements $a = 1$ of the former centers of force, yet the $R_c = 10$ surface of section shown in Fig. 14 shows no significant qualitative differences from the $b = 0$ case of Fig. 11. In fact much of the microstructure persists. For example, the small island around $(2.21, 1.49)$ in Fig. 14 and that around $(2.11, 1.46)$ in Fig. 11 are due to similarly shaped $7:10$ orbits, and the narrow tadpoles around $(1.11, 1.71)$ in Fig. 14 and $(0.92, 1.62)$ in Fig. 11 are due to similar $5:6$ orbits.

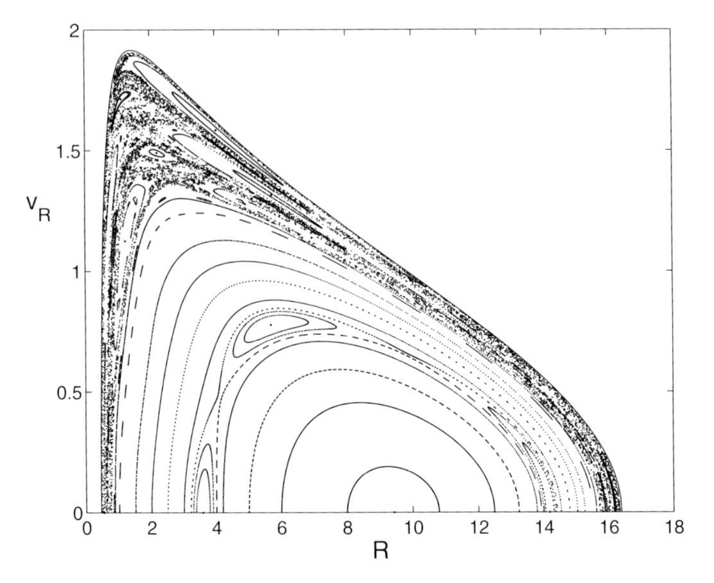

Fig. 14. Surface of section for axisymmetric orbits for $R_c = 10$ and $k = 0.1$, as in Fig. 11 but for a thickened potential with $b = 0.5$

6 Discussion and Conclusions

6.1 Related Work

This work adds another example to those already known in which chaos is induced when orbits pass between two or more regions, in each of which the motion is integrable, but in which different integrals apply. One that has attracted considerable attention in recent years is that of a black hole at the center of an otherwise smooth-cored triaxial galaxy. Shortly after de Zeeuw [14] had shown how Stäckel potentials account nicely for the major orbit families which Schwarzschild [23] had found in his numerical integrations of a smooth-cored triaxial galaxy, Gerhard and Binney [24] showed how destructive a central black hole, or cusp, would be for box orbits which came close to it. Motion near the black hole is nearly Keplerian, as has been discussed in detail by Sridhar and Touma [25].

Transitions between regions in which different integrals apply are basic to the mapping constructed by Wisdom [26] to explain the Kirkwood gaps in the asteroid belt. His mapping concentrates high-frequency perturbations by Jupiter into a periodic sequence of impulses which modify the motion periodically. Between impulses, the motion is determined entirely by secular terms which describe an integrable system. Touma and Tremaine [27] have applied a similar mapping in galactic dynamics. They study eccentric orbits in a plane non-axisymmetric potential using a map obtained by concentrating the non-axisymmetry so that it acts impulsively at apocenters of orbits. Between successive apocenters the non-axisymmetry is ignored and the orbital motion is integrable.

From a mathematical perspective, this work adds another instance in which there is evidence that KAM curves occur, even though the conditions for which the KAM theorem has been proved are not satisfied. Moser's [28] proof of that theorem has been refined to the case of continuous derivatives through to third order. Yet Kuzmin-like potentials have discontinuous first derivatives. Like the example of Varvoglis [29], ours comes directly from a significant physical context. Many studies in which KAM curves have been found for Hamiltonians which lack the smoothness required by Moser's proof, such as that of Benettin and Strelcyn [30] have studied variants of the billiard ball problem introduced by Birkhoff [31].

6.2 Conclusions

Our largely numerical study of orbits in Kuzmin-like potentials has shown that Kuzmin's disk is not typical. Its out-of-plane orbits are all integrable, despite the abrupt changes in the angular momentum vector which occur whenever the disk is crossed. Other Kuzmin-like models (other than the simple case in which the potential is harmonic) are not integrable. The presence of the disk causes many more resonances than have been found in similar smooth potentials [32] [17]. Those resonances grow at relatively modest values of energy, overlap, and give rise to many stochastic orbits for which the disk crossings may scatter the angular momentum vector fairly randomly. The extent of the stochasticity grows with increasing energy and decreasing angular momentum. However, there are many orbits which remain regular despite the impulsive changes due to crossing the disk. We have investigated only a very limited class of Kuzmin-like models, and Kuzmin-like models are themselves special because they do not permit the disk density to be adjusted independently of the volume density. Yet it is hard to see why similar mechanisms will not induce stochasticity in disk-crossing orbits in galaxies with disks of substantial size. The similar results from our even more limited tests with thickened Kuzmin-like potentials, for which the disk is not a sharp discontinuity and for which the motion is not integrable outside the disk, is evidence of the likely robustness of our findings. Martinet et al [4]–[7] investigated quite different models. Some of their surfaces of section show large stochastic regions, more extensive than any we display, and even instances in which the central periodic orbit of an axisymmetric case has become unstable. Because the models are so different, no detailed comparison of our results with theirs is possible.

Although we drew a distinction in Sect. 1 between our interest in orbits and that of Ostriker et al [3] on the internal structure of globular clusters, that distinction may be somewhat fuzzy. The occurrence of chaos is commonly ascribed to sensitivity to initial conditions, when small changes in initial conditions lead to greatly different outcomes [20]. The calculations in [3] are based on changes in the relative motion of two nearby points of a cluster caused by disk shocking. Orbits of nearby points in a stable cluster are nearby when viewed on the galactic scale. They grow far apart if and when the cluster disrupts. Our work finds that there are substantial regions of phase space in which orbits remain regular. In them nearby orbits diverge only gradually from each other, whereas they diverge rapidly from each other in stochastic regions. This implies that the vulnerability of a globular cluster to disk shocking is likely to depend considerably on its orbit This matches what Aguilar, Hut, and Ostriker [33] indeed find; that disk shocking is much more destructive of globular clusters on highly elongated radial orbits.

Acknowledgements

It is a pleasure to thank George Contopoulos, and Nikos Voglis and his committee, for organizing a stimulating workshop, and for their hospitality. This work has been supported in part by the National Science Foundation through grant DMS-0104751.

References

1. R. Bender, S. Döbereiner, C. Möllenhoff, A&AS, **74**, 385 (1988)
2. C. Scorza, R. Bender, A&A, **293**, 20 (1995)
3. J.P. Ostriker, L. Spitzer, R.A. Chevalier, ApJ, **176**, L51 (1972)
4. L. Martinet, A. Hayli, A&A, **14**, 103 (1971)
5. F. Mayer, L. Martinet, A&A, **27**, 199 (1973)
6. L. Martinet, A&A, **32**, 329 (1974)
7. L. Martinet, F. Mayer, A&A, **44**, 45 (1975)
8. M. Schmidt, Bull. Astron. Inst. Neth., **13**, 15 (1956)
9. M. Schmidt: 'Rotation parameters and distribution of mass in the Galaxy'. In: *Stars and Stellar Systems, Vol. 5* ed. by A. Blaauw, M. Schmidt (University of Chicago Press, Chicago, 1965) pp. 513-530
10. J.E. Tohline, K. Voyages, ApJ, **555**, 524 (2001)
11. G.G. Kuzmin, Astron.Zh., **33**, 27 (1956)
12. J. Binney, S. Tremaine: *Galactic Dynamics* (Princeton University Press, Princeton 1987)
13. H. Goldstein: *Classical Mechanics*, 2nd edn. (Addison-Wesley, Reading MA 1980)
14. P.T. de Zeeuw, MNRAS, **216**, 273 (1985)
15. J. Binney, MNRAS, **201**, 1 (1982)
16. J. Miralda-Escudé, M. Schwarzschild, ApJ, **339**, 752 (1989)
17. C. Hunter, B. Terzić, A.M. Burns, D. Porchia, C. Zink, Ann.N.Y. Acad. Sci., **867**, 61 (1998)
18. O.E. Gerhard, A&A, **151**, 279 (1985)
19. N.H. Packard, J.P. Crutchfield, J.D. Farmer, R.S. Shaw, Phys. Rev. Lett., **45**, 712 (1980)
20. K.T. Alligood, T.D. Sauer, J.A. Yorke: *Chaos: An Introduction to Dynamical Systems* (Springer, New York 1996)
21. M. Schwarzschild, ApJ, **409**, 563 (1993)
22. M. Miyamoto, R. Nagai, PASJ, **27**, 533 (1975)
23. M. Schwarzschild, ApJ, **232**, 236 (1979)
24. O.E. Gerhard, J.J. Binney, MNRAS, **216**, 467 (1985)
25. S. Sridhar, J. Touma, MNRAS, **303**, 483 (1999)
26. J. Wisdom, AJ, **87**, 577 (1982)

27. J. Touma, S. Tremaine, MNRAS, **292**, 905 (1997)
28. J. Moser: *Stable and Random Motions in Dynamical Systems* (Princeton University Press, Princeton 2001)
29. H. Varvoglis, J. Physique, **46**, 495 (1985)
30. G. Benettin, J.-M. Strelcyn, Phys. Rev. A **17**, 773 (1978)
31. G.D. Birkhoff: *Dynamical Systems* (American Mathematical Society, Providence 1927)
32. D.O. Richstone, ApJ, **252**, 496 (1982)
33. L. Aguilar, P. Hut, J.P. Ostriker, ApJ, **335**, 720 (1988)

Chaos and Chaotic Phase Mixing in Galaxy Evolution and Charged Particle Beams

Henry E. Kandrup

University of Florida, Gainesville, FL 32611, USA

Abstract. This paper discusses three new issues that necessarily arise in realistic attempts to apply nonlinear dynamics to galaxy evolution, namely: (i) the meaning of chaos in many-body systems, (ii) the time-dependence of the bulk potential, which can trigger intervals of *transient chaos*, and (iii) the self-consistent nature of any bulk chaos, which is generated by the bodies themselves, rather than imposed externally. Simulations and theory both suggest strongly that the physical processes associated with galactic evolution should also act in nonneutral plasmas and charged particle beams. This in turn suggests the possibility of testing this physics in real laboratory experiments, an undertaking currently underway.

1 Introduction

As recently as 1990, most galactic dynamicists ignored completely the possible role of chaos in galaxies. However, the past decade has seen a growing recognition that chaos can be important in determining the structure of real galaxies. Still, much recent work involving chaos in galactic astronomy has been simplistic in that it has involved naive applications of standard techniques from nonlinear dynamics developed to analyse two- and three-degree-of-freedom time-independent Hamiltonian systems. The object here is to discuss some of the additional complications which arise if nonlinear dynamics is to be applied to real galaxies, which are *many-body systems* comprised of a large number of interacting masses and characterised by a *self-consistently determined bulk potential* which, during their most interesting phases, can be *strongly time-dependent.*

2 Transient Chaos and Collisionless Relaxation

2.1 Transient Chaos Induced by Parametric Resonance

It is well known to nonlinear dynamicists that the introduction of an oscillatory time-dependence into even an otherwise integrable potential can trigger an interval of *transient chaos*, during which many orbits exhibit an exponentially sensitive dependence on initial conditions. Physically, this transient chaos arises from a resonance overlap between the frequencies $\sim \Omega$ of the unperturbed orbits and the frequency or frequencies $\sim \omega$ of the time-dependent perturbation.

In the past, the possible effects of such chaos have been considered for both nonneutral plasmas [1] and charged particle beams [2]. More recently, such transient chaos has also begun to be considered in the context of galactic astronomy [3]. That work has shown that, for large fractional amplitudes, > 0.1 or so, this resonance can be very broad, triggering significant amounts of chaos for $10^{-1} \leq \omega/\Omega \leq 10$; and that the existence of the phenomenon is robust, comparatively insensitive to details. It will, for example, persist if one allows for damped oscillations and/or modest drifts in frequency (generated, *e.g.*, by making ω a random variable sampling an Ornstein-Uhlenbeck colored noise process).

The breadth of the resonance and the insensitivity to details suggest that transient chaos could well prove common, if not generic, in violent relaxation [4], the collective process whereby a (nearly) collisionless galaxy or galactic halo evolves towards an equilibrium or near-equilibrium state. Violent relaxation typically involves damped oscillations triggered, *e.g.*, by interactions with another galaxy or, in the early Universe, by the cosmological details preceding galaxy formation. However, when considering collective effects there is only one natural time scale, the dynamical time $t_D \sim 1/\sqrt{G\rho}$, with ρ a typical mass density, which determines both the characteristic orbital time scale and (at least initially) the oscillation time scale. The exact numerical values of these time scales will involve numerical coefficients which will in general be unequal and vary as a function of location within the galaxy. If, however, one need only demand that the oscillation and orbital time scales agree to within an order of magnitude, it would seem likely that this resonance could trigger transient chaos through large parts of the galaxy. In real galaxies the oscillations will presumably damp and the frequencies drift as the density changes and, presumably, power cascades from longer to shorter scales. To the extent, however, that the details are unimportant such variations should not obviate the basic effect.

2.2 Chaotic Phase Mixing and Collisionless Relaxation

But why might such transient chaos prove important in galactic evolution? Detailed numerical simulations indicate that violent relaxation can be a very rapid and efficient process, but simple models involving regular orbits, such as Lynden-Bell's [4] balls rolling in a pig-trough, do not approach an equilibrium nearly fast enough [5]. The important point, however, is that allowing for the effects of chaos can in principle dramatically accelerate both the speed and efficacy of violent relaxation. An initially localised ensemble of regular orbits evolved into the future in a time-independent potential will begin by diverging as a power law in time and, only after a very long period, slowly evolve towards a time-averaged equilibrium, *i.e.*, a uniform population of the KAM tori to which it is restricted. By contrast, a corresponding ensemble of chaotic orbits will begin by diverging exponentially at a rate that is comparable to a typical value of the largest finite time Lyapunov exponent for the orbits in the ensemble; and then converge exponentially towards an equilibrium or near-equilibrium state at a somewhat smaller, but still comparable, rate. The exponential character of this *chaotic phase mixing* means that the time scale associated with this process is typically far shorter than the time scale associated with *regular phase mixing* [6–8].

It is evident that chaotic phase mixing in a time-independent Hamiltonian system can trigger a very rapid approach towards an equilibrium, but this does not necessarily 'explain' violent relaxation. If, *e.g.*, most of the orbits in the system are regular, it would seem unlikely that chaotic phase mixing could be sufficiently ubiquitous to explain an approach towards a (near-)equilibrium for the galaxy as a whole. Indeed, one would expect that, for a galaxy that is in or near equilibrium the relative measure of chaotic orbits should be comparatively small: If the galaxy exhibits nontrivial structures like a bar or a cusp, the types of structures which one has come to associate with chaos, one would also expect large numbers of regular orbits must be present to serve as a 'skeleton' to support that structure [9]. Moreover, it is evident that, although chaotic mixing in a time-independent potential can be very efficient in mixing orbits on a constant energy surface, the energy of each particle remains conserved, so that there can be no mixing in energies. The extent to which chaotic phase mixing in a time-dependent potential will trigger an efficient shuffling of energies is not completely clear.

The important point, then, is that chaotic phase mixing associated with transient chaos in a time-dependent potential is likely to explain these remaining lacunae. At least for large amplitude perturbations, (say) 10% or more, this parametric resonance can trigger a huge increase in the relative abundance of chaotic orbits so that, for pulsation frequencies near the middle of the resonance, virtually all the orbits exhibit substantial exponential sensitivity. Moreover, given that this chaos involves a resonant coupling, it tends typically to cause a substantial shuffling of energies: those frequencies which are most apt to trigger lots of chaos are also apt to induce the largest shuffling of energies.

Still it should be noted that one *can* get a 'near-complete' shuffling of orbits on different constant energy surfaces even if the orbital energies are not especially well shuffled. This, however, is not necessarily a problem. Simulations of systems exhibiting efficient collective relaxation do not necessarily involve masses which completely 'forget' their initial conditions. Rather, comparatively efficient and complete violent relaxation is completely consistent with an evolution in which masses 'remember' (at least partially) their initial binding energies, *i.e.*, in which masses that start with comparatively large (small) binding energies end up with comparatively large (small) binding energies [10].

That it may be possible to achieve efficient chaotic phase mixing in an oscillating galactic potential while still relaxing towards a nearly integrable state within $10t_D$ or so is illustrated in Fig. 1. This Figure was generated from orbits evolved in a time-dependent potential of the form

$$V(x,y,z,t) = -\frac{m(t)}{(1+x^2+y^2+z^2)^{1/2}}, \qquad m(t) = 1 + \delta m \frac{\sin \omega t}{(t_0+t)^p}, \tag{1}$$

with $\delta m = 0.5$, $t_0 = 100$ and $p = 2$, which represents a galaxy damping towards an integrable Plummer sphere. The four curves in the top panel exhibit the x-component of the phase space *emittance*, $\epsilon_i = (\langle r_i^2\rangle\langle v_i^2\rangle - \langle r_i v_i\rangle^2)^{1/2}$ $(i = x, y, z)$, all computed for the same localised ensemble of initial conditions, but allowing for four different frequencies ω. The curves exhibit considerable structure but, at least for early times, the overall evolution is exponential. The bottom panels exhibit the x and y coordinates at five different times for the ensemble represented

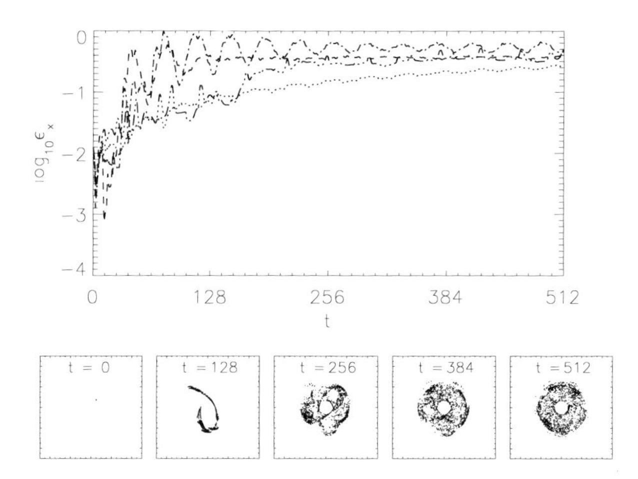

Fig. 1. (top) The emittance ϵ_x for an ensemble of initial conditions evolved in an integrable Plummer potential subjected to damped oscillations with four different frequencies: $\omega = 0.035$ (triple-dot-dashed), 1.40 (dashed), 3.50 (dot-dashed), and 7.00 (dotted). (bottom) x-y scatter plots corresponding to the uppermost curve with $\omega = 3.50$.

by the uppermost of the four curves. Here $t_D \sim 20$, so that $t = 256$ corresponds to roughly $12t_D$.

Intuitively, one might expect that strong oscillations, which trigger the largest finite time Lyapunov exponents and the largest number of chaotic orbits, would yield the fastest chaotic phase mixing and, hence, the most rapid and most complete violent relaxation. A time-dependence with a weaker oscillatory component, *e.g.*, a time-dependence corresponding to a near-homologous collapse, might instead be expected to yield less chaos and, hence, less efficient and less complete violent relaxation. There is, therefore, an important need to determine the extent to which, in real simulations of violent relaxation, many/most of the orbits (or phase elements) are strongly chaotic, and the degree to which the rate and completeness of the observed violent relaxation correlate with the size of the largest finite time Lyapunov exponents and/or the relative measure of chaotic orbits. Investigations of these issues are currently underway.

3 The Role of Discreteness Effects

3.1 Microchaos and Macrochaos

The discussion in the preceding section, like most applications of nonlinear dynamics to galactic astronomy, neglects completely discreteness effects associated with the 'true' many-body potential, assuming that masses in a galaxy can be approximated as evolving in a smooth, albeit time-dependent, three-dimensional potential and that 'chaos' has its usual meaning. That this is justified is not completely obvious. The gravitational N-body problem for a large number of bodies of comparable mass is strongly chaotic in the sense that individual orbits have large positive Lyapunov exponent χ_N *even when there is absolutely no chaos in the continuum limit!* If, *e.g.*, a smooth density distribution corresponding to an integrable potential is sampled to generate an N-body density distribution, one finds that orbits evolved in this N-body distribution will be strongly chaotic, even for very large N, despite the fact that characteristics in the smooth potential generated from the same initial condition are completely integrable. Moreover, there is no sense in which the exponential sensitivity decreases with increasing N: if anything χ_N is an increasing function of N [11]. In this sense, *larger N implies more chaos, not less!*

This situation has led some astrophysicists to question, either implicitly or explicitly, the reliability of the entire smooth potential approximation. Thus, *e.g.*, it has been suggested [12] that "the approximation of a smooth potential is useful for studying orbits, but not for studying their divergence." This is of course a problematic statement in that the distinction between exponential and power law divergence, emblematic of the differences between regular and chaotic behaviour, lies at the heart of applications of nonlinear dynamics to galactic dynamics. If the Lyapunov exponents associated with the bulk potential have nothing to do with the N-body problem, one must perforce reject completely all conventional applications of nonlinear dynamics to galactic astronomy.

The crucial point, then, is that there *does* appear to be a well-defined continuum limit, even at the level of individual orbits [13–15]. Suppose that a smooth density distribution, corresponding either to an integrable potential or to a potential admitting large measures of regular orbits, is sampled to generate a fixed, *i.e.* frozen in time, N-body density distribution, and that the trajectories of test particles evolved in this frozen distribution are compared with smooth potential characteristics with the same initial conditions. In this case, there is a precise sense in which, as N increases, the frozen-N trajectories converge towards the smooth potential characteristics. Both visually and in terms of the *complexity* [16] of their Fourier spectra, the frozen-N trajectories come to more closely resemble the smooth potential characteristics; and, *viewed mesoscopically*, the frozen-N and smooth potential orbits remain closer in phase space for progressively longer times. In particular, a frozen-N orbit corresponding to an integrable characteristic will have a large Lyapunov exponent χ_N even if, visually, it is essentially indistinguishable from the regular characteristic!

But how can this be? The key recognition here is that two 'types' of chaos can be present in the N-body problem, characterised by *two different sets of Lyapunov exponents associated with physics on different scales.* Close encounters between particles trigger *microchaos*, a generic feature of the N-body problem, which leads to large positive Lyapunov exponents χ_N. If, however, the bulk smooth potential is chaotic, one will also observe *macrochaos*, which is again characterised by positive, albeit typically much smaller, Lyapunov exponents χ_S. Suppose, for example, that one compares the evolution of two nearby chaotic initial conditions in a single frozen-N background or the same chaotic initial condition evolved in two different frozen-N realisations of the same bulk density. In this case, one typically observes a three-stage evolution, namely: (1) a rapid exponential divergence at a rate χ_N set by the true Lyapunov exponents associated with the N-body problem, which persists until the separation becomes large compared with a typical interparticle spacing; followed by (2) a slower exponential divergence at a rate comparable to the (typically much smaller) smooth potential Lyapunov exponent χ_S, which persists until the separation becomes macroscopic; followed by (3) a power law divergence on a time scale $\propto (\ln N)t_D$. For regular initial conditions, the second stage is absent and the time scale for the third stage scales instead as $N^{1/2}t_D$.

Microchaos becomes stronger as N increases in the sense that the value of χ_N increases with increasing N [17]. Despite this, however, it becomes progressively less important macroscopically in that the *range* of the chaos, *i.e.*, the scale on which the microchaos-driven exponential divergence of nearby orbits terminates, decreases with increasing N. In the limit $N \to \infty$ microchaos will become completely irrelevant but, for finite N, it does have an effect, at least on sufficiently short scales; and it is possible from an N-body simulation to extract estimates of both χ_N and the typically much smaller χ_S [15].

3.2 Modeling Discreteness Effects as Friction and Noise

It has been long recognised that, for sufficiently small N and/or over sufficiently long times, discreteness effects will not be completely negligible. Systems like galaxies are 'nearly collisionless' in the sense that the stars interact primarily via collective macroscopic forces associated with the bulk density distribution; but, at least in principle, if one waits long enough discreteness effects should have an appreciable effect.

Astronomers are accustomed to modeling discreteness effects in the context of a Fokker-Planck description analogous to that formulated originally in the context of plasma physics [18]. However, it is not completely clear to what extent this is really justified. The conventional Fokker-Planck description was formulated originally to extract statistical properties of orbit ensembles and distribution functions over long time scales, assuming implicitly that the bulk potential is regular. To what extent, then, can Langevin realisations of a Fokker-Planck equation yield reliable information about individual orbits over comparatively short time scales, particularly if the orbits are chaotic?

Analyses of flows in frozen-N systems indicate [14] that, at the level of both orbit ensembles and individual orbits, discreteness effects can in fact be modeled *extremely* well by Gaussian white noise in the context of a Fokker-Planck description, allowing for a dimensionless diffusion constant $D \propto 1/N$, consistent with the predicted scaling $D \propto \ln \Lambda/N$, with Λ the so-called Coulomb logarithm [18]. For localised ensembles of initial conditions corresponding to both regular and chaotic orbits, phase mixing in frozen-N systems and phase mixing in smooth potentials perturbed by Gaussian white noise yield virtually identical behaviour, both in terms of the evolution of various phase space moments such as the emittance and the rate at which individual orbits in the ensemble exhibit nontrivial 'transitions', *e.g.*, passing through some *entropy barrier* from one phase space region to another. And similarly, a comparison of frozen-N orbits and noisy smooth potential orbits with the same initial condition reveals that their Fourier spectra typically exhibit comparable complexities. Gaussian white noise is even successful in mimicking some of the effects of microchaos. If, *e.g.*, one tracks the divergence of two noisy orbits with the same chaotic initial condition evolved in a smooth potential, one observes the same three-stage evolution as for a pair of frozen-N orbits evolved in two different frozen-N potentials.

An example of this agreement is illustrated in Fig. 2, which exhibits data generated by averaging over 100 pairs of orbits evolved in frozen-N density distributions which correspond in the continuum limit to a triaxial homogeneous ellipsoid with axis ratios 1.95 : 1.50 : 1.05, perturbed by a spherically symmetric central mass spikes ('black hole'). The top two solid curves represent (from top to bottom) results for $N = 10^{4.5}$ and $N = 10^{5.5}$. The four dotted curves represent analogous results derived for pairs of noisy orbits evolved from the same initial conditions in the smooth potential with (from top to bottom) diffusion constant $D = 10^{-4}$, 10^{-5}, 10^{-6}, and 10^{-7}. The near-coincidence of the top two solid and dotted curves indicates that discreteness effects for $N = 10^{p+1/2}$ are well-mimicked by Gaussian white noise with $D = 10^{-p}$.

Such striking agreement suggests strongly that investigations of how orbits in smooth potentials are impacted by the introduction of friction and noise can provide important insights into the role of graininess in real galaxies. It is customary to assert that, in a system as large as a galaxy, discreteness effects reflecting close encounters between stars are unimportant because the relaxation

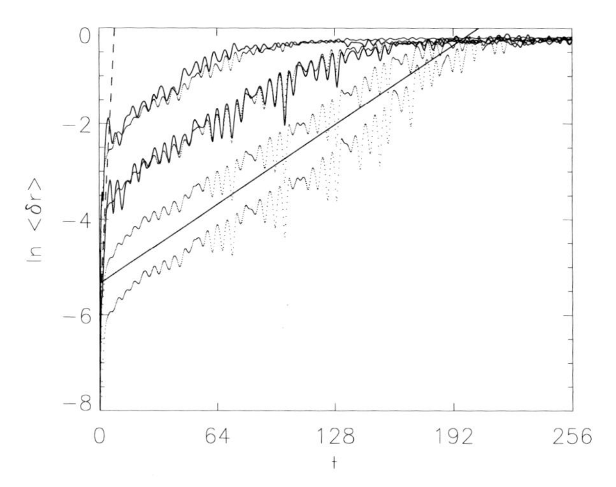

Fig. 2. The mean spatial separation between the same initial conditions evolved in two different frozen-N backgrounds (solid curves) and different noisy orbits evolved in the smooth potential from the same initial condition (dots). The solid line has a slope 0.022, equal to the mean value of the smooth potential Lyapunov exponent χ_S. The dashed curve has a slope 0.75, equal to the mean value of the N-body Lyapunov exponent χ_N.

time t_R on which they can induce appreciable changes in quantities like the energy is orders of magnitude longer than the age of the Universe [19]. This is likely to be true if the galaxy is an exact equilibrium, especially an equilibrium characterised by an integrable potential. However, the assertion is suspect if (as must usually be the case) the system is only 'close to' an equilibrium or near-equilibrium, especially if the bulk potential is characterised by a phase space admitting a complex coexistence of regular and chaotic orbits.

Over the past decade, analyses of flows in time-independent Hamiltonian systems have revealed that even very weak perturbations, idealised as friction and white noise corresponding to $t_R \sim 10^6 - 10^9 t_D$ and, hence, $D \sim 10^{-6} - 10^{-9}$, can have significant effects within a time as short as $100 t_D$ or less by facilitating phase space diffusion through cantori or along the Arnold web [20–22]. The basic point is that the motions of chaotic orbits in a complex potential can be constrained significantly by topological obstructions like cantori or the Arnold web which, albeit not completely preventing motions from one phase space region to another, serve as an *entropy barrier* to impede such motions. In many respects, the physical picture is similar to the elementary problem of effusion of gas through a tiny hole in a wall. There is nothing in principle to prevent a gas molecule from passing through the hole and, hence, escaping from the region to which it is originally confined; but, if the hole is very small, the time scale associated with this effusion can be extremely long.

In the same sense, and for much the same reason, chaotic orbits trapped in one phase space region may, in the absence of perturbations, remain stuck in that region for a very long time. However, subjecting the orbits to noise will 'wiggle' them in such a fashion as to increase the rate at which they pass through the entropy barrier, thus accelerating phase space transport. Numerical simulations indicate that, in at least some cases, this escape process can be well approximated by a Poisson process, with the number of nonescapers decreasing exponentially at a rate Λ that is determined by the perturbation [23,24]. This effect appears to result from a resonant coupling between the orbits and the noise. White noise is characterised by a flat power spectrum and, as such, will couple to more or less anything. If, however, the noise is made coloured, *i.e.*, if instantaneous kicks are replaced by impulses of finite duration, the high frequency power is reduced; and, if the autocorrelation time becomes sufficiently long that there is little power at frequencies comparable to the orbital frequencies, the effect of the noise decreases significantly. Significantly, it appears that, overall, the details of the perturbation may be largely irrelevant: additive and multiplicative Gaussian noises tend to have comparable effects and the presence or absence of friction does not seem to matter. All that appears to matter are the amplitude and the autocorrelation time upon which there is a relatively weak, roughly logarithmic, dependence.

But what does all this imply for a real galaxy? Given that collisionless near-equilibria must be more common than true equilibria, it would seem quite possible that, during the early stages of evolution, a galaxy might settle down towards a near-equilibrium, rather than a true equilibrium, *e.g.*, involving what have been termed [25] 'partially mixed' building blocks. If discreteness effects and all other perturbative effects could be ignored, such a quasi-equilibrium might persist without exhibiting significant changes over the age of the Universe. If, however, one allows for discreteness effects or, alternatively, other perturbations reflecting, *e.g.*, a high density cluster environment, the orbits could become shuffled in such a fashion as to trigger significant changes in the phase space density and, consequently, a systematic secular evolution [26].

Such a scenario could, for example, result in the destabilisation of a bar. Many models of bars (*e.g.* [27]) incorporate 'sticky' [28] chaotic orbits as part of the skeleton of structure, replacing crucial regular orbits which can be absent near corotation and other resonances. Making these 'sticky' orbits become unstuck could cause the bar to dissipate. Similar effects could also cause an originally nonaxisymmetric cusp to evolve towards a more nearly axisymmetric state. To the extent that the triaxial Dehnen potentials are representative, one can argue that chaotic orbits may be extremely common near the centers of early-type galaxies, but that many of these chaotic orbits are extremely sticky [29] and, as such, could help support the nonaxisymmetric structure. Perturbations that make these sticky orbits wildly chaotic could *de facto* break the bones of the skeleton supporting the structure and trigger an evolution towards axisymmetry.

4 Experimental Tests of Galactic Dynamics

4.1 Similarities Between Galaxies and Nonneutral Plasmas

Even though electrostatics and Newtonian gravity both involve $1/r^2$ forces, electric neutrality implies that the physics of neutral plasmas is very different from the physics of self-gravitating systems. Viewed over time scales $> t_R$, nonneutral plasmas and charged-particle beams are also very different from self-gravitating systems: the attractive character of gravity leads to phenomena like evaporation and core-collapse which cannot arise in a beam or a plasma. If, however, one restricts attention to comparatively short times $\ll t_R$, much of the physics should be the same. Theoretical expectations, supported by numerical simulations, suggest that it is the existence of long range order, not the sign of the interaction, which is really important; but, to the extent that this be true, collisionless nonneutral plasmas and collisionless self-gravitating systems should be quite similar.

Typical sources of charged-particle beams configure the beams in trains of 'packets' or 'bunches', as they are termed by accelerator dynamicists. The objective of a good high-intensity accelerator is to generate bunches comprised of a large total number of charges confined to a small phase space volume and then accelerate those bunches to very high energies while minimising any growth in emittance. As one example, modern photocathode-based sources of electron beams routinely generate bunches comprised of some $10^{10} - 10^{11}$ electrons with transverse 'emittance' $\tilde{\epsilon}$ of a few microns. (Here $\tilde{\epsilon} = \epsilon/v_0$, where v_0 is the mean axial velocity of the particle distribution.) The energy relaxation time t_R associated with such bunches typically corresponds to the time required for a bunch to travel a distance ~ 1 km or so which, in many cases, is much longer than any distance of experimental interest, so that the bunches are 'nearly collisionless.'

Models of equilibrium configurations of nonneutral plasmas and charged-particle beams confined by electromagnetic fields can be characterised by a complex phase space quite similar to that associated with models of elliptical galaxies and, as such, have orbits with very similar properties. For example [9], the so-called 'thermal equilibrium model' [32] of beam dynamics, which involves a self-interacting nonneutral plasma in thermal equilibrium confined by an anisotropic harmonic oscillator potential, is strikingly similar [29] to the non-spherical generalisations of the Dehnen potential of galactic astronomy in terms of such properties as the degree of 'stickiness' manifested by chaotic orbits or how the relative measure of chaotic orbits and the size of the largest Lyapunov exponent vary with shape.

As in galactic dynamics, questions have been raised regarding the validity of the continuum approximation for nearly collisionless charged particle beams [31]. However, comparatively short time integrations ($t \ll t_R$) involving discreteness effects and the nature of the continuum limit in nonneutral plasmas [33] yield results essentially identical to what is observed for gravity – although the behaviour associated with neutral plasmas is *very* different. In particular, the macroscopic manifestations of phase mixing, for both regular and chaotic orbits, are indistinguishable, and the coexistence of microchaos and macrochaos persists unabated.

Nontrivial effects associated with a time-dependent potential have also been predicted for both nonneutral plasmas [1] and charged particle beams [2]. Although the time-dependence that is envisioned in a beam is typically less violent than that anticipated in violent relaxation within a galaxy, such is not always the case. Indeed, there is compelling experimental evidence that, in a beam, such a time-dependence can have the undesireable effect of ejecting particles from the core into an outerlying halo [34].

Perhaps most interesting, however, is the fact that numerical simulations that reproduce successfully 'anomalous relaxation' observed in real laboratory experiments involving accelerator beams have shown compelling evidence of chaotic phase mixing. One classic example involves the propagation of five nonrelativistic high-intensity beamlets in a periodic solenoidal transport channel, where self-consistent space-charge forces are extremely important [35]. Ideally, these beamlets should exhibit coherent periodic oscillations (quite literally disappearing and reappearing) which might be expected to decay only on a relaxation time scale t_R that corresponds to a propagation distance ~ 1 km. However, regardless of how well the beam was matched to the transport channel, the beamlets were seen to reappear only once, at a point ~ 1 m from the source, disappearing completely within 2 m or so (*cf.* Fig. 6.10 in [35]). Their failure to reappear again would seem to reflect some collisionless process that, in effect, causes the particles to 'forget' their initial conditions.

Detailed simulations using the particle-in-cell code *WARP* [36], which do an extremely good job of reproducing what is actually seen, demonstrate seemingly unambiguously that, because of the time-dependent space-charge potential, a

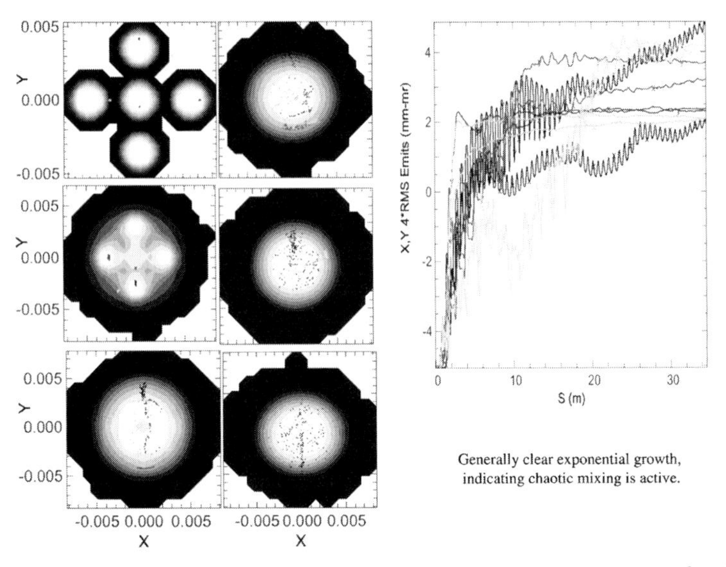

Fig. 3. Evolution of five representative ensembles of test particles in the five-beamlet simulation. The left hand panel shows snapshots of the ensembles at (top-to-bottom left column) 0 m, 0.98 m, 2.88 m, and (top-to-bottom right column) 5.24 m, 11.52 m, and 31.68 m, with the x- and y-axes labeled in meters. The right panel shows the evolution of the logarithm of the emittances ϵ_x and ϵ_y as a function of distance $S(z)$ along the accelerator.

large fraction of the particles in the beam experience the effects of strong, possible transient, macrochaos [37,38]. This is, *e.g.*, evident from Fig. 3, which illustrates the evolution of representative test particles which interact with the bulk potential but not with each other. Here the left hand panel shows snapshots of the beam after it has travelled distances 0 m, 0.98 m, 2.88 m, 5.24 m, 11.52 m, and 31.68 m, with the representative ensembles superimposed. The right hand panel exhibits the evolution of the emittances ϵ_x and ϵ_y for these ensembles. It is evident that the initially localised ensembles are diverging exponentially so as to fill much of the accessible phase space, and that this exponential divergence coincides with the beamlets losing their individual identities. Also evident is the fact that the behaviour observed here is very similar to that exhibited in Fig. 1 which, recall, was generated for orbits in a perturbed Plummer potential exhibiting damped oscillations.

4.2 Testing Galaxy Evolution with Charged Particle Beams

The aforementioned similarities between galactic astronomy and charged particle beams suggest the possibility of using accelerators as a laboratory for astrophysics in which one can perform experimental tests of galactic dynamics, a possibility currently being developed by a University of Florida – Fermilab/Northern Illinois University – University of Maryland collaboration. This collaboration, which has the dual aims of (1) obtaining an improved understanding of the applicability of nonlinear dynamics to nearly collisionless systems interacting via long range forces and (2) using that understanding to generate more sharply focused bunches by minimising undesirable increases in emittance, is currently planning concrete experiments which can, and presumably will, be performed on the University of Maryland Electron Ring (*UMER*) currently under construction. Here a number of obvious issues, all experimentally testable, come to mind:

How ubiquitous is chaotic phase mixing as a source of anomalous relaxation? Older experiments with lower-intensity beams, where the space-charge forces were comparatively unimportant, tended not to manifest extreme examples of anomalous relaxation. Anomalous relaxation appears more common in high intensity beams, especially in settings where a time-dependent density distribution generates a strongly time-dependent potential; and it is obvious to ask whether chaos is the principal culprit. The idea here is to identify the types of scenaria that tend generically to yield anomalous relaxation and to determine, *e.g.*, whether such scenaria tend typically to be associated with a bulk potential that incorporates a strong, roughly oscillatory component. Do numerical simulations of orbits ensembles evolved in such systems exhibit evidence of chaotic phase mixing? And do individual orbits in those ensembles exhibit strong exponential sensitivity, associated, *e.g.*, with transient chaos?

Do instabilities tend to trigger transient chaos? Instabilities in collisionless systems can exhibit behaviour qualitatively similar to that associated with turbulence in collision-dominated systems, but it is well known that turbulence is a strongly chaotic phenomenon. This possibility is especially interesting in that turbulence is another setting where different 'types' of chaos, characterised by wildly different time scales, can act on different length scales.

What types of geometries, both strongly time-dependent and nearly time-independent, tend to yield the most efficient chaotic phase mixing and the largest measures of chaotic orbits? Do time-dependent evolutions involving strongly convulsive oscillations tend generically to exhibit especially fast relaxation? And do they tend to yield especially large amounts of chaos, as probed by the relative measure of chaotic orbits and/or the sizes of the largest (finite time) Lyapunov exponents? To the extent that bulk properties of such 'accelerator violent relaxation' correlate with the degree of chaos exhibited in the evolving beam, and that the degree of chaos correlates with the form of the macroscopic time dependence, one will have a physically well-motivated explanation of which sorts of scenaria would be expected to exhibit complete and efficient violent relaxation and which would not!

Is it, *e.g.*, true that, for nearly axisymmetric configurations, prolate (or oblate) bunches tend generically to exhibit especially large amounts of chaos? And do any such trends that are observed coincide with trends observed in models of galactic equilibria [29]? Even if a beam bunch remains nearly axisymmetric during its evolution, the acceleration mechanism can – and in general will – change its shape as it passes down an accelerator in a fashion that depends on the accelerator design. The obvious question, then, is whether the oblate or prolate phase tends generically to be especially chaotic.

Addressing this and related issues could provide important insights as to *why* galaxies have the detailed shapes that they do, a general question for which, at the present time, no compelling dynamical explanation exists. One knows, *e.g.*, that elliptical galaxies tend to have isophotes that are slightly boxy or disky, and that this boxiness or diskiness correlates with such properties as the rotation rate, the steepness of the central cusp, and the size of any deviations from axisymmetry [39]. Must all these effects be attributed to the detailed form of the formation scenario, or is there a clear dynamical explanation? Is it, *e.g.*, true that the observed deviations from perfect ellipsoidal symmetry conspire to reduce the relative number of chaotic orbits or to increase the numbers of certain regular orbit types required as a skeleton to support the observed structure?

One might also use accelerator experiments to probe the role of discrete substructures and the extent to which they can be modeled as friction and noise in the context of a Fokker-Planck description [40]. If the injection of a beam involves a large mismatch, a significant charge redistribution will occur, resulting in violent relaxation, 'turbulent' behaviour, and the formation of substructures ('lumps') on a variety of scales. To the extent that such a time-dependent evolution can be described in a continuum approximation, one might then expect that the bulk potential will correspond to a highly complex time-dependent phase space and that the substructures could act as a 'noisy' source of extrinsic diffusion, facilitating both transitions between 'sticky' and 'wildly chaotic' behaviour and, in some cases, transitions between regularity and chaos. Given the evidence (*cf.* [33]) that, at least over short times, discreteness effects act similarly for attractive and repulsive $1/r^2$ forces, such insights could be directly related to such issues as the destabilisation of bars in spirals and/or the secular evolution of nonaxisymmetric ellipticals towards more nearly axisymmetric states.

Do systems tend to evolve in such a fashion as to minimise the amount of chaos? There is an intuitive expectation amongst many galactic astronomers (*cf.* [41]) that galaxies tend to evolve towards equilibria which incorporate few if any chaotic orbits, *i.e.*, that nature somehow favors nearly-regular equilibria. It would certainly appear true that a model must incorporate significant numbers of regular orbits to support interesting structures like bars and/or triaxiality, but this does not *a priori* preclude the possibility of chaotic orbits also being present. Generic time-independent three-degree-of-freedom potentials are neither completely regular nor completely chaotic, admitting instead a complex coexistence of regular and chaotic phase space regions. The obvious question then is: are galactic equilibria or near-equilibria typically well-represented by potentials which are generic in this sense; or are they, for reasons unknown, special in that they tend to be rather nearly regular?

5 Conclusions

This paper has focused on several fundamental issues that arise in attempts to apply nonlinear dynamics to real galaxies, many-body systems characterised by a self-consistently determined bulk potential which, during their most interesting phases, can be strongly time-dependent. As recently as a decade ago these issues would have been considered of largely academic, rather than practical, interest. However, recent observational advances – which facilitate improved high resolution photometry of individual objects as well as statistical analyses of large samples with varying redshift – and improved computational resources – which allow unparalleled explorations of multi-scale structure –, together with the recognition that the basic physics can be also probed in the context of charged particle beams, imply that theoretical predictions regarding these 'academic' issues can in fact be tested observationally, numerically, and experimentally.

Thanks to Ioannis Sideris for providing Fig. 1 and to Rami Kishek for providing Fig. 3. Thanks also to Court Bohn for his comments on a preliminary draft of the manuscript. Partial financial support was provided by NSF AST-0070809.

References

1. S. Strassburg and R. C. Davidson: Phys. Rev. **E61**, 5753 (2000)
2. R. Gluckstern: Phys. Rev. Lett **73**, 1247 (1994)
3. H. E. Kandrup, I. M. Vass, and I. V. Sideris: Mon. Not. R. astr. Soc.: submitted (2002) (astro-ph/0211056)
4. D. Lynden-Bell: Mon. Not. R. astr. Soc. **136**, 101 (1967)
5. H. E. Kandrup: 'Collisionless Relaxation of Stellar Systems'. In: *Galaxy Dynamics, A Rutgers Symposium*, ed. by D. Merritt, J. N. Sellwood, and M. Valluri (Sheridan, San Francisco 1999)
6. H. E. Kandrup and M. E. Mahon: Phys. Rev. **E49**, 3735 (1994)
7. M. E. Mahon, R. A. Abernathy, B. O. Bradley, and H. E. Kandrup: Mon. Not. R. astr. Soc. **275**, 443 (1995)
8. D. Merritt and M. Valluri, Astrophys. J. **471**, 82 (1996)
9. J. Binney: Comments Astrophys. **8**, 27 (1978)
10. P. J. Quinn and W. H. Zurek: Astrophys. J. **331**, 1 (1988)
11. M. Hemsendorff and D. Merritt: astro-ph/0205538 (2002)
12. D. Heggie: 'Chaos in the N-Body Problem of Stellar Dynamics'. In: *Predictability, Stability, and Chaos in N-Body Dynamical Systems*, ed. A. E. Roy (Plenum, New York 1991)
13. H. E. Kandrup and I. V. Sideris: Phys. Rev. **E64**, 056209-1 (2001)
14. I. V. Sideris and H. E. Kandrup: Phys. Rev. **E65**, 066203-1 (2002)
15. H. E. Kandrup and I. V. Sideris: Astrophys. J. in press (2003) (astro-ph/0207090)
16. H. E. Kandrup, B. L. Eckstein, and B. O. Bradley: Astron. Astrophys. **320**, 65 (1997)
17. I. V. Pogorelov: Phase Space Transport and the Continuum Limit in Nonlinear Hamiltonian Systems. Ph. D. Thesis, University of Florida, Gainesville (2000)
18. M. N. Rosenbluth, W. M. MacDonald, and D. L. Judd: Phys. Rev. **107**, 1 (1957)
19. J. Binney and S. Tremaine, *Galactic Dynamics* (Princeton University, Princeton, 1987)

20. H. E. Kandrup: Astrophys. J. **480**, 155 (1997)
21. C. Siopis and H. E. Kandrup: Mon. Not. R. astr. Soc. **391**, 43 (2000)
22. H. E. Kandrup and S. J. Novotny: Celestial Mechanics, submitted (2002) (astro-ph/0204019)
23. I. V. Pogorelov and H. E. Kandrup: Phys. Rev. **E60**, 1567 (1999).
24. H. E. Kandrup, I. V. Pogorelov, and I. V. Sideris: Mon. Not. R. astr. Soc. **311**, 719 (2000)
25. D. Merritt and T. Fridman: Astrophys. J. **460**, 136 (1996)
26. H. E. Kandrup: Space Science Reviews, in press (2003) (astro-ph/0011302)
27. P. A. Patsis, E. Athanassoula, and A. C. Quillen: Astrophys. J. **483**, 731 (1997)
28. G. Contopoulos: Astron. J. **76**, 147 (1971)
29. H. E. Kandrup and C. Siopis: Mon. Not. R. astr. Soc.: submitted (2002)
30. C. L. Bohn and I. V. Sideris: Phys. Rev. **ST AB**: submitted (2002)
31. J. Struckmeier, Phys. Rev. **E54**, 830 (1996)
32. M. Brown and M. Reiser: Phys. Plasmas **2**, 965 (1995)
33. H. E. Kandrup, I. V. Sideris, and C. L. Bohn: Phys. Fluids, submitted (2002)
34. R. W. Garnett *et al*: In *Linac 2002: Proceedings of the XXI International Linac Conference*, in press.
35. M. Reiser: *Theory and Design of Charged Particle Beams* (Wiley, New York, 1994)
36. D. P. Grote, A. Friedman, I. Haber, and S. Yu: Fusion Eng, Design **32-33**, 193 (1996)
37. R. A. Kishek, P. G. O'Shea, and M. Reiser: Phys. Rev. Lett. **85**, 4514 (2000)
38. R. A. Kishek, C. L. Bohn, P. G. O'Shea, M. Reiser, and H. E. Kandrup: In: *Proceedings of the IEEE 2001 Particle Accelerator Conference*, ed P. Lucas and S. Weber (IEEE, Chicago, 2001)
39. J. Kormendy and R. Bender: Astrophys. J. Lett. **464**, 119 (1996)
40. C. L. Bohn and J. R. Delayen: Phys. Rev. **E50**, 1516 (1994)
41. M. Schwarzschild: Astrophys. J. **232**, 236 (1979)

Motions of a Black Hole near the Center of a Galaxy

Richard H. Miller

University of Chicago, Astronomy Department,
5640 S. Ellis Ave., Chicago 60637-1433, U. S. A.

Abstract. Some years ago we published an account of experiments which indicated that the nucleus of a galaxy orbits around the mass centroid. This can be viewed as an orbiting density wave which grows near the center in a galaxy model that starts without such motions. While these experiments were run without a massive particle, we suggested that similar physical effects might cause a massive particle near the center to oscillate with larger amplitudes than indicated by simple Brownian motion arguments. Results from recent experiments will be reported to clarify some of the issues raised by a massive particle (a black hole) near the center.

Motions of a massive black hole near the center of a galaxy have excited considerable interest lately, since a typical galaxy may harbor a massive black hole near its center [1]. More recent work is described in a recent review by Ferrarese and Merritt [2].

Dynamical effects cause a cusp to build up around a black home. A few of the obvious questions follow:

- the amount of mass in the cusp,
- the density profile of the cusp,
- black hole motions within the cusp, and
- larger scale motions as the black hole–cusp combination move together.

Our methods are best suited to the larger scale motions of the combined black hole and cusp, and those aspects are the subject of this paper. The other questions require a different computational approach. Ferrarese and Merritt discuss some of these features.

Our approach is experimental. It is based on numerical experiments carried out in a computer, using $n-$body programs. Results to be presented in this note are best understood with some appreciation of the methods used to obtain them.

1 Experimental Details

The $n-$body approach produces an initial value problem, which involves one set of computer programs to establish the starting condition (the loader) and another set to advance the system in time (the integrator). Of course, the state of the system at any stage along the way can be regarded as a new initial condition.

1.1 Generalities

We refer to the black hole throughout this paper simply as a "massive particle," abbreviated "MP," to stress the fact that it is being treated as if it were a simple Newtonian particle. Unusual features of a black hole caused by general–relativistic effects are confined to spatial regions which are much too small to resolve in many n–body studies, including this one. Field particles feel the black hole only as if it were a massive particle acting through Newtonian gravitation. Forces due to the MP are softened to avoid excessive forces.

Boundary Conditions

Boundary conditions are important aspects of any self–consistent self–gravitating problem like that considered here. Periodic boundary conditions provide a way to avoid having to follow the dynamics of an entire galaxy. The idea is to provide a suitable representation of the region surrounding the MP and its associated cusp that at once is computationally manageable and provides an appropriate physical surrounding. The immediate environs of the MP are also treated as being isotropic.

Equipartition

The notion of "equipartition" is often used in this note. It comes from statistical mechanics, where it refers to equal mean energies per degree of freedom, and it is used in that sense. There is no reason to believe that equipartition should hold under the present circumstances, but the concept provides a convenient terminology to discuss the phenomenon that MP motions do not quite fit the expected pattern.

A workaround is to define an "effective" MP mass. Call the "actual" MP mass $M_{\rm act}$ and the "effective" mass $M_{\rm eff}$, where the mean MP energy would be the same as the mean field particle energy if the MP mass were $M_{\rm eff}$. The relation between $M_{\rm eff}$ and $M_{\rm act}$ is of interest.

But a warning is called for: constant reference to $M_{\rm eff}$ tends to suggest that there is some "mass" that so moves, while the physical reality is likely to be a more abstract concept like a "density wave" or a "mode." Certainly there is no identifiable "thing" associated with the excess of $M_{\rm eff}$ over $M_{\rm act}$.

1.2 Loader

An isothermal configuration is generated by the loader. Some design considerations follow.

Isothermal Equation Around a Massive Particle

The standard recipe for generating an isothermal is to use an isothermal distribution function [19], $f = f(\mathcal{E}) = \mathcal{N} \exp(-\mathcal{E}/E_0)$, where $\mathcal{E}$ is the energy of a particle moving within the system, E_0 is the (one dimensional) velocity dispersion of field particles, and $\mathcal{N}$ is a normalization factor. Use energies per unit mass, so $\mathcal{E} = \frac{1}{2}v^2 + U_{\rm tot}$, where $U_{\rm tot}$ is the *total* potential within which the particle moves. For present purposes, $U_{\rm tot}$ is made up of three parts, which happily add linearly. Those parts are (1) $U_{\rm sc}$, the self–consistent potential generated by the complexion of field particles, (2) $U_\bullet$, the potential around the massive particle, and (3) $U_{\rm ext}$ an external potential. The external potentials that will be considered are (isotropic) harmonic, and the potential due to the MP is Keplerian.

Densities are computed in the usual way by integrating over velocities at the selected point in configuration space.

$$\rho = \int f d^3v = \rho_0 \exp\left(-\frac{U_{\rm tot}}{E_0}\right) = \rho_0 \exp\left(-\frac{U_{\rm sc} + U_\bullet + U_{\rm ext}}{E_0}\right), \qquad (1)$$

where constants from the integration and the normalization factor have been absorbed into a single constant, ρ_0.

Remark. An analytic self–consistency problem can be formulated for spherical symmetry by incorporating this density into the Poisson equation to yield

$$\frac{1}{r^2}\frac{d}{dr}\left(r^2 \frac{dU_{\rm sc}}{dr}\right) = 4\pi G \rho_0 \exp\left(-\frac{U_\bullet + U_{\rm ext}}{E_0}\right) \exp\left(-\frac{U_{\rm sc}}{E_0}\right). \qquad (2)$$

Save for the first exponential, this is the standard isothermal equation [4]. Polytropic systems follow the same line of argument, with $f_{\rm poly} = \mathcal{N}_{\rm poly}(-\mathcal{E}/E_0)^p$. Because of the extra terms resulting from $U_\bullet$, neither the polytropic nor the isothermal cases have the usual homology invariances [4].

The singularity in $U_\bullet$ as $r \to 0$ requires special treatment in the integration of (2). Integrating once, (2) becomes

$$r^2 \frac{dU_{\rm sc}}{dr} = 4\pi G \rho_0 \int_{r0}^{r} \exp\left(-\frac{U_\bullet + U_{\rm sc}}{E_0}\right) r^2\, dr,$$

if $U_{\rm ext}$ is set to zero. Huntley and Saslaw [5,6] discuss treatment of the lower limit, $r0$, on the integral to work around the singularity in $U_\bullet$. The contribution of $U_\bullet$ is more complicated in stellar dynamical problems than they indicate.

A Practical Loader

Fortunately, computational solutions to the Poisson equation usually work from the integral form and get around issues of singular potentials by softening the forces. In the absence of analytic solutions to the self–consistency problem in periodic boundary conditions, we use iterative solutions built with densities in the form provided by (1).

The iterative procedure runs as follows. Guess a density distribution, then find the Newtonian potential generated by that density distribution using the same Poisson solver as for the integrations. Add any external potential and the potential due to the massive particle, and then take the exponential of that summed potential at every tabulated point to generate a new density distribution. Taking that new density distribution as a new starting point, repeat the process until it converges in some sense. Happily, this process converges nicely.

Care is required on a few points: to obtain the desired number of particles, to determine the final value for the constant that replaces the physical gravitational constant, to obtain the desired ratio of maximum to minimum densities, to handle scaling and additive constants in the three kinds of potential consistently, and so on.

This builds a density from which each field particle can be assigned a position by a quasi–Monte Carlo process. Once given a location, each particle is assigned a velocity sampled from an isotropic Maxwellian distribution with E_0 set to ensure self–consistency. The MP was always loaded on the origin.

Loads so generated are usually very near self–consistency, as tested by integrating them forward in time and noting that most properties are constant to within a percent or so. Polytropic models differ only in detail, but none were used in the experiments discussed here.

1.3 Integrator

A time–centered leapfrog integrator is used. It is symplectic, and so guarantees a Liouville theorem, which is an essential ingredient of the physical properties of a stellar dynamical system. Particles can cross a "periodic boundary" freely.

Forces are derived from a potential, which is worked out anew at each integration step. The self–consistent part is computed a Poisson solver, and any external potential is then added. Forces due to the massive particle are handled in the integrator to avoid having to treat large forces in the Poisson solver. Tabulated densities serve as input to the Poisson solver, and the output is tabulated at the same set of points. Our set of tabulation points forms a cubic cartesian lattice with lattice spacing L. The periodicity length is N tabulation points, so a the edge of a periodic cell is NL.

Field particles have mass, m, and T is our integration time step. Physically consistent units are related by the dimensionless constant $W = GmT^2/L^3$, which replaces the physical gravitational constant in the calculations. Numerical values quoted later in this paper are given in dimensionless form, so distances are in units of L, masses in units of m, and velocities in units of L/T, and so on. MP masses are quoted in units of m, so an MP mass of 100K is $100\,000m$. There is no simple way to relate these units to observations.

2 Results from Experiments

Experiments reported in this paper came from 3 sets: (1) Periodicity length $N = 128$, (2) Periodicity length $N = 64$, and (3) the entire configuration embedded in a strong isotropic harmonic potential, again with periodicity length $N = 128$. All experiments ran with $P = 800\,768$ particles. The MP was initially at the center of a periodic cell, so the even values of N used here place the MP at equal distances from eight neighboring tabulation points, at the common corner of the small cubic cells centered on those 8 tabulation points. Even very small motions cause the MP to move through various ones of those small cells.

Experiments with Periodicity Length $N = 128$ are described first, followed by the remaining sets, which begin with Sect. 2.4.

2.1 Nature of MP Motions

MP motions look like a growing oscillation which levels off at later times. The Y–component of the MP position is shown as a function of time for six runs in Fig. 1, which are stacked for clarity. The numbers at the end of each track indicate the mass of the MP. Amplitudes for the 100K track are magnified 5 times and they are doubled for the 10K track.

While the tracks at lower mass (< 10K) look like a growing oscillation that levels off at some amplitude, all with nearly the same period, tracks for the more massive 10K and 100K experiments look different. Both the amplitudes and periods are irregular. It is tempting to guess that they may become chaotic, but we cannot make any definitive claims in that regard.

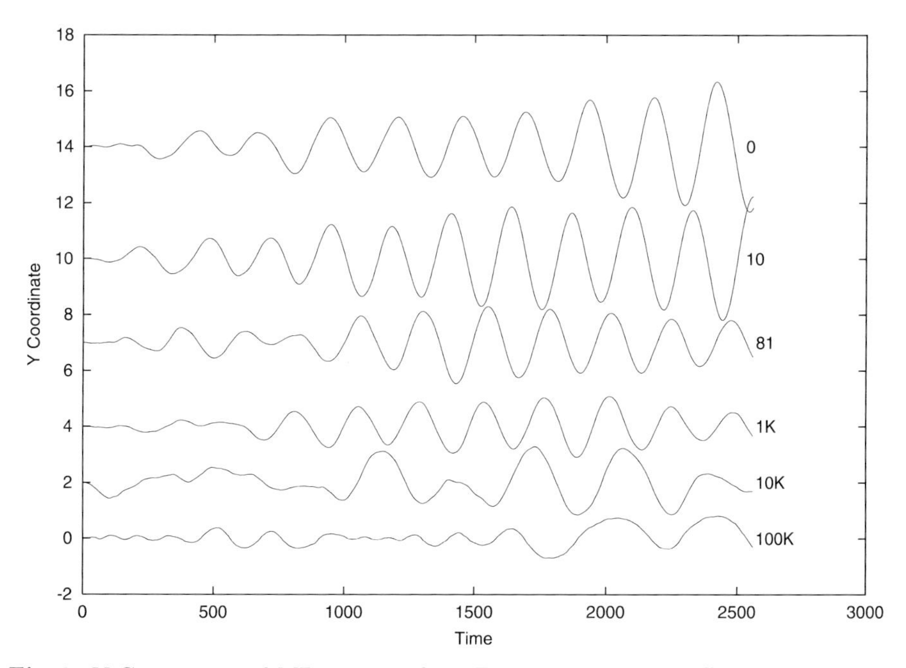

Fig. 1. Y-Component of MP motions from Experiments with Different MP Masses. Tracks are spaced vertically for clarity. Labels at the end of each track indicate the mass of the MP

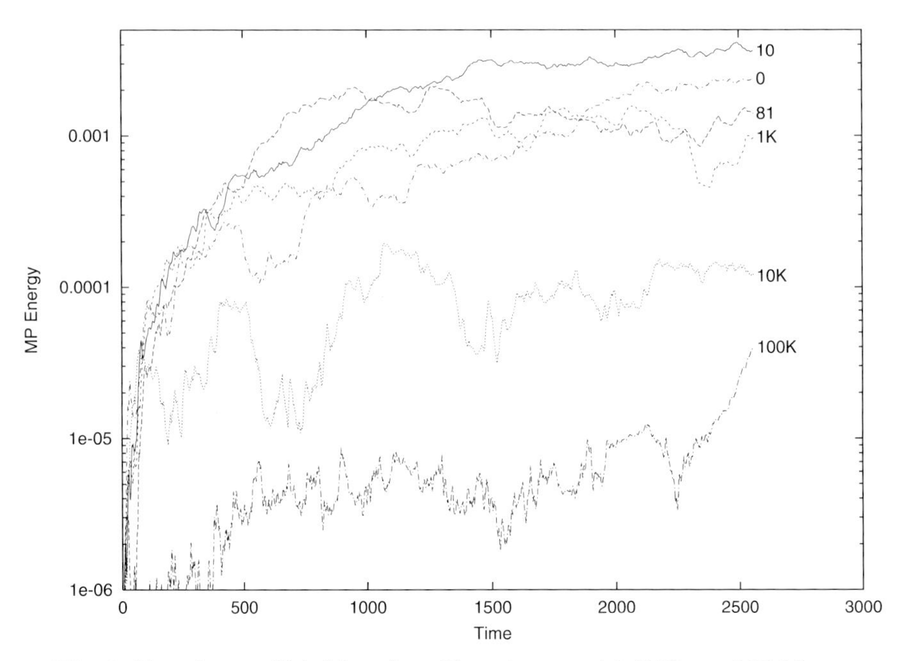

Fig. 2. Energies per Unit Mass from Experiments with Different MP Masses

2.2 MP Energies and Effective Masses

Orbital periods are sufficiently constant throughout these runs that a total MP energy per unit mass can be estimated from $\frac{1}{2}((\omega x)^2 + v^2)$. A convenient way to estimate $1/\omega$ is to scale the displacements so they match the velocities throughout the runs, which they do pretty well. When energies so determined are plotted as functions of time, all on the same page, the Fig. 2 is generated.

Log (energy per unit mass) is plotted in Fig. 2. At smaller MP masses the curves nearly overlap, but for the greatest MP masses, they "equilibrate" at lower values. The near overlap is of interest here.

The track for MP mass of 100K climbs rapidly at the end. The MP had wandered off to a large distance on a nice smooth orbit.

Tracks in Fig. 2 that are associated with runs with smaller MP masses are reasonably smooth, while those with the largest MP masses become increasingly irregular, even apart from the high–frequency "grass" along the track for 100K. Part of that "grass" arises from roundoff.

Weight Energies by "Effective" MP Mass

Tracks can each be scaled so that they all settle down at about the 0.75 level, the mean energy of a field particle, giving Fig. 3. In the picture provided by equipartition, the scale factor represents an "effective" mass.

The overlap is by no means exact. The irregularity in a given track may be regarded as typical of the scatter of MP energy values due to the statistical

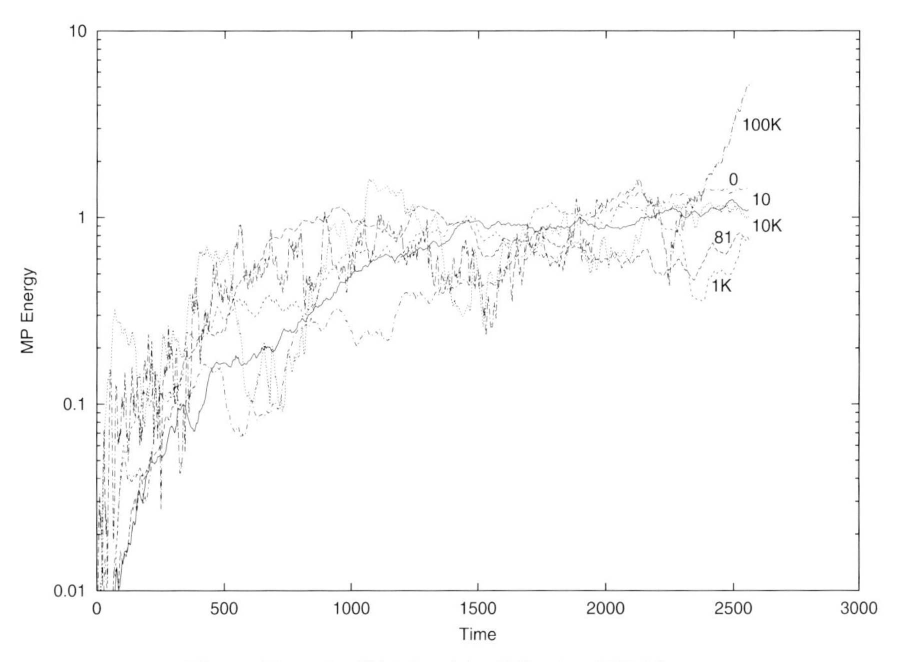

Fig. 3. Energies Weighted by Effective MP Masses

processes inherent in an equipartition picture. The scatter from experiment to experiment is about the same as the scatter within a given experiment in this plot. Call the "actual mass" M_{act} and the "effective" mass M_{eff}.

Compare Effective and Actual Masses

When the two kinds of "mass" are compared in Fig. 4, an interesting pattern emerges. The heavy diagonal line in that plot indicates equality of the effective mass and the actual mass, $M_{\mathrm{eff}} = M_{\mathrm{act}}$. Filled black squares indicate experiments from this $N = 128$ sequence. The other points come from the other sequences. The lines connecting these points are there simply to guide the eye.

There is a bit of a cheat in the abscissa of Fig. 4: $M_{\mathrm{act}} = 0$ has been entered as about 1.4, to prevent the plotted points from disappearing off to the left. Otherwise both scales are logarithmic.

For small M_{act}, M_{eff} is nearly constant at about 600. Once M_{act} exceeds that value of 600, the two are nearly equal: $M_{\mathrm{eff}} \approx M_{\mathrm{act}}$.

To zero order, this set of experiments fits a pattern, $M_{\mathrm{eff}} \approx \max(600, M_{\mathrm{act}})$. There may be a dip partway along the constant M_{eff} portion, but that hangs on just one experiment. The constant portion reflects the near–overlap of tracks noted in Fig. 2. It is not clear what physical features set the effective mass at about 600 for low M_{act}.

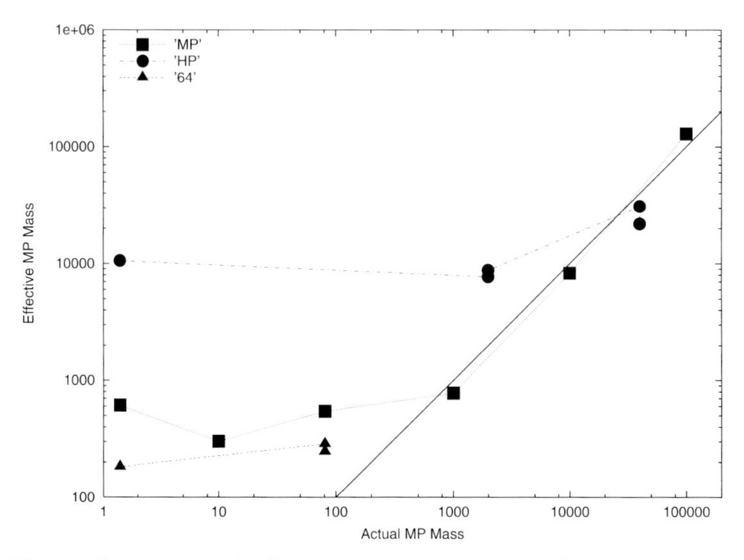

Fig. 4. Comparison of Effective MP Masses against Actual MP Masses

2.3 The Run with MP = 100K

The run with the very large $M_{\mathrm{act}} = 100\,000$ showed several features that set it apart from the other runs in the $N = 128$ sequence.

Density Cusp Around the MP

Field particles near the MP form a definite density cusp. The cusp is shown as the solid track in Fig. 5. The dashed track comes from the run in which the MP mass is zero, for comparison. Both loads had the same number of particles and the same initial density contrast, $\rho_{\max}/\rho_{\min} = 100$, so the profile with 100K mass has much more mass at great distance.

This cusp is quite robust. It has a near power–law slope, $\rho \sim 17.5r^{-1.50}$, and the cusp remains substantially unchanged to the end of the experiment. However, it does not have much mass – only about 5000 field particles, some 5% of the mass of the MP.

The principal reason there is so little mass in the cusp is that so little phase volume is available. The cusp is confined to a small configuration volume, and the MP force field, which is softened with a softening length around 1.5 of our length units, restricts the velocities that can be bound. In analytic solutions (1), more mass accumulates around the MP with shorter softening in an isothermal configuration, finally diverging in the limit as the softening goes to zero.

The cusp seen in this run fits the expected density to fill the potential of the MP according to (1). There is no evidence of a nonlinear increase in cusp mass due to the mass in the cusp itself. Doubtless that increase must be present for more massive cusps.

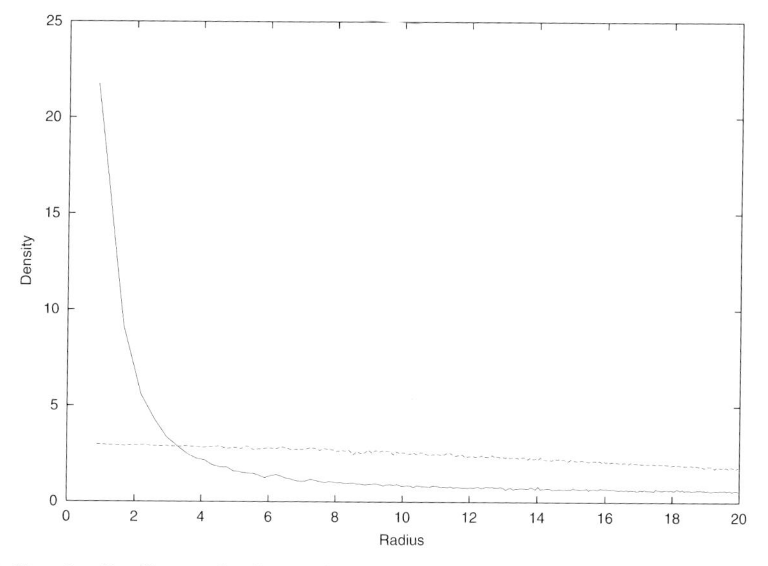

Fig. 5. Density Profiles at the Start of Experiments with MP mass = 100K (*solid line*), and with MP mass = 0 (*solid line*)

The Cusp Moves With the MP. A second property is that the cusp moves with the MP. It remains pretty well centered on the MP. This is shown in Fig. 6, in which the MP position and cusp position are plotted together at the same times. Figure 6 extends the time scale beyond that of Fig. 2. Data were filtered with a low–pass box filter because the raw data for the cusp position is noisy. Both the MP position and the cusp center positions were filtered by the *same* filter for this plot to avoid problems with filter calibration. Only the $x-$component of position is shown in Fig. 6, but agreement in the other two components is equally good. We regard the agreement as spectacular.

Particles Bound to the Cusp. A third property is that there seem to be NO particles permanently bound to the cusp. Passing field particles dwell there a bit longer, probably because their trajectories are curved, and thus account for what appears to be an enhanced density. This property can be seen from (1) if a density well is placed in an otherwise uniform isothermal sea of particles.

The Large MP Orbit

Energy plots were cut off $T = 2560$ to avoid a large drift of the MP toward the end of the run. This drift turned out to be useful, even though it is harmless, because it helped to nail down the case that the cusp moves with the MP (Fig. 6).

Features of MP Motion in the 100K Run

MP motions looked nearly oscillatory in runs with smaller MP masses, but they look quite irregular (albeit smooth) in the 100K run. There is noise in the MP

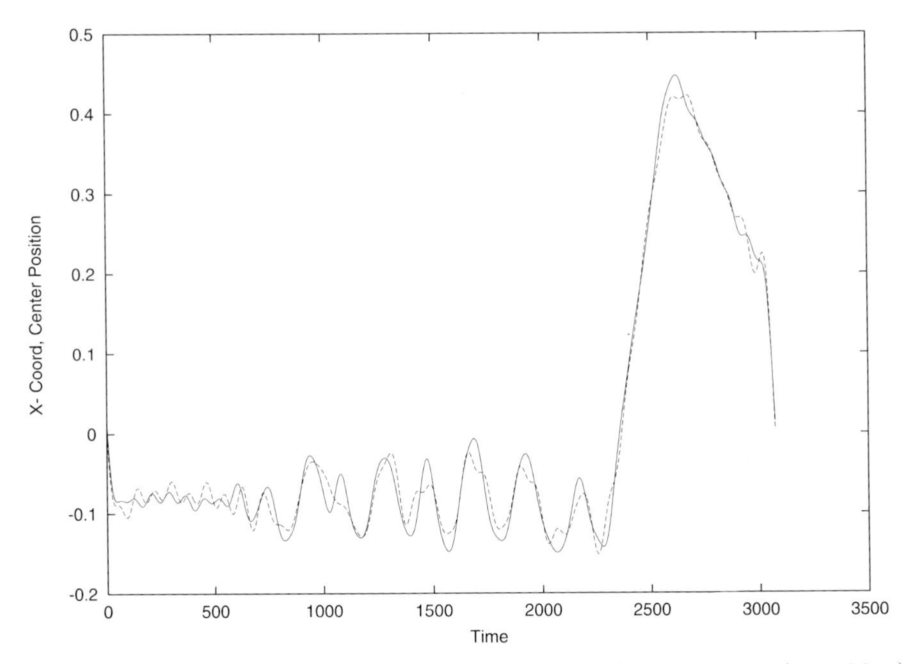

Fig. 6. Comparison of Cusp and MP Positions. (*solid line*) MP position, (*dotted line*) Cusp position

energy plots and in the MP velocity plots, but this is related to roundoff because of the small MP velocities in this run. Velocities are small because of the large mass.

The similarity of displacement and velocity plots broke down for the 100K run when the MP took off on its large drift, so the rule for estimating MP energy was no longer valid once that drift started. It was pretty good up to that time.

2.4 External Harmonic Potential

MP motions appear nearly harmonic, at least in the low-mass portions of Fig. 4, leading to the question how a batch of self–gravitating particles would act if placed within a fixed, fairly strong, harmonic potential.

This set of experiments was undertaken to investigate that question. The quick answer, as indicated by the filled black circle points in Fig. 4, is that systems in a harmonic background potential act much like systems without the harmonic potential, save that the effective mass along the low–mass portion is significantly higher at around 8 000. The double point at $M_{\mathrm{act}} = 2000$ represents two experiments. Points at $M_{\mathrm{act}} = 40\,000$ lie near the $M_{\mathrm{eff}} = M_{\mathrm{act}}$ line, just as the heavy points do in other experiments.

Tracks of the MP energy weighted by M_{eff} *vs.* time for this series of experiments are quite similar to those of Fig. 3, except that they are not as noisy. This is especially true of tracks with large M_{act}.

An unusual aspect of the experiments in a strong external harmonic potential is that they all show a strong oscillation in the total kinetic energy (sum of all field particles plus the MP). The fractional amplitude of these oscillations ((peak to peak difference)/mean) was 9 to 12%, which is several times as great as the worst case encountered with other kinds of external potential. Fractional total KE oscillations were typically under 1% for other experiments, often mere fractions of a percent. Oscillations in the harmonic potential continue undamped for the duration of any experiment. The period of these oscillations was quite accurately half the period of the external harmonic potential, just the expected value. The design period of the external potential is set at 77 of our time units in all these experiments.

Lagrangian radii show the oscillation to have an amplitude proportional to the mean radius for that mass, a pattern that we have called the "fundamental" or "breathing" mode a different study [9]. The phase of the oscillation is the same at all masses.

These oscillations defied all efforts to get rid of them, through many trials and many checks. This suggests that there is something peculiar to harmonic potentials that makes it difficult to attain a Vlasov equilibrium. Of course, *everything* goes at the same frequency in a harmonic potential, so there are many resonances.

The period of MP oscillation in each of these experiments is around 74, measurably less than the period of 77 set by the external potential. Difference periods in the two kinds of oscillation are around 2000 of our time units. Our integrations extend beyond a period of the difference frequency, so MP motions are demonstably *not* in resonance with the frequency of the background potential.

The shorter period arises from the fact that the collection of field particles contributes a potential that itself looks harmonic, and which modifies the total frequency down near the center of the configuration. It increases the frequency, which decreases the period. Again, this is reasonable and expected.

2.5 Other Features

Some other results applied across the entire set of experiments reported here.

Trajectory Separation

One feature was noticed in every one of the experiments of the three sets described here. As noted earlier, the MP was started from rest at the center of a "periodic cube." One field particle among the full load also started from rest at the same location. One of the routine post–run analyses consisted in tabulating the difference between the location of that one field particle and the MP throughout each experiment. The remarkable feature is that the field particle and the MP showed identical motions. Their position never deviated by more than roundoff in the accuracy with which the field particle position was tabulated: 0.002 of our length units at $N = 128$ and 0.001 at $N = 64$.

In a normal n–body problem, trajectories of these two particles would separate, and their distance (in phase space) would grow exponentially with time, like a Lyapunov exponent. We checked that this property holds for experiments run with grid codes like that used for the present experimental sequences, and found that exponential separation held in those cases as well. These tests were run some years ago.

These two trajectories do not separate. They remain very accurately together for the entire experiment. It is even more remarkable that they do not separate even in cases where $M_{\mathrm{act}} = 0$ – when, in effect, there is no MP.

This seems to indicate that motions near the MP are not chaotic, since otherwise the trajectories should separate. There seems to be a patch near the center that is locally harmonic, and harmonic potentials are not chaotic. With large values of M_{act} the softened potential of the MP is locally harmonic, and it produces a benign environment within which that particular field particle might move. In principle, that field particle might orbit around the MP with small amplitude, but if so, its amplitude is too small to be detected.

MP Started with Nonzero Velocity

The MP started with nonzero velocity in experiments from another sequence. It settled down to the same amplitude of motion (same MP energy) as was attained by an MP starting from rest. This is consistent with any picture in which the steady state is described by equipartition. That other sequence is not described in this note. There was no external potential in that series of experiments.

3 Discussion

A few conclusions follow from the experimental results presented here.

3.1 The Relation Between Actual and Effective MP Masses

Experiments with low M_{act} show limited amplitude of MP motion which follows

$$M_{\mathrm{eff}} \approx \max(M_0, M_{\mathrm{act}}) \tag{3}$$

approximately, with differing M_0 depending on circumstances. We find $M_0 \approx 600$ for the $N = 128$ sequences, 250 for the $N = 64$ sequences, and 8 000 for the sequences with a strong harmonic external potential. So far, we have no way to estimate M_0, although trends seem reasonable.

The pattern at low M_{act} ($M_{\mathrm{act}} < M_0$) requires that the field particles and the MP be treated as a self–consistent whole. Smooth galaxy centers have a locally harmonic potential near the center, where a kind of collective effect takes over. This pattern confirms our earlier result [7,8] since those experiments all had $M_{\mathrm{act}} = 0$. An interesting example was included in [10], where a group of 1000 particles, each with mass equal to that of a field particle, showed coherent motion that grew in a manner similar to that seen here. But it looks as if one early speculation is not valid: we guessed that a massive black hole might show displacements well in excess of those consistent with equipartition. It doesn't.

The break in (3) at high mass where $M_{\text{eff}} > M_{\text{act}}$ and the different character of MP motions with large M_{eff} may be caused by a cusp, which could modify the force law in the neighborhood of the MP. The cusp moves with the MP, indicating that an effective mass for the combined MP and cusp should include the mass of the cusp. That enhancement is small in the experiments reported here, attaining a value at most around 5%, well below our ability to detect it.

An important aspect of these results is that the plateau at M_0 holds right down to $M_{\text{act}} = 0$ According to equipartition, the MP velocity should diverge in that case, but nothing of the sort happens.

Remark. The question how a batch of particles interact in a strong harmonic background potential might be illustrated by the following perturbation–type problem.

A batch of particles moves in a harmonic potential. They do not interact in the unperturbed state, but have a weak interaction is turned slowly. How do they respond?

The interesting case with $1/r$ Keplerian interaction is impossible to handle, so we seek an interaction with similar features that can be handled analytically. That leads to the following problem, which can be solved completely. Complete analytic solutions like this often suggest features which a more complex system must share.

Imagine a set of n particles in a harmonic potential, but with a harmonic interaction between *absolutely every* pair of particles. Start with a one–dimensional problem. The formulation looks a lot like the normal modes problems discussed in mechanics texts.

Let the coordinate of the ith particle be x_i. All particles have the same mass and the interaction between the ith and j^{th} particles is $\frac{1}{2}b(x_j - x_i)^2$. Characterize the background harmonic potential by its frequency, Ω. Divide out the mass of each particle so the equation is per unit mass. The equation of motion for the ith particle (of a set of N particles) is

$$\ddot{x}_i = -\Omega^2\, x_i + b\sum_{j\neq i}^{n} (x_j - x_i).$$

The sign on the interaction term puts the acceleration in the positive direction if $x_j > x_i$, which is what we want. The $j = i$–term has been omitted, but it can be included with no harm since it is always zero. Include it, and assume harmonic motion for all particles (with the explicit exponential, $e^{i\omega t}$ cancelled throughout).

Then the equation of motion (always per unit mass) can be written

$$(\omega^2 - \Omega^2 - nb)\, X_i + b \sum_{j=1}^{n} X_j = 0, \tag{4}$$

where X_i and X_j are the (complex) coefficients that multiply those $e^{i\omega t}$ terms. ω^2 is an eigenvalue of the system.

In matrix notation,

$$MX = 0 \qquad \text{with} \qquad M = aI + bJ, \tag{5}$$

where $a = \omega^2 - \Omega^2 - nb$ and b is dimensionally a frequency squared. The matrix I is an $n \times n$ identity matrix and the matrix J is the $n \times n$ matrix all of whose entries are 1's.

The matrix J is the dyad, $f\,f^T$, of a $1 \times n$ column vector all of whose entries are 1's, so its rank is one. Superscript T denotes the transpose of a matrix. f is an eigenvector of J with eigenvalue n. The eigenvalues of J are all zero save for one whose value is n. It is straightforward to construct a complete basis for J, but it is not needed for present purposes.

Since $M = aI + bJ$ is real symmetric by construction, it can be reduced to diagonal form with real diagonal elements. There are $n - 1$ matrix diagonal elements of a and one whose value is $a + bn$. These statements are more or less evident, but the argument can be facilitated by the following considerations. Let R be the ($n \times n$ orthogonal) matrix that diagonalizes J. Then

$$R^T\,M\,R = a\,(R^T\,I\,R) \;+\; b\,(R^T\,J\,R).$$

But $R^T\,J\,R$ is diagonal, and $R^T\,I\,R$ remains diagonal (since the identity remains an identity under any rotation), so this transformation also diagonalizes M. The diagonal elements add, term–by–term, so M has $n - 1$ diagonal elements of a and one of $a + bn$.

This solves our problem, since (5) requires each diagonal element to be zero. When the defining values are plugged in, we have one eigenvalue with $\omega^2 = \Omega^2$ and $n - 1$ eigenvalues of $\omega^2 = \Omega^2 + n\,b$.

The single eigenvalue, $\omega = \pm\Omega$ goes with the eigenvector f. All particles are in the same place and move together, oscillating with the frequency of the background harmonic potential. This solution, with a single large blob, should have been expected.

The set of $n - 1$ degenerate eigenvalues goes at a higher frequency. At a given interaction strength their frequency grows without bound as n becomes very large. Particles move in any of a variety of manners, save that they do not move as one large blob. "Modes" in the degenerate set feel the full acceleration due to all the particles, irrespective of where those particles might be.

The linear analysis singles out the property that all the particles, piled up at a single location and moving together, oscillate with the frequency of the background potential and don't feel the interactions. Any other motions feel the combined interaction of all other particles. This property may well be characteristic of this problem, but it might break down when the interactions become singular at zero separation.

An unpublished 1962 note by Joel L. Brenner [11], of Stanford Research Institute, pointed out how simple it is to manipulate matrices of the class encountered, $aI + bJ$. He also pointed out that the argument generalizes to block matrices. Sadly, his note is no longer available.

Because the harmonic oscillator problem is separable, the three dimensional extension is easy to construct. It gives rise to a block diagonal matrix, with three blocks each $n \times n$. The same arguments apply to each block. They would even apply for an anisotropic external oscillator potential if the problem is discussed in a coordinate system in which the potential is diagonal. Anisotropic b's would have to be diagonal in the same frame, which is a bit artificial.

Unfortunately, the approach described here does not generalize easily to particles with different masses.

3.2 Theoretical Approaches

The problem of a massive particle in a sea of other particles, all embedded within a strong harmonic potential, looks a lot like the problem of Brownian motion in a harmonic potential. This problem has been discussed in the literature, apparently first by Ornstein. It is best known to astronomers through a 1943 paper by Chandrasekhar [12] but it was also included in [13]. That solution is one-dimensional, but a linearized three-dimensional extension appears in [14]. All three [12–14] are reprinted in [15]. The argument was recently retraced by Chatterjee [16]. All these approaches show a distribution of velocities and positions for the Brownian particle as a function of time. In a steady state, after transients have died out, expectation values are consistent with equipartition.

A Langevin equation is written in these treatments, with force on the Brownian particle treated as a series of impulses governed by a probability distribution. Those impulses are caused by close encounters, and field particles exert no forces on each other or on the Brownian particle other than the impulses. This picture is not completely applicable to the stellar dynamical problem because the underlying harmonic potential is generated by the field particles themselves, and the field particles interact, making the formulation of the theoretical problem somewhat inconsistent in the manner for self-gravitating particles.

Chatterjee *et al.* [16] included an n−body calculation and reported agreement with the Brownian motion picture. Because their field particles could not feel each other, they did not move self-consistently, so these authors did not find the plateau at low MP mass.

Calculations for this program were carried out at the NASA-Ames Research Center's NAS Systems Division under a grant of computing resources, which is gratefully acknowledged. We thank Dr. Bruce F. Smith of NASA-Ames Research Center for arranging the access. It is a pleasure to acknowledge helpful discussions with Prof. Kevin Prendergast.

References

1. John Kormendy and Douglas Richstone. *Ann.Rev.Astronomy & Astrophysics*, **33**, 581–620, (1995)
2. L. Ferrarese and D. Merritt. *Phys. World*, **15, No. 6**, 41–46, (2002)
3. J. Binney and S. Tremaine. *Galactic Dynamics*. (Princeton University Press, Princeton, NJ., 1987)

4. S. Chandrasekhar. *An Introduction to the Study of Stellar Structure.* (Dover, New York, NY, 1957)
5. James M. Huntley and W. C. Saslaw. *Astrophys.J*, **199**, 328–335, (1975)
6. W. C. Saslaw. *Gravitational Physics of Stellar and Galactic Systems.* (Cambridge University Press, New York, 1985)
7. R. H. Miller and B. F. Smith. 'Goings-on at the center of a galaxy'. In D. J. Benney, F. H. Shu, and Chi Yuan, editors, *Applied Mathematics, Fluid Mechanics, Astrophysics, a Symposium to Honor C. C. Lin*, (World Scientific, Singapore 1988) pp. 366–372.
8. R. H. Miller and B. F. Smith. *Astrophys.J.*, **393**, 508–515, (1992)
9. R. H. Miller and B. F. Smith. *Celest. Mech. & Dyn. Astron*, **59**, 161–199, (1994)
10. R. H. Miller, G. R. Roelofs, and B. F. Smith. 'An experimental study of counter-rotating cores in elliptical galaxies'. In J. W. Sulentic, W. C. Keel, and C. M. Telesco, editors, *Paired and Interacting Galaxies, Proceedings of IAU Colloquium 124*, NASA Conference Publication 3098, (NASA GPO Washington DC, 1992) pp. 549–553.
11. J. L. Brenner. 'A set of matrices for testing computer programs'. Stanford Research Institute, Menlo Park, CA., 1962.
12. S. Chandrasekhar. *Reviews of Modern Physics*, **15**, 1–89, (1943)
13. G. E Uhlenbeck and L. S. Ornstein. *Phys.Rev.*, **36**, 823, (1930)
14. M. C. Wang and G. E. Uhlenbeck. *Reviews of Modern Physics*, **17**, 323–342, (1945)
15. Nelson Wax, editor. *Selected Papers on Noise and Stochastic Processes*, (Dover Publications, Inc. New York, 1954)
16. P. Chatterjee, Lars Hernquist, and A. Loeb. *Astrophys.J.*, **572**, 371–381, (2002)

Weak Homology of Bright Elliptical Galaxies

Giuseppe Bertin

Università degli Studi, Dipartimento di Fisica, via Celoria 16, I-20133 Milano, Italy

Abstract. Studies of the Fundamental Plane of early-type galaxies, from small to intermediate redshifts, are often carried out under the guiding principle that the Fundamental Plane reflects the existence of an underlying mass-luminosity relation for such galaxies, in a scenario where elliptical galaxies are homologous systems in dynamical equilibrium. Here I will re-examine the issue of whether empirical evidence supports the view that significant systematic deviations from strict homology occur in the structure and dynamics of bright elliptical galaxies. In addition, I will discuss possible mechanisms of dynamical evolution for these systems, in the light of some classical thermodynamical arguments and of recent N-body simulations for stellar systems under the influence of weak collisionality.

1 Introduction

This article focuses on three main questions: (1) Are elliptical galaxies structurally similar to each other? (2) Which detailed dynamical mechanisms can make elliptical galaxies evolve? (3) Are there general trends to be anticipated for the evolution of these stellar systems?

Here I will report on a long-term research project aimed at providing answers to the above questions. Some interesting clues have been discovered only very recently [5], [6], [12]. Most of the paper refers to the class of bright ellipticals only; low-luminosity ellipticals are known to be characterized by different dynamical properties.

2 Structure of Bright Elliptical Galaxies

The answer to whether elliptical galaxies can be considered to be structurally similar to each other depends on the specific context in which the question is posed and addressed. Below, I will focus on the context of the physical interpretation of the Fundamental Plane ([26], [23]).

As demonstrated by a number of investigations (e.g., see [29], [30] for a study based on a sample of more than 200 early-type galaxies), the observed correlation that defines the Fundamental Plane, $\log R_e = \alpha \log \sigma_0 + \beta SB_e + \gamma$ (with $\alpha = 1.25 \pm 0.1$, $\beta = 0.32 \pm 0.03$, $\gamma = -8.895$ in the B band; the effective radius being measured in kpc, the central velocity dispersion in km/sec, the mean surface brightness in $mag/arcsec^2$ [2], [29]), is remarkably tight, with a scatter on the order of 15% in R_e.

The following simple physical argument has been put forward as an interpretation of this important physical scaling law. If we note that (1) the observed luminosity law of bright elliptical galaxies appears to be universal (the so-called $R^{1/4}$ law; [22]) and (2) the kinematical structure of these systems is regular and uniform ([4], [28]), it is natural to conclude that elliptical galaxies should be considered as homologous dynamical systems, in the sense that the relevant virial coefficient K_V should be taken to be approximately constant from galaxy to galaxy. Then, (3) given the existence of the virial constraint, the Fundamental Plane can be seen as the manifestation of a mass–luminosity relation for galaxies (see [27], [46]). In fact, the virial theorem can be written as $GM_\star/R_e = L(G/R_e)(M_\star/L) = K_V\sigma_0^2$, where $M_\star$ is the mass of the luminous component and L is the total luminosity. By eliminating σ_0 from the Fundamental Plane relation, one finds:

$$\left(\frac{M_\star}{L}\right)\frac{1}{K_V} \propto R_e^{(2-10\beta+\alpha)/\alpha} L^{(5\beta-\alpha)/\alpha} \sim L^{(5\beta-\alpha)/\alpha} . \tag{1}$$

The latter relation follows from the *empirical* fact that $2 - 10\beta + \alpha \approx 0$.

Unfortunately, there are empirical and theoretical findings that work against the hypotheses at the basis of the previous argument. First of all, significant deviations from the $R^{1/4}$ law have long been noted (see [17], [40]), and found to correlate systematically with the galaxy luminosity (see also [25]). Second, studies that have measured the amount and distribution of dark matter in ellipticals (see [4]) have shown that the presence of dark matter is more prominent in brighter and spatially larger galaxies, thus demonstrating that the virial coefficient may vary significantly from galaxy to galaxy. A curious theoretical point adds further caution to the perception that ellipticals should be considered homologous systems. This derives from direct inspection of the so-called f_∞ sequence of models [9]. As demonstrated in [5], models that appear to be all (see Fig. 1, for Ψ in the range $7 - 10$) very well fitted by the $R^{1/4}$ law, over a luminosity range of more than ten magnitudes, may be characterized by significantly different values of the relevant virial coefficient (see Fig. 2, the triangles representing the virial coefficient for the f_∞ sequence of models), as a result of the impact of a more and more concentrated nucleus.

In [5] we have further confirmed, by close inspection of four cases (NGC 1379, NGC 4458, NGC 4374, NGC 4552; studied in great detail by comparing the performance of a number of fitting procedures on data taken from [19], [18]), that the Sersic [41] index n for the $R^{1/n}$ photometric profiles can indeed be very different from 4 (in particular, for NGC 4552 we find $n \approx 11$, with residuals on the order of 0.2 magnitudes; a fit performed with the $R^{1/4}$ law would lead to residuals up to one magnitude, while a fit based on an $R^{1/4}$+ exponential profile would have residuals up to half a magnitude). On the other hand, we have checked that, if the luminosity range where the fit to the photometric profile is performed is reduced to less than 5 magnitudes, then (see [16]) the profiles are indeed well fitted by a "universal" $R^{1/4}$ law.

In conclusion, while we find it necessary to dismiss strict homology as a viable description of elliptical galaxies in relation to the interpretation of the

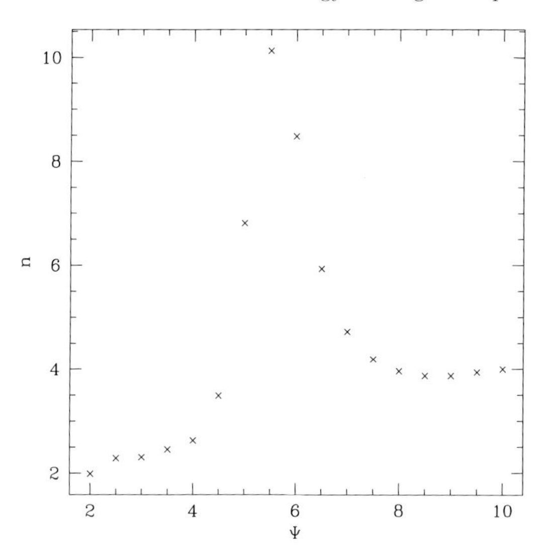

Fig. 1. The best-fit $n(\Psi)$, obtained by fitting the f_∞ models, projected along the line of sight, with $R^{1/n}$ profiles. Note the plateau at $n = 4$ reached by concentrated (high-Ψ) models, for which the radial range adopted in the fit is $0.1 \leq R/R_e \leq 10$ (from [5])

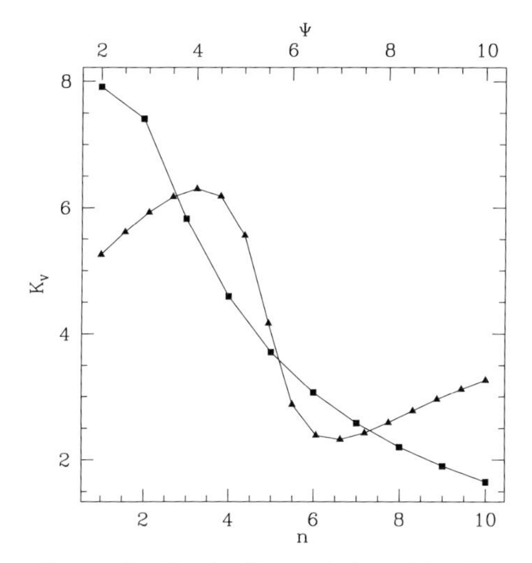

Fig. 2. The virial coefficient for the f_∞ (triangles) and for the isotropic $R^{1/n}$ (squares) models, based on an aperture of radius $R_e/8$ (from [5])

Fundamental Plane, the existence of the empirical scaling law suggests that some kind of *weak homology* must be enforced (expressed by (1)), as a correlation between structural properties and total luminosity. In [5] we have also proved that a large scatter in the dynamical correlations (e.g., in the $n \sim -19 + 3 \log L$

relation noted in [17], [25]) may well be compatible with the observed tightness of the Fundamental Plane.

3 Mechanisms of Dynamical Evolution

Given the conclusion that elliptical galaxies have to be considered only weakly homologous systems, it is natural to ask whether and how individual galaxies may change their internal structure via dynamical processes. This general issue is especially important, if we recall that typically, in the study of the cosmological evolution of the Fundamental Plane (see [44] and references therein), strict homology and thus a mass–luminosity relation is assumed for the observed galaxies and an interpretation of the data (see Fig. 3) is made in terms of pure *passive evolution* (through the evolution of the luminosity resulting from the evolution of the properties of stellar populations).

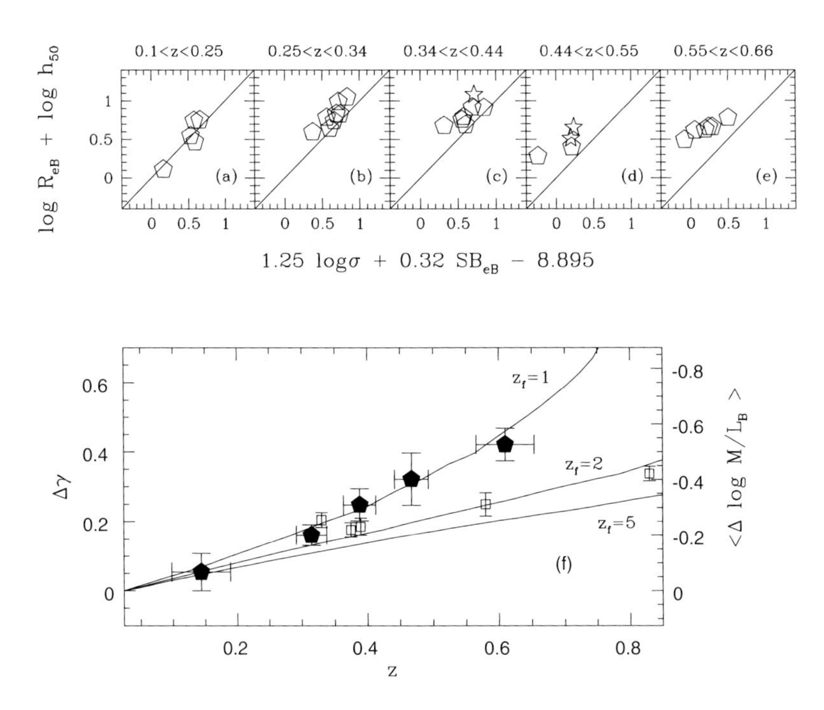

Fig. 3. The Fundamental Plane in the rest frame B band. In panels (a) to (e), field E/S0 galaxies are shown, binned in redshift, and compared to the Fundamental Plane found in the Coma Cluster by [2]. Panel (f) shows the average offset of the intercept of field galaxies from the local Fundamental Plane relation as a function of redshift (large filled pentagons) compared to the offset observed in clusters (open squares). The solid lines represent the evolution predicted for passively evolving stellar populations formed in a single burst at $z = 1, 2, 5$ (from top to bottom). This figure is taken from [44] where full references are given to the sources for cluster data points and stellar synthesis models

Besides the possibility of major merger events, are there significant sources of dynamical evolution for elliptical galaxies to be considered? As noted recently [3], the traditional approach to the study of elliptical galaxies, in terms of equilibrium and stability for the solutions of the collisionless Boltzmann equation, supplemented by the Poisson equation, may be misinterpreted. Given the very large values of typical star-star relaxation times in elliptical galaxies (see [20], [42]) it is generally taken for granted that, unless a system happens to be in a dynamically unstable state (for example, a condition of excessive radial anisotropy; see [39]), its state is basically "frozen" into an equilibrium distribution function. Thus the only task left to the dynamicist would be to decipher which distribution function best describes the observed states (a task that is particularly difficult for non-spherical systems) taken to be strictly stationary.

In our opinion, the above picture is oversimplified and may lead to an improper perception of the dynamics of real stellar systems. If, for simplicity, we take the view that elliptical galaxies have formed via collisionless collapse (see [45]), we should realize that splitting past and present conditions (that is formation processes and a collisionless equilibrium state) is just an idealization that the theory makes in order to define a basic state and to study its properties. In reality, stellar systems evolve continually and we should check to what extent the evolution processes change the internal structure of galaxies.

There are several specific mechanisms and causes for dynamical evolution that could be studied: (i) "Granularity" in phase space left over from the initial collapse; (ii) Presence of gas in various phases, especially of the hot X-ray emitting interstellar medium; (iii) Interactions with a compact central object; (iv) Interactions between the galaxy and its own globular cluster system; (v) Interactions with external satellites and effects of tides and minor mergers.

In a recent paper [6] we have tried to quantify the role of items (iv) and (v) above by means of N-body simulations. The idea at the basis of these studies is that heavy objects can suffer dynamical friction and then be dragged in toward the galaxy center, as studied earlier, for example, in [15], [14], [48]; in fact, the parallel momentum transport relaxation time T_{fr} is related to the deflection relaxation time T_D by a factor that can be very small when a heavy test particle moves through a field of lighter particles: $T_{fr} = 2T_D m_f/(m_t + m_f)$. We have thus revisited the problem of simulating the orbital decay of a satellite, placed initially on a circular orbit at the periphery of a galaxy, and basically confirmed the general findings presented in [14]; note that our simulations have been made with about one million particles, while the earlier simulations had been carried out with a few thousand particles. Then we have proceeded to address a quasi-spherical problem in which the satellite is fragmented into many smaller objects (several runs were made with either 20 or 100 fragments), distributed on a spherical shell. The quasi-spherical symmetry that characterizes this study has the important advantage of allowing for a smoother framework, basically free from other effects unrelated to dynamical friction, such as those associated with lack of equilibrium in the initial configuration. Furthermore, with respect to the earlier studies of the orbital decay of a single satellite, our attention here

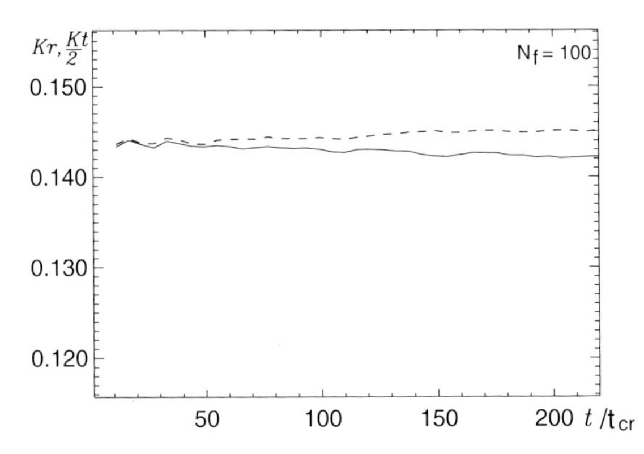

Fig. 4. The development of pressure anisotropy in a galaxy as a result of the interaction with a shell of $N = 100$ fragments dragged in toward the galaxy center by dynamical friction. The broken line represents the evolving value of $K_T/2$, where K_T is the total kinetic energy associated with the star motions in the tangential directions; the solid line represents the evolving value of K_r, the total kinetic energy associated with the star motions in the radial direction (from [6])

is mostly shifted to measuring the evolution of the underlying structure of the hosting galaxies. One effect observed, while the fragments are slowly dragged in toward the center, is a general change in the stellar density distribution with respect to the initial polytropic basic state. Another expected effect that we have been able to quantify, starting from an initially isotropic distribution of stellar orbits, is the slow growth of a tangentially biased pressure anisotropy (see Fig. 4). All these slow dynamical evolution effects appear to be genuinely associated with the process of dynamical friction exerted by the stars on the minority component of heavier objects. We are planning a survey of cases that should allow us to identify general properties of dynamical evolution in elliptical galaxies resulting from the interaction between the stars and a significant population of globular clusters or of the merging of a large number of small satellites.

4 General Trends from Thermodynamical Arguments

In order to study possible general trends that may be anticipated for the evolution of elliptical galaxies, we refer to the general framework that has been successfully applied to the context of the evolution of globular clusters. Globular clusters appear to be well represented by King [34] models (see [24]). They are recognized to be non-homologous stellar systems, subject to dynamical evolution resulting from internal effects (such as weak collisionality and evaporation) and external perturbations (such as disk-shocking, when, in our Galaxy, their orbits happen to cross the disk). It has been noted that these mechanisms of dynamical evolution make a globular cluster evolve approximately along the King equilibrium sequence (see [47] and references therein).

For globular clusters, an important paradigm is provided by the *gravothermal catastrophe* [37], which offers interesting applications and physical interpretation (for a review, see [42]). Here we recall that, starting from the study of isothermal gas spheres [13], the gravothermal catastrophe is expected to occur also in stellar systems (see [1], [37]). The instability is interpreted as due to the curious property of self-gravitating systems of being characterized by a negative effective specific heat. Although for stellar systems a rigorous proof has been provided only for idealized models where an isothermal set of stars is confined by a spherical box, the paradigm is generally believed to be sufficiently robust to be applicable to real stellar systems, provided that they possess a sufficient level of internal collisionality. An independent element that strengthens the view that the paradigm is indeed robust has been added by an analysis that has shown, for an isothermal gas, that spherical symmetry is not a necessary ingredient [35].

Following some arguments initially put forward by Lynden-Bell (see [36], [37]), would there be a way to lay out a similar scenario for elliptical galaxies as partially relaxed stellar systems? If so, we would gather powerful "thermodynamical" arguments to determine general trends for evolution, beyond the specific paths produced by a given dynamical mechanism.

In our view, there are two aspects of the problem that require clarification. A first point is that we would like to start from a physically justified equilibrium sequence, much like King models for globular clusters, able to describe the general properties of elliptical galaxies. A second point is that, formally, the origin of the gravothermal catastrophe can be traced to the Poincaré stability of linear series of equilibria (see [31], [32]). For a proper mathematical derivation, one would thus like to start from a sequence of collisionless models derived rigorously from the Boltzmann entropy. In the absence of such a sequence, a derivation of the gravothermal catastrophe has been based on either an *unjustified ansatz* (see [33], [38]), that the global temperature of the system would be associated with the coefficient multiplying the energy in the distribution function, or the use of non-standard entropies [21] (but for unrealistic models).

In order to address the first point, we may refer to a sequence of models that have been found to be very promising for a realistic description of elliptical galaxies (the so-called f_∞ models; [9], see the review [11]). These models have been inspired by the characteristics of the products of collisionless collapse, as derived from N-body simulations [45]. In the simple spherical case, they are based on the distribution function $f_\infty = A(-E)^{3/2} \exp{(-aE - cJ^2/2)}$, with A, a, c positive constants, and define a one-parameter equilibrium sequence, which, much like King models, can be parameterized in terms of the dimensionless central potential $\Psi = -a\Phi(0)$. For positive values of E the distribution function is taken to vanish. When Ψ increases beyond a certain value, around $\Psi = 7$, the models have a projected mass density profile that is well fitted by the $R^{1/4}$ law and indeed they turn out to be an excellent tool to fit the observations. From the point of view of statistical mechanics, they have been found [43] to be compatible with a derivation based on a partition of phase space in terms of the star energy and the star angular momentum square, under the assumption that detailed conservation of the star angular momentum is required at large values of angular momentum. This closely follows our understanding of the process of partial violent relaxation [36]. Unfortunately, the derivation is based on heuristic arguments and the distribution function does not follow from a straightforward exact mathematical extremization of the Boltzmann entropy; in particular, the orbit time that acts as a weight to the cells in phase space is replaced, for simplicity, by a factor $1/(-E)^{3/2}$, which is approximately correct only for weakly bound orbits. Therefore, attempts at using this equilibrium sequence to study the gravothermal catastrophe in the context of elliptical galaxies, while definitely appealing from the physical point of view (see also [8]), would remain less satisfactory from the formal point of view.

Now we have shown [12] that we can carry out a program that is satisfactory not only from the physical point of view (because it is based on an equilibrium sequence, also inspired by studies of collisionless collapse [45], that is able to match the properties of observed galaxies), but also from the mathematical point of view (because the distribution is derived rigorously from the Boltzmann entropy by requiring the conservation of a third global quantity Q, in addition to total energy and total mass). The program is made possible by the second option explored in [43] for the construction of models of partially relaxed stellar systems. This option leads to the so-called $f^{(\nu)}$ models. It was already noted [43] that the general physical properties of the $f^{(\nu)}$ models are close to those of the f_∞ models and, in particular, that for ν in the range $0.5 - 1$ their projected mass distribution, for concentrated models, follows the $R^{1/4}$ law.

Let f be the single-star distribution function, E the single-star specific energy, and J the magnitude of the single-star specific angular momentum. Consider the standard Boltzmann entropy:

$$S = -\int f \ln f d^3v d^3x \tag{2}$$

and extremize it under the constraint that the total mass

$$M = \int f d^3v d^3x, \tag{3}$$

the total energy

$$E_{tot} = \frac{1}{3}\int E f d^3v d^3x, \tag{4}$$

and a third global quantity

$$Q = \int J^\nu |E|^{-3\nu/4} f d^3v d^3x \tag{5}$$

are assigned. Then the resulting distribution function is

$$f^{(\nu)} = A \exp\left(-aE - dJ^\nu |E|^{-3\nu/4}\right). \tag{6}$$

In the above expression, the quantities A, a, and d are positive constants. The parameter ν is a free (positive) parameter, which was argued [43] to be in the range 0.5 – 1.0. In the following we refer to the case $\nu = 1$. Note that the three constants appearing in the distribution function define two scales and one dimensionless parameter, which we take to be $\gamma = ad^{2/\nu}/(4\pi GA)$.

Self-consistent models generated by such distribution function are computed from the Poisson equation, solved under the boundary conditions of regular potential at the center and of Keplerian potential at very large radii. For positive values of E the distribution function is taken to vanish. If we introduce the dimensionless central potential $\Psi = -a\Phi(0)$, the outer boundary condition defines a sort of eigenvalue problem that is solved by the relation $\gamma = \gamma(\Psi)$. The self-consistent models thus make a one-parameter equilibrium sequence.

By careful numerical integration, one may then proceed to calculate the functions $S = S(M, Q, \Psi)$ and $E_{tot} = E_{tot}(M, Q, \Psi)$ on the equilibrium sequence (see Fig. 5) and from here the inverse *global temperature*

$$\zeta = \left(\frac{\partial S}{\partial E_{tot}} \right)_{M,Q} . \tag{7}$$

The onset of the gravothermal catastrophe is thus determined by inspection of the equilibrium sequence studied in the (E_{tot}, ζ) plane (following [31]).

In [12] we have implemented the above program and shown that for the $f^{(\nu)}$ models the gravothermal catastrophe is expected to set in at $\Psi \approx 9$. Surprisingly, around this value of the concentration, the projected mass distribution turns out

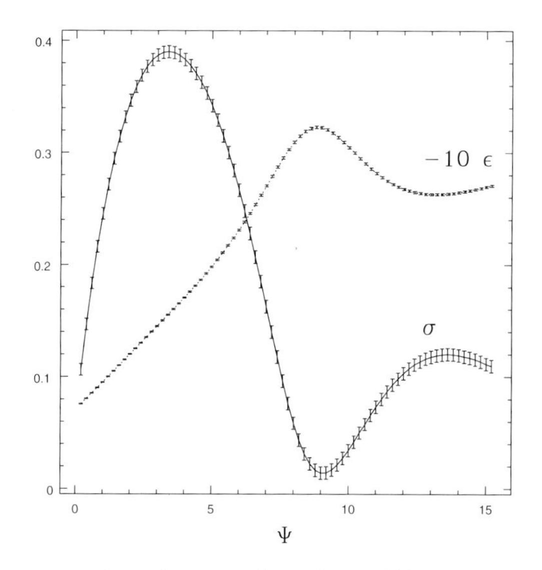

Fig. 5. Specific entropy and total energy along the equilibrium sequence of $f^{(\nu)}$ models with $\nu = 1$ (as a function of the concentration parameter Ψ, at constant M and Q, and thus expressed by means of the dimensionless functions $\sigma(\Psi)$ and $\epsilon(\Psi)$). Note that for $\Psi < 3.5$ the models are characterized by a negative global temperature, because the derivatives of S and E_{tot} have opposite signs. This figure is taken from [12]

to be very well fitted by the $R^{1/4}$ law (this general point had already been noted in [43], but outside the context of the gravothermal catastrophe). For values of Ψ close to and beyond 9, the general properties of the instability "spiral" in the (E_{tot}, ζ) plane, based on the proper thermodynamical definition of the global temperature, are the same as in the $(E_{tot}, \hat{a})$ plane, based on the *ansatz* that the temperature of the models is determined by the coefficient a (see Fig. 6).

One important point noted in [12] is a qualitative departure of the behavior of the instability "spiral" at low values of Ψ. For the original gas sphere and for the stellar dynamical analogue of a stellar system confined by a box with reflecting walls, the limit of low concentration was identified as that of a *non-gravitating ideal gas*, subject to Boyle's law. In our case, the analogy breaks down. In fact, the global temperature turns out to *change sign* at $\Psi \approx 3.5$ (see Fig. 5). Such a drastic event should be accompanied by some physical counterpart in the dynamical behavior of the system. Surprisingly, the value of $\Psi \approx 3.5$ coincides with that for the threshold of the radial-orbit instability [39] (for the context of f_∞ models, see [7] and [10]). In other words, by undertaking a thermodynamical description of the equilibrium sequence of models defined by the $f^{(\nu)}$ distribution function, we have found arguments that lead us naturally not only to the interpretation of the observed $R^{1/4}$ law, but also to one clue for the interpretation of the radial-orbit instability of collisionless stellar systems. Besides the proper-

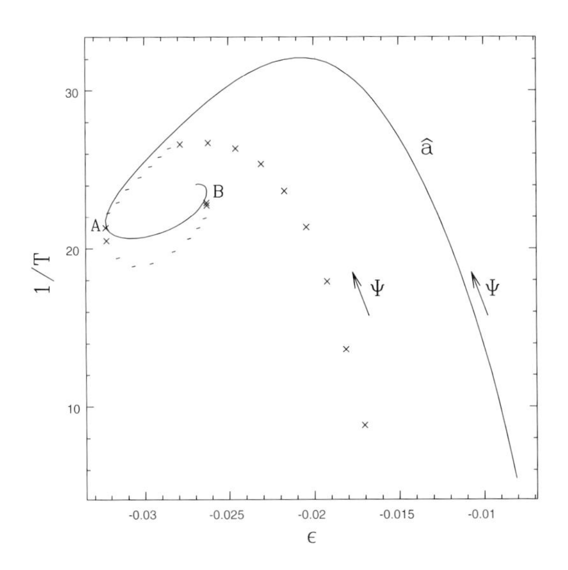

Fig. 6. The instability "spiral" of $f^{(\nu)}$ models with $\nu = 1$. The solid line refers to the results obtained from the *ansatz* that the coefficient a represents the inverse global temperature. Crosses represent the inverse global temperature from the definition $\partial S/\partial E_{tot}$; other symbols indicate estimated points for which the adopted numerical differentiation is less reliable. Point A marks the onset of the gravothermal catastrophe (from [12])

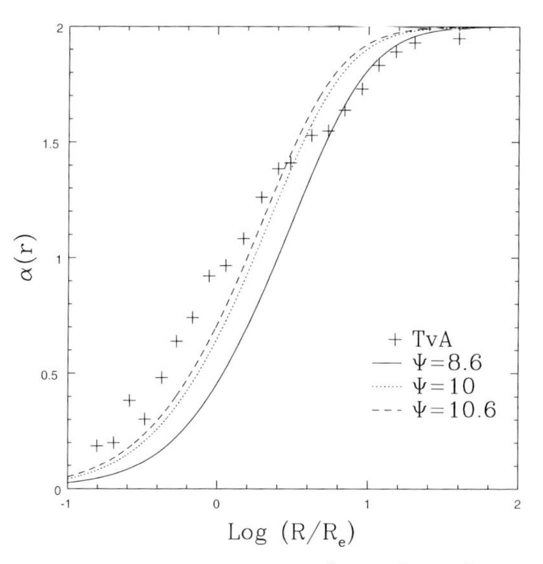

Fig. 7. Pressure anisotropy profiles $\alpha = 2 - (\langle v_\phi^2 \rangle + \langle v_\theta^2 \rangle)/\langle v_r^2 \rangle$ as a function of radius for selected $f^{(\nu)}$ models ($\nu = 1$) compared to the pressure anisotropy profile found [45] in numerical simulations of collisionless collapse. This figure has been prepared by M. Trenti

ties just outlined, one important additional aspect that makes the $f^{(\nu)}$ models, at this point, more appealing than the f_∞ models is their anisotropy level. We had noted (e.g., see [11]) that the f_∞ models are actually too isotropic, when compared with the final products of simulations of collisionless collapse [45]. The present models turn out to be much more interesting even in this respect. We have indeed checked that their characteristic anisotropy profile, for values of Ψ close to the onset of the gravothermal catastrophe, is very similar to that observed in the numerical simulations (see Fig. 7).

5 Conclusions

For a physically justified family of equilibrium models, representing the result of incomplete violent relaxation, and derived rigorously from the Boltzmann entropy, we have shown that, at high concentration values, the onset of the gravothermal catastrophe is found to occur at $\Psi \approx 9$, in the parameter domain where models are characterized by an $R^{1/4}$ projected density distribution. At low concentration values, the equilibrium sequence presents a drastic departure from the limit of the classical isothermal sphere, because models become associated with a negative global temperature. The transition point, $\Psi \approx 3.5$, turns out to coincide with the point of the sequence where the radial-orbit instability sets in. In the intermediate concentration regime, $3.5 < \Psi < 9$, the structural properties of the models change, much like those of models along the

King equilibrium sequence, a family of models that is known to capture the non-homologous properties of globular clusters. It is our hope that, in this domain of intermediate concentration values, the $f^{(\nu)}$ models may be used to describe the characteristics of weak homology of elliptical galaxies.

Acknowledgements

I would like to thank L. Ciotti, M. Del Principe, T. Liseikina, M. Lombardi, F. Pegoraro, M. Stiavelli, M. Trenti, T. Treu, and T. van Albada, for their collaboration in this work, and the organizers of the workshop"Galaxies & Chaos", for their warm hospitality in Athens. This work has been partially supported by MIUR of Italy.

References

1. V.A. Antonov: Vestnik Leningr. Univ. no. 19, 96 (1962) (Engl. Transl.: in *Structure and Dynamics of Elliptical Galaxies*, ed. by T. de Zeeuw (Reidel, Dordrecht 1986) pp. 531-548
2. R. Bender, R.P. Saglia, B. Ziegler et al.: Astrophys. J. **493**, 529 (1998)
3. G. Bertin: 'Gravitational plasmas'. In *Plasmas in the Universe, 142nd Course of the International School of Physics "Enrico Fermi"*, ed. by B. Coppi, A. Ferrari, E. Sindoni (Società Italiana di Fisica, Bologna 2000) pp. 373-393
4. G. Bertin, F. Bertola, L.M. Buson et al.: Astron. Astrophys. **292**, 381 (1994)
5. G. Bertin, L. Ciotti, M. Del Principe: Astron. Astrophys. **386**, 149 (2002)
6. G. Bertin, T. Liseikina, F. Pegoraro: paper submitted (2002)
7. G. Bertin, F. Pegoraro, F. Rubini, E. Vesperini: Astrophys. J. **434**, 94 (1994)
8. G. Bertin, R.P. Saglia, M. Siavelli: 'Spiraling into the $R^{1/4}$ law of ellipticals'. In *New ideas in Astronomy*, ed. by F. Bertola, J.W. Sulentic, B.F.Madore (Cambridge University Press, Cambridge 1988) pp. 93-95
9. G. Bertin, M. Stiavelli: Astron. Astrophys. **137**, 26 (1984)
10. G. Bertin, M. Stiavelli: Astrophys. J. **338**, 723 (1989)
11. G. Bertin, M. Stiavelli: Rep. Prog. Phys. **56**, 493 (1993)
12. G. Bertin, M. Trenti: Astrophys. J. **584**, in press (2003)
13. W.B. Bonnor: Mon. Not. Roy. Astron. Soc. **116**, 351 (1956)
14. Tj.R. Bontekoe: Orbital decay of satellite galaxies. PhD Thesis, Groningen University, Groningen (1988)
15. Tj.R. Bontekoe, T.S. van Albada: Mon. Not. Roy. Astron. Soc. **224**, 349 (1987)
16. A. Burkert: Astron. Astrophys. **278**, 23 (1993)
17. N. Caon, M. Capaccioli, M. D'Onofrio: Mon. Not. Roy. Astron. Soc. **265**, 1013 (1993)
18. N. Caon, M. Capaccioli, M. D'Onofrio: Astron. Astrophys. Suppl. **106**, 199 (1994)
19. N. Caon, M. Capaccioli, R. Rampazzo: Astron. Astrophys. Suppl. **86**, 429 (1990)
20. S. Chandrasekhar: Astrophys. J. **97**, 251 (1943)
21. P.H. Chavanis: Astron. Astrophys. **386**, 732 (2002)
22. G. De Vaucouleurs: Ann. d'Astrophys. **11**, 247 (1948)
23. S. Djorgovski, M. Davis: Astrophys. J. **313**, 59 (1987)
24. S. Djorgovski, G. Meylan: Astron. J. **108**, 1292 (1994)

25. M. D'Onofrio, M. Capaccioli, N. Caon: Mon. Not. Roy. Astron. Soc. **271**, 523 (1994)
26. A. Dressler, D. Lynden-Bell, D. Burstein et al.: Astrophys. J. **313**, 42 (1987)
27. S.M. Faber, A. Dressler, R.L. Davies et al.: 'Global scaling relations for elliptical galaxies and implications for formation'. In *Nearly normal galaxies: From the Planck time to the present*, ed. by S.M. Faber (Springer, New York 1987) pp. 175-183
28. O. Gerhard, A. Kronawitter, R.P. Saglia, R. Bender: Astron. J. **121**, 1936 (2001)
29. I. Jørgensen, M. Franx, P. Kjærgaard: Astrophys. J. **411**, 34 (1993)
30. I. Jørgensen, M. Franx, P. Kjærgaard: Mon. Not. Roy. Astron. Soc. **280**, 167 (1996)
31. J. Katz: Mon. Not. Roy. Astron. Soc. **183**, 765 (1978)
32. J. Katz: Mon. Not. Roy. Astron. Soc. **189**, 817 (1979)
33. J. Katz: Mon. Not. Roy. Astron. Soc. **190**, 497 (1980)
34. I.R. King: Astron. J. **71**, 64 (1966)
35. M. Lombardi, G. Bertin: Astron. Astrophys. **375**, 1091 (2001)
36. D. Lynden-Bell: Mon. Not. Roy. Astron. Soc. **136**, 101 (1967)
37. D. Lynden-Bell, R. Wood: Mon. Not. Roy. Astron. Soc. **138**, 495 (1968)
38. M. Magliocchetti, G. Pucacco, E. Vesperini: Mon. Not. Roy. Astron. Soc. **301**, 25 (1998)
39. V.L. Polyachenko, I.G. Shukhman: Sov. Astron. **25**, 533 (1981)
40. P. Prugniel, F. Simien: Astron. Astrophys. **321**, 111 (1997)
41. J.L. Sersic: *Atlas de galaxias australes* (Observatorio Astronomico, Cordoba 1968)
42. L. Spitzer: *Dynamical evolution of globular clusters* (Princeton University Press, Princeton 1987)
43. M. Stiavelli, G. Bertin: Mon. Not. Roy. Astron. Soc. **229**, 61 (1987)
44. T. Treu, M. Stiavelli, S. Casertano et al.: Astrophys. J. Lett. **564**, 13 (2002)
45. T.S. van Albada: Mon. Not. Roy. Astron. Soc. **201**, 939 (1982)
46. T.S. van Albada, G. Bertin, M. Stiavelli: Mon. Not. Roy. Astron. Soc. **276**, 1255 (1995)
47. E. Vesperini: Mon. Not. Roy. Astron. Soc. **287**, 915 (1997)
48. M.D. Weinberg: Mon. Not. Roy. Astron. Soc. **239**, 549 (1989)

Part III

Observations

Observing Chaos in Disk Galaxies*

Preben Grosbøl

European Southern Observatory, Karl-Schwarzschild-Str. 2, 85748 Garching, Germany

Abstract. Regions in disk galaxies where one would expect to find chaotic behavior are likely to be associated with major stellar resonances such as the co-rotation. The possible identification of such locations in real galaxies is illustrated by examples of four spiral galaxies observed in the K band. Observational issues related to the detection of chaotic regions are discussed. Although surface photometry may suggest chaotic regions, it is essential to compare detailed velocity profiles with dynamic models to estimate the probability of such claims. Finally, the feasibility of performing observations of chaos with current state-of-the-art facilities such as the VLT is considered. It is found that it should be possible down to a surface brightness level of I $\approx$ 20 mag/arcsec^2 corresponding to the end of bars in typical disk galaxies whereas access to detailed studies of chaos in the main spiral pattern would require more powerful facilities.

1 Introduction

Chaotic behavior of orbits in galactic potentials is frequently seen in analytical models and numerical experiments (e.g. N-body simulations). By increasing perturbations in models of spiral galaxies, one can observe a transition of stable orbits to chaotic ones [4]. Numeric techniques make it simple to identify such orbits by calculating their dynamic spectra [29]. The existence of chaotic behavior in models does not automatically mean that it is an important phenomena in real galaxies. It is also possible that growing spiral modes are damped by non-linear effects, causing an increased velocity dispersion, before a significant fraction of the stars becomes chaotic. Thus, it is of high interest to estimate the level of chaotic behavior in disk galaxies.

The current paper considers the possibility of observing chaos in real galaxies. The next section looks on the importance of the environment while the main regions where one may expect chaotic behavior are identified in Sect. 4. Different indicators for chaos in disk galaxies, including our own, are discussed in the following sections. The feasibility of actually observing chaos is then considered in the last section using VLT instrumentation as a reference.

2 Environment

Current disk galaxies would have had time enough to reach a relative stable and relaxed state if they were formed in a single collapse at an early epoch of the

* Based on observations collected at the European Southern Observatory, La Silla, Chile.

universe. In a Cold Dark Matter scenario where galaxies are formed in hierarchical mergers over a longer time, it is less obvious that present time galaxies are relaxed, isolated systems as frequently assumed in models. Thus, it is of interest to verify whether typical nearby spiral galaxies have been able to achieve a quasi-stationary state or not.

The history of star formation in field galaxies provides some information on the environment at earlier epochs since a higher frequence of encounters between galaxies may yield an increased star formation rate. Madau et al. [21] used data from the Hubble Deep Field (HDF) to estimate the past star formation rate and found that it increased back to around a redshift of $z \approx 1.5$. At higher redshifts, the star formation rate is either close to constant or still monotonically increasing [19] depending on the exact corrections applied for dust attenuation and cosmological surface brightness dimming which are significant at these redshifts. Although this does not exclude a monolithic collapse scenario, it would predict a higher metal mass density at high redshifts than observed for absorber in quasi-stellar objects.

Another approach is to study the morphology of galaxies as function of their redshift as done by van den Berg [10] who analyzed the HDFs. He found that galaxies with $z<1$ appear largely as disk-like while those with $z>2$ either had chaotic knots or were centrally concentrated which may be precursors to elliptical galaxies or bulges in spirals. For the higher redshifts, a larger fraction of galaxies showed evidence of recent mergers. Considering the star formation rate at earlier epochs discussed above, van den Bergh [10] suggested that a change of slope around $z \approx 1.5$ may be cause in a transition from a merger driven star formation at early epochs to one mainly occurring in galactic disks.

The Hubble morphological classification scheme was found satisfactory for redshifts $z<0.5$ while it at higher redshift became increasingly difficult to apply as the fraction of peculiar galaxies got larger [10]. The frequency of barred spiral galaxies also became smaller at redshifts $z>0.5$ which could be caused by their disks being hotter and therefore more stable against bar instabilities. Spiral structure observed at these higher redshifts also appear more chaotic and less well developed.

The influence of the environment on the internal structure of disk galaxies was investigated by van den Bergh [8,9] who looked on a set of 930 Northern galaxies in the Revised Shapley-Ames catalog [24]. From this sample, he found no statistical significant change in the morphology of bars or spiral structures as function of their environment which suggests that such structures primary depend on the general properties of the parent galaxy. The merger rate of local disk galaxies was estimated by Keel & Wu [16] to be 4.2 per Hubble time for pairs and 0.33 per Hubble time extrapolated to all spirals.

This indicates that a majority of nearby disk galaxies have had upto 5 Gyr (i.e. roughly corresponding to $z \approx 0.5$) to reach a quasi-stationary state. The central potential well of typical spiral galaxies seems to be deep enough to avoid strong influence from the environment. It is therefore reasonable to believe that chaotic behavior observed in the central part of most disk galaxies is due to the dynamics properties of the galaxies rather then a signature of a violent past.

3 Regions of Chaos

The stability of stellar orbits in disk galaxies was studied by Contopoulos [5] who used a 2D isochrone potential with a superimposed bar or spiral perturbation rotating with a constant pattern speed. For weak and intermediate bar perturbations, the main family x_1 of periodic orbits is stable for values of the Hamiltonian $h < h(L_1)$ where L_1 is the unstable Lagrangian point. Close to this point, the x_1 family breaks up into an infinity of families. When the perturbation is increased, the main family becomes unstable for lower values of h at a bifurcation of an important resonant family and remains unstable. Typical invariant curves for orbits in a bar potential show stable regions around the families x_1 and x_4 (corresponding to retrograde orbits) representing non-periodic orbits trapped around stable periodic orbits. Chaotic regions surround the stable ones which decrease in size at stronger perturbations. The transition to chaotic motions in spiral galaxies occurs at lower values of h than for bars with the same amplitude. Contopoulos [5] concluded that the two main mechanisms for producing this sudden increase of chaotic behavior are the breaking up of the x_1 family into an infinity of families close to $h(L_1)$, and that it becomes unstable at a resonance and remains unstable for higher h. The former mechanism is also responsible for the assumption that bars end just inside their co-rotation (CR) due to an increase of chaotic orbits [4].

Models for the response density in a realistic galactic potential with an imposed two armed spiral perturbation were made by Contopoulos & Grosbøl [6,7] in the case of normal spirals and by Kaufmann & Contopoulos [15] for barred galaxies. The response was in phase and supported the imposed spiral between Inner Lindblad Resonance (ILR) and Outer Lindblad Resonance (OLR) for weak perturbations in agreement with the linear density wave theory [20]. For strong spiral potentials, the chaotic region around CR increased and the spiral was only nearly self-consistent between ILR and the 4:1 resonance (or -4:1 and OLR) due to the 45° phase shift of the stable periodic orbits just outside the 4:1 resonance. Nonlinear effects started to become important for spirals with a relative radial force perturbation $F_r \geq 5\%$ [13]. It was found that satisfactory models which matched imposed and response densities could be constructed for barred galaxies even in the regions close to CR where a significant fraction of the orbits was chaotic.

In many N-body simulation fast bars are formed while a slower rotating, more transient spiral pattern is seen outside [25]. This suggests that a spiral galaxy may have several patterns with different angular speeds. One possible explanation was given by Tagger et al. [28] who considered nonlinear coupling of such spiral modes. Bars ending at their CR could in this way be coupled to or drive outer spirals through their ILR. This mechanism provide a sharp selection of possible pattern speeds and can excite both harmonics and sub-harmonics such as m=1 waves [22]. Sellwood & Sparke [26] showed that even if bar and spiral pattern in a galaxy had different pattern speeds its appearance would most of the time suggest that the bar was connected to the inner part of the spiral as seen most frequently. Less than 10% of the time, bar and spiral in such

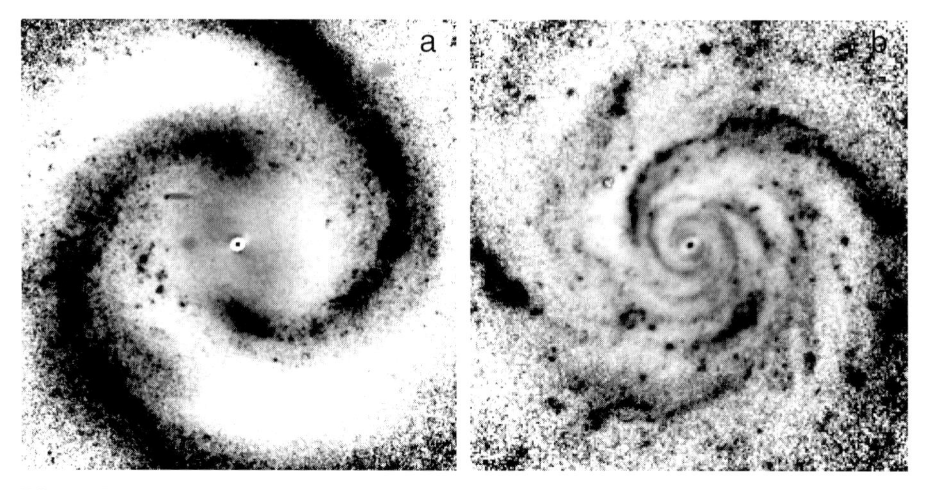

Fig. 1. Relative maps in the K' band of the central parts of two spiral galaxies: (**a**) NGC 1566 and (**b**) NGC 4030. Spherical bulges were subtracted before the images were de-projected. The maps were divided by their average radial profiles and presented in negative i.e. dark indicates higher than average intensity. The full range from white to black corresponds to ±30%.

systems would look fully separated such as in NGC 1566 (see Fig. 1a). Barred galaxies with rings at the end of their bar may also be candidates for systems with double pattern speeds. Although no detailed studies of the stability of orbits in the interface region between a fast bar and a slow spiral have been made, it is likely that chaotic behavior will develop there.

Whereas nonlinear response to a growing spiral perturbation will lead to increased velocity dispersion [3] and therefore possibly to more stochastic orbits, the most important locations where one would expect chaotic behaviors are the main resonances and in particular the CR region. For strong perturbation, chaos may be present for lower h at an important resonance. A region of special interest is the termination of bars as it is likely to be close to CR, and in the case of an outer spiral pattern rotating with a different angular speed, be even more proven to chaotic behavior.

4 Chaos in the Galaxy

We have a unique opportunity to study the distribution of space velocities of individual stars in our Galaxy after the Hipparchus satellite observed proper motions and parallaxes for a large number of stars in the solar neighborhood. The local velocity distribution was recently analyzed by Dehnen [11] and Fux [12]. Velocities in the Galactic plane shows a significant structure beside its general ellipsoidal shape. Whereas features associated to stars with (B–V)<0.4 (i.e. relative young) are still likely to trace their initial conditions at formation (e.g. moving groups or open clusters), older stars will be more relaxed and can therefore be used to probe the Galactic potential. The distribution of older stars with (B–V)>0.6 displays clear velocity features like the Hercules stream which is absent in the diagram for the younger stars.

Using a realistic model of the Galaxy including a central bar, Fux [12] calculated orbits of test particles and performed 3D N-body simulation to find the possible origin of the major stream in the velocity distribution. He estimated Liapunov divergence timescales for orbits representing the local velocity distribution to quantify their stability with various parameters for the bar. Also comparing the velocity ellipsoid derived from N-body simulation, he concluded that the Hercules stream could have several origins such as being induced by cooler chaotic orbits from in the bar region or hot chaotic orbits with $h > h(L_1)$. It was found possible but less likely that quasi-x_1 orbits could contribute to streams like the Hercules and Hyades.

The analysis of the velocities in the solar neighborhood provides a detail view of a possible typical distribution in the disk of spiral galaxies. The number of stars associated with the Hercules stream could suggest that up to 15% of the stars just outside a weak bar in a disk galaxy could be chaotic. The size and shape of features like the Hercules stream also show the difficulty in observing similar structures in external galaxies due to projection effects.

5 Tracers of Chaos in External Galaxies

For external galaxies, one can in general not observe individual stars but only integrated properties such as surface brightness distribution and line-of-sight-velocity-profiles (LOSVP). This makes it significantly more difficult to distinguish between collections of stars following non-periodic orbits trapped around stable ones and such which exhibit a chaotic behavior. One must rely on detailed dynamic models of the galaxies to determine to what degree chaotic behavior is present in a specific region.

Chaotic regions will typically have a phase space with less structure than those occupied by stable motions although it is easy to construct models with smooth appearance consisting of stable orbits (e.g. axisymmetric disks). In disk galaxies, radial regions occupied by bars or spirals must have a high fraction of ordered motions to support such structures. If more than one non-axisymmetric mode exists in a galaxies (e.g. a bar and a spiral, or two spiral patterns), the interface between them may indicate the location of a resonance and therefore possibly a larger amount of chaotic motions than in the regions dominated by a single mode. More chaotic motions would be expected in areas between modes with different pattern speeds (e.g. in the region between the end of a bar and the start of a spiral pattern). Thus, a zone with relative small azimuthal perturbations between regions with significant bar or spiral modes would be a likely candidate for increased chaotic behavior.

A more subtle indicator for an increased fraction of chaotic motions is a radial variation of the amplitude of spiral modes. Although such variation also could be caused by the interaction of different spiral modes, their details shape may be

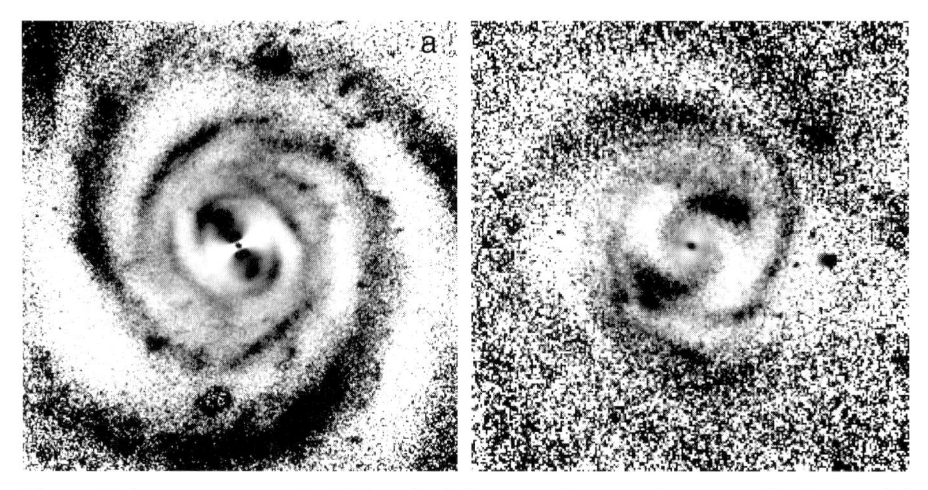

Fig. 2. Relative maps in the K' band of the central parts of two spiral galaxies: (**a**) NGC 4939 and (**b**) NGC 6902. The representation is identical to that of Fig. 1.

used as a diagnostic. An interaction between spiral modes would yield sinusoidal variation to the first approximation while an amplitude change due to a higher fraction of chaotic orbits in resonance regions would appear as a decrease at specific radii corresponding to major resonances in the galaxy. Attenuation of dust and patchy star formation would make it very difficult to detect such variations within the main spiral structure but radial amplitude changes in relative strong bars may display such features.

To illustration the way one may interpret the surface brightness distribution in disk galaxies, the K′ images of four spiral galaxies are shown in Figs. 1 and 2. The K band was chosen since it better represents the mass distribution of old disk stars although some population effects are still present [23]. A Sérsic $r^{1/n}$ profile [27] was fitted to the bulge and subtracted together with foreground stars before the images were de-projected. The bulge fitting was in some cases not fully satisfactory and left residuals in the very central parts. The figures show face-on, relative intensity maps normalized to the average radial profiles of the galaxies.

The first example, NGC 1566, shown in Fig. 1a, is a grand-design spiral galaxy classified as Sc(s)I in [24]. After the bulge was subtracted, a weak bar with an amplitude of ∼7% became visible. Its position angle is offset with more than 30° with respect to the start of the two armed spiral pattern and resembles the N-body simulations with different pattern speeds for bar and spiral [26]. The patchy nature of high intensity regions in the arms suggests that a significant fraction of the light in the arms originates from young objects. Especially at the start of the spiral arms just outside the bar, the star formation rate seems to be enhanced. This could be caused by gas clouds following more stochastic orbits in this region and therefore more likely collide when they encounter the spiral arms. Thus, the region between the bar and the start of the spiral is likely to exhibit some chaos.

The galaxies NGC 4030 (see Fig. 1b) has the type Sbc(r)I in [24] and show a more irregular arm structure in its inner regions while a two-armed pattern prevails at larger radii. This could indicate the existence of separate spiral modes as the inner pattern do not smoothly join the outer one. However, it is difficult to judge whether all the arm sections, seen in the central region, are associated with mass perturbations or some are mainly tracing recent star formation. If several spiral modes do exist in this galaxy, one may expect an increase of chaotic behavior where they interact.

In the case of NGC 4939 classified as a Sbc(rs)I, a strong bar is present with three sets of symmetric arcs just outside as seen on Fig. 2a. The first set of arcs is located just outside the bar but slightly offset relative to the orientation of the bar. The next set is situated almost parallel to the bar and shifted $\sim 90°$ with respect to the first arcs. Finally, a third set is again offset by $\sim 90°$ with the main grand design two-armed spiral pattern starting at the same radius. The arcs have a relative smooth appearance which suggests that they are density enhancements in the disk although significant star formation are likely to be present. The symmetry, alignment and shape of the arcs point to a stellar dynamical origin associated to specific resonances and families of periodic orbits. The exact relation can only be made after a detailed dynamic model is compared to the intensity distribution. The radial regions between the arcs (more noticeable for the two outer ones) have significantly smaller azimuthal variations than for the arcs themselves and are possibly related to a higher amount of chaotic orbits.

The last sample galaxy NGC 6902 of type Sa(r) is shown in Fig. 2b. This galaxy has two spiral pattern where the inner and outer spirals are winding with different orientations. This suggests that a major resonance is located at the radius where the two patterns join each other. An increased star formation is also observed at this location. Although the presence of spiral perturbations excludes strong chaotic behavior, an increased fraction of chaotic orbits is expected.

A more indirect way to see the results of non-linear dynamic effects and possibly increased chaotic behavior is to consider the distribution of the mean relative amplitude of the main spiral arms as function of their pitch angle for normal spirals [14] as shown in Fig. 3. It shows a lack of strong, tight spiral which could be explained by non-linear effects starting to damp growing spiral modes [3] when the relative radial force perturbation becomes large enough [13].

Two main features in the velocity distribution may be expected for regions with a substantial fraction of chaotic motion, namely: a) a general increase in the velocity dispersion and b) a non-Gaussian distribution. The chaotic orbits are not trapped around stable, periodic ones and will typically have a wider distribution function depending on the actual potential. Since it is only possible to observe one velocity component for external galaxies, the viewing angle is important as seen in the case of the velocity distribution in the solar neighborhood [11,12] where features like the Hercules stream only could be observed at certain

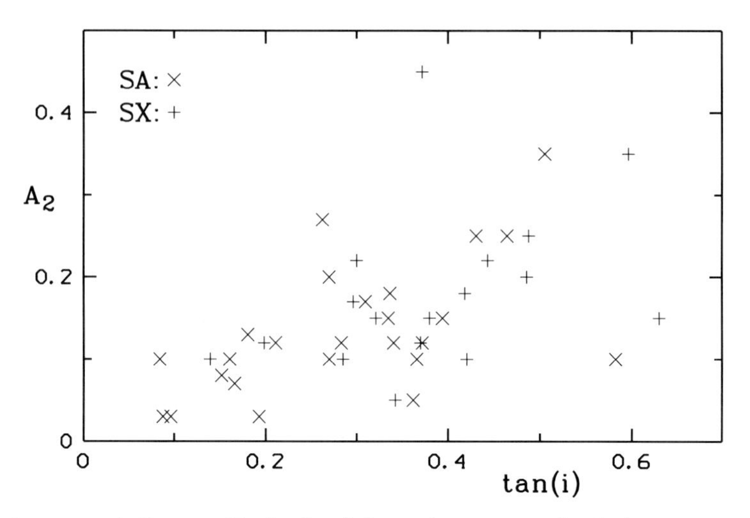

Fig. 3. Average relative amplitude A_2 of the main two-armed spiral pattern measured in K for 53 normal spiral galaxies as function of the mean pitch angle i of their arms.

projections. Also the integration along the line of sight may mask velocity structures which originates from chaotic behavior. One may also be able to detect velocity features associated with the existence of multiple families of periodic orbits near resonance regions where chaotic motion do not dominate.

6 Observational Considerations

Although the study of surface brightness distribution of disk galaxies may yield some indications on possible locations of chaotic regions, it is essential to obtain detailed kinematic data in order to support a claim of chaotic behavior. It is clear from the analysis of the stellar velocity distribution in the solar neighborhood [12] that a unique interpretation may be very difficult even with high quality data.

The need for an accurate dynamic model demands that both the mass distribution in the disk and the total potential including a possible dark matter component are estimated. The main problems in deriving a mass distribution from surface photometry are population effects and attenuation by dust. These effects are significantly reduced when using the near-infrared K band [23] as can be seen in Fig. 4 where both B and K maps of NGC 2997 are shown. Strong dust lanes are seen in the B band along the major arms but also in the inter-arm regions. Further, and the bulge appears significantly more prominent in the K band. Although attenuation by dust is strongly reduced in K, population effects must be considered as one notices strings of knots along the arms. Their compactness and location close to the main dust lanes suggest that they are associated to young objects (e.g. star forming regions). Besides these knots, it is

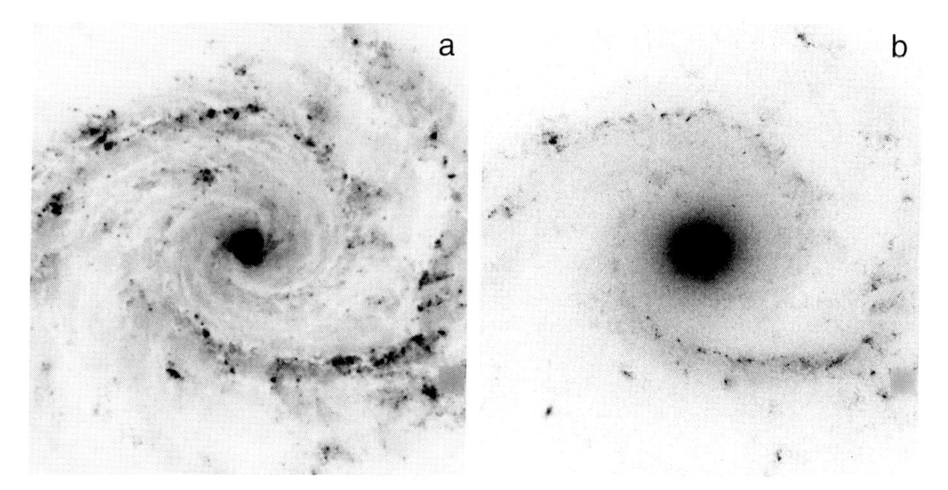

Fig. 4. Images of NGC 2997 observed in **a**) B band image and (**b**) K′ band. Foreground stars were removed.

likely that a more defuse component of young stars contributes to the K band luminosity in the arm regions.

The rotation curve of a galaxy can be obtained through long slit spectroscopy (LSS) along the major axis or using an integral field unit (IFU) which yields a full velocity map. It is simpler to used emission lines to measure the velocity field, however, since they measure the gas kinematics corrections for possible effects due to shocks, streaming motions and differences in velocity dispersion compared to the stellar component must be applied. A safer approach is to measure stellar absorption lines (e.g. MgI at 518 nm or CaII at 854 nm). They still have to be corrected for velocity perturbations in the disk (e.g. spiral or bar modes) before the average potential can be derived. Systematic effects due to attenuation by dust [1] and asymmetric, non-Gaussian velocity profiles [18] should also be considered.

It is also important to choose the region suspected to exhibit chaotic behavior carefully including its position relative to its parent galaxy. If it is close to the major axis of the galaxy, it is necessary to subtract the velocity component due to the general rotation of the galaxy. At the minor axis, a contribution to the LOSVP from the bulge may be significant. In all cases, the integration over the finite thickness of the disk will introduce systematic effects.

Taking as an example the local stellar velocity distribution, one would have to detect velocity features with a separation of $\sim$50 km/sec and an amplitude contrast of less than 10%. This would require a signal-to-noise ratio (SNR) in the range of 20-50 depending of number of free parameters in the model.

7 Feasibility of Observing Chaos

Whereas it is trivial to obtain both deep K band surface photometry and long slit spectra for deriving a general potential model of a disk galaxy with 4m class telescopes, the observation of detailed LOSVP's with sufficient SNR and spectral resolution is significantly more challenging. To estimate the feasibility of such observations with current state-of-the-art instrumentation, the ESO Very Large Telescope (VLT) facility was taken as an example. Its four 8m unit telescopes are located at Paranal in the Atacama desert, Chile, and provide excellent conditions for this type of project. At present, four VLT instruments could be considered for obtaining LOSVP, namely:

FORS1/2 have both imaging and spectroscopic modes in the visual part of the spectrum. The maximum spectral resolution is ~1700 for long slit mode.

VIMOS is a visual multiple object spectrograph with imaging modes. It has several IFU modes including one with a field of almost 1 arcmin^2 and a spectral resolution of ~2200.

FLAMES/GIRAFFE is a multi-fiber, high resolution spectrograph for visual wavelengths. There are several small IFU's and one $7' \times 11'$ IFU head with a lower resolution of ~9000.

ISAAC is an infrared instrument with both imaging and long slit spectroscopic modes. It's higher spectral resolution in the K band is ~3000 which is just sufficient to resolve the OH lines and thereby give access to the low background inter-line regions.

The typical surface brightness at the end of a bar in a spiral galaxy is K $\approx$ 18 mag/arcsec^2 with a color index (I-K) $\approx$ 2 mag. Strong spiral arms are on average at least 1 mag. fainter while inter-arm regions typically are 1 mag. fainter than the arms. To estimate the feasibility of observing chaos with the VLT, the ESO Exposure Time Calculator(version 2.8.3) [2] was used assuming a surface brightness in K = 18 mag/arcsec^2 or I = 20 mag/arcsec^2 with a seeing of $0.8''$ and an airmass of 1.2. The results are given in Table 1 where spectral resolution and typical SNR are listed for a 1 hour exposure. The SNR estimate for ISAAC may vary significantly depending on the exact location of the lines to be measured relative to OH lines.

By averaging over several spectral and spatial channels, a somewhat higher SNR can be obtained. Even so, it is clear from the estimates in Table 1 that it is just feasible to obtain an acceptable SNR and spectral resolution at the end of the bar while regions in the spiral structure would require even larger facilities.

8 Conclusions

Studies of the past star formation rate and the morphology of galaxies in the HDF's indicate that a majority of local disk galaxies was formed at least 5 Gyr ago and therefore has had time enough to reach a relaxed, quasi-stable state. There are no evidence that bar and spiral structure depend on the environment

Table 1. Performance of VLT instruments in spectroscopy mode for a 1 hour exposure with 0.8″ seeing. The wavelength ($\Delta\lambda$) and velocity (Δv) resolution per detector pixel together with the Point Spread Function (PSF) are given for each configuration. Finally, the number of electron from the source and the corresponding SNR are listed as calculated by the ESO Exposure Time Calculator for an extended source with a surface brightness of I = 20 mag/arcsec^2.

			$\Delta\lambda$	Δv	PSF	Source	
Instrument	Mode		(nm/pix)	(km/s/pix)	(pix)	(e^-)	SNR
FORS	LSS	600I	0.132	50	4	329	27
	LSS	1200R	0.075	34	4	274	26
	LSS	1400V	0.063	36	4	154	15
ISAAC	LSS	MR	0.121	16	7	195	3:
VIMOS	IFU	R2150	0.061	29	5	248	10
	IFU	R1000	0.273	116	5	1531	31
GIRAFFE	IFU	LR04	0.020	11	3	62	5

of the parent galaxy. Thus, chaotic behavior observed in disk galaxies is likely to have an internal dynamic origin if there is no evidence of recent mergers.

The most likely regions to find chaos in spiral galaxies are the major stellar resonances in the disk especially CR. If a galaxy hosts several spiral modes with different angular speeds (e.g. a fast bar and a slower rotating spiral), one would expect increased chaotic behavior in the interface region between them.

The analysis of the stellar velocity distribution in the solar neighborhood show that some streams (e.g. Hercules) may originate from a population of stars with chaotic orbits. Even with access to high quality data for individual stars, the interpretation is ambiguous and rely on detailed comparisons with dynamic models.

For external galaxies where only integrated properties can be observed, it is possible to identify regions where chaotic behavior may be expected but it is essential to compare detailed LOSVP with dynamic model to access the probability of chaos. Candidate regions are major resonances and interface zones between different spiral modes. It is important to consider possible contamination of measured velocity profiles by galactic rotation, disk thickness and attenuation by dust.

Whereas surface photometry and basic kinematics data can be obtained with 4m class telescopes, velocity profiles with sufficient spectral resolution and SNR are much more demanding. With current state-of-the-art facilities like VLT, it is just feasible to access a surface brightness of I $\approx$ 20 mag/arcsec^2 corresponding to the end of the bar in a typical disk galaxy. Observation to search for chaotic behavior in the disk related to the main spiral structure would only be possible with significantly larger facilities.

References

1. M. Baes, H. Dejonghe: Mon. Not. R. Astron. Soc. **335**, 441 (2002)
2. P. Ballester, A. Disarò, A. Dorigo et al.: ESO Messenger **96**, 19 (1999)
3. Z. Bin, Y. Zeng-yuan: Appl. Math. Mech. (Eng. Ed.) **11**, 901 (1990)
4. G. Contopoulos: Astrophys. Astron. **81**, 198 (1980)
5. G. Contopoulos: Astrophys. Astron. **117**, 89 (1983)
6. G. Contopoulos, P. Grosbøl: Astrophys. Astron. **155**, 11 (1986)
7. G. Contopoulos, P. Grosbøl: Astrophys. Astron. **197**, 83 (1988)
8. S. van den Bergh: Astron. J. **124**, 782 (2002)
9. S. van den Bergh: Astron. J. **124**, 786 (2002)
10. S. van den Bergh: Pub. Astron. Soc. Pacific **114**, 797 (2002)
11. W. Dehnen: Astron. J. **119**, 800 (2000)
12. R. Fux: Astrophys. Astron. **373**, 511 (2001)
13. P. Grosbøl: Pub. Astron. Soc. Pacific **105**, 651 (1993)
14. P. Grosbøl, E. Pompei, P.A. Patsis: ASP Conf. Ser. **275**, eds. E.Athanassoula, A.Bosma, R.Mujica, 305 (Astron. Soc. Pacific, San Francisco 2002)
15. D.E. Kaufmann, G. Contopoulos: Astrophys. Astron. **309**, 381 (1996)
16. W. Keel, W. Wu: Astron. J. **110**, 129 (1995)
17. S.M. Kent: Astron. J. **91**, 1301 (1986)
18. K. Kuijken, M.R. Merrifield: Mon. Not. R. Astron. Soc. **264**, 712 (1993)
19. K.M. Lanzetta, N. Yahata, S. Pascarelle et al.: Astrophys. J. **570**, 492 (2002)
20. C.C. Lin, F.H. Shu: Astrophys. J. **140**, 646 (1964)
21. P. Madau, L. Pozzetti, M. Dickinson: Astrophys. J. **498**, 106 (1998)
22. F. Masset, M. Tagger: Astrophys. Astron. **322**, 442 (1997)
23. H.-W. Rix, M.J. Rieke: Astrophys. J. **418**, 123 (1993)
24. A. Sandage, G.A. Tammann: *A Revised Shapley-Ames Catalog of Bright Galaxies* (Carnegie Inst., Washington 1981)
25. J.A. Sellwood: Mon. Not. R. Astron. Soc. **217**, 127 (1985)
26. J.A. Sellwood, L.S. Sparke: Mon. Not. R. Astron. Soc. **231**, 25P (1988)
27. J.L. Sérsic: *Atlas de galaxias australes* (Obs. Astron. de Cordoba, Cordoba 1968)
28. M. Tagger, J.F. Sygnet, E. Athanassoula, R. Pellat: Astrophys. J. **318**, L43 (1987)
29. N. Voglis, G. Contopoulos, C. Efthymiopoulos: Cel. Mech. Dyn. Astron. **73**, 211 (1999)

Observational Determination of the Gravitational Potential and Pattern Speed in Strongly Barred Galaxies

Per A.B. Lindblad and Per Olof Lindblad

Stockholm Observatory, AlbaNova, SE–10691 Stockholm, Sweden

Abstract. In order to compute stellar orbits in spiral galaxies the gravitational potential and its pattern speed must be known. Observationally, these parameters are difficult to determine, in particular for strongly barred galaxies. We will briefly review different methods and illustrate in more detail the case where the problem has been approached by numerical gasdynamical simulations.

1 Introduction

When computing orbits in real galaxies one needs to know the gravitational potential and its pattern speed, i.e. the angular velocity of its non-axisymmetric component. In the quasi-steady density wave picture of galaxy dynamics these two quantities are considered to be more or less constant throughout the system and over a certain period of time.

To derive the gravitational potential from photometry requires very accurate multicolour photometry to very faint levels with corrections for extinction and assuming mass-luminosity ratios for a mixture of stellar populations. To this must be added the potential of an unknown amount of dark matter. Where the symmetry plane of the galaxy is suitably inclined to the plane of the sky, radial velocities will give information of the kinematics in the plane of the galaxy, complementing the information from photometry.

The axisymmetric part of the potential in the plane of the galaxy is generally described by the rotation curve, whereby we mean the set of circular orbital velocities as a function of the distance from the centre, given by the axisymmetric Fourier component of the density distribution.

The interplay between the orbital motions and the pattern speed gives rise to various resonance phenomena, of which we in particular note the corotation resonance (CR) and the Inner (ILR) and Outer (OLR) Lindblad Resonances. The CR occurs where the orbital angular velocity is close to the pattern speed. Any regular non-circular particle orbit can be described as a closed orbit with n pericentra, rotating with a certain angular velocity. The rotation can be slower than the mean orbital velocity, so that the particle describes the rotating closed orbit mainly in the forward direction, or faster, so that the particle describes the closed orbit mainly in the retrograde direction. Where the rotation of the orbit, for $n = 2$, is close to the pattern speed, we have an ILR in the former case and an OLR in the latter. For one and the same pattern ILR and OLR fall on different sides of CR. There remains the possibility that one and the same system may contain different structural components each with its own pattern speed [10].

In the case of a weak bar with small deviations from circular motion we have a circular corotation region, where the circular angular velocity is close to the pattern speed. This case is well demonstrated by England, Hunter and Contopoulos [2] by hydrodynamical model computations of motions in various bar potentials. Figure 1 shows the case of a rather weak bar perturbation. To the left is shown the gas density response pattern when the gas has been settled into a quasi-steady state, and to the right the gas velocity vectors in a frame rotating with the pattern.

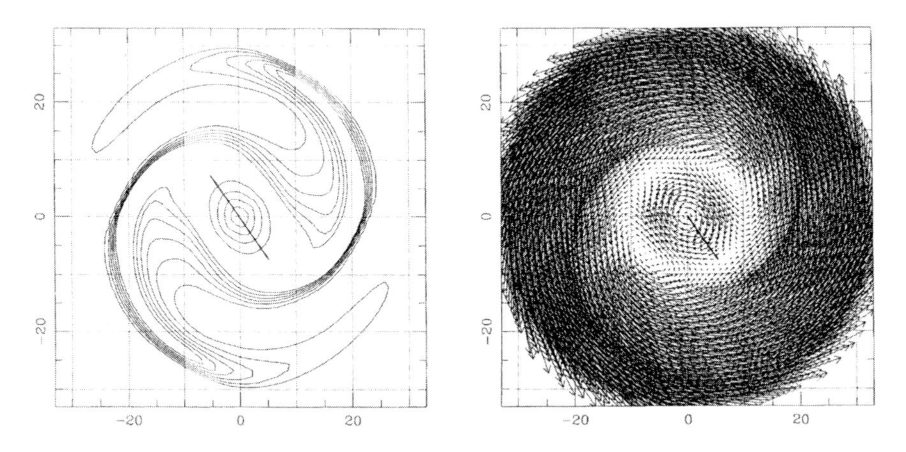

Fig. 1. Gas density response (left) and velocity vectors (right) of a rotating bar perturbation of moderate strength. The system rotates clock-wise. A straight line in the figures shows the position and extent of the bar. From [2]

The circular corotation region of low velocities is clearly seen in this figure. It contains two vortex regions, corresponding to the Lagrangian L4 and L5 equilibrium points of the restricted 3-body problem. In these regions the gas rotates counter clock-wise and expands from the vortex centre.

The case of a very strong bar, as shown in Fig. 2, however, is different. In the bar region there is no continuous set of nearly circular orbits and no circle of corotation resonance. Here the corotation region has broken down. There are still the two vortices placed at right angels to the bar, now very pronounced. In addition, there are two vortices close to the gas density maxima in the bar, around which the gas now flows in the same direction as the bar rotation and inwards towards the vortex centre. There also seem to be low velocity corotation regions outside the ends of the bar. The entire bar is close to corotating.

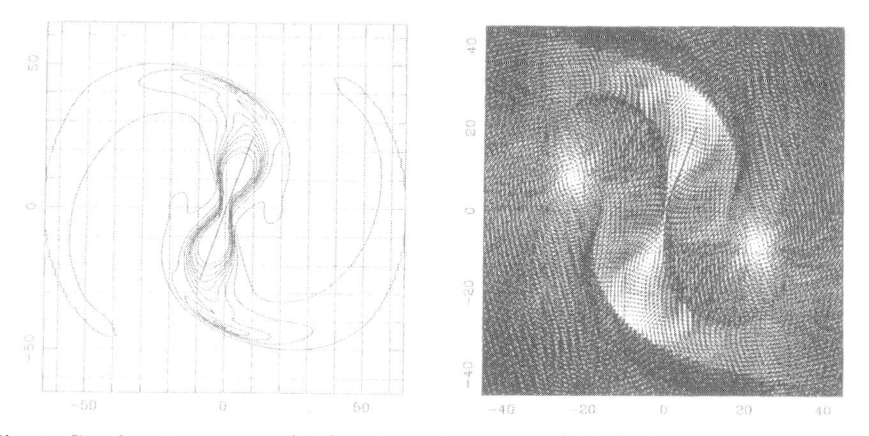

Fig. 2. Gas density response (left) and velocity vectors (right) of a rotating very strong bar perturbation. The system rotates clock-wise. A straight line in the figures shows the position and extent of the bar. From [2]

To derive a rotation curve in the bar region from the observed radial velocities in such a galaxy would be a very difficult task and its usefulness might be questioned.

2 Methods to Determine the Pattern Speed

A sophisticated scheme to derive the pattern speed for a non-circular-symmetric pattern in open spiral and barred galaxies was suggested by Tremaine and Weinberg [11]. The scheme is based on the continuity equation and is rather model independent. It is assumed that the disk has a well defined pattern speed, that the surface brightness of the tracer obeys the continuity equation, and that there is no streaming velocity normal to the disk plane.

The continuity equation is integrated over a strip parallel to the apparent major axis of the system. The gain and loss of matter across the strip, due to the rotation of the pattern, is related to the radial velocities and luminosity distribution along the strip.

Figure 3 illustrates the version of this method designed by Merrifield and Kuijken [9], here applied to the early type SB galaxy NGC 4596 [3]. The spectra are added all along the slit, and the right side shows the Doppler broadening function of this single absorption line spectrum determined by means of a templet star. This is done for three different slit offsets along the minor axis. On the bottom we see the continuum luminosity distribution along the same slits. Each pair gives a point in the diagram. The slope of the line is proportional to the pattern speed multiplied with the sin i of the inclination of the plane of the galaxy to the plane of the sky. This can be extended to any number of slits.

The method has been applied to a handful of early type barred galaxies. Its potential usefulness, however, for H I observations in barred galaxies is doubtful. Figure 4 shows an optical image as well as the total neutral hydrogen map of the

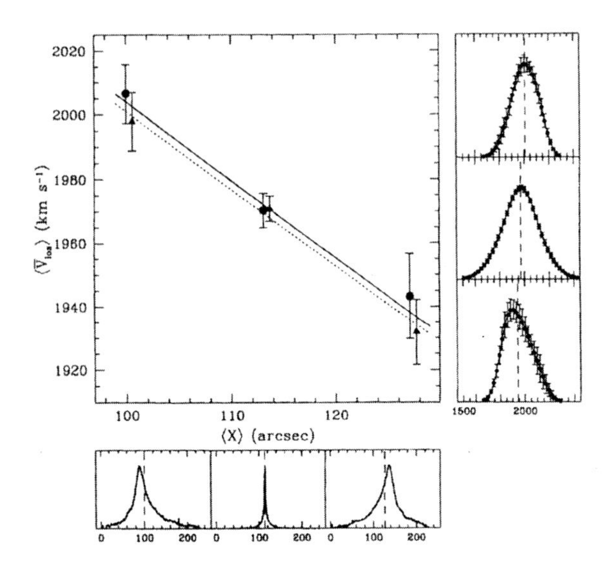

Fig. 3. Mean line-of-sight velocities versus luminosity centroid position for three slits parallel to the major axis of NGC 4596. From [3]

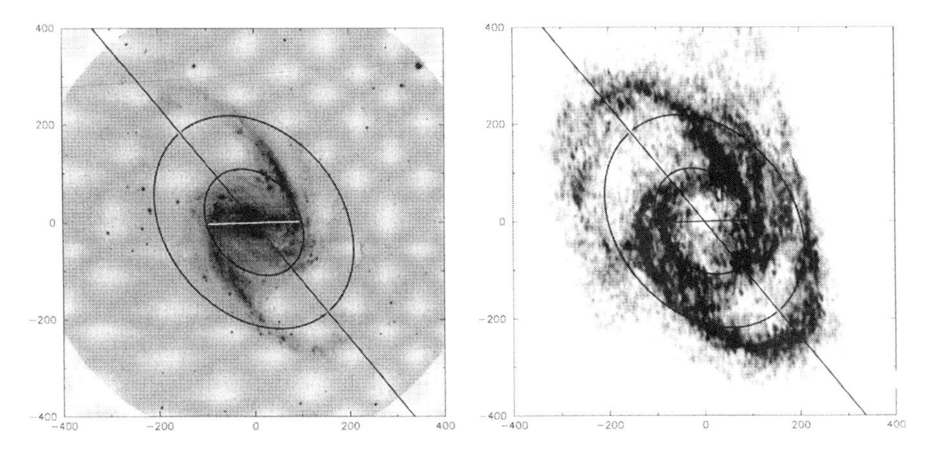

Fig. 4. An optical ESO 3.6 m prime focus plate (left) and total H I column density map (right) of NGC 1365. The ellipses enclose the region where the rotation curve is believed to give a reliable description of the axisymmetric forces. The bar major axis and the line of nodes are marked as straight lines running through the centre. From [5]

barred galaxy NGC 1365. Typically, H I disappears in the bar region, where the deviation from circular symmetry is the largest and the method is most useful. The continuity equation is not valid. In the spiral arms H I is forming stars and along the bar H I is transferred to molecular hydrogen streaming towards the centre. Our efforts have shown that the method cannot be meaningfully applied to the H I data for NGC 1365.

Several authors have tried to identify the positions of a corotation circle, or other specific resonances, in real galaxies and from the rotation curve get the pattern speed. In some cases this procedure seems to start from assumptions that should be proven in the end.

A method, based on the linear density wave theory, has been suggested by Canzian [1]. According to this theory the residual radial velocities along the line of sight, when the rotation curve has been subtracted, should be proportional to $\sin \Phi$ close to the ILR, where Φ is the central angle in the plane of the galaxy, and proportional to $\sin 3\Phi$ close to the OLR. At corotation the amplitudes of the two functions should have a specific ratio, dependent of the shape of the rotation curve. The method is very sensitive to the proper derivation of the rotation curve and the residual velocities. Being based on the linear density wave theory it should not be applicable to barred spirals.

If there is no easy way out, the ultimate method is to make a simulation of the entire galaxy, taking into consideration all available observational information. Several authors have done this in different ways. In the course of such a simulation procedure one should get a rotation curve, perturbing potential and pattern speed, consistent with the observed velocity field and giving resonance regions compatible with the observed morphology.

3 The Case of NGC 1365

To illustrate the procedure, let us consider the case of NGC 1365 which was simulated by Per A.B. Lindblad in a project at Stockholm Observatory, in which we collaborated with E. Athanassoula in Marseille [5]. A similar analysis has been performed, among others, by Weiner, Sellwood, Williams, and van Gorkom in the case of the SB galaxy NGC 4123 [13] [12], using the same gasdynamical code.

NGC 1365 (Fig. 4) is one of the more thoroughly studied nearby isolated barred galaxies [6]. Its inclination to the plane of the sky of $40°$ is suitable for radial velocity studies of the kinematics, and the inclination of the bar to the line of nodes close to optimal for a study of streaming both across and along the bar. The distance is 18 Mpc, which gives a scale where $1''$ corresponds to 100 pc. With a diameter of $11'$, or 66 kpc, it is a supergiant galaxy.

Detailed VLA observations in H I have been presented by Jörsäter and van Moorsel [4]. The total H I density map is given in Fig. 4. As was mentioned, H I is very scarce in the bar region. In the very nucleus H I is seen in absorption.

However, the central region was filled in with velocities from long slit emission line spectra, and a complete radial velocity map for the interstellar gas was constructed [7]. In Fig. 5 we see the characteristic twist along the bar and wiggles along spiral arms.

In Figs. 4 and 5 the ellipses separate the *bar region*, *intermediate region* and *outer region*. As the basis for the rotation curve we adopt the azimuthally averaged rotation curve of Jörsäter and van Moorsel. However, in the bar region

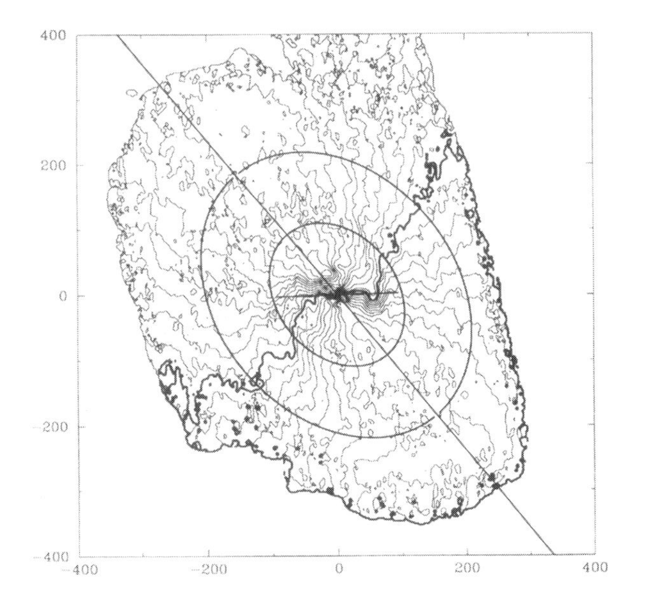

Fig. 5. The observed radial velocity field of NGC 1365. The contour interval is 20 km/s, and the systemic velocity is drawn as a thick line. The ellipses are the same as in Fig. 4. From [5]

such an azimuthally averaging method does not give a good approximation to the rotation curve, and in the outer region the system is warped, as seen from the kinematics in Fig. 5, which again makes the rotation curve uncertain.

To estimate the perturbing potential, we use infrared photometry, as the bar region shows a multitude of dust. We choose an infrared J-band image obtained as part of the Ohio-State University Bright Galaxy Imaging Survey (Fig. 6). This image was analysed in terms of even azimuthal Fourier components. As the spiral arms seem firmly attached to the ends of the bar, we assume that the spiral part of the structure, at least for a considerable time, has the same pattern speed as the bar.

At the start of the fitting procedure we let the axisymmetric part of the potential be represented by the Jörsäter–van Moorsel rotation curve. The perturbing potential is derived from the infrared surface photometry, where the mass/luminosity ratio M/L (in arbitrary units, as the photometry is not absolutely calibrated) is kept as a free parameter. This free parameter also compensates for effects of the unknown thickness of the bar.

The model fit in the intermediate region is rather insensitive to the exact shape of the rotation curve in the bar region. Thus, we can now make simulations with a sequence of different pattern speeds and M/L ratios to get the best fit to the structure in the *intermediate region*. The detailed structure in this region is particularly sensitive to the choice of pattern speed, and this speed can now be fixed to within a few km s^{-1} kpc^{-1}, at a value of 18 km s^{-1} kpc^{-1} for NGC 1365.

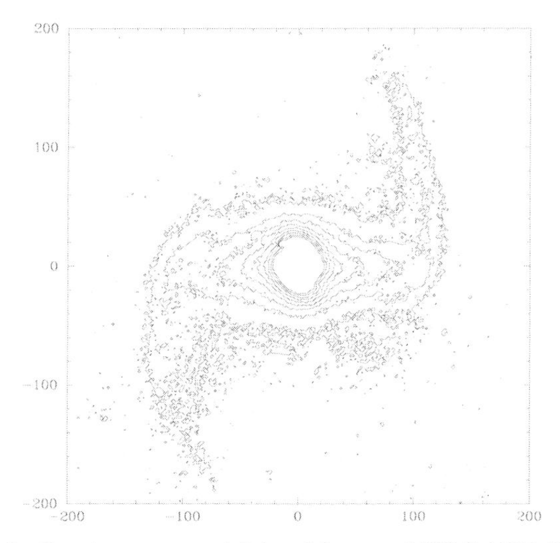

Fig. 6. Inclination corrected J-band image of NGC 1365. From [5]

With the chosen value for the pattern speed we now adjust the Jörsäter–van Moorsel rotation curve in the *bar region* as well as the final choice of the perturbing M/L ratio, until the simulation reproduces the observed radial velocities from slit spectra as well as the morphology of the dust lanes along the bar, both of which lay constraints on the position of the ILR.

The result is seen in Fig. 7, where density contours of the model are overlaid the H I total column density map. The slight mismatch in the outer region is due to the steep decline of the Jörsäter–van Moorsel rotation curve. If raised about 10 km/s in the outer region, the OLR and spiral arms move outward (i.e. a less drastic warp is assumed) and the match is improved.

Figure 8 compares the observed radial velocity field with that given by the model as observed from the same angle. The forced bisymmetry of the model limits the fit in both Figs. 7 and 8. In spite of this, the main features are reproduced fairly well. The velocity pattern in Fig. 8 shows the characteristics of orbits elongated along the bar. The shocks along the bar are smoothed in frame (**a**) due to the procedure with which the velocity field was constructed from randomly positioned slits. Thus, here a comparison should be made directly with slit measurements. Figure 9 shows this for a slit placed perpendicular to the bar 29″ East of the nucleus. The jump across the shock of 300 km/s is well reproduced.

Thus, we arrive at a possible variation of the axisymmetric forces and a pattern speed which are in agreement with the observed morphology and velocity field. In contrast, Weiner et al. [12] through accurate photometry and by the perturbations required, assuming the same M/L for disk and bar, derive by simulations similar to ours this M/L and deduce the mass of the halo required to reproduce the outer rotation curve. This means that they use the perturbations from the bar to infer a M/L for the disk and bar and get the mass of the dark

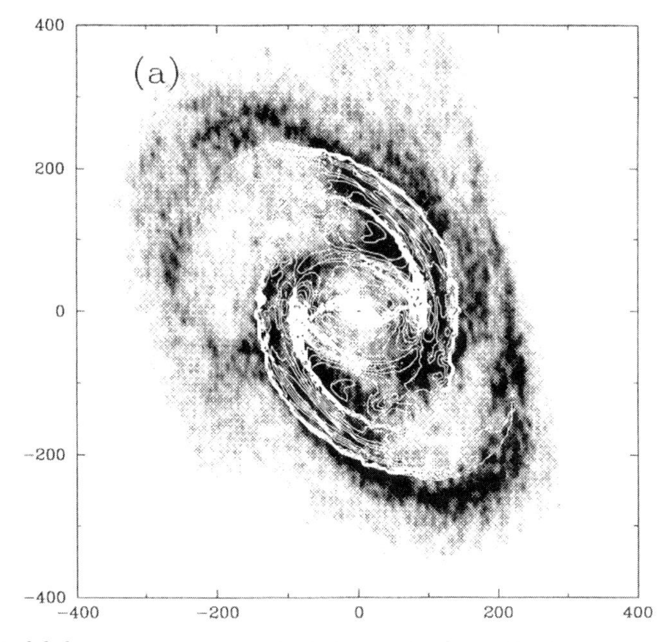

Fig. 7. Model density contours overlaid the total column H I density map. From [5]

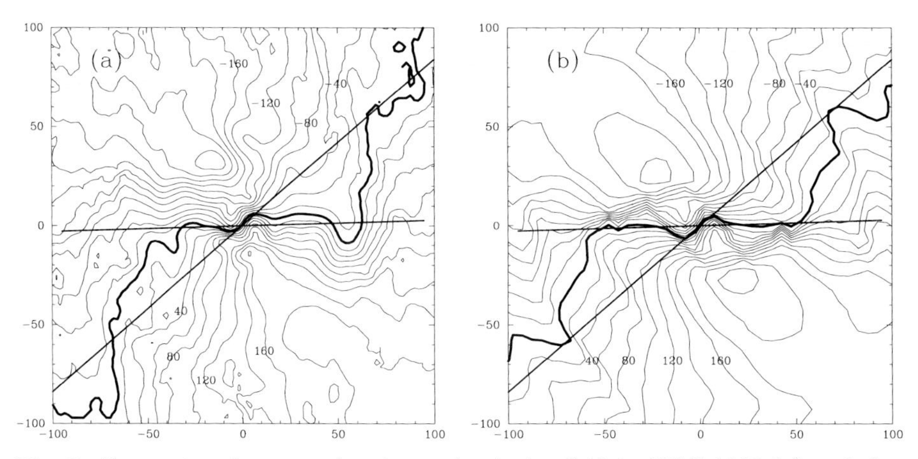

Fig. 8. Comparison between the observed velocity field in NGC 1365 (**a**) and that given by the model (**b**). The orientations of the bar axis in PA 92° and the minor axis of the galaxy in PA 130° are shown as straight lines. From [5]

halo component. For NGC 1365 this would not be safe due to the warp, which makes the outer rotation curve uncertain.

The gas flow in a frame rotating with the pattern is shown in Fig. 10, where the circles mark the ILR, CR and OLR resonance positions. Only half of the velocity field is shown. The orbits show the familiar twist around the ILR [8] as well as the change of direction of the flow at the shock fronts along the bar. The

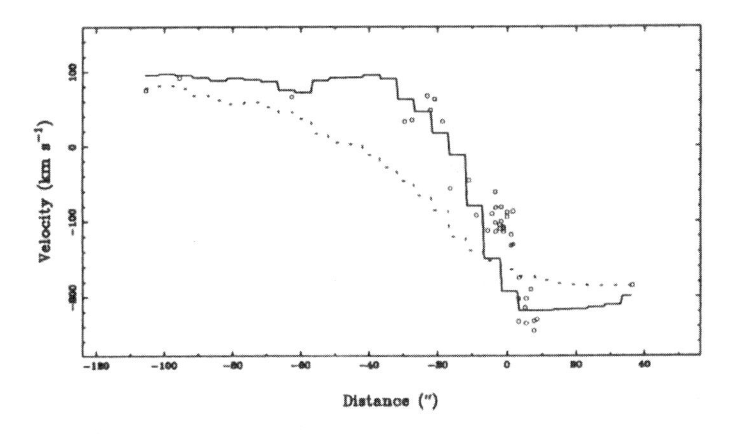

Fig. 9. Radial velocities along a slit placed perpendicular to the bar 29″ East of the nucleus. Open circles: observed velocities. Solid line: model velocities. Dotted line: pure rotational motion according to the rotation curve. From [5]

spiral arms extending from the ends of the bar appear also for a purely barred perturbation, but the spiral part of the potential is necessary to drive the arms through corotation. The vortex regions at CR are apparent.

Thus, through detailed comparison between observations and models, estimates of the pattern speed as well as total gravitational potential can be obtained for individual galaxies. This will permit us to compute stellar orbits in rather realistic, but in our case bi-symmetrical, galaxy potentials.

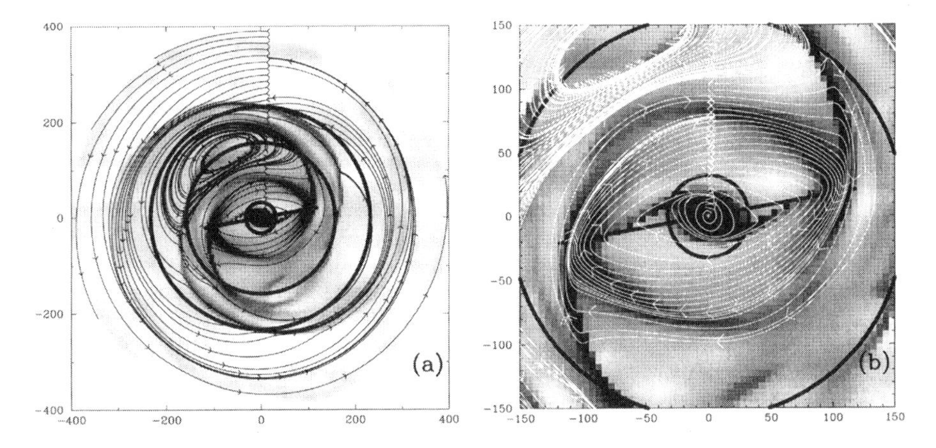

Fig. 10. Gas flow lines from the model overlaid gray scale maps of the same model. Full scale (**a**), bar region (**b**). From [5]

References

1. B. Canzian: Astrophys. J. **414**, 487 (1993)
2. M.N. England, J.H. Hunter Jr., G. Contopoulos: Astrophys. J. **540**, 154 (2000)
3. J. Gerssen, K. Kuiken, M.R. Merrifield: Month. Not. Roy. Astr. Soc. **306**, 926 (1999)
4. S. Jörsäter, G. van Moorsel: Astron. J. **110**, 2037 (1995)
5. P.A.B. Lindblad, P.O. Lindblad, E. Athanassoula: Astron. Astrophys. **313**, 65 (1996)
6. P.O. Lindblad: Astron. Astrophys. Rev. **9**, 221 (1999)
7. P.O. Lindblad, M. Hjelm, J. Högbom, S. Jörsäter, P.A.B. Lindblad, M. Santos-Lleó: Astron. Astrophys. Suppl. **120**, 403 (1996)
8. P.O. Lindblad, P.A.B. Lindblad: Publ. Astr. Soc. Pacific Conf. **66**, 29 (1994)
9. M.R. Merrifield, K. Kuijken: Month. Not. Roy. Astr. Soc. **274**, 933 (1995)
10. J.A. Sellwood, L.S. Sparke: Month. Not. Roy. Astr. Soc. **231**, 25p (1988)
11. S. Tremaine, M.D. Weinberg: Astrophys. J. **282**, L5 (1984)
12. B.J. Weiner, J.A. Sellwood, T.B. Williams: Astrophys. J. **546**, 931 (2001)
13. B.J. Weiner, T.B. Williams, J.H. van Gorkom, J.A. Sellwood: Astrophys. J. **546**, 916 (2001)

Observational Manifestation of Chaos in Grand Design Spiral Galaxies

Alexei M. Fridman[1,2], Roald Z. Sagdeev[3],
Oleg V. Khoruzii[1,4], and Evgenii V. Polyachenko[1]

[1] Institute of Astronomy RAS, 48 Pyatnitskaya st., Moscow 117647, Russia
[2] Sternberg Astronomical Institute, Moscow State University, University prospect, 13, Moscow, 119899, Russia
[3] University of Maryland, College Park, MD. 20742, USA
[4] Troitsk Institute for Innovation and Thermonuclear Research, National Research Center of Russian Federation, Troitsk, Moscow Reg. 142092, Russia

Abstract. To study dynamic properties of the gaseous disk of the grand design spiral galaxy NGC 3631 we calculate the Lyapunov characteristic numbers (LCN) for different families of streamlines in the disk. For the trajectories near separatrices of the giant vortices and near saddle points presenting in the velocity field, the LCN turned out to be positive. The result is insensitive to the method of the calculation. Both methods — using two trajectories and based on linearized equations — give the identical results. The values of the LCN in the gaseous disk of NGC 3631 are independent on the Riemannian metric used for the calculations in agreement with the classical mathematical theorem. The spectra of the 'short-time' LCN (stretching numbers) evaluated for the same trajectories turned out to be non-invariant. We confirmed this result obtained for the real galactic disk on classical model examples.

1 Introduction

The main topic of our paper consists in the demonstration of a fact that results of analysis of the observed velocity field of the galactic disk can serve as a source of our knowledge of the stochastic galactic dynamics.

This topic has relation to a gaseous disk rather than to stellar one. In spite of the evident recent progress in the measurements of the line-of-sight velocity field of stellar disks, our knowledge of stochastic stellar dynamics of external galaxies is based for the most part on theoretical investigations. Information in this respect from observational data on stellar velocity fields is moderate for the following reason.

As a rule, the external galaxies are not resolved into individual stars. Using the analysis of absorption lines we can measure the line-of-sight velocity field of the stellar population averaged over some local spatial region. The trajectory of this region of the stellar disk may differ qualitatively from the trajectories of the stars in the same region. For example, a stellar bar rotates as a solid body, while the stellar orbits in the bar may be complicated and far from a simple rotation. Such a behaviour is typical for collisionless selfgravitating systems.

From observational data we can construct gravitational potential as a function of coordinates. But some variations of the potential within errors of observations often result in the transformation of regular stellar orbits into chaotic ones and vice versa. This may lead to artefacts.

At first glance the line-of-sight velocity field of a gaseous disk also can not be used directly to study the stochastic dynamics of the disk. However, the use of the method of restoration of 3D velocity field from the observed line-of-sight velocity field [1], [2] enables to determine regular and chaotic trajectories by the calculation of the Lyapunov characteristic numbers (LCN) [3].

In this case a natural question may appear. It is well known (see, e.g. [4]), that the LCN is calculated for the trajectories in the phase space, while the restoration method [1], [2] gives 3D velocity field in the coordinate space[1]. Hence it allows to see the behaviour of the trajectories in the coordinate space rather than in the phase one. The book [5] may help to resolve this question. In this book the hydrodynamical equations of 3D stationary incompressible flows are reduced to nonstationary dynamical equations in 2D phase space. In other words, in this book it is shown, that the problem of analysis of the properties of trajectories in 3D stationary incompressible flows is equivalent to the problem for nonstationary dynamical systems with 2D phase space. These systems, evidently, can demonstrate both regular and chaotic motions. The same is done in [6] for compressible 3D stationary flows. The idea of the reduction is the following.

A steady-state 3D flow is described by the set of equations:

$$\frac{\mathrm{d}x}{\mathrm{d}t} = v_x(x,\, y,\, z)\,, \quad \frac{\mathrm{d}y}{\mathrm{d}t} = v_y(x,\, y,\, z)\,, \quad \frac{\mathrm{d}z}{\mathrm{d}t} = v_z(x,\, y,\, z)\,, \tag{1}$$

which can be rewritten in the following form:

$$\frac{\mathrm{d}x}{v_x} = \frac{\mathrm{d}y}{v_y} = \frac{\mathrm{d}z}{v_z} = \mathrm{d}t\,. \tag{2}$$

For our aim, a more convenient notation is

$$\frac{\mathrm{d}x}{\mathrm{d}z} = \frac{v_x}{v_z} \equiv f_1(x,\, y,\, z)\,, \quad \frac{\mathrm{d}y}{\mathrm{d}z} = \frac{v_y}{v_z} \equiv f_2(x,\, y,\, z)\,. \tag{3}$$

The latter equations show that we are dealing with the "nonstationary" problem for a dynamical system in 2D phase space $(x,\, y)$. The variable z is playing the role of time τ:

$$\frac{\mathrm{d}x}{\mathrm{d}\tau} = f_1(x,\, y,\, \tau)\,, \quad \frac{\mathrm{d}y}{\mathrm{d}\tau} = f_2(x,\, y,\, \tau)\,. \tag{4}$$

These equations describe the dynamical systems where stable and unstable trajectories may coexist. As it follows from above (and [5], [6]), that corresponds to the coexistence of stable and unstable streamlines in 3D coordinate space.

[1] In more details see [1]–[3].

2 Restored Velocity Field of the Grand Design Spiral Galaxy NGC 3631

In Fig. 1 one can see the reconstructed velocity field of NGC 3631[2] in the galactic plane with superimposed lines of constant phase of the vertical (perpendicular to the galactic plane) velocity. Squares mark the maxima of the absolute values of the vertical velocity of gas at each radius. Asterisks show the locations of the zeros of the vertical velocities. The vertical velocity amplitude is not shown. Thin lines mark the location of the vortices — anticyclones (upper left and lower right) and cyclones.

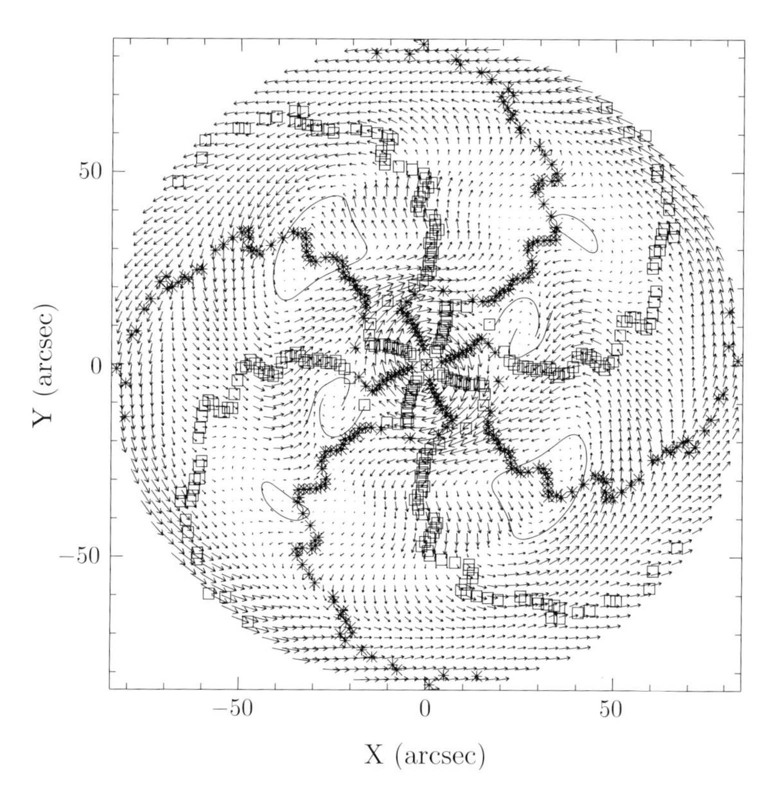

Fig. 1. The reconstructed velocity field of NGC 3631 in the galactic plane and superimposed lines of constant phases of the vertical motions. Squares mark the maxima of the absolute values of the vertical velocity of gas at each radius. Asterisks show the locations of the zeros of the vertical velocities. Thin lines show the location of the vortices — anticyclones (upper left and lower right) and cyclones.

[2] A solution of the ill-posed problem of the reconstruction of three component velocity field from the observed line-of-sight velocity field of gaseous disk of grand design spiral galaxies is described in papers [1], [2], [7], [8] and in the review [9].

From observations it follows that gaseous disks of spiral galaxies – the set of their main parameters – lie close to the boundary of their dynamic instability [10]. That might seem natural: in the course of developing the instability, the velocity dispersion grows, and the disk approaches the boundary of the instability [9]. Since the instability generating both the spiral arms and vortices is saturated, then the 3D motion of the gas should be quasi-stationary in the reference frame corotating with the density wave. It means that trajectories coincide with streamlines in 3D coordinate space. In the regions close to those where V_z equals to zero, every fluid particle participates in 2D motion only and hence its trajectory should coincide with its 2D streamlines.

If vortex lie in the regions of the 2D motion, the fluid particles trapped in the vortex are separated from the transiting (untrapped) ones by the separatrix – the last closed streamline around the vortex center. As we can see in Fig. 1, the vortices with centers close to the zeros of the vertical motions ($V_z = 0$) are surrounded by closed separatrices. Two cyclones located far from the zeros do not demonstrate the presence of a clear separatrix in the 2D streamlines. These facts can be considered as an evidence of the real three-dimensionality of the velocity field of the gaseous disk. The whole structure of the reconstructed velocity field in Fig. 1 agrees with the assumption of its quasi-stationarity.

Besides the separatrices surrounding vortices the 2D velocity field in the disk plane contains saddle points (marked by crosses in Fig. 2). Choosing the beginning of streamlines near separatrices or near the saddle points, one can see that these streamlines diverge. We would like to know, if this divergence is exponential or not, in other words, if the streamlines are chaotic or regular. To clarify it we need to calculate the LCN.

3 The Calculation of the Lyapunov Characteristic Numbers

In the case of exponential divergence of the trajectories we have

$$d(t) \sim d_0 e^{\lambda t}\,, \tag{5}$$

where d_0 is the initial separation between neighbouring trajectories, $d(t)$ is the separation for the time t, λ is a rate of the exponential divergence and is equal to the maximum LCN.

The rigorous definition of the LCN is [4]

$$\lambda = \lim_{t\to\infty,\, d_0\to 0} \frac{1}{t} \ln \frac{d(t)}{d_0}\,. \tag{6}$$

In our case of the gaseous disk of NGC 3631, it is difficult to use the definition of the LCN (6) for the following reasons:

1) the duration of observations is much smaller than the characteristic time λ^{-1} of the exponential divergence of two points moving along nearby trajectories;

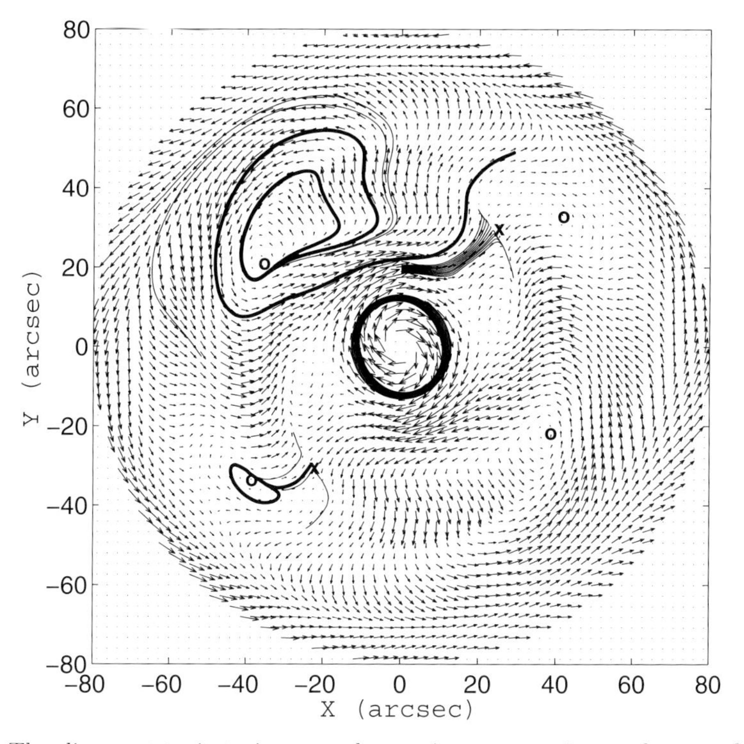

Fig. 2. The divergent trajectories near the vortices separatrices and a set of trajectories near the saddle point superimposed on the restored two dimensional velocity field. Thick solid lines show the trajectories which are used to calculate the stochastic characteristics in the vicinity of the vortices. "o"-signs mark the centers of the vortices. "x"-signs mark the saddle points. Also, non-divergent trajectories near the center of the disk are shown.

2) the presence of a minimal distance $d_{\min}$ between two trajectories owing to a finite spatial resolution of measurements δ, $d_{\min} \geq \delta$;
3) the ratio of the characteristic scale R_{ch} of the velocity field variations to the resolution δ is not too large, moreover there are some regions where $R_{ch} \simeq \delta$;
4) we cannot measure the velocity field of the overall disk but only of a part of the disk.

To overcome the first difficulty we use the mentioned above property of the stationarity of the velocity field.

The second and the third difficulties restrict an allowable maximal length of the trajectory (and thus a maximal time T of the calculation of the LCN). First, a trajectory can eventually leave the area, for which the velocity field is defined. Second, a trajectory may come to the region where the characteristic scale of the velocity field variations is of the order of the spatial resolution of the velocity data, $R_{ch} \simeq \delta$ (according to the second restriction), that contradict to the linear

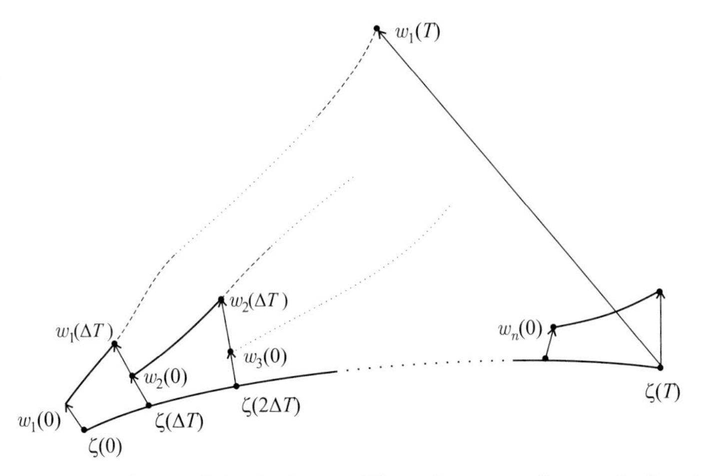

Fig. 3. The method of the LCN calculation. The reference trajectory is denoted by the longest solid line and some auxiliary trajectories by dashed curves with short regions of solid curves.

approximation condition $d/R_{ch} \ll 1$. Hence, these regions are "forbidden" for the method used to find the LCN.

In the previous paper [3] (hereafter Paper I) we tried to overcome the difficulty, connected with the limitation on the maximal length of the trajectory, using a method proposed by Casartelli *et al.* [11] and described in the well-known monograph [4] (see also [12], [13], [14]).

In our case we specify the number of steps n and divide the given time interval $[0, T]$ into n time intervals $\Delta T = T/n$. Choosing the initial deviation vector $\boldsymbol{w}_1(0)$ we evaluate two trajectories from $\boldsymbol{\zeta}_0$ and $\boldsymbol{\zeta}_0 + \boldsymbol{w}_1(0)$ and determine $\boldsymbol{w}_1(\Delta T)$ (see Fig. 3). After each step, following [11], we will renormalize the deviation vector to the initial length d_0, preserving its direction

$$\begin{aligned} \boldsymbol{\zeta}_{i+1}(0) &= \boldsymbol{\zeta}(i\Delta T) + \boldsymbol{w}_{i+1}(0), \\ \boldsymbol{w}_{i+1}(0) &= \frac{d_0}{d_i}\boldsymbol{w}_i(\Delta T), \end{aligned} \tag{7}$$

where $d_i \equiv d(\boldsymbol{w}_i(\Delta T))$.

According to [11] and [15], for sufficiently large T the reliable estimate of the LCN (6) is the following:

$$\lambda \approx \lambda^{(n)} = \frac{1}{T}\sum_{i=1}^{n} \ln \frac{d_i}{d_0}. \tag{8}$$

The described method implies integration over two trajectories — the reference trajectory, shown in Fig. 3 by the thick solid line, and an auxiliary trajectory.

The limited length of the trajectory used in the calculations poses a question on the accuracy of the LCN determination. We consider the results to be reliable when:

$$\xi \equiv \lambda^{(n)} T \gg 1 \,. \tag{9}$$

Sometimes, the trajectories are so short that the condition (9) is not fulfilled. According to [13] and [14] the reliability can be improved, if one takes a set of trajectories instead of one. In this case the LCN is calculated as follows:

$$\lambda^{(n)} = \frac{1}{N} \sum_{k=1}^{N} \lambda_k^{(n)} \,, \tag{10}$$

where $\lambda_k^{(n)}$ is calculated according to the formula (8), using the k-th trajectory as a reference one, N is the total number of the basic trajectories. The reliability condition turns into

$$\eta \equiv \xi N = \lambda^{(n)} T N \gg 1 \,. \tag{11}$$

All details of the calculations of the LCN for different families of streamlines in the gaseous disk of NGC 3631 are contained in Paper I. Our calculations of the LCN led to the conclusion, that the gaseous disk of NGC 3631 contained both the regular and chaotic streamlines. The formers are located near the disk center, the latters – in the vicinity of separatrices of vortices and near the saddle points.

To perform the calculations of the LCN one needs to define the Riemannian metric d. The general form of the metric used in the present work is

$$d(\boldsymbol{w}) = \sqrt{g_1 x^2 + g_2 y^2}, \tag{12}$$

where x and y are the Cartesian coordinates of the vector $\boldsymbol{w}$. Metrics d differ by the positive metric coefficients g_1 and g_2. For the sake of simplicity we refer to different metrics as follows: (g_1, g_2). For example, we often use metric (1,1), that implies $d = \sqrt{x^2 + y^2}$.

The Oseledec theorem [16] claims that the result of the LCN calculation should not depend on the Riemannian metric. It would be interesting to note that for the considered real system the LCN, calculated using different metrics, also deviate very little (see Fig. 4). Namely, near the separatrix of the anticyclone for the metric (1,1) $\lambda^{(n)} = 0.8768 T_0^{-1}$, for the metric (1,0) $\lambda^{(n)} = 0.8931 T_0^{-1}$, for the metric (0,1) $\lambda^{(n)} = 0.8513 T_0^{-1}$ ($T_0 = 7.5 \cdot 10^7$ years)[3].

In parallel with the use of the mentioned above technique, which employs two trajectories for evaluation of the LCN, we recalculated LCN for the same regions using the linearized equation for the deviation vector w (see [4]). It turns out that for the sufficiently smooth vector field interpolation[4] the results obtained in both cases are identical.

[3] According to the metric definition if x^2 or y^2 is positive, we should have $d > 0$. Rigorously speaking, metrics $(g_1, g_2) = (1,0)$ and $(0,1)$ do not fulfil this requirement. To overcome this obstacle one can assume that the corresponding coefficients have infinitely small positive values.

[4] Here we use fourth order tensor product spline interpolation.

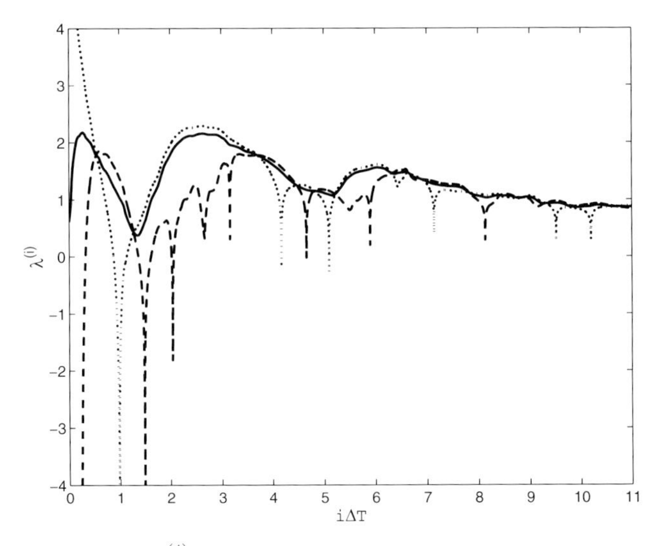

Fig. 4. The behavior of $\lambda^{(i)}$ with $i\Delta T/T_0$ for trajectory near the separatrix of the anticyclone, calculated for different metrics: (1,1) – solid line, (1,0) – dashed line, (0,1) – dotted line. Number of time steps $n = 1000$.

4 Spectrum of the Stretching Numbers

The expansion coefficient (in terminology of Oseledec [16]) or the stretching number [17] [13] [14] is defined as follows:

$$a_i = \frac{1}{\Delta T} \ln \frac{d_i(\Delta T)}{d_0} . \tag{13}$$

Using (13) one can find the spectra of the stretching numbers for real objects. According to the definition [13], [14], the spectrum of the stretching numbers is

$$S(a, x_0, y_0) = \lim_{N \to \infty} \frac{1}{N} \frac{\mathrm{d}N(a)}{\mathrm{d}a} , \tag{14}$$

where $\mathrm{d}N(a)$ is the number of appearances of the stretching number a_i in the interval $(a,\ a + da)$.

The spectra, calculated for chaotic and regular trajectories using different metrics are shown in Figs. 5–8. In all cases, the spectra are not invariant to the metric change.

This is very interesting fact, since as it follows from (8), (13) and (14) the LCN is the first moment of the spectrum of the stretching numbers

$$\lambda = \int aS(a)da, \tag{15}$$

but the LCN itself preserve the mentioned invariance.

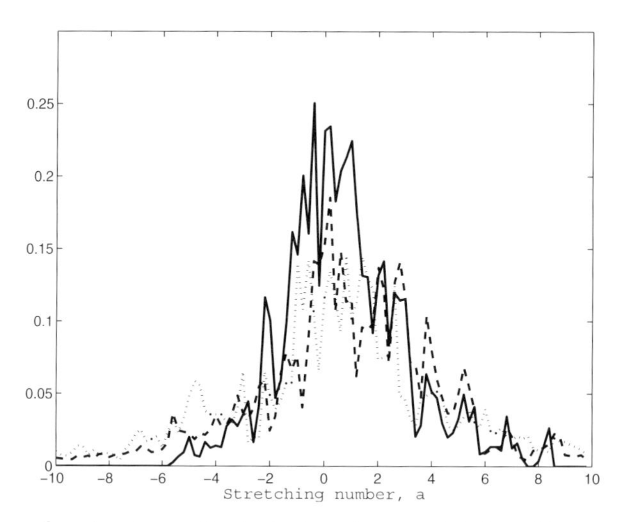

Fig. 5. Spectra for the stochastic trajectory in the vicinity of the anticyclone (see Fig. 2), calculated for different metrics: (1,1) – solid line, (1,0) – dashed line, (0,1) – dotted line. Number of time steps $n = 1000$.

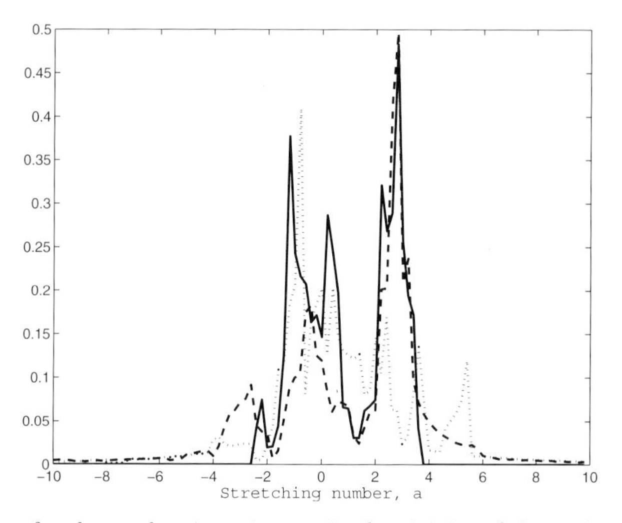

Fig. 6. Spectra for the stochastic trajectory in the vicinity of the cyclone (see Fig. 2), calculated for different metrics: (1,1) – solid line, (1,0) – dashed line, (0,1) – dotted line. Number of time steps $n = 1000$.

The dependence of the spectrum form on the metric has not been known previously, although it can be checked that it holds either in theoretical and real observable systems. Figures 9, 10 show the LCN and the spectra, calculated for different metrics for the well-known standard (Chirikov) map and the Lorenz attractor. One can see that in both cases the LCN, calculated using different metrics are equal, whereas the forms of the spectra are different.

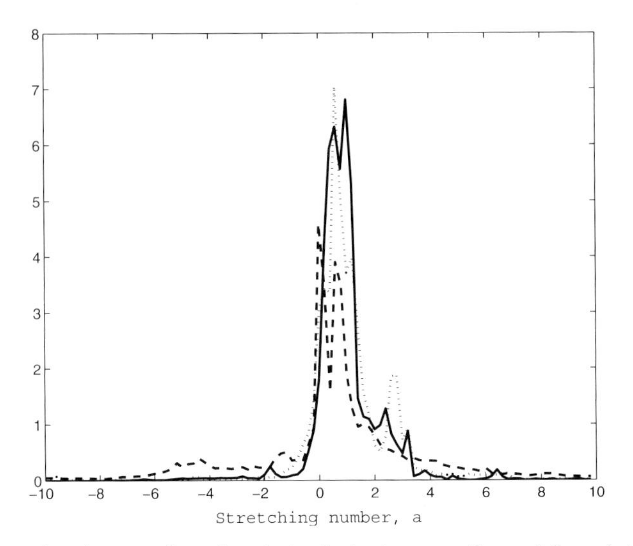

Fig. 7. Spectra for the set of stochastic trajectories near the saddle point (see Fig. 2), calculated for different metrics: (1,1) – solid line, (1,0) – dashed line, (0,1) – dotted line. Number of time steps $n = 1000$.

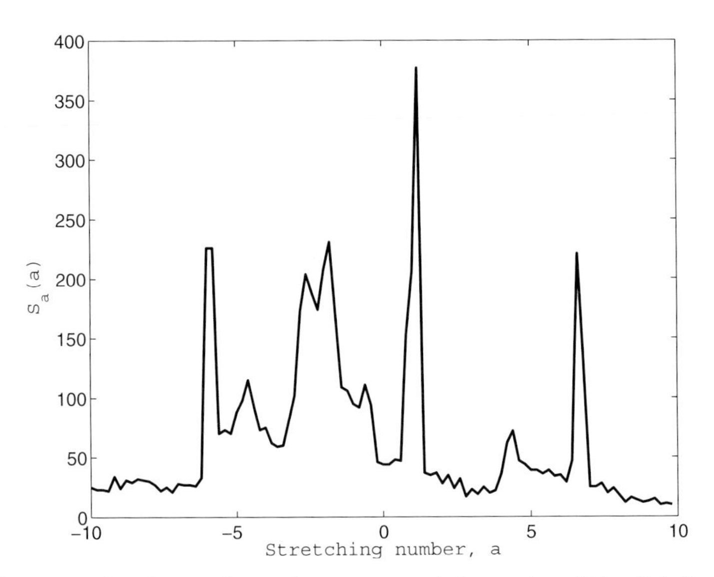

Fig. 8. Spectrum for the regular trajectory around the center of the disk (see Fig. 2), calculated for the metric (1,1).

5 Conclusions

In conclusion let us summarize the dynamical properties of the gaseous disk of NGC 3631 revealed in this paper and the Paper I.

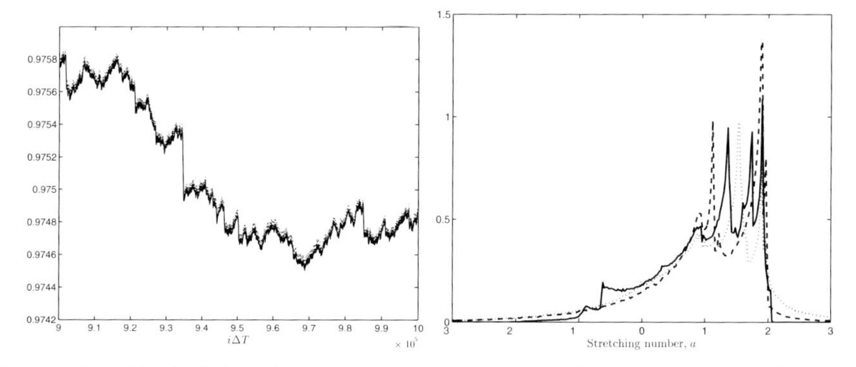

Fig. 9. The LCN (left figure) and the spectra (right figure) for the standard (Chirikov) model, calculated for different metrics: (1,1) – solid line, (0,1) – dashed line, (1,0) – dotted line.

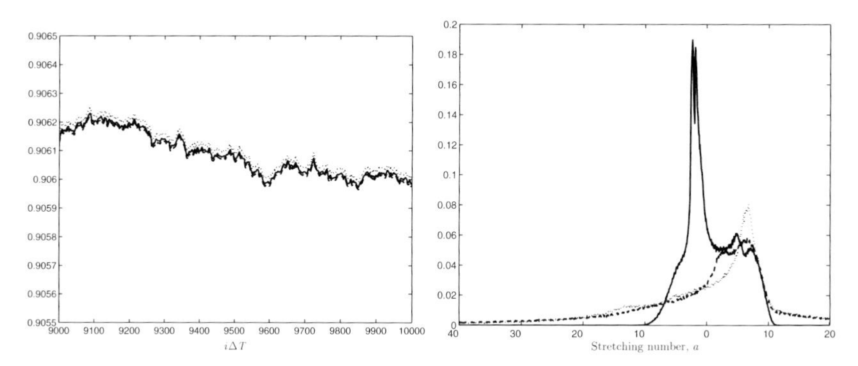

Fig. 10. The LCN (left figure) and the spectra (right figure) for the Lorenz attractor) model, calculated for different metrics: (1,1) – solid line, (0,1) – dashed line, (1,0) – dotted line.

1. The three component velocity field restored from the observed line-of-sight velocity field of the gaseous disk of the galaxy NGC 3631 is stationary and demonstrates the presence of both regular and stochastic trajectories of the gas.
2. The regular trajectories are observed near the center of the disk, the stochastic ones — near the saddle points and the separatrices of giant vortices presenting in the velocity field.
3. The type of the divergence of the trajectories was determined by two different methods of the calculation of the Lyapunov characteristic numbers (LCN): using two neighbouring trajectories and based on the linearised equations. Both methods gave identical results.
4. The LCN obtained for the real galaxy are turn out to be invariant to the metric change, in full agreement with the Oseledec mathematical theorem [16].
5. For the first time it was demonstrated, that the form of the spectra of the 'short-time' LCN (stretching numbers) varies with the metric change.

6. The correctness of two latter conclusions was confirmed also for the classical model dynamical systems — the standard map and the Lorenz attractor.

Acknowledgments

Authors are grateful to V. Arnold, B. Chirikov, G. Contopolous, Y. Fridman, V. Oseledec, M. Rabinovich, and Ja. Sinai for numerous and very fruitful discussions. This work was performed with partial financial support of RFBR grants N 02-02-16878, N 02-02-06603, "Leading Scientific Schools" grant N 00-15-96528, and "Fundamental Space Researches. Astronomy" grants: N 1.2.3.1, N 1.7.4.3.

References

1. V.V. Lyakhovich, A.M. Fridman, O.V. Khoruzhii, A.I. Pavlov: Astronomy Report **41**, 447 (1997)
2. A.M. Fridman, O.V. Khoruzhii, V.V. Lyakhovich, V.S. Avedisova, O.K. Sil'chenko, A.V. Zasov, A.S. Rastorguev, V.L. Afanasiev, S.N. Dodonov, J. Boulesteix: Astroph. and Space Sci. **252**, 115 (1997)
3. A.M. Fridman, O.V. Khoruzhii, E.V. Polyachenko: 'Observational Manifestation of Chaos in the Gaseous Disk of the Grand-Design Spiral Galaxy NGC 3631'. In: *Observational Manifestation of Chaos in Astrophysical Objects, International Workshop, Moscow, August 28-29, 2001*, ed. by A.M. Fridman, M.Ya. Marov, R.H. Miller, Space Science Rev. **102**, 51 (2002)
4. A.J. Lichtenberg, M.A. Lieberman: *Regular and Stochastic Motion* (Springer-Verlag, New York, Heidelberg, Berlin 1983)
5. G.M. Zaslavsky, R.Z. Sagdeev, D.A. Usikov, A.A. Chernikov: *Weak Chaos and Quasi-Regular Patterns* (Cambridge Univ. Press, Cambridge 1991)
6. Govorukhin et al. (1999)
7. A.M. Fridman, O.V. Khoruzhii, V.V. Lyakhovich, O.K. Sil'chenko, A.V. Zasov, V.L. Afanasiev, S.N. Dodonov, J. Boulesteix: Astron. & Astroph. **371**, 538 (2001)
8. A.M. Fridman, O.V. Khoruzhii, E.V. Polyachenko, A.V. Zasov, O.K. Sil'chenko, A.V. Moiseev, A.N. Burlak, V.L. Afanasiev, S.N. Dodonov, J.H. Knapen: Mon. Not. R. Astr. Soc. **323**, 651 (2001)
9. A.M. Fridman, O.V. Khoruzhii: Space Science Rev., accepted for publication in 2002.
10. A.V. Zasov, S. Simakov: Astrofizika **29**, 190 (1988)
11. M. Casartelli, E. Diana, L. Galgani, A. Scotti: Phys. Rev. A **13**, 1921 (1976)
12. B.V. Chirikov, F.M. Izrailev, V.A. Tayurski: Comput. Physics Commun. **5**, 11 (1973)
13. N. Voglis, G. Contopoulos: J. Phys. **A27**, 4899 (1994)
14. G. Contopoulos, N. Voglis: A&A, 317, 73 (1997)
15. G. Bennetin, L. Galgani, J.-M. Strelcyn: Phys. Rev. **A14**, 2338 (1976)
16. V.I. Oseledec: Tr. Mosk. Mat. Obsch., **19**, 179, 1968 (Trans. Mosc. Math. Soc.**19**, 197, 1968).
17. C. Froeschle', Ch. Froeschle, E. Lohinger: Cel. Mech. Dyn. Astron. **56**, 307 (1993)

Quarter-Turn Spirals in Galaxies

Evgenii Polyachenko

Institute of Astronomy RAS, 48 Pyatnitskaya st., Moscow 117647, Russia

Abstract. Observations in the optical show that grand design spirals consist of a set of principle arms and characteristic near-circular extensions that can be described as quarter-turn spirals. Arguments are presented in favor of the idea that the latter set of spirals is caused by the response of the material of the disk to the gravitational potential of the main spiral arms. The peculiarities of the potential in the narrow transitional annulus between the regions of spiral and multipole behavior can explain basic characteristic features of the quarter-turn spirals (their angular length and small pitch angles).

1 Introduction

Optical observations give many examples of grand design spiral galaxies in which arms consist of two parts: strong symmetric primary spirals, and adjacent faint secondary spirals. Figure 1 shows three images of such galaxies. The transition between the parts is marked by steep brightness gradients and by changing in the pitch angle of the arms.

As it is shown in many papers, for many spiral galaxies the last part of the optical grand design spirals almost vanishes in the near infrared wavelengths (see, e.g., [1], [2], [3]). Such a discrepancy between optical and near IR data suggests possible different formation mechanisms of these parts and allow one to consider the secondary spirals as a specific part of a whole spiral structure.

Basic characteristic features of the secondary spirals can be established by studying the azimuthal Fourier spectra of brightness maps for such galaxies:

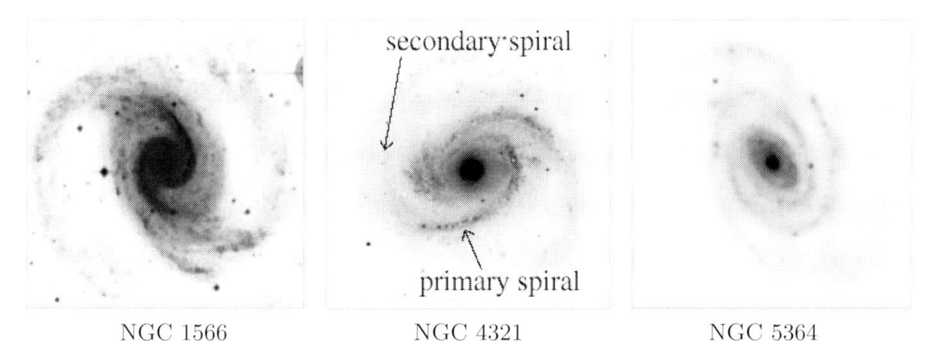

Fig. 1. Examples of galaxies with QTS

1. Their pitch angles are small (compared to those of the primary spirals);
2. Their angular length is of about 90°.

Due to the first two features one may refer to the secondary arms as the quarter-turn spirals (QTS). Lack of the QTS in the near IR implies that

3. The secondary spirals formed mainly from the cold component of the galactic disk.

Below, the explanation of the QTS phenomena is presented.

2 The Simplest Theory

In this section the simplest theory is described, in which QTS are treated as the disk response to the gravitational potential of the primary spirals. The calculations are carried out under the following basic assumptions:

- the galaxies under consideration have well defined spiral structures with certain pattern velocities;
- the simplest approximation of a cold 2-dimensional disk can be used (it is expected that this approximation contains the main effects, and the corrections for the velocity dispersion are small);
- linear theory of the disk response can be used;
- QTS are located sufficiently far from the main resonances.

Accordingly, the density response to the gravitational potential can be calculated by employing the linearized hydrodynamical equations with the pressure equal to zero

$$\frac{\partial \tilde{\sigma}}{\partial t} + \Omega \frac{\partial \tilde{\sigma}}{\partial \varphi} + \frac{1}{r}\frac{\partial}{\partial r}(r\sigma_0 v_{r1}) + \frac{\sigma_0}{r}\frac{\partial v_{\varphi 1}}{\partial \varphi} = 0,$$
$$\frac{\partial v_{r1}}{\partial t} + \Omega \frac{\partial v_{r1}}{\partial \varphi} - 2\Omega v_{\varphi 1} = -\frac{\partial \Phi_1}{\partial r},$$
$$\frac{\partial v_{\varphi 1}}{\partial t} + \Omega \frac{\partial v_{\varphi 1}}{\partial \varphi} + \frac{\kappa^2}{2\Omega} v_{r1} = -\frac{1}{r}\frac{\partial \Phi_1}{\partial \varphi},$$

where v_{r1} and $v_{\varphi 1}$ — the perturbed velocities, κ is the epicyclic frequency, $\kappa^2 = 4\Omega^2 + r d\Omega^2/dr$, $\Omega = \Omega(r)$ is the disk angular velocity, $\tilde{\sigma}$ and Φ_1 are the disk (density) response and the potential of the primary spirals, respectively. Assuming that all perturbations are proportional to the exponent $e^{i(m\varphi - \omega t)}$, one can obtain the response in the form

$$\tilde{\sigma} = -\frac{1}{r}\frac{\partial}{\partial r}\left(r\varepsilon \frac{\partial \Phi_1}{\partial r}\right) + \varepsilon \frac{m^2}{r^2}\Phi_1 + \frac{2m}{r\omega_*}\frac{\partial}{\partial r}(\varepsilon \Omega)\Phi_1, \tag{1}$$

where

$$\varepsilon \equiv \frac{\sigma_0(r)}{\omega_*^2 - \kappa^2} \tag{2}$$

is the gravitational analog of dielectric permittivity [4], [5]; $\omega_* = \omega - m\Omega(r)$; $\omega = m\Omega_p$ is the perturbation frequency, Ω_p is the pattern speed.

2.1 Qualitative Considerations

In the case when Φ_1 varies quickly with radius (it is proved below by numerical calculations), one can neglect all but one term in (1) with the highest derivative of Φ_1. Thus, formula (1) turns to

$$\tilde{\sigma} \simeq -\varepsilon \Phi_1''.$$

It is also implied here that the contributions from the resonance terms is of no significance. As it is clear from (2), $\varepsilon(r) < 0$ in the region between the inner and outer Lindblad resonances, at which $\Omega_p = \Omega - \kappa/m$ and $\Omega_p = \Omega + \kappa/m$, respectively. As a rule, the spiral structure is localized within this region (see, e.g. [6]). Then we obtain the qualitative formula for the response in the form:

$$\tilde{\sigma} \propto \Phi_1''. \tag{3}$$

To use the formula (3) one need to know the potential of the primary spirals. In the general case, this potential should be calculated numerically. However, the behavior of the potential can be predicted from qualitative considerations in two regions: in the spiral region and in the region sufficiently far from the spirals.

1. Spiral Region. For tightly-wound spirals, the curves of minima of the potential coincide with the curves of maxima of the surface density of the primary spirals. It follows from the well-known relation between the potential Φ_1 and the surface density σ_1 in Toomre's theory of the tightly-wound spirals [7]:

$$\Phi_1(r) = -\frac{2\pi G}{|k(r)|}\sigma_1(r),$$

where $k(r)$ is the wavenumber, G is the gravitational constant. The same correspondence approximately holds for open spirals, as it is demonstrated below on the model examples (see also [8]).

Applying the formula (3) for such a spiral-like potential $\Phi_1(r) \propto e^{-iF_\Phi}$ (here the phase $F_\Phi = \int^r k(r')dr'$), one can obtain

$$\tilde{\sigma}(r) \propto \Phi_1''(r) \simeq -k^2\Phi_1(r), \tag{4}$$

i.e. in this region, the curves of maxima of the density response follow the curves of minima of the gravitational potential (see Fig. 2).

2. Region Beyond the Primary Spirals. Well away from the spirals, the potential Φ_1 tends to its asymptotic quadrupole form [9]

$$\Phi_1(r,\varphi) \to -r^{-3}\cos 2(\varphi - \varphi_0), \quad r \to \infty, \quad \varphi_0 = const. \tag{5}$$

Applying the formula (3) to the quadrupole potential (5) one can obtain:

$$\tilde{\sigma}(r) \propto \Phi_1''(r) \simeq +12\Phi_1(r), \tag{6}$$

i.e. the curves of maxima of the response follow the curves of maxima of the gravitational potential (see Fig. 2).

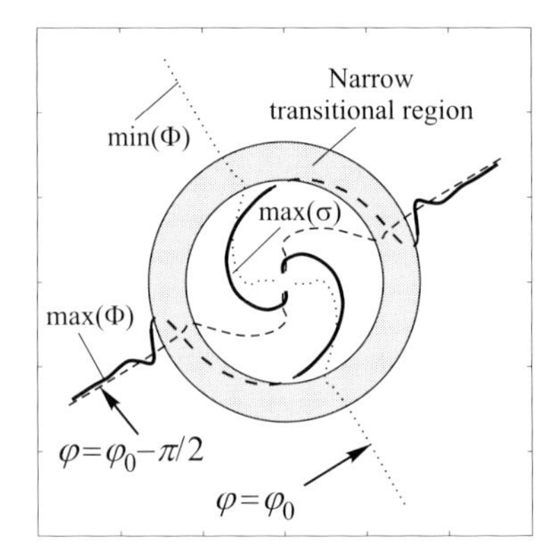

Fig. 2. Spiral, multipole, and transitional behavior of the gravitational potential.

It is notable, that the curves of the potential minima take its asymptotic direction $\varphi = \varphi_0$ just after departing from the primary spirals near its ends. It follows from numerical calculations of the potential. This implies that the effective region, that determines the asymptotic direction of the curves of the potential minima is small, compared to the size of the primary spirals (see [10] for some qualitative explanations based on the multipole expansion of the potential). Thus, the potential can be presented in the multipole form

$$\Phi_1(r, \varphi) = A_\Phi(r) \cos 2(\varphi - \varphi_0) \simeq -r^n \cos 2(\varphi - \varphi_0)$$

not only at sufficiently large radii, but almost up to the primary spirals. Evidently, the mentioned correspondence between maxima of the response and the potential holds in the whole multipole region.

In the narrow transitional region between the spiral and the multipole regions (see Fig. 2) the response switches between minima and maxima of the potentials. The pitch angle of the spiral response is small due to the narrowness of the transitional region, and the angular length is apparently about $\pi/2$. Thus, the response in the transitional region is QTS.

QTS cannot contain any significant amount of old stars of the disk, since it contradicts to the observed small pitch angles of QTS. Indeed, QTS occur in the gas and young stars. The velocity dispersion of young stars increases with time and stars should simply leave the QTS region, otherwise the latter would be much wider.

2.2 Model Examples

In this section, several model examples will be considered. Figure 3 presents the response to the logarithmic spirals $\sigma_1 \propto e^{iB \log r}$ derived numerically using the exact formula (1) and the general expression for the simple layer potential.

The parameter B defines the pitch angle i of a spiral: $\tan i = 2/B$. In the first two figures, the primary spirals are open, while the last figure shows a tightly wound primary spiral. The responses in the figures follow the maxima of the

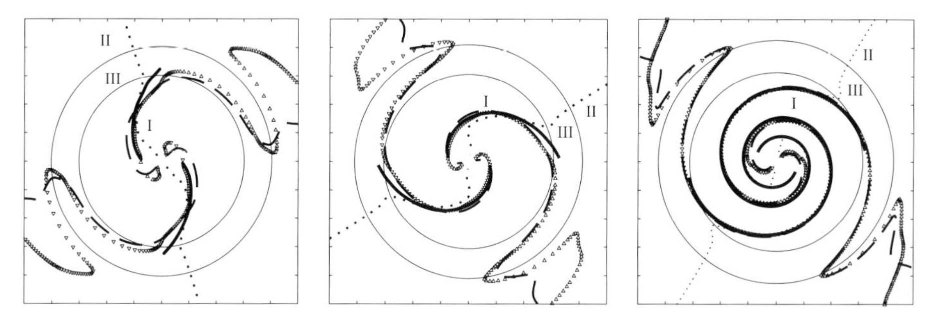

Fig. 3. The response of the galactic disk to the gravitational potential of the logarithmic spirals $\sigma_1 \propto e^{iB \log r}$. The spiral, multipole, and transitional regions are divided by the circles. Solid lines show the maxima of the primary spirals, dotted lines — minima of the gravitational potential, dashed lines — maxima of the second derivative of the gravitational potential, triangles — maxima of the response, calculated using (1).

second derivatives of the potential in accordance with the elementary theory. For open primary spirals, the QTS are evident. On the contrary, for the tightly wound primary spirals, it is more difficult to reveal QTS.

The pitch angle of the QTS seems to be roughly independent of the pitch angle of the primary spirals. For all these cases the pitch angles are about 10°.

3 Example of QTS in the Galaxy NGC 3631

NGC 3631 is rather bright non-barred galaxy of type Sc with well-defined spiral structure, observed nearly face-on.

As it was shown in [11], the second Fourier harmonic dominates over the others for this galaxy. It can be seen from Fig. 4a, where the contribution of the individual Fourier harmonics to the brightness deviation from axial symmetry is presented. Figure 4b shows the maximum of the R-map second Fourier harmonic superimposed on the image of the galaxy. The quarter turn spirals are clearly seen (this is also true for other bands — B, V, R, Hα). In the Fig. 4b the inner QTS are clearly seen. They arise just from the same mechanism as the outer QTS due to the decreasing of the primary spiral amplitude to the center of the disk.

The axisymmetric disk density profile and the primary spiral density can be inferred from the analysis of a brightness map, assuming the light to mass ratio to be constant. The obtained amplitude and phase of the second Fourier harmonic is given in Fig. 5. Thick curves show the smoothed functions that describe the primary spiral. The potential is restored by using the general formula of a simple layer. The rotation curve and the pattern velocity of the spiral structure is taken from [11], [12].

Figure 6 shows the response to the gravitational potential of the primary spirals for the grand design galaxy NGC 3631. It is seen that the response repeats the curves of the second derivative of the potential. The comparison of the

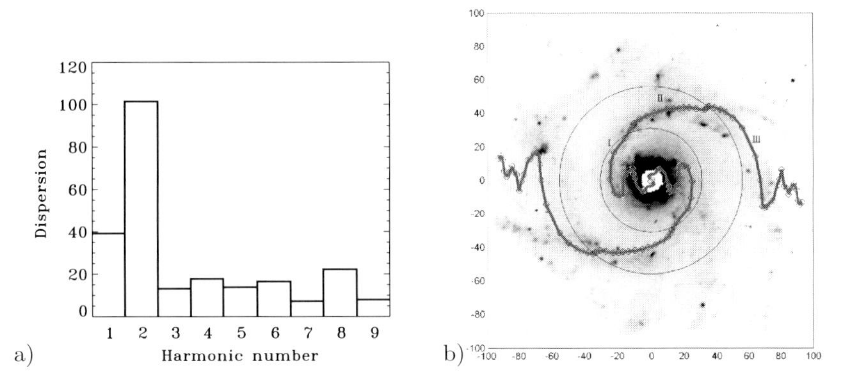

Fig. 4. a) Contributions of the individual Fourier harmonics to the brightness deviation from axial symmetry; b) maxima of the $m = 2$ Fourier harmonic superimposed on the R-map of the galaxy (ING archive). Circles divide the regions of inner QTS (I), spiral (II), and outer QTS (III).

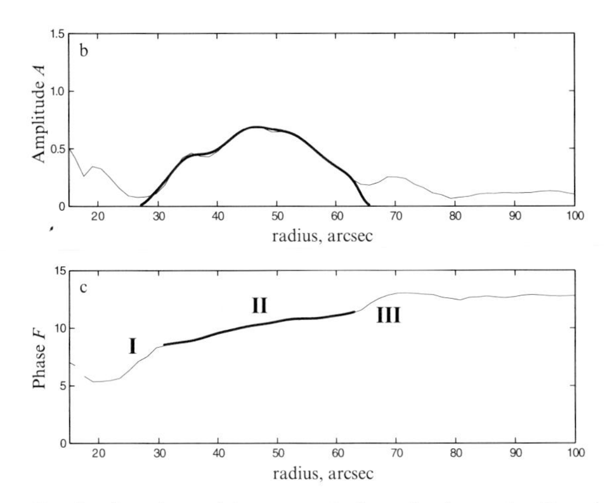

Fig. 5. The amplitude (in the arbitrary units) and phase (radians) of the $m = 2$ Fourier harmonic of the R-band image of the galaxy NGC 3631. Thick curves show the amplitude and phase of the primary spirals used in calculations.

response with the second Fourier harmonic (solid line) demonstrates the qualitative agreement of theory with observations. In the outer part, the response is somewhat longer, but the pattern structure of the QTS is reproduced. In the center, the response practically coincide with the Fourier harmonic: the length, the pitch angle and even the radially aligned structure in the very center is reproduced. Thus for the inner part, the good agreement is observed.

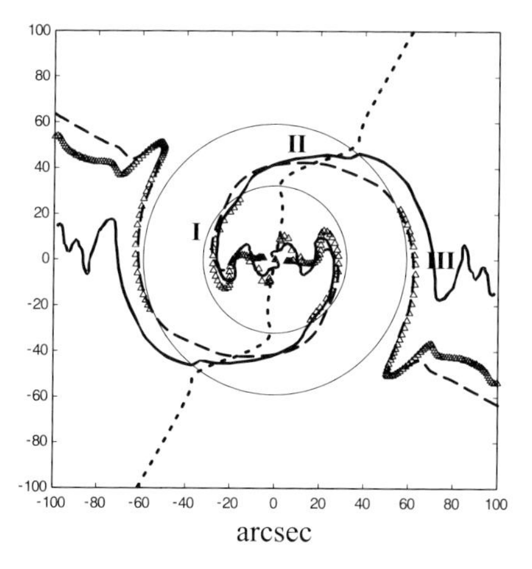

Fig. 6. The response of the grand design galaxy NGC 3631. Circles divide inner QTS, spiral, and outer QTS regions. Dotted lines show minima of the gravitational potential, dashed lines — maxima of the second derivative of the gravitational potential, triangles — maxima of the response. Solid lines show maxima of the $m = 2$ Fourier harmonics as in the Fig. 4b.

4 Conclusions

1. The phenomenon of QTS at the end of the primary arms of normal grand design galaxies has been analyzed. It is shown that:

- QTS are the universal nonresonance response of the galactic disk to the potential of the primary spirals;
- characteristic observed features of the QTS are explained by the peculiarities of the potential behavior near the end of the primary spirals.

2. QTS occurs not only in normal spiral galaxies. The example of the barred galaxy NGC 1365 is given in [10], [13]. When the bar is formed, the primary spiral must appear as the disk response to the potential of the bar. For the resonance excitation [15], the length of this spiral does not exceed $\pi/2$ for the standard fast bar, and π, for the Lynden-Bell slow bar [14], [15]. If the primary spiral is strong enough to change the multipole potential of the bar to the spiral-type potential, then the first quarter-turn spirals should form that elongate the primary spirals over $\pi/2$.

3. QTS provides a natural explanation for the sequential elongation of spirals both in normal and barred galaxies. The development of the QTS can make sufficiently powerful so that the total potential will turn into the spiral form once again. The secondary quarter-turn spirals will then appear, and the spiral structure elongates further over $\pi/2$.

Acknowledgments

This research has made use of the NASA/IPAC Extragalactic Database (NED) which is operated by the Jet Propulsion Laboratory, California Institute of Technology, under contract with the National Aeronautics and Space Administration and was partially based on data from the ING Archive. The work was supported in part by grants No. 99-02-18432 and 00-15-96528 provided by RFBR.

References

1. D.L. Block, G. Bertin, A. Stockton, P. Grosbøl, A.F.M. Moorwood, R.F. Peletier: Astron. & Astrophys. **288**, 365 (1994)
2. P.J. Grosbøl, P.A. Patsis: Astron. & Astrophys. **336**, 840 (1998)
3. D.M. Elmegreen, F.R. Chromey, B.A. Bissell, K. Corrado: Astron. J. **118**, 2618 (1999)
4. A.M. Fridman, A.B. Mikhailovskii: JEPT **61**, 457 (1971)
5. G.S. Bisnovatii-Kogan, A.B. Mikhailovskii: Sov. Astron., **17**, 205 (1973)
6. C.C. Lin, F.H. Shu: Astrophys. J. **140**, 646 (1964)
7. A. Toomre: Astrophys. J. **139**, 1217 (1964)
8. G. Contopoulos, P. Grosbøl: Astron. & Astrophys. **155**, 11 (1986)
9. L.D. Landau, E.M. Lifshitz: *The classical theory of fields* (Pergamon Press, London, New York 1976), sect. 41
10. V.L. Polyachenko, E.V. Polyachenko: Astron. Rep., **46**, 1 (2002)
11. A.M. Fridman, O.V. Khoruzhii, E.V. Polyachenko, et al.: Mon. Not. R. astr. Soc. **323**, 651 (2001)
12. A.M. Fridman, O.V. Khoruzhii, E.V. Polyachenko, et al.: Phys. Letters, **A264**, 85 (1999)
13. E.V.Polyachenko: Mon. Not. R. astr. Soc. **331**, 394 (2002)
14. D. Lynden-Bell: Mon. Not. R. astr. Soc. **187**, 101 (1979)
15. I.I.Pasha, V.L. Polyachenko: Mon. Not. R. astr. Soc. **266**, 92 (1994)

Dynamics of Galaxies: From Observations to Distribution Functions

Herwig Dejonghe and Veronique De Bruyne

Astronomical Observatory, Ghent University, Krijgslaan 281, S9, 9000 Gent, Belgium

Abstract. An overview is given of the currently most used dynamical modelling methods, with an emphasis on those methods that allow to derive a distribution function from observations of elliptical galaxies. Special attention is paid to the applications of distribution functions in the study of the internal dynamical structure of galaxies. It is indicated how existing modelling methods can be improved to correspond better with state-of-the-art observations and computation facilities.

1 Introduction

In general terms, a dynamical model for a stellar system provides a statistical description of a gravitational system, that is based as much as possible or practical on dynamical theory. More specifically, the best one can hope for is the determination of the distribution function (hereafter DF), which provides the probability to find a star with a given position and a given velocity. Such a model is supposed to provide a (good) approximation of the observed morphology and kinematics of a galaxy. In many cases, a dynamical model is only a vehicle to determine global parameters (e.g. total mass) or special characteristics (e.g. the presence of a central black hole). However, if dynamical modelling also involves the determination and interpretation of a DF, more detailed and structural information about the galaxy can be obtained. It is the purpose of this contribution to show that this goal is coming within reach.

2 General Considerations

2.1 Scope

Dynamical modelling always implies a simplification of some sort because in a realistic situation dynamical theory is virtually absent or impractical to apply. An obvious way out is the use of N-body simulations, which can righteously claim a generality and applicability that theory cannot match. Moreover, the sampling of a DF from an N-body simulation is rather trivial in principle. On the other hand, our capabilities to interpret huge amounts of data remains rather limited, though the amount of information one can handle keeps increasing. One should also keep in mind that (the possibility of) the presence of dark matter, the nature of which still eludes us, may vitiate many a, otherwise state-of-the-art, simulation, because it is by no means obvious if and how dark matter can be modelled by means of particles. We will not discuss the N-body technique in this contribution.

Many dynamical analyses restrict themselves to the study of the Jeans equations, which are integrals of the Liouville equation and thus are differential equations involving moments of the DF. The Jeans equations therefore provide the tool to study galaxies as hydrodynamical systems, filled with a "stellar gas". However, they do not generally form a closed set of equations, implying that they generally require a priori assumptions on the anisotropy of the dispersion tensor. They also cannot account for the fact that stars are on orbits, that link rather distant regions of the galaxy, by virtue of the fact that the Jeans equations are local because they are differential equations. As a consequence, more hydrodynamical models can be made than models based on DFs, not all of them however with a positive DF. Despite the extensive use that has been made of the Jeans equations in the past (e.g. [9], [46], [56]), we will not discuss them in this review.

2.2 Spiral Galaxies

As of now, relatively little has been done to determine DFs in spiral galaxies, at least compared to elliptical galaxies. There are many reasons for this. To begin with, there is generally a lot of extinction in spiral galaxies, which makes observations difficult. Spirals also tend to be less massive than ellipticals, and therefore high spectral resolution is needed to determine the relatively small velocity dispersions. The observational capabilities to accomplish this have only recently become practical. Moreover, there is the presence of the spiral arms. While it is well established that these are only perturbations on an otherwise rather smooth distribution, their presence, together with the rather patchy dust distribution associated with them, does not directly make the analysis any simpler. Last but not least, because of the spiral structure that must be rotating, spiral galaxies are not in a photometric dynamical equilibrium, and therefore the potential must be time dependent. This makes the modelling quite hard (though not impossible, see e.g. [78], [79]). Therefore, we will also limit our focus to elliptical galaxies. In the end, elliptical galaxies may prove to be equally complicated as spiral galaxies, but at least, from a photometric point of view, they look rather relaxed and smooth, and it does not seem therefore to be a gross simplification to assume that their underlying gravitational potential is time independent. Even if it is not (and surely it isn't), it is unlikely to evolve on timescales that are comparable to the crossing times of an individual orbit in an elliptical galaxy, as seems to be indicated by the presence of rather standard-looking ellipticals up to high redshift.

There is a rather extensive literature on the dynamical structure of our Galaxy. In many respects, it involves all issues that are relevant for spiral galaxies (e.g. the study of the central bar, the 3 kpc ring...), and many that are more typical for ellipticals (the structure of bulge and halo). It is clearly not within the scope of this review to cover all aspects of Galaxy research, and we will only mention attempts to determine the orbital structure of populations in our Galaxy for as far as equilibrium DFs are involved.

2.3 The Gravitational Potential

A dynamical theory implies somehow a model for the gravitational potential that is supposed to govern the motion of baryonic matter. Unfortunately, little is known about the exact form of the underlying gravitational potential of an elliptical galaxy. There are several philosophies for including a potential into the modelling.

One of the basic assumptions concerns the symmetries in the total mass distribution (and therefore the gravitational potential) of elliptical galaxies. It must be that the potential is somehow connected to the luminous matter distribution, which can be estimated from deprojecting the projected surface brightness. The luminous matter distribution can be given as a parameterized function (e.g. Nuker density, or [21], [61], [76]), or as a sum of basis densities (MGE [8],[36]) or completely numerically. Similarly, the potential can be a parameterized expression (e.g. power law models [37], [38]) or a sum of basis functions ([17],[57]), or a purely numerical function.

Triaxial systems provide the most general description for elliptical galaxies. Due to their complexity they are not widely used to fit observations (but see [57], [58]), and their characteristics are more often studied on theoretical grounds (e.g. [28], [51], [60], [74]). We will not discuss this case any further.

Since the symmetries in the potential largely dictate the modelling flexibility that dynamical theory will allow, the sequel of this review is structured accordingly.

3 Spherical Potentials

In the spherical world view, the simplest models for elliptical galaxies have DFs that depend only on the energy E and therefore have an isotropic dispersion tensor. For the construction of this type of models, it suffices to have an expression for the spherical spatial density (and hence the potential through Poisson's equation), and to solve an integral equations (Eddington's formula). There is a close connection between such models and spherical isotropic models based on the Jeans equations. In spite of the fact that the DF is a more fundamental quantity than the velocity dispersion, this stellar dynamical description of elliptical galaxies, in the spherical and isotropic framework at least, and for as far as it is a derivative of observational information, has failed to make important contributions in addition to what has been obtained with the Jeans equations.

Spherical models can be made anisotropic by introducing the modulus L of the angular momentum in the DF $F(E, L)$. There are many anisotropic models possible that reproduce the same mass density and velocity dispersion ([25], [26]). While the anisotropic generalization is fairly obvious from a mathematical point of view, it is quite unclear why a stellar system would form and settle in a form that is completely blind for the orientation of the orbital planes of its orbits (orbits in spherical potentials are planar). Nevertheless, such models learn us that there is a rather bewildering variety of dynamical models that could fit the photometry in a spherical geometry, though the observed velocity dispersion could be quite different. It remains a mystery why nature chooses to realize only a subset of them, so as to make the fundamental plane thin.

Finally, the presence of dark matter offers the possibility to consider non-spherical models within spherical potentials, the idea being that any inconsistency will be taken care of by the dark matter. It is clearly hard to defend this assumption in any rigorous way, but it is certainly not improbable that the dark matter is rounder than the luminous matter. Such models could therefore be seen as a useful approximation of the real thing [20], [58].¡

4 Axisymmetric Potentials

4.1 Two-Integral Dynamical Models

In the axisymmetric paradigm, the component of the angular momentum parallel to the symmetry axis L_z is in general the only conserved integral besides the energy. Two-integral models with a DF of the form $F(E, L_z)$ have equal radial and vertical velocity dispersion, but these are different from the tangential dispersion. Such DFs consist of an even part in L_z and and odd part in L_z. The spatial density does only depend on the even part of the DF. As a consequence, if a DF generates a system with given spatial density, an unlimited number of DFs can be constructed that generate that spatial density by adding odd functions in L_z to the DF. The odd part of the DF is important for the rotation in the galaxy.

The analytical theory concerning the determination of the DF from the mass density and the mean rotation has been initiated in [49], [25]. Not many of these analytical techniques allow to include the observed dynamics in an easy way, and therefore they have not been applied very often. A contour integral formula for the calculation of two-integral DFs for axisymmetric systems, derived directly from the density has been developed by Hunter & Qian [50]. The method requires an analytical expression for the potential for the density in terms of the potential. The DF can be calculated as a contour integral. Applications can be found in e.g. [63], [82].

Kuijken ([54]) developed a completely numerical technique for the construction of DFs. The numerical inversion of the integral equations connecting the spatial density and the streaming velocity with the DF, requires some smoothing which is achieved by a parametrization of the DF in continuous bilinear segments in this case.

When inhomogeneous data sets containing photometric and kinematic informations are considered, a quadratic programming algorithm [27] is a practical tool to obtain a dynamical model. For applications, see [35], [68], [69].

Two-integral DFs can also be constructed using the Richardson-Lucy algorithm ([22]). Furthermore, there are also a number of papers that describe classes of analytical models, but that are not really used for modelling data so far ([10], [11], [12], [52], [65]).

A dynamical system with two integrals of motion does not provide enough freedom for the modelling of our galaxy: observations have shown that in the solar neighbourhood the radial and vertical velocity dispersion are clearly different; hence the need for an additional degree of freedom. Moreover, numerical integration of orbits in axisymmetric systems has revealed that most orbits obey a third integral.

The introduction of such a third integral into dynamical models makes the modelling considerable more complicated, mainly because a third integral can only be determined approximately. There are several ways to approximate a third integral of motion. Some of these techniques use perturbation methods, e.g. [23], [44], [59] or derive a so-called 'partial integral', e.g. [34], [39]. Another option is the use of Stäckel potentials, where the expression for a third integral is analytically known. Some commonly used techniques in three-integral dynamical modelling are mentioned here.

4.2 Three-Integral Dynamical Models

Schwarzschild Methods. Schwarzschild [67] proposed a method that relies on the numerical calculation of a large library of orbits in a given potential. During integration, the characteristics of the orbits (e.g. spatial and projected density, line profiles, velocity moments) are stored on a grid. The weighted orbits are then combined in order to approximate the observed quantities as good as possible. The orbits are parameterized by their starting points, that are connected to integrals of motion. Hence, there is no need for an expression for an effective third integral.

This method is very general because there are no prerequisites for the expression for the potential or the DF. On the other hand it is a computationally demanding method and it is not straightforward to obtain a sufficiently smooth DF. In most applications the main issue is to reproduce the velocity moments and line profiles in order to estimate a (central) mass for the galaxy. Applications of this method can be found in [7], [8], [13], [14], [15], [41], [42], [64], [76], [81].

Whereas the method is independent of a priori assumptions, it does require a substantial degree of regularization. One way to make this easier is to use building blocks that are smoother than individual orbits [60]. One step further is to use two integral DF components whose observables can be calculated as integrals of the DF without going through the step of orbit integration, but also here regularization is required in order to derive smooth DFs [80]. Models that are created in this way give up a great deal of generality of the Schwarzschild method and are closely related to another widely used type of models, those that consist of a combination of basis function DFs.

Basis Function DFs. Independently of how the third integral of motion is approximated, DFs can be created as a sum of basis functions. This technique relies on the linearity of the equations and on a minimization algorithm to find the best fitting DF (e.g. [16], [18], [23], [24], [30], [31], [36], [45], [53], [55], [58], [59], [66], [71]).

Within this class of dynamical models a range of different types of basis functions can be used. Depending on the specific applications, the number of assumptions may be modest. Expansion methods offer the possibility to obtain non-parametric fits. The most important advantage however is that they can deliver smooth and positive DFs that are explicitly known, thereby offering possibilities that are hard to retrieve from models that rely on pure numerical integration of orbits. These properties make this kind of modelling rather appealing if the goal is to get insight in the internal structure of elliptical galaxies (e.g. [18]).

The Use of Separable Potentials. Potentials for which the Hamilton-Jacobi equations are separable in a given orthogonal coordinate system are called separable Stäckel potentials. For this family of potentials, three integrals of motion are explicitly known for all initial conditions, and the expressions for them are analytical and simple. Moreover, the orbital families in Stäckel potentials are easily classifiable and correspond to orbits that are found in more general potentials. In many cases they provide a good general description of elliptical galaxies. Despite the fact that these are not the most general potentials, studying models based on them may provide valuable insight in the internal structure of elliptical galaxies ([32], [29], [30]). However, there are no irregular orbits in these potentials [33]. The explicit knowledge of a third integral of motion allows to write down the moments of judiciously chosen DFs in an analytical way. This is very practical in combination with the use of basis function DFs (e.g. [18], [30], [36], [62], [71]).

5 What Can We Learn from Distribution Functions?

In most of the cases where three-integral DFs are explicitly shown (e.g. [8], [30], [59], [81]) this is done to illustrate that there is indeed structure in the third integral and to draw conclusions concerning the degree of anisotropy of the stellar system. However, not all of the presented DFs are smooth enough to get more detailed information on the internal dynamical structure of the galaxy.

Figure 1 gives an illustration of how information in a three-integral DF can be visualized, in this case by means of intersections with a plane of constant third integral in integral space or in turning point space. A representation in turning point space seems to be the most intuitive, since it can be most easily linked to physical orbits and their spatial extent. However, when a modelling procedure is adopted that makes explicitly use of the integrals of motion, a representation in integral space gives more insight in how the model has used the freedom available in the basis DFs. Both representations are complementary, in the sense that radial orbits are highlighted in a representation in the (E, L_z)-plane, while a representation in turning point space gives a clearer view on circular orbits.

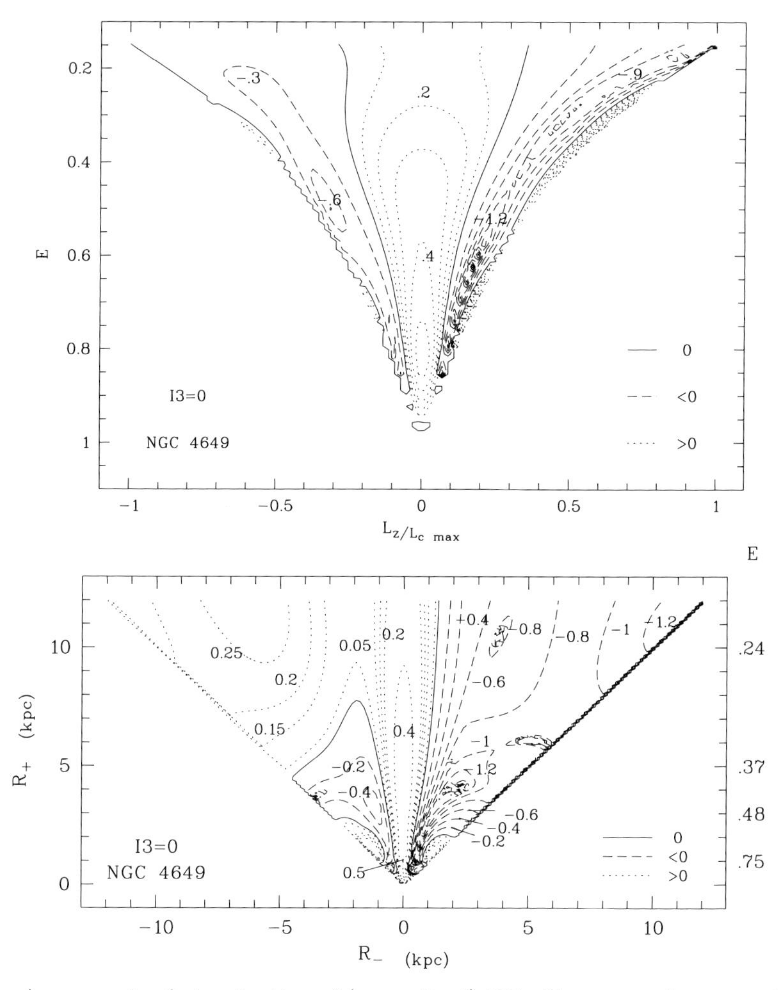

Fig. 1. An example of visualization of (normalized) DFs. Upper panel: representation in integral space of an intersection of a three-integral DF with a plane of constant third integral ($I_3 = 0$). Lower panel: representation of the same quantity in turning point space. Here, R_+ is the apocentre in the equatorial plane, R_- is the pericentre in the equatorial plane, the sign of R_- is equal to the sign of L_z.

5.1 Disentangle Photometric and Kinematic Information

Since a dynamical model is based on a mix of photometric and kinematic observables, the resulting DF carries information on both. When dynamical models are calculated for elliptical galaxies that behave kinematically different, the DFs should also be significantly different. When these galaxies have different photometric properties, a key question is whether there are significant differences between the DFs that can be attributed to the kinematic information alone. To answer such a question, it is necessary to eliminate from the DFs the signature of the photometry.

Such a normalization can be realized by dividing the three-integral DF, obtained by fitting the photometry and kinematics, by the even two-integral DF that is determined completely by fitting the photometry only. In addition, such a normalized DF is dimensionless and coordinate independent, which makes it a suitable quantity for comparison between different dynamical models.

Figure 1 shows a logarithmic representation of such a normalized three-integral DF for NGC 4649. A detailed description of the modelling and the observations can be found in [18]. The main conclusion from the figure is that orbits with small $|L_z|$ seem to play an important role in this model (largest values for the contours in a vertically stretched region in the middle of the contour plot), and this is an effect attributable to the kinematical information only. In the representation in turning point space, a small region with positive contours for circular or near-circular orbits with negative L_z becomes clear, although the observed mean velocity is positive. This shows that small amounts of counterrotation in elliptical galaxies may not leave their imprint on the observed kinematics.

5.2 Additional Classification Parameter

The class of elliptical galaxies is generally seen as a twodimensional manifold, characterized by flattening and boxyness/diskyness, which are photometric parameters. One may think of refining the classification by using kinematic information. This could be done by means of comparing normalized DFs as described in the previous part. In order to quantify this idea, a diagnostic could be used:

$$\nu = \frac{log(M_1) - log(M_2)}{2\sqrt{\sigma_{M_1}^2 + \sigma_{M_1}^2}}, \tag{1}$$

with $\log(M_i)$ the logarithm of the normalized DF for model i and σ_{M_i} the error on the normalized DF for model i.

An illustration of what can be concluded from this diagnostic can be found in Fig. 2, where a comparison between normalized DFs of NGC 4649 and NGC 7097 indicates that it is indeed possible to distinguish between galaxies based on their kinematics alone. The figure shows that the orbital density of radial orbits and circular orbits is significantly high for NGC 4649, while orbits with moderate to high $|L_z|$ are preferred in the model for NGC 7097. These differences in orbital densities are due only to differences in kinematic parameters.

5.3 Counterrotation in NGC 7097

NGC 7097 is a disky elliptical where the mean velocity in the inner 1.76 kpc indicates a counterrotating core (lowest $\langle v \rangle$ is -20 km/s at 0.88 kpc). This galaxy was modeled with a three-integral model based on a Stäckel potential [18].

A representation in turning point space of the DF where the number of orbits with $L_z < 0$ is subtracted from the number of orbits with $L_z > 0$ (see left panel

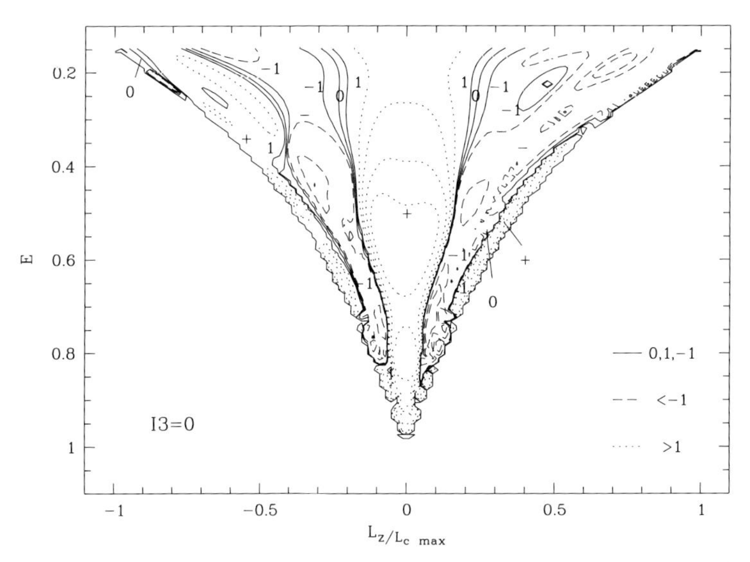

Fig. 2. Comparison of normalized DFs in a plane with zero third integral, by means of the diagnostic ν, as defined in equation 1. This figure compares DFs for NGC 4649 and NGC 7097. Negative contours are in dashed lines and indicate that ($DF_{\mathrm{NGC\ 7097}} > DF_{\mathrm{NGC\ 4649}}$), positive contours are in dotted lines and indicate that ($DF_{\mathrm{NGC\ 4649}} > DF_{\mathrm{NGC\ 7097}}$). Solid lines display contours for -1,0,1.

in Fig. 3) revealed that the counterrotation in this galaxy is not necessarily caused by a compact group of stars. Stars contributing to the counterrotation can be found as far as 4 kpc from the centre, while the mean velocity profile shows only counterrotation up to 1.76 kpc. For a toy galaxy where the amount of rotation and counterrotation is doubled with respect to the counterrotation present in NGC 7097, the counterrotating orbits are confined to a much smaller spatial region (right panel in Fig. 3). This indicates that the amount of signal in the $\langle v \rangle$-profile for NGC 7097 is not enough to constrain the counterrotating stars to the central part of the galaxy. This has also implications for possible formation scenarios for ellipticals with counterrotating cores.

6 Continuous Improving on Modelling Techniques

Dynamical modelling is a vivid research field, where considerable efforts are spent on improving existing modelling methods. These continuous developments are triggered by practical considerations such as ever improving computational capabilities and/or observational facilities, that deliver data of ever increasing quality and quantity.

6.1 Spectra as Source of Information

The shape of the absorption lines in a galaxy spectrum depends on the composition of the stellar mix and the dynamics of the galaxy. The most frequently used

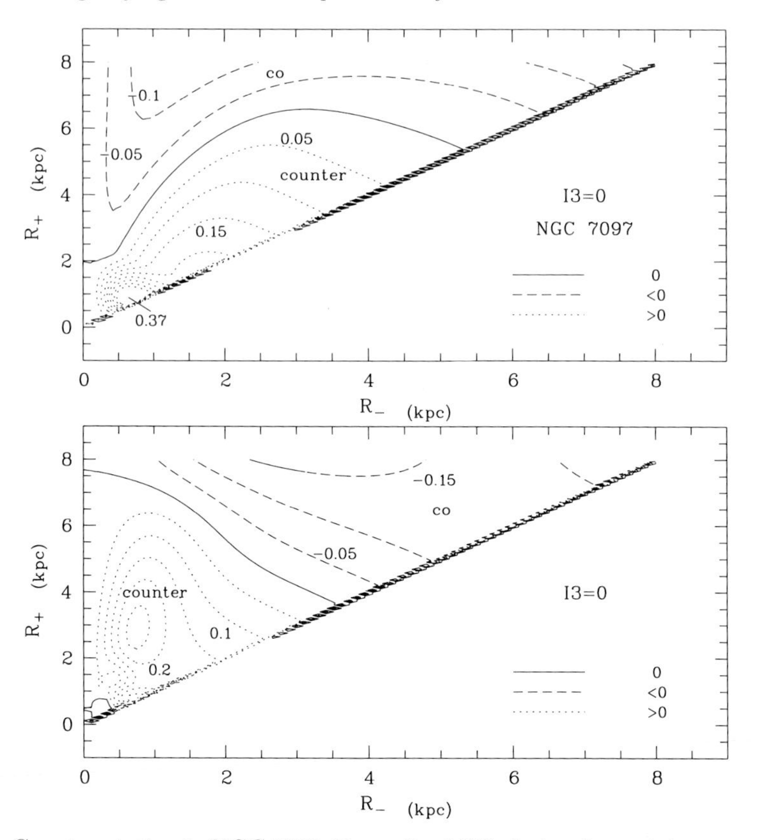

Fig. 3. Counterrotation in NGC 7097. Normalized DFs in turning points space, where the number of orbits with negative angular momentum is subtracted from the number of orbits with positive angular momentum. For NGC 7097 in the left panel: this galaxy contains counterrotating orbits with radii up to 4 kpc. The right panel shows the case for a toy galaxy, where the amount of rotations was doubled with respect to the case of NGC 7097.

approach to dynamical modelling is a two-step process: first the line-of-sight velocity distribution (hereafter LOSVD) is determined from the observations and kinematic parameters are derived. In a second step these parameters are used to constrain a DF.

A new strategy uses the spectra as they come to constrain (1) a DF and (2) a template mix in one modelling process [31]. This is equivalent to the construction of a detailed dynamical model and a population synthesis at the same time. However, fitting directly to spectra implies a considerable increase in the number of data points. It seems that only when analytically tractable basis functions are used to construct the DF, in which case the contribution of the dynamical components to the spectra can be calculated analytically, this is a feasible setup.

With this method, a three-integral dynamical model has been constructed for NGC 3258 [20]. Spectral features from the Ca II triplet were modeled using

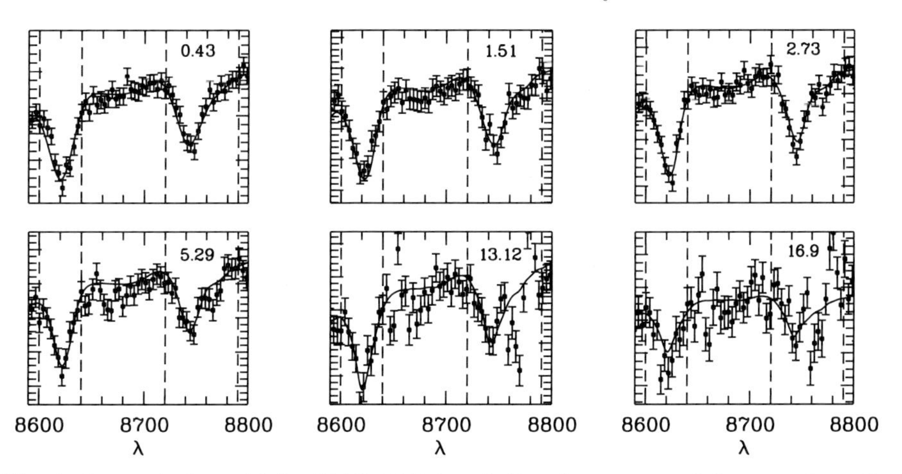

Fig. 4. Dynamical model for NGC 3258 fitted directly to spectral features from the Ca II triplet (data points with error bars) and the fit (in solid lines). The fit regions are located between the intervals indicated by dashed lines with centre around 8620 Å and 8750 Å. Projected radii are indicated in the top right corners.

two different template stars (a G5 dwarf and a K4 giant). Figure 4 shows the fit to the two strongest lines of the Ca II triplet at different projected radii.

Figure 5 shows some results of the modelling: the number density of the different populations (upper panel) and the kinematic parameters (projected and spatial) for the different populations. The obtained DF can be used to study the dynamics of the separate stellar populations in the galaxy. This can be done by comparing their spatial kinematics (see Fig. 5), their LOSVDs (see Fig. 6) or representations of their DFs (see Fig. 7).

The plot with relative densities in Fig. 5 shows that the centre contains mainly K4 giants, between 0.5 and 2 kpc there is almost an equal amount of both stellar types and for larger radii the K4 giants are again dominating the stellar light distribution. As for the kinematics, it is clear that the G5 dwarfs rotate more than the K4 giants do (illustrated in Figs. 5 and 6). The anisotropy parameter shows that the model is isotropic in the centre, becomes radial anisotropic soon and becomes tangential anisotropic at 2 kpc.

The shapes of the line profiles for the total DF (solid lines in Fig. 6) are not Gaussian but have a large variety of shapes. These shapes can be better understood when the LOSVDs for the total DF and the ones for the separate populations are compared. The LOSVDs for the total DF are composed of the LOSVDs of the G5 dwarfs (dotted profiles) and the K4 giants (dashed profiles). The relative heights of these profiles indicate the relative number densities of the stellar populations.

Intersections for the DFs with the equatorial plane are shown in Fig. 7, presented in the (E, L_z)-plane and in turning point space. Also here, the characteristics of the total DF (upper row) are a mix of the characteristics of the DFs for the separate stellar populations.

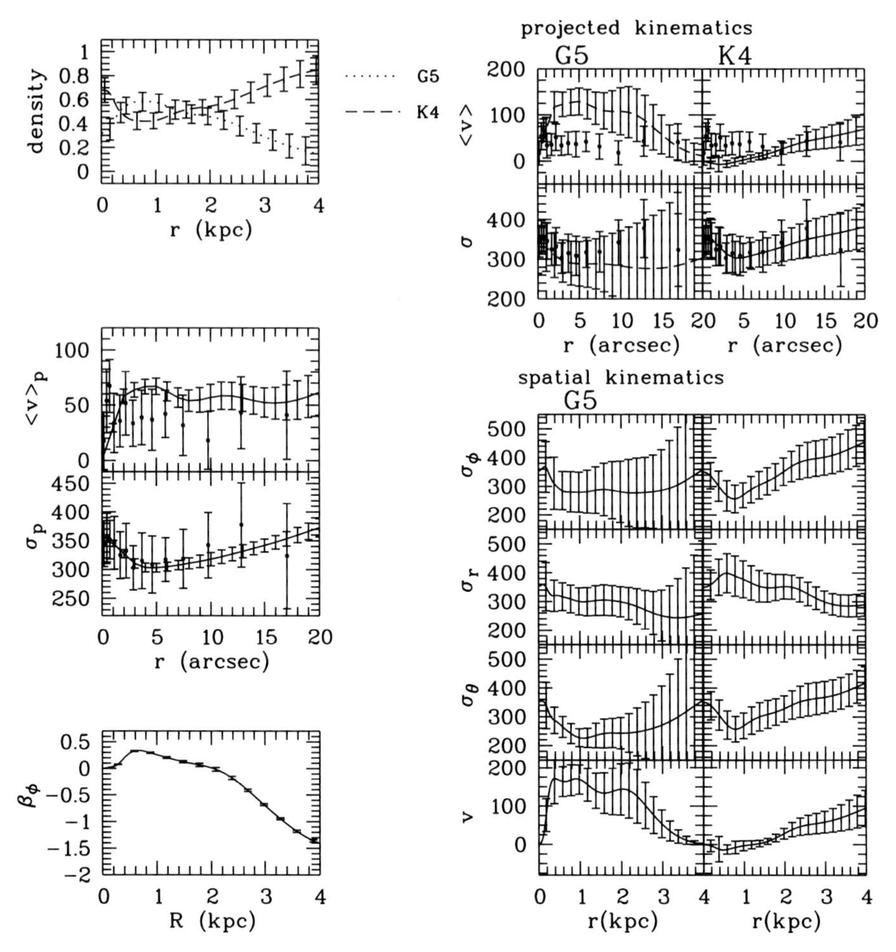

Fig. 5. Dynamical model for NGC 3258. Upper left panel: densities of the stellar populations. Middle left panels: projected mean velocity and projected velocity dispersion for the total DF in solid line and with error bars. The data points presented by dots are kinematic parameters obtained from fitting a Gauss-Hermite series up to fourth order to the observed spectra. Lower left panel: The anisotropy parameter for the model. Upper right panels: Projected kinematics with error bars for the stellar templates (G5 dwarfs on the left, K4 giants on the right), from top to bottom: projected mean velocity, projected velocity dispersion. The data points presented by dots are kinematic parameters obtained from fitting a Gauss-Hermite series up to fourth order to the observed spectra. Lower right panels: Spatial kinematics with error bars for the stellar templates (G5 dwarfs on the left, K4 giants on the right), from top to bottom: spatial dispersions and spatial velocity.

6.2 Construction of a Third Integral

Most dynamical models with DFs currently found in the literature use Stäckel potentials and the explicit expression for the third integral offered by these sys-

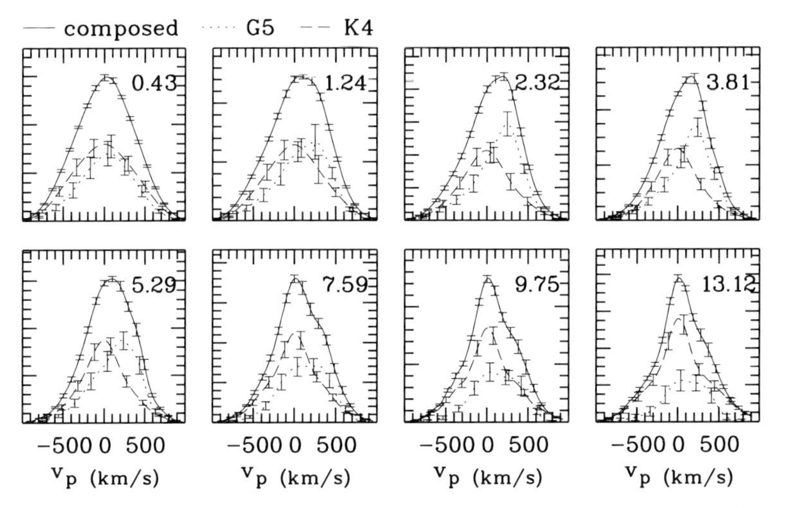

Fig. 6. LOSVDs and error bars calculated from the dynamical model for NGC 3258, for the total DF in solid lines, for the G5 dwarfs in dotted lines and for the K4 giants in dashed lines.

tems. To improve on the generality of this approach, one may think of approximating a galaxy potential by a set of Stäckel potentials instead of one single Stäckel potential.

This can be done in an elegant way, since creating a grid in energy and angular momentum in integral space creates a sequence of surfaces in space so that for fixed angular momentum, orbits with higher energy fill a volume that is completely embedded in the volume filled by an orbit with lower energy. Similarly, for fixed energy, orbits with larger angular momentum are completely within the volume filled by orbits with smaller angular momentum. Hence, dividing the (E, L_z)-plane into a number of rectangles is equivalent to dividing space into a number of bounded domains, see Fig. 8. For each of these domains, a Stäckel potential can be determined that is locally a good approximation to the original potential. The validity of the approximation can be checked in several ways, based on numerical integration of orbits in the original potential and fitted Stäckel potential. Various criteria to judge the agreement between the orbits in the original potential and local Stäckel potential are surfaces of section, conservation of I_3, orbital densities, topology of orbits. An application of this procedure can be found in [17].

7 Some Concerns Before Modelling Takes of

7.1 Influence of Dust

In the context of dynamical modelling, elliptical galaxies are traditionally considered as dust free stellar systems, but it is well known that in nature they do

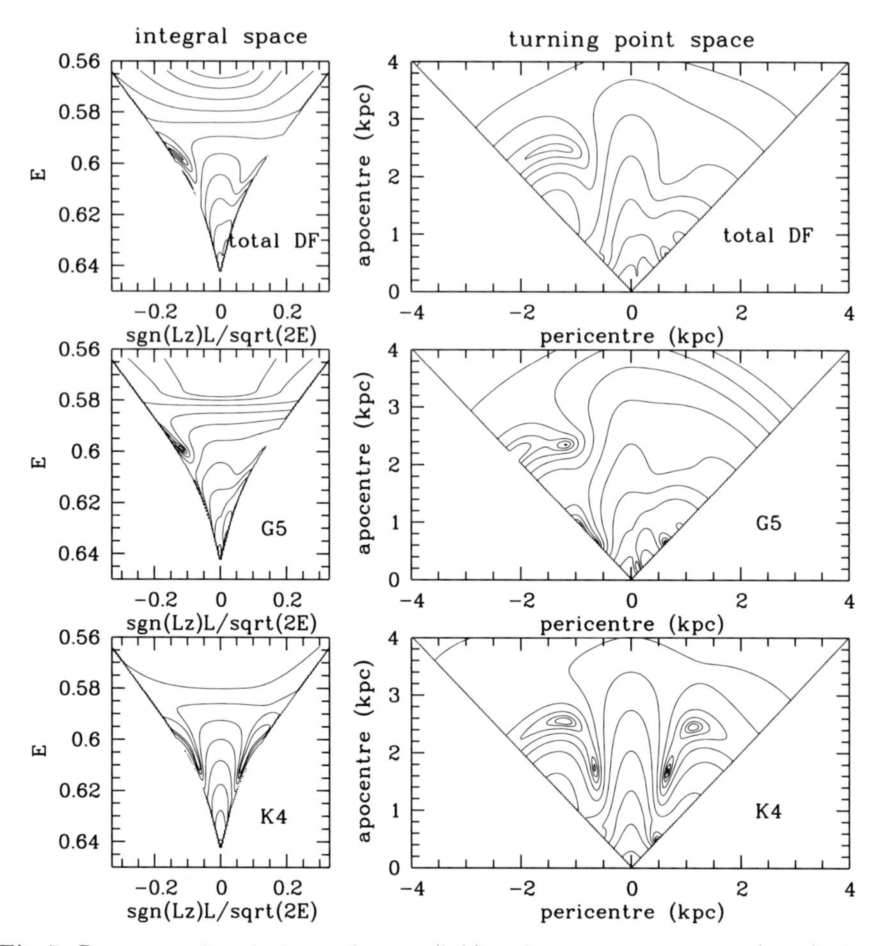

Fig. 7. Representations in integral space (left) and turning point space (right) of the DF for NGC 3258 in the equatorial plane. Here, R_+ is the apocentre in the equatorial plane, R_- is the pericentre in the equatorial plane, the sign of R_- is equal to the sign of L_z. Contours are equally spaced in log(phase space density). From top to bottom: for the total DF, for the G5 dwarfs, for the K4 giants.

contain various amounts of dust. Moreover, observed photometry and kinematics can be seriously effected by the dust.

There are numerous observations that prove the existence of dust in elliptical galaxies. Discrete optical dust features are seen in a number of ellipticals as dust lanes ([6], [48]) or in central regions ([40], [72], [77], [73]). Thermal far-infrared emission has revealed, unexpectedly, large IRAS fluxes and a comparison with optical dust features indicated that most of the dust has to exist as a diffuse component ([47]).

The dust grains present in ellipticals can cause absorption and scattering of the light. The result is that photons disappear from the line of sight and reappear on other lines of sight. The effect of this on the observed photometry and

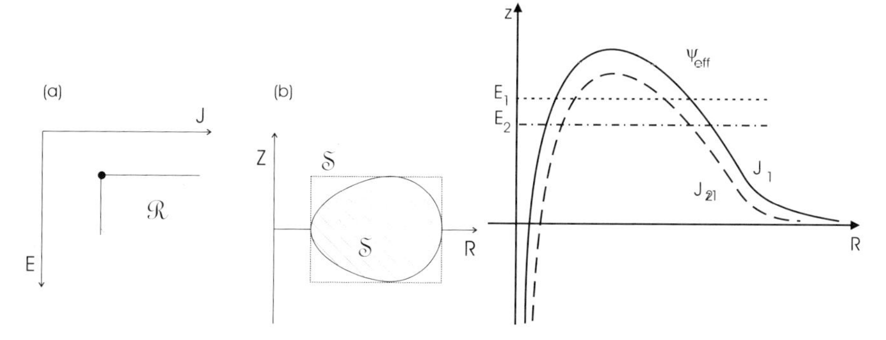

Fig. 8. Left: A grid in integral space (a) is a subdivision in space (b). Right: Intersections of lines of constant energy with the curve for the effective potential (solid line) illustrate that orbits with higher energy (E_1) fill a volume that is completely embedded in the volume filled by an orbit with lower energy (E_2). Similarly, for fixed energy (e.g. E_1), orbits with larger angular momentum (J_1)are completely within the volume filled by orbits with smaller angular momentum (J_2).

kinematics can be studied by solving the radiative transfer equation, taking dust attenuation into account and calculating the light profile and projected kinematics numerically ([1], [3], [4]). In practice however, a Monte Carlo routine that generates millions of photons turns out to be most cost-effective and flexible. It is able to simulate quite naturally the scattering process and it includes velocity information in an elegant way. Moreover, there are no restrictions on the geometry of stars and dust and a decent error analysis is possible.

This modelling showed that absorption and scattering by dust grains affect all observables. As for the kinematic profiles, there are no large effects in the central regions of the galaxy. On the other hand, there are serious effects on the kinematics in the outer regions: the projected velocity dispersion decreases more slowly while the h_4 profile increases quite dramatically (see Fig. 9). These effects are identical to what is generally considered as the kinematic signature of a dark matter halo. Hence, dust attenuation may reduce or even eliminate the need for a dark matter halo, giving rise to a new mass-dust degeneracy. This is investigated in more detail in [2]. Possible ways to break this degeneracy is to include dust attenuation in dynamical modelling and/or to use near-infrared kinematics which are less influenced by dust attenuation.

7.2 Error Bars on Data

Undoubtly, it is important to have photometry and kinematic data of good quality, but it is of equal importance to have a reliable error estimate on kinematic parameters. For dynamical modelling, these estimates are often used in goodness of fit indicators, and the decision whether to accept or reject a particular dynamical model often depends critically on the error bars of the data (e.g. [13], [64], [53]).

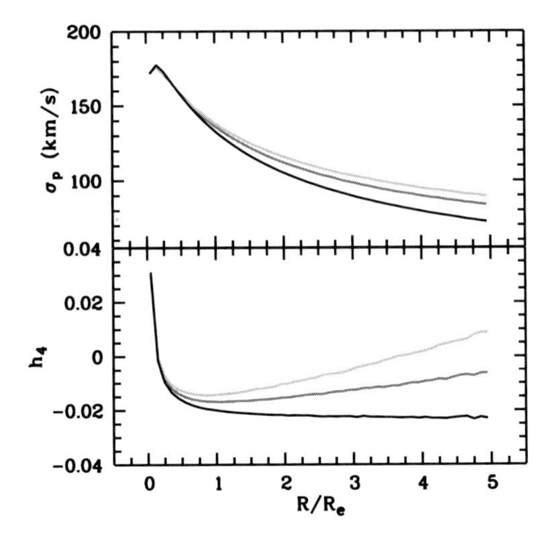

Fig. 9. Effect of dust attenuation on the projected kinematics, in particular the velocity dispersion and the h_4 profiles. The profiles are shown for a model without dust attenuation and for models with optical depths of V = 0.5 and V = 1.

An observed galaxy spectrum is the sum of individual stellar spectra of stars moving along the line of sight that have been redshifted by the rotation of the galaxy and smeared out by the velocity dispersion of the stars. Since the seventies, much effort has been spent on retrieving the LOSVD from observations [19]. In this context, it is convenient to express the information in that LOSVD in a limited number of parameters.

The higher order moments of the LOSVD depend critically on the wings of the profiles. Instead of expressing the characteristics of a LOSVD through its moments, a parameterized version of the LOSVD is often used in the form of a truncated Gauss-Hermite series ([43], [75]).

These parameters and an error estimate on them are often obtained using a least-squares minimization. The statistical interpretation of this method relies on the assumption that the noise is independent and Gaussian distributed on the input data. These conditions are generally not met after standard data reduction steps. As a consequence, the errors derived from standard statistics will in many cases differ from the real errors on the kinematic parameters.

Sometimes, Monte Carlo simulations are used ([5], [70]) to estimate the uncertainties on the derived parameters. For the realization of synthetic galaxy spectra, a Gaussian noise distribution with given S/N is used. So also this approach will show a similar tendency to underestimate the errors. People are aware of this, but it is mostly left unclear how large the differences between the estimated and real error bars are.

Recently, a method was proposed that makes a diagnosis of the characteristics of the real noise on the spectra and that allows to calculate more realistic error bars on kinematic parameters [19]. An illustration is given in Fig. 10. Panel (a) shows a galaxy spectrum and a fit to this spectrum, panel (b) shows the fitting

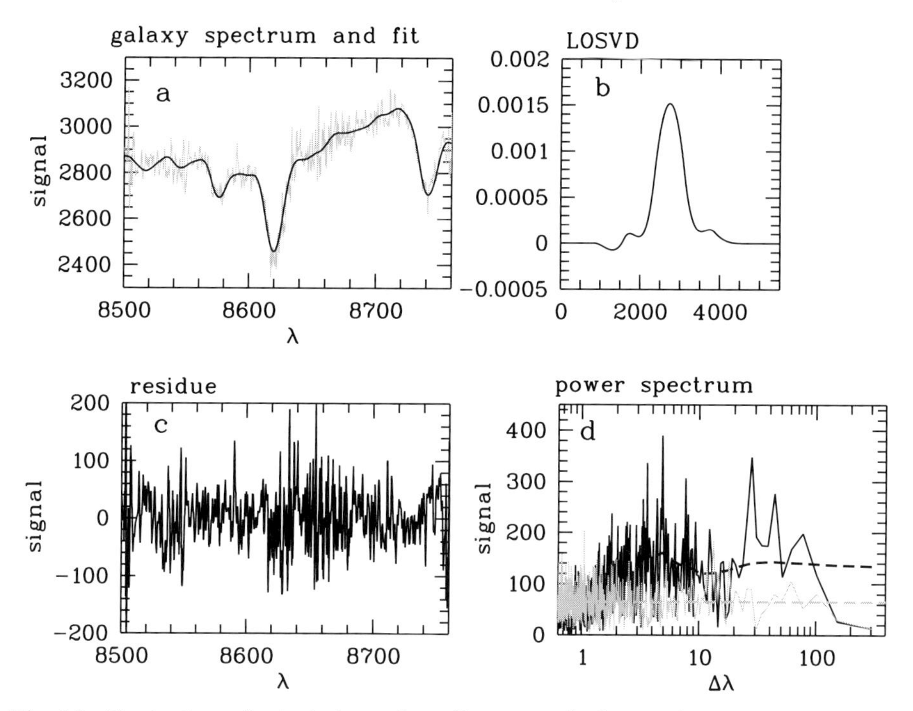

Fig. 10. Illustration of a technique that allows to calculate realistic error estimates. Panel (a): observed spectrum and fit. Panel (b): fitting LOSVD. Panel (c): residue of the fit. Panel (d): power spectrum of residue (black) and power spectrum with S/N of input spectrum (grey).

LOSVD and panel (c) shows the residual of the fit, that is considered as the real noise involved in the problem. Panel (d) shows the power spectrum of the residual in black and the power spectrum of a Poisson noise profile according to the S/N of the spectrum in panel (a) in grey. It is clear that the real noise involved is not Poissonian. The characteristics of the noise can be represented by the smoothed representation (dashed line). Realistic error estimates are obtained from Monte Carlo simulations with synthetic galactic spectra where the noise distribution has the same power spectrum as the real noise.

In the test case of NGC 3258, the realistic errors appeared to be almost a factor of 2 larger than the errors based on least squares statistics. Moreover, for the first time is was shown that the way the spectra are sampled in the data reduction has a non-negligible influence on the quality of the derived kinematics.

8 Conclusions

In this contribution we have given an overview of the most frequently used dynamical modelling methods. If detailed and structural information about a galaxy is to be obtained, the dynamical modelling should involve the determination of a DF. By means of a number of case studies it is shown how structural information contained in DFs can be visualized and interpreted. It is clear that this requires the DF to be a smooth function. However, not all dynamical modelling methods that are currently used deliver a DF that can be used for this purpose. If one agrees to use analytical approximations to the third integral (axisymmetric case), an expansion of the DF in analytic basis functions seems to offer only advantages.

We finally discuss a few recent developments that have become possible thanks to improved observational and computational capabilities, and we also critically reexamine some established paradigms. In particular, new astrophysical issues can be addressed if a direct fit to the spectra can be constructed. There are now possibilities to perform dynamical modelling and population synthesis in one go. We also explore what would need to be revised if one of the basic premises of dynamical modelling of elliptical galaxies is wrong: that there is little or no dust in elliptical galaxies. It turns out that the presence of dust has a non-negligible influence on the observable kinematics, especially in the outer regions that are crucial in our assessment of the presence of dark matter. Finally, we revisit the usual error analyses on the kinematic parameters, and show that in general, error bars of higher order kinematic parameters are underestimated by almost a factor of 2.

References

1. M. Baes, H. Dejonghe, MNRAS, **313**, 153 (2000)
2. M. Baes, H. Dejonghe, ApJ, **563**, L19 (2001)
3. M. Baes, H. Dejonghe, MNRAS, **318**, 798 (2000)
4. M. Baes, H. Dejonghe, MNRAS, **335**, 441 (2002)
5. R. Bender, R.P. Saglia, O.E. Gerhard, MNRAS, **269**, 785 (1994)
6. F. Bertola, G. Galetta, ApJ, **226**, L15 (1978)
7. G.A. Bower, R.F. Green, R. Bender, K. Gebhardt, T.R. Lauer, J. Magorrian, D.O. Richstone, A. Danks, T. Gull, J. Hutchings, C. Joseph, M.E. Kaiser, D. Weistrop, B. Woodgate, C. Nelson, E.M. Malumuth, ApJ, **550**, 75 (2001)
8. M. Cappellari, E.K. Verolme, R.P. van der Marel, G.A. Verdoes Kleijn, G.D. Illingworth, M. Franx, C.M. Carollo, P.T. de Zeeuw, ApJ, **578**, 787 (2002)
9. P. Cinzano, R.P. van der Marel, MNRAS, **270**, 325 (1994)
10. L. Ciotti, ApJ, **520**, 574 (1999)
11. L. Ciotti, ApJ, **471**, 68 (1996)
12. L. Ciotti, B. Lanzoni, A&A, **321**, 724 (1997)
13. N. Cretton, F. C. van den Bosch F., ApJ, **514**, 704 (1999)
14. N. Cretton, P.T. de Zeeuw , R.P. van der Marel, H.-W. Rix, ApJS, **124**, 383 (1999)
15. N. Cretton, H.-W. Rix, P.T. de Zeeuw, ApJ, **536**, 319 (2000)
16. J. de Brijne, R. P. van der Marel, P. T. de Zeeuw, MNRAS, **282**, 909 (1996)
17. V. De Bruyne, F. Leeuwin, H. Dejonghe, MNRAS, **311**, 297 (2000)
18. V. De Bruyne, H. Dejonghe, A. Pizzella, M. Bernardi, W.W. Zeilinger, ApJ, **546**, 903 (2001)
19. V. De Bruyne, P. Vauterin, S. De Rijcke, H. Dejonghe, MNRAS, in publication
20. V. De Bruyne, S. De Rijcke, H. Dejonghe, W.W. Zeilinger, submitted to MNRAS
21. W. Dehnen, MNRAS, **265**, 250 (1993)

22. W. Dehnen, MNRAS, **274**, 919 (1995)
23. W. Dehnen, O.E. Gerhard, MNRAS, **261**, 311 (1993)
24. W. Dehnen, O.E. Gerhard, MNRAS, **268**, 1019 (1994)
25. H. Dejonghe, Phys. Rep, **133**, 217 (1986)
26. H. Dejonghe, MNRAS, **224**, 13 (1987)
27. H. Dejonghe, ApJ, **343**, 113 (1989)
28. H. Dejonghe, D. Laurent, MNRAS, **252**, 606 (1991)
29. H. Dejonghe, P.T. de Zeeuw, ApJ, **333**, 90 (1988)
30. H. Dejonghe, V. De Bruyne, P. Vauterin, W.W. Zeilinger, A&A, **306**, 363 (1996)
31. S. De Rijcke, H. Dejonghe, MNRAS, **298**, 677 (1998)
32. P.T. de Zeeuw, MNRAS, **216**, 273 (1985)
33. P.T. de Zeeuw, M. Franx, Annu. Rev. Astron. Astrphys., **29**, 239 (1991)
34. P.T. de Zeeuw, N.W. Evans, M. Schwarzschild, MNRAS, **280**, 903 (1996)
35. S. Durand, H. Dejonghe, A. Acker, A&A, **310**, 97 (1996)
36. E. Emsellem, H. Dejonghe, R. Bacon, MNRAS, **303**, 495 (1999)
37. N.W. Evans, MNRAS, **260**, 191 (1993)
38. N.W. Evans, MNRAS, **267**, 333 (1994)
39. N.W. Evans, R.M. Häfner, P.T. de Zeeuw, MNRAS, **286**, 315 (1997)
40. Ferrari F., Pastoriza M.G., Macchetto F.D., Bonatto C., Panagia N., Sparks W.B., A&A, **389**, 355 (2002)
41. K. Gebhardt, D. Richstone, J. Kormendy, T.R. Lauer, E.A. Ajhar, R. Bender, A. Dressler, S.M. Faber, C. Grillmair, J. Magorrian, S. Tremaine, AJ, **119**, 1157 (2000)
42. K. Gebhardt, T.R. Lauer,J. Kormendy, J. Pinkney, G.A. Bower, R. Green, T. Gull, J.B. Hutchings, M.E. Kaiser, C.H. Nelson, D. Richstone, D. Weistrop, AJ, **122**, 2469 (2001)
43. O.E. Gerhard, MNRAS, **265**, 213 (1993)
44. O.E. Gerhard, P. Saha, MNRAS, **251**, 449 (1991)
45. O.E. Gerhard, G. Jeske, R.P. Saglia, R. Bender, MNRAS, **295**, 197 (1998)
46. J. Gerssen, R.P. van der Marel, K. Gebhardt, P. Guhathakurta, R.C. Peterson,C. Pryor, AJ, **124**, 3270 (2002)
47. P. Goudfrooij, R. de Jong, A&A, **298**, 784 (1995)
48. T.G. Hawarden, A.J. Longmore, S.B. Tritton, R.A.W. Elson, H.G. Corwin, MNRAS, **196**, 747 (1981)
49. C. Hunter, AJ, **80**, 783 (1975)
50. C. Hunter, E. Qian, MNRAS, **262**, 401 (1993)
51. C. Hunter, P.T. de Zeeuw, ApJ, **389**, 79 (1992)
52. Z. Jiang, D. Moss, MNRAS, **331**, 117 (2002)
53. A. Kronawitter, R.P. Saglia, O. Gerhard, R. Bender, A&AS, **144**, 53 (2000)
54. K. Kuijken, ApJ, **446**, 194 (1995)
55. J. Magorrian, D. Ballantyne, MNRAS, **322**, 702 (2001)
56. J. Magorrian, S. Tremaine, D. Richstone, R. Bender, G. Bower, A. Dressler, S.M. Faber, K. Gebhardt, R. Green, C. Grillmair, J. Kormendy, T. Lauer, AJ, **115**, 2285 (1997)
57. A. Mathieu, H. Dejonghe, MNRAS, **303**, 455 (1999)
58. A. Mathieu, H. Dejonghe, X. Hui, A&A, **309**, 30 (1996)
59. M. Matthias, O. Gerhard, MNRAS, **310**, 879 (1999)
60. D. Merritt, T. Fridman, MNRAS, **460**, 136 (1996)
61. D. Merritt, M. Valluri, ApJ, **471**, 82 (1996)
62. A. Pizzella, P. Amico, F. Bertola, L.M. buson, I.J. Danziger, H. Dejonghe, E.M. Sadler, R.P. Saglia, P.T. de Zeeuw, W.W. Zeilinger, A&A, 323, 349 (1997)

63. E.E. Qian, P.T. de Zeeuw, R.P. van der Marel, C. Hunter, MNRAS, **274**, 602 (1995)
64. H.-W. Rix, P.T. de Zeeuw, N. Cretton, R. van der Marel, C. Carollo, ApJ, **488**, 702 (1997)
65. F.H.A. Robijn, P.T. de Zeeuw, MNRAS, **279**, 673 (1996)
66. R.P. Saglia, A. Kronawitter, O. Gerhard, R. Bender, AJ, **119**, 153 (2000)
67. M. Schwarzschild, ApJ, **232**, 236 (1979)
68. M.N. Sevenster, H. Dejonghe, H. Habing, A&A, **299**, 689 (1995)
69. M.N. Sevenster, H. Dejonghe, K. Van Caelenberg, H. Habing, A&A, **355**, 537 (2000)
70. T.S. Statler, AJ, **109**, 1371 (1995)
71. T.S. Statler, H. Dejonghe, T. Smecker-Hane, AJ, **117**, 126 (1999)
72. A. Tomita, K. Aoki, M. Watanabe, T. Takata, S. Ichikawa, AJ, **120**, 123 (2000)
73. H.D. Tran, Z. Tsvetanov, H.C. Ford, J. Davies, W. Jaffe, F.C. van den Bosch, A. Rest, AJ, **121**, 2928 (2001)
74. I. Trujillo, A. Asensio Ramos, J.A. Rubino-Martin, A.W. Graham, J.A.L. Aguerri, J. Cepa, C.M. Gutierrez, MNRAS, 333, 510 (2002)
75. R.P. van der Marel, M. Franx, ApJ, **407**, 525 (1993)
76. R.P. van der Marel, N. Cretton, P.T. de Zeeuw, H.-W. Rix, ApJ, **493**, 613 (1998)
77. P.G. van Dokkum, M. Franx, AJ, **110**, 2027 (1995)
78. P. Vauterin, H. Dejonghe, A&A, **313**, 465 (1996)
79. P. Vauterin, H. Dejonghe, MNRAS, **286**, 812 (1997)
80. E.K. Verolme, P.T. de Zeeuw, MNRAS, **331**, 959 (2002)
81. E.K. Verolme, M. Cappellari, Y. Copin, R.P. van der Marel, R. Bacon, M. Bureau, R.L. Davies, B.M. Miller, P.T. de Zeeuw, MNRAS, **335**, 517 (2002)
82. F. Wernli, E. Emsellem , Y. Copin, A&A, **396**, 73 (2002)

Arp 158: A Study of the HI

Mansie G. Iyer[1], Caroline E. Simpson[1],
Stephen T. Gottesman[2], and Benjamin K. Malphrus[3]

[1] Florida International University, Miami, FL, USA
[2] University of Florida, Gainesville, FL, USA
[3] Morehead State University, Morehead, KY, USA

Abstract. We present here 21 cm observations of Arp 158. We have performed a study of the HI to help us understand the overall formation and evolution of this system and its role in galaxy evolution.

According to the evolutionary sequence described by Hibbard & van Gorkom (1996), Arp 158 appears to be an intermediate stage merger. There seem to be three distinct knots connected by a bar embedded in luminous material. There is also a diffuse spray to the south-east.

As noted by A. Toomre (Chincarini & Heckathorn, 1973), Arp 158 bears a certain optical resemblance to NGC 520 (Arp 157). We agree with this. From our 21 cm observations of Arp 158, we also see a comparable HI content with NGC 520. Such similarities between the two galaxies could imply like formation processes. However, the HI morphologies and kinematics for both systems are quite dissimilar.

1 Introduction

Galaxy-galaxy interactions and mergers can cause major perturbations to the structure and content of galaxies. The gas can lose angular momentum and collapse into the core, triggering nuclear starbursts and the formation of young star clusters (Duc & Mirabel 1997). On the other hand, strong tidal forces resulting from these interactions can produce features like plumes, shells, rings, tidal tails and bridges. This was shown in work done by Toomre and Toomre (1972). They also identified a series of galaxies that they believed represented galaxies at different stages of merging ("The Toomre Sequence"). Hibbard & van Gorkom (1996) presented H-alpha, R-band and 21cm observations of several galaxy systems from the Toomre sequence and showed that they most likely did form an evolutionary chain. They identified three stages in this chain: (i) the early stage, in which the disks are well separated and only marginally distrupted, (ii) the intermediate stage, which exhibit distinct nuclei embedded in luminous material and, (iii) late stage mergers in which tidal appendages are seen to emerge from a single nucleus.

This paper is the first part of a study to examine the HI morphologies and kinematics of a series of perturbed galaxies. These galaxies, Arp 213, Arp 78, Arp 135, Arp 31 and Arp 158 are all in various stages of the evolutionary sequence described above. Our main goal is to better understand and interpret this sequence. With this aim in mind, we present 21 cm Very Large Array[1] (VLA) observations of Arp 158.

Arp 158 is located at $\alpha = 01^h25^m20.9^s$ and $\delta = +34d01m29s$ (J2000). It is described in Zwicky's Catalogue of Selected Compact Galaxies and Post-Eruptive Galaxies as "post-eruptive, blue, with fan shaped jets, and three compact knots connected by a bright bar." The galaxy also appears to be in an intermediate stage of merging as defined by Hibbard & van Gorkom (1996), as it has two distinct nuclei embedded in an envelope of luminous material.

The optical image, Fig. 1 (Arp's Catalogue, 1966), shows that there are three visible knots, with a central bar and tail extending to the west and an extensive, faint diffuse tail to the south-east. The westernmost nucleus was examined by Chincarini & Heckathorn (1973, hereafter referred to as C&H) and was claimed to be a foreground star. From a visual examination of Arp's image (1966), this knot appears to be an unresolved point source, seeming perfectly round and scattered in the image. Dahari (1985), however, argued that his measured redshift for this knot was similar to that of the south-east one and thus could not be a foreground star. Without confirming spectra, we cannot determine the nature of this knot.

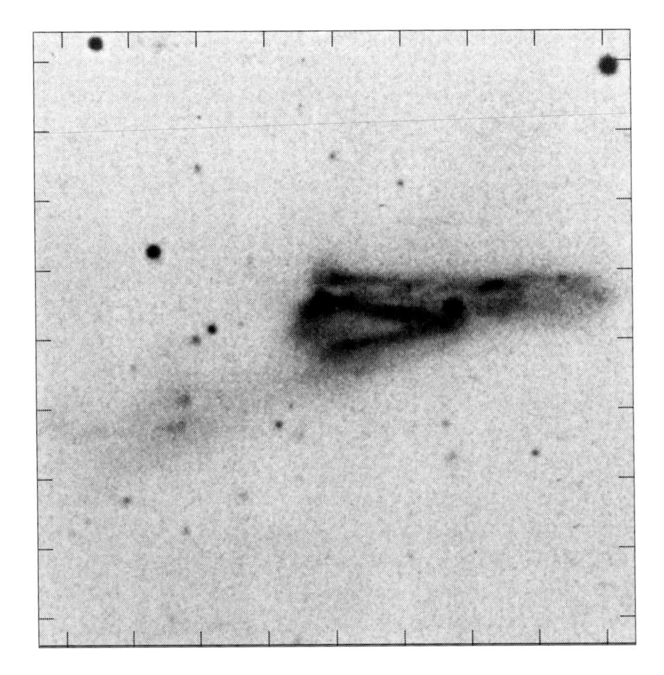

Fig. 1. Optical Image of Arp 158 from the Arp's Catalogue

---

[1] The VLA is operated by the Radio Astronomy Observatory, which is a facility of the National Science Foundation operated under the cooperative agreement by Associated Universities, Inc.

2 Observations and Data Reduction

We obtained high senstivity D-configuration observations of the 21 cm line of atomic hydrogen (HI) for the galaxy, Arp 158. This was done using the VLA in its spectral line mode. The galaxy was observed for a total of five hours, including move time and calibration time, using a 128 channel spectrometer with a total bandwidth of 6.25 Mhz. The channel separation of 48.8 kHz provides a velocity resolution of 12.4 km s^{-1}. The observations were made at a central velocity of 4758 km s^{-1} and the data was collected for a single polarization. The observational parameters are given in Table 1.

Table 1. Observational Parameters

Date	14 and 17 May 1999
Total Number of Visibilities	65344
Total Bandwidth	6.25 MHz
Total Number of Channels	127
Channel Seperation	48.8 KHz
Heliocentric Velocity	4758 km s^{-1}
Beam (FWHM)	47.22" × 40.88"
rms noise in channel maps	0.97 mJy
Equivalent Brightness Temperature	0.29K

The data reduction and calibration were carried out using the AIPS data reduction package available from the National Radio Astronomy Observatory. Owing to the fact that our observations were performed during the day, the data suffered from heavy solar contamination. The D-array is very susceptible to solar contamination, particularly at short baselines. To rectify this problem, the "CLIP" routine in AIPS was used to remove data points greater than 1 Jy for all baselines less than 315m (uv-range less than 1.5kλ).

After editing, the data was then calibrated, with the gain and phase solutions having very few or no closure errors. The CLEAN routine was applied and images were then made by implementing the fast Fourier transform to the UV data plane and the moment maps were obtained.

3 Results

Through this paper we use a value of Hubble's constant of 75 km $s^{-1}Mpc^{-1}$, and a heliocentric velocity of 4758 km s^{-1} (NASA IPAC Extragalactic Database, NED) for Arp 158. These values imply a distance of 63 Mpc. Figure 2 shows the MOM0 map. At this resolution, we can see that there are three distinct HI concentrations. In Fig. 3, which shows the HI contours on an optical grayscale, we

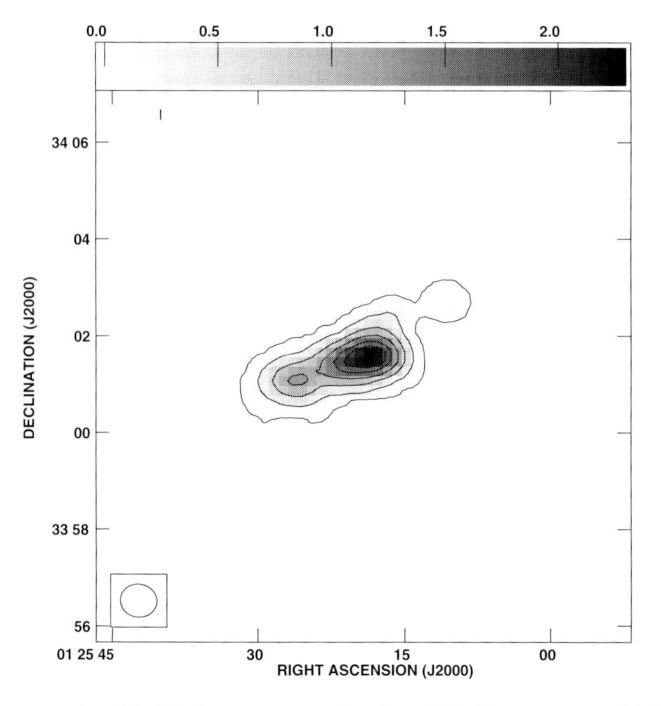

Fig. 2. HI greyscale with HI flux contours for Arp 158. Contours are (2 (2σ), 18, 42, 66, 90, 108) $\times 10^{19}$ atoms cm^2.

can see that the concentration with the highest column density ($N_H \leq 1.0 \times 10^{21}$ atoms cm^{-2}), named system A, coincides with the three-optical-knot bar system. System B ($N_H \leq 6.9 \times 10^{20}$ atoms cm^{-2}) is the second distinct concentration and appears to coincide with the diffuse tail to the south-east. There is also a very interesting gas knot to the north-west of the system which appears to have absolutely no optical counterpart in the DSS image. This is system C, with a peak column density of 1.2 $\times 10^{20}$ atoms cm^{-2}. A deeper optical image is needed to see if there is any stellar component to this object.

Figure 4 shows the MOM1 map. There again appear to be three distinct kinematic areas coinciding with systems A, B and C, which can also be seen from their velocity profiles (Fig. 5). Systems A and B are separated by almost 400 km s^{-1}, whereas systems A and C are separated by about 100 km s^{-1}. The velocity widths for systems A, B, and C are approximately 215 km s^{-1}, 82 km s^{-1}, and 29 km s^{-1} respectively. There is an area of steep velocity gradient between A and B, which overlies the bright optical emission in the system. The channel maps for Arp 158 show that there is a continuity in velocity, i.e. there is emission in every channel implying that there is gas connecting the two systems, which is not visible in the moment maps due to superposition. Furthermore, in these channel maps, there is no HI connecting A and C, although they share the same velocity space.

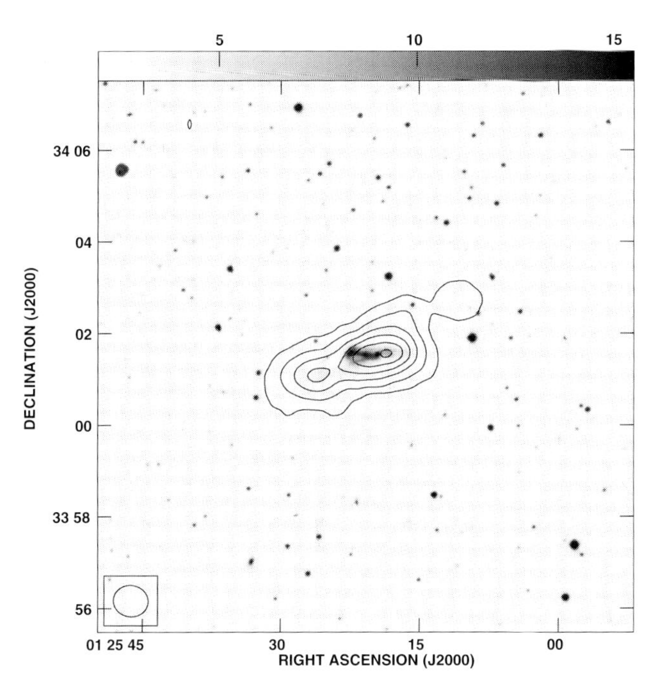

Fig. 3. Optical Digitized Sky Survey (Lasker et al, 1990) image of Arp 158 with HI flux contours. Contours are ((2 (2σ), 18, 42, 66, 90, 108) $\times 10^{19}$ atoms cm^2.

The isovelocity contours in the MOM1 map are very different for systems A and B. The apparent kinematical axes of these two systems, as indicated by the isovelocity contours, are oriented in different directions. The position angle of the kinematical axis for system A is at approximately 25 deg on the western side, rotating to a position angle of approximately 90 deg on the eastern side. This almost 90 deg kink in the isovels in system A show that this is a fairly abrupt change. For system B, the position angle is approximately 150 deg. (These position angles are measured north through east.) The contours for system B also show a smooth run of velocity, with widely separated isovels indicating a shallow gradient in velocity. Thus system B, which optically resembles a tail, also stretches out in the HI and with its narrow spread in velocity and the smooth run of velocities is consistent with this feature being a tidal tail (Hibbard & van Gorkom, 1996).

4 Discussion

In the optical image for Arp 158, there are bright components which include the knots and the bar, as well as a diffuse component (the southeastern tail). C&H estimated the optical extent of the bright components to be approximately 21 kpc; with the diffuse components extending to about 41-52 kpc. There are a few small optical condensations in the faint south east tail; they may be HII

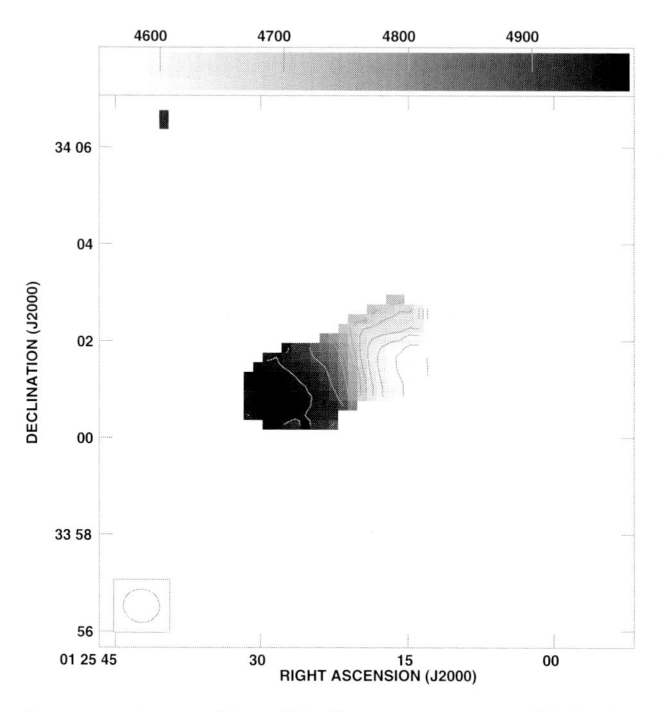

Fig. 4. Velocity moment map of Arp 158. Contours represent (4560, 4590, 4650, 4710, 4740, 4780, 4810, 4890, 4950, 4980) km s^{-1}.

regions (C&H), which would not be unusual in a tidal tail (Hibbard & van Gorkom, 1996). Using Arp's image, we estimate the knots and the bar to extend to approximately 27 kpc. The bright tail, defined as the central knot to the beginning of the diffuse tail, is about 12 kpc. The extensive, faint diffuse tail to the south east spreads out to approximately another 26 kpc.

C&H presumed that the eastern knot and the central knot were two separate galactic nuclei from two interacting galaxies. We have estimated their separation to be approximately 6 kpc. From their spectra, they assumed a similar composition and mass-to-light ratio for the knots, and attributed the optical brightness of the eastern knot to a higher central star density. Using this information, they calculated the mass ratio of the two nuclei to be 5:1. This mass ratio implies that the gravitational interaction between the two nuclei is weak as compared to the gravitational attraction between two bodies of equal mass. To get a large amount of disruption in a weak encounter, the objects need to pass very close to each other. Since Arp 158 is very disrupted, the encounter must have been close. According to A. Toomre, as reported by C&H, the best proof that such an encounter had occurred was the optical tail in the south east direction. Toomre also noted the optical similarity between NGC 520 (Arp 157) and Arp 158. We agree with this and discuss it further in this section.

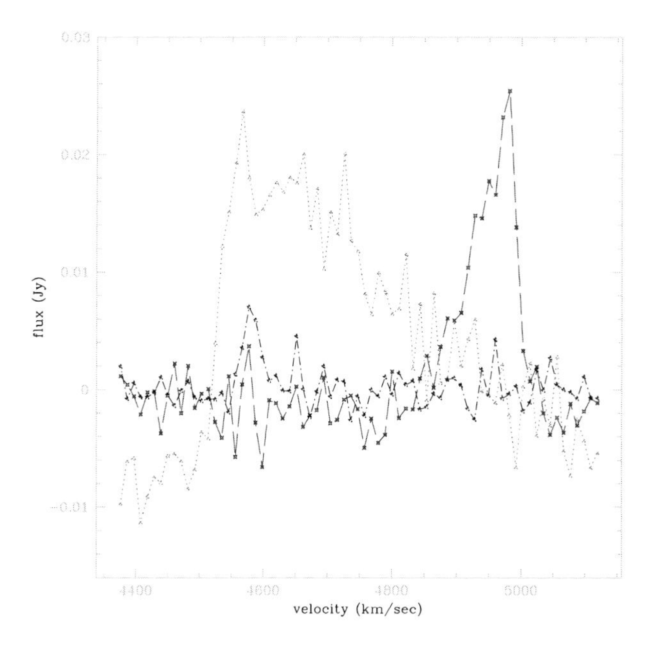

Fig. 5. Velocity profiles of Arp 158. The dots represent system A, long dashes represent system B. System C is represented by the dot and dashes.

NGC 520, classified as an intermediate stage merger by Hibbard & van Gorknom (1996) consists of two parent nuclei separated by $\approx$ 6 kpc. There is also a tail to the south which extends to $\approx$ 23 kpc. There is a plume which emanates from this tail, bends east, and then stretches to the dwarf galaxy UGC 957 at a distance of $\approx$ 52 kpc. We note here the optical similarities between NGC 520 and Arp 158; their tails extend to almost the same lengths and the separation between the parent nuclei for NGC 520 as well as the separation between the eastern and central nuclei for Arp 158 are comparable.

Just as the HI for Arp 158 extends beyond the optical image, so too is the case for NGC 520. The total HI mass for Arp NGC 520 was measured to be 7.1 $\times 10^9$ $M_\odot$ (Hibbard & van Gorkom, 1996), roughly similar to that of Arp 158, which is 6.19 $\times 10^9$ $M_\odot$. Even though both the galaxies contain large amounts of HI, the distribution is very different. Most of the gas in NGC 520 is distributed in an extensive gas disk and shows differential rotation (Hibbard & van Gorkom, 1996). This is not the case with Arp 158. There is a large amount of HI in the galaxy, but it is not concentrated over a disk and shows no signs of differential rotation. So although there are many optical similarities between NGC 520 and Arp 158, their HI morphologies and kinematics are quite different.

5 Conclusion

Although there is considerable debate on the origin of NGC 520, recent studies of the galaxy are leaning towards a merger hypothesis for it (Stanford 1990, 1991; Bernlöhr, 1993). Hibbard and van Gorkom (1996), suggest that NGC 520 formed out of an encounter between an extensive gas rich disk system and a gas poor system, although the southernmost feature, an extension to the southern tail, is a drawback to this idea. Such a feature would require two gas rich progenitors as opposed to one gas rich and one gas poor.

As we have seen, there are various optical similarities between Arp 158 and NGC 520 and comparable HI content. This could lead us to believe that their formation processes are also the same. However, the fact that their HI morphologies and kinematics are considerably dissimilar must be accounted for.

M. Iyer wishes to acknowledge an International Travel Grant from the A.A.S.

This research has made use of the NASA/IPAC Extragalactic Database (NED) which is operated by the Jet Propulsion Laboratory, California Institute of Technology, under contract with the National Aeronautics and Space Administration.

References

1. H. Arp : Atlas of Peculiar Galaxies, Washington D.C, Carnegie Institution, (1966)
2. K. Bernlöhr: Astronomy and Astrophysics, **270**, 20 (1993)
3. G. Chincarini, M.H. Heckathorn: PASP, **85**, 568, (1973)
4. O. Dahari: Astrophysical Journal Supplement, **57**, 643, (1985)
5. P.A. Duc, I.F. Mirabel: Messenger, **89**, 14, (1997)
6. J.E. Hibbard, J.H. van Gorkom: Astronomical Journal, **111**, 655, (1996)
7. B.M. Lasker, C.R. Sturch, B.J. Mclean, J.L. Russell, H. Jenker and M.M. Shara: Astrophysical Journal, **99**, 2019, (1990)
8. S.A. Stanford: Astrophysical Journal, **358**, 153, (1990)
9. S.A. Stanford: Astrophysical Journal, **381**, 409, (1991)
10. A. Toomre, J. Toomre: Astrophysical Journal, **178**, 263, (1972)
11. F. Zwicky: Catalogue of Selected Compact Galaxies and of Post-Eruptive Galaxies, Guemligen, Switzerland (1971)

Dark Matter in Spiral Galaxies

Albert Bosma

Observatoire de Marseille, 2 Place Le Verrier, 13248 Marseille Cedex, France

Abstract. In this talk I will discuss several issues concerning the dark matter problem in spiral galaxies. I will give first my version of the state of the debate about cuspy halos in low surface brightness galaxies, and then discuss the situation in bright spirals. I conclude that the dark matter profiles of low surface brightness late-type galaxies do not show much evidence for cuspy halos with NFW type profiles, and that the inner parts of high surface brightness spirals are unlikely to be dominated by dark matter.

1 Introduction

Recent developments in cosmology show a rather coherent picture on large scales, in favor of a ΛCDM model of structure formation, with Ω_Λ about 0.7, and $\Omega_{\rm matter}$ about 0.3. Since $\Omega_{\rm baryon}$ is about 0.04, there is room for dark matter in the form of a WIMP, but, with $\Omega_{\rm stars}$ about 0.005 at $z = 0$, there are a lot of baryons unaccounted for at the present epoch (see e.g. [45] and references therein).

Nevertheless, there is a dark matter "crisis" on small (galaxy) scales, since numerical simulations of cosmological structure formation seem to predict:

- cuspy halos, in apparent conflict with data on low surface brightness (hereafter LSB) galaxies and the inner parts of the Milky Way
- a lot of substructure, in apparent conflict with the number and size of the Milky Way's companions
- an offset in the Tully-Fisher relation, referred to as the angular momentum problem.

Since the majority of people working in the field think that the ΛCDM model of structure formation is the best picture describing the large scale structure, great care has to be exercised to ascertain the validity of the conclusions on the galaxy scale, before concluding about the validity of the ΛCDM model. The reactions to this situation are of several kinds: proponents of the ΛCDM model question the validity of the observations, others examine the role of additional physical processes which could remove material in a dark halo cusp while not questioning the validity of ΛCDM, and yet others examine other theories, such as warm dark matter, self-interacting dark matter, or even MOND.

In this paper I will examine the two issues concerning dark matter on galactic scales on which a lot of work has been done recently, i.e. the dark matter profiles of LSB galaxies, and the question of the dominance of dark matter in the central parts of bright spirals.

2 No Cuspy Halos in LSB Galaxies

2.1 Expectations vs. Observations

Dwarf and low surface brightness galaxies are thought to be dark matter dominated, and thus provide a crucial test for the current cosmological numerical simulations, which are predicting the density profiles of dark matter halos. Inner power law slopes for dark halos produced in cosmological simulations of CDM and ΛCDM models come out to be -1.5 ([33], [17]) or -1.0 (Navarro et al. 1996 [35], hereafter NFW). The latter argue that their NFW profile is universal, and thus can be scaled down to the dwarf galaxy scale (even though these scales are not yet fully modelled directly up till now). For warm dark matter models, similar slopes are found (e.g. [22]).

If one neglects the minor contribution of the stellar and gaseous components in these galaxies, an upper limit can be found for the slope of the density profile of the dark halo directly from inverting the observed rotation curves into a density distribution. It is crucial to get rotation data at the highest spatial resolution possible, but some claims in the literature have been made based not on real data, but on HI rotation curves corrected for beam smearing. Since a clear measurement is better than several arguments, we have collected a large data set using long slit Hα-spectra, and find that we can exclude the cuspy halos predicted by the current cosmological numerical simulations ([13], [14], [32], [11]). In the last paper, we concentrated on nearby dwarf galaxies, so as to have the highest linear resolution possible. Our results clearly show that the high resolution data favour models with a core, and exclude the steeper slopes required by the NFW and other CDM models.

2.2 Reply to Some Criticism

Primack ([39], [40]) severely criticises our work as published in De Blok et al. [13]. Like e.g. Van den Bosch & Swaters [46], he orients the discussion towards asking whether the data are still consistent with NFW profiles, rather than towards trying to find which power law slope fits the data best. In any case, he prunes the data, and eliminates galaxies which have data at poor resolution, edge-on galaxies, etc., so as to retain only a restricted sample. While I disagree with his statistics in the end, I will follow here Primack's arguments, select from our data the galaxies which are well resolved according to his criteria, and see whether his conclusions are justified.

a) *resolution.* Though in [13] and [11] we do discuss the effects of resolution, let me consider only galaxies with **two** independent points inside 1 kpc, the radius at which the discrimination with the NFW model prediction becomes significant. This strict criterion allows me to retain 16 galaxies from the original sample discussed in [13].

b) *edge-on galaxies.* Contrary to Primack's assertions, there is nothing mysterious about their kinematics. In [8] we show that small edge-on galaxies like NGC 100 are transparent, a conclusion confirmed e.g. by [31]. Moreover, our

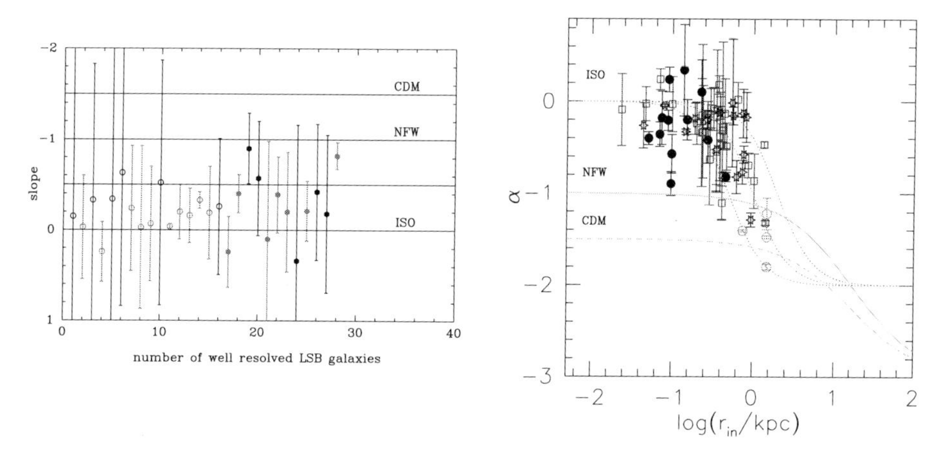

Fig. 1. Left: Plots of the slopes for the 16 best resolved galaxies of the De Blok et al. [13] sample (open circles) and the additional 12 galaxies in the De Blok & Bosma [11] sample (filled circles), with 3 σ error bars. 17 galaxies are 3 σ away from the NFW prediction. **Right:** Value of the inner slope α of the mass-density profiles plotted against the radius of the innermost point on the rotation curve in kpc, for a number of dwarf and low surface brightness galaxies observed by De Blok et al. Open circles and squares are from the sample in [13], stars and filled circles are from the sample in [11]. Overplotted are the theoretical slopes of a pseudo-isothermal halo model (dotted lines) with core radii of 0.5 (left-most), 1 (centre) and 2 (right-most) kpc. The full line represents a NFW model, the dashed line a CDM $r^{-1.5}$ model. Both of the latter models have parameters $c = 8$ and $V_{200} = 100$ km s^{-1}, which were chosen to approximately fit the data points in the lower part of the diagram.

spectra do not show low radial velocity wings expected for a non solid body rotation in the central parts, which by itself rules out NFW profiles for these edge-on galaxies. So I retain the 5 edge-on galaxies in the final sample, but note that my conclusions do not depend on their inclusion.

c) *more data.* I add new data from the February 2001 run of [11]. Primack [39] did not dispose of these data, but in [40] chose to ignore them. This brings the total number of galaxies in the sample considered here to 28 galaxies.

With these new selection criteria, I rule out NFW slopes at the 3σ-level for 17 galaxies, of which 5 edge-on (see Fig. 1). Primack's statement that of the dozen cases probed in [13], about half appear consistent with the cuspy NFW profile is simply not true (I find only 2 out of 12 galaxies he selected having slopes $\leq$ -0.5), and his final conclusion that this data set may be consistent with an inner density profile $\alpha \sim$ -1 but probably not steeper, is not warranted, and not corroborated by the newer data.

2.3 Further Observations and Modelling OF Possible Errors

In [11] we already show that data taken by independent observers agree for F561-1, F563-1, F568-3, and UGC 5750. Our results have been corroborated by an independent study by Marchesini et al. [28, 29], who did a further comparison of raw data for 4 other galaxies. Fabry-Perot data from [18] for a few galaxies in common are also in good agreement.

During a February 2002 observing run, I checked again several galaxies previously observed by others, and find good agreement between the raw data. Moreover, I experimented for UGC 4325 with deliberate slit offsets of ± 5", and find very good agreement between the position-velocity curves thus obtained. Such agreement is expected for velocity fields dominated by solid-body rotation. These observations are reported in [12].

Furthermore, to counter objections that systematic errors play a significant role in our results, and always bias the result away from cusps, and in favour of cores, we have simulated the observational analysis using Monte Carlo simulations (cf. [12]. Part of the scatter in the slope α could come from the fact that the "no disk" solution is not strictly true, but this gives a lower limit on α. Part of the scatter could come from lopsidedness, part could come from streaming motions, and part from slit offsets, and/or mismatches between the optical and the dynamical centers.

In all these cases, it seems very difficult to erase a peak in the distribution around slope -1 expected if the NFW profile is really valid in LSB galaxies. Only extreme values of the possible biasing effects have to be postulated, i.e. slit offsets of 5" or more in all cases, etc. On the other hand, the good resolution data favour a slope of -0.2, with a scatter of 0.2 which can naturally be explained by a modest amount of every effect considered.

3 Spirals

As already remarked by Kalnajs [19], there is not much need for a dark halo in order to explain optical rotation curves of bright spiral galaxies. This has been corroborated for many more optical rotation curves by Kent [20, 21], Athanassoula et al. [2], Buchhorn [9] as reported in Freeman [15], Moriondo et al. [34], and Palunas & Williams [37]. It is thus the extended HI rotation curves which provide the strongest evidence for the need of a dark halo. However, there is no simple way to determine how important the disk is in the central parts, and whether it dominates the potential there. Even so, models based on cosmological simulations (e.g. [36]) predict that the inner parts of spirals are also dark matter dominated. Hence the importance to study spiral galaxies in such a way so as to discriminate between maximum disk or sub-maximum disks.

3.1 Importance of Good Data in the Inner Parts

As for the LSB galaxies discussed before, it is very important to get good rotation curve data in the inner parts, so that the decomposition into a mass model with disk (and bulge sometimes), gas and dark halo components can be done correctly. This means in practice that the HI data for the outer parts need to be supplemented with data in the inner parts, if the spatial resolution of the data at 21-cm is inadequate. In the latter case, beam smearing will lower systematically the derived rotation curve. This problem, while thoroughly understood in principle, is still playing an unnecessary large role in the discussion. Partly this is due to the unfortunate circumstance that for late type spirals the differences between the Hα and HI rotation curves are not very large, and sometimes so small within the admittedly large errors of both datasets, that some people are misled into assuming that they can be ignored. This is sometimes justified *a posteriori* by taking a high resolution Hα spectrum along the major axis, which agrees with the HI rotation curve in some cases, but not in others. Systematic programs are now underway to remedy this situation, using several techniques: long slit spectroscopy of emission lines such as Hα (e.g. [13]), 3D imaging using Fabry-Pérot techniques (e.g. [10, 5]), or CO data from e.g. the BIMA interferometer [50].

3.2 Importance of Disk Self-gravity

For brighter spirals, the problem of decomposition into mass components and the degeneracy of this procedure becomes crucial. I have discussed this problem several times elsewhere, e.g. [6]. There are several arguments pleading for the disk mass being dominant in the central parts of spiral galaxies, i.e. the constraints posed by spiral structure theory, the constraints imposed by fitting systematic peculiar velocities due to motions induced by spiral arms, and the amplitude of the radial velocity jumps across shocks in barred spirals.

Swing Amplifier Criteria: Athanassoula et al. [2] have discussed the application of spiral structure constraints to composite mass models. For each model they examine if swing amplification, based on the mechanism discussed by Toomre [43], is possible. A more graphical description is given in [7]. The physics of the swing amplifier depend on the shape of the rotation curve and on a characteristic X parameter, which in turn depends on the epicyclic frequency κ, the number of arms m, and the active surface mass density of the disk.

By requiring that the swing amplification of the $m = 2$ perturbations is possible, one can thus limit the range of mass-to-light ratios to a factor of 2: the lower limit is set by requiring that the disk is massive enough so that amplification of the $m = 2$ perturbations is just allowed, and the upper limit is set by requiring that amplification of the $m = 1$ perturbations is just prohibited. Usually the latter condition is fulfilled if one requires a model with maximum disk and a halo with non-hollow core. See [2] for more details.

Non Circular Motions in Spiral Galaxies: Already in the M81 HI data obtained with the Westerbork telescope ([41]) the effects of peculiar motions due to the spiral arms are clearly seen. These were modeled with a response calculation by Visser [47], who did not include a dark halo in his models. Lately, Alfaro et al. [1] show a single long-slit spectrum of the galaxy NGC 5427, where the presence of "wiggles" in the position-velocity curve are quite clearly associated with the spiral arms. This is not new, of course, since wiggles were already seen in long slit data of quite a number of galaxies (e.g. [42, 30]), and taken to imply maximum disk (e.g. [15]). It is clear that the presence of such wiggles indicates that the disk is self-gravitating enough to produce them.

Kranz et al. [23] present long-slit data for NGC 4254, a spiral galaxy in the Virgo cluster for which also HI data are available [38]. They try to reproduce the observed velocity perturbations with a stationary gas flow model using the K-band image of the galaxy as input to the evaluation of the disk part of the galactic potential. They find that a maximum disk model produces too large velocity perturbations, and put an upper limit on the disk mass fraction (the mass ratio between a given disk model and the maximum disk model) of 0.8. However, this galaxy is lopsided in the HI, the spiral may be evolving, the small bar in the center of the galaxy might have a different pattern speed than the main spiral pattern, and the inclination may be higher than the authors take it.

In his thesis, Kranz ([24, also 25], reports on a similar analysis for four more cases, and finds a trend that the brightest spirals (those with the highest rotational velocities), seem to have maximum disks, but that towards lower luminosity spirals the relative influence of the dark matter in the inner parts increases. A comparison with the data from Athanassoula et al. [2] shows in fact good agreement with this trend.

Non Circular Motions in Barred Spirals: Weiner et al. [48, 49] model in detail the gas flow in the barred spiral NGC 4123, and find that the best fit to the velocity data requires a maximum disk model for the mass distribution. Note that here again the modelling is done as a response calculation for a stationary flow in the potential derived from an optical image. As in the case of NGC 4254, no time evolution has been considered.

Lindblad, Lindblad & Athanassoula [26] have analyzed similar data for the bright barred spiral NGC 1365, and find a relatively good fit with a maximum disk model. Since the rotation curve of NGC 1365 is declining, no dark halo has been included in these models.

3.3 Lensing Data

Gravitational lensing is a promising tool for the derivation of mass distributions, since there are no assumptions to make about the messy stellar population contents of the object which does the lensing. Maller et al. [27] discuss the possibility to determine in the case of strong lensing the flattening of the gravitational potential from eventual misalignments of the lensed images and the object doing the lensing. Trott & Webster [44] discuss the case of the lens 2237+0305, for which they combine the lens model with data on the outer rotation curve from VLA HI observations. From their models there is clear evidence for little mass due to the dark halo in the central parts, which are dominated by a bulge-bar system (also [51]). Their interpretation that the disk is not maximal is partly influenced by their inclusion of the bar into the bulge, even though bars are thought to originate in the disk.

For our own Galaxy data on the microlensing towards the bulge-bar system has been used to estimate whether there could be a NFW-like dark matter cusp in the central parts ([3, 4]). Since the lensing data has to be explained by the stars in the bar-bulge system, one can calculate how much the dark matter contributes to the inner rotation curve of the Galaxy, while keeping the constraints set by the situation in the solar neighbourhood. The models for the stellar distribution of the lensing sources slightly underpredict the lensing rate, so that there is hardly any room for dark matter in the central parts of our Galaxy.

4 Conclusion

In conclusion, our results show that the predicted steep slopes in the density profiles of LSB galaxies are not observed. The evidence for brighter spirals is that also there dark matter may not be dominant in the central parts of galaxies. This means that the current description by cosmological numerical models is incomplete at small scales, if not incorrect.

Acknowledgements

Thanks are due to Lia Athanassoula for frequent discussions, and Erwin de Blok and Stacy McGaugh for a fruitful collaboration on LSB galaxies.

References

1. E.J. Alfaro, E. Perez, R.M. Gonzalez Delgado, M.A. Martos, J. Franco: Astrophys. J. **550**, 253 (2001)
2. E. Athanassoula, A. Bosma, S. Papaiannou: Astron. Astrophys. **179**, 23 (1987)
3. N. Bessantz, O. Gerhard: Mon. Not. R. Astron. Soc. **330**, 591 (2002)
4. J.J. Binney, N.W. Evans: Mon. Not. R. Astron. Soc. **327**, L27 (2001)
5. S. Blais-Ouellette, P. Amram, C. Carignan: Astron. J. **121**, 1952 (2001)
6. A. Bosma: astro-ph/0112080 (2001)
7. A. Bosma: in ASP Conf. Ser. Vol. **182**, Galaxy dynamics, eds. D.R. Merritt, M. Valluri & J.A. Sellwood, (San Francisco: ASP), 339 (1999)
8. A. Bosma, Y.I. Byun, K.C. Freeman, E. Athanassoula: Astrophys. J. **400**, L21 (1992)
9. M. Buchhorn: Ph.D. thesis, Australian National University (1992)
10. R. Corradi, J. Boulesteix, A. Bosma, M. Capaccioli, P. Amram, M. Marcelin: Astron. Astrophys. **244**, 27 (1991)
11. W.J.G. de Blok, A. Bosma: Astron. Astrophys. **385**, 816 (2002)
12. W.J.G. de Blok, A. Bosma, S.S. McGaugh: Mon. Not. R. Astron. Soc.(in press), astro-ph/0212102 (2002)
13. W.J.G. de Blok, S.S. McGaugh, V.C. Rubin: Astron. J. **122**, 2396 (2001)
14. W.J.G. de Blok, S.S. McGaugh, A. Bosma, V.C. Rubin: Astrophys. J. **552**, L23 (2001)
15. K.C. Freeman: in Physics of Nearby Galaxies, Nature or Nurture ?, eds. T.X. Thuan, C. Balkowski & J. Tran Thanh Van (Gif-sur-Yvette: Editions Frontières), 201 (1992)
16. K.C. Freeman: in IAU. Coll. 171, The low surface brightness Universe, eds. J.I. Davies, C. Impey & S. Phillips, (San Francisco: ASP), 3 (1998)

17. T. Fukushige, J. Makino: Astrophys. J. **557**, 533 (2001)
18. O. Garrido, M. Marcelin, P. Amram, J. Boulesteix: Astron. Astrophys. **387**, 821 (2002)
19. A. Kalnajs: in IAU Symp. 100, Internal Kinematics and Dynamics of Galaxies, ed. E. Athanassoula, (Dordrecht: Reidel), 87 (1983)
20. S.M. Kent: Astron. J. **91** 1301 (1986)
21. S.M. Kent: Astron. J. **93** 816 (1987)
22. A. Knebe, J.E.G. Devriendt, A. Mahmood, J. Silk: Mon. Not. R. Astron. Soc. **329**, 813 (2002)
23. T. Kranz, A. Slyz, H.-W. Rix: Astrophys. J. **562**, 164 (2001)
24. T. Kranz: Ph.D. thesis, Heidelberg (2002)
25. T. Kranz, A. Slyz, H.-W. Rix: astro-ph/0212290 (2002)
26. P.A.B. Lindblad, P.O. Lindblad, E. Athanassoula: Astron. Astrophys. **313**, 65 (1996)
27. A.H. Maller, L. Simard, P. Guhathakurta, J. Hjorth, A.O. Jaunsen, R.A. Flores, J.R. Primack: Astrophys. J. **533**, 194 (2000)
28. D. Marchesini, E. D'Onghia, G. Chincarini, C. Firmani, P. Conconi, E. Molinari, A. Zaachei: astro-ph/0107424 (2001)
29. D. Marchesini, E. D'Onghia, G. Chincarini, C. Firmani, P. Conconi, E. Molinari, A. Zaachei: astro-ph/0202075 (2002)
30. D.S. Mathewson, V.L. Ford, M. Buchhorn: Astrophys. J. Suppl. **81**, 413 (1992)
31. L.D. Matthews, K. Wood: Astrophys. J. **548**, 150 (2001)
32. S.S. McGaugh, V.C. Rubin, W.J.G. de Blok: 2001, Astron. J. **122**, 2381 (2001)
33. B. Moore, T. Quinn, F. Governato, J. Stadel, G. Lake: Mon. Not. R. Astron. Soc. **310**, 1147 (1999)
34. G. Moriondo, R. Giovanelli, M.P. Haynes: Astron. Astrophys. **346**, 415 (1999)
35. J.F. Navarro, C.S. Frenk, S.D.M. White: Astrophys. J. **462**, 563 (1996)
36. J.F. Navarro: astro-ph/9807084 (1998)
37. P. Palunas, T.B. Williams: Astron. J. **120**, 2884 (2000)
38. B. Phookun, S.N. Vogel, L.G. Mundy: Astrophys. J. **418**, 113 (1993)
39. J.R. Primack: astro-ph/0112255 (2001)
40. J.R. Primack: astro-ph/0205391 (2002)
41. A.H. Rots, W.W. Shane: Astron. Astrophys. **31**, 245 (1974)
42. V.C. Rubin, D. Burstein, W.K. Ford Jr, N. Thonnard: Astrophys. J. **289**, 81 (1985)
43. A. Toomre: 'What amplifies the spirals'. In *The Structure and Evolution of Normal Galaxies*, eds. S.M. Fall & D. Lynden-Bell (Cambridge: Cambridge Univ. Press), pp. 111-136, (1981)
44. C.M. Trott, R.L. Webster: Mon. Not. R. Astron. Soc. **334**, 621 (2002)
45. M.S. Turner: astro-ph/0207297 (2002)
46. F. van den Bosch, R.A. Swaters: Mon. Not. R. Astron. Soc. **325**, 1017 (2001)
47. H.C.D. Visser: Astron. Astrophys. **88**, 159 (1980)
48. B. Weiner, J.A. Sellwood, J.H. van Gorkom, T.B. Williams: Astrophys. J. **546**, 916 (2001)
49. B. Weiner, J.A. Sellwood, T.B. Williams: Astrophys. J. **546**, 931 (2001)
50. T. Wong: Ph.D. Thesis, University of Berkeley (2000)
51. H.K.C. Yee: Astron. J. **95**, 1331 (1988)

A SAURON View of Galaxies

Ellen K. Verolme[1], Michele Cappellari[1], Glenn van de Ven[1], P. Tim de Zeeuw[1], Roland Bacon[2], Martin Bureau[3], Yanick Copin[4], Roger L. Davies[5], Eric Emsellem[2], Harald Kuntschner[6], Richard McDermid[1], Bryan W. Miller[7], and Reynier F. Peletier[8]

[1] Sterrewacht Leiden, Leiden, The Netherlands
[2] Centre de Recherche Astronomique de Lyon, Saint–Genis–Laval, Lyon, France
[3] Department of Astronomy, Columbia University, New York, USA
[4] Institut de Physique Nucléaire de Lyon, Villeurbanne, France
[5] Physics Department, University of Oxford, Oxford, UK
[6] European Southern Observatory, Garching, Germany
[7] Gemini Observatory, La Serena, Chile
[8] Department of Physics and Astronomy, University of Nottingham, Nottingham, UK

Abstract. We have measured the two-dimensional kinematics and line-strength distributions of 72 representative nearby early-type galaxies, out to approximately one effective radius, with our panoramic integral-field spectrograph SAURON. The resulting maps reveal a rich variety in kinematical structures and linestrength distributions, indicating that early-type galaxies are more complex systems than often assumed. We are building detailed dynamical models for these galaxies, to derive their intrinsic shape and dynamical structure, and to determine the mass of the supermassive central black hole. Here we focus on two examples, the compact elliptical M32 and the E3 galaxy NGC4365. These objects represent two extreme cases: M32 has very regular kinematics which can be represented accurately by an axisymmetric model in which all stars rotate around the short axis, while NGC4365 is a triaxial galaxy with a prominent kinematically decoupled core, with an inner core that rotates about an axis that is nearly perpendicular to the rotation axis of the main body of the galaxy. Our dynamical models for these objects demonstrate that two-dimensional observations are essential for deriving the intrinsic orbital structure and dark matter content of galaxies.

1 The SAURON Project

The formation and evolution of galaxies is one of the most fundamental research topics in astrophysics. A key question in this field is whether early-type galaxies form very early in the history of the universe or are gradually built up by mergers and the infall of smaller objects. The answer to this problem is closely tied to the distribution of intrinsic shapes, the internal dynamics and linestrength distributions, and the demography of supermassive central black holes.

In the few past decades, it has become clear that ellipticals, lenticulars, and spiral bulges display a variety of velocity fields and linestrength distributions. Two-dimensional spectroscopy of stars and gas is essential when attempting to derive information on the intrinsic structure. For this reason, we have built a panoramic integral-field spectrograph, SAURON ([1]), which provides large-scale two-dimensional kinematic and linestrength maps in a single observation.

We commissioned SAURON on the 4.2m William Herschel Telescope on La Palma in 1999. In low-resolution mode, the spectrograph combines a large field-of-view ($33'' \times 41''$) with a pixel size of $0\farcs94$. When the seeing conditions are good, the high-resolution mode, with a pixel size of $0\farcs28$, allows zooming in on galactic nuclei. SAURON observes in the spectral range of 4810–5340 Å, which contains the gaseous emission lines Hβ and [OIII] and [NI], as well as a number of stellar absorption features (Mgb, Fe, Hβ). The instrumental dispersion is $\sim$100 km/s. Between 1999 and 2002, we have used SAURON to observe a carefully-selected representative sample of 72 ellipticals, lenticulars and Sa bulges, distributed over a range of magnitudes, ellipticities, morphologies and environments ([29]).

We have finalized the data reduction, have accurately separated the emission- and absorption lines, have calibrated the line-strength measurements, and have in hand maps of the stellar and gaseous kinematics and linestrengths for all 48 E and S0 objects, with those for the spirals to follow soon. The maps reveal many examples of minor axis rotation, decoupled cores, central stellar disks, non-axisymmetric and counter-rotating gaseous disks, and unusual line-strength distributions ([2,29]). We have also developed new methods to spatially bin the data cubes to a given signal-to-noise ([4]), and to quantify the maps with Fourier methods ([5,17]). This allows accurate measurements of, e.g., the opening angle of the isovelocity contours and of the angle between the direction of the zero-velocity contour and the minor axis of the surface brightness distribution ([5]), enabling various statistical investigations of the entire sample of objects.

2 Dynamical Models

We are constructing detailed dynamical models which fit all kinematics and eventually even observations of the stellar line-strengths of the galaxies in the SAURON survey. We do this by means of Schwarzschild's ([18]) orbit superposition method, which was originally developed to reproduce theoretical density distributions (e.g., [12,14,18–20]), and was subsequently adapted to incorporate observed kinematic data in spherical and axisymmetric geometry ([6,9,11,16]). We have implemented a number of further extensions including the ability to deal with a Multi-Gaussian Expansion of the surface brightness distribution ([3,8,13]). We have also shown that the large data sets that are provided by instruments such as SAURON can be modelled without any problems ([26]).

Recently, we completed the non-trivial extension to the software that allows inclusion of kinematic measurements in triaxial geometry ([27]). As in the axisymmetric case, observational effects such as pixel binning and point-spread-function convolution are taken into account. The chaotic orbits are dealt with in the 'standard' way (see [24]), and the line-of-sight velocity profile is used to constrain the models. In the next two sections, we describe two applications in more detail, one in axisymmetry and the other for a triaxial intrinsic shape.

3 Axisymmetric Models for M32

We applied our axisymmetric modeling software to the nearby compact E3 galaxy M32 ([26]). By complementing the SAURON maps (Fig. 1) with high-resolution major axis stellar kinematics taken with STIS ([10]), the models are constrained at both small and large radii, which allows us to measure an accurate central black hole mass M_{BH}, stellar mass-to-light ratio M/L, *and* inclination i. The left panels of Fig. 2 show the dependence of $\Delta\chi^2$, which is a measure of the discrepancy between model and data, on M_{BH}, M/L (in solar units, for the I-band) and i. The inner three contours show the formal 1, 2 and 3σ-confidence levels for a distribution with three degrees of freedom. The black hole mass and mass-to-light ratio are constrained tightly at $M_\bullet = 2.5 \times 10^6 M_\odot$ and $M/L = 1.8 M_\odot/L_\odot$, and the inclination is constrained to a value near $70° \pm 5°$. The right panels of the same figure show similar contours, but now for a data-set consisting of the STIS-kinematics together with four slits extracted from the SAURON-data. In this case the constraints on all three parameters, but most notably on the inclination, are much less stringent. This demonstrates that two-dimensional observations are essential to gain insight into the intrinsic structure of galaxies.

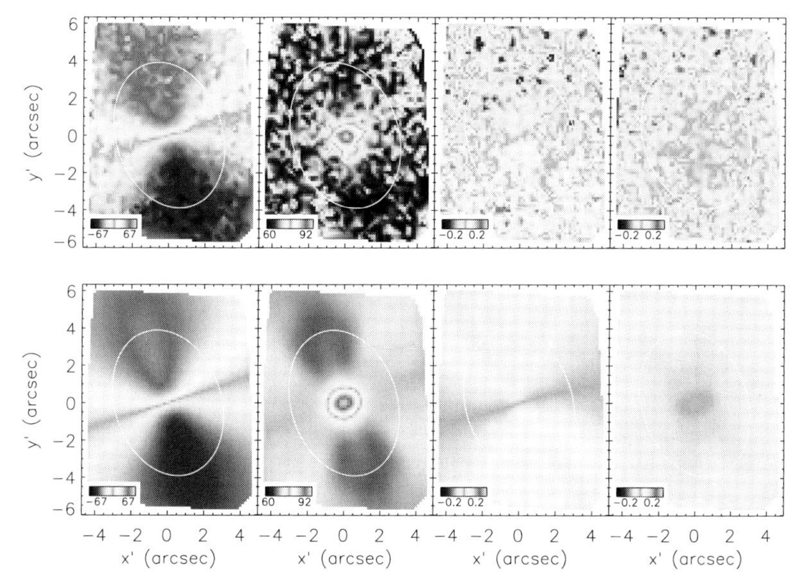

Fig. 1. Top panels: the SAURON kinematic maps for M32. From left-to-right: the mean velocity, velocity dispersion and Gauss–Hermite parameters h_3 and h_4, which measure the first and second order deviations of the line-of-sight velocity distribution from a Gaussian shape. Bottom panels: idem, but now for the best-fit axisymmetric dynamical model with I-band $M/L = 1.8 M_\odot/L_\odot$, $M_{\mathrm{BH}} = 2.5 \times 10^6 M_\odot$, and $i = 70°$.

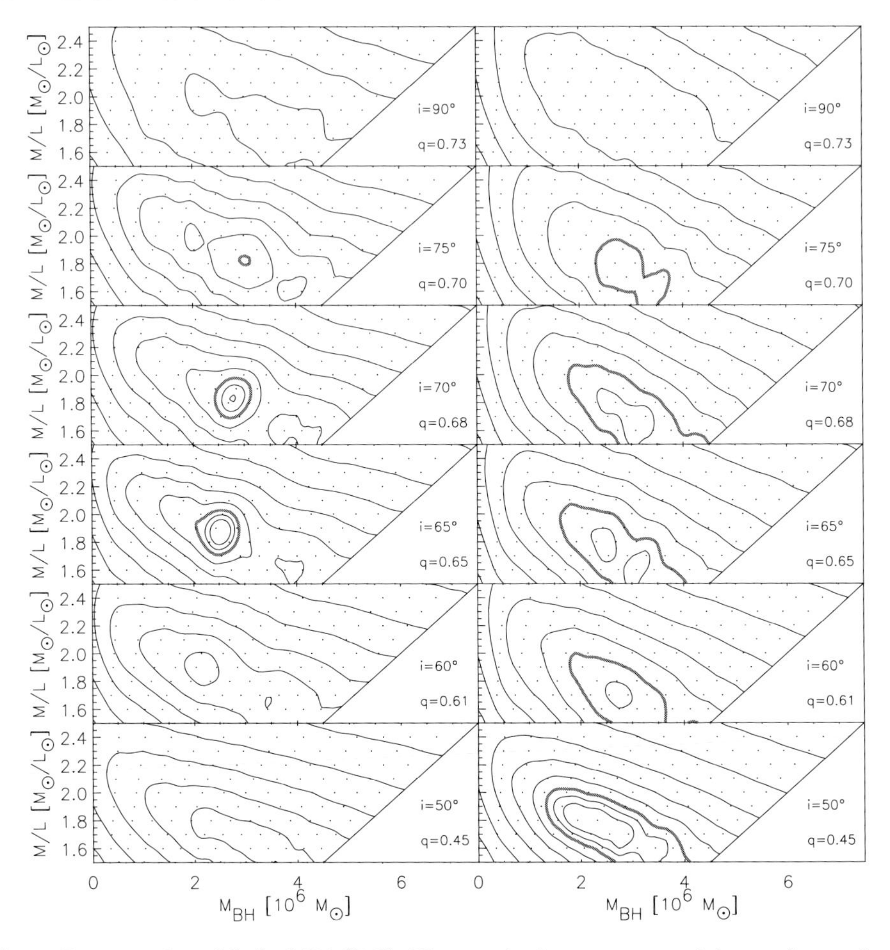

Fig. 2. Dynamical models for M32 ([26]). The panels show contours of the goodness-of-fit parameter $\Delta\chi^2$ as a function of the central black hole mass M_{BH}, the stellar mass-to-light ratio M/L and the inclination i. Each dot represents a specific axisymmetric dynamical model. The intrinsic flattening q of the models is indicated in the lower-right corner of each panel. The models are constrained by `STIS` kinematics along the major axis ([10]) together with two-dimensional observations obtained with `SAURON` in its high resolution mode ([29]). The inner three contours represent the formal $1, 2$ and 3σ-confidence levels for a distribution with three degrees of freedom. *Left panels*: model fits to a data set consisting of the `STIS`-data and the full `SAURON` field. Tight constraints are placed on the central black hole mass and mass-to-light ratio, as well as on the allowed range of inclinations. *Right panels*: the $\Delta\chi^2$ for models that were constrained by four extracted slits from the $9'' \times 11''$ `SAURON` field (major and minor axis, and at $\pm 45°$, as in [11]) and the `STIS` data. This shows that the traditional kinematic coverage provides almost no constraint on i, and that the resulting uncertainties on the inferred values of M/L and M_{BH} are correspondingly larger.

4 The Triaxial Galaxy NGC 4365

The upper panels of Fig. 3 show the stellar kinematics in the central $30'' \times 60''$ of the giant elliptical galaxy NGC4365, derived from two SAURON pointings ([7]). The velocity field clearly shows a prominent decoupled core in the inner $3'' \times 7''$

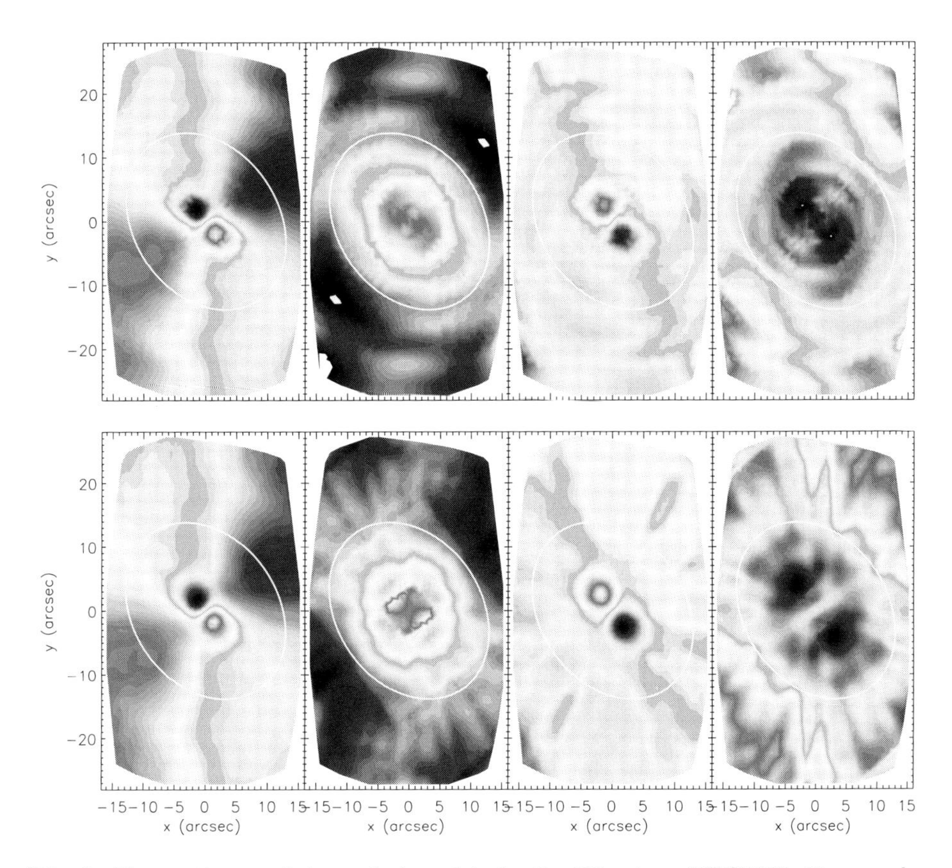

Fig. 3. Observations and dynamical models for the E3 galaxy NGC4365. *Top panels*: from left to right, the stellar velocity field, velocity dispersion, and Gauss–Hermite moments h_3 and h_4, as observed with SAURON. The maps are based on two semi-overlapping pointings, sampled at $0\farcs8 \times 0\farcs8$, and were constructed via a kinemetric expansion to provide the best representation of the data that is consistent with an intrinsically triaxial geometry (e.g., point-antisymmetry for the V and h_3 maps, cf. [5]). The original maps can be found in ([7]). The amplitude of the velocity field is about 60 km/s, the peak velocity dispersion is 275 km/s, and the contours in the h_3 and h_4 maps range between ± 0.10. The decoupled core measures $3'' \times 7''$. *Bottom panels*: idem, but now for a dynamical model with average intrinsic axis ratios $p = 0.93$ and $q = 0.69$ (triaxiality parameter $T = (1 - p^2)/(1 - q^2) \sim 0.22$), observed from a direction defined by the viewing angles $\vartheta = 85°$ and $\varphi = 15°$. This model reproduces all the main characteristics of the observations ([28]).

(cf. [23]). It has a rotation axis which lies $82^\circ \pm 2^\circ$ away from that of the body of the galaxy, which rotates around its long axis. Such a structure is possible when the shape is intrinsically triaxial because of the presence of orbits that have net mean streaming around either the long or the short axis.

The globular cluster system of NGC 4365 shows evidence for an intermediate age population ([15]). The SAURON linestrength maps, however, indicate a predominantly old stellar population ([7]), suggesting that the observed kinematic structure may have been in place for over 12 Gyr and the galaxy is in stable triaxial equilibrium. We therefore applied our developed modeling software to this case, to investigate whether it is possible to reproduce all the kinematic data in detail, and to constrain M/L and the intrinsic shape and orbital structure.

We represented the observed surface brightness distribution of NGC 4365 by a Multi-Gaussian Expansion which accurately fits the observed radial variation of ellipticity, the boxyness of the isophotes, and the modest isophotal twisting. We derived the deprojected density by assuming that each of the constituent Gaussian components is stratified on similar concentric triaxial ellipsoids. The three Euler angles that specify the orientation of the ellipsoids can be chosen freely. For each choice, we computed a library of 4000 orbits, obtained from 20 energy shells with 200 orbits each, covering the four major orbit families, and including orbits from minor families and chaotic orbits. As the spatial resolution of the SAURON measurements is modest, we did not consider the effect of a central black hole. The preliminary results indicate that the quality-of-fit parameter $\Delta\chi^2$ varies quite significantly with M/L and the parameters defining the intrinsic shape. The lower panels of Fig. 3 show the predictions of one model that fits the data well. This illustrates that the software works, and shows that NGC 4365 is indeed consistent with a triaxial equilibrium shape.

In principle, best-fit values of the shape parameters, the direction of observation, and the mass-to-light ratio can be determined by a systematic investigation of the parameter space, just as was done for M32. For triaxial systems this is a very time-consuming effort, but a first-order guess of the galaxy parameters can be obtained by using other, simpler, schemes (see, e.g., [21,22,25]). Our detailed dynamical modeling software can then be used to explore this more restricted parameter range. Work along these lines is in progress, and will make it possible to deduce, e.g., the intrinsic properties of the kinematically decoupled cores seen in many of these systems. Inclusion of higher spatial resolution data will allow accurate measurement of the mass of the central black hole.

5 Concluding Remarks

We have presented two examples of recent results from our program to construct detailed axisymmetric and triaxial dynamical models for galaxies in the SAURON representative survey of nearby ellipticals, lenticulars and Sa bulges. The panoramic SAURON observations tighten the constraints on the possible orientation of a galaxy considerably. The extension of the modeling software to triaxial shapes including kinematic constraints works, and that it will help us gain significant insight into the structure of early-type galaxies.

References

1. Bacon R., et al., 2001, MNRAS, 326, 23
2. Bureau M., et al., 2002, in ASP Conf. Ser., 273, 53
3. Cappellari M., et al., 2002, ApJ, 578, 787
4. Cappellari M. & Copin Y., 2003, MNRAS, in press
5. Copin Y. et al., 2001, EDPS Conf. Ser. in Astron. & Astrophys., eds F. Combes, D. Barret, F. Thévenin, 289 (astro-ph/0109085)
6. Cretton N., de Zeeuw P.T., van der Marel R.P., Rix H.–W., 1999, ApJS, 124, 383
7. Davies R.L., et al., 2001, ApJ, 548, L33
8. Emsellem E., Monnet G., Bacon R., 1994, A&A, 285, 739
9. Gebhardt K., et al., 2003, ApJ, in press (astro-ph/0209483)
10. Joseph C.L., et al., 2001, ApJ, 550, 668
11. van der Marel R.P., Cretton N., de Zeeuw P.T. & Rix H.-W., 1998, ApJ, 493, 613
12. Merritt D., Fridman T., 1996, ApJ, 460, 136
13. Monnet G., Bacon R., Emsellem E., 1992, A&A, 253, 366
14. Poon M.Y., Merritt D., 2002, ApJ, 568, 89
15. Puzia T., Zepf S.E., Kissler–Patig M., Hilker M., Minniti D., Goudfrooij P., 2002, A&A, 391, 453
16. Rix H., de Zeeuw P.T., Cretton N., van der Marel R.P., Carollo C.M., 1997, ApJ, 488, 702
17. Schoenmakers R.H.M., Franx M., de Zeeuw P.T., 1997, MNRAS, 292, 349
18. Schwarzschild M., 1979,, ApJ, 232, 236
19. Schwarzschild M., 1993, ApJ, 409, 563
20. Siopis C., Kandrup H. E., 2000, MNRAS, 319, 43
21. Statler T.S., 1994a, ApJ, 425, 458
22. Statler T.S., 1994b, ApJ, 425, 481
23. Surma P. & Bender R., 1995, A&A, 298, 405
24. Terzic B., Hunter C., de Zeeuw P.T., 2001, in *Stellar Dynamics: from Classic to Modern*, eds L.P. Osipkov & I.I. Nikiforov, (St. Petersburg State University, Russia), 303
25. van de Ven G., Hunter C., Verolme E.K., de Zeeuw P.T., 2003, MNRAS, submitted
26. Verolme et al., 2002, MNRAS, 335, 517
27. Verolme E.K., Cappellari M., van de Ven, G., de Zeeuw P.T., 2003, MNRAS, submitted
28. Verolme E.K., et al. 2003, in prep.
29. de Zeeuw et al., 2002, MNRAS, 329, 513

Photometric Properties of Karachentsev's Mixed Pairs of Galaxies

Alfredo Franco-Balderas, Deborah Dultzin-Hacyan, and Héctor M. Hernández-Toledo

Instituto de Astronomía, UNAM. Apartado Postal 70-264, C.P. 04510 México D.F., México.

Abstract. We present multicolor broad band BVRI photometry for a sample of 42 mixed morphology binary galaxies taken from the "Karachentsev Catalogue of Isolated Pairs of Galaxies" (KPG). Images were obtained with 0.84m and 1.5m telescopes of the Observatorio Astronómico Nacional, San Pedro Mártir, Baja California, México, operated by the Instituto de Astronomía, UNAM. Our goal is to identify and isolate some structural and photometric properties of disk galaxies and elliptical galaxies at different stages of interaction.

1 Review

The mid to late 70's saw an astronomical debate that led to the recognition that gravitational interaction is an important factor in galactic evolution affecting directly properties such as size, morphological type, luminosity, star formation rate, and mass distribution (Sulentic, [20]; Larson & Tinsley,[14] ; Stocke, [19]). According to current popular models of galaxy formation, galaxies are assembled through a hierarchical process of mass aggregation, dominated either by mergers (Kauffmann, White & Guiderdoni, [11]; Baugh, Cole & Frenk, [3]), or by gas accretion (Avila-Reese, Firmani & Hernández [2]; Avila-Reese & Firmani, [1]). In the light of these models, the influence of environment factors and interaction phenomena in the shaping and star formation of the disks is natural, at least for a fraction of the present-day galaxy population.

Pairs of galaxies occupy an initial position in the spectrum of galaxy populations and are used to measure, in approximate way, the mass of the components as well as to know the form of the gravitational potential on the basis of the morphology of the components and its evolution in time. For binary galaxies, current ideas suggest that most physical pairs are morphologically concordant, that is, with components showing similar initial star formation and angular momentum properties. However, the number of (E+S) pairs ($\sim$128) in The Catalogue of Isolated Pairs of Galaxies in the Northern Hemisphere (KPG, Karachentsev, [8]) means that, for a flux limited sample ($m < 15.7$), more than two out of every ten pairs are of the (E+S) type, suggesting that a considerable number of them must be physical binaries. The KPG was done under a criterion of strict isolation that excluded, as far as possible the optical pairs. Redshift information, available for the whole (E+S) sample, suggests that most of them are likely to be physically proximate. Digital Sky Survey images show that most of them have visible signs of disturbance; bridges, tails, common envelopes and distortions that are regarded as evidence for gravitational interaction. In addition, statistical studies indicate that a high fraction (~ 65 %) show an enhancement in the optical and FIR emission (Xu & Sulentic, [21]; Hernández-Toledo et al., [6]). This enhancement is interpreted as a by-product of interaction-induced star formation activity in physical binaries (Rampazzo & Sulentic, [16]). This is at odds with the current models of galaxy formation.

We present multicolor broad band BVRI photometry for a sample of 40 elliptical-spiral (E+S) binary galaxies taken from the "Karachentsev Catalogue of Isolated Pairs of Galaxies" (KPG). Images were obtained with 0.84m and 1.5m telescopes of the Observatorio Astronómico Nacional, San Pedro Mártir, Baja California, México, operated by the Instituto de Astronomía, UNAM. Images were calibrated using standard stars from Landolt's [13] list.

Our goal is to identify and isolate some structural and photometric properties of disk galaxies and elliptical galaxies at different stages of interaction.

2 Interactions, Mergers and Evolution

Most E+S pairs show signs of disturbance like bridges, tails, geometric and morphologic distortion, and in some cases nuclear activity, which are evidence of gravitational interactions.

Karachentsev [8] identified three basic interaction classes (AT, LI and DI) that describe the pairs which show obvious signs of interaction. AT class identifies pairs with components in a common luminous halo with a symmetric, amorphous or shredded, asymmetric (sh) structure (Fig. 1a). LI pairs show evidence of tidal bridges (br), tails (ta) or both (br+ta) (Fig. 1b). DI pairs show evidence of structural distortion in one (1) or both (2) components (Fig. 2a). We add to this sequence NI for pairs with no obvious morphological distortion (Fig. 2b).

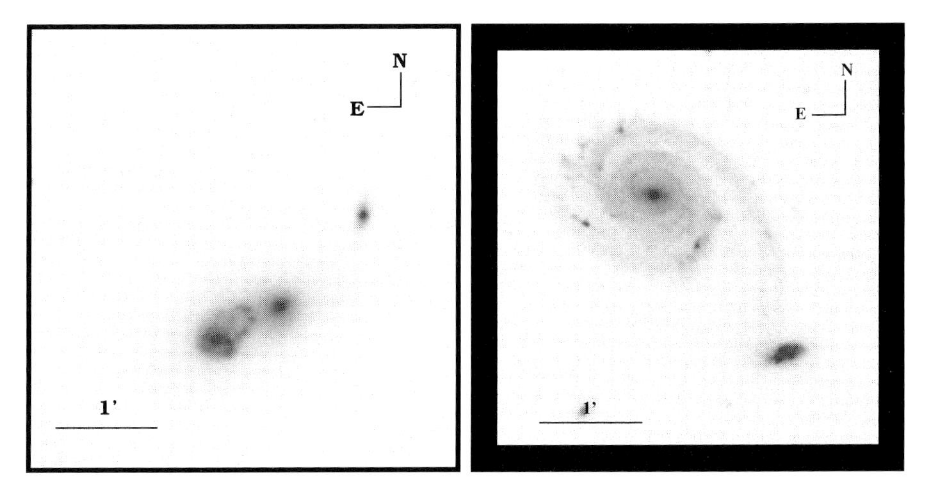

Fig. 1. Interaction classes. a)Left: KPG83 (AT pair). b)Right: KPG591 (LI pair)

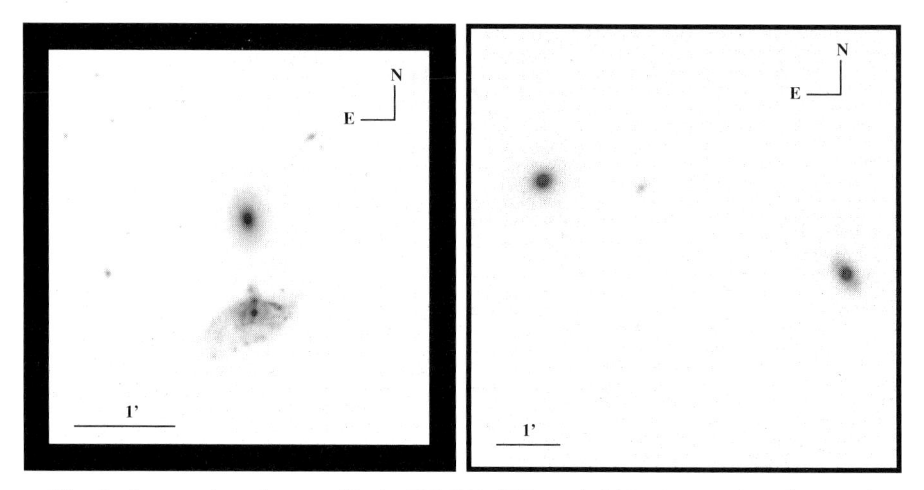

Fig. 2. Interaction classes. a)Left: KPG29 (DI pair). b)Right: KPG38 (NI pair)

The order AT-LI-DI-NI can be regarded as a sequence from strongest to weakest evidence for tidal distortion or, alternatively, most to least dynamically evolved (interpreting a common envelope as a sign of extensive dynamical evolution in pairs).

While mergers between comparable mass galaxies may be responsible for some of the most dramatic extragalactic events, minor mergers of small mass satellites may play a subtle but nonetheless critical role in the evolution of galaxies. Satellites are commonly found in the vicinity of normal galaxies (Zaritsky et al., [22]).

3 Surface Photometry

Mixed pairs of galaxies are excellent laboratories for the study of interaction in galaxies because they represent a set of objects in which we see the effects of the interaction of a rich gas object (S member) on the presence of a relatively clean disturber (E member).

The surface brightness profile of a galaxy is produced by the spatial distribution of the stars as well as the spatial distribution of the dust. To discuss the optical morphology (that could be modified by the presence of bars, spiral arms, rings, etc) and its relationship to the global photometric properties, the final results for each pair are presented in the form of a mosaic (see Fig. 3). Each mosaic includes: 1)A B-band image, 2) A B-band filtered image, 3) Surface brightness and color profiles, and 4) Its correspondent geometric profiles (radial ϵ, P.A. and A_4 coefficient). In most of the cases, foreground stars in the field have been removed.

We are proposing to use our filtered images in combination with the estimated geometric and surface brightness profiles to look for morphological features in more detail. The morphologic evaluation can be done in three parts: 1.- Visual

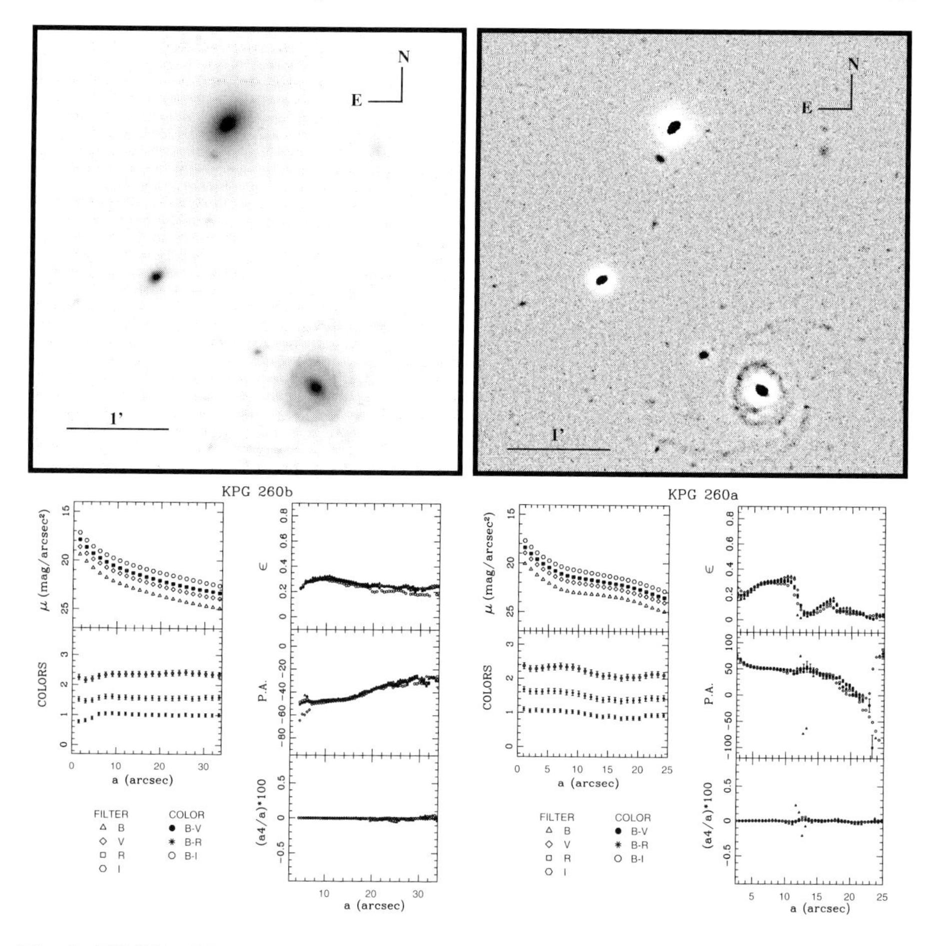

Fig. 3. KPG260 Mosaic. Top left: B-band image. Top right: B-band filtered image. Bottom left: Surface Brightness Profile and Geometric Parameters for KPG260A (west). Bottom right: Surface Brightness Profile and Geometric Parameters for KPG260B (east).

identification of false pairs E+S. 2.- Evaluation of the geometric parameters for each component. 3.- Evaluation of all characteristics, mainly those related to structures like spiral arms, shells, rings, bars and large regions of stellar formation, presumably associated with the interactions.

HST has revealed that the cores of some ellipticals (as well as the central regions of some spirals) have an excess of light relative to the de Vaucouleurs law fit. The excess light and rapid decline are evidence for central black holes. Some bright ellipticals also deviate from the de Vaucouleurs law in their outer regions, with a surface brightness in excess of the expected fit. There is evidence that mergers, capture of material and tidal disturbances from other galaxies play an important role in the final structure of elliptical galaxies.

On the other hand, spiral galaxies have bulges that are very similar to elliptical galaxies. In the disk, the brightness decreases approximately exponentially, with a characteristic scale of length for each galaxy. Barred galaxies, like non-barred galaxies, have exponential disks. There are two types of bars: Flat bars have almost constant surface brightness along the bar, that is, they have a much shallower decline than the disk. Exponential bars, in contrast, have the same scale of length as the disk.

Measurements of surface brightness profiles are essential for quantitative investigations of galaxy morphology, decomposition of bulge and disc, studies of galaxy structure and stellar populations, and measurements of dust distribution. Important parameters like the size, the distribution of mass, the star formation rate, as well as the magnitude of nuclear activity can be determined with the help of our observations.

4 Deviations from Perfect Ellipticity

The intensity of an elliptical galaxy can be expressed as:

$$I = I_0 + A_n * \sin(n * \phi) + B_n * \cos(n * \phi) \tag{1}$$

where ϕ measures the position angle of the major axis. The A_n and B_n coefficients for n > 1 represent the amplitudes of the deviations form perfect ellipticity, which are typically around 1 %. The quantity :

$$\frac{a_4}{a} = \frac{\sqrt{1-\epsilon} * B_4}{a * |\frac{dI}{da}|} \tag{2}$$

(see Bender & Möllenhoff, [4]), were a is the length of the isophote's semimajor axis, forms a dimensionless measure of the diskiness of the isophote, indicate whether a galaxy is boxy ($a_4/a < 0$) or disky ($a_4/a > 0$). See Fig. 4.

The traditional view on the formation and evolution of giant elliptical galaxies is that they are very old stellar systems and all formed very early at a redshift of more than two (Searle, Sargent & Bagnuolo [18]). Alternatively, hierarchical theories of galaxy formation predict that massive galaxies were assembled relatively late in many generations of mergers of disk-type galaxies or smaller subunits and mass accretion. It has been argued by Kauffman [9] and Kauffman & Charlot [10] that this merger scenario is consistent with observations of elliptical galaxies at different redshifts. Naab & Burkert [15] performed a large set of collisionless N-body simulations (taking into account the stellar and dark matter component) of merging disk-galaxies with mass ratios of 1:1, 2:1, 3:1, and 4:1. They show that 1:1 merger remnants rotate slowly, are supported by anisotropic velocity dispersions, have significant minor-axis rotation, and show predominantly boxy isophotes in good agreement with observations of bright giant ellipticals. 3:1 and 4:1 remnants are isotropic, fast rotators, show a small amount of minor-axis rotation, and have disky isophotes in perfect agreement with observations of faint fast rotating giant elliptical galaxies. 2:1 remnants show intermediate properties.

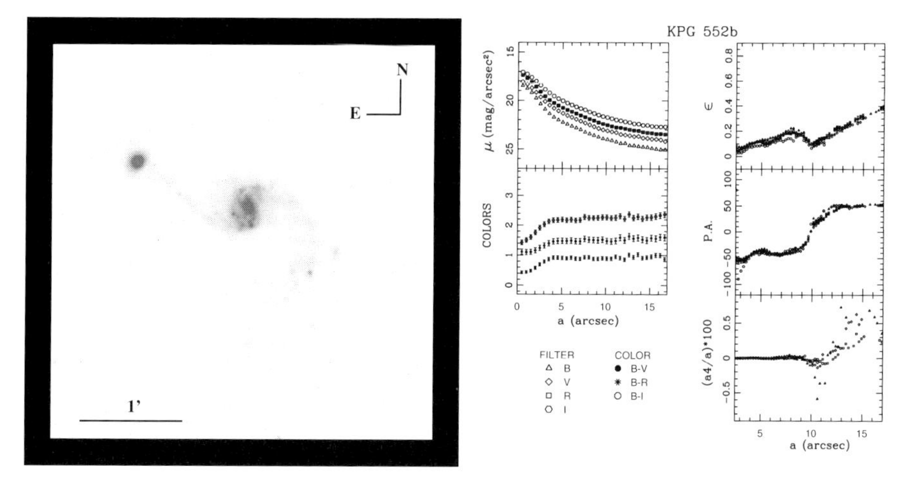

Fig. 4. Example of an elliptical galaxy with isophotal twist and external disky structure. Left: B-band image of KPG552. Right: Surface Brightness Profile and Geometric Parameters for KPG552B (west).

Projection effects lead to a large spread in the data in good agreement with our observations.

The frequency and spatial distribution of disky and boxy ellipticals in pairs E+S could provide interesting information on the frequency of equal- and unequal mass mergers in different environments

5 Isophotal Twist

The major axis position angle (PA) gives the orientation of the galaxy in the sky: it is measured counterclockwise from north to the major axis. Variation of PA with radius, or twist, may be an indication of triaxiality for elliptical galaxies with no axis of rotational symmetry. Isophotal twist may be common in spirals. Such twist may also be interpreted as an indication of the presence of small bars, rings or nonaxisymmetric bulges in the central regions. Variations in ellipticity are associated with a isophotal twist.

6 The Holmberg Effect

Holmberg [7] compared the photographic colors of paired galaxies and found a significant correlation between the colors of pair components. The physical explanation of the Holmberg effect is complex, it has been interpreted as reflecting a tendency for similar types of galaxies to form together (morphological concordance), a possible reflection of the role of local environment in determining galaxy morphology, but alternatively, it can presumably also reflect mutually induced star formation (Kennicutt et al., [12]) in physical pairs.

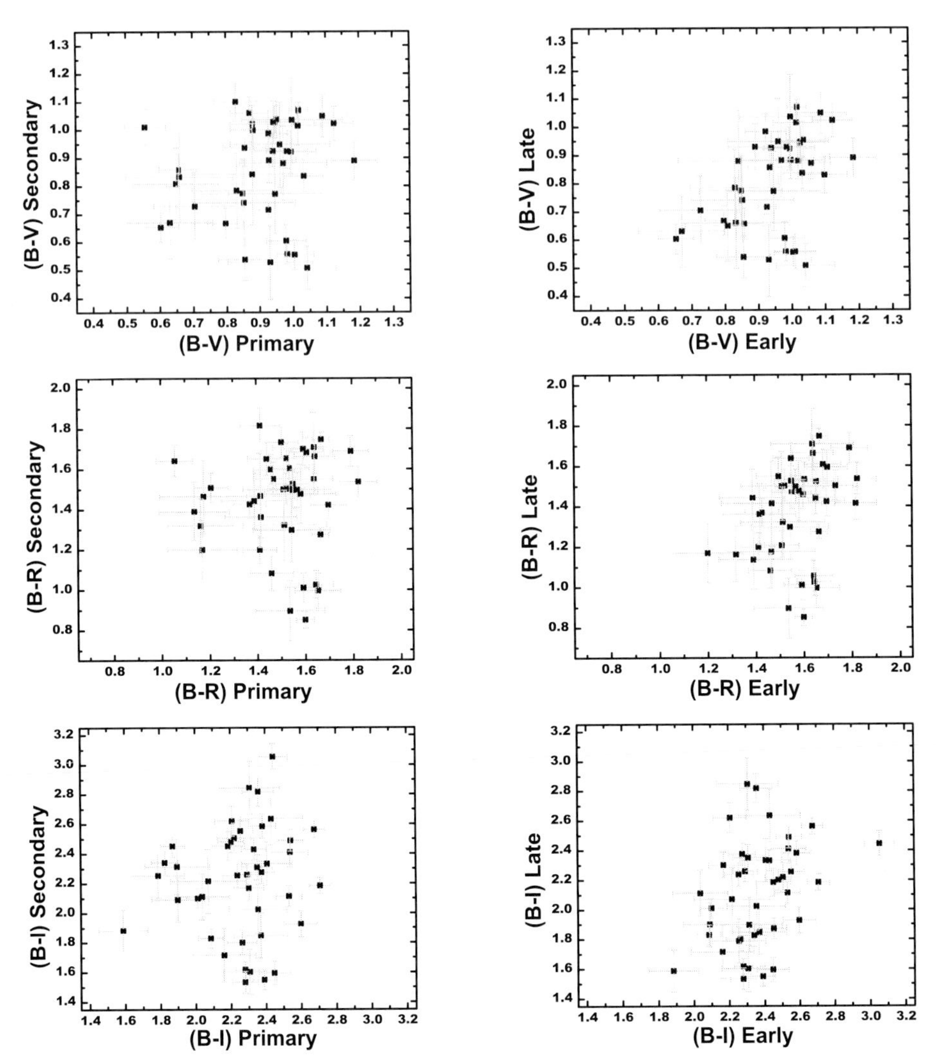

Fig. 5. Holmberg Effect. Left column:color index of the fainter member in a E+S pair (secondary component) referred to the color index of the brighter (primary component). Right column: color index of the S member versus the color index of the E member.

Figure 5 show the Holmberg Effect for our sample. The color correlation between pair components in mixed pairs is poor in any plot, contrary to the results of Demin [5]. Any tendency, if present, could be explained by the intrinsic scatter in the $(B-V)^0_T$ - Morphological Type correlation as reported by Roberts & Haynes [17]. However the evidence is not conclusive due to the magnitudes are not corrected by galactic extinction, K-correction and inclination correction.

7 Summary

Until this moment, we conclude that:

• An important number of E+S pairs are misclassified ($\sim$ 25 %). The pairs not classified as true (E+S) pairs are primarily of two types: 1) disky pairs composed of spiral and lenticular components and 2) early pairs composed of elliptical and lenticular components. Both of these classes raise interesting questions about galaxy formation because of the discordant star formation and angular momentum properties of the components.

• Most spirals in mixed pairs have redder colors than normal spirals [(B-V) > 0.6] which could be related with formation of dust by an event of starburst.

• On the other hand, the ellipticals in mixed pairs have redder colors than normal ellipticals too [(B-V) > 0.9] which could be related with transference of gas and dust belonging to the spiral galaxy.

• In E+S pairs, "grand design" spirals are common and luminous, whereas a few low-luminosity S members exhibit flocculent spiral structure.

• Boxy ellipticals tend to be more luminous than disky ones, in E+S pairs.

• Possibly boxy ellipticals are formed by mergers of (mostly) ancestral objects. On the other hand, disky ellipticals may have been formed from a single proto-galaxy, or from the merger of mainly gaseous ancestral objects.

• Morphological distortion is common in all the sample.

References

1. Avila-Reese V. & Firmani C., 2000, Rev. Mex. Astron. Astrofís. 36, 23.
2. Avila-Reese V., Firmani C. & Hernández X., 1998, ApJ 505, 37.
3. Baugh C. M., Cole S. & Frenk C. S., 1996, MNRAS 283, 1361.
4. Bender R. & Möllenhoff C., 1987, A&A 177, 71.
5. Demin V. V., Zasov A. V., Dibai E.A., Tomov A.N., 1984, SvA 28, 367.
6. Hernández-Toledo H. M., Dultzin-Hacyan D., González J. J. & Sulentic J. W., 1999, AJ 118, 108.
7. Holmberg E., 1958, Lund Medd. Astron. Obs. Ser. II, 136, 1.
8. Karachentsev I. D., 1972, Catalogue of Isolated Pairs of Galaxies in the Northern Hemisphere, Comm. Spec. Ap. Obs. 7, 1.
9. Kauffman, L.H., 1996, magr.meet.
10. Kauffman, L.H., & Charlot 1998, pfg..conf.
11. Kauffmann G., White S. D. M. & Guiderdoni B., 1993, MNRAS 264, 201.
12. Kennicutt R. C., Roettiger K. A., Keel W. C., van der Hulst J. M. & Hummel E., 1987, AJ 93, 1011.
13. Landolt A. U., 1992, AJ 104, 340.
14. Larson R. B. & Tinsley B. M., 1978, ApJ 219, 46.
15. Naab T. & Burkert A., 2001, ApJ 555, 91.
16. Rampazzo R. & Sulentic J. W., 1992, A&AS 259,43.
17. Roberts M. S. & Haynes M. P. 1994, ARA&A 32, 115.
18. Searle L., Sargent W. L. W. & Bagnuolo W. G., 1973, ApJ 179, 427.
19. Stocke J. T., 1978, AJ 83, 348.
20. Sulentic J. W., 1976, ApJS 32, 171.
21. Xu C. & Sulentic J. W., 1991, ApJ 374, 407.
22. Zaritsky D., Kennicutt R.C. & Huchra J. P., 1994, ApJ 420, 87.

Spline Histogram Method for Reconstruction of Probability Density Functions of Clusters of Galaxies

Dmitrijs Docenko[1] and Kārlis Bērziņš[2]

[1] Institute of Astronomy, University of Latvia, Raina blvd 19, Riga LV-1586, Latvia; dima@latnet.lv
[2] Ventspils International Radio Astronomy Center, Akademijas laukums 1-1503, Riga LV-1050, Latvia; kberzins@latnet.lv

Abstract. We describe the spline histogram algorithm which is useful for visualization of the probability density function setting up a statistical hypothesis for a test. The spline histogram is constructed from discrete data measurements using tensioned cubic spline interpolation of the cumulative distribution function which is then differentiated and smoothed using the Savitzky-Golay filter. The optimal width of the filter is determined by minimization of the Integrated Square Error function.

The current distribution of the TCSplin algorithm written in f77 with IDL and Gnuplot visualization scripts is available from www.virac.lv/en/soft.html.

1 Introduction

Whenever one makes a physical measurement one obtains a discrete result, starting from spatial measurements and ending with classification of some set of objects by some quantity. Particular measured value follows from the statistical properties of the system strictly following the probability distribution function, hereafter PDF. The PDF, in its turn, is determined by the physical properties of the system. If a measured data set is statistically complete then its PDF contains information about the system's physical properties. The PDF shows a character of unimodal or multimodal systems. It is natural to assume that the PDF of unimodal physical systems contain only one global maximum and several maxima indicate the multimodality of a data set. Therefore the shape of the PDF allows one to classify the measured data points, e.g. to find structure in case of positional measurements.

In statistics it is widely accepted to use histograms as the PDF approximations. It is also well known that ordinary histograms being dependent on two free parameters (bin size and its location) give very subjective results. Many methods have been developed trying to solve this problem [5]. However, most of them are still dependent on some parameters in a non-objective manner.

Generally, the probability density estimation methods can be divided into two main groups: parametric and non-parametric. The first ones assume some definite type of the PDF function (e.g. Gaussian or their superposition) and try to find the best-fit parameters for it. A good such example is the KMM mixture modelling algorithm [2]. A main disadvantage of these methods is that not all data sets can be well fitted with any chosen function. Rather often the real PDF of physical system has significant difference from a chosen best-fit function, and in many cases it is not known *a priori* what function it should be at all.

Non-parametric methods try to construct PDF estimates as compromise of two opposite demands. First, the estimate should be as close as possible to the measured PDF. Second, statistical noise due to a finite volume of the selection should be filtered out. There are several ways how to do it.

It is possible to minimize a functional that is a sum of two terms – the statistical noise and the one increasing with a difference between data points and the PDF estimate (Vondrak's method) [21]. Unfortunately, there is still one free coefficient responsible for the smoothing degree. This coefficient is not determined in any automated way and usually is found from good-looking conditions. Another method is to convolve the initial guess of the PDF with some kernel (kernel methods) for data smoothing. Also in this case there remains a free coefficient – the kernel width, that is responsible for smoothing of the function in an "optimal way", besides the result is weakly dependent on the chosen kernel shape [5], [20].

There is, however, a method that allows one to choose an optimal smoothing width: the PDF should not be over-smoothed and lose its true features, and the noise level should be diminished as far as possible on the other hand. This method is described e.g. in [20]. Its main idea is to define an Integrated Square Error (ISE) function that shows the difference between the real PDF and its estimate, and then to minimize it. The ISE function method is implemented for kernel methods in e.g. [15], and results are encouraging. However, the ISE function itself is often rather noisy.

In this paper we propose another approach to estimate PDF of a given one-dimensional data set in automatic and optimal way. This is the spline histogram method. We have found that the tensioned cubic splines are suitable for this task and the corresponding algorithm has been called TCSplin. The current version of the TCSplin code is freely available in the internet, it is also included in the CD-ROM.

For demonstration purposes in this article we will use the spline histogram algorithm to find a "well determined" redshift structure of galaxies within clusters Abell 2256 and Abell 3626.

This paper has the following structure. The spline histogram algorithm will be discussed in section 2. Bootstrapping simulations, discussed in section 3, help to evaluate errors of the PDF estimates. In section 4 the spline histogram application to data sets of clusters of galaxies A2256 and A3526 will be shown as examples. Finally, some concluding remarks will be given in section 5.

2 The Spline Histogram Algorithm

The spline histogram method is a non-parametric approach for reconstruction of probability density function underlying statistical selection. It was first discussed in [4] as one of possible methods to detect substructure in clusters of galaxies.

Recently it was further developed in [9] and these results are summarized in this paper.

From spectroscopic observations we obtain redshifts of galaxies in clusters. Let us denote redshift of the ith galaxy by z_i and order them ascendentally ($z_i \leq z_{i+1}$). Next step is to construct a step-like cumulative distribution function (CDF) obtained purely from observational data: $F_{obs}(cz) = N(z_j < z)/N_{gal}$, where $N(z_j < z)$ is a number of galaxies with redshift smaller than z, and N_{gal} is a total number of detected galaxies, c is the speed of light. At this stage we assume that the data set is statistically complete being representative of the physical situation. The PDF $f(x)$ by definition is the derivative of $F_{obs}(cz)$ in respect to cz. If CDF is constructed as shown before then $f(x)$ is a sum of Dirac δ-functions.

In the spline histogram method the points z_i with ordinates $F_{obs}(cz_i)$ are consequently connected by non-decreasing smooth analytical spline $S(cz)$. After construction of $S(cz)$ the latter is analytically differentiated leading to the PDF estimate $\hat{f}(cz)$. This procedure guarantees that the obtained continuous PDF is in agreement with the discrete distribution of the data points. The PDF contains all initially observed information about the cluster and it has a lot of noise as a consequence. To diminish the noise, $\hat{f}(cz)$ has to be smoothed.

Trying to utilize usual cubic splines for interpolation of the CDF, one encounters the problem that they will generally have negative derivative intervals if both first and second derivatives in the data points are put to be equal. Although there is an infinite amount of possibilities how to construct a smooth continuous spline that has non-smooth first derivative at data points.

We have found that tensioned cubic splines (hereafter TCS) nicely fit all spline histogram needs. They are defined such that the cubic polynomial spline length between two data points is minimal, and only the interpolating function and its first derivative are continuous in data points. Also in this case sometimes a derivative of the TCS is negative. Then in order to exclude a non-physical decreasing of the CDF estimate, we use non-tensioned splines increasing accordingly the spline length within these regions.

To reduce the statistical noise, the algorithm has been symmetrized. For the same purpose there was added a possibility to unite close points in the data set, that would otherwise give unphysical high PDF peaks. Nevertheless the resulting PDF is noisy and due to this the next step of the algorithm is a smoothing procedure.

In our case the noise is seen as narrow high peaks in the PDF arising from high CDF derivatives between close data points. The Savitzky-Golay filters [16] have been chosen for the smoothing remembering that PDF construction without any *a priori* knowledge about the system character was one of the main reasons for developing the spline histograms. These filters locally conserve first moments of the smoothed function. The remaining problem is to define the optimal width of the filter such that it reduces the noise but not over-smoothes the real PDF features.

This is done using the Integrated Square Error (ISE) function [20]:

$$ISE(\hat{f}(cz)) = \int\limits_{cz_{min}}^{cz_{max}} \left(\hat{f}(cz) - f(cz)\right)^2 d(cz), \tag{1}$$

where $f(cz)$ is a true PDF underlying the observed selection, and $\hat{f}(cz)$ is a PDF estimated from the observed data, i.e. the smoothed spline histogram in our case. As we are using digital filters to smooth the data, this should be rewritten for case of discrete points. It follows from the theory [18], [9], [20] that quantity $P(h)$ will have minimum for the same smoothing width h as $ISE(\hat{f}(x))$:

$$P(h) = \sum_{i=1}^{N} \left(\hat{f}(cz_i)^2 - 2\hat{f}(cz_i) + 2C_0^{(h)}\right), \tag{2}$$

where $C_0^{(h)}$ are the smoothing filter zeroth coefficients, and it was taken into account that $\sum_{i=1}^{N} \hat{f}(cz_i) = N$. In contrary to the equation defining the ISE function, $P(h)$ can be easily calculated from the data.

The filter width that gives the minimal $P(h)$ value is the optimal one because the corresponding deviation between the true and estimated PDFs is also minimized. As a result the spline histogram is obtained but it says nothing about the remaining statistical noise level in it. To find it out we use a bootstrapping technique described in the next section.

3 Simulated Data Analysis

Simulated data are produced and analysed as follows. Using the obtained spline histogram as a true PDF, we generate the same amount N of random numbers. Then from this selection we compute another smoothed spline histogram. Repeating this sufficient number of times (say 100), one can calculate the average of the simulated spline histograms and its scattering. It is useful to characterize the scattering by the distribution quartiles. The upper quartile shows that the estimated PDF has 75% probability to be below it. For the lower quartile, accordingly, this probability is 25% (see Fig. 1).

To estimate the quality of the approximation, the first moments of several simulated distributions were computed and compared with the original values (see Table 1). It can be seen that the average values are the same within statistical error bars (1σ), whereas the standard deviations are about 10% larger than the original values because of the smoothing effect. For Gaussian distributions the asymmetry and excess are significantly different from zero, although in non-Gaussian cases they are rather close to original values.

Dependence of the smoothing size on the selection volume was also analysed. From theoretical considerations [18], [20], [15], we know that the optimal smoothing size depends on the selection volume N in the following way: $h_{opt} \propto N^{-1/5}$. Analysing different volume random number selections for the same initial distribution, we have empirically found that for our algorithm $h_{opt} \propto N^{-0.195}$, that shows an excellent agreement with the theoretical prediction.

Table 1. Moments of the initial distribution from simulations of the 500 point selection

Gaussian distribution	Average	St.Dev.	Asymmetry	Excess
General distribution	0.500	0.089	0.000	-0.006
Average from simulations	0.499	0.093	-0.486	2.287
St.Dev. from simulations	0.004	0.003	0.109	0.355
2 equal dispersion Gaussians	Average	St.Dev.	Asymmetry	Excess
General distribution	0.475	0.190	0.190	-0.718
Average from simulations	0.474	0.192	0.153	-0.676
St.Dev. from simulations	0.009	0.005	0.068	0.104
2 different dispersion Gaussians	Average	St.Dev.	Asymmetry	Excess
General distribution	0.565	0.190	-0.112	-0.836
Average from simulations	0.564	0.192	-0.165	-0.704
St.Dev. from simulations	0.009	0.004	0.064	0.105

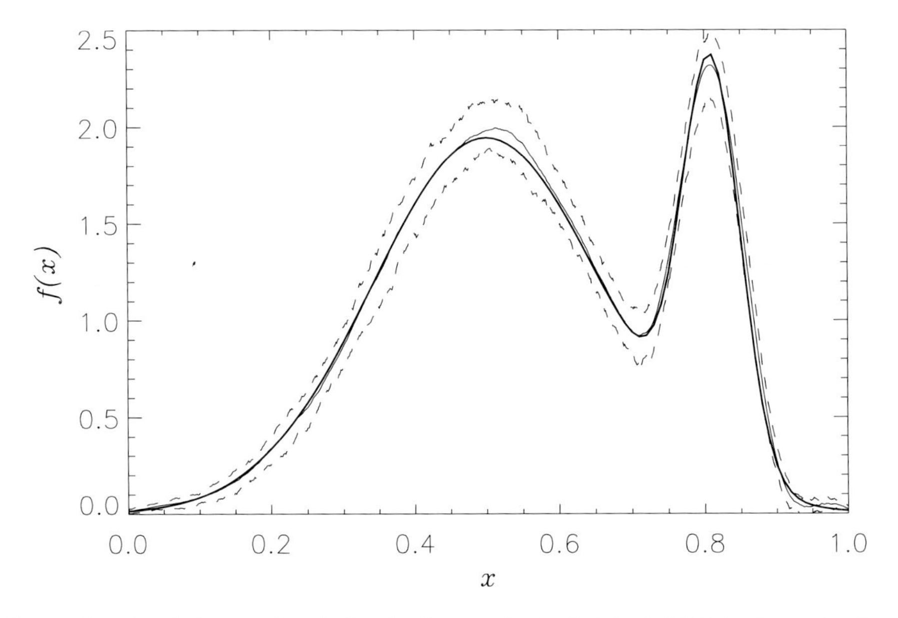

Fig. 1. Result of the simulated distribution analysis. Original PDF is shown by the thick solid line, the thin solid line represents the average value of 100 smoothed spline histograms using 500 point selection each, and the lower and upper dashed lines are the first and third quartiles, respectively.

4 Galaxy Cluster Data Analysis

As example we show the implementation of the algorithm on two clusters of galaxies – Abell 2256 and 3526.

Abell 2256 is a rich regular cluster at $z \approx 0.06$ ($\alpha \approx 17^h 03.7^m$, $\delta \approx +78°43'$, equinox 2000.0 [1]). It has similar properties to the Coma cluster (similar X-ray luminosities, both have optical and X-ray substructure and a radio halo), but is situated approximately 2.5 times farther.

A2256 has been previously studied in x-rays, optical and radio by several authors, e.g. [10], [6], [7]. It is accepted and understood, that Abell 2256, being one of the best studied clusters of galaxies, exhibit complex inner structure.

Result of the implementation of the TCSplin algorithm to the data of [10], consisting of 89 galaxy redshift measurements, is shown in Fig. 2. From the figure one can obviously see that the cluster is unrelaxed and has strongly non-Gaussian velocity PDF. Most likely it consists of two or more merging parts that currently are undergoing a final stage of unification.

Centaurus cluster A3526 ($z \approx 0.011$, $\alpha \approx 12^h 48.9^m$, $\delta \approx -41°18'$, equinox 2000.0 [1]) has been extensively studied, as it is a nearby rich cluster of galaxies. It is intermediate between Coma and Virgo clusters in richness and in distance and has richness class 1 or 2 (e.g. [17]). Centaurus is irregular in appearance, like Virgo. The cluster core has two apparent centres of concentration, one being centred on NGC 4696 and the other being 0.5° further east ([12], [3]).

Extensive study of this cluster is made in [8], [13] and [14]. The research included determination of redshifts for 259 galaxies and photometry for 329 galaxies within 13° field centred on the cluster, and the following analysis of data. The bimodal galaxy velocity distribution and extensive substructure in both subclusters have been found. Mean heliocentric velocities and line-of-sight dispersions of two main cluster components, within 3° of the cluster centre, are 3041 and 586 km sec^{-1} (denoted Cen30), and 4570 and 262 km sec^{-1} (denoted Cen45), respectively. The projected distributions of members of each component overlap on the sky. Other small galaxy groups also have been found in this study.

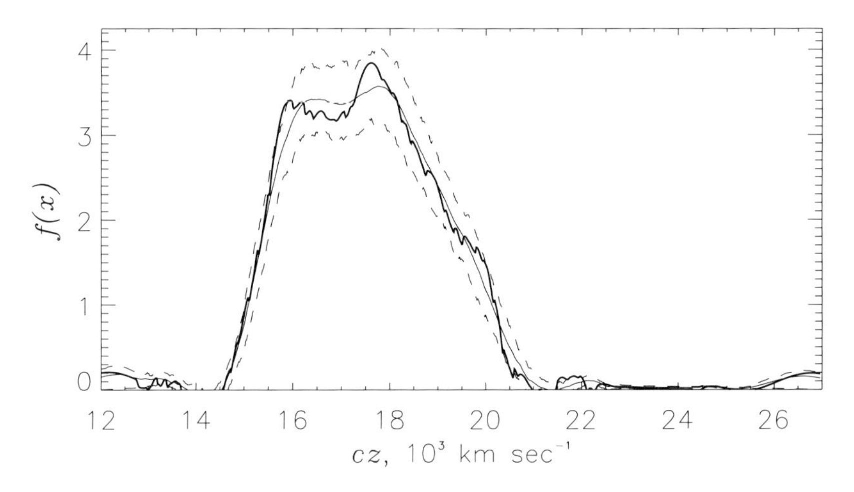

Fig. 2. The spline histogram of A2256. The meaning of lines is the same as in Fig. 1

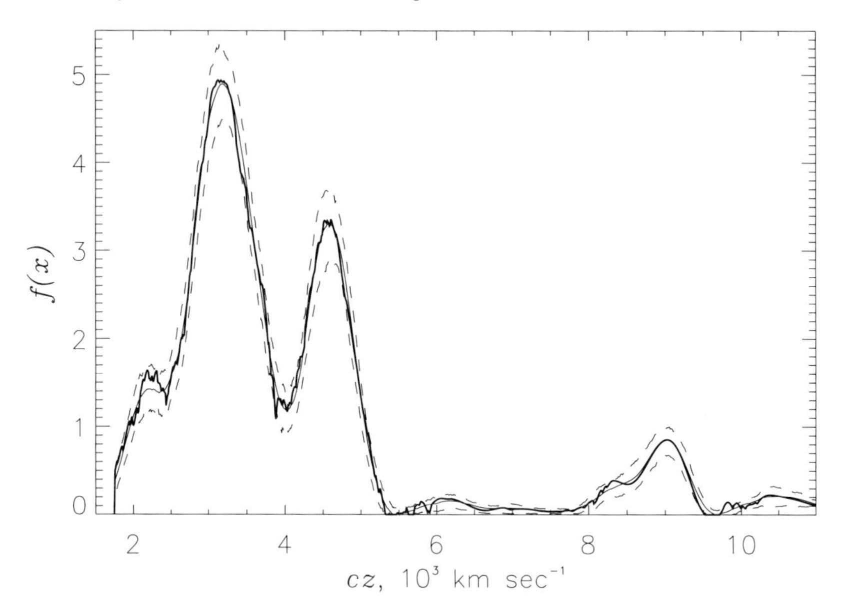

Fig. 3. The spline histogram of A3526. The meaning of lines is the same as in Fig. 1

Recently bimodality of the cluster has been confirmed in [19]. The authors used it to test a non-parametric method of the PDF estimation proposed by [11] and the same two main features of the cluster were noticed.

Our result of processing the data set of [8] is shown in Fig. 3. We find the same two main structures as in the original analysis. Clearness of these features demonstrates the quality of the algorithm. Shape of each of these components is close to Gaussian indicating their relatively relaxed state. Besides that the spline histogram shows additional left "shoulder" of the Cen30 group at around $cz \approx 2100$ km sec^{-1}. This probably is one of separate groups noticed by [14]. Possibly this as well as those features around 6200 and 8300 km sec^{-1} are not spatially real but just the redshift space caustics artefacts.

We see that a direct implementation of the algorithm leads to a good estimate of the PDF of clusters of galaxies. The only difference is the dispersions of the group velocities that are overestimated due to our PDF smoothing. One should keep that in mind and calculate the dispersion directly from the original data if needed.

5 Concluding Remarks

This paper has demonstrated the usefulness of the spline histogram algorithm in statistical studies of 1D data sets. It has all advantages over the well known ordinary histogram approach estimating the probability density functions. In principle the spline histograms may be expanded to higher dimensional cases but that introduces higher effect of the sampling noise. Unfortunately enlargement of a data set size does not necessarily guarantee larger signal to noise ratio. More generally it is dependent on the distribution character.

The latest version of the spline histogram algorithm TCSplin code is freely available online from http://www.virac.lv/en/soft.html. Presently it is written in f77 with IDL and Gnuplot visualization scripts.

Acknowledgments

DD is grateful to the European Physical Society EWTF that has provided a travel grant and the LOC of Research centre of Astronomy of Academy of Athens for possibility to participate in the Workshop. DD and KB also thank Bernard Jones for the fruitful discussions. The work of KB was supported by grant No.01.0024.4.1 of the Latvian Council of Sciences.

References

1. G.O. Abell, H.G. Corwin, Jr, and R.P. Olowin: ApJSS, **70**, 1 (1989)
2. K.A. Ashman, C.M. Bird, and S.E. Zepf: AJ, **108**, 2348 (1994)
3. N.A. Bahcall: ApJ, **193**, 529 (1973)
4. K. Berzins: in *Generation of Cosmological Large-Scale Structure* ed. by D.N. Schramm and P. Galeotti (Kluwer Academic Publishers, 1997) pp.283-288
5. K. Berzins: *Substructure of clusters of galaxies: Methodology*, Master Thesis, University of Copenhagen and University of Latvia, Copenhagen, Riga (1998), http://www.virac.lv/papers/kberzins_msc.tar.gz
6. U.G. Briel, et al.: A&A, **246**, L10 (1991)
7. T.E. Clarke, and T.A. Ensslin: astro-ph/0106137 (2001)
8. R.J. Dickens, M.J. Currie and J.R. Lucey, MNRAS **220**, 679 (1986)
9. D. Docenko: *Spline histograms and their application to analysis of clusters of galaxies* (in Latvian), Master Thesis, University of Latvia, Riga (2002), http://www.virac.lv/papers/ddocenko_msc.pdf
10. D.G. Fabricant, S.M. Kent, and M.J. Kurtz: ApJ, **336**, 77 (1989)
11. D. Fadda, E. Slezak, and A. Bijaoui: A&ASS **127**, 335 (1998), astro-ph/9704096
12. A.R. Klemola: AJ, **74**, 804 (1969)
13. J.R. Lucey, M.J. Currie and R.J. Dickens: MNRAS, **221**, 453 (1986)
14. J.R. Lucey, M.J. Currie and R.J. Dickens: MNRAS, **222**, 427 (1986)
15. A. Pisani: MNRAS **265**, 706 (1993)
16. W.H. Press, S.A. Teukolsky, W.T. Vetterling, and B.P. Flannery: *Numerical Recipes in FORTRAN*, Cambridge University Press, Cambridge (1992)
17. A. Sandage: ApJ, **183**, 731 (1973)
18. B.W. Silverman: *Density Estimation for Statistics and Data Analysis*, Chapman & Hall, London (1986)
19. P. Stein, H. Jerjen, and M. Federspiel: A&A **327**, 952 (1997), astro-ph/9707211
20. R. Vio, G. Fasano, M. Lazzarin, and O. Lessi: A&A, **289**, 640 (1994)
21. J. Vondrak: Bulletin of Astronomical Institute of Czechoslovakia, **20**, 349 (1969)

Stars Close to the Massive Black Hole at the Center of the Milky Way

Nelly Mouawad[1], Andreas Eckart[1], Susanne Pfalzner[1], Christian Straubmeier[1], Rainer Spurzem[2], Reinhard Genzel[3], Thomas Ott[3], and Rainer Schödel[3]

[1] I. Physikalisches Institut, University of Cologne, Zülpicher Str.77, 50939 Cologne, Germany
[2] Astronomisches Rechen-Institut, Mönchhofstr. 12 bis 14, 69120 Heidelberg, Germany
[3] Max-Planck-Institut für extraterrestrische Physik, Giessenbachstr., 85748 Garching, Germany

Abstract. Recent measurements of stellar velocities ([5], [7]) and variable X-ray emission [3] near the center of the Milky Way have already provided the strongest case for the presence of a super-massive black hole in our Galaxy. Information on the enclosed mass and stellar number density counts, in the central stellar cluster of the Galaxy, now allows to derive realistic potentials to study stellar orbits. We present the results of calculations using a 4th-order Hermite integrator. They provide valuable additional information on the three dimensional distribution and dynamics of the He-Stars. We also discuss the importance of Newtonian peri-astron shifts for stellar orbits in the central cluster and how future observations with infrared interferometers (LBT, VLTI, Keck) [6] will help to improve our understanding of the dynamics and distribution of the stars in this region.

1 Introduction

Stellar proper motions, radial velocities and accelerations obtained with high angular resolution techniques over the past decade have convincingly proven the presence of 3 million solar masses in the center of our Galaxy. This mass is associated with the compact radio source Sagittarius A* and currently represents the best candidate for a super massive Black Hole. In this gravitational potential at the center of the Milky Way, the stars show large orbital velocities. In the central arcsecond those can be observed as proper motions via repeated imaging at the highest possible resolution.The largest velocities in the vicinity of SgrA* are several 1000 km/s with a maximum $\geq$5000 km/s for the early type star S2. The location at the maximum stellar velocity dispersion and the center of gravitational force, as determined from the orbital accelerations of the two stars S1 and S2 agree to within less than 100 mas with the position of the compact radio source SgrA*. These accelerations were consistent with bound orbits but still allowed for a wide range of possible orbits ([9], [7]).

For the first time, and after 10 years of observations with the MPE speckle camera SHARP at the ESO NTT and the new adaptive optics CONICA/NAOS at the UT4 of the Very Large Telescope (VLT), we are now able to determine an

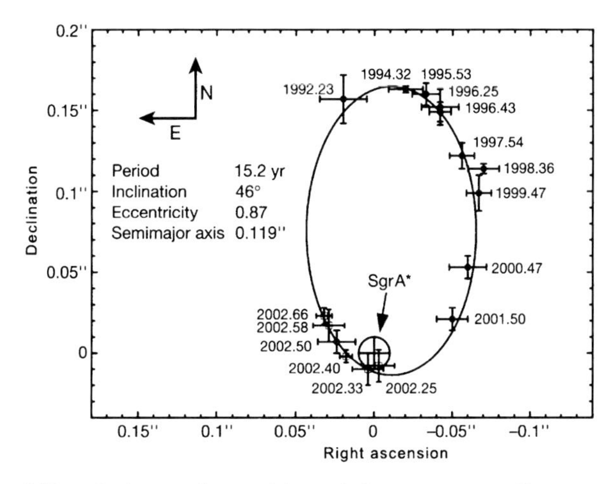

Fig. 1. Orbit of S2, relative to the position of the compact radio source SgrA* (large cross and circle, denoting the ±10 mas uncertainties of the infrared-radio astrometry). The data obtained with the MPE SHARP camera at the NTT and the VLT UT4 NACO adaptive optics are shown. The projection of the best-fitting Kepler orbit is shown as a continuous curve with its main parameters listed adjacent to the orbit.

orbit for the star S2, currently closest to the compact radio source SgrA* (semi-major axis = 4.62 mpc), these data trace an almost complete orbit (Fig. 1) and are well fitted by a bound, highly elliptical Keplerian orbit (ϵ=0.87), with an orbital period of 15.2 years, requiring an enclosed point mass of $(3.7\pm1.5)\times10^6$ $M_\odot$ [15]. This star passed through peri-center in April 2002, at a distance of only 17 light hours from the radio source, when it moved at $\geq$5000 km/s.

2 Mass Estimate for the Inner Cusp

2.1 Stellar Density

With the highest spatial resolution observations presently available in the near-infrared (50 - 60 mas), spatial scales from light hours to a few light years can be probed. The new CONICA/NAOS data were estimated from direct and crowding corrected $K_s\leq18$ stellar counts of in annuli centered on the position of SgrA*. They clearly confirm the presence of a local stellar concentration, a cusp centered on SgrA*, indicated earlier by the SHARP/NTT and KECK data ([4], [1]). To estimate the cusp mass, we were able to fit the combined SHARP and CONICA stellar count data with a superposition of several Plummer models of the form: $\rho(r)=\rho(0)[1+(r/R)^2]^{-\alpha/2}$ (α=5), with different densities $\rho(0)$, and different core radii R. Fig. 2 shows three different fits, where the dotted curve shows a fit for the inner cusp with a PLummer model of a core radii R=0.55" and a spatial density $\rho(r)=4.35*10^7$ $M_\odot$ pc^{-3}. The solid line shows the sum of that initial model with a further inner Plummer model of R=0.135" and $\rho(r)=6.5*10^8$ $M_\odot$ pc^{-3}. The average fit is shown in the large dotted curve. It is similar to the solid

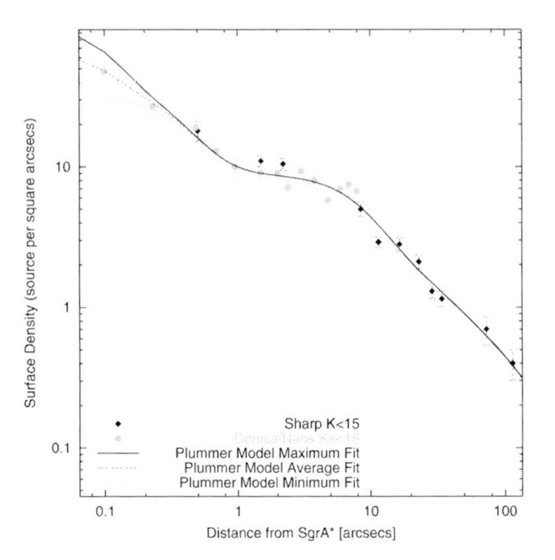

Fig. 2. A Plummer Model fit to the surface density of stars as a function of distance. The grey, filled circles represent the CONICA/NAOS data for $K_s \leq 18$. The darker, filled diamonds represent the SHARP/NTT data for stars with $K_s \leq 15$, and scaled upward by a factor of 5 in order to match the fainter CONICA/NAOS counts [8]. The small dotted curve and the solid curve represent the minimum and maximum fit to the data with our model, respectively. The big dotted middle curve represents the average fit.

curve but with a smaller density of $\rho(r)$=3.25*10^8 $M_\odot$ pc^{-3} of the additional R=0.135" component.

2.2 The Enclosed Mass

From these parameters, we were able to derive the enclosed mass as a function of the separation from the Galactic Center. Under the assumption that the mass to light ration of the inner cluster is comparable to that of the outer cluster (M/L∼ 2μm), the mass present at a distance of 0.55" from the BH was evaluated to be between 5000 $M_\odot$ and 6100 M⊙. Figure 3 shows 3 different curves in solid, dash-dotted and grey dotted. Similar to Fig. 2 they correspond to maximum, mean, and minimum fit, respectively. Using a fourth order Hermite integrator derived from the one used in high-accuracy N-body simulations ([2], [13] for the first introduction of the Hermite scheme see [10]), and adapting the mid-value of our models, we computed ,for an S2 like orbit [15] , the trajectory of a star through the extended mass and around the BH . The Hermite scheme allows a fourth order accurate integration based on only two time steps. For that it requires the analytic computation of the time derivative of the gravitational force; therefore the use of Plummer model superpositions as they are used here is very convenient. Further studies with more general density distributions should be undertaken , because they may influence the precise value of the periastron shift. Also it should be noted that this integration is purely classical so any relativistic periastron shifts are not taken into account.

The resulting retrograde periastron shift amounts to a value of ~1.7 arcmins per orbital revolution which is few times smaller than the relativistic prograde periastron shift [12]. Figure 4 below gives us different periastron shift values for different cusp masses, i.e. mass to light ratios.

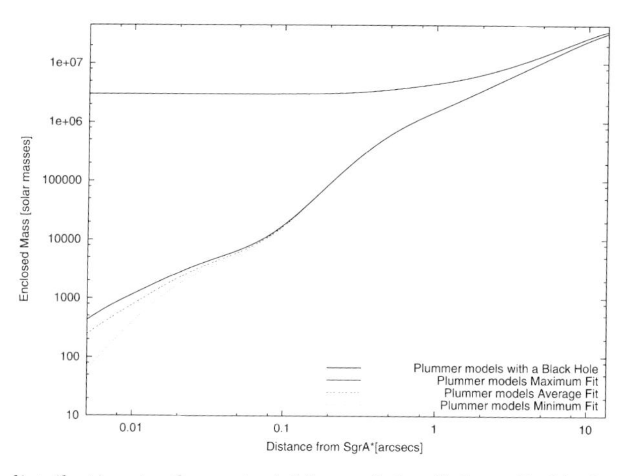

Fig. 3. Mass distribution in the central 10 pc of the Galaxy. In black a model fit to the data [8] resulting from the Plummer models fit to the stellar number density data and a 3×10^6 $M_\odot$ black hole. The 3 lower curves show a the stellar enclosed mass only. They clearly show the contribution of the inner cusp to the overall mass. The grey dotted curve represents a minimum of 5000 $M_\odot$ present inside a sphere of 0.55". The dash-dotted curve is the same as for the first curve with an additional mass of 550 $M_\odot$ present in 0.135". The solid curve is similar to the dash-dotted curve but with a larger additional mass of 1100 $M_\odot$

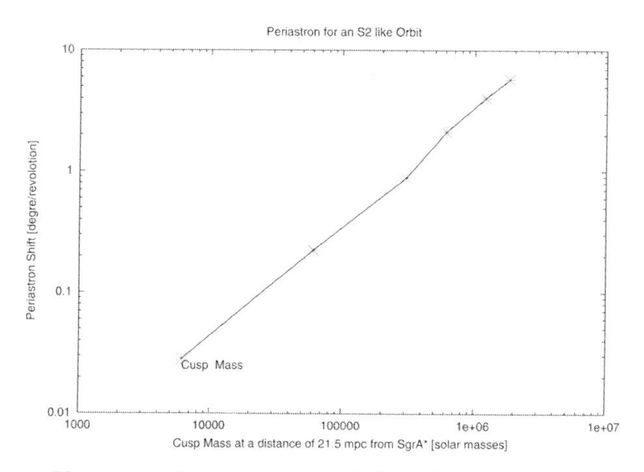

Fig. 4. Variation of the periastron shift with the inner cusp mass.

3 Estimate of the Line of Sight Positions of 13 Helium Stars

An enigmatic case in the central cluster of our Galaxy is the presence of the He I emission line stars and other early type stars (e.g in IRS16/13 complexes). They are confined in a projected radius on the sky not larger than about 400 mpc. Genzel et al. [8] show that most of them are on tangential orbits and seem to have a projected clockwise rotation on the sky. A determination of the spatial positions of these stars, will help to clarify the way they were formed, how they heat the gas and dust in the central parsec, and - potentially - how they influence the accretion stream onto the central arcsecond. We studied 13 (of the 29 known) He I emission-line stars, for which Genzel et al. [8] had given 3D velocities, and 2D projected separations from SgrA* (see also the list in [11]). With these parameters we compute the trajectories of the stars using a fourth order Hermite integrator, and assuming a potential derived from the the plummer model fit described above (Fig. 3).

Assuming that these stars are orbiting inside a sphere of 400 mpc or inside a sphere of 200 mpc, we can then determine the maximum line of sight ranges for the different stars. Figure 5 shows the results obtained for the 400 mpc case (dotted) and the 200 mpc case (solid black). For some of the stars the line of

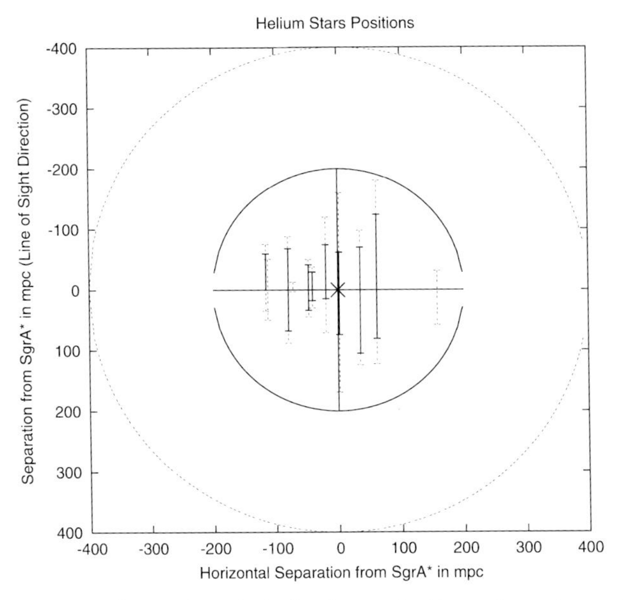

Fig. 5. The dash-dotted exterior circle is the projection of a sphere of 400 mpc of radius, Black solid interior circle is the projection of a sphere of 200 mpc. Dotted red bars are possible ranges along the line of sight positions of the He-stars with orbits that stay well within a sphere of 400 mpc, and for the case of the black solid bars well within a sphere of 200 mpc of radius. The crass at the center represents the position of SgrA*.

sight position is not very well constrained and the range can be quite large with respect to the radius of the sphere.

However, the result can be tested by making use of the fact the early type stars are mostly on tangential orbits and therefore show a significant anisotropy [8]. We computed the anisotropy parameter

$$\gamma = (V_T^2 - V_R^2)/(V_T^2 + V_R^2) \tag{1}$$

for 13 early type stars, covering one or more orbital time scale. For the 2 cases, 13 different random distributions of our sample in the possible ranges of line of sight positions were chosen, and the γ-value for 11 different snapshots in times calculated. Figure 6 results in the summation of 130 different combinations for the first case shown in the black solid line, and for the second case in a dotted line. The dotted histogram (200 mpc radius) shows clear anisotropy behaviour towards tangential orbits. This trend is less pronounced for the black curve (400 mpc radius).

We can conclude that the He stars of our sample are present mostly inside a sphere of about 200 mpc. A further constraint can be introduced via the presence of the mini-spiral, especially the northern arm. Here we can take into account that inspite of its presence some of the stars are not reddened (e.g IRS16 complex), and we deduce that these stars should be located in front of the mini-spiral. Vollmer & Duschl [14] give a model of the mini-spiral and describe the northern arm via gas motion within a plane and give its orientation. In Fig. 7, we show the final possible ranges in the line of sight direction obtained for our sample of stars taking al constraints into account that we mentioned above. The table gives the appropriate values of spatial positions for each star.

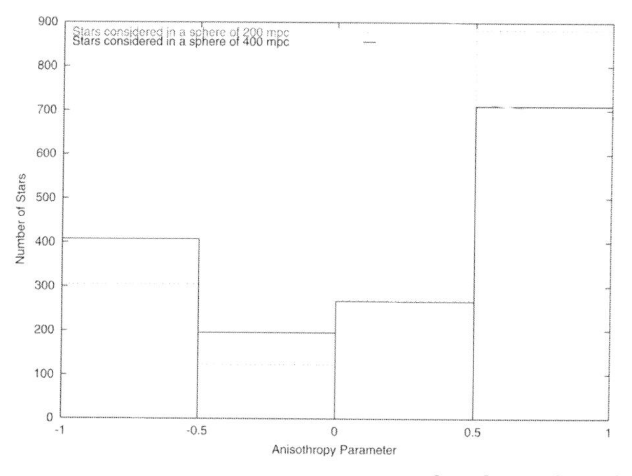

Fig. 6. Histogram of the anisotropy parameter γ=($V_T{}^2$-$V_R{}^2$)/($V_T{}^2$+$V_R{}^2$). Summation of 26 arbitrary distributions of the 13 Helium stars. The distribution covers a complete range in the line of sight positions of our sample. In dotted and black solid histogram lines, values are plotted considering that the He stars orbit inside a sphere of 200 mpc and 400 mpc of radius, respectively

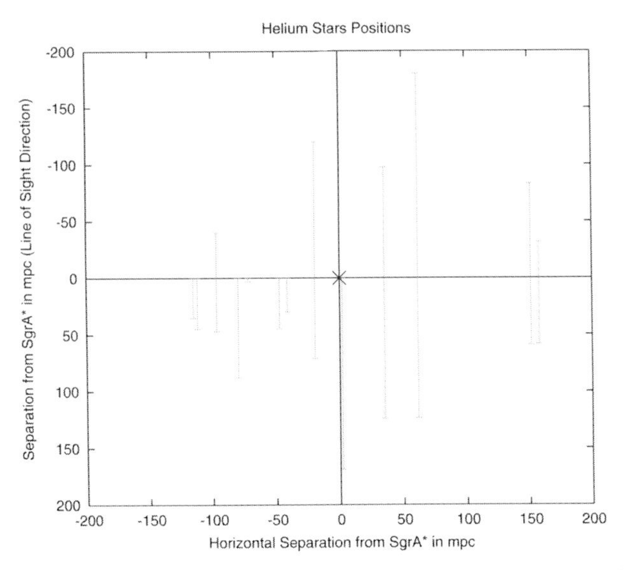

Fig. 7. The final possible ranges for the line of sight positions of the helium stars are represented by bars, the crass at the center represents the position of SgrA*

Table 1. Positions of the 13 He-stars including the ranges of possible for their minimum and maximum line of sight positions with respect to the plane of the sky that includes SgrA*.

He Stars Names	R.A[mpc]	Dec[mpc]	Max Z[mpc]	Min Z[mpc]
IRS16NW	+1.99	+46.05	169	0.0
IRS16C	-47.21	+17.03	44.5	0.0
IRS16SW	-41.00	+37.92	29	0.0
IRS16CC	-79.72	+19.74	88	0.0
IRS29N	+ 61.50	+54.57	123	-180.2
IRS16SE1	-71.98	-44.89	3	0
IRS29NE1	+35.22	+78.95	124	-98.0
IRS16NE	-111.8	+42.6	45	0.0
IRS16SE2	-114.94	-46.4	35	0.0
IRS33E	-19.35	-127.7	71	-120.0
IRS7SE	-96.75	+104.89	47	-40
IRS34W	+157.89	+ 62.69	58.5	-31.5
IRS7W	+151.32	+193.5	59	-83

Acknowledgements

Support by the SFB494 at the University of Cologne and partial support by the SFB439 at the University of Heidelberg is acknowledged.

References

1. Alexander, T., The Distribution of Stars near the Supermassive Black Hole in the Galactic Center. *The Astrophysical Journal*, **527**, Issue 2, 835-850 (1999)
2. Aarseth, S.J., From NBODY1 to NBODY6: the Growth of an Industry. *Proc. Astron. Soc. Pac*, **111**, 1333-1346
3. Baganoff, F.K., et al., Rapid X-ray flaring from the direction of the supermassive black hole at the Galactic Centre. *Nature*, **413**, 45–48 (2001)
4. Eckart, A., Genzel, R., Hofmann, R., Sams, B. J.& Tacconi-Garman, L. E.,High angular resolution spectroscopic and polarimetric imaging of the galactic center in the near-infrared. *Astrophys. J.*, **445**, L23-L26 (1995)
5. Eckart, A. & Genzel, R., Observations of stellar proper motions near the Galactic Centre. *Nature*, **383**, 415-417 (1996)
6. Eckart,A., Mouawad, N., Krips, M., Straubmeier, C., and Bertram, T., 2002, SPIE 4835-03, Cnf. Proc. of the SPIE Meeting on 'Astronomical Telescopes and Instrumentation', held in Waikoloa, Hawaii, 22-28 August 2002
7. Eckart, A., Genzel, R., Ott, T. & Schödel, R., Stellar orbits near Sagittarius A*. *Mon.Not.R.Soc.*, **331**, 917-934 (2002)
8. Genzel, R., Pichon, C., Eckart, A., Gerhard, O. & Ott, T., Stellar dynamics in the Galactic Centre: proper motions and anisotropy. *Mon.Not.R.Soc.*, **317**, 348-374 (2000)
9. Ghez, A., Morris, M., Becklin, E.E., Tanner, A. & Kremenek, T., The accelerations of stars orbiting the Milky Way's central black hole. *Nature*, **407**, 349-351 (2000)
10. Makino, J. & Aarseth, S.J., On a Hermite integrator with Ahmad-Cohen scheme for gravitational many-body problems. *Proc. Astron. Soc. Japan*, **44**, 141-151
11. Ott, T., Eckart, A., Genzel, R., Ap.J. **523**, 2480 (2000)
12. Rubilar, G. F.& Eckart, A. Periastron shifts of stellar orbits near the Galactic Center. *Astronomy and Astrophysics*, **374**, 95-104 (2001)
13. Spurzem, R., Direct N-Body Simulations. *Jl. Comp. Appl. Math*, **109**, 407-432
14. Vollmer, B. & Duschl,W.J., The Minispiral in the Galactic Center - Exploring a Data Cube With a Three Dimensional Method. *The Central Parsecs of the Galaxy, ASP Conference Series*, Ed by Heino Falcke, Angela Cotera, Wolfgang J. Duschl, Fulvio Melia, and Marcia J. Rieke, **186**, ISBN: 1-58381-012-9, p. 265 (1999)
15. Schödel, R., Ott, T., Genzel, R., Hofmann, R.; Lehnert, M.; Eckart, A., Mouawad, N., Alexander, T., Reid, M. J., Lenzen, R., Hartung, M., Lacombe, F., Rouan, D., Gendron, E., Rousset, G., Lagrange, A.-M., Brandner, W., Ageorges, N., Lidman, C., Moorwood, A. F. M., Spyromilio, J., Hubin, N., Menten, K. M.,A star in a 15.2-year orbit around the supermassive black hole at the centre of the Milky Way. *Nature*, **419**, 694-696 (2002)

Part IV

Formation and Evolution of Galaxies

Angular Momentum Redistribution and the Evolution and Morphology of Bars

Lia Athanassoula

Observatoire de Marseille, 2 Place Le Verrier, 13248 Marseille cedex 04, France

Abstract. Angular momentum exchange is a driving process for the evolution of barred galaxies. Material at resonance in the bar region loses angular momentum which is taken by material at resonance in the outer disc and/or the halo. By losing angular momentum, the bar grows stronger and slows down. This evolution scenario is backed by both analytical calculations and by N-body simulations. The morphology of the bar also depends on the amount of angular momentum exchanged.

1 Introduction

Bars are common features of disc galaxies. De Vaucouleurs ([11]), using a classification based on images at optical wavelengths, found that roughly one third of all disc galaxies are barred (family SB), while yet another third have small bars or ovals (family SAB). Observations in the near infrared have shown that galaxies that had been classified as non-barred from images at optical wavelengths may have a clear bar component when observed in the near infrared. Thus Eskridge et al. ([17]) classified more than 70% of all disc galaxies as barred, while Grosbøl, Pompei & Patsis ([18]) found that only $\sim 5\%$ of all disc galaxies are definitely non barred.

Bars come in a large variety of strengths, lengths, masses, axial ratios and shapes. Great efforts have been made in order to obtain some systematics on bar structure and important advances have been made. Elmegreen & Elmegreen ([16]) have shown that earlier type bars are relatively longer (i.e. relative to the disc diameter) on average than bars in later type galaxies. They also find that early type bars have flat intensity profiles along the bar major axis, while late type bars have exponential-like profiles. A correlation has been found ([5], [27]) between the length of bars and the size of bulges. This is in good agreement with the trend found in [16], since earlier type galaxies have larger bulges than late types. Important differences between early and late type bars are also found with the Fourier decomposition of the surface density. Indeed the relative $m = 2$ and 4 components are much stronger in early than in late type bars. Moreover, the higher order components ($m = 6$ and 8), which for the late type bars are negligible, are still important for early types. Finally, the shape of the bar isodensities differ and Athanassoula et al. ([7]), using a sample of strongly barred early type galaxies, showed that their bar isophotes are rectangular-like, particularly in the region near the end of the bar.

The first trials of N-body simulations (e.g. [28]) show that bars grow spontaneously and are long-lived. Yet it is only recently that simulations have achieved sufficient quality to provide information on the morphology of N-body bars and on the mechanisms that govern bar formation and bar evolution. I will here discuss some of the latest results of N-body simulations. I will argue that it is the exchange of angular momentum within the galaxy that will determine the bar strength and the rate at which the pattern speed decreases after the bar has formed, as well as the bar morphology.

2 Angular Momentum Exchange and Bar Evolution

Exchange of energy and angular momentum between stars at resonance with a spiral density wave has been first discussed by Lynden-Bell & Kalnajs ([26]). Using linear perturbation theory, these authors showed that, for a steady forcing, stars emit, or absorb, angular momentum only if they are at resonance. Stars at the inner Lindblad resonance (hereafter ILR) lose angular momentum, while stars at the outer Lindblad resonance (hereafter OLR) gain it. This groundbraking work has to be extended in order to be applied to bars in general and N-body bars in particular. HI observations, basically starting with [10], have now established that, if Newton's law of gravity is valid, then disc galaxies are embedded in a dark matter component, called the halo, whose mass exceeds that of the disc. This component should now be taken into account as an extra partner in the angular momentum exchange process. Furthermore, bars are strongly non-linear features, since they contain a large fraction of the mass in the inner parts of the disc and a considerable part of the total disc mass. Thus any linear theory should be thought of as a guiding line, to be supplemented by and tested against adequate N-body simulations. It is obvious that such simulations should be fully self-consistent, since rigid components can not exchange energy and angular momentum.

In [4] I extended the analytical work of [26] to include spheroidal components, like a halo and/or a bulge, and also supplemented it with fully self-consistent N-body simulations. In the analytical part I showed that, if the distribution function of the spheroidal component is a function only of the energy, then at all resonances the halo and bulge particles can only gain angular momentum. Also, since the bar is a strongly nonlinear feature, higher multiplicity resonances should be taken into account. Thus angular momentum is emitted by particles (stars) at the resonances in the inner disc, mainly the ILR, but also the inner -1:m resonances nearer to corotation (hereafter CR). It is absorbed by disc particles (stars) at the OLR, or the outer 1:m resonances, outside corotation, or by the resonant particles in the halo and/or bulge components. Since the bar is a negative angular momentum perturbation ([26]), by losing angular momentum it will grow. This clearly outlines a scenario for the evolution of barred galaxies.

3 The Effect of the Halo on Bar Evolution

As the bar loses angular momentum, it grows stronger. This, however, can only happen if there are absorbers that can absorb the angular momentum that the bar region emits. Thus the existence of a massive halo component, whose resonances can absorb considerable amounts of angular momentum, will help the bar grow. At first sight this may be thought to go against old claims that haloes stabilise bars. In fact, a more precise wording is necessary. Indeed, the halo slows down the bar growth in the initial stages of the evolution. At later stages, however, the situation can be reversed, since the halo may absorb the angular momentum emitted by the bar, and thus it may allow the latter to grow further. Thus bars that grow in halo-dominated discs can be stronger than bars that grow in disc-dominated surroundings. This effect was not seen till recently, since the older studies were either 2D (e.g. [33], [8]), or 3D but with few particles (e.g. [31]), or had rigid haloes. In all these cases the halo was denied from the onset its destabilising influence. Its effect becomes clear in fully self-consistent N-body simulations, with an adequate particle number and resolution. Thus [6] showed that stronger bars can grow in cases with more important halo components.

The influence of the halo is also illustrated in Fig. 1, where I compare the results of two N-body simulations. Initially their disc is exponential, with unit mass and scale-length ($M_d = 1$, $R_d = 1$) and its Q parameter ([32]) is independent of radius and equal to 1.2. Since $G = 1$, taking the mass of the disc equal to 5×10^{10} $M_{\odot}$, and its scale-length equal to 3.5 kpc implies that the unit of velocity is 248 km/sec and the unit of time is 1.4×10^7 yrs. This calibration is reasonable, but is not unique, so in the following I will give all quantities in computer units. The reader can then convert the values to astronomical units according to his/her needs. The halo component is spherical, non-rotating and has an isotropic velocity distribution. It follows a pseudo-isothermal radial density profile ([19]) and has a total mass $M_h = 5$, a core radius $\gamma = 0.5$ and a cutoff radius of $r_c = 10$. Its density is truncated at 15 disc scale-lengths, i.e. at a radius containing more than 96% of its mass. In building the initial conditions I loosely followed [19] and [6], and the simulations were run on the Marseille GRAPE-5 systems (for a description of the GRAPE-5 boards see [22]). The only difference between the initial conditions of the two simulations is that for simulation LH, illustrated in the left panels, the halo is live and represented by roughly 10^6 particles, while for simulation RH, illustrated in the right panels, it is rigid, i.e. represented by an analytical potential and thus can neither emit nor absorb angular momentum. Although their initial conditions are so similar, the two simulations evolve in a very different way. After some initial multi-spiral episodes, LH forms a bar which grows stronger with time. Its morphology at $t = 700$ is shown in the left panels. The bar is long and strong and has ansae-type features near the end of its major axis. It is surrounded by a ring, which can be compared to the inner rings often observed in barred galaxies. The bar formation entails considerable redistribution of the disc matter, both radially and azimuthally. On the other hand the disc in simulation RH stays close to axisymmetric, except for some multi-armed spirals which die out with time. Only at the latest stages of the evolution does it form an oval distortion, and even that is weak and short and is confined to the innermost parts of the disc, as can be seen for $t = 900$ in the right

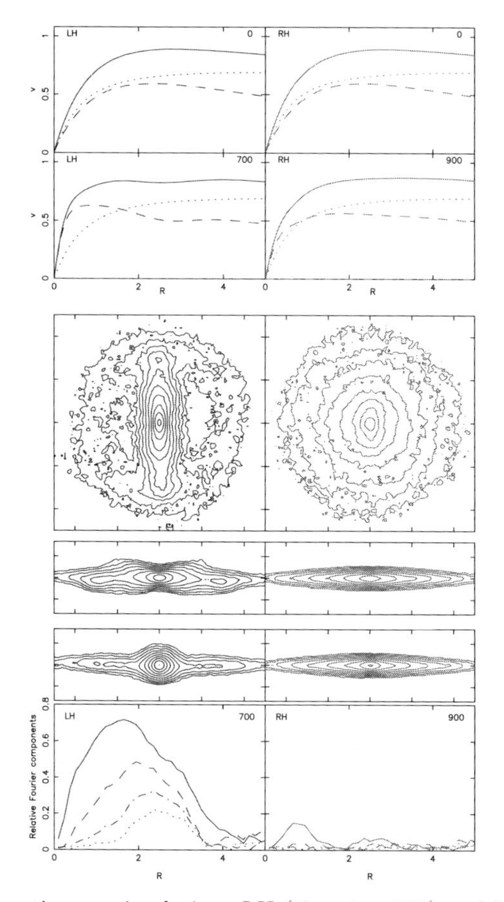

Fig. 1. Basic information on simulations LH (time $t = 600$) and RH (time $t = 900$). The two upper rows give the circular velocity curves at time 0 and t. The dashed and dotted lines give the contributions of the disc and halo respectively, while the thick full lines give the total circular velocity curves. The third row of panels gives the isocontours of the density of the disc particles projected face-on and the fourth and fifth row give the side-on and end-on edge-on views, respectively. The side of the box for the face-on views is 10 length units and the height of the box for the edge-on views is 3.33. The isodensities in the third row of panels have been chosen so as to show best the features in the bar and in the inner disc. No isodensities for the outer disc have been included, although the disc extends beyond the area shown in the figure. The sixth row of panels gives the $m = 2$, 4, 6, and 8 Fourier components of the mass.

panels of Fig. 1. I show this simulation at a later time than that for simulation LH because at earlier times there is very little structure visible.

Seen edge-on with the bar seen side-on (i.e. with the line of sight along the bar minor axis), simulation LH exhibits a very strong peanut, which is totally absent from simulation RH (fourth row of panels). Seen edge-on with the bar seen end-on (i.e. with the line of sight along the bar major axis), the peanut in LH resembles a bulge (left panel on fifth row). This underlines the hazards involved in classifying edge-on galaxies, since the classifier may in such cases easily misinterpret the bar for a bulge.

A useful way of quantifying the bar strength is with the help of the Fourier components of the mass, or density. These can be defined as

$$F_m(r) = \sqrt{A_m^2(r) + B_m^2(r)}/A_0(r), \qquad m = 0, 1, 2, ... \tag{1}$$

where

$$A_m(r) = \frac{1}{\pi} \int_0^{2\pi} \Sigma(r, \theta) cos(m\theta) d\theta, \qquad m = 0, 1, 2, ... \tag{2}$$

and

$$B_m(r) = \frac{1}{\pi} \int_0^{2\pi} \Sigma(r, \theta) sin(m\theta) d\theta, \qquad m = 1, 2, ... \tag{3}$$

For runs LH and RH, these components for $m = 2$, 4, 6 and 8 are shown in the lower panels of Fig. 1. For run LH all four components are important, due to the strength of the bar. Their amplitude decreases with increasing m, while the location of the maximum shifts outwards. On the other hand, for model RH only the $m = 2$ component stands out from the noise, but its amplitude is rather small, smaller than e.g. that of the $m = 8$ for model LH.

Since the only difference between the initial conditions of models LH and RH is that the halo of the one is responsive, while that of the other is rigid, we can conclude that the halo response is crucial for determining the evolution of barred galaxies.

4 Bar Slow-Down

I ran a large number of simulations similar to those described in the previous sections. I noted that, as it loses angular momentum, the bar grows longer, and more massive, thus stronger. Angular momentum loss, however, is not only linked to an increase in the bar strength. It is also linked to a slow-down, i.e. to a decrease of the bar pattern speed Ω_p with time. Such a slow-down has indeed been seen in a number of simulations and has also been predicted analytically ([34], [36], [24], [25], [20], [1], [12], [13]). It can also be seen in Fig. 2, which shows the run of the bar pattern speed with time for simulation LH, whose morphology at time $t = 700$ is shown in the left panels of Fig. 1. Note that the bar slows down considerably with time.

5 Resonances

In order for haloes to be able to absorb angular momentum, they need to have a considerable fraction of their mass at resonance. This was shown to be true in ([2]). I will illustrate it here for model LH. The procedure is the same as that followed in [2]. I calculate the potential from the mass distribution in the

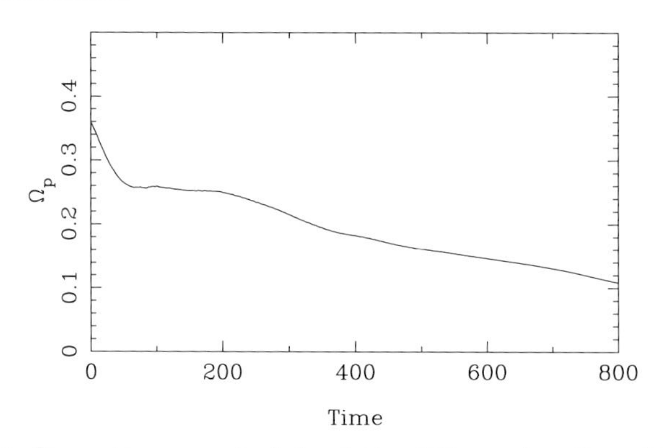

Fig. 2. Bar pattern speed of simulation LH as a function of time.

disc and halo component at time $t = 800$, by freezing all motion except for the bar, to which I assign bulk rotation with a pattern speed equal to that found in the simulation at that time. I then pick at random 100 000 disc and 100 000 halo particles and, using their positions and velocities as initial conditions, I calculate their orbits for 40 bar rotations. Using spectral analysis ([9], [23]), I then find the principal frequencies of these orbits, i.e. the angular velocity Ω, the epicyclic frequency κ and the vertical frequency κ_z. An orbit is resonant if there are three integers l, m and n, such that

$$l\kappa + m\Omega + n\kappa_z = -\omega_R = m\Omega_p \tag{4}$$

Orbits on planar resonances fulfill

$$l\kappa + m\Omega = -\omega_R = m\Omega_p \tag{5}$$

The ILR corresponds to $l = -1$ and $m = 2$, CR to $l = 0$, and OLR to $l = 1$ and $m = 2$.

The upper panels of Fig. 3 show, for time $t = 800$, the mass per unit frequency ratio M_R of particles having a given value of the frequency ratio $(\Omega - \Omega_p)/\kappa$ as a function of this frequency ratio[1]. The distribution is not uniform, but has clear peaks at the location of the main resonances, as was already shown in [2] and [4]. The highest peak for the disc component is at the ILR, followed by $(-1, 6)$ and CR. In all simulations with strong bars the ILR peak is strong. The existence of peaks at other resonances as well as their importance varies from one run to another and also during the evolution of a given run. For example the CR peak is, in many other simulations, much stronger than in the example shown here. For the spheroidal component the highest peak is at CR, followed by peaks at the ILR, OLR and $(-1, 4)$.

The lower panels show the angular momentum exchanged. For this I calculated the angular momentum of each particle at time 800 and at time 500, as described in [4], and plotted the difference as a function of the frequency ratio

[1] See [4] for more information on M_R and on how it is derived.

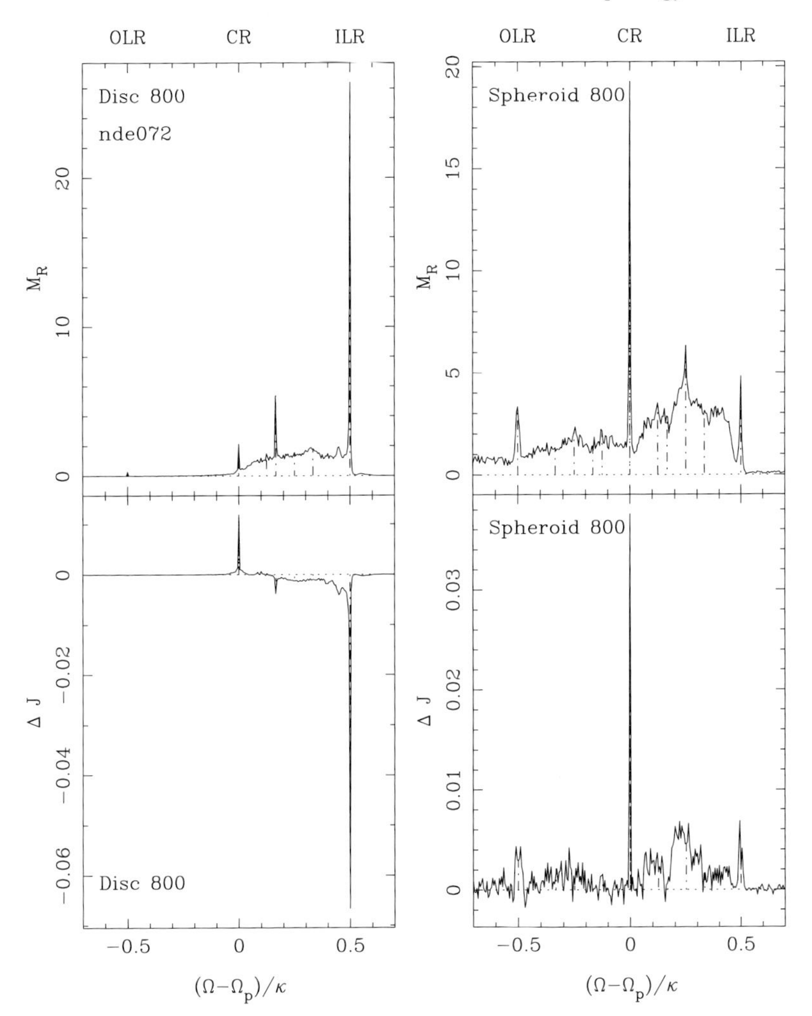

Fig. 3. The upper panels give, for time $t = 800$, the mass per unit frequency ratio, M_R, as a function of that ratio. The lower panels give ΔJ, the angular momentum gained or lost by the particles between times 800 and 500, plotted as a function of their frequency ratio $(\Omega - \Omega_p)/\kappa$, calculated at time $t = 800$. The left panels correspond to the disc component and the right ones to the halo. The component and the time are written in the upper left corner of each panel. The vertical dot-dashed lines give the positions of the main resonances.

of the particle at time 800. It is clear from the figure that disc particles at ILR and at the $(-1, 6)$ resonance lose angular momentum, while those at CR gain it. There is a also a general, albeit small, loss of angular momentum from particles with frequencies between CR and ILR. This could be partly due to particles trapped around secondary resonances, and partly due to angular momentum taken from particles which are neither resonant, nor near-resonant, but can still lose a small amount of angular momentum because the bar is growing. The corresponding panel for the spheroid is, as expected, more noisy, but shows that particles at all resonances gain angular momentum. Thus this plot, and similar ones which I did for other simulations, confirm the analytical results of [4], and show that the linear results concerning the angular momentum gain or loss by resonant particles, qualitatively at least, carry over to the strongly nonlinear regime.

6 What Determines the Strength of Bars and Their Slow-Down Rate?

I have shown in the previous sections that the halo can take angular momentum from the bar, thus making it stronger and slower. However, for this effect to be important, the amount of angular momentum exchanged must be considerable. For the latter to happen the halo must

- be sufficiently massive in the regions containing the principal resonances.
- not be too hot, i.e. not have too high velocity dispersion. Indeed, hot haloes can not absorb much angular momentum, even if they are massive (e.g. [4]).

Thus the length and the slow-down rate of bars are naturally limited by the mass and velocity distribution of the halo. Examples of this can be found in [4].

7 Trends and Correlations

In [4] I found trends and correlations between the angular momentum absorbed by the spheroid (i.e. the halo plus, whenever existent, the bulge), the bar strength and the bar pattern speed. They are based on a set of simulations analogous to those described in the previous sections. Such plots are given also in Fig. 4, based on a somewhat larger sample of simulations. About three quarters of them were run with the Marseille GRAPE-5 systems, and roughly one quarter was run on PCs using Dehnen's treecode ([14], [15]). Each point represents one simulation and the trends are the same as those found in [4]. The upper panels show the results for the whole sample, the middle panels contain only simulations where the halo has a small core radius ($\gamma < 2$), $M_h = 5$ and does not extended beyond 15 disc scale-lengths, and the lower ones contain only simulations where the halo has a large core radius ($\gamma > 2$), $M_h = 5$ and again does not extended beyond 15 disc scale-lengths.

The right panel shows that there is a correlation between the angular momentum of the spheroid and the bar strength. This correlation holds also when I restrict myself to simulations with large (or small) core radii as seen in the second and third row. A trend also exists between the spheroid angular momentum and the bar pattern speed. In this case, however, simulations with large core radii behave differently from those with small radii. Indeed, for simulations with a small core radius (i.e. centrally concentrated halos) I find a very strong correlation

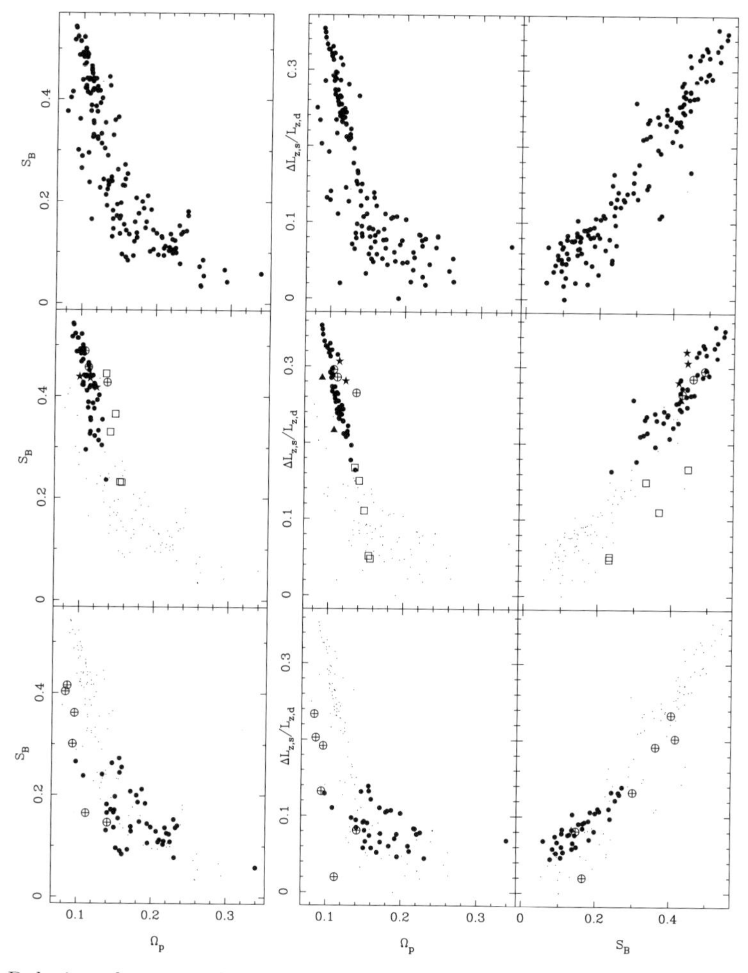

Fig. 4. Relations between the bar strength and the pattern speed (left panels), the spheroid angular momentum and the pattern speed (middle panels) and the spheroid angular momentum and the bar strength (right panels), at times $t = 800$. The spheroid angular momentum is normalised by the initial disc angular momentum ($L_{z,d}$). The simulations under consideration in each panel are marked with a filled circle and the rest by a dot. The upper row includes all simulations, the middle and the lower ones subsamples, as described in the text. In the middle panel simulations with a bulge are marked with a $\oplus$, simulations with $\gamma = 0.01$ with a filled star, simulations with $2 > \gamma \geq 1$ with a filled triangle and simulations with $Q_{init} \geq 2$ with an open square. In the lower panel simulations with $Q_{init} \geq 1.4$ and $z_0 \geq 0.2$ are marked with an $\oplus$.

between the spheroid angular momentum and the pattern speed, particularly if I restrict myself to one value of γ. In such simulations the angular momentum is exchanged primarily between the bar region and the spheroid, thus accounting for the very tight correlation. Simulations with large cores behave differently (lower middle panel). They show only a rough trend, except for simulations with a hot disc, which show a tight correlation. This is easily explained in the scenario of evolution via angular momentum exchange. The outer parts of hot discs absorb only little angular momentum, so that the exchange is basically between the bar region and the spheroid, thus accounting for the tight correlation. On the other hand, if the outer disc is cold, then it can participate more actively in the exchange. Since the angular momentum absorbed by the spheroid (plotted in Fig. 4) is not the total angular momentum exchanged, but only a fraction of it, I find only a trend.

8 Comparing the Morphology of *N*-Body and of Real Bars

The correlations discussed in section 7 show clearly that models that have exchanged more angular momentum have stronger bars than models that have exchanged little. By examining the results of the individual simulations, I could see that, in cases where large amounts of angular momentum have been exchanged, the bars are long, relatively thin and have rectangular-like isodensities, particularly in their outer parts. A typical example of such a case is given in the left panels of Fig. 5 (see also [6]). Note also the existence of ansae at the ends of the bar, a feature sometimes observed in early type barred galaxies. On the other hand, models that have exchanged little angular momentum have less homogeneous properties. For example they can have either ovals, or short bars. Typical examples of such cases are given in the middle and right panels of Fig. 5, respectively. The model in the left panel has exchanged about 15 percent of the disc angular momentum by the time shown in Fig. 5, while the other two models only of the order of a percent.

The edge-on morphology also is strongly influenced by the amount of angular momentum exchanged. The strong bar, when seen edge-on, displays a clear peanut morphology, as often observed. On the other hand the oval has a boxy edge-on appearance, while the small bar has not changed significantly the edge-on morphology of the galaxy.

The difference in bar strength is also illustrated in the lower panels of Fig. 5, which show the relative Fourier components of the density for m = 2, 4, 6 and 8 for the three simulations. The simulation that exchanged a lot of angular momentum has a very strong $m = 2$ component, with a secondary maximum roughly at the position of the ansae. The remaining components, even the $m = 6$ and 8 ones, are also important. The location of their maximum moves outwards with increasing m. The oval has much lower Fourier components, and only the $m = 2$ stands out from the noise. For small radii all components are nearly zero, which means that the oval must be nearly axisymmetric in its innermost parts. On the other hand, the $m = 2$ amplitude drops slowly with radius in the outer parts, thus extending to large radii. The small bar has Fourier components which drop rapidly with radius, i.e. they are noticeable only in the central region, as expected since the bar is confined there. Only the m = 2 and 4 components stand out from the noise.

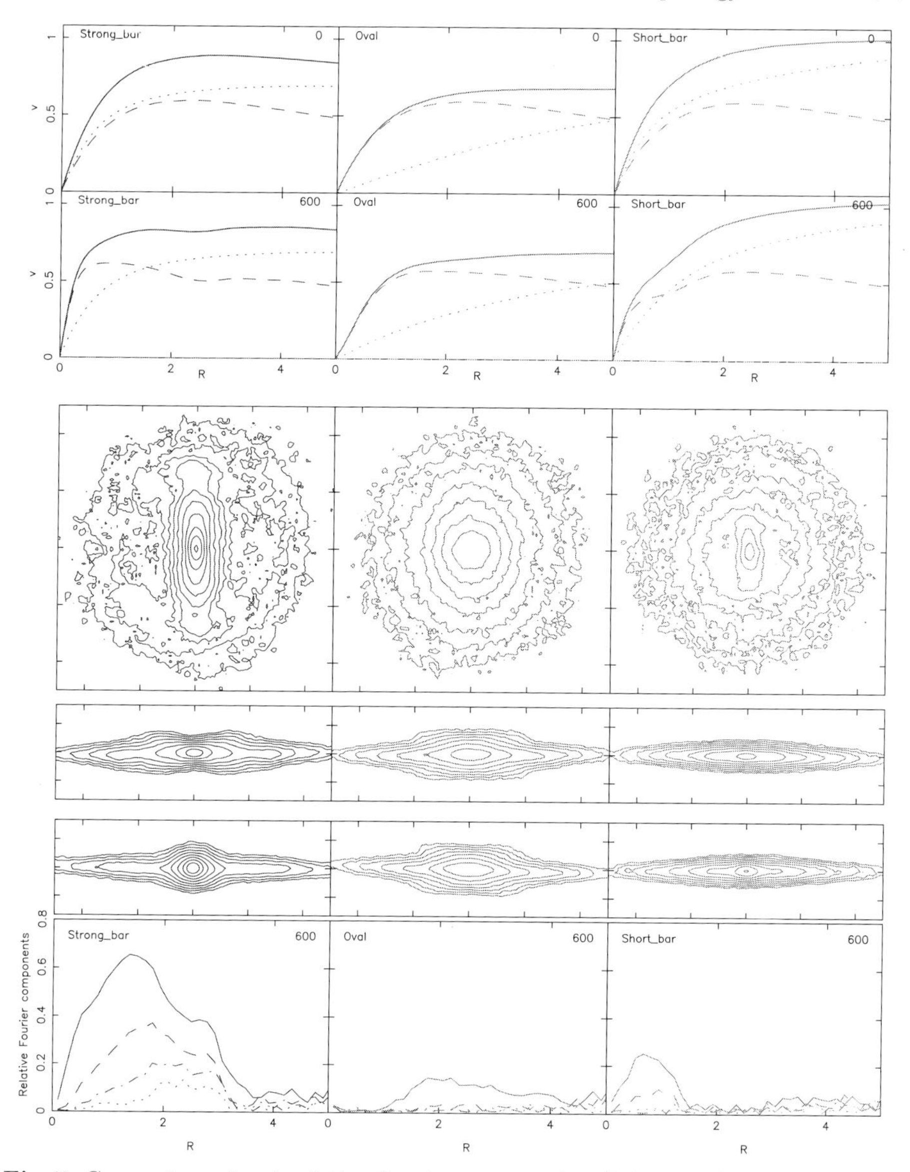

Fig. 5. Comparison of a simulation forming a strong bar (left panels), one forming an oval (middle panels) and one forming a short bar (right panels). The layout is as for Fig. 1.

The radial rearrangement of the disc material due to the bar can be inferred by comparing the initial with the current circular velocity curves, given in the first and second rows of panels. The strong bar has entailed a substantial radial rearrangement, the final disc mass distribution being considerably more centrally concentrated than the initial one. On the other hand, in the other two simula-

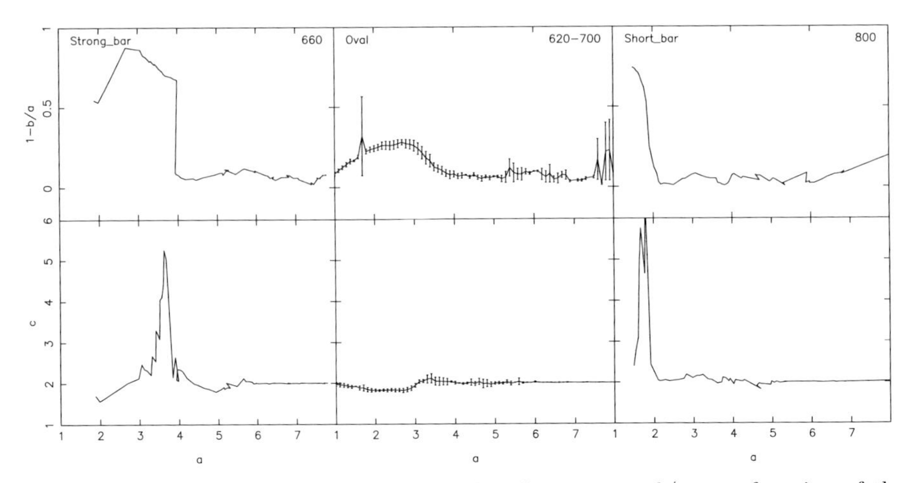

Fig. 6. The upper panels show the run of the ellipticity 1 - b/a as a function of the semi-major axis a. The lower panels show the run of the shape parameter c, also as a function of a. The left panels corresponds to a model with a strong bar, the middle ones to model with an oval and the right ones to a model with a short bar. To improve the signal-to-noise ratio for the model with the oval I took an average over a time interval, namely [620-700]. The dispersion during that time is indicated by the error bars. The times are given in the upper right corner of the upper panels.

tions, and particularly in the one producing the oval, there is very little radial rearrangement of the disc material. Since there is also hardly any radial rearrangement of the halo material ([6], [3], [35]), this means that the disc-to-halo mass ratio changes most in the simulations where more angular momentum has been exchanged.

Quantitative comparison of the bar form of the three models is given in Fig. 6. The values of the bar semi-major and semi-minor axes (a and b, respectively) and of the shape parameter (c) were obtained by fitting generalised ellipses of the form

$$(|x|/a)^c + (|y|/b)^c = 1, \tag{6}$$

to the bar isodensities. The shape parameter c is 2 for ellipses, larger than 2 for rectangular-like generalised ellipses, and smaller than 2 for diamond-like ones. From this figure one can note that both the strong and the short bar are thin, and in general see how their axial ratios vary with the semi-major axis. The shape parameter is given in the lower panels. We see that both the strong and the short bar have rectangular-like isodensities in the outer regions of the bar, while the oval has a shape very close to elliptical. In fact the strong bar has axial ratios and shapes very similar to those found in [7] by applying the same type of analysis to a sample of early type barred galaxies.

Plotting the run of the density along the bar major axis ([6]) I find for the strong bar a profile which is rather flat within the bar region, similar to what was found in [16] for early type bars.

It is thus clear that the amount of angular momentum exchanged influences the morphology of the bar. In my first example, where a lot of angular momentum was transferred from the bar to the outer halo (mainly), the result is a long, strong bar, with some rectangular-like isophotes and ansae at its ends. The examples where little angular momentum was exchanged have a very different morphology, one forming an oval and the other a short bar. What determines which one of the two it will be? In the examples shown here, and in a rather large sample of similar cases, the oval was formed in an initially hot disc, while the short bar grew in a hot halo. The existing theoretical framework, however, gives no predictions on this point and work is in progress to elucidate this further.

9 Summary

In this paper I reviewed evidence that shows that angular momentum is exchanged between the bar region in the one hand, and the outer disc and the spheroid on the other. This exchange determines the slow-down rate of the bar, as well as its strength and its overall morphology.

Acknowledgments

I thank the organisers for inviting me to this interesting conference. I thank M. Tagger, A. Bosma, W. Dehnen, A. Misiriotis, C. Heller, I. Shlosman, F. Masset, J. Sellwood, O. Valenzuela, A. Klypin and P. Teuben for many stimulating discussions. I thank J. C. Lambert for his help with the GRAPE software and with the administration of the simulations and W. Dehnen for making available to me his tree code and related programs. I also thank the INSU/CNRS, the University of Aix-Marseille I, the Region PACA and the IGRAP for funds to develop the GRAPE and Beowulf computing facilities used for the simulations discussed in this paper and for their analysis.

References

1. E. Athanassoula: 'Evolution of bars in isolated and in interacting disk galaxies'. In *Barred Galaxies*, eds. R. Buta, D. Crocker and B. Elmegreen (Astron. Soc. of the Pacific Conference series, 91), pp. 309–321 (1996)
2. E. Athanassoula: Astrophys. J. **569**, L83 (2002)
3. E. Athanassoula: 'Formation and Evolution of Bars in Disc Galaxies'. In *Disks of Galaxies: Kinematics, Dynamics and Perturbations*, eds. E. Athanassoula, A. Bosma, R. Mujica (Astron. Soc. of the Pacific Conference Series, 275) pp. 141–152 (2002)
4. E. Athanassoula: Mon. Not. R. Astron. Soc., in press (2003)
5. E. Athanassoula, L. Martinet: Astron. Astrophys. **87**, L10 (1980)
6. E. Athanassoula, A. Misiriotis: Mon. Not. R. Astron. Soc. **330**, 35 (2002)
7. E. Athanassoula, S. Morin, H. Wozniak, et al.: Mon. Not. R. Astron. Soc. **245**, 130 (1990)
8. E. Athanassoula, J. A. Sellwood: Mon. Not. R. Astron. Soc. **221**, 213 (1986)

9. J. Binney, D. Spergel: Astrophys. J. **252**, 308 (1982)
10. A. Bosma: Astron. J., **86**, 1825 (1981)
11. G. de Vaucouleurs: Astrophys. J. Suppl. **8**, 31 (1963)
12. V. P. Debattista, J. A. Sellwood: Astrophys. J. **493**, L5 (1988)
13. V. P. Debattista, J. A. Sellwood: Astrophys. J. **543**, 704 (2000)
14. W. Dehnen: Astrophys. J. **536**, L39 (2000)
15. W. Dehnen: J. Comp. Phys. **179**, 27 (2002)
16. B. G. Elmegreen, D. M. Elmegreen: Astrophys. J. **288**, 438 (1985)
17. P. B. Eskridge, J. A. Frogel, R. W. Pogge, et al.: Astron. J., **119**, 536 (2000)
18. P. Grosbøl, E. Pompei, P. Patsis: 'Spiral Structure Observed in Near-Infrared'. In *Disks of Galaxies: Kinematics, Dynamics and Perturbations*, eds. E. Athanassoula, A. Bosma, R. Mujica (Astron. Soc. of the Pacific Conference Series, 275) pp. 305–310 (2002)
19. L. Hernquist: Astrophys. J. Suppl. **86**, 389 (1993)
20. L. Hernquist, M. D. Weinberg: Astrophys. J. **400**, 80 (1992)
21. F. Hohl: Astrophys. J. **168**, 343 (1971)
22. A. Kawai, T. Fukushige, J. Makino, M. Taiji: Pub. Astron. Soc. Japan **152**, 659 (2000)
23. J. Laskar: Icarus, **88**, 266 (1990)
24. B. Little, R. G. Carlberg: Mon. Not. R. Astron. Soc. **250**, 161 (1991a)
25. B. Little, R. G. Carlberg: Mon. Not. R. Astron. Soc. **251**, 227 (1991b)
26. D. Lynden-Bell, A. J. Kalnajs: Mon. Not. R. Astron. Soc. **250**, 161 (1972)
27. P. Martin: Astron. J. **109**, 2428 (1995)
28. R. H. Miller, K. H. Prendergast, W. J. Quirk: Astrophys. J. **161**, 903 (1979)
29. K. Ohta: 'Global Photometric Properties of Barred Galaxies'. In *Barred Galaxies*, eds. R. Buta, D. A. Crocker, B. G. Elmegreen (Astron. Soc. of the Pacific Conference Series, 91) pp. 37–43, (1996)
30. K. Ohta, M. Hamabe, K. Wakamatsu: Astrophys. J. **357**, 71 (1990)
31. J. P. Ostriker, P. J. E. Peebles: Astrophys. J. **186**, 467 (1973)
32. A. Toomre: Astrophys. J. **139**, 1217 (1964)
33. A. Toomre: 'What amplifies the spirals' In: *The Structure and Evolution of Normal Galaxies*, ed. S. M. Fall, D. Lynden-Bell (Cambridge University Press) pp. 111–136 (1981)
34. S. Tremaine, M. D. Weinberg: Mon. Not. R. Astron. Soc. **209**, 729 (1984)
35. O. Valenzuela, A. Klypin: in preparation (2003)
36. M. D. Weinberg: Mon. Not. R. Astron. Soc. **213**, 451 (1985)

Major Mergers and the Origin of Elliptical Galaxies

Andreas Burkert[1] and Thorsten Naab[2]

[1] Max-Planck-Institut für Astronomie, Königstuhl 17, D-69117 Heidelberg, Germany
[2] Institute of Astronomy, Madingley Road, Cambridge CB3 0HA, UK

Abstract. The formation of elliptical galaxies as a result of the merging of spiral galaxies is discussed. We analyse a large set of numerical N-Body merger simulations which show that major mergers can in principle explain the observed isophotal fine structure of ellipticals and its correlation with kinematical properties. Equal-mass mergers lead to boxy, slowly rotating systems, unequal-mass mergers produce fast rotating and disky ellipticals. However, several problems remain. Anisotropic equal mass mergers appear under certain projections disky which is not observed. The intrinsic ellipticities of remnants are often larger than observed. Finally, although unequal-mass mergers produce fast rotating ellipticals, the remnants are in general more anisotropic than expected from observations. Additional processes seem to play an important role which are not included in dissipationless mergers. They might provide interesting new information on the structure and gas content of the progenitors of early-type galaxies.

1 Introduction

Giant elliptical galaxies are believed to be very old stellar systems that formed by a major merger event preferentially very early at a high redshift of more than two ([59], [61]). The merger triggered an intensive star-formation phase which turned most of the gas of the progenitors into stars. Some fraction of the gas was heated to temperatures of order the virial temperature, producing X-ray coronae which are still visible today. The stellar disks of the progenitors were destroyed as a result of the strong tidal forces during the merger, leading to kinematically hot, spheroidal stellar remnants. Subsequently, the systems experienced very little accretion and merging with negligible star formation [22]. This scenario is supported by many observations which indicate that ellipticals contain stellar populations that are compatible with purely passive evolution ([21], [1], [27], [65]). or with models of an exponentially, fast decreasing star formation rate [66].

An alternative scenario which is based on hierarchical theories of galaxy formation predicts that massive galaxies are assembled relatively late in many generations of mergers through multiple mergers of small subunits, with additional smooth accretion of gas ([38], [39]). In this case, ellipticals might form either if the multiple subunits are already preferentially stellar or if star formation was very efficient during the protogalactic collapse phase [42].

The idea that ellipticals form from major mergers of massive disk galaxies has been originally proposed by Toomre & Toomre [61]. Their "merger hypothesis" has been explored in details by many authors, using numerical simulations. Gerhard [31], Negroponte & White [52], Barnes [2] and Hernquist [33] performed the first fully self-consistent merger models of two equal-mass stellar disks embedded in dark matter halos. The remnants are slowly rotating, pressure supported and anisotropic. They generally follow an $r^{1/4}$ surface density profile for radii $r \geq 0.5r_e$, where r_e is the effective radius. However it turns out that due to phase space limitations [23], an additional massive central bulge component is required [35], to fit the observed de Vaucouleurs profile [19] also in the inner regions. All simulations demonstrated consistently that the global properties of equal mass merger remnants resemble those of ordinary slowly rotating massive elliptical galaxies.

More recently it has become clear that ellipticals have quite a variety of fine structures with peculiar kinematical properties which, in contrast to their universal global properties, can give a more detailed insight into their formation history. It is interesting to investigate whether the merging hypothesis can explain these observations and, if yes, whether they provide more information on the validity of this scenario, the orbital parameters of the mergers and the structure and gas content of the progenitors from which the ellipticals formed.

Elliptical galaxies can be subdivided into two major groups with respect to their structural properties ([9], [7], [8], [41]). Faint ellipticals are isotropic rotators with small minor axis rotation and disky deviations of their isophotal shapes from perfect ellipses. Their isophotes are peaked in the rotational plane and a Fourier analyses of the isophotal deviation from a perfect ellipse leads to a positive value of the fourth order coefficient a_4. These galaxies might contain secondary, faint disk components which contribute up to 30% to the total light in the galaxy, indicating disk-to-bulge ratios that overlap with those of S0-galaxies ([55], [58]). Disky ellipticals have power-law inner density profiles ([43], [28]) and show little or no radio and X-ray emission [10]. Most massive ellipticals have boxy isophotes, with negative values of a_4. They also show flat cores ([43], [28] Faber et al. 1997) and their kinematics is more complex than that of disky ellipticals. Boxy ellipticals rotate slowly, are supported by velocity anisotropy and have a large amount of minor axis rotation. Like the secondary disks of disky ellipticals, the boxy systems occasionally reveal kinematically decoupled core components, that most likely formed from gas that dissipated its orbital energy during the merger, accumulated in the center and subsequently turned into stars ([29], [57], [8]). The cores inhibit flattened rapidly rotating disk- or torus-like stellar structures that dominate the light in the central few hundred parsecs ([56], [45]), but they contribute only a few percent to the total light of the galaxy. The fact that the stars are metal-enhanced confirms that gas infall and subsequently violent star formation, coupled with metal-enrichment must have played an important role in the centers of merger remnants ([11], [25], [12], [26]). Boxy ellipticals show strong radio emission and high X-ray luminosities, resulting from emission from hot gaseous halos [16] that probably formed from gas heating during the merger. These hot gaseous bubbles are however absent in disky ellipticals. The distinct physical properties of disky and boxy elliptical galaxies indicate that both types of ellipticals experienced different formation histories.

In order to understand the origin of boxy and disky ellipticals the isophotal shapes of the numerical merger remnants have been investigated in detail. It has been shown that the same remnant can appear either disky or boxy when viewed from different directions (Hernquist 1993b) with a trend towards boxy isophotes ([36], [58]). Barnes [4] and Bendo & Barnes [15] analysed a sample of disk-disk mergers with a mass ratio of 3:1 and found that the remnants are flattened and fast rotating in contrast to equal mass mergers. Naab et al. [47] studied the photometrical and kinematical properties of a typical 1:1 and 3:1 merger remnant in details and compared the results with observational data . They found an excellent agreement and proposed that fast rotating disky elliptical galaxies can originate from purely collisionless 3:1 mergers while slowly rotating, pressure supported ellipticals form from equal mass mergers of disk galaxies.

Despite these encouraging results no systematic high-resolution survey of mergers has yet been performed to explore the parameter space of initial conditions and specify the variety of properties of merger remnants that could arise. Recently, Naab & Burkert [50] completed a large number of 112 merger simulations of disk galaxies adopting a statistically unbiased sample of orbital initial conditions with mass ratios η of 1:1, 2:1, 3:1, and 4:1. This large sample allows a much more thorough investigation of the statistical properties of merger remnants in comparison with observed disky and boxy ellipticals.

2 The Merger Models

Cosmological simulations currently are not sophisticated enough to predict initial conditions of major spiral mergers. Some insight can however be gained by investigating the typical conditions under which dark matter halos merge in standard cold dark matter models. Such a detailed analysis was done by Khochfar, Burkert & White [40]. The first encounter is in most cases a parabolic orbit with an impact parameter of order the scale radius of the more massive dark halo, with random orientation of the net spin axes of the progenitors. Unequal mass mergers with mass ratios η of 3:1 to 4:1 are as likely as equal-mass mergers with $\eta = 1:1-2:1$. The cold dark matter simulations however do not provide information on the internal structure and gas content of the merging spirals. In fact, simulations of hierarchical structure formation including gas lead to disk galaxies which do not fit the zero point of the Tully-Fisher relation with disk scale radii that are up to a factor of 10 smaller than observed [51]. Unless these problems are solved we cannot study the subsequent merging of disk galaxies self-consistently, including the large-scale evolution of the Universe. In the meantime, the best strategy is to construct plausible equilibrium models of disk galaxies and investigate their merging in isolation.

Equilibrium spirals were generated using the method described by Hernquist [34]. The following units are adopted: gravitational constant G=1, exponential scale length of the larger disk $h = 1$ and mass of the larger disk $M_d = 1$. For a typical spiral like the Milky Way these units correspond to $M_d = 5.6 \times 10^{10} M_{\odot}$, h=3.5 kpc and a unit time of 1.3×10^7 yrs. Each galaxy consists of an exponential disk, a spherical, non-rotating bulge with mass $M_b = 1/3$ and a Hernquist density profile [32] with a scale length $r_b = 0.2$. The stellar system is embedded in a spherical pseudo-isothermal halo with a mass $M_d = 5.8$, cut-off radius $r_c = 10$ and core radius $\gamma = 1$.

The mass ratios η of the progenitor disks were varied between $\eta = 1$ and $\eta = 4$. For equal-mass mergers ($\eta = 1$) in total 400000 particles were adopted with each galaxy consisting of 20000 bulge particles, 60000 disk particles, and 120000 halo particles. Twice as many halo particles than disk particles are necessary in order to reduce heating and instability effects in the disk components [47]. For the mergers with $\eta = 2, 3, 4$ the parameters for the more massive galaxy were as described above. The low-mass companion however contained a fraction of $1/\eta$ less mass and number of particles in each component, with a disk scale length of $h = \sqrt{1/\eta}$, as expected from the Tully-Fisher relation [54].

The N-body simulations for the equal-mass mergers were performed by direct summation of the forces using the special purpose hardware GRAPE6 [44]. The mergers with mass ratios $\eta = 2, 3, 4$ were followed using the newly developed treecode WINE [64] in combination with the GRAPE5 [37] hardware. WINE uses a binary tree in combination with the refined multipole acceptance criterion proposed by Warren & Salmon (1996). This criterion enables the user to control the absolute force error which is introduced by the tree construction. We chose a value of 0.001 which guarantees that the error resulting from the tree is of order the intrinsic force error of the GRAPE5 hardware which is 0.1%. For all simulations we used a gravitational softening of $\epsilon = 0.05$ and a fixed leap-frog integration time step of $\Delta t = 0.04$. For the equal-mass mergers simulated with direct summation on GRAPE6 the total energy is conserved. The treecode in combination with GRAPE5 conserves the total energy up to 0.5%.

For all mergers, the galaxies approached each other on parabolic orbits with an initial separation of $r_{sep} = 30$ length units and a pericenter distance of $r_p = 2$ length units. Free parameters are the inclinations of the two disks relative to the orbital plane and the arguments of pericenter. In selecting unbiased initial parameters for the disk inclinations we followed the procedure described by Barnes [2]. To determine the spin vector of each disk we define four different orientations pointing to every vertex of a regular tetrahedron. These parameters result in 16 initial configurations for equal mass mergers and 16 more for each mass ratio $\eta = 2, 3, 4$ if the initial orientations are interchanged. In total we simulated 112 mergers.

In all simulations the merger remnants were allowed to settle into equilibrium approximately 8 to 10 dynamical times after the merger was complete. Then their equilibrium state was analysed.

3 Photometric and Kinematical Properties of the Remnants

To compare our simulated merger remnants with observations we analysed the remnants with respect to observed global photometric and kinematical properties of giant elliptical galaxies, e.g. surface density profiles, isophotal deviation from perfect ellipses, velocity dispersion, and major- and minor-axis rotation. Defining characteristic values for each projected remnant we followed as closely as possible the analysis described by Bender et al. [8].

3.1 Isophotal Shape

An artificial image of the remnant was created by binning the central 10 length units into 128 × 128 pixels. This picture was smoothed with a Gaussian filter of standard deviation 1.5 pixels. The isophotes and their deviations from perfect ellipses were then determined using a data reduction package kindly provided by Ralf Bender. Following the definition of Bender et al. [8] for the global properties of observed giant elliptical galaxies, we determined for every projection the effective a_4-coefficient $a4_{\mathbf{eff}}$ as the mean value of a_4 between $0.25r_e$ and $1.0r_e$, with r_e being the projected spherical half-light radius. Like for observed ellipticals we find two types of remnants. Disky systems show a positive characteristic peak of a_4 roughly at $0.5r_e$. In boxy ellipticals, the a_4 coefficient might be positive in the innermost regions. It decreases however systematically outwards with a mean value that is negative. The characteristic ellipticity $\epsilon_{\mathbf{eff}}$ for each projection was defined as the isophotal ellipticity at $1.5r_e$. To investigate projection effects we determined for each simulation $a4_{\mathbf{eff}}$ and $\epsilon_{\mathbf{eff}}$ for 500 random projections of the remnant. These values were used to calculate the two-dimensional probability density function for a given simulated remnant to be "observed" in the $a4_{\mathbf{eff}}$-$\epsilon_{\mathbf{eff}}$ plane.

Figure 1 shows the ellipticities and a_4-coefficients of mergers with $\eta = 1, 2, 3,$ and 4. The contours indicate the areas of 50% (dashed line), 70% (thin line) and 90% (thick line) probability to detect a merger remnant with the given properties. Observed data points from Bender et al. [11] or [13] are over-plotted. Filled boxes are observed boxy ellipticals with $a4_{\mathbf{eff}} \leq 0$ while open diamonds indicate observed disky ellipticals with $a4_{\mathbf{eff}} > 0$. The error bar in determining a_4 from the simulations is shown in the upper left corner and was estimated applying the statistical bootstrapping method [36]. Ellipticity errors are in general too small to be visible.

We find that the isophotal shapes of ellipticals and their ellipticities are affected by the initial mass ratio of the merger and by projection effects. The area covered by 1:1 remnants with negative $a4_{\mathbf{eff}}$ is in very good agreement with the observed data for boxy elliptical galaxies. In particular the observed trend for more boxy galaxies to have higher ellipticities is reproduced. However we also find configurations of 1:1 mergers which under certain projection angles appear disky with $0 \geq a4_{\mathbf{eff}} \leq 1$. In addition, note that the remnants with $a4_{\mathbf{eff}}$ around zero can have higher ellipticities than observed.

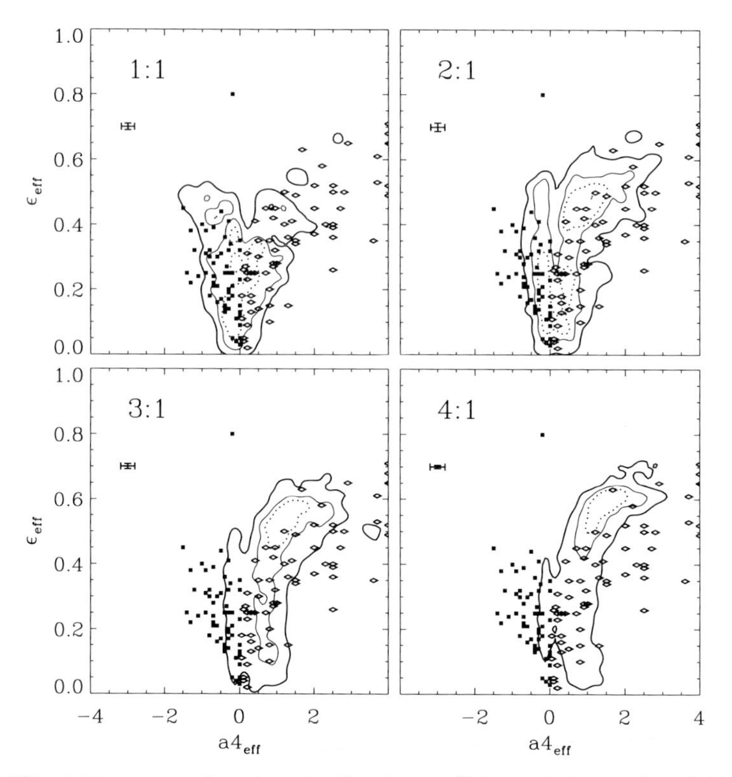

Fig. 1. Ellipticities versus fourth-order Fourier coefficient of the isophotal shape deviations is shown for simulations with different initial mass ratios. The contours indicate the 50% (dotted line), 70% (thin solid line) and the 90% (thick solid line) probability to find a merger remnant in the enclosed area. Black squares indicate values for observed boxy ellipticals, open diamonds show observed disky ellipticals.

The distribution function of isophotal shapes for 1:1 merger remnants peaks at $a4_{\mathbf{eff}} \approx -0.5\%$ (dashed curve in Fig. 2). It declines rapidly for more negative values and has a broad wing towards positive $a4_{\mathbf{eff}}$ values. Almost half of the projected remnants are disky. In contrast, remnants of mergers with higher mass ratios shift in the direction of positive $a4_{\mathbf{eff}}$. 2:1 remnants peak at $a4_{\mathbf{eff}} \approx 0$. Now, 75% of the projected remnants show disky isophotes. For these cases, the observed trend of more disky ellipticals to be more flattened is also clearly visible in Fig. 1. 3:1 and 4:1 mergers peak at $a4_{\mathbf{eff}} \approx 1$. Their fraction of boxy projections is only 11% and 7%, respectively. The very high positive values of $a4_{\mathbf{eff}} \geq 4\%$ observed in some ellipticals cannot be reproduced. One might argue that these objects formed from mergers with even higher mass ratios of $\eta \geq 5:1$. However, in this case, test simulations show that the merger remnants do not look like typical ellipticals anymore with characteristic de Vaucouleurs profiles as the more massive disk is not destroyed. Their surface brightness profiles instead remain exponential.

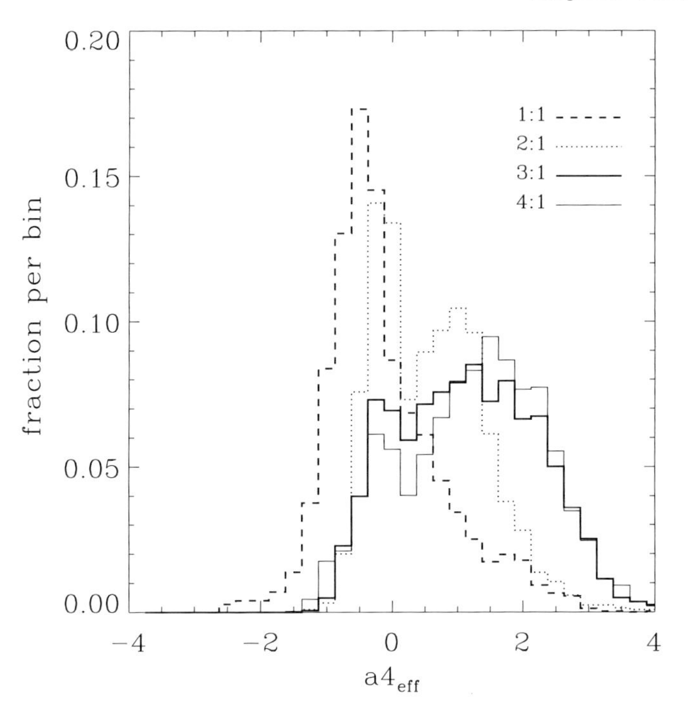

Fig. 2. Normalized histograms of the shape parameter $a4_{\mathbf{eff}}$ for mergers with various mass ratios.

In summary, there is a clear trend for unequal-mass mergers to produce more disky remnants. Responsible for the disky appearance of the 3:1 and 4:1 remnants is the distribution of the particles of the massive disk [3]. The particles originating from the small progenitor accumulate in a torus-like structure with peanut-shaped or boxy isophotes while the luminous material of the larger progenitor still keeps its disk-like appearance. In combination, the contribution from the larger progenitor – since it is three to four times more massive – dominates the overall appearance of the remnant. This result holds for all 3:1 and 4:1 merger remnants. For equal mass mergers however both disks are destroyed efficiently during the merger. No dominant disk-like structure remains after the merger and the system looses the information about the initial configuration.

3.2 Kinematics

The central velocity dispersion σ_0 of every remnant is determined as the average projected velocity dispersion of the stars inside a projected galactocentric radius of $0.2r_e$. The characteristic rotational velocity v_{maj} along the major axis is defined as the projected rotational velocity determined around $1.5r_e$. Like for the isophotal shape we constructed probability density plots for the kinematical properties of the simulated remnants and compared them with observational

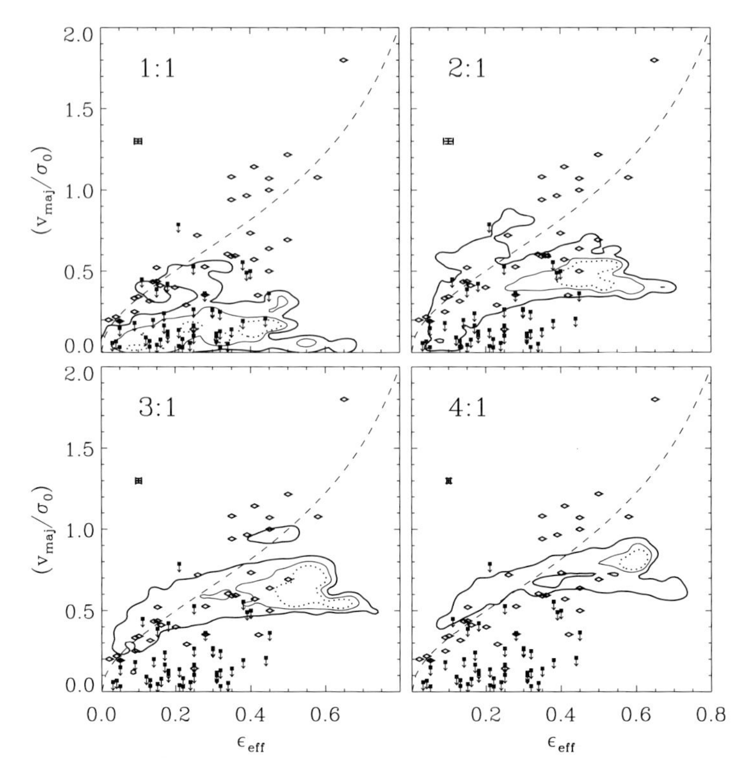

Fig. 3. Rotational velocity over velocity dispersion versus characteristic ellipticity for mergers with various mass ratios. Values for observed ellipticals are overplotted. The dashed line shows the theoretically predicted correlation for an oblate isotropic rotator.

data from elliptical galaxies. Figure 3 shows the distribution function in the (v_{maj}/σ_0)-$\epsilon_{\mathbf{eff}}$ plane.

The region of slowly rotating boxy ellipticals (filled squares) is almost completely covered by the data of 1:1 mergers. Unequal-mass merger remnants are clearly fast rotating. They can be associated with disky ellipticals. Although the simulated remnants are in good agreement with observations there is again the trend for the ellipticities to be higher than observed, especially when the system is seen edge-on.

The anisotropy parameter $(v_{maj}/\sigma_0)^*$ is defined as the ratio of the observed value of (v_{maj}/σ_0) and the theoretical value for an isotropic oblate rotator $(v/\sigma)_{theo} = \sqrt{\epsilon_{\mathbf{obs}}/(1-\epsilon_{\mathbf{obs}})}$ with the observed ellipticity $\epsilon_{\mathbf{obs}}$ [17]. This parameter is frequently used by observers to test whether a given galaxy is flattened by rotation ($(v_{maj}/\sigma_0)^* \geq 0.7$) or by velocity anisotropy ($(v_{maj}/\sigma_0)^* < 0.7$) ([25], [7], [53], [58]). Figure 4 shows the normalized histograms for the $(v_{maj}/\sigma_0)^*$ values of the simulated remnants. 1:1 remnants peak around $(v_{maj}/\sigma_0)^* \approx 0.3$ with a more prominent tail towards lower values. They are consistent with being supported by anisotropic velocity dispersions. As these systems also have

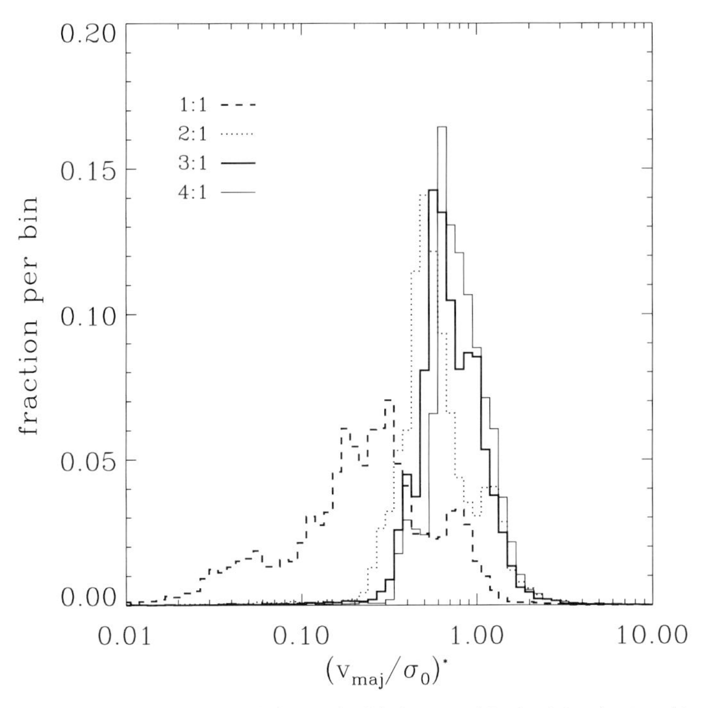

Fig. 4. Normalized histograms of $(v_{maj}/\sigma_0)^*$ for 1:1 (dashed line), 2:1 (dotted line), 3:1 (thick line) and 4:1 (thin line) mergers.

preferentially negative a_4-values they agree with observations of boxy ellipticals (Fig. 5). Unequal mass mergers peak at $(v_{maj}/\sigma_0)^* \approx 0.7$, as expected for oblate isotropic rotators. Since especially the 3:1 and 4:1 remnants also have predominantly disky isophotes they cover the area populated by observed disky ellipticals in the $log(v_{maj}/\sigma_0)^*$ - $a4_{\mathbf{eff}}$ diagram which is shown in Fig. 5.

We also investigated the minor-axis kinematics of the simulated remnants by determining the rotation velocity along the minor axis at $0.5r_{eff}$. The amount of minor axis rotation was characterized by $(v_{min}/\sqrt{v_{maj}^2 + v_{min}^2})$ [18]. Minor axis rotation in elliptical galaxies, in addition to isophotal twist, has been suggested as a sign for a triaxial shape of the main body of elliptical galaxies ([62], [30]). In general, 1:1 mergers show a significant amount of minor-axis rotation, whereas 3:1 and 4:1 remnants have only small minor axis rotation (for details see [50]).

4 Conclusions

The analysis of a large set of mergers with different mass ratios and orbital geometries shows that their properties are in general in good agreement with the observational data for elliptical galaxies.

Only equal mass mergers can produce boxy, anisotropic and slowly rotating remnants with a large amount of minor axis rotation. However, in the more un-

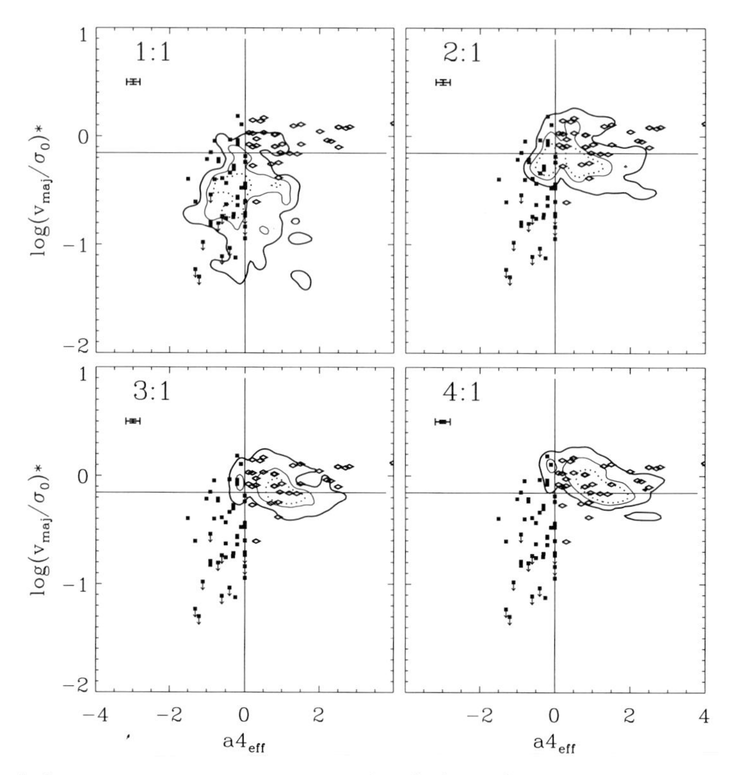

Fig. 5. Anisotropy parameter versus isophotal shape for mergers with various mass ratios. Values for observed ellipticals are overplotted.

likely case that the initial spins of the progenitor disks are aligned, the remnants appear isotropic and disky or boxy depending on the orientation. In contrast, 3:1 and 4:1 mergers form a more homogeneous group of remnants. They have preferentially disky isophotes, are always fast rotating and show small minor axis rotation independent of the assumed projection. 2:1 mergers have properties intermediate between boxy or disky ellipticals, depending on the projection and the orbital geometry of the merger.

There still exist problems which are not solved up to now. Certain projections of 1:1 mergers lead to anisotropic, disky remnants which are not observed. Edge-on projections of merger remnants often show very high ellipticities $\epsilon > 0.6$ which are larger than observed. Finally, some 2:1 to 4:1 remnants are more anisotropic than expected from their rotation. Their values of $(v/\sigma)^*$ are smaller and their ellipticities are larger than observed. A problem arises especially for very low luminosity giant ellipticals which are characterized by exceptionally high rotational velocities in the outer regions that cannot be reproduced [24]. A detailed analyses of the intrinsic kinematics of disky, fast rotating merger unequal-mass remnants which are called isotropic due to their high $(v/\sigma)^* \approx 0.7$ demonstrates that in most cases the velocity dispersion tensor is as anisotropic as for equal-mass, boxy and anisotropic mergers with $(v/\sigma)^* = 0.1$ [20]. The anisotropy parameter therefore is not necessarily a good indicator of anisotropy. It rather measures the amount of rotation in the systems.

The present simulations were purely dissipationless, taking into account only the stellar and dark matter components. The importance of gas in determining the structure of merger remnants is not clear up to now. Kormendy & Bender [41] proposed a revised Hubble sequence with disky ellipticals representing the missing link between late type systems and boxy ellipticals. They noted that gas infall into the equatorial plane with subsequent star formation could explain the origin of diskyness. Scorza & Bender [58] demonstrated that ellipticals with embedded disks would indeed appear disky when seen edge-on and boxy otherwise. Although this scenario appears attractive, it cannot explain why X-ray halos are found only in boxy ellipticals. As the detection of hot gas around galaxies should be independent of their orientation, the isophotal shapes of ellipticals would not correlate with their X-ray emission, if these shapes are merely a result of projection effects.

Our simulations indicate that it is preferentially the initial mass ratio which determines the isophotal shapes of merger remnants. Still, gas could have played an important role in affecting the final structure and stellar population of ellipticals ([5], [6], [46]), not only in their central regions and might solve the problem of dissipationless mergers. Naab & Burkert [48] have shown that extended gas disks can form as a result of a gas rich unequal mass mergers (see also [4]). Naab & Burkert [49] investigated line-of-sight velocity distributions of dissipationless merger remnants and found a velocity profile asymmetry that is opposite to the observations. They concluded that this disagreement can be solved if ellipticals would indeed contain a second disk-like substructure that most likely formed through gas accretion. The situation is however not completely clear, as another study by Bendo & Barnes [15] found a good agreement of the observed asymmetries for some cases. More simulations, including gas and star formation will be required to understand the role of gas in mergers and to answer the question of how early-type galaxies formed.

Acknowledgement

A. Burkert thanks the organizers of the workshop "Galaxies and Chaos" for the invitation and a very stimulating conference.

References

1. Aragon-Salamanca, A. , Ellis, R. S., Couch, W. J. & Carter, D. 1993, MNRAS, 262, 764
2. Barnes, J. E. 1988, ApJ, 331, 699
3. Barnes, J. E. 1998, Galaxies: Interactions and Induced Star Formation: Lecture Notes 1996 / Saas Fee Advanced Course 26, eds. D. Friedli, L. Martinet, and D. Pfenniger, Springer, 275
4. Barnes, J. E. 2002, MNRAS, 333, 481

5. Bekki, K. 1998, ApJL, 502, L133
6. Bekki, K. 1999, ApJ, 513, 108
7. Bender, R. 1988a, A&A, 193, L7
8. Bender, R. 1988b, A&A, 202, L5
9. Bender, R., Döbereiner, S., & Möllenhoff, C. 1988, A&AS, 74, 385
10. Bender, R., Surma, P., Döbereiner, S., Möllenhoff, C. & Madejsky, R. 1989, A&A, 217, 35
11. Bender, R. & Surma, P. 1992, A&A, 258, 250
12. Bender, R. 1996, IAU Symp., 171, New Light on Galaxy Evolution, ed. R. Bender & R. L. Davies (Dordrecht: Kluwer), 181
13. Bender, R., Burstein, D. & Faber, S.M. 1992, ApJ, 399, 462
14. Bender, R., Ziegler, B. & Bruzual, G. 1996, ApJL, 463, L51
15. Bendo, G. J. & Barnes, J. E. 2000, MNRAS, 316, 315
16. Beuing, J., Döbereiner, S., Böhringer, H. & Bender, R. 1999, MNRAS, 302, 209
17. Binney, J. 1978, MNRAS, 183, 779
18. Binney, J. 1985, MNRAS, 212, 767
19. Burkert, A. 1993, A&A, 278, 23
20. Burkert, A., Binney, J. & Naab, T. 2003,in preparation
21. Bower, R. G., Lucey, J. R. & Ellis, R. S. 1992, MNRAS, 254, 589
22. Bruzual A., G. & Charlot, S. 1993, ApJ, 405, 538
23. Carlberg, R. G. 1986, ApJ, 310, 593
24. Cretton, N., Naab, T., Rix, H., & Burkert, A. 2001, ApJ, 554, 291
25. Davies, R. L., Efstathiou, G., Fall, S. M., Illingworth, G. & Schechter, P. L. 1983, ApJ, 266, 41
26. Davies, R. L. 1996, IAU Symp., 171, New Light on Galaxy Evolution, ed. R.Bender & R. L. Davies (Dordrecht: Kluwer), 37
27. Ellis, R. S., Smail, I., Dressler, A. , Couch, W. J., Oemler, A. , Jr., Butcher, H. & Sharples, R. M. 1997, ApJ, 483, 582
28. Faber, S. M., et al. 1997, AJ, 114, 1771
29. Franx, M. & Illingworth, G. D. 1988, ApJL, 327, L55
30. Franx, M., Illingworth, G. & de Zeeuw, T. 1991, ApJ, 383, 112
31. Gerhard, O.E. 1981, MNRAS 197, 179
32. Hernquist, L. 1990, ApJ, 356, 359
33. Hernquist, L. 1992, ApJ, 400, 460
34. Hernquist, L. 1993a, ApJS, 86, 389
35. Hernquist, L. 1993b, ApJ, 409, 548
36. Heyl, J. S., Hernquist, L. & Spergel, D. N. 1994, ApJ, 427, 165
37. Kawai, A., Fukushige, T., Makino, J., & Taiji, M. 2000, PASJ, 52, 659
38. Kauffmann, G. 1996, MNRAS, 281, 487
39. Kauffmann, G. & Charlot, S. 1998, MNRAS, 297, L23
40. Khochfar, S., Burkert, A. & White. S. 2003, in preparation
41. Kormendy, J. & Bender, R. 1996, ApJL, 464, L119
42. Larson, R.B. 1974, MNRAS, 166, 585
43. Lauer, T. R., et al. 1995, AJ, 110, 2622
44. Makino, J., Fukushige, T. & Namura, K. 2003, to be submitted to PASJ.
45. Mehlert, D., Saglia, R. P., Bender, R. & Wegner, G. 1998, A&A, 332, 33
46. Mihos, J. C. & Hernquist, L. 1996, ApJ, 464, 641
47. Naab, T. , Burkert, A., & Hernquist, L. 1999, ApJL, 523, L133
48. Naab, T. & Burkert, A. 2001a, ASP Conf. Ser. 230: Galaxy Disks and Disk Galaxies, 451

49. Naab, T. & Burkert, A. 2001b, ApJL, 555, L91
50. Naab, T. & Burkert A., submitted to ApJ
51. Navarro, J.F. & Steinmetz, M. 2000, ApJ, 538, 477
52. Negroponte, J. & White, S. D. M. 1983, MNRAS, 205, 1009
53. Nieto, J.-L., Capaccioli, M. & Held, E. V. 1988, A&A, 195, L1
54. Pierce, M. J. & Tully, R. B. 1992, ApJ, 387, 47
55. Rix, H. -W. & White, S. D. M. 1990, ApJ, 362, 52
56. Rix, H. -W. & White, S. D. M. 1992, MNRAS, 254, 389
57. Jedrzejewski, R. & Schechter, P. L. 1988, ApJL, 330, L87
58. Scorza, C. & Bender, R. 1995, A&A, 293, 20
59. Searle, L., Sargent, W. L. W. & Bagnuolo, W. G. 1973, ApJ, 179, 427
60. Steinmetz, M. & Buchner, S. 1995, Galaxies in the Young Universe, Proceedings of a Workshop held at Ringberg Castle, eds. H. Hippelein, K. Meisenheimer & H.-J. Röser, Springer, p. 215
61. Toomre, A. & Toomre, J. 1972, ApJ, 178, 623
62. Wagner, S. J., Bender, R. & Moellenhoff, C. 1988, A&A, 195, L5
63. Warren, M. S., & Salmon, J. K. 1996, ApJ, 460, 121
64. Wetzstein, M., Nelson, A., Naab, T., & Burkert, A. 2003, in preparation
65. Ziegler, B. L. & Bender, R. 1997, MNRAS, 291, 527
66. Ziegler, B. L., Saglia, R. P., Bender, R., Belloni, P., Greggio, L. & Seitz, S. 1999, A&A, 346, 13

Dynamical Evolution of Galaxies: Supercomputer N-Body Simulations

Edward Liverts, Evgeny Griv, Michael Gedalin, and David Eichler

Ben-Gurion University of the Negev, P.O. Box 653, Beer-Sheva 84105, Israel

Abstract. The time evolution of a computer model for an isolated disk representing a flat galaxy is studied. The method of direct integration of Newton's equations of motion of particles–"stars" is applied. Using the modern 128-processor SGI Origin 2000 supercomputer in Israel, we make long simulation runs with a large number of particles, $N = 100\,000$. One of the goals of the simulation is to test the validities of the modified Safronov–Toomre criterion for stability of arbitrary but not only axisymmetric Jeans-type gravity disturbances (e.g., those produced by a spontaneous perturbation and/or a companion system) in a self-gravitating, thin, and almost collisionless stellar disk. We are also interested in how model particles diffuse in chaotic (residual) velocity space. This is of considerable interest in the nonlinear theory of stellar disks.

1 Introduction

One can learn much about the properties of stellar systems of galaxies experimentally by computer simulation of N-body systems. In this work, we analyze the evolution and stability of structures in an N-body model of an isolated and rotating stellar disk representing a flat galaxy by integration of Newton's equations of motion of N identical particles. Use of the 128-processor SGI Origin 2000 computer, enabled us to make long simulation runs using a large number of particles, $N = 100\,000$, in the direct summation code and thus simulate phenomena not previously studied numerically. The essential difference between the present and previous simulations is the comparison between the results of N-body experiments and the kinetic stability theory as developed in [1–8].

2 N-Body Simulations

Different methods are currently employed to simulate the evolution of collisionless point-mass systems of flat galaxies by N-body experiments. See, e.g., [9] as a review. For instance, one can use an algorithm for a simulation code, which is an analog of plasma particle-mesh (PM) codes. It is believed that simulating many billions of stars in actual galaxies by using only several ten or hundred thousands particles in PM experiments will be enough to capture the essential physics, which includes wave-like collective motions. In other fields, such as the simulation of spiral structures, PM codes may be used with moderate success. This is because these fine-scale $\lesssim$ 1 kpc structures can basically be governed by collisionless processes. By increasing the number of cells to reproduce the microstructures, one reduces the average number of particles per cell, and thus increases the undesirable effect of particle encounters. The problem is serious because there are regions in the phase space which are important to the problem but in which the distribution function is small. In contrast, our model is based on the direct numerical integration of the equations of motion in three dimensions for N mutually gravitating particles.

The numerical procedure is used first to seek stationary solutions to the Boltzmann equation in the self-consistent field approximation, and then to determine the stability of those solutions to small gravity perturbations. At the start of the N-body integration, our similation initilizes the particles on a set of 100 concentric circular rings with a circular velocity $\boldsymbol{V}_{\text{rot}}$ of galactic rotation in the equatorial plane; the system is isolated. Then the position of each particle was slightly perturbed by applying a pseudorandom number generator. The Maxwellian-distributed chaotic (residual) velocities $\boldsymbol{v}$ were added to the initial circular velocities $\boldsymbol{V}_{\text{rot}}$, and $|\boldsymbol{v}| \ll |\boldsymbol{V}_{\text{rot}}|$. The acceleration of the ith particle is

$$\boldsymbol{a}_i = \sum_{j \neq i}^{N} \frac{(\boldsymbol{r}_j - \boldsymbol{r}_i)}{(r_{ij}^2 + r_{\text{cut}}^2)^{3/2}}. \tag{1}$$

In (1), r_{cut} is the so-called cutoff radius. This "softening" of the gravitational potential is a device used in N-body simulations to avoid numerical difficulties at very close but rare encounters. Units are chosen such that the mass of each particle is $10^5 M_\odot$ so that the total mass of the disk galaxy is $10^{10} M_\odot$. The initial radius of the disk is 10 kpc. When N is large, the main computational problem is the large number, $\propto N(N-1)$, of operations required to determine $\boldsymbol{a}_i$.

We consider a rotating model disk of stars of thickness h with a surface mass density variation given by $\sigma_0(r) = \sigma(0)\sqrt{1 - r^2/R^2}$, where $\sigma(0)$ is the central surface density and R is the radius of the initial disk. As a solution of a time-independent collisionless Boltzmann equation, to ensure initial equilibrium, the angular velocity to balance the zero-velocity dispersion disk, $\Omega_0 = \pi\sqrt{G\sigma(0)/2R}$, was adopted [10]. For this uniformly rotating disk, the Maxwellian-distributed chaotic velocities with radial c_r and azimuthal c_φ dispersions in the plane $z = 0$, according to the Safronov–Toomre criterion [11], $c_{\text{T}} = 3.4G\sigma_0/\kappa = 0.341\Omega_0\sqrt{R^2 - r^2}$, may be added to the initial circular velocities $V_{\text{rot}} = r\Omega_0$. Here κ is the epicyclic frequency. According to [1–8,12–14], it is crucial to realize that such a spatially inhomogeneous disk is Jeans-stable only against the axisymmetric (radial) gravity perturbations but unstable against the nonaxisymmetric (spiral) perturbations. The initial vertical velocity dispersion was chosen $c_z = 0.15c_r$. Finally, the angular velocity Ω_0 was replaced by [10]

$$\Omega = \left\{ \Omega_0^2 + \frac{1}{r\sigma(r)} \frac{\partial}{\partial r} \left[\sigma(r) c_r^2(r) \right] \right\}^{1/2}.$$

The sense of disk rotation was taken to be counterclockwise, the cutoff radius was $r_{\text{cut}} = 0.004R$, and the initial disk thickness was $h = 0.006R$, that is, $h > r_{\text{cut}}$.

Slight corrections have been applied to the resultant velocities and coordinates of the model stars so as to ensure the equilibrium between the centrifugal and gravitational forces, to preserve the position of the disk center of gravity at the origin, and to include the weak effect of the finite thickness of the disk to the gravitational potential. Thus, the initial model is very near the dynamical equilibrium for all radii. A time $t = 1$ was taken to correspond to a single revolution of the initial disk. In the experiment the simulation was performed up to a time $t = 10$. It should be noted here that after about three rotations the picture is practically stabilized and no significant changes in gross properties of the model over this time are observed. Tests indicated that the results were insensitive to changes in the number of particles in the range $N = 10\,000 - 100\,000$, the cutoff parameter in the range $r_{\rm cut} = (0.001 - 0.01)R$, the initial velocity dispersion in the range $c_r = (1 - 1.3)c_{\rm T}$, and the initial disk thickness in the range $h = (0 - 0.08)R$. We argue that structures observed in our N-body simulations originate from the collective modes of oscillations — the classical Jeans-type gravitational modes and firehose-type bending modes.

3 Results of Simulations

Figure 1 displays a series of snapshots from a simulation run. The figures include only a face-on view of the simulation region. In accordance with the theoretical explanation [1–8,12–14], the effects of the Jeans instability of spontaneous non-axisymmetric gravity perturbations appear quickly in the simulation. One can see at first a strong multi-armed spiral structure. It is interesting to notice that in a sample of 654 optical spiral galaxies [15], two-armed (grand design) galaxies like M 51 are roughly a factor of six times rare than such many-armed galaxies like NGC 613, an SBb galaxy in Sculptor.

At a later time, $t > 0.5$, the multi-armed structure disappears quickly and is replaced by a weak spiral structure with three main spiral arms, $m = 3$, or sometimes two, $m = 2$, or only one, $m = 1$, spiral arms. These spirals are evidently gravitationally (Jeans-)unstable Lin–Shu density waves [16–19] and not material arms, since test particles pass right through them. The $m = 1$ mode shifts the point with highest density from the center of mass [9]. Interestingly, in many disk-shaped galaxies, e.g., in the spiral galaxies M 101 and NGC 1300,

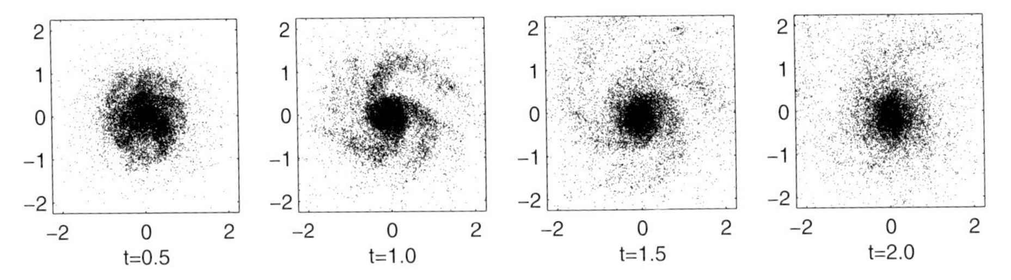

Fig. 1. The time evolution (face-on view) of an initially equilibrium, Toomre-stable disk of $N = 100\,000$ stars. The effects of the Jeans-type gravitational instability of spontaneous spiral disturbances appear quickly in the simulation. A moderately tightly wound, low–m spiral structure develops in the plane of the system at a time $t \approx 1.0$.

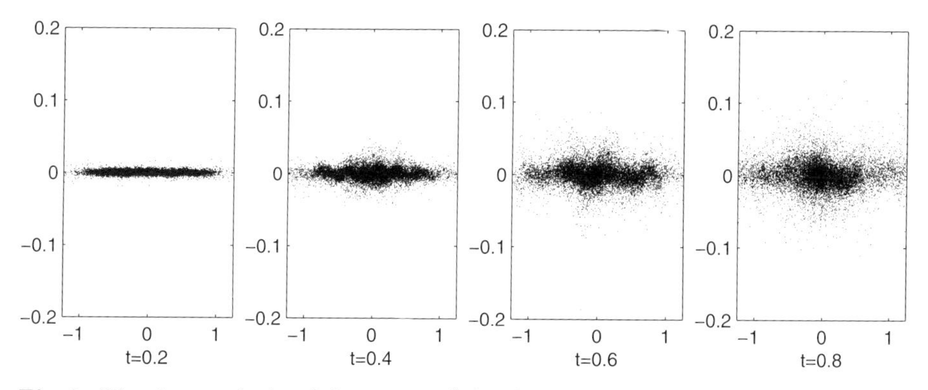

Fig. 2. The time evolution (edge-on view) for the simulation run shown in Fig. 1. At $t \approx 0.4$, the firehose-type bending instability fiercely develops in the central, almost nonrotating parts of the system (and is switched off at $t \approx 2.0$ [9,21]).

there appears to be a deviation from rotational symmetry. In principle, such a deviation may be due to the one arm Jeans instability. Also note that two unusual single-arm galaxies turn up in [20] sample of some 54 differentially rotating spiral galaxies. In one of them, namely NGC 4378, the spiral arm can be traced over most $\frac{1}{4}$ revolutions.

From the edge-on view pictured in Fig. 2, one can see that a fully three-dimensional disk develops immediately at $t \approx 0.2$. A straightforward estimate shows that a mean height Δ of the disk above the plane corresponds to the force balance between the gravitational attraction in the plane and the "pressure" due to the velocity dispersion c_z (i.e., "temperature") in the z-direction [9,19]. Clearly, this pressure-supported (in the z-direction) three-dimensional structure seen to form very rapidly on the time scale of a single vertical epicyclic oscillation, $< \Omega^{-1}$, with rather sharp edges. After a time $t \approx 0.2$ there is no change in the edge-on structure until at $t \approx 0.4$. It is noteworthy that at $t \approx 0.4$, the firehose-type bending instability rapidly develops in the central parts of the system (and is switched off at $t \approx 2.0$ [9,21]). At later times, $t > 2.0$, no dramatic evolution is observed in our simulations (see [9]). To emphasize, the bending instability develops in the *central*, almost *nonrotating* region of the system under study [9]. New spectroscopic optical and H I observations constitute a strong case in favour of this bar-buckling mechanism for the formation of boxy/peanut-shaped bulges in spiral galaxies [22]. Apparently, the authors of [23–25] first found the firehose-type bending instability as a precursor of galactic bulge formation in the central, almost nonrotating regions of a warm in the plane N-body disk, which initially developed planar bars.

At somewhat later times, $t > 2.0$, a "box-shaped" or sometimes "peanut-shaped" bar structure is developed [9]. The simulations show that, soon after a central bar develops in the equatorial plane, it buckles and settles with an increased thickness and vertical velocity dispersion, appearing boxy-shaped when seen end-on and peanut-shaped when seen side-on [9,24,25]. The projected den-

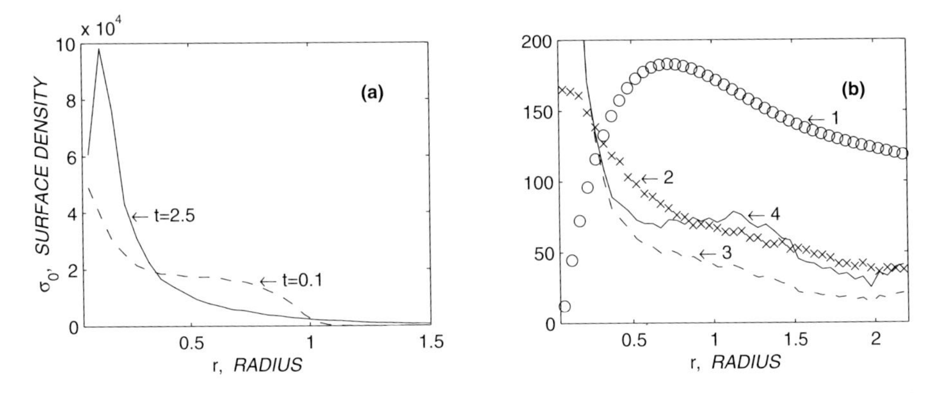

Fig. 3. In Fig. 3*a*, we show the evolution of the surface density for a stellar disk ($\times 2M_{\odot}/\text{kpc}^2$) shown in Fig. 1. In Fig. 3*b*, rotation curve V_{rot} (km/sec) (*1*), radial velocity dispersion (km/sec) (*2*), Safronov–Toomre critical velocity dispersion (km/sec) (*3*), and modified dispersion (km/sec) (*4*) at the time $t = 2.5$.

sities of this central bar resemble the bulge light distribution measured by the COBE satellite in the Milky Way's Galaxy [26]. Moreover, a significant fraction of edge-on spiral galaxies, and therefore presumably of all spirals, show boxy or peanut-shaped isophotes in the bulge region [22]. The firehose-type bar-buckling instability is the currently favored mechanism for the formation of boxy/peanut-shaped bulges in spiral galaxies.

Figure 3*a* shows the evolution of the azimuthally averaged surface density σ_0 as a function of radius. It can be seen that the mass density is redistributed by the Jeans-unstable waves on the dynamical time scale, $\lesssim \Omega^{-1}$; the surface density of the quasi-steady state system at $t > 1$ falls off exponentially.

3.1 Modified Stability Criterion

To emphasize it again, even though the initial velocity dispersion is equal to the Safronov–Toomre [11] stabilizing one c_{T}, the model is still violently Jeans-unstable. The reason for such a behaviour of a disk has been explained in [1–8]. See also [12–14,27–30] for a discussion. Accordingly, the presence of the differential rotation (or shear) results in quite different dynamical properties of the axisymmetric and nonaxisymmetric gravity perturbations. In differentially rotating disks the azimuthal force resulting from azimuthal displacements is more important in determining the stability than is the radial force resulting from radial displacements [28,31]. In a nonuniformly rotating disk for nonaxisymmetric perturbations the modified dispersion c_{M} of a marginally Jeans-stable system is larger than c_{T} (although still of the order of c_{T}), and is approximately

$$c_{\text{M}} \approx (2\Omega/\kappa)c_{\text{T}}. \tag{2}$$

In disk-shaped galaxies, $2\Omega/\kappa = 1.5 - 1.8$. It is obvious that in differentially rotating galaxies, disks manage to keep their local stability parameter close to the critical value, $c_r \approx (2\Omega/\kappa)c_T \approx 2c_T$ or Toomre's Q-stability parameter $Q \equiv c_r/c_T \approx 2\Omega/\kappa \approx 2$, respectively. In this case, once the entire differentially rotating disk has been heated to values $c_r \approx 2c_T$ (or $Q \approx 2$), no further spiral waves can be sustained by virtue of the Jeans instability — unless some "cooling" mechanism is available leading to Toomre's Q-value under approximately 2 or to the value of c_r smaller than approximately $2c_T$, respectively (e.g., by the dissipation in the gas and/or by the star formation in an interstellar medium [9]).

Thus, the Jeans-unstable perturbations can be stabilized by the chaotic velocity spread. The critical Safronov–Toomre [11] velocity dispersion should stabilize only radial perturbations of the Jeans type. The differentially rotating and spatially inhomogeneous disk is still unstable against spiral Jeans perturbations. The modified stability criterion against arbitrary but not only axisymmetric gravity perturbations is given by (2). The spiral arms in nonuniformly rotating systems are a mechanism for angular momentum transfer [31].

We compare the radial velocity dispersion values c_r predicted by the Safronov–Toomre criterion c_T and by the modified criterion (2) with values obtained in the numerical experiment. Many investigators have remarked that the experimental c_r significantly exceeds c_T. This effect is also apparent in Fig. 3*b*, which represents the quasi-steady state for the computer experiment with a disk shown in Fig. 1. On the other hand, the quantity c_M calculated from (2) satisfactorily fits the experimental c_r (in the central, pressure-supported parts of the system where $V_{rot} \lesssim c_r$, both c_T and c_M fail to employ). Thus, the experiment yields a radial velocity dispersion for the particles significantly greater than predicted by the Safronov–Toomre criterion, but it is nearly equal to the modified dispersion.

3.2 Chaotic Velocity Diffusion

One of the important problems of stellar disks is the determination of the chaotic velocity diffusion. Such a velocity diffusion can be caused by gravitational instabilities of a disk. To compute the velocity diffusion we calculate the mean-square spread in the planar chaotic velocity c^2 as a function of time for different radii.

As is seen in Fig. 4, along with the growth of the oscillation amplitude (planar spiral density waves): (a) chaotic velocities increase, and eventually in the disk a quasi-stationary distribution is established at times $t \gtrsim 1$ so that the Jeans stability sets in, and (b) during the first rotation the squared plane velocity dispersion of particles increases with time as roughly $c^2 \equiv c_r^2 + c_\varphi^2 \propto t$. The results (a) and (b) are in good agreement with the predictions of weakly nonlinear (quasilinear) kinetic theory as developed in [7,8]. Interestingly, observations already showed about the same law of "heating" (increase of the velocity dispersion of the young stellar population with age of stars t) in the Solar vicinity, $c^2 \propto t$ [32]. We conclude that both the quasilinear theory [7,8] and the N-body simulation presented here are able to account for the form of the age–velocity dispersion law in the plane of the Galaxy.

In turn, there are numerous observations showing that there exists ongoing dynamical relaxation on the time scale of < 10 rotation periods ($< 2 \times 10^9$ yr) in

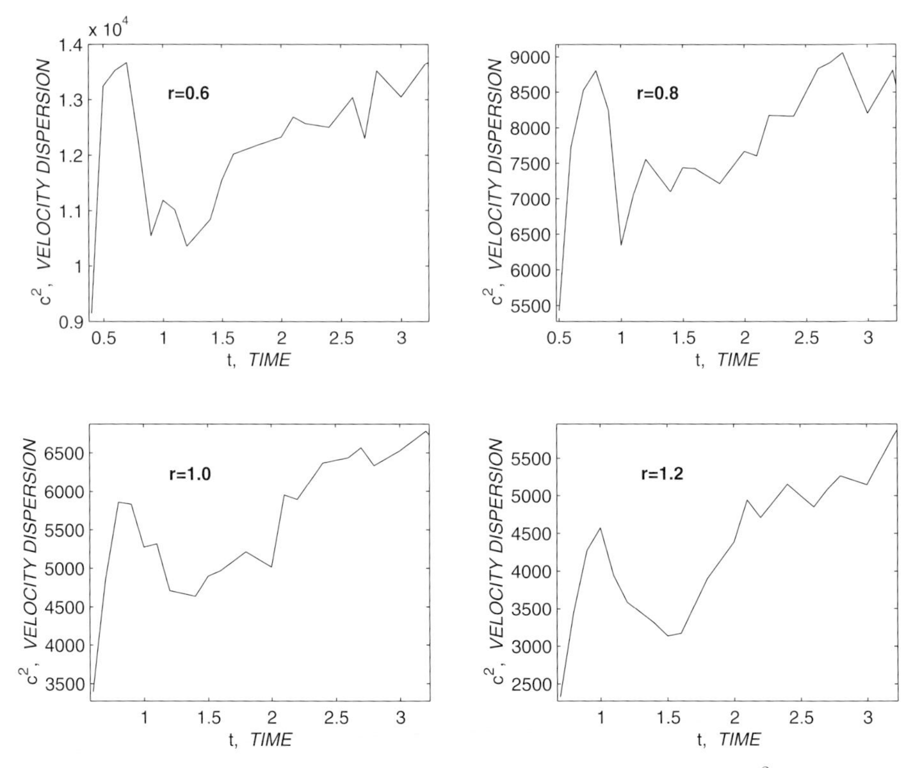

Fig. 4. The time evolution of the squared planar velocity dispersion c^2 for different radii r ($\times 10$ kpc). Initially, the dispersion increases with with time rapidly, $c^2 \propto t$. Later in a quasi-steady state, $t \gtrsim 1$, the dispersion grows only slightly.

the collisionless disk of the Milky Way's Galaxy [19,32–34]. It was observed that in the Solar neighborhood the velocity distribution function of stars with an age $t \gtrsim 10^8$ yr is close to a Schwarzschild distribution — a set of Gaussian distributions along each coordinate in chaotic velocity space, i.e., close to equilibrium along each coordinate. Also, older stellar populations are observed to have a higher velocity dispersion than younger ones. Thus, this dynamical relaxation of the distribution of young stars which were born in the equilibrium disk of the Galaxy results in a randomization of the velocity distribution and a monotonic increase of the chaotic velocity dispersion. The latter indicates a significant irregular gravitational field in the Galactic disk [32–34]. The irregular field causes a diffusion of stellar orbits in velocity [7] (and positional [8]) space. Various mechanisms for the relaxation have been proposed. In the present work, we suggest the idea of the collective *collisionless* relaxation: planar Jeans-unstable gravity perturbations affect effectively the averaged velocity distribution of young stars in the equatorial plane (see [7,8,35,36] for a discussion).

We thank Tzi-Hong Chiueh, Frank H. Shu, Irina Shuster, Raphael Steinitz, and Chi Yuan for valuable discussions. This work was supported in part by the Israel Science Foundation and the Israeli Ministry of Immigrant Absorption.

References

1. E. Griv, W. Peter: ApJ **469**, 84 (1996)
2. E. Griv, M. Gedalin, C. Yuan: A&A **328**, 531 (1997)
3. E. Griv, B. Rosenstein, M. Gedalin, D. Eichler: A&A **347**, 821 (1999)
4. E. Griv, C. Yuan, M. Gedalin: MNRAS **307**, 1 (1999)
5. E. Griv, M. Gedalin, D. Eichler, C. Yuan: Phys. Rev. Lett. **84**, 4280 (2000)
6. E. Griv, M. Gedalin, D. Eichler, C. Yuan: Ap&SS **271**, 21 (2000)
7. E. Griv, M. Gedalin, D. Eichler: ApJ **555**, L29 (2001)
8. E. Griv, M. Gedalin, C. Yuan: A&A **383**, 338 (2002)
9. E. Griv, T. Chiueh: ApJ **503**, 186 (1998)
10. F. Hohl: J. Comput. Phys. **9**, 100 (1972)
11. A. Toomre: ApJ, **139**, 1217 (1964)
12. A.G. Morozov: SvA **24**, 391 (1980)
13. A.G. Morozov: SvA **25**, 421 (1981)
14. E. Liverts, E. Griv, D. Eichler, M. Gedalin: Ap&SS **274**, 315 (2000)
15. B.G. Elmegreen, D.M. Elmegreen: ApJ **342**, 677 (1989)
16. C.C. Lin, F.H. Shu: Proc. Natl. Acad. Sci. **55**, 229 (1966)
17. C.C. Lin, C. Yuan, F.H. Shu: ApJ **155**, 721 (1969)
18. F.H. Shu: ApJ **160**, 99 (1970)
19. J. Binney, S. Tremaine: *Galactic Dynamics* (Princeton Univ. Press, Princeton, NJ 1987)
20. J. Kormendy, C.A. Norman: ApJ **233**, 539 (1979)
21. E. Griv, M. Gedalin, C. Yuan: ApJ **580**, L27 (2002)
22. M. Bureau, K.C. Freeman: AJ **118**, 126 (1999)
23. F. Combes, R.H. Sanders: A&A **96**, 164 (1981)
24. F. Combes, F. Debbasch, D. Friedli, D. Pfenniger: A&A **233**, 82 (1991)
25. N. Raha, J.A. Sellwood, R.A. James, F.D. Kahn: Nature **352**, 411 (1991)
26. E. Dwek, R.G. Arendt, M.G. Hauser et al.: ApJ **445**, 716 (1995)
27. Y.Y. Lau, G. Bertin: ApJ **226**, 508 (1978)
28. C.C. Lin, Y.Y. Lau: SIAM Stud. Appl. Math. **60**, 97 (1979)
29. G. Bertin: Phys. Rep. **61**, 1 (1980)
30. C.C. Lin, G. Bertin: Adv. Appl. Mech. **24**, 155 (1984)
31. D. Lynden-Bell, A.J. Kalnajs: MNRAS **157**, 1 (1972)
32. R. Wielen: A&A **60**, 263 (1977)
33. G. Gilmore, I.R. King, P.C. van der Kruit: *The Milky Way as a Galaxy* (Univ. Science Book, Mill Valley, CA 1990)
34. J. Binney: 'Secular Evolution of the Galactic Disk'. In: *Galaxy Disks and Disk Galaxies, ASP Conf. Series, **230***, ed. J.G. Funes, E.M. Corsini (PASP, San Francisco 2001), pp. 63–70
35. E. Griv, M. Gedalin, D. Eichler, C. Yuan: 'The Dynamical Relaxation of the Milky Way'. In: *Astrophysical Ages and Times Scales, ASP Conf. Series, **245***, ed. T. von Hippel, C. Simpson, N. Manset (PASP, San Francisco 2001), pp. 280–287
36. E. Griv, M. Gedalin, C. Yuan: A&A (2003), in press

Formation of the Halo Stellar Population in Spiral and Elliptical Galaxies

Tetyana Nykytyuk

Main Astronomical Observatory of Academy of Sciences of Ukraine, 03680 Kyiv, Zabolothnoho 27, Ukraine

Abstract. A scenario of galactic halo formation through mergers of fragments has been considered. In the framework of the scenario sets of fragments have been obtained from the observed halo metallicity distribution function of the Milky Way Galaxy and others. Our results allow us to conclude that 1) in our Galaxy a halo field star formation can be occured in the fragments evolved as closed system 2) the formation of the bulk of halo field stars of M31 and NGC 5128 perhaps is not associated with the formation of the halo globular clusters in these galaxies 3) in our Galaxy the formation of the halo field stars could be associated with the halo globular cluster formation

1 Introduction

The oldest stellar systems carry information about the processes that took place in galaxies during the early epoch of their formation. Studies of ages and metallicities of these systems allow us to put restrictions on the theoretical description of processes of galaxy formation. Therefore the halo stellar population is good test for models of protogalaxy formation.

Studying globular clusters of our Galaxy Searle and Zinn [18] have concluded that the halo globular clusters were formed in fragments which have merged with the main body on timescales of more than 1 Gyr. Our work is based on this scenario. It is supposed that the halo stellar population represents a mixture of stars that were formed in fragments originally evolved separately from the main protogalactic cloud. Hence, there should be a set of fragments which will reproduce the observed halo metallicity distribution of stars. The aim of this work is to identify a set of fragments containing stars which when mixed give rise to to the observed metallicity distribution of halo field stars and of halo globular clusters.

2 The Model

Let the mass of fragments from which the halo is formed equal the sum of the masses of the stellar population and of gas fallen onto the disk at the present epoch. Each fragment is supposed to evolve as a closed system; mergers among fragments are not taken into account. Star formation process in different fragments can begin at different times. A fragment may evolve up to given astration level s ($s = 1 - \mu$, μ - the fraction of gas in a fragment) before it falls on a protogalaxy. The stars formed up to this moment add to the halo stellar population and the gas falls onto the disk. Fragments can be captured by the protogalaxy untill they begin to form an internal stellar population, i.e. only gas fragments are captured. The contribution of elements, synthesized by a stellar population that formed in one fragment, to interstellar medium depends on 1) synthesis of elements by stars of various masses 2) the number of stars formed in a given interval of stellar masses i.e. on the initial mass function. The initial mass function in a mass range from m up to $m + dm$ is

$$N(m) = \phi_0 m^{-A} dm, \tag{1}$$

where ϕ_0 is the normalization coefficient determined from the condition

$$\phi_0 \sum_{j=1}^{n} m_j^{-A+1} \Delta m = 1 M_\odot. \tag{2}$$

It is taken that the initial mass function is described by the Salpeter law with A=2.35 [15]. The upper and lower mass limits of formed stars are taken to equal $m_U = 120 M_\odot$ and $M_L = 0.1 M_\odot$ accordingly. The mass of matter which has been ejected by the stellar population at the moment t is

$$Q_m(t) = \int_0^t \int_{m_L}^{m_U} \dot{Q}(m,\tau) \phi_0 m^{-A} dm d\tau, \tag{3}$$

where $\dot{Q}(m,\tau)$ is rate of mass loss by a star with mass m and lifetime τ. The mass of a synthesized element i ejected by all stars of a population at the moment t:

$$Q_i(t) = \int_0^t \int_{m_L}^{m_U} \dot{Q}(m,\tau)(Z_i(m,\tau) - Z_i(0)) \phi_0 m^{-A} dm d\tau, \tag{4}$$

where $Z_i(0)$ is abundance of element i in gas from which the stars were formed, $Z_i(m,\tau)$ is abundance of element i in matter ejected by stars with mass m and lifetime τ [2].

We assume that each fragment evolves as a closed system and that its evolution is considered within the framework of simple model of chemical evolution of galaxies. Star formation process in a fragment is considered as sequence of bursts with a population of stars is formed during each burst. The mass of gas m_g, the mass of element i m_i and the mass converted into stellar remains m_s at the start of the star formation burst t_{b_j}:

$$m_g(t_{b_j}) = m_g(t_{b_{j-1}}) - m_{b_{j-1}} - \sum_{k=1}^{j-1} m_{b_k} [Q_m(\tau_{j,k}) - Q_m(\tau_{j-1,k})] \tag{5}$$

$$m_i(t_{b_j}) = m_i(t_{b_{j-1}}) - m_{b_{j-1}} z_i(t_{b_{j-1}}) - \\ - \sum_{k=1}^{j-1} m_{b_k} [Q_m(\tau_{j,k}) - Q_m(\tau_{j-1,k})] z_i(t_{b_k}) + \\ + \sum_{k=1}^{j-1} m_{b_k} [Q_i(\tau_{j,k}) - Q_i(\tau_{j-1,k})] \quad (6)$$

$$m_s(t_{b_j}) = m_s(t_{b_{j-1}}) + m_{b_{j-1}} - \sum_{k=1}^{j-1} m_{b_k} [Q_m(\tau_{j,k}) - Q_m(\tau_{j-1,k})] \quad (7)$$

where

$$\tau_{j,k} = t_{b_j} - t_{b_k}$$
$$\tau_{j-1,k} = t_{b_{j-1}} - t_{b_k},$$

where m_{b_j} is the mass of a star formation burst j , $z_i(t_{b_j})$ is the abundance of element i at the moment t_{b_j}, $\tau_{j,k}$ is the age of burst k at the moment t_{b_j} [2]. The second part on the right hand side of equation 6 takes into account a mass change of element i in the interstellar medium as a result of conversation of a gas into stars. The first sum on the right hand side of equation 6 describes the contribution of matter ejected by stars as if the stars would eject unconverted matter. The second sum represents the contribution of syntesized elements. Having solved numerically the equations 5 – 7 we obtain a metallicity distribution function for the stars formed in a fragment of unit mass (with 1 $M_\odot$) with an astration level $s = 1$, see Fig. 1a. Since the evolution of all fragments is described by the simple model the metallicity distribution function of stars in all fragments will have same shape (Fig. 1a) but the upper metallicity limit of the stars in fragments with different astration levels will be different.

In order to test the validity of the building-up numerical model we carried out a comparison of numerical results with analytical. The comparison of stellar metallicity distributions calculated within the framework of the numerical model

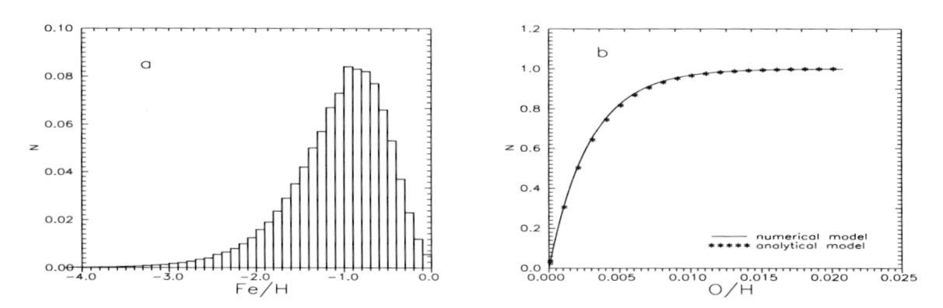

Fig. 1. (**a**) the metallicity distribution function calculated in the framework of the numerical model for a fragment with a level astration $s = 1$. (**b**) A comparison of the results from numerical (*solid line*) and analytical (*asterisks*) modelling of the cumulative stellar metallicity distribution

and with the help of the analytical expression of the simple model taken from [3] is shown in Fig. 1b.

Let us now consider the method by which it is possible to obtain masses of fragments where the halo stellar population was formed. Let a_j be the fraction of stars found in a range of metallicities Z_j, $Z_j + \Delta Z$ in a fragment with mass m (in our case $m = 1M_{sun}$). Then an observed amount of halo stars finding in a range of metallicities Z_j, $Z_j + \Delta Z$ will be represent the total amount of stars $N_{Z_j} = a_j \sum_{i=1}^{n} m_i$ of a given metallicity Z_j from all fragments whose maximum metallicity exceeds Z_j. For given set of metallicities we have

$$\begin{aligned}
a_1 \cdot m_1 + a_1 \cdot m_2 + a_1 \cdot m_3 + \ldots + a_1 \cdot m_n &= N_{Z_1} \\
a_2 \cdot m_2 + a_2 \cdot m_3 + \ldots + a_2 \cdot m_n &= N_{Z_2} \\
\vdots \quad &\vdots \\
a_n \cdot m_n &= N_{Z_n}
\end{aligned}$$

The fragment with the highest astration level s determines the number of stars with the greatest value of a metallicity Z_n. Having solved the set of equations (8a) it is possible to obtain the masses of fragments with maximum metallicities of stars from Z_1 up to Z_n (i.e. the number of fragments with unit mass falling on given Z_j and responsible for the halo stars with such metallicity). Thus, using an observed metallicity distribution of a halo stars and a modelled metallicity distribution function for a fragment of unit mass we shall obtain a value of the total mass of the unit mass fragments evolved up to each given value of astration level s. It is necessary to note that we can obtain the total mass of fragments evolved up to a given value of metallicity Z_j but not the number of fragments that are included in this total mass.

3 Results and Discussion

3.1 The Halo Field Stars of Our and Some Other Galaxies

Let us now compare the obtained set of fragments for the halo field stars of our Galaxy (Fig. 2a), M31 (Fig. 2c), NGC 5128 (Fig. 2e). It is necessary to mention that obtained values of masses of fragments are conditional since in order to obtain the precise masses of fragments the mass of the stellar halos of the investigated galaxies were necessary. This value was only available for our Galaxy. Therefore the value of a halo mass $5*10^{10}M_{sun}$ for a halo field stars and $10^9 M_{sun}$ for a halo globular clusters was accepted for all galaxies. The obtained results for our Galaxy show that the halo stellar population was formed in fragments with low astration levels and high metallicity dispersion. The theoretical distribution obtained from a mixture of stars from different fragments of the set quite well reproduces the observable distribution (Fig. 2b). It allows us to consider that the halo field stars of our Galaxy were formed in a fragments with a low astration levels and high metallicity dispersion.

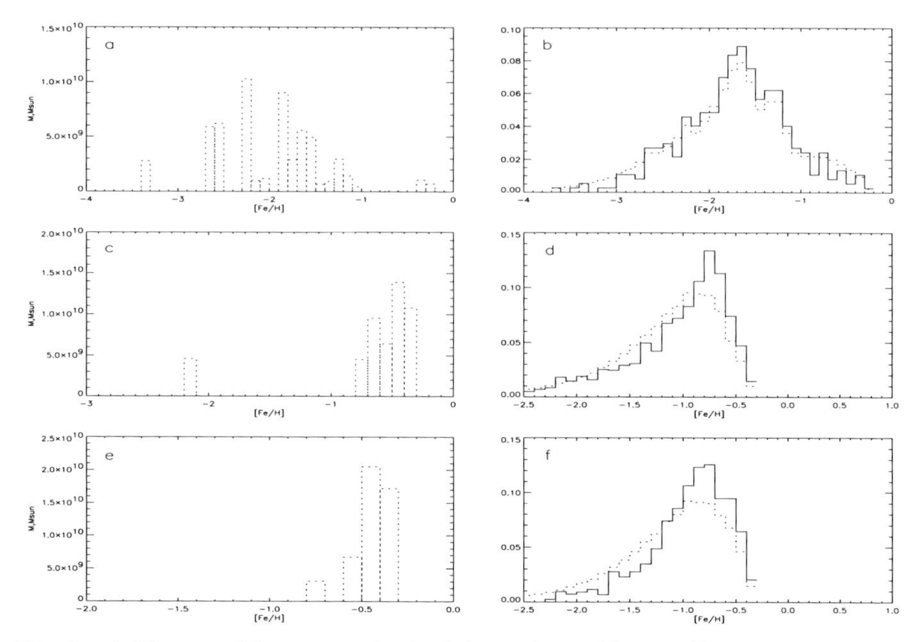

Fig. 2. a) The set of fragments obtained from observable metallicity distribution of a halo field stars of our Galaxy **b)** The comparison of observable (*solid line*)[14] and theoretical (*dotted line*) metallicity distributions obtained from a mixture of the stellar population of fragments in a Fig. 2 **c)** the set of fragments obtained from the observable metallicity distribution of halo field stars of M31 **d)** The comparison of the observable (*solid line*)[8] and the theoretical (*dotted line*) metallicity distributions of M31 halo field stars **e)** The set of fragments obtained from observable metallicity distribution of a halo field stars of NGC 5128 **f)** The comparison of the observed (*solid line*)[11] and the theoretical (*dotted line*) metallicity distributions of NGC 5128 halo field stars

Although the metallicity distribution of stars from the obtained set of fragments for M31 reproduces a general wiev of observed distributions it shows more metal-poor stars (Fig. 2d) than it is really observed. The similar pattern is also gained for NGC 5128 (Fig. 2e, 2f). Harris and Harris[12] who investigated the observed metallicity distributions of halo stars of these galaxies concluded that the halo of NGC 5128 and M31 (opposite to the halo of our Galaxy) was formed by the merger of large satellites (similar to LMC, Small Magellanic Cloud and M32 by sizes) rather than by an accretion of smaller stellar systems. Bekki with collaborators[5] have considered a formation of an ellipticals (in particular NGC 5128) by merging of a spiral galaxies and they also have found out that the stellar halo of elliptic galaxies formed by this way, is populated mainly by stars with rather high metallicity ($[m/H] \sim -0.4$) which came from a exterior parts of disks of merging spiral galaxies.It is interesting that if a halo mass fraction of their merging spirals is more then 0.2 the simulated metallicity distribution function shows more metal-poor stars than observable metallicity distribution function. As suggests [5], similarity between distributions of NGC 5128 and of

M31 can be explained in the case when bulge of M31 also was formed by merging of two spirals. Thus, it is possible to assume that if the halo field stars of M31 and NGC 5128 were really formed in mergers the closed box approximation of accreted fragment does not suit this case.

3.2 A Globular Cluster System of Our and Some Other Galaxies

The part of a globular cluster metallicity distribution which includes halo globular clusters was only used for computation of masses of fragments. The distribution of halo globular clusters of galaxies was obtained by Gauss approximation of observed distributions in all cases except our Galaxy. In Fig. 3a the set of fragments obtained from observable metallicity distribution of halo globular clusters of our Galaxy is shown. The comparison of observable distribution and distributions obtained from a mixture of stars of fragments in Fig. 3a shows (Fig. 3b) that the obtained set of fragments not so well reproduces observable metallicity distribution of halo globular clusters of our Galaxy as it was expected. Although metallicity distribution for halo globular clusters is more narrow than for field stars (Fig. 3) peaks of observable metallicity distributions of halo field stars of and halo globular clusters of our Galaxy coincide. The Fig. 3d shows age - metallicity relation for globular clusters of our Galaxy. The comparison of observations with tracks of fragments of a unit mass (evolved as closed system) shows that the stellar population of fragments whose star formation began in a different times can reproduce declination and scatter of observable values of ages and metallicities in Fig. 3d. Thus it is quite possible that the halo globular clusters could be formed together with a major part of halo field stars in the same fragments and consideration of formation of halo globular clusters of our Galaxy as an isolated subsystem is not meaningful.

The comparison of observed and theoretical (obtained from a mixture of stars of the fragments (Fig. 3e)) distributions shows (Fig. 3f) that the theoretical distribution don't well reproduce the metallicity distribution of halo globular clusters of M31. Some surplus of metal-poor stellar population takes place here. The peaks of observed metallicity distributions of halo field stars and of halo globular clusters of M31 do not coincide (the observations give $[Fe/H]$ ~-0.6 [6] and $[Fe/H]$ ~-1.4[4] accordingly). Peak of distribution of halo field stars of M31 coincides with peak of distribution of bulge globular clusters which value [Fe/H] also makes ~ -0.6 [4]. Apparently, in the case of M31 we really deal with the population of a spheroid, as it was already mentioned in [8]. In such case, it is possible to consider the evolution of halo globular clusters isolately. However, in any case the surplus of the metal-poor stellar population in theoretical distribution of M31 halo globular clusters (as well as halo field stars) does not allow to assume that the halo globular clusters were formed in fragments evolved as the closed system.

The theoretical metallicity distribution o NGC 5128 (Fig. 3h) which was obtained from a mixture of the stellar population of fragments in a Fig. 3g shows the surplus of the metal- poor stellar population.

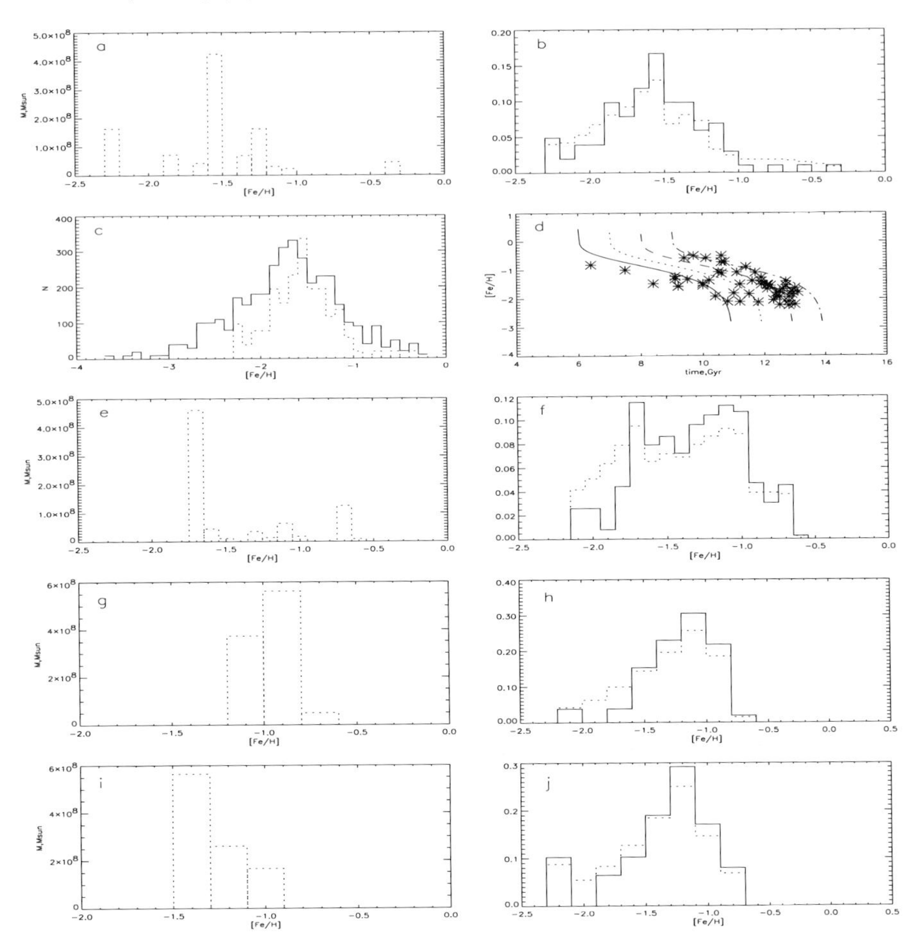

Fig. 3. a) The set of fragments obtained from the observed halo globular cluster metallicity distribution of our Galaxy **b)** The comparison of observable (*solid line*)[1] and theoretical (*dotted line*) metallicity distribution of Galaxy halo globular clusters. **c)** The comparison of observed metallicity distributions of halo field stars (*solid line*)[14] and halo globular clusters (*dotted line*) [1] for our Galaxy **d)** Age - metallicity relation for globular clusters of our Galaxy. The evolutionary tracks of unit mass fragments (with the time of evolution of 5 Gyr) whose star formation began 11, 12, 13 and 14 Gyr ago is shown by lines. Observational data are taken from [17,16] **e)** The set of fragments obtained from observable metallicity distribution of halo globular clusters of M31 **f)** The comparison of observable (*solid line*)[4] and theoretical (*dotted line*) metallicity distributions of M31 halo globular clusters. **g)** The set of fragments obtained from observable metallicity distribution of halo globular clusters of NGC 5128 **h)** The comparison of observable (*solid line*)[10] and theoretical (*dotted line*) metallicity distribution of NGC 5128 halo globular clusters **i)** The set of fragments obtained from observable metallicity distribution of halo globular clusters of M87 **j)** The comparison of observable (*solid line*)[7] and theoretical (*dotted line*) metallicity distribution of M87 halo globular clusters

The comparison of metallicity distributions of halo field stars and halo globular clusters of NGC 5128 shows an discrepancy of distribution peaks (the observations give $[Fe/H]$ ~-0.75 [13] and $[Fe/H]$ ~-1.11[9] accordingly).

The theoretical distribution for a case M87 (Fig. 3j) generally reproduces an overall view of observed metallicity distribution though the obtained number of globular clusters in each bin is not corresponding with the number of globular clusters in observed metallicity distribution of halo globular clusters for M87.

4 Conclusions

Our results allow us to conclude that halo field star formation in our Galaxy can be occured in the fragments evolved as closed system. If the formation of halo field stars of M31 and NGC 5128 has occured by mergers of massive fragments the closed box model of merging fragments doesn't suit this case. The formation of halo field stars of these galaxies perhaps is not associated with the formation of the halo globular clusters in these galaxies In our Galaxy the formation of the halo field stars could be associated with the halo globular cluster formation. The formation of halo globular cluster subsystem as an isolated subsystem in ellipticals (NGC 5128 and M87) is more probable then in spirals (Galaxy and M31). If the halo globular clusters were formed in fragments, a consideration of fragment's evolution as an closed system is not enough to reproduce the observed metallicity distribution of halo globular clusters of ellipticals as spirals.

The autor would like to thank the Organizing Commitee for their kind hospitality and partial financial support. The autor also thanks Pilyugin L.S. and the anonymous referee for a series of valuable notes and help in process of preparation of the paper, Khomenko E.V. for her invaluable help in the development of the necessary programs, Dr.Cardwell and Ms.Rebecca from the IAC for the reading and correcting of this paper. This work was supported by the Ukrainian Fund of Fundamental Researches (grant 02.07/00132)

References

1. T.V. Borkova, V.A. Marsakov: AZh **77**, 750 (2000)
2. L.S. Pilyugin: AZh **71**, 825 (1994)
3. R.J. Tayler: *Galaxies: Structure and Evolution* (Mir, Moscow 1981)
4. P. Barmby, J.P. Huchra, J.P. Brodie, D.A. Forbes, L.L. Schroder, C.L. Grilmair: Astron. J. **119**, 727 (2000)
5. K. Bekki, W.E. Harris, G.L.H. Harris : astro-ph/0212545
6. M. Bellazzini, C.Cacciari, L.Federici, F. Fusi Pecci, M. Rich:astro-ph/0212531
7. J.G. Cohen, J.P. Blakeslee, A. Ryzhov: Astrophys. J. **496**, 808 (1998)
8. P.R. Durrell, W.E. Harris, C.J. Pritchet: Astron. J. **121**, 2557 (2001)
9. H.Eerik, P.Tenjes:astro-ph/0212522
10. G. Harris, D. Geisler, H. Harris, J. Hesser: Astron. J. **104**, 613 (1992)
11. G. Harris, W. Harris, G. Poole: Astron. J. **117**, 855 (1999)
12. G. Harris, W. Harris: Astron. J. **120**, 2423 (2000)

13. W.E.Harris, G.L.H. Harris: Astron.J. **122**, 3065 (2001)
14. S.G. Ryan, J.E. Norris: Astron. J. **101**, 1865 (1991)
15. E. Salpeter: Astrophys. J. **121**, 161 (1955)
16. M.Salaris, A. Weiss: astro-ph/9704238
17. M. Salaris, A.Weiss A.: astro-ph/0204410
18. L. Searle, W. Zinn: Astroph. J. **225**, 357 (1978)

Model of Ejection of Matter from Dense Stellar Cluster and Chaotic Motion of Gravitating Shells

Maxim V. Barkov[1], Vladimir A. Belinski[2], Genadii S. Bisnovatyi-Kogan[1], and Anatoly I. Neishtadt[1]

[1] Space Research Institute, Russian Academy of Sciences, 117997, 84/32 Profsoyuznaya Str, Moscow, Russia;
[2] National Institute of Nuclear physics (INFN) and International Center of Relativistic Astrophysics (ICRA), Rome, Italy

Abstract. It is shown that during the motion of two initially gravitationally bound spherical shells, consisting of point particles moving along ballistic trajectories, one of the shell may be expelled to infinity at subrelativistic speed $v_{exp} \leq 0.25c$. The problem is solved in Newtonian gravity. Motion of two intersecting shells in the case when they do not runaway shows a chaotic behaviour. We hope that this simple toy model can give nevertheless a qualitative idea on the nature of the mechanism of matter outbursts from the dense stellar clusters.

1 Introduction

Dynamical processes around supermassive black holes in quasars, blazars and active galactic nuclei (AGN) are characterised by violent phenomena, leading to formation of jets and other outbursts. Here we consider the possibility of a shell outburst from a supermassive black holes (SBH) surrounded by a dense massive stellar cluster, basing on a pure ballistic interaction of gravitating shells oscillating around SBH.

Investigation of spherical stellar clusters using shell approximation was started by Hénon [6], and than have been successfully applied for investigation of the stability [7], violent relaxation and collapse [10,6,5], leading to formation of a stationary cluster. Investigation of the evolution of spherical stellar cluster with account of different physical processes was done on the base of a shell model in the classical series of papers of L. Spitzer and his coauthors [15–22], see also [12].

Numerical calculations of a collapse of stellar clusters in a shell approximation [23,3,4] had shown, that even if all shells are initially gravitationally bound, after a number of intersections some shells obtain sufficient energy to become unbound, and to be thrown to the infinity. In the Newtonian gravity the remnant is formed as a stationary stellar cluster, and in general relativity SBH may be formed as a remnant.

Important example of a quasi-spherical mass ejection is a relativistic collapse of a spherical stellar system, which is considered [24,9,8,14] as the main mechanism of a formation of supermassive black holes in the galactic centers.

Approximation of such collapse by consideration of spherical shells is the simplest approach, which reflects all important features of such collapse [5,3,4]. Our consideration is related to the motion of stars (shells) which remain outside the newly formed supermassive black hole.

Here we consider a simplified problem of a motion of two massive spherical shells, each consisting of stars with the same specific angular momentum and energies, around SBH. In a more complicated case of numerous shell intersections this elementary act is a key process of the energy exchange between stars and of matter ejection. This is also the elementary process leading to the violent relaxation of the cluster [10], studied in the shell approximation in [5,3,4].

Development of chaos during the motion of two gravitating intersecting shells had been found first by Miller and Youngkins [11] in the oversimplified case with a pure radial motion and reflecting inner boundary. We have found [2] a chaotic behaviour in a more realistic model where stars with the same energy and angular momentum are dispersed isotropically over the spherical shell, and each star moves along its ballistic trajectory in the averaged gravitational field, with account of the shell self-gravity. We find conditions at which one of two shells is expelling to infinity taking energy from another shell. We find a maximum of the velocity of the outbursting shell as a function of the ratio of its mass m to the mass M of SBH using Newtonian theory of the shell's motion.

We show, that for equal masses of two shells the expelling velocity reach the value $v_{max} \approx 0.3547 v_p$ at $m/M = 1.0$, and $v_{max} \geq 0.3 v_p$ was obtained at the masses of shells $0.25 \div 1.5M$, where the parabolic velocity of a shell in the point of a smallest distance to the black hole v_p may be of a considerable part of c.

In Sects. 2,3 we describe outburst effect and in Sects. 4 we present the evidence of the chaos in the system of intersecting shells. The exact solution of these problems in the context of General Relativity have been found in [1].

2 Two Shells Around SBH

Physically the nature of the ballistic ejection is based on the following four subsequent events. The outer shell is accelerated moving to the center in a strong gravitational field of a central body and inner shell. Somewhere near the inner minimal radius of the trajectory shells intersect. After that the former outer shell is decelerating moving from the center in the weaker gravitational field of only one central body. The second intersection happens somewhere far from the center. That may result in the situation when the total energy (negative) of the initially gravitationally bound outer shell is becoming positive as a result of two subsequent intersections with another shell. The quantitative analysis of this process is done in [2].

Equation of motion of a shell with mass m and total conserved energy E in the field of a central body with mass M is

$$E = \frac{mv^2}{2} - \frac{Gm(M + m/2)}{r} + \frac{J^2 m}{2r^2}, \tag{1}$$

where $v = dr/dt$ is the radial velocity of the shell and $J^2 m/2r^2$ is the total kinetic energy of tangential motions of all particles, which the shell is made up from. The constant $J > 0$ has that interpretation that Jm is the sum of the absolute values of the angular momenta of all particles.The term $m/2$ in (1) is due to the self-gravity of the shell.

Let us consider two shells with parameters m_1, J_1 and m_2, J_2 moving around SBH with mass M. Let the shell "1" be initially outer and the shell "2" be the inner one. Then equations of motion are:

$$E_{1(0)} = \frac{m_1 v_{1(0)}^2}{2} - \frac{Gm_1(M + m_1/2 + m_2)}{r} + \frac{J_1^2 m_1}{2r^2}, \tag{2}$$

$$E_{2(0)} = \frac{m_2 v_{2(0)}^2}{2} - \frac{Gm_2(M + m_2/2)}{r} + \frac{J_2^2 m_2}{2r^2}. \tag{3}$$

By the index (0) we mark the initial evolution stage before the first intersection of the shells. Assume that both shells are moving to the center. Such shells intersect each other at a some radius $r = a_1$ and at some time $t = t_1$ after first intersection the shell "1" becomes inner and shell "2" outer and at a some radius $r = a_2$ and at some time $t = t_2$ they have second intersection. The motion of the shells after first intersection is designated by the index (1), and after second intersection is designated by the index (2):

$$E_{1(2)} = E_{1(1)} - \frac{Gm_1 m_2}{a_2} = E_{1(0)} + Gm_1 m_2 \left(\frac{1}{a_1} - \frac{1}{a_2} \right), \tag{4}$$

$$E_{2(2)} = E_{2(1)} + \frac{Gm_1 m_2}{a_2} = E_{2(0)} - Gm_1 m_2 \left(\frac{1}{a_1} - \frac{1}{a_2} \right). \tag{5}$$

Let us describe the situation, when one shell is ejected to infinity after intersection of to initially bound shells. We consider a case when a_2 is larger then a_1 so, that the second term in (4) has larger absolute value, then the first one, the first shell gains a positive energy and goes to infinity. Both shells have initial negative energies $E_{1(0)}$ and $E_{2(0)}$, but with small enough absolute values. The first shell takes the energy from the second one, which is becoming more bound with larger absolute value of the negative energy $E_{2(2)}$, according to (4).

3 Numerical Solution

Let us illustrate the foregoing scenario by an exact particular example of two shells of equal masses. We choose parameters in the following way:

$$m_1 = m_2 = m, \qquad E_{1(0)} = E_{2(0)} = 0, \qquad J_1 < J_2, \tag{6}$$

In fact such exact solution represents the first approximation to the more general situation when $E_{1(0)}$ and $E_{2(0)}$ are non-zero (negative) but small in that sense that both modulus $|E_{1(0)}|$ and $|E_{2(0)}|$ are much less than Gm^2/a_1.

We assume that initially both shells are moving towards the center. It follows from (2), (3) that under condition (6) such shells will intersect inescapably at the point $r = a_1$, $t = t_1$. After the second intersection at $r = a_2$, $t = t_2$ the shell "1" will be thrown to infinity with expelling velocity v_{1exp}. It follows from (4) that

$$v_{1exp} = v_{1(2)}\Big|_{r\to\infty} = \sqrt{2Gm\left(\frac{1}{a_1} - \frac{1}{a_2}\right)}. \tag{7}$$

In order to construct a solution with maximal possible v_{1exp} we consider a case when initially inner shell "2" reaches the inner turning point at minimal possible radius $r = r_m$ of the order of few $r_g = 2GM/c^2$, and intersects with the initially outer shell after, during its outward motion (see Fig. 1).

Between the first and second intersection there exists the inner turning point of the shell "1" (now inner shell). We take that additional restriction that the shell "1" reaches this turning point also at the minimal possible radius $r = r_m$. It is easy to show that a wide class of solutions with such restriction really exists (see Fig. 1).

We now introduce the "parabolic" velocity v_p of the any outer shell at the point of its minimal distance to the center $r = r_m$ as:

$$v_p = \sqrt{\frac{2G(M + 3m/2)}{r_m}}. \tag{8}$$

Then we have

$$\frac{v_{1exp}}{v_p} = \sqrt{\frac{m}{M + 3m/2}\left(\frac{r_m}{a_1} - \frac{r_m}{a_2}\right)}. \tag{9}$$

Consequently this ratio is also a function of a_1/r_m and m/M. We find numerically the value of a_1/r_m maximizing the ratio v_{1exp}/v_p which value is a function

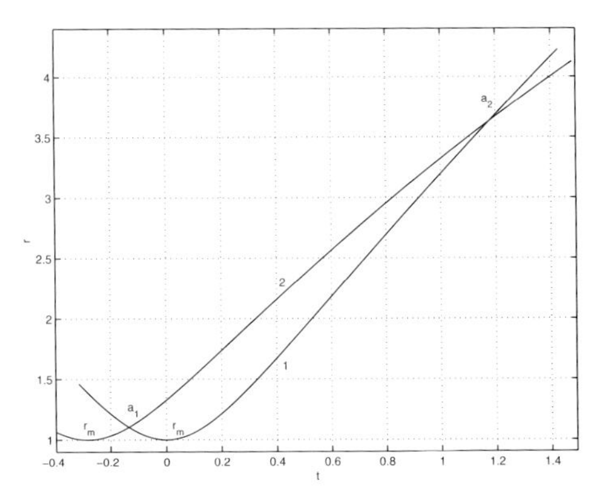

Fig. 1. Time dependence of radii r of two shells (in units r_m) on time t (in units r_m/v_p, $v_p = \sqrt{2G(M + 3m/2)/r_m}$). The shells intersect at points $a_1 = 1.1401$, $a_2 = 3.6316$. Here $m/M = 0.1$, and after the second intersection the first shell is running away with velocity at infinity $v_{1exp} = 0.23415 v_p$.

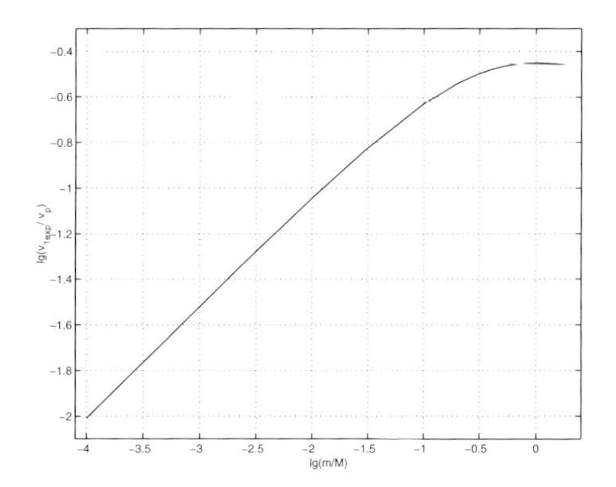

Fig. 2. Dependence of the maximum runaway velocity v_{1exp} in units v_p on the ratio of the shell mass to the mass of central body m/M. The maximum value of $v_{1exp} = 0.3547$ corresponds to $m/M = 1.0$.

of m/M only. Dependencies for v_{1exp}/v_p as functions of the parameter m/M for these maximizing solutions are given in Fig. 2.

The numerical calculations (illustrated by the Fig. 2) shows that the runaway velocity v_{1exp} reaches its maximal possible value at $m/M \approx 1.0$ and it is $v_{1exp} \approx 0.3547 v_p$. If we consider the shells around SBH then the minimal radius r_m of the shell "1" orbit cannot be less then the two gravitational radii $2r_g = 4G(M + 3m/2)/c^2$. In the extreme case when $r_m = 2r_g$ we have $v_p \sim c/\sqrt{2}$ and for the maximal possible runaway velocity we get $v_{1exp} \sim 0.25c$.

4 Chaos in the Shell Motion

The first evidence that the motion of two intersecting shells can show chaotic character was given by B.N. Miller & V.P. Youngkins [11]. They investigated the special case when the central body is absent and particles making up the shells are moving only in radial direction (in our notation $M = 0$ and $J_1 = J_2 = 0$) with an artificial reflection at a given inner radius. This situation, however, cannot model astrophysical cluster with massive nuclei, and also the problem of the influence of central Newtonian singularity arises, which need some additional care. In any case a study of more physically realistic models with nonzero M, J_1 and J_2 from the point of view of possible chaotic behaviour represents an essential interest. We report here some numerical results for such more general two-shells model which were investigated in the previous sections but again for the shells with equal masses.

We fix here the initial specific angular momenta and energies, different for both shells, initial radii of the shells, and vary only the mass ratios of the shells and the central body. It was found that at our choice of parameters the motion of the shells becomes chaotic at $m/M = 2\%$.

For more clear understanding of the problem let us consider the following simplification. Put one of the shell mass equal to zero. We fix the specific angular momenta of the shells (the angular momentum of the light shell is less than the angular momentum of the heavy one) and initial specific energy of the heavy shell. The initial specific energy of the light shell is chosen to satisfy the following condition: specific energy of the heavy shell is less than that of the light one. Due to these conditions on every turnover the light shell has two intersections with the heavy shell.

We implement the approach from [13] for analysis of this model. Namely, we study Poincaré section and Poincaré return map of the problem. We define in the phase space the surface for Poincaré section by conditions that the radial velocity of the light shell is 0, and the radius of the light shell has a local minimum. Let E, E' and η, η' (defined modulus 2π) be specific energies of the light shell and phases of the heavy shell at two subsequent time moments when the radius of the light shell has minima. (Phase of the heavy shell is just the mean anomaly of Keplerian motion of this shell). The Poincaré map sends pair E, η to the E', η'. This map is area-preserving. The trajectories of several initial points under the action of the Poincaré map are shown in Fig. 3. While for the problem under consideration the forces are discontinuous, the Poincaré return map is analytic, and therefore KAM theory can be applied for small values of m/M, where $m = m_2$. Note that $E - E' \sim m/M$, $\eta - \eta' \sim (-E')^{-3/2}$, similarly to [13]. Following [13], we get the estimate for the border value E^* of the chaotic region in E-space (see Fig. 3) $E^* \sim (m/M)^{2/5}$, i.e. the chaotic behavior is possible at any value of m/M (see Figs. 3-4). Our numerical simulation suggests relation $E_1^* \sim (m/M)^{0.42}$; at $E > E_1^*$ runaway of the light shell is possible and at $E < E_1^*$ the light shell is not be able to escape. The region $E < E_1^*$ up to a residue of a small measure is filled with invariant curves of the Poincaré map in accordance with KAM theory. Another relation $E_2^* \sim (m/M)^{0.46}$ was found for energy value separating the region $E > E_2^*$ in where the shell motion is chaotic and the region $(E_1^* < E < E_2^*)$ in which the regular motion is possible. As one can see, the numerical simulations and analytical estimation give close results.

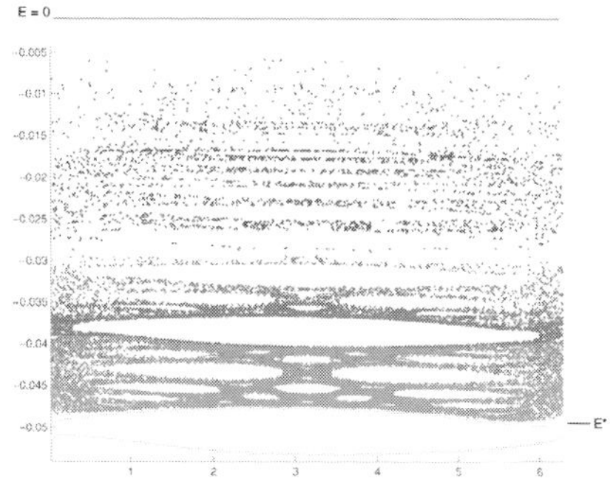

Fig. 3. Distribution of points in (E,η)-space for $m/M = 0.005$. The upper line indicates the border of capture of a shell, i.e. E=0. E^* is the lower limit of the chaotic motion, related to the breakdown of the invariant curve in the KAM theory. The structure is similar [13] for different values of the m/M.

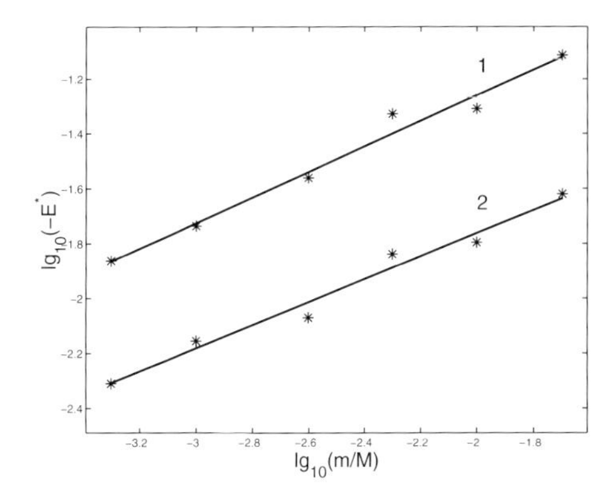

Fig. 4. The lines divide the plane into 3 regions. Above the first line ($E_1^* \sim -(m/M)^{0.42}$) escape to infinity is impossible for any initial parameter values. Below the second line ($E_2^* \sim -(m/M)^{0.46}$) escape to infinity is possible for any initial parameter values. For region between lines "1" and "2" the possibility of escaping to infinity depends on initial conditions.

In the case when the light shell does not have two intersections on every turnover, the Poincaré return map is continuous but not smooth. In this case KAM theory is not applicable. Invariant sets become belt-like and have non zero width.

Acknowledgement

The work of G.S.B.-K. and M.V.B. was partly supported by RFBR grant 99-02-18180, and INTAS grant 00-491. The work of A.I.N. was partly supported by RFBR grant 03-01-00158, and INTAS grant 00-221.

References

1. Barkov M. V., Belinski V. A., and Bisnovatyi-Kogan G. S., 2002, JETP, 95, 3, 371–391.
2. Barkov M. V., Belinski V. A. and Bisnovatyi-Kogan G. S.,2002, MNRAS 334, 338–344
3. Bisnovatyi-Kogan, G. S., Yangurazova, L. R. 1987, Asrtofizika 27, 79
4. Bisnovatyi-Kogan, G. S., Yangurazova, L. R. 1988, Ap. Sp. Sci. 147, 121
5. Gott, J. P. 1975, ApJ 201, 296
6. Hénon, M. 1964, Ann. d'Ap. 27, 83
7. Hénon, M. 1973, Astron. Ap. 24, 229
8. Ipser, J. . 1980, ApJ 238, 1101
9. Lightman, A. P., Shapiro, S. L. 1978, Rev. Mod. Phys. 50, 437
10. Lynden-Bell, D. 1967, MNRAS 136, 101
11. Miller B. and Youngkins V. 1997, Chaos 7, 187.
12. Palmer P. and Voglis N., 1983, MNRAS 205, 543.
13. Petrosky T.Y. 1986, Phs. Lett. A, 117, 7, 328-332

14. Rasio, F., Shapiro, S., Teukolsky, S. 1989, ApJ 336, L63
15. Spitzer, L. Jr., Hart, H. M. 1971a, ApJ 164, 399
16. Spitzer, L. Jr., Hart, H. M. 1971b, ApJ 166, 483
17. Spitzer, L. Jr., Shapiro, S. L., 1972, ApJ 173, 529
18. Spitzer, L. Jr., Thuan, T. X. 1972, ApJ 175, 31
19. Spitzer, L. Jr., Chevalier, R., 1973, ApJ 183, 565
20. Spitzer, L. Jr., Shull, J. M., 1975a, ApJ 200, 339
21. Spitzer, L. Jr., Shull, J. M., 1975b, ApJ 201, 773
22. Spitzer, L. Jr., Mathieu, R. D., 1980, ApJ 241, 618
23. Yangurazova, L. R., Bisnovatyi-Kogan, G. S. 1984, Ap. Sp. Sci. 100, 319
24. Zeldovich, Ya. B., Podurets, M. A. 1965, AZh 42, 963 (transl. Soviet Astron. -AJ, 29, 742 [1966])

Direct vs Merger Mechanism Forming Counterrotating Galaxies

Maria Harsoula and Nikos Voglis

Academy of Athens, Research Center for Astronomy, 14 Anagnostopoulou Str., GR-10673, nvogl@cc.uoa.gr, mharsoul@phys.uoa.gr

Abstract. A comparison of the formation of counterrotating elliptical galaxies is made in two alternative scenarios: (a) The scenario of merging of a primary and a satellite galaxy, and (b) the direct scenario, in which counterrotating galaxies are formed directly from cosmological initial conditions. We conclude that, although both scenarios might have worked in parallel, the scenario of merging has a large number of parameters that should be well tuned to form a counterrotating galaxy in contrast to the direct scenario that is controlled by only two or three parameters. This could give more chances to the direct scenario.

Introduction

It is known, nowadays, from observations, that almost 1/3 of the elliptical galaxies present a counterrotating core (CRCs hereafter) with respect to the rest of the galaxy [4], [14], [15], [5], [22], [23]. A number of cases of spiral counterrotating core galaxies are also reported [8].

The scenarios proposed in the literature explaining the formation of such galaxies are split into two main categories: a) the merger scenarios and b) the non-merger scenarios. According to the first scenarios CRCs ellipticals are the result of the merging between two galaxies. The merging can be either dissipationless, i.e. between a main galaxy and a satellite one [2], [1], or between two spiral galaxies [26], [3]. Similar scenarios consider episodic gas infall, or accretion of a gas rich companion [6], [7]. Most of these scenarios fail to explain the reddening and the increased metallicity of the core of the CRC galaxy.

A non-merger scenario has been proposed by Hau and Thomson [18] according to which tidal torquing on the main body of the galaxy is caused by another passing galaxy. This torque can reverse the sense of rotation of the outer parts of the galaxy and therefore can in general produce a kinematically decoupled core.

An alternative non-merger scenario, is the one that we call direct scenario and it was proposed by Voglis et al [28], [17], [29], where a bar-like density excess in the early Universe (i.e. Decoupling) that makes bound the material of a protogalactic cloud, can work as a seed to form a CRC galaxy. The great advantage of this scenario is that it explains the almost total lack of any population differences between the core and the main body of the galaxy as well as between the CRC galaxies and other normal galaxies.

A brief discussion of the direct scenario is given in Sect. 1. In Sect. 2 the initial conditions and the results of a numerical experiment are described, in which a counterrotating galaxy is formed from cosmological clumpy initial conditions. In Sect. 3 we describe the initial conditions and the results of another experiment in which a counterrotating galaxy is formed by the merging process. The parameters in the latter case were arranged so that the two galaxies have almost the same rotation velocity curve. This allows a comparison between the two scenarios discussed in Sect. 4.

1 Basic Features of the Direct Scenario

Let's consider a density perturbation at the early post-decoupling Universe which has a bar-like shape and is formed by small clumps of mass. The most tightly bounded particles have a quadrupole moment with major axis A1 along a random direction (Fig. 1). Then the distribution of the more loosely bound particles has a quadrupole moment with major axis along the direction A2. This protogalactic cloud is surrounded by an anisotropic environment which creates a tidal field with random orientation of its main axis (direction of larger forces).

If the axis A1 is along the direction of 2nd-4th quadrant, then these particles will acquire positive angular momentum while the more loosely bound particles being mainly along the direction 1st-3rd-quadrant (along A2 in Fig. 1), acquire negative angular momentum.

Therefore, this configuration has a certain distribution of angular momentum along the radius, that can initiate a counterrotating system. After the collapse and the mixing of the particles this initial distribution of the angular momentum can survive under some conditions.

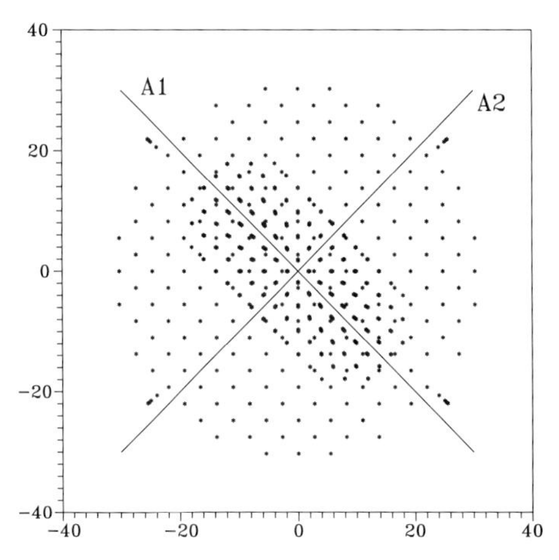

Fig. 1. A bar like distribution with major axis A1. Loosely bound particles have major axis A2

The violent relaxation of the system cannot erase all the memory of the initial conditions and therefore the particles remember their initial energies (and as a consequence their initial positions) in a statistical sense, even for a Hubble time.

This scenario is verified by N-body simulations in two different cases of initial conditions, namely quiet initial conditions [28], [17], [16] and clumpy initial conditions [29].

In the case of quiet initial conditions we found that a counterrotating galaxy is formed when the bar-like initial density perturbation has an axial ratio around the value 0.5. If this ratio is changed to 0.4, negative rotation dominates all along the radius, while for a ratio 0.8, positive rotation dominates all along the radius [17].

A second parameter that influences the appearance of counterrotation is the strength of the external tidal field. It was found that a modest external tidal field favors the formation of counterrotating galaxies, while a strong or a weak tidal field favor the formation of galaxies with simple rotation [16].

2 Counterrotation from Clumpy Cosmological Initial Conditions

This scenario can be realized in the case of clumpy cosmological initial conditions as follows. We used two different mass scales: the galactic mass scale named G-scale which contains 2 M_u (where M_u is the mass unit, i.e. the mass of a galaxy) and the environment scale named E-scale which contains 664 M_u.

The G-scale distribution of particles is composed of a number of $N_g = 5616$ particles of equal mass initially arranged in Lagrangian coordinates (q_1, q_2, q_3) in a cubic grid limited by a sphere. The size of the system is arranged in such a way that when it expands with the Universe it has total energy equal to zero, simulating the expansion of an Einstein-de Sitter Universe (i.e. $\Omega = 1$, all distance increase with time t as $t^{2/3}$).

The positions and velocities of particles, when the system is perturbed, are evolved by the Zeldovich approximation [31], as explained analytically in [29].

The E-scale is resolved into 664 particles put initially in a cubic grid limited by a sphere. Eight central particles of this grid have been removed to make room for the location of the G-scale. Each of the remaining particles of the E-scale has mass equal to 1 M_u. Their grid is slightly deformed in such a way that undesirable non-cosmological torques acting of the G-scale due to the discreteness of the E-scale is almost eliminated (see for details [27]). Then the positions and the velocities of the E-scale particles are evolved in such a way so that it simulates correctly the initial strength and the time behavior of the cosmological tidal torque in agreement with the analytical calculations during the linear phase of evolution of the density perturbation. The Y-axis is the axis of strongest forces of the tidal field. In Fig. 2a the overall projection of the initial configuration is shown at a time=0, when the N-body calculations start. The G-scale is shown in magnification in Fig. 2b at the same time.

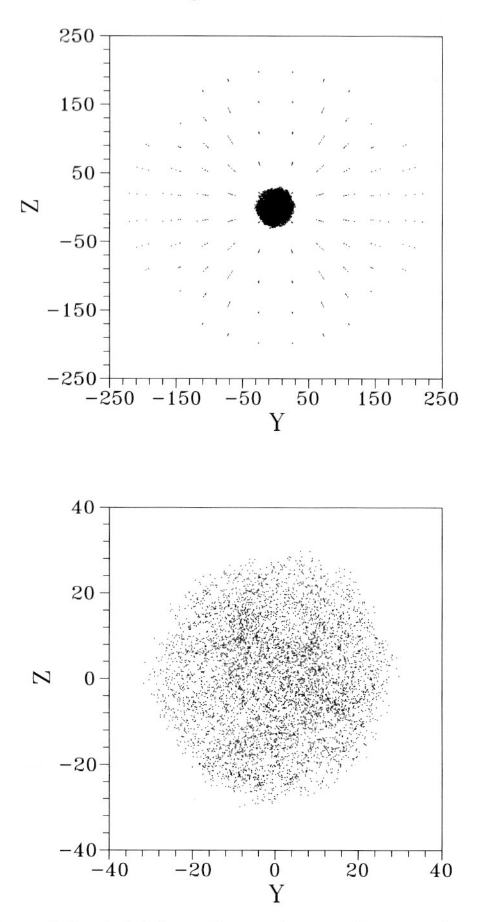

Fig. 2. The projection of the initial configuration on the y-z plane at time=0. (**a**) the whole configuration and (**b**) the g-scale only

The evolution of the whole system (of both scales) is followed simultaneously by Aarseth's N-body2 code. The units of time t_{un}, length r_{un} and velocity v_{un} in our unit system are expressed in terms of real units as:

$$t_{un} = 0.875\beta^{3/2} Myears \tag{1}$$

$$r_{un} = 1.5\beta(\frac{M_{un}}{M_{12}})^{1/3}(Kpc) \tag{2}$$

$$v_{un} = 1677\beta^{-1/2}(\frac{M_{un}}{M_{12}})^{1/3}(km/sec) \tag{3}$$

where $M_{12} = 10^{12}M_{\odot}$ and β is a re-scaling parameter.

The G-scale configuration evolves so that a part of about 4800 particles of it reaches a maximum radius of expansion and then detaches from the general

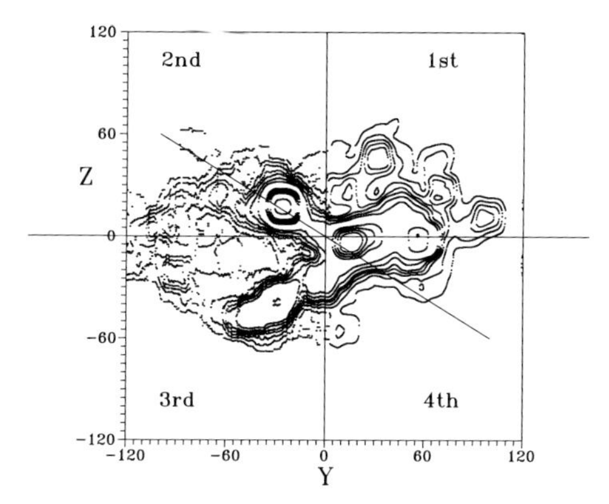

Fig. 3. The isodensity contours plotted on the Y-Z plane at time=$400t_{un}$ for the experiment with clumpy initial conditions

expansion. During this period clumps grow inside it, and gradually merge to form a single main clump i.e. a galaxy. 'Violent relaxation' occurs by the process of gradual merging.

For a certain range of the initial parameters i.e., consistent to the values used in the case of quiet initial conditions in Harsoula and Voglis 1998, we detected counterrotation in the relaxed system.

In Fig. 3 the isodensity contours are plotted on the Y-Z plane at a time corresponding to $400t_{un}$. It's obvious that the most tightly bound particles form a bar-like density perturbation with a major axis of about 50° with respect to the tidal axis. These particles acquire therefore positive angular momentum, while the less tightly bound particles have a major axis almost perpendicular to the axis of the most bound ones and acquire negative angular momentum.

Beyond a time of $4000t_{un}$ the bar and a considerable part of the G-scale around the bar has collapsed and been relaxed to a bound system i.e. a galaxy. The secondary infall is no more important.

Figure 4 shows the rotational velocity profile of an experiment with counterrotation. The dots upon the rotation curve correspond to fractions 10%, 20% etc. of the total mass inside the respective radius. We see that about 60% of the bound mass has positive rotation, while the rest of the bound mass has negative rotation. As it was mentioned above the strength of the cosmological tidal field is important for counterrotation to appear. For example, if we repeat this experiment having the G-scale exposed to a tidal field, 30% stronger counterrotation disappears. Angular momentum of negative sign dominates throughout the whole configuration. The corresponding rotational velocity profile is shown in Fig. 5.

Another parameter that could play a role is the initial orientation of the bar formed by the most tightly bound particles. However, the results are not expected to be particularly sensitive on this orientation, unless the bar is nearly parallel or perpendicular to the tidal axis.

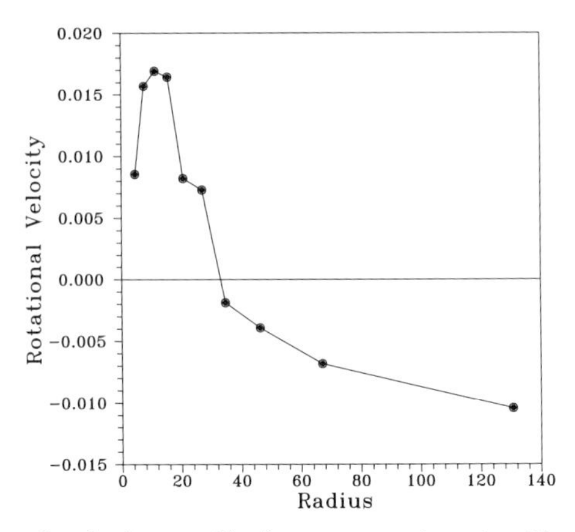

Fig. 4. (The rotational velocity profile for an experiment with clumpy cosmological initial conditions giving a counterrotating elliptical galaxy

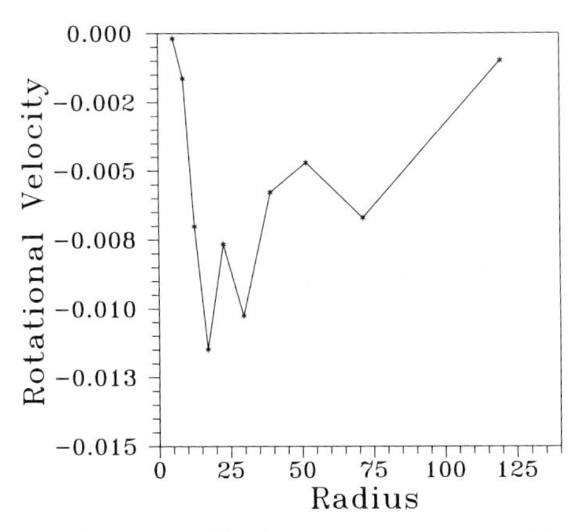

Fig. 5. The rotational velocity profile for an experiment with clumpy cosmological initial conditions and a tidal field 30% stronger that the one of Fig. 3

3 Counterrotation from Merging with a Satellite Galaxy

Mergers can also produce CRCs. We have already performed some experiments with mergers in order to compare the two scenarios [30]. Here we present a deeper investigation studying also the possibility to distinguish CRCs formed by mergers from those formed by cosmological initial conditions via some observable quantities.

We therefore have performed several experiments of dissipationless merging of a primary galaxy and a satellite galaxy. The primary is designed so that all the orbits rotate in one direction only i.e. contains no retrograde orbits at all.

To form such a primary galaxy, we start with the relaxed configuration of the experiment with the rotational velocity profile of Fig. 5. Then all the velocity components u_{yz}, parallel to the Y-Z plane are turned perpendicular to their position vector component R_{yz}, with common direction of rotation. Thus the system has no retrograde orbits. It remains in the same virial equilibrium and has a value of the spin parameter $\lambda \approx 0.1$ [25].

The satellite galaxy is produced by isolating a central spherical part of the initial state of the primary galaxy. We have performed three numerical experiments namely A,B and C. In the experiment A the mass of the satellite galaxy is about 20% of the primary's mass, while in the experiment B and C it is about 10% of the primary's mass. In particular, in C the velocities of the particles of the satellite have been multiplied by a factor $0.5^{1/2}$ so that its kinetic energy is reduced by 50% and the satellite becomes quite compact in this experiment. We run the satellite separately for several dynamical times (4000 t_{un}) so that it relaxes to a more compact system (with higher central density) before the start of the merging process.

Then, in all three experiments, the satellite galaxy is placed on a retrograde orbit on the plane of rotation of the primary (i.e. on the Y-Z plane) at a distance of $150r_{un}$ in Y-direction and $50r_{un}$ in Z-direction from the center of mass of the primary galaxy. The initial velocity of the center of mass of the satellite galaxy for all the experiments is $U_s \approx Vp/3$ where Vp is the velocity of the parabolic orbit around the primary galaxy.

We notice here that prograde orbits of the satellite galaxy fail to produce counterrotating cores [2]. This is expected for thermodynamical reasons. Furthermore, simulations with orbital inclination different than zero have very little possibility to give counterrotating systems [1]. The reason is that the dynamical friction between the satellite and the rotating primary causes the satellite's orbit to pivot such as to be oriented almost upright in relation to the rotation plane of the primary. A rough idea of the expected change in the orbit's inclination was first given by Chandrasekhar with his dynamical friction formula [10].

In Fig. 6 we can see the initial configuration of the primary and of the satellite galaxy at the start of the N-body run, and the orbit of the center of mass of the satellite galaxy around the common center of mass of the two galaxies for the ex. C. The satellite galaxy describes a spiral orbit around the common center of mass being subject to dynamical friction as it moves towards the primary galaxy. The most compact part of it relaxes at the center.

In Fig. 7 the line with stars shows the rotation velocity of the primary before merging. The line with dots in this figure shows the new form of the rotation velocity of the relaxed system after merging for exp A. This form is similar to the rotational velocity of Fig. 4 obtained from cosmological initial conditions, with a little greater maximum value. It is clear therefore that the satellite sinking inside the primary is able to reverse the rotation near the center, leaving a signature of its proper angular momentum.

In Fig. 8 we present the rotational velocity profiles for all the three experiments. In ex. A (stars) the central region seems to have greater positive values of rotation and this is due to the fact that the satellite has twice the number of particles than in the other two experiments (mass ratio 1:5). Moreover the absolute values of the rotational velocities for the outer parts of the galaxy are smaller than in the other two experiments. This is again due to the mass ratio

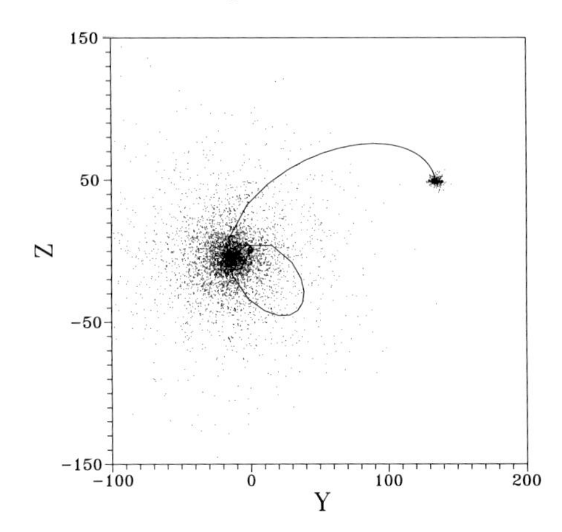

Fig. 6. The initial configuration of the primary and the satellite galaxy at time=0 for ex. C and the spiral orbit of the center of mass of the satellite until the whole system is relaxed

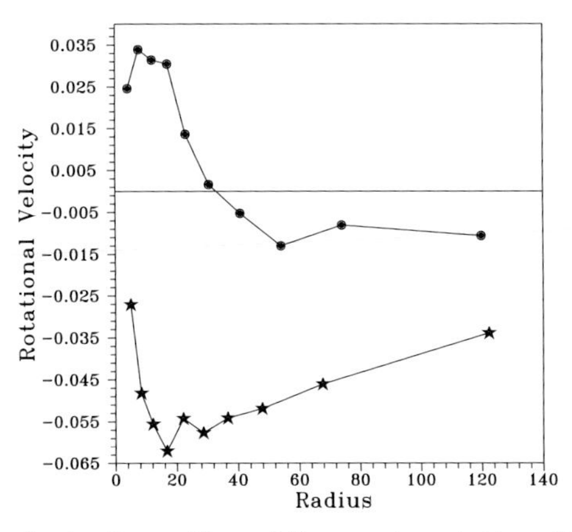

Fig. 7. (The rotational velocity profiles, of the counterrotating elliptical galaxy formed by merging of a primary and a satellite galaxy for ex. A (dots) and of the primary galaxy before the merging (stars)

of the two galaxies which is higher is in ex. A (1:5) and therefore the particles of the satellite galaxy, having positive angular momentum with respect to the target galaxy, affect a greater number of particles of the target and can reverse their rotation. In ex. B (dots) where the mass ratio of the two galaxies is 1:10, we observe that the whole galaxy rotates with one sign, except of a small fraction of 10% of the remnant's mass which has a small value of positive rotation. Therefore the mass ratio can seriously affect the rotational velocity profile and the ratio 1:10 is an approximate lower threshold for having counterrotating galaxies. In ex. C (open squares) the mass ratio is still 1:10, but the satellite galaxy is initially more compact and therefore it can preserve better it's orbital angular momentum. Therefore a fraction of 40% of the remnant's mass has positive ro-

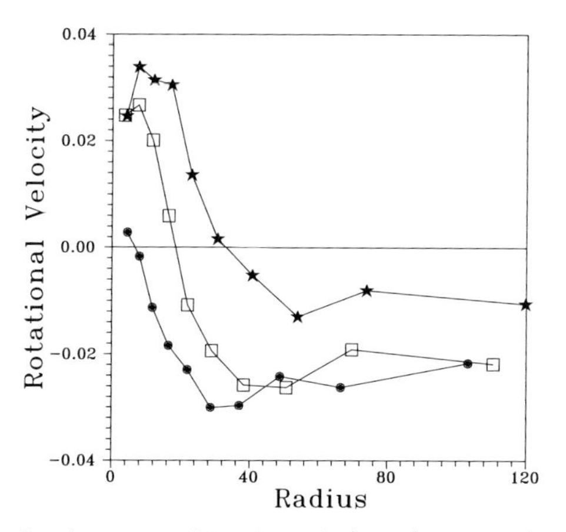

Fig. 8. The rotational velocity profile of ex. A (stars), ex. B (dots) and ex. C (open squares) in the case of merger

tation. It seems that in this case the mixing of the particles is less than in the other two cases.

4 Comparison of the Results and Discussion

It is useful here to convert the values of the rotational velocities and the radius of Figs. 4 and 7 in real units and compare with the values observed in counterrotating galaxies. Barcells and Quinn [2] proposed a typical counterrotation velocity $40km/sec$ for $M_{un} = 5\text{x}10^{11} M_{\odot}$ and half mass radius r_o=$10Kpc$, or at most a factor of 2 higher than this limit. By comparing with our results where $r_o \approx 18 r_{un}$ we get from (2) $\beta \approx 0.42$. From equation (3) we get $v_{un} = 1942km/sec$. The maximum of the counter rotation velocity of Figs. 3a and 6a is $v_{max} = (0.018 \; to \; 0.033)v_{un} \approx \; 35 \; to \; 64km/sec$. Therefore our results, either from cosmological initial conditions or from merging, are in agreement with the observational data.

We have tested the merging scenario for greater values of the initial velocity of the satellite galaxy. We found that for $V_p > U_s > Vp/3$, the final velocity profile shows greater values of positive rotation in the center.

The scenario of merging events and the direct scenario can occur independently. Thus, it is expected that some of the counterrotating galaxies observed today were formed in one way, while others may have formed in the other way. In both cases, our simulations have shown that the distribution of the angular momentum can survive for a Hubble time and therefore it can be observed.

A question that arises is whether, via some observable quantity, counterrotating elliptical galaxies formed by mergers can be distinguished from those formed by cosmological collapses.

Hausman and Ostriker [19] proposed that once a typical victim has a higher central density than the cannibal, its core survives ingestion to produce a core-within-a-core structure. According to Barcells and Quinn [2], if the mass ratio of the merger event is $1:5$ or higher no obvious trace in the surface brightness profile is left apart from the fact that, in general, there is no flattening in the center. This seems to be confirmed in the simulations of Duncan et al [12] where they have performed mergers with several mass ratios $1:k$ with $1 \leq k \leq 5$. The surface density profiles of these experiments show no hint of a core-within-a-core structure.

According to our results a hump in the surface density profile is obvious when the mass ratio is 1:10 and the initial reduction of the satellite's initial kinetic energy is greater that 50%. This is seen in Fig. 9 where the surface density profile is plotted for all the three experiments. In Fig. 9a the surface profile of ex. A (solid line) and ex. B (dashed line) is plotted. The two profiles seem quite similar with no obvious hump. The outer part of the remnant galaxy in both experiments

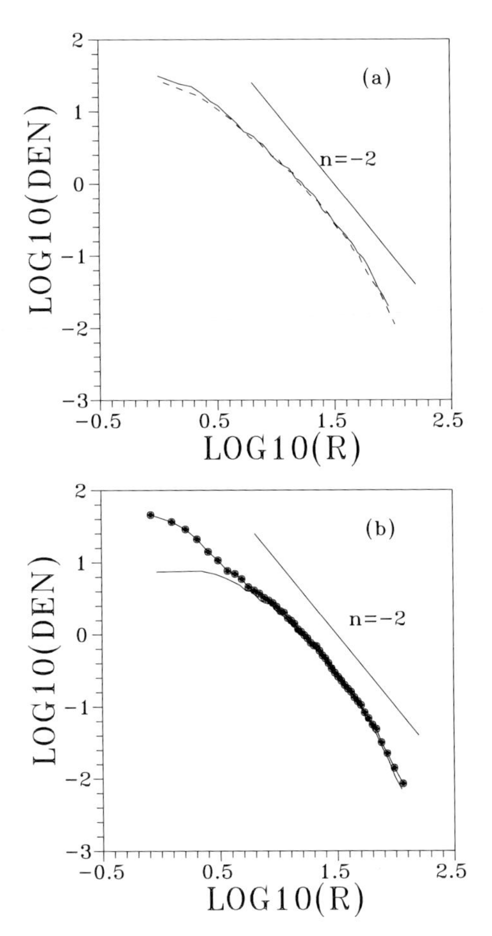

Fig. 9. The surface density as a function of the radius in logarithmic scale. (**a**) the solid line corresponds to ex. A and the dashed line to ex. B (**b**) the solid line with dots corresponds to ex. C and the solid line without dots to the primary galaxy before merger

follows approximately Hubbles' power low profile with exponent $n \approx -2$, while it becomes steeper beyond a radius of $\approx 50r_{un}$. In Fig. 9b, the solid line with dots corresponds to the remnant of ex. C and the solid line without dots to the primary galaxy before merging. While the primary profile before merging flattens inside a radius $r \approx 10r_{un}$, the remnant profile continue to rise inward due to the addition of the secondary material and flattens only at the softening length.

In the curve corresponding to ex. C the hump is obvious in the region where the satellite dominates the surface density. A similar density profile is presented, for the first time, in Efstathiou at al [13] for the observed Kinematical decoupled core galaxy NGC5813 and as Kormendy [21] suggested this can be a good candidate for a merger remnant.

In Fig. 10 we compare the surface density profile of ex. C with the one derived from the experiment with cosmological initial conditions. The solid line with dots corresponds to ex. C and the solid line without dots corresponds to the counterrotating system created from cosmological initial conditions. The latter one seems to flatten near the core, while the profile corresponding to the merger continues to rise towards the center, due to the addition of the satellite's material.

A parameter that is always affected from the merging is the galaxy's spin angular momentum. The mixing of particles between the two galaxies and the tidal torques exerted during the merging can cause the reduction of the spin of the primary galaxy, as it was remarked already from 1982 by Negroponte and White [24]. In their simulations the initial spin of the primary galaxy has reduced up to a factor of 3. This is confirmed in Fig. 11 where the spin angular momentum parallel to the x-axis (which is the axis of the rotation of the two

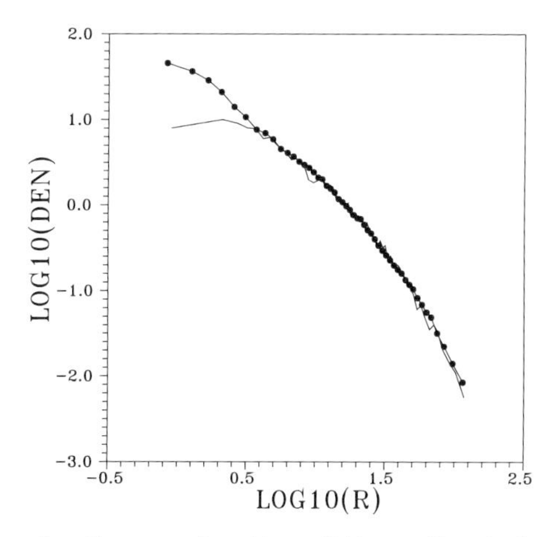

Fig. 10. The surface density as a function of the radius in logarithmic scale. The solid line with dots corresponds to the galaxy formed by merger (ex. C) and the solid line (without dots) corresponds to the galaxy formed by clumpy cosmological initial conditions

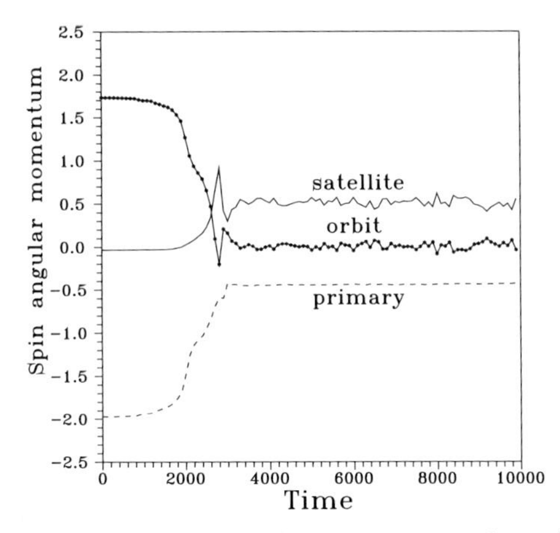

Fig. 11. The spin angular momentum of the primary galaxy (dashed line), of the satellite (solid line) and the orbital angular momentum (line with stars) as a function of time for ex. A.

galaxies)is plotted as a function of time for ex. A. The dashed line corresponds to the spin angular momentum of the primary galaxy, the solid line to the spin of the satellite and the solid line with stars to the angular momentum of the orbit with respect to the common center of mass. The spins along the other two directions are very close to zero. From this figure it is obvious that the orbit of the satellite is retrograde with respect to the rotation of the primary galaxy and that angular momentum is transferred from the orbit to the individual particles of the two galaxies. The spin of the primary galaxy starts from a negative value and is reduced by a factor of 4, until the remnant of the merging has relaxed (at a time of about 3000 t_{un}) and then remains constant. The satellite, on the other hand, starts with an almost zero initial value of spin angular momentum and after the relaxation acquires a positive spin angular momentum with approximately the same mean value with the value of the primary. These results are in agreement with Barcells and Quinn [2], who had mentioned that the survival of the satellite does not imply the survival of its spin. The spin of the satellite becomes always aligned with the orbital angular momentum, after the merger, irrespective of its initial orientation or sign. Therefore, the major effect for the presence of counterrotation is due to the orbital angular momentum of the satellite. Another parameter that could be useful to check and can be thought as a trace of a past merger event, is the ellipticity profile of the galaxy. Bender [4] pointed out that in some observed counterrotating galaxies, the radial transition between the core and the main body is clearly visible in their ellipticity profiles. He proposed that this should be a sign of a past merger event. However, the ellipticity profiles derived from our simulations (Fig. 12), both from merger and from cosmological initial conditions, look very much alike with the ones presented in Bender's paper. In Fig. 12 the line with dots correspond to the ellipticity of the merger remnant of ex. A as a function of it's longest axis, while the line with stars corresponds to the ellipticity of the galaxy with cosmological initial conditions.

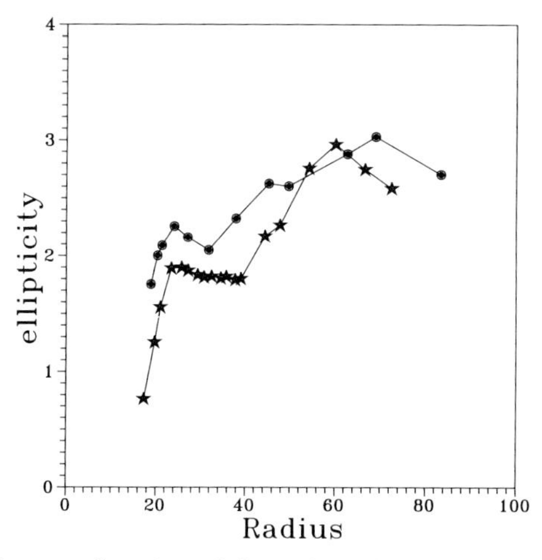

Fig. 12. The ellipticity as a function of the radius along the major axis of the remnant galaxy of ex. A.(line with dots) and for the galaxy formed by cosmological initial conditions (line with stars)

A peak of the radial profile of the ellipticity appears in both experiments and therefore one cannot use the ellipticity profile of an observed counterrotating galaxy to decide whether it is the product of a merger or it is produced from a single primordial collapse.

A theoretical argument below gives more chances to the direct scenario. In the case of cosmological initial conditions there are two main parameters that have to be arranged in order to produce counterrotating galaxies, namely, the axial ratio of the bar-like density excess, and the strength of the initial tidal field. Statistical estimations of these parameters can be directly derived from the power spectrum of the initial density perturbations (and its various moments).

On the other hand, only a small number of merger events can guarantee the formation of counterrotating remnants. Those in which a considerably large number of parameters are well tuned.

For example:

a) The mass ratio of the satellite and the primary as well as the density profile of the satellite must be inside a certain range of values, so that the satellite not only survives into the center but also dominates the core of the remnant.

b) the satellite's orbit must be on the rotation plane of the primary, otherwise the dynamical friction between the satellite and the rotating primary causes the satellite to pivot such as to be oriented almost upright in relation to the rotation plane of the primary,

c) the orbit of the satellite must be retrograde with respect to the rotation of the primary and have initial kinetic energy such that it does not escape, disrupted, or totally reverse the sign of the rotation throughout the radius of the remnant.

These arguments considerably reduce the possibilities of successful merger events. A question therefore is how to assign the formation of so many observed counterrotating ellipticals in such a demanding mechanism.

Recently a study has been made for the environment of galaxies in which there is gas or stellar counterrotation by Bettoni et al [9]. They concluded that no significant differences appear between the environments of counterrotating and normal galaxies. Therefore the hypothesis that counterrotating galaxies and polar rings derived from a recent interaction with a small satellite or a galaxy of similar size seems to be disproved and these galaxies seem to follow the idea that all galaxies are born through a merger process of smaller objects occurring very early in their life (cosmological initial conditions), or that they have been derived from a continuous infall of gas that formed stars later.

Acknowledgements

This work was supported by the Research Committee of the Academy of Athens (grant 200/517). We wish to thank Dr. Sv. Aarseth for his code.

References

1. Bak, J.: Thesis: "Retrograde minor mergers and counterrotating cores" (2000)
2. Barcells, M., and Quinn, P.J.: ApJ **361**, 381 (1990)
3. Barcells, M. and Gonzalez, A.C.: ApJ **505**, L109 (1998)
4. Bender R.: A&A **202**, L5 (1988)
5. Bender, R., and Surma P.: A&A **258**, 250 (1992)
6. Bertola, F. and Bettoni, D.: ApJ **329**, 102 (1988)
7. Bertola, F., Buson, L.M., and Zeilinger, W.W.: Nature, **335**, 705 (1988)
8. Bettoni, D., Fasano, G., and Galletta, G.: Astron. J. **99**, 1789 (1990)
9. Bettoni, D., Galleta, G., and Prada F.: A&A **374**, 83 (2001)
10. Chandrasekhar, S.: ApJ **97**, 255 (1943)
11. Ciri, R., Bettoni, D., and Galletta, G.: Nature **375**, 661 (1995)
12. Duncan, M., Farouki R., and Shapiro S.: ApJ **271**, 22 (1983)
13. Efstathiou, G., Ellis R.S. and Carter D.: MNRAS **201**, 975 (1982)
14. Franx, M., and Illingworth G.: ApJ **327**, L55 (1988)
15. Franx, M., Illingworth G. and Heckman T.: ApJ **344**, 613 (1989)
16. Harsoula, M.: Thesis: "Cosmological collapses and formation of counterrotating galaxies" (1999)
17. Harsoula, M. and Voglis, N.: A&A **335**, 431 (1998)
18. Hau, G.K.T. and Thomson R.C.: MNRAS **270**, L23 (1994)
19. Hausman, M., and Ostriker J.: ApJ **224**, 320 (1978)
20. Hernquist, L., Barnes, J.: Nature **354**, 210 (1991)
21. Kormendy, J.: ApJ **287**, 577 (1984)
22. Kuijken, K., Fisher D., Merrifield M.R.: MNRAS, **283**, 543 (1996)
23. Mehlert, D., Saglia R.P., Bender R. and Wegner G.: A&A **332**, 33 (1998)
24. Negroponte, J., White, S.: MNRAS **205**, 1009 (1983)
25. Peebles P.J.E.: ApJ **155**, 393 (1969)
26. Rubin, V.C., Graham, J.A., and Kenney J.D.P.: ApJ **394**, L9 (1992)
27. Voglis, N., Hiotelis, N.: A&A **218**, 1 (1989)

28. Voglis, N., Hiotelis, N., Hoeflich, P.: A&A **249**, 5 (1991)
29. Voglis., N., Harsoula, M., Efthymiopoulos, Ch.: Cel. Mech. Dyn. Astr. **78**, 265-278 (2000)
30. Voglis, N. and Harsoula, M.: in *Modern Theoretical and Observational Cosmology*, Plionis M. and Kotsakis S. (eds), Kluwer Academic Publishers, 276 (2001).
31. Zel'dovich, Ya.: A&A **5**, 84 (1970)

Pitch Angle of Spiral Galaxies as Viewed from Global Instabilities of Flat Stellar Disks

Shunsuke Hozumi

Max-Planck-Institut für Astronomie, Königstuhl 17, D-69117 Heidelberg, Germany

Abstract. We investigate whether the behavior of the pitch angle, predicted by a local dispersion relation derived from the density wave theory, can be applied to that of the fastest growing, globally unstable modes of flat stellar disks. We pay attention to two-armed modes, and obtain such global modes by numerically integrating the linearized collisionless Boltzmann equation. The results show that the pitch angle of the fastest growing modes has a tight correlation with the ratio of the square of the radial velocity dispersion to the surface density, as indicated by the local theory, over a sufficiently wide range of radii. The correspondence between the global modes and global properties of spiral galaxies is briefly discussed.

1 Introduction

Spiral structure is one of the most prominent features in disk galaxies. In fact, Hubble [6] classified spiral galaxies on the basis of the observed appearance of spiral arms. However, what his classification scheme, known as the Hubble sequence, represents physically remains still unclear.

One of the most successful theories to understand spiral structure is the density wave theory, which was originally conceived by Lindblad [10], and was first formulated by Lin and Shu [9]. This theory was based on local analysis, so that owning to the inhomogeneous density distributions in galaxy disks, density waves are propagated with a group velocity, and disappear in a few dynamical times [20]. In addition, it is found that spiral shocks [15] induced by a gaseous component in the disk damp the underlying spiral potential, again on a dynamical time scale [8].

The difficulties confronted with the density wave theory mentioned above may be overcome by the concept of global modes. Indeed, global analysis of thin disks has revealed that many unstable modes exist in self-gravitating disks treated with a fluid approximation [1] and in those composed of stars [7]. In particular, Bertin et al. [2] have shown that all Hubble morphological types can be realized by global modes using fluid disk models with appropriate basic states. In their modal calculations, it is demonstrated that the pitch angle of global modes depends on a fraction of the active disk mass, the distribution of Toomre's Q parameter [19], the typical Q value, and so on. Since their major concern was to specify the basic state to generate and support a given spiral structure, they did not describe explicitly what determined the pitch angle of spiral arms.

Since the density wave theory gives, in general, a good description of real spiral galaxies [17], it can help us understand the physics of spiral structure in spite of a local theory based on a WKBJ approximation. However, it is not necessarily clear whether the properties of global modes can be fully described by local density waves (see [21]). In this paper, we first show what is predicted for the pitch angle of unstable modes on the basis of the density wave theory. Next, this prediction is examined for the fastest growing two-armed modes of stellar disks in a global context.

2 Prediction by the Density Wave Theory

For tightly wrapped spiral waves in a cool disk, the dispersion relation for a one-component fluid model [3] is given by

$$(\omega - m\Omega)^2 = c^2k^2 - 2\pi G\Sigma|k| + \kappa^2, \tag{1}$$

where ω, Ω, κ, k, c, G, Σ, and m are the wave frequency, angular speed of a star, epicyclic frequency, wavenumber, sound speed, gravitational constant, surface density of the disk, and number of arms in spiral patterns, respectively. Equation (1) is rewritten as

$$(\omega - m\Omega)^2 = c^2(|k| - \pi G\Sigma/c^2)^2 + \kappa^2\left(1 - 1/Q_{\rm g}^2\right), \tag{2}$$

where $Q_{\rm g} = \kappa c/(\pi G\Sigma)$ is the Toomre stability parameter for gas disks. When $Q_{\rm g} > 1$, all waves are stable, so that there is no specific wave to be selected. However, taking into account unstable waves, the wave mode with the highest growth rate should appear in the disk by overwhelming the others with time. Since the Toomre parameter is a criterion valid only for local axisymmetric Jeans instabilities, we neglect the stability condition that $Q_{\rm g} > 1$ in the case of non-axisymmetric instabilities like those two-armed modes focused on here, and will find what is expected if (2) includes unstable waves.

Equation (2) indicates that the most unstable wave has the wavenumber $|k_{\rm m}| = \pi G\Sigma/c^2$, which is converted to the wavelength $\lambda_{\rm m}$ given by

$$\lambda_{\rm m} = 2c^2/G\Sigma. \tag{3}$$

Thus, we can see that the wavelength of the most unstable mode depends on c^2/Σ. As the wavelength is larger, the pitch angle of spiral arms becomes larger. Therefore, the density wave theory predicts, as is well-known, that the pitch angle increases with increasing c^2/Σ.

3 Models and Method

Since we aim not at the reproduction of real spiral structures but at demonstrating how well and to what extent the local theory gives a representation of the features of global modes, we use disk models appropriate for our concern, irrespective of the deviation from realism.

We adopt infinitesimally thin Toomre disks [18] without bulges, whose surface density distributions, Σ, and potentials, Ψ, are given, respectively, by

$$\Sigma(R) = \frac{Mq}{2\pi a^2}\left(1 + \frac{R^2}{a^2}\right)^{-3/2}, \tag{4}$$

and

$$\Psi(R) = -\frac{GM}{\sqrt{R^2 + a^2}}, \tag{5}$$

where R is the distance from the center of the disk, and a is the length scale. Here, q $(0 < q \leq 1)$ represents an active disk mass fraction of the total mass. Since the Toomre disks are the flat version of Plummer's models [14] in three-dimensional configurations, for each disk model the fraction, $1 - q$, of the total mass is considered to reside in a surrounding dark halo represented by a Plummer model with the same length scale a as that of the disk. Thus, in our models, the total mass distribution including a disk and a halo is identical, and the mass fraction of the disk is different.

For this mass profile, we can use Miyamoto's distribution functions (DFs) [12] with respect to directly rotating stars, $F^+(\varepsilon, j)$, where ε and j are the energy and angular momentum of a star per unit mass, respectively. Retrograde stars are introduced in the same manner as that adopted by Nishida et al. [13]. Then, the equilibrium DFs, F_0, are given by

$$F_0(\varepsilon, j) = \begin{cases} q[(1/2)F_0^+(\varepsilon) + F_1^+(\varepsilon, j)], & (j \geq 0) \\ (1/2)qF_0^+(\varepsilon), & (j < 0), \end{cases} \tag{6}$$

where the functions, $F_0^+(\varepsilon)$ and $F_1^+(\varepsilon, j)$, are derived from the expansion of $F^+(\varepsilon, j)$ as

$$F^+(\varepsilon, j) = F_0^+(\varepsilon) + F_1^+(\varepsilon, j). \tag{7}$$

The Miyamoto DFs [12] are characterized by the parameter n that specifies the distribution of radial velocity dispersion, c_{r}, such that

$$c_{\mathrm{r}}^2(R) = -\frac{1}{2n + 4}\Psi(R). \tag{8}$$

This equation indicates that c_{r} can be provided independently of the disk surface density.

If disk models satisfy the condition such that $(n + 2)q = \text{constant}$, they have the same c_{r}^2/Σ distribution throughout the disk, as derived from (4), (5), and (8). We take $(n + 2)q = 5$ with the pairs of $(n, q) = (3, 1)$, $(4, 5/6)$, $(5, 5/7)$, $(6, 5/8)$, $(7, 5/9)$, and $(10, 5/12)$, and $(n + 2)q = 3$ with the pairs of $(n, q) = (3, 3/5)$, $(4, 1/2)$, $(5, 3/7)$, and $(6, 3/8)$. In Fig. 1, the distributions of the Q parameter for both model sequences are presented. Here, Q is defined for stellar disks as

$$Q = \frac{\kappa c_{\mathrm{r}}^2}{3.36 G \Sigma}. \tag{9}$$

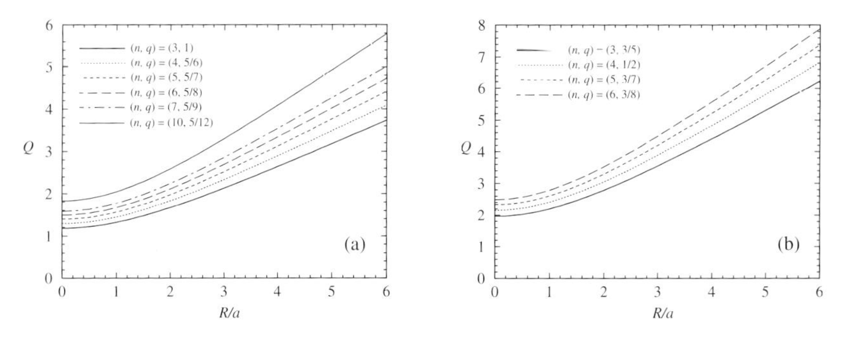

Fig. 1. Distribution of the Toomre stability parameter Q for (a) the models with $(n+2)q = 5$, and for (b) the models with $(n+2)q = 3$

It is found from this figure that all models are locally stable, i.e., $Q > 1$.

The most unstable global modes are obtained by numerically integrating the linearized collisionless Boltzmann equation given by

$$\frac{df_m}{dt} = \frac{\partial \psi_m}{\partial R}\frac{\partial F_0}{\partial u} + im\psi_m \frac{\partial F_0}{\partial j}, \tag{10}$$

where f_m and ψ_m are the perturbed distribution function and perturbed potential, respectively, with respect to the m-armed mode, and i is the unit of the imaginary part. Equation (10) is solved as an initial value problem by evolving an arbitrary form of perturbation, imposed initially, forward in time until it has reached the exponential growth in f_m and ψ_m. We restrict ourselves only to $m = 2$ modes, which are usually the most unstable modes among various m-armed modes. In fact, most of grand-design spirals show two-armed features. The numerical details are described in [5].

The units of mass and length, and the gravitational constant are taken so that $M = a = 1$ and $G = 1$, respectively. Then, the unit of time is $(a^3/GM)^{1/2}$.

4 Results

We have obtained the fastest growing two-armed modes of the models with $(n+2)q = 5$ and those with $(n+2)q = 3$. The growth rate, pattern speed, and corotation radius of each model are summarized in Table 1 for $(n+2)q = 5$, and in Table 2 for $(n+2)q = 3$.

We calculate the pitch angle, $i_{\rm p}$, given by

$$\cot i_{\rm p} = \left| mR\frac{\partial \phi}{\partial R} \right|, \tag{11}$$

where for a specified arm in the disk, ϕ is the phase angle of the density crest of the mode along R [3]. Here, we employ $m = 2$.

The pitch angle of each mode, measured in degrees, is shown in Fig. 2, which illustrates that the pitch angle profile does not change substantially from model

Table 1. Results for the fastest growing modes of the models with $(n+2)q = 5$

n	q	Growth Rate	Pattern Speed	Corotation Radius
3	1	0.224	0.333	1.83
4	5/6	0.167	0.328	1.85
5	5/7	0.119	0.313	1.92
6	5/8	0.0851	0.294	2.03
7	5/9	0.0652	0.275	2.15
10	5/12	0.0516	0.228	2.49

Table 2. Results for the fastest growing modes of the models with $(n+2)q = 3$

n	q	Growth Rate	Pattern Speed	Corotation Radius
3	3/5	0.0309	0.226	2.51
4	1/2	0.0158	0.213	2.62
5	3/7	0.00744	0.201	2.74
6	3/8	0.00449	0.190	2.86

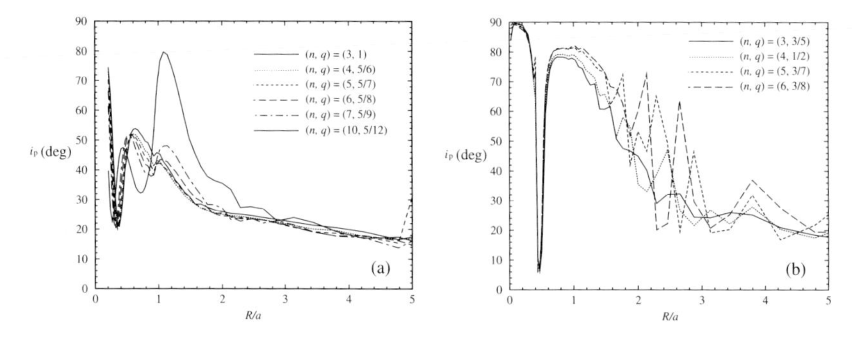

Fig. 2. Pitch angle of the fastest growing two-armed modes for (a) the models with $(n+2)q = 5$, and for (b) the models with $(n+2)q = 3$

to model for each model sequence. For the $(n+2)q = 5$ sequence, all but the model with $(n, q) = (10, 5/12)$ have nearly the same pitch angle profile within $R = 5$. Even if the model with $(n, q) = (10, 5/12)$ is included in this sequence, the difference in pitch angle is rather small from $R \sim 2$ to $R \sim 5$. For the $(n+2)q = 3$ sequence, the pitch angles are not well determined because the very small growth rates do not enhance the density contrast of the unstable modes conspicuously. In spite of the large scatters, the pitch angles of these models appear to fluctuate around a certain profile. Therefore, also in this sequence, it can be found that the pitch angle profiles are approximately similar to one another. These results

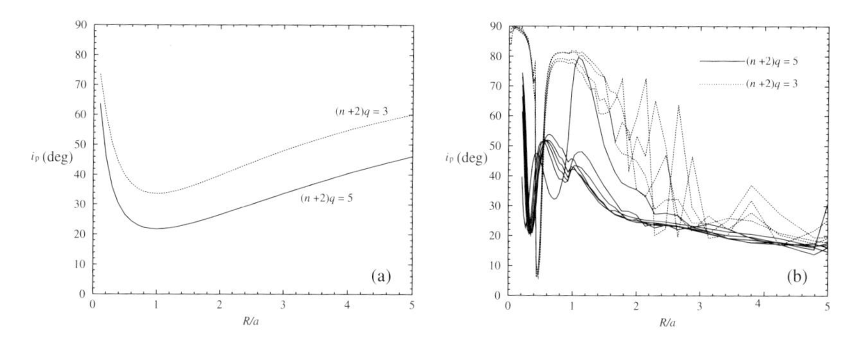

Fig. 3. Pitch angle of unstable two-armed modes for (a) expected profiles calculated from the local dispersion relation for $(n+2)q = 5$ (solid line), and for $(n+2)q = 3$ (dotted line), and for (b) obtained profiles of all the models with $(n+2)q = 5$ (solid lines) and $(n+2)q = 3$ (dotted lines)

are remarkable in that models with different radial velocity dispersion profiles and different degrees of self-gravity show almost the same pitch angle profile over a sufficiently wide range of radii.

According to the local theory prediction, the increase in c_{r}^2/Σ leads to the increase in pitch angle, as found from (3) and (11). In our models, c_{r}^2/Σ is inversely proportional to $(n+2)q$, and so, the pitch angles for the $(n+2)q = 3$ sequence should be systematically larger than those for the $(n+2)q = 5$ sequence, as indicated by Fig. 3 (a), in which the pitch angles are calculated from (3) and (11). We can see this expected behavior in Fig. 3 (b), where the pitch angle profiles of all the models are put together. However, the difference in pitch angle for both model sequences becomes very small at larger radii than $R = 3$ unlike those values obtained from the local dispersion relation which are presented in Fig. 3 (a).

5 Discussion and Conclusions

We have found that the pitch angles of the fastest growing modes obey the prediction derived from the density wave theory even in a global context. We conclude from our global mode calculations that the pitch angles of spiral arms are regulated to a considerable degree by a typical value of c_{r}^2/Σ, if spiral structure is viewed as a manifestation of the fastest growing modes of stellar disks. In addition, our results indicate that the density wave theory is helpful to understand the properties of global modes qualitatively. In particular, although the theory is developed for tightly wound spirals, it is still useful for loosely wound spirals because the pitch angles obtained here are typically 20°.

Roberts and Haynes [16] have summarized the global properties of spiral galaxies along the Hubble sequence. According to their Fig. 2, there is no practically systematic change in total mass at least from Sa to Sc. Furthermore, their Fig. 3 illustrates that the total surface density including gaseous and stellar components decreases slightly from Sa to Sc. If the mass distributions of spiral galaxies would not change substantially, the decrease in total surface density and the increase in pitch angle, together with our results, suggest that $c_{\rm r}$ would be nearly constant or increase from Sa to Sc. Therefore, observations of $c_{\rm r}$ along the Hubble sequence will clarify the relevance of our argument.

Recently, Ma [11] has revealed that the pitch angle increases as the disk surface density decreases by analyzing all Hubble type galaxies altogether. Ma's finding is consistent with our results of global mode calculations. His analysis shows a large scatter in the relation between the pitch angle and the surface density. If a typical $c_{\rm r}^2/\Sigma$ instead of Σ was chosen for each spiral galaxy, this scatter might be reduced.

From the arguments mentioned above, real spiral structures appear to reflect the characteristics of the fastest growing, global modes of stellar disks. In fact, the old stellar population observed at near-infrared wavelengths forms large-scale smooth symmetric arms in the disk (e.g., [4]). These features may be a manifestation of globally unstable modes.

References

1. S. Aoki, M. Noguchi, M. Iye: Publ. Astron. Soc. Jpn. **31**, 737 (1979)
2. G. Bertin, C. C. Lin, S. A. Lowe, R. P. Thurstans: Astrophys. J. **338**, 78 (1989)
3. J. Binney, S. Tremaine: *Galactic Dynamics* (Princeton University Press, Princeton 1987) Chap. 6
4. D. L. Block, G. Bertin, A. Stockton, P. Grosbøl, A. F. M. Moorwood, R. F. Peletier: Astron. Astrophys. **288**, 365 (1994)
5. S. Hozumi, T. Fujiwara, M. T. Nishida: Publ. Astron. Soc. Jpn. **39**, 447 (1987)
6. E. P. Hubble: Astrophys. J. **64**, 321 (1926)
7. A. J. Kalnajs: Astrophys. Lett. **11**, 41 (1972)
8. A. J. Kalnajs: Astrophys. J. **175**, 63 (1972)
9. C. C. Lin, F. H. Shu: Astrophys. J. **140**, 646 (1964)
10. B. Lindblad: Stockholm Observ. Ann. **22**, 3 (1963)
11. J. Ma: Astron. Astrophys. **388**, 389 (2002)
12. M. Miyamoto: Publ. Astron. Soc. Jpn. **23**, 21 (1971)
13. M. T. Nishida, Y. Watanabe, T. Fujiwara, S. Kato: Publ. Astron. Soc. Jpn. **36**, 27 (1984)
14. H. C. Plummer: Mon. Not. R. Astro. Soc. **71**, 460 (1911)
15. W. W. Roberts: Astrophys. J. **158**, 123 (1969)
16. M. S. Roberts, M. P. Haynes: Annu. Rev. Astron. Astrophys. **32**, 115 (1994)
17. W. W. Roberts, Jr., M. S. Roberts, F. H. Shu: Astrophys. J. **196**, 381 (1975)
18. A. Toomre: Astrophys. J. **138**, 385 (1963)
19. A. Toomre: Astrophys. J. **139**, 1217 (1964)
20. A. Toomre: Astrophys. J. **158**, 899 (1969)
21. A. Toomre: 'What amplifies the spirals'. In: *The Structure and Evolution of Normal Galaxies*, ed. by S.M. Fall, D. Lynden-Bell (Cambridge University Press, Cambridge 1981) pp. 111–136

Collisionless Evaporation from Cluster Elliptical Galaxies

Veruska Muccione[1] and Luca Ciotti[2]

[1] Geneva Observatory, 51. ch. des Maillettes, 1290 Sauverny, Switzerland
[2] Dipartimento di Astronomia, Università di Bologna, via Ranzani 1, 40127 Bologna, Italy

Abstract. We describe a particular aspect of the effects of the parent cluster tidal field (CTF) on stellar orbits inside cluster Elliptical galaxies (Es). In particular we discuss, with the aid of a simple numerical model, the possibility that *collisionless stellar evaporation* from elliptical galaxies is an effective mechanism for the production of the recently discovered intracluster stellar populations (ISP). A preliminary investigation, based on very idealized galaxy density profiles (Ferrers density distributions), showed that over an Hubble time, the amount of stars lost by a representative galaxy may sum up to the 10% of the initial galaxy mass, a fraction in interesting agreement with observational data. The effectiveness of this mechanism is due to the fact that the galaxy oscillation periods near equilibrium configurations in the CTF are comparable to stellar orbital times in the external galaxy regions. Here we extend our previous study to more realistic galaxy density profiles, in particular by adopting a triaxial Hernquist model.

1 Introduction

Observational evidences of an Intracluster Stellar Population (ISP) are mainly based on the identification of *intergalactic* planetary nebulae and red giant stars (see, e.g., [1],[2],[3],[4],[5]). Overall, the data suggest that approximately 10% (or even more) of the stellar mass of clusters is contributed by the ISP [6]. The usual scenario assumed to explain the finding above is that gravitational interactions between cluster galaxies, and interactions between the galaxies with the gravitational field of the cluster, lead to a substantial stripping of stars from the galaxies themselves.

Here, supported by a curious coincidence, namely by the fact that *the characteristic times of oscillation of a galaxy around its equilibrium position in the cluster tidal field (CTF) are of the same order of magnitude of the stellar orbital periods in the external part of the galaxy itself*, we explore the effects of interaction between stellar orbits inside the galaxies and the CTF. In fact, based on the observational evidence that the major axis of cluster Es seems to be preferentially oriented toward the cluster center, N-body simulations showed that model galaxies tend to align, as observed, reacting to the CTF as rigid bodies [7] . By assuming this idealized scenario, a stability analysis then showed that this configuration is of stable equilibrium, and allowed to calculate the oscillation periods in the linearized regime [8]. In particular, oscillations around two stable equilibrium configurations have been considered, namely: 1) when the center of mass of the galaxy is at rest at center of a triaxial cluster, and the galaxy inertia ellipsoid is aligned with the CTF principal directions, and 2) when the galaxy center of mass is placed on a circular orbit in a spherical cluster, and the galaxy major axis points toward the galaxy center while the galaxy minor axis is perpendicular to the orbital plane.

Here, prompted by these observational and theoretical considerations, we extend a very preliminary study of the problem [9], by evolving stellar orbits in a more realistic galaxy density profile: for simplicity we restrict to case 1) above, while the full exploration of the parameter space, together with a complete discussion of case 2), will be given elsewhere [10]. It is clear, however, that both cases are rather exceptional. Most cluster galaxies neither rest in the cluster center nor move on circular orbits, but they move on elongated orbits with very different pericentric and apocentric distances from the cluster's center; in a triaxial cluster many orbits are boxes and some orbits can be chaotic. These latter cases can be properly investigated only by direct numerical simulation of the stellar motions inside the galaxies, coupled with the numerical integration of the equations of the motion of the galaxies themselves.

2 The Physical Background

Without loss of generality we assume that in the (inertial) Cartesian coordinate system C, with the origin on the cluster center, the CTF tensor $\boldsymbol{T}$ is in diagonal form, with components T_i $(i = 1, 2, 3)$. By using three successive, counterclockwise rotations (φ around x axis, ϑ around y' axis and ψ around z'' axis), the linearized equations of the motion for the galaxy near the equilibrium configuration can be written as

$$\ddot{\varphi} = \frac{\Delta T_{32} \Delta I_{32}}{I_1} \varphi, \qquad \ddot{\vartheta} = \frac{\Delta T_{31} \Delta I_{31}}{I_2} \vartheta, \qquad \ddot{\psi} = \frac{\Delta T_{21} \Delta I_{21}}{I_3} \psi, \tag{1}$$

where ΔT is the antisymmetric tensor of components $\Delta T_{ij} \equiv T_i - T_j$, and I_i are the principal components of the galaxy inertia tensor. In addition, let us also assume that $T_1 \geq T_2 \geq T_3$ and $I_1 \leq I_2 \leq I_3$, i.e., that $\Delta T_{32}, \Delta T_{31}$ and ΔT_{21} are all less or equal to zero (see, e.g., [8], [10]). Thus, the equilibrium position associated with (1) is *linearly stable*, and its solution is

$$\varphi = \varphi_{\rm M} \cos(\omega_\varphi t), \quad \vartheta = \vartheta_{\rm M} \cos(\omega_\vartheta t), \quad \psi = \psi_{\rm M} \cos(\omega_\psi t), \tag{2}$$

where

$$\omega_\varphi = \sqrt{\frac{\Delta T_{23} \Delta I_{32}}{I_1}}, \quad \omega_\vartheta = \sqrt{\frac{\Delta T_{13} \Delta I_{31}}{I_2}}, \quad \omega_\psi = \sqrt{\frac{\Delta T_{12} \Delta I_{21}}{I_3}}. \tag{3}$$

For computational reasons the best reference system in which calculate stellar orbits is the (non inertial) reference system C' in which the galaxy is at rest, and its inertia tensor is in diagonal form. The equation of the motion for a star in C' is

$$\ddot{\boldsymbol{x}}' = \mathcal{R}^{\rm T} \ddot{\boldsymbol{x}} - 2\boldsymbol{\Omega} \wedge \boldsymbol{v}' - \dot{\boldsymbol{\Omega}} \wedge \boldsymbol{x}' - \boldsymbol{\Omega} \wedge (\boldsymbol{\Omega} \wedge \boldsymbol{x}'), \tag{4}$$

where $\boldsymbol{x} = \mathcal{R}(\varphi, \vartheta, \psi)\boldsymbol{x}'$, and

$$\boldsymbol{\Omega} = (\dot{\varphi}\cos\vartheta\cos\psi + \dot{\vartheta}\sin\psi, -\dot{\varphi}\cos\vartheta\sin\psi + \dot{\vartheta}\cos\psi, \dot{\varphi}\sin\vartheta + \dot{\psi}). \tag{5}$$

In (4)

$$\mathcal{R}^{\mathrm{T}}\ddot{\boldsymbol{x}} = -\nabla_{\boldsymbol{x}'}\phi_{\mathrm{g}} + (\mathcal{R}^{\mathrm{T}}\boldsymbol{T}\mathcal{R})\boldsymbol{x}', \tag{6}$$

where $\phi_{\mathrm{g}}(\boldsymbol{x}')$ is the galactic gravitational potential, $\nabla_{\boldsymbol{x}'}$ is the gradient operator in C', and we used the tidal approximation to obtain the star acceleration due to the cluster gravitational field.

3 Galaxy and Cluster Models

For simplicity we assume that the galaxy and cluster densities are stratified on homeoids. In particular, the galaxy density belongs to ellipsoidal generalization of the widely used γ-models ([11],[12]):

$$\rho_{\mathrm{g}}(m) = \frac{M_{\mathrm{g}}}{\alpha_1\alpha_2\alpha_3}\frac{3-\gamma}{4\pi}\frac{1}{m^{\gamma}(1+m)^{4-\gamma}}, \tag{7}$$

where M_{g} is the total mass of the galaxy, $0 \leq \gamma \leq 3$ and

$$m^2 = \sum_{i=1}^{3}\frac{(x_i')^2}{\alpha_i^2}, \qquad \alpha_1 \geq \alpha_2 \geq \alpha_3. \tag{8}$$

The inertia tensor components of a generic homeoidal density distribution (in the natural reference system adopted in (8)), are given by

$$I_i = \frac{4\pi}{3}\alpha_1\alpha_2\alpha_3(\alpha_j^2 + \alpha_k^2)h_{\mathrm{g}}, \tag{9}$$

where $h_{\mathrm{g}} = \int_0^{\infty}\rho_{\mathrm{g}}(m)m^4 dm$, and so $I_1 \leq I_2 \leq I_3$. Note that, from (3) and (9) it results that the frequencies for homeoidal stratifications *do not depend on the specific density distribution assumed*, but only on the quantities $(\alpha_1, \alpha_2, \alpha_3)$. We also introduce the two ellipticities

$$\frac{\alpha_2}{\alpha_1} \equiv 1 - \epsilon, \qquad \frac{\alpha_3}{\alpha_1} \equiv 1 - \eta, \tag{10}$$

where $\epsilon \leq \eta \leq 0.7$.

A rough estimate of *characteristic stellar orbital times* inside m is given by $P_{orb}(m) \simeq 4P_{\mathrm{dyn}}(m) = \sqrt{3\pi/G\overline{\rho_{\mathrm{g}}}(m)}$, where $\overline{\rho_{\mathrm{g}}}(m)$ is the mean galaxy density inside m. We thus obtain

$$P_{orb}(m) \simeq 9.35 \times 10^6\sqrt{\frac{\alpha_{1,1}^3(1-\epsilon)(1-\eta)}{M_{\mathrm{g},11}}}m^{\gamma/2}(1+m)^{(3-\gamma)/2} \quad \text{yrs}, \tag{11}$$

where $M_{\mathrm{g},11}$ is the galaxy mass normalized to $10^{11}M_{\odot}$, $\alpha_{1,1}$ is the galaxy “core” major axis in kpc units (for the spherically symmetric $\gamma = 1$ Hernquist model [13],

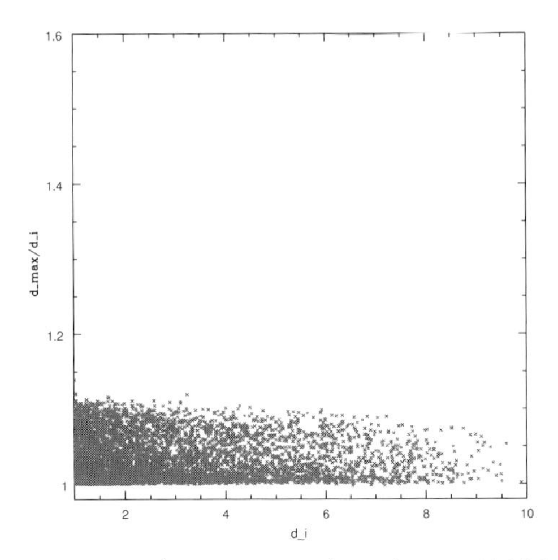

Fig. 1. Distribution of the $d_{\rm max}/d_{\rm i}$ ratio vs. $d_{\rm i}/\alpha_1$ after an Hubble time for the model galaxy at rest. $d_{\rm i}$ is the initial distance of the star from the galaxy center, while $d_{\rm max}$ is the maximum distance from the galaxy center reached during the simulation.

$R_{\rm e} \simeq 1.8\alpha_1$); thus, in the outskirts of normal galaxies orbital times well exceed 10^8 or even 10^9 yrs. For the cluster density profile we assume

$$\rho_{\rm c}(m) = \frac{\rho_{\rm c,0}}{(1+m^2)^2}, \tag{12}$$

where m is given by an identity similar to (8), with $a_1 \geq a_2 \geq a_3$, and, in analogy with (10) we define $a_2/a_1 \equiv 1-\mu$ and $a_3/a_1 \equiv 1-\nu$, with $\mu \leq \nu \leq 1$. It can be shown (see, e.g., [8],[10]) that the CTF components at the center of a non-singular homeoidal distribution are given by

$$T_i = -2\pi G \rho_{\rm c,0} w_i(\mu,\nu), \tag{13}$$

where the dimensionless quantities w_i are independent of the specific density profile, $w_1 \leq w_2 \leq w_3$ for $a_1 \geq a_2 \geq a_3$, and so the conditions for stable equilibrium in (1) are fulfilled ([8],[10]). The quantity $\rho_{\rm c,0}$ is not a well measured quantity in real clusters, and for its determination we use the virial theorem, $M_{\rm c}\sigma_{\rm V}^2 = -U$, where $\sigma_{\rm V}^2$ is the virial velocity dispersion, that we assume to be estimated by the observed velocity dispersion of galaxies in the cluster. Thus, we can now compare the galactic oscillation periods:

$$P_\varphi = \frac{2\pi}{\omega_\varphi} \simeq \frac{8.58\times 10^8}{\sqrt{(\nu-\mu)(\eta-\varepsilon)}} \frac{a_{1,250}}{\sigma_{\rm V,1000}} \,\text{yrs}\,,$$

$$P_\vartheta = \frac{2\pi}{\omega_\vartheta} \simeq \frac{8.58\times 10^8}{\sqrt{\nu\eta}} \frac{a_{1,250}}{\sigma_{\rm V,1000}} \,\text{yrs}\,,$$

$$P_\psi = \frac{2\pi}{\omega_\psi} \simeq \frac{8.58\times 10^8}{\sqrt{\mu\varepsilon}} \frac{a_{1,250}}{\sigma_{\rm V,1000}} \,\text{yrs}\,.$$

(for small galaxy and cluster flattenings, where $a_{1,250} = a_1/250$ kpc and $\sigma_{V,1000} = \sigma_V/10^3$ km/s, [10]) with the characteristic orbital times in galaxies. Thus, from (11) and (14abc), it follows that *in the outer halo of giant Es, stellar orbital times can be of the same order of magnitude as the oscillatory periods of the galaxies themselves near their equilibrium position in the CTF*. For example, in a relatively small galaxy of $M_{g,11} = 0.1$ and $\alpha_{1,1} = 1$, $P_{orb} \simeq 1$ Gyr at $m \simeq 10$ (i.e., at $\simeq 5R_e$), while for a galaxy with $M_{g,11} = 1$ and $\alpha_{1,1} = 3$ the same orbital time characterizes $m \simeq 7$ (i.e., $\simeq 3.5R_e$).

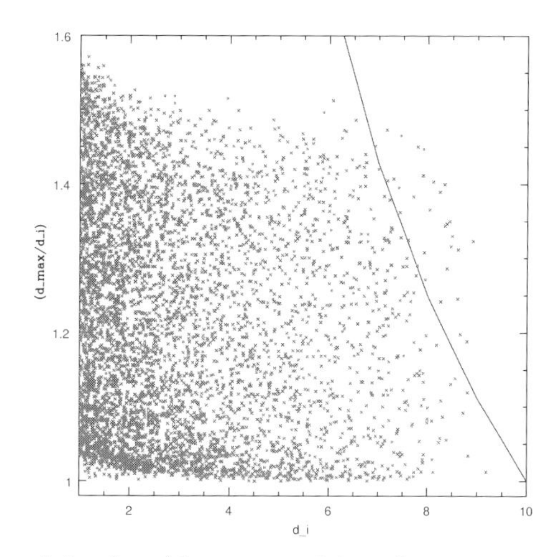

Fig. 2. Distribution of the $d_{\rm max}/d_{\rm i}$ ratio vs. $d_{\rm i}/\alpha_1$ after an Hubble time for the same galaxy model as in Fig. 1, when oscillating around its equilibrium position in the CTF.

In order to understand the effects of the galaxy oscillations on the stellar orbits, we performed a set of Monte-Carlo simulations, in which we followed the evolution of 10^4 - 10^5 "1-body problems" over the Hubble time by integrating numerically (4). At variance with [9], where we used simple and easy-to-integrate Ferrers density profiles, here we study orbital evolution in a more realistic (but also more demanding from the numerical point of view) galaxy density profile, namely a triaxial Hernquist model, obtained by assuming $\gamma = 1$ in (7). The gravitational potential inside the galaxy density distribution, in a form suitable for the numerical integration, was obtained by using an expansion technique useful in case of small density flattenings ([10],[14]). The initial conditions are generated by using the Von Neumann rejection method in phase-space (for details see [10]): note that, at variance with the analysis [9], now "stars" are characterized by initial velocities that can be different from zero. The code, a double-precision fortran code based on a Runge-Kutta scheme, runs on GRAVITOR, the Geneva Observatory 132 processors Beowulf cluster (**http://obswww.unige.ch/~pfennige/gravitor/gravitor_e.html**). The computation of 10^4 orbits usually requires 2 hours when using 10 nodes.

4 Preliminary Results and Conclusions

We show here, as an illustrative case, the behavior of the ratio $d_{\rm max}/d_{\rm i}$ as a function of $d_{\rm i}/\alpha_1$, for a moderately flattened galaxy model ($\epsilon \simeq 0.2$ and $\eta \simeq 0.3$), with $M_{\rm g} = 10^{11} M_{\odot}$, semi-mayor axis $\alpha_1 = 3$ kpc, and maximum oscillation angles equals to 0.1 rad. The cluster parameters are $a_{1,250} = \sigma_{\rm V,1000} = 1$, $\mu = 0.2$, $\nu = 0.4$, and the total number of explored orbits is $N_{\rm tot} = 10^4$. In order to show the effect of oscillations, in the following simulations we artificially eliminated the *direct* contribution of the CTF, as given by the second term in the r.h.s. of (6).

In Fig. 1 we show the result of a first simulation in which the galaxy is *not* oscillating: obviously, the ratio $d_{\rm max}/d_{\rm i}$ is in general (slightly) larger than unity, due to the initial velocity of each star. In Fig. 2 we show the result for the same galaxy model, when oscillating around the equilibrium position: the effects of the galaxy oscillations are clearly visible as a global "expansion" of the galaxy. As a reference, the solid line indicates the expansion ratio required to reach the representative distance of $10R_{\rm e}$ from the galaxy center. Thus, it is clear that the galaxy oscillations are certainly able to substantially modify the galaxy density profile. In particular, it will be of interest the study of the (more realistic) case in which the galaxy is in rotation around the cluster center. In this case we expect a different behavior of stellar orbits as a function of the distance of the galaxy center of mass from the cluster center: in fact, while inside the cluster core the CTF is *compressive* (see, e.g., [7],[8]), outside the CTF is *expansive* along the cluster radial direction, and in this latter case its direct effect should increase the expansive effect due to the galaxy oscillations. These cases are discussed in detail in [10].

References

1. T.Theuns, S.J. Warren: MNRAS, 284 (1996)
2. R.H. Méndez et al: ApJ, 491 (1998)
3. J.J. Feldmeier , R. Ciardullo, G.H. Jacoby: ApJ, 503 (1998)
4. M. Arnaboldi, J.A.L. Aguerri, N.R. Napolitano, O. Gerhard, K.C. Freeman, J. Feldmeier , M. Capaccioli, R.P. Kudritzki, R.H. Méndez: AJ, 123 (2002)
5. P.R. Durrell, R. Ciardullo, J.J. Feldmeier, G.H. Jacoby, S. Sigurdsson: ApJ, 570 (2002)
6. H.C. Ferguson, N.R. Tanvir, T. von Hippel: Nature, 391 (1998)
7. L. Ciotti, S.N. Dutta: MNRAS, 270 (1994)
8. L. Ciotti, G. Giampieri: Cel. Mech. & Dyn. Astr., 68 (1998)
9. V. Muccione, L. Ciotti: Mem. S.A.It., in press (2003)
10. L. Ciotti, V. Muccione: in preparation
11. W. Dehnen: MNRAS, 256 (1993)
12. S. Tremaine et al.: AJ, 107 (1994)
13. L. Hernquist: ApJ, 356, 359 (1990)
14. L. Ciotti, G. Bertin: in preparation

Part V

Solar System Dynamics

Chaos in Solar System Dynamics

Rudolf Dvorak

Institute of Astronomy, University of Vienna, Türkenschanzstrasse 17, A-1180 Vienna, Austria

Abstract. The effect of solar system chaos on small bodies, such as comets and asteroids, is quite different from that experienced by the major planets. It is more obvious in the motion of the large number of small bodies in the planetary region. Thus, the existence of different groups of comets and asteroids is due to the different qualities of the various resonances – mean motion resonances, secular resonances and three-body resonances – but especially because of resonance overlap. Moreover, chaotic motion has also been found in the motion of the planets and appears to be present on even a larger level in extra-solar planetary systems.

1 Introduction

When dealing with the celestial mechanics of solar system bodies we have to be aware that there is a fundamental difference from galactic dynamics even though the governing physical laws – primarily gravitation and also in special cases relativity – are the same. We can directly observe the motion of solar system bodies on time scales as short as years, months and even days. Furthermore, there is one special planetary system we are dealing with although now we have knowledge of about 100 extrasolar planets. For these systems the observational conditions are very different from observations in our planetary environment. In galactic dynamics we cannot observe directly the motion, for centuries the picture does not change for the observer on the Earth; we thus have a "snapshot" of the dynamics of our galaxy. On the contrary we have observations of galaxies in different stages of their age and can therefore deduce information on the dynamics of our own Galaxy. Furthermore, although a galaxy consists of billions of stars, we can treat their motions in the corresponding galactic potential as a motion in a "simple" Hamiltonian using an averaged gravitational potential. In solar system dynamics we have to deal with an n-body system where n is small (depending on the number of planets we wish to take into account). As we will see in this text the model of an n-body system has the advantage that it can be treated as an integrable dynamical system (namely the Keplerian motion of any body around the Sun, for a satellite around the parent planet) perturbed by the other bodies with significantly smaller masses than the Sun.

There are very good recent reviews on this subject ([15], [17]) which point out quite well the fundamental rôle of chaos in the dynamics of the solar system. In the present text we will, besides repeating more or less the fundamental ideas given in these two papers, critically give light to some of the results presented therein and report on new results concerning the rapidly increasing number of dynamical studies of extrasolar planetary systems.

One of the fundamental questions of astrodynamics is whether the solar systems is stable or not. Perturbation theory cannot give a final answer because of the limited time scales over which such theories are valid. It is also clear that straightforward numerical integration of the equations of motion cannot give a definitive answer because the inevitable accumulation errors lead to a deviation of the computed orbits from the "real" ones. An additional point of weakness is the incomplete dynamical model, which cannot perfectly represent the "real" solar system, even when all known effects, such as the gravitational forces of the largest minor planets [1], the oblateness of the Sun and the planets, the galactic field and the relativistic effects are included. Additionally there exists an intrinsic effect in nonlinear dynamical systems which makes the solutions uncertain owing to extreme sensitivity with respect to the initial conditions, known to physicists as *deterministic chaos*. This was pointed out for galactic dynamics in the seminal paper of Hénon and Heiles [9] and was found later through the work of Wisdom in the chaotic motion of asteroids near the 3:1 resonance with Jupiter [25].

The following sections are devoted to basic considerations of the dynamics of an n-body system with a dominating central mass, the appearance of chaotic motion due to the nonlinearity of the equations of motion and the action of different kinds of resonances, the motion of comets and asteroids, the long-term evolution of the orbits of the planets and the special orbit of Mercury. The conclusion summarizes the rôle of chaotic motion in solar systems dynamics. In an epilog some interesting results for extrasolar planetary systems are presented and show how our own system may serve as a "toy model" for planetary systems.

2 The n-Body Solar System

In heliocentric coordinates the equations of motion for the planets (modelled as mass points m_i) may be written as

$$\ddot{\boldsymbol{q}}_\mathrm{i} = k^2 \left(-\frac{m_1 + m_\mathrm{i}}{r_\mathrm{i}^3} \boldsymbol{q}_\mathrm{i} + \sum_{j=2, j \neq i}^{\mathrm{n}} m_\mathrm{j} \left(\frac{\boldsymbol{q}_\mathrm{j} - \boldsymbol{q}_\mathrm{i}}{r_\mathrm{ji}^3} - \frac{\boldsymbol{q}_\mathrm{j}}{r_\mathrm{j}^3} \right) \right) . \tag{1}$$

Here the first term describes two-body motion around the Sun, and the second term is a small perturbing acceleration due to the presence of the other planets (with masses m_j of the order of 10^{-3} to 10^{-7} of the Sun's dominating mass). The appearance of the 3rd power of the distances $\mid \boldsymbol{q}_\mathrm{j} - \boldsymbol{q}_\mathrm{i} \mid = r_\mathrm{ji}$ in the denominator may cause large accelerations, even though the masses of the planets are small, when these distances become small themselves. In planetary theories this is not a practical problem because planets move on well separated orbits. It is different

[1] which are in fact taken into account for computing short time ephemerides of the planets

for comets, which often come quite close to planets (especially Jupiter) which can change their orbits significantly. Consequently comets having initially parabolic orbits may end up on elliptic orbits with moderate eccentricities. Besides the step-by-step numerical integration method (e.g. with a classical Runge-Kutta or a Lie series method [8]) a special method of solving the equations of motion has been developed in the last centuries by mathematicians and astronomers (Lagrange, Laplace, Brown, etc.). In perturbation theory one works with complicated series expansions including perhaps thousands of terms. The solutions are such that substituting the time in the series (also in form of a Fourier series) immediately yields the position in space (and in the sky). This method will be discussed briefly, as it gives deep insights into the nature of the dynamics of the planetary system and reveals the fundamental rôle of resonances.

2.1 Outline of the Classical Perturbation Method

For an approximate solution of the orbit of a planet the theory of two-body-motion can be applied which leads to six constants known as the orbital elements of a planet $\boldsymbol{\sigma} = (a, e, i, \omega, \Omega, T)^{\mathrm{T}}$. Under the attraction of the other planets these elements change slowly in time; because the other masses are orders of magnitudes smaller than the Sun these changes are small. One can thus describe the motion of the planet i by a 1st order differential equation of the form

$$\frac{d\boldsymbol{\sigma}_{\mathrm{i}}}{dt} = \sum_{j=1, j\neq i}^{n} F_{\mathrm{ij}}(\boldsymbol{\sigma}_{\mathrm{i}}, \boldsymbol{\sigma}_{\mathrm{j}}, \frac{\partial R_{\mathrm{ij}}}{\partial \boldsymbol{\sigma}_{\mathrm{i}}}) \tag{2}$$

where R_{ij} is the so-called perturbing function, which can be written as a Fourier-series with respect to the time. For a single planet of mass m_1 perturbed by another planet with mass m_2 the perturbing function is the following

$$R_{12} = m_2 \sum_{j_1=-\infty}^{\infty} \sum_{j_2=-\infty}^{\infty} C_{\mathrm{j_1,j_2}} \cos[(j_1 n_1 + j_2 n_2)t + D_{\mathrm{j_1,j_2}}]. \tag{3}$$

The perturbations of the other planets may just be added, leading to a perturbing function $R_1 = R_{12} + R_{13} + \ldots + R_{1\mathrm{n}}$ depending on the number of planets taken into account; this simplification holds only in a first order theory. In a rather comprehensive formulation the Delaunay elements can be used

$$\begin{aligned} L_1 &= \kappa_1 \sqrt{a} & l_1 &= M_1 = n_1 t \\ G_1 &= L_1 \sqrt{(1-e^2)} & g_1 &= \omega_1 \\ H_1 &= G_1 \cos i & h_1 &= \Omega_1 \end{aligned} \tag{4}$$

where $\kappa_1 = k\sqrt{m_1 + M}$, M is the mass of the Sun and n_1 stands for the mean motion [2]. Each canonical pair obeys the canonical equations

$$\frac{d\boldsymbol{\Gamma}_1}{dt} = \frac{\partial R_{12}}{\partial \gamma_1}, \qquad \frac{d\gamma_1}{dt} = -\frac{\partial R_{12}}{\partial \boldsymbol{\Gamma}_1} \tag{5}$$

[2] which relates to the semimajor axes a_1 via the 3rd Kepler law $n^2 a^3 = \kappa^2$

where $\boldsymbol{\Gamma}_1 = (L_1, H_1, G_1)^{\mathrm{T}}$ has the conjugate vector $\boldsymbol{\gamma}_1 = (l_1, h_1, g_1)^{\mathrm{T}}$. As an example we now show how the perturbations act on the Delaunay element H_1. To derive the respective perturbations of the first order we need to compute $\delta H_1 = \int \frac{\partial R_{12}}{\partial h_1} dt$. Inserting the perturbing function (3) one can construct the partial derivative with respect to the conjugate variable

$$\delta H_1 = m_2 \int \sum_{j_1=-\infty}^{\infty} \sum_{j_2=-\infty}^{\infty} \frac{\partial C_{j_1,j_2}}{\partial h_1} \cos[(j_1 n_1 + j_2 n_2)t + D_{j_1,j_2}] dt, \tag{6}$$

which integrates to

$$\delta H_1 = m_2 \sum_{j_1=-\infty}^{\infty} \sum_{j_2=-\infty}^{\infty} \frac{\partial C_{j_1,j_2}}{\partial h_1} \frac{\sin[(j_1 n_1 + j_2 n_2)t + D_{j_1,j_2}]}{(j_1 n_1 + j_2 n_2)} \tag{7}$$

and hence so-called *small divisors* appear. Note that the C_{j_1,j_2} depend only on the action variables $\boldsymbol{\Gamma}$ and the D_{j_1,j_2} only on the angle variables $\boldsymbol{\gamma}$. Whenever the two mean motions involved, n_1 and n_2, are in resonance, the small divisor – being a number close to zero – makes the perturbation for that element very large. A good example is the *great inequality* between Jupiter and Saturn. The two giant planets are in the *mean motion resonance* 5:2 which means that $2n_{\text{jup}} - 5n_{\text{sat}} \sim 0$ for the summation indices $j_1 = 2$ and $j_2 = 5$. Inserting the values of the mean motions for Jupiter and Saturn one gets in degrees per day $2 \cdot 0.^{\circ}08309 - 5 \cdot 0.^{\circ}03346 = -0.^{\circ}00112$ and thus for the great inequality a period of ~ 880 years. A more detailed computation shows that the amplitude of this perturbation in longitude reaches values up to almost $0.^{\circ}5$ for Jupiter due to Saturn (and vice versa for Saturn due to the perturbation of Jupiter almost 1°)[3].

2.2 High Order Resonances, Secular Resonances and the Fundamental Frequencies

In the summation of the Fourier expansion of the perturbing function only a finite number of terms for j_1 and j_2 are to be taken into account, because beyond these terms the amplitudes are rather small even for "small divisors". The main point is, that it is known since Poincaré [23], that the series expansion of the perturbation function is not convergent for high orders and has to be truncated at some point. Nevertheless high order resonances (like $j_1 = 29$ and $j_2 = -72$ for Jupiter and Saturn) or secular resonances (see below) may appear inside the main resonance and occasionally produce chaotic motion, due to overlapping separatrix layers for larger eccentricities. This is less important for planetary motion than for the motion of the asteroids. Of special interest for the stability of the planetary system are long-period perturbations, which can be computed by averaging over the fast variable $l = M = nt$. The system described in (5) then decouples into two sets of equations, where one involves only the eccentricity and

[3] The necessary double integration for the element l_1 leads to a divisor of the form $(j_1 n_1 + j_2 n_2)^2$ and makes the perturbation especially large.

the perihelion, while the other involves only inclinations and nodes (there are no equations for L and l). The solutions of the so–called Lagrange system describe the motion of the nodes and the perihelia of the orbits. The determination of the secular frequencies is of great interest in connection with the chaotic behaviour of the planetary system (see Sect. 4.2).

When one takes into account that the nodes and the perihelia of the orbits change slowly with time ($\omega = \omega_o + \omega_1 t$ and $\Omega = \Omega_o + \Omega_1 t$) the aforementioned "phase coefficient" in (3) becomes also time dependant. Therefore, in the integration of the secular evolution another small divisor, (e.g. $j_1\omega_1 + j_2\omega_2 \approx 0$) may appear which leads again to strong perturbations due to a *secular resonance.*

In the so–called *Kozai resonance* [12] the eccentricities and the inclinations of small bodies perturbed by the planets are coupled in the sense that the eccentricity of the orbit has a maximum when the inclination has a minimum and vice-versa. This can be explained by the fact that for a constant semimajor axis a the Delaunay element $H = \kappa\sqrt{a(1-e^2)}\cos i$ is a constant of motion (e.g. [20]).

The recently discovered *three-body mean motion resonances* take into account in the series expansion (3) the mean motions of the perturbed body and two perturbing planets ($j_1 n_1 + j_2 n_2 + j_3 n_3 \approx 0$), and seem to act especially in the motion of asteroids [20]. Without going into the details we point out that overlap of these resonances can lead to chaotic motion (for an extensive description see [4]).

3 Dynamics of Comets and Asteroids

Our solar system is populated by a large number of small bodies orbiting the Sun in more or less eccentric orbits. The distinction between the two physically different populations – the comets and the asteroids – is already evident because of the different apparition for the observer on Earth. We can divide the comets, according to their orbits, into short–period comets with orbital periods smaller than 200 years and the long–period ones. The latter very probably come from the Oort Cloud, where on the order of 10^8 comets may be present with aphelion distances up to 0.4 pc. Their hyperbolic orbits, which are not confined to small inclinations, led them to the inner part of the solar system due to perturbations from passing stars, interstellar clouds and galactic tides. Within the group of short–period comets[4] we can distinguish the Jupiter family with periods less than that of Jupiter and the ones with orbits similar to that of comet Halley [5]. The orbital inclinations of comets of the Jupiter family are small and most of them are supposed to come from the second reservoir for comets, the *Edgeworth-Kuiper* belt outside Neptune's orbit. The orbits of short-period comets are often deformed due to close encounters with planets, particularly with Jupiter.

---

[4] we now know about 150 of this group

[5] with the orbital elements $a = 17.94\ AU, e = 0.967, i = 162.^o2, \omega = 112^o, \Omega = 58.^o1$ and $T = 1986\ 02\ 19.0$

These close encounters can be modeled via scattering and are the source of unpredictability of the orbit for longer time intervals. Using numerical integrations of real and fictitious objects [5] and simplified mapping methods [3] the statistical properties for these encounters were determined and the respective Poincaré surfaces of sections unveiled the chaotic structure of these orbits. In Fig. 1 we see the orbits of comet Halley for two slightly different initial conditions, which lead, after several close approaches to Jupiter, to completely different dynamical behavior. Whereas the upper orbit leads to hyperbolic escape after 2.10^5 years, the lower orbit shows the typical character of jumping from one mean motion resonance with Jupiter to another. In the lowest graph we can see that the comet's orbit is in the Kozai resonance, where the inclination and the eccentricity move oppositely (well visible especially between 170 and 190 kyrs).

We can also estimate dynamical lifetimes of these objects [19], which are on the order of 10^6 years. This lifetime is determined by: (i) collision with a planet (e.g. crash of SL9 on Jupiter in 1994), (ii) breakup in a sun-grazing encounter or (iii) escape from the solar system after a close encounter with a giant planet.

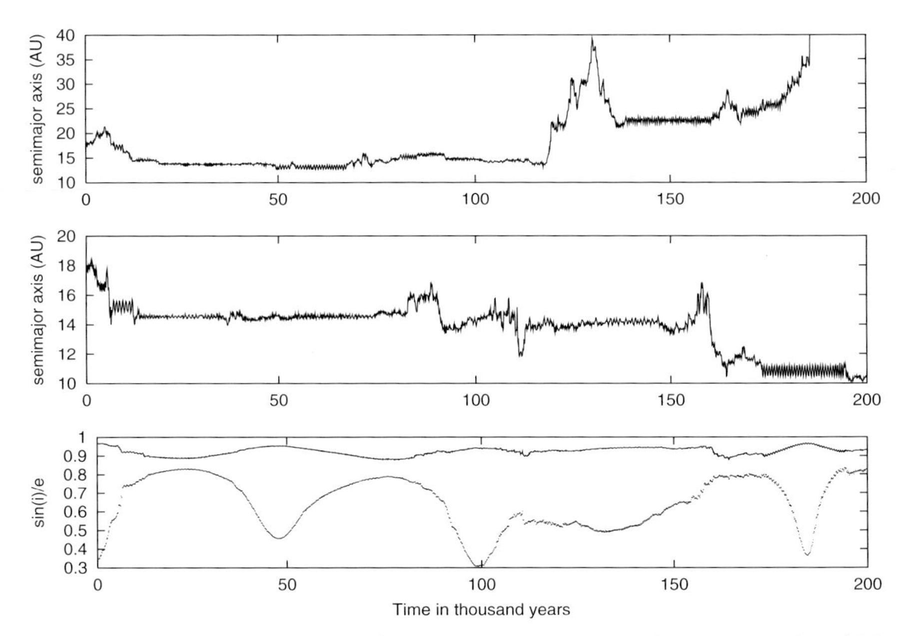

Fig. 1. The time evolution for two fictitious objects in Halley like orbits with a $\Delta M = 0.^{\circ}01$. The upper graphs plot the semimajor axis versus the time, the lower graph shows the eccentricity (upper curve) and the sin(i) (lower curve) to illustrate that the 2nd object is in the Kozai resonance

For the asteroids we can distinguish 4 different groups[6] (i) the Edgeworth-Kuiper-Objects (KBO, moving outside Neptune) (ii) the cloud of Jupiter Trojans (moving close to the Lagrange equilibrium points L_4 and L_5), (iii) the main belt asteroids (with semimajor axes between Mars and Jupiter) and (iv) the Near Earth Asteroids (NEAs), with orbits which bring them close to the Earth. Only the Trojans form a well defined group, the other ones are not so well separated, respectively they change their membership to a specific group.

The members of the NEAs are usually divided into three subgroups: the ATENS, with a semimajor axis smaller than that of the Earth and an aphelion distance $Q = a(1 + e) > 0.983$ AU (mean perihelion distance of Earth), the APOLLOS, with a semimajor axis larger than that of the Earth and a perihelion distance $q = a(1 - e) < 1.017$ AU (mean aphelion distance of Earth) and the AMORS, with a semimajor axis larger than that of the Earth and a perihelion distance $1.017 < q < 1.3$ AU (they do not cross Earth's orbit, but they stay inside the orbit of Mars)

The number of known asteroids is growing rapidly because during the observing programs of the NEAs more and more small and also large objects[7] are discovered. The actual numbers of asteroids in addition to the ≈ 20000 main belt asteroids – are Atens (172), Apollos (1043), Amors (1013), Centaurs (125), Jupiter Trojans around the preceding Lagrange point L_4 (962), L_5 (602) and KBOs (664).

In Fig. 2 one can see the sculpting of the main belt inside Jupiter's orbit which is mainly due to mean-motion resonances, as well as three-body resonances and secular resonances (inner main belt) and also by the chaotic behavior of separatrix crossing (e.g. 3:1 mean motion resonance). We can order the importance of the acting resonances [22] as follows: (1) three–body resonances with Jupiter and Saturn, (2) resonances with Mars, (3) mean motion resonances with Jupiter, and (4) three–body resonances involving both Mars and Jupiter. In the inner part of the main belt Mars is – besides the secular resonances with Jupiter and Saturn [21] – an important perturbing planet; the relatively large eccentricity of its orbit compensates for its small mass. The middle part $2.5 \leq a \leq 3$ AU is the most stable one. In the outer belt Jupiter is acting with its resonances and sub–resonances.

An important step further in the knowledge of solar system dynamics occurred, when Wisdom [25] discovered that the depletion of the main belt in the 3:1 mean motion resonance with Jupiter is caused by the chaotic behaviour of asteroids located there. The asteroid suffers from a sudden increase in the eccentricity which may bring it into a resonance overlap area, which in turn leads to an additional increase in eccentricity and finally to a close approach with Mars. This process of separatrix crossing is a basically chaotic behavior known

[6] we exclude the Centaurs as a group which are asteroids outside Jupiter but inside the 3:2 resonance with Neptune

[7] Quaoar, a recently discovered KBO, has a semimajor axis a=43.4 AU, an eccentricity of e=0.03 and a diameter of 1250km (which is half the size of Pluto).

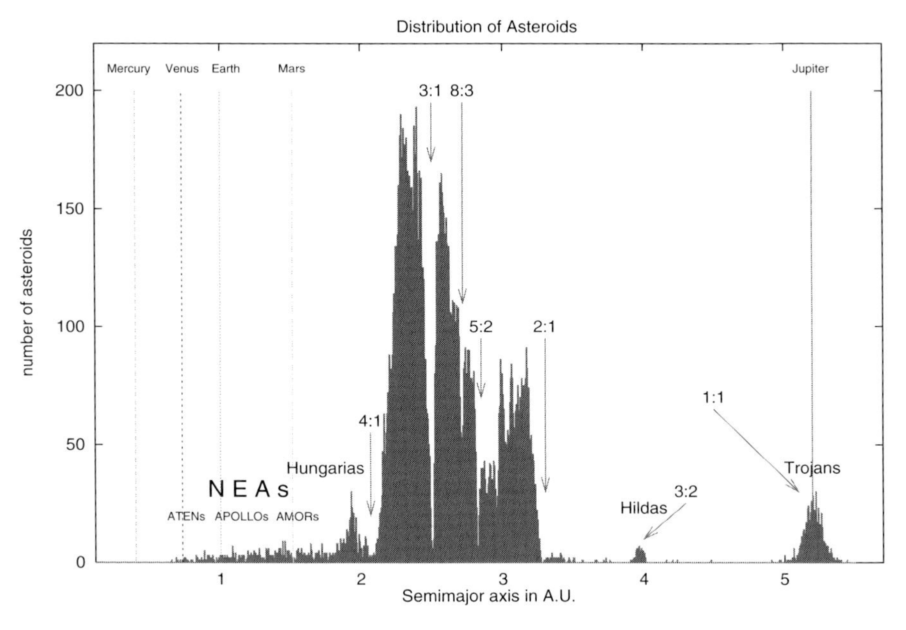

Fig. 2. Distribution of asteroids inside Jupiter's orbit

from the perturbed pendulum, where the qualitative change from libration to circulation is due to the sensitivity on the initial conditions.

The riddle of the different dynamical behaviour of the 2:1 resonance, which is almost depleted from planetoids, and the 3:2 Hilda family of asteroids has been solved recently by the work of different authors (e.g. Ferraz-Mello and Nesvorny [6]) and is primarily due to an early depletion of asteroids in the 2:1 resonance because of the different location of Jupiter and Saturn in the early days of the solar system, where they were somewhat closer. A detailed description of the structure of these two resonances can be found in the previously mentioned book by Morbidelli [20] (p. 286–294).

In Fig. 3 one can see how a NEA jumps from one mean motion resonance with the Earth to another (upper panel), which is caused by more or less close encounters with this planet (lower panel). The detailed transport of asteroids from the Kuiper belt to the Centaurs and from there to the inner regions of the solar system as well as the transport from the main belt to Earth–crossing orbits, is governed mainly by secular resonances [24].

4 The Planets

The problem of the stability of our planetary system was first studied by Laplace in the 18th century, who found that the semimajor axes of the planets suffer only periodic changes up to the first order; 50 years later Poisson extended this theorem up to second order. As already mentioned Poincaré [23] discovered

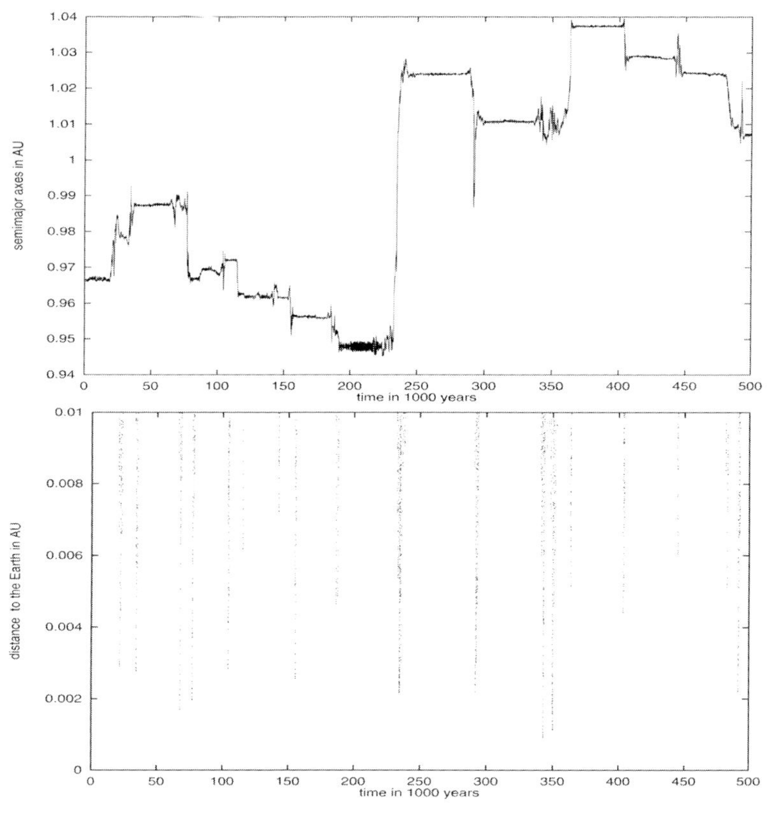

Fig. 3. The jumping orbit of 2262 Aten

that the usual perturbation techniques used to represent the solution in form of power series in small parameters such as the eccentricities, inclinations and/or the planetary masses are not convergent, because of small divisors. Using the method of averaging over the mean motions of the planets (and thus not being able to determine the longitude of the planet) one can solve the secular part of the perturbing function. To lower orders one can find good approximations to the solutions because the system decouples into an inclination–node independent and an eccentricity–perihelion independent system. For a qualitatively good solution over millions of years one can introduce high order terms which result in a shifting of the proper mode frequencies and combinations of them [1]. Using numerical techniques we can nowadays – with the aid of very fast computers – integrate the full equations of motion up for times scales comparable to the age of the solar system [10].

4.1 Numerical Solutions

We have integrated the full equations of motion for a time interval of 200 million years (±100 million years) for a dynamical model consisting of the Sun and 8

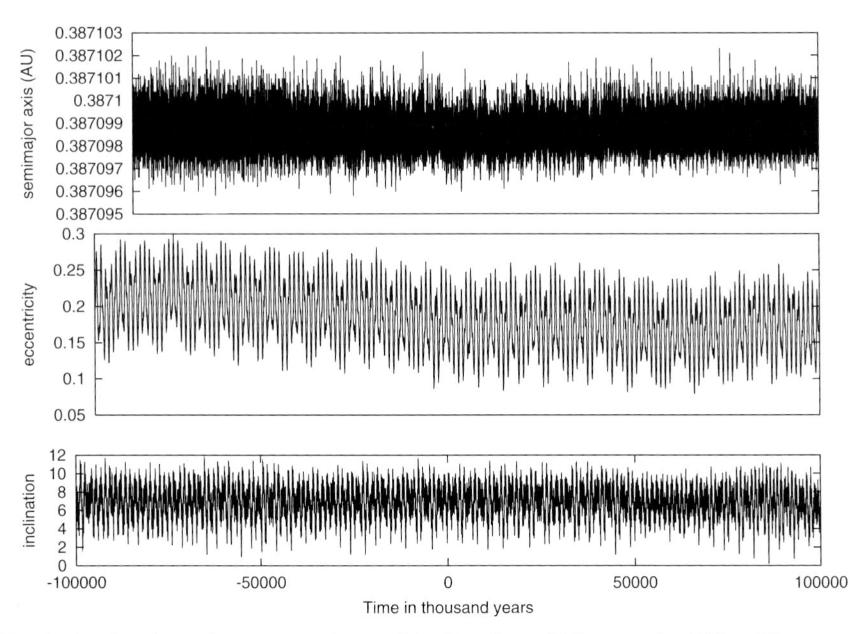

Fig. 4. Semimajor axis, eccentricity and inclination of Mercury for 200 million years

planets (treated as point masses, Earth + Moon as one body, relativistic effects added afterwards) with the Lie series method ([18], [8]). In Fig. 4 one can see the relative large variations in the eccentricity and inclination for Mercury, while the semimajor axis is very constant even for such a long integration. These results illustrate the precision of the method because the first sign of the lack of precision of an integration is a secular drift of the semimajor axes of the innermost planet in a simulation of the motion of the planetary system [7].

The coupling between Earth and Venus, which have almost the same mass, is visible in Fig. 5. We show – besides the time interval -2.5 to +2.5 million years around today – the time evolution of their eccentricities for 5 million years in the past and in the future. It is remarkable that no differences at all can be seen there!

For the interval of 200 million years the minimum and maximum values of the semi-major axes, the eccentricities and the inclinations of the planets are shown in Table 1.

4.2 Determination of the Fundamental Frequencies

The analysis of the data to determine the so-called fundamental frequencies (the motion of the nodes and of the longitudes of the perihelia) was done using a program provided by [2] which uses a Chebycheff approximation of the data, determines the largest amplitude, subtracts it and repeats the analysis (etc.)

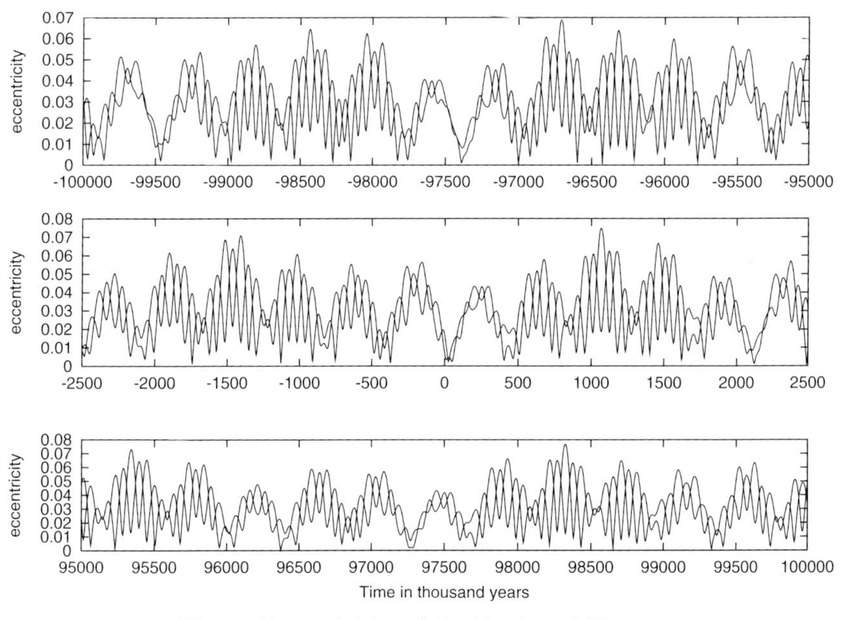

Fig. 5. Eccentricities of the Earth and Venus

Table 1. Extreme values of the action variables for 2.10^8 years

planet	a_{min}	a_{max}	e_{min}	e_{max}	i_{min}	i_{max}
Mercury	0.38710	0.38710	0.07874	0.29988	0.17600	11.72747
Venus	0.72332	0.72336	0.00002	0.07709	0.00076	4.91515
Earth	0.99997	1.00004	0.00002	0.06753	0.00075	4.49496
Mars	1.52354	1.52386	0.00008	0.13110	0.00291	8.60320
Jupiter	5.20122	5.20504	0.02513	0.06191	1.09172	2.06598
Saturn	9.51281	9.59281	0.00742	0.08959	0.55867	2.60187
Uranus	19.09807	19.33511	0.00008	0.07835	0.42170	2.73888
Neptune	29.91013	30.32452	0.00001	0.02317	0.77977	2.38597

This new determination ([7]) is compared to the values published by previous authors in Table 2. One can see that these frequencies are quite close to the already published ones using other methods.

Laskar [13] used a different approach to model the long term evolution of the planetary orbits: he integrated numerically the secular system, (truncated up to 2nd order in the masses and to 5th order with respect to the eccentricities and the inclinations). The surprising result was the discovery that a secular resonant term, namely $\theta = 2(\dot{\tilde{\varpi}}_{M} - \dot{\tilde{\varpi}}_{E}) - (\dot{\Omega}_{M} - \dot{\Omega}_{E})$, where E stands for the Earth and M stands for Mars, is alternatively librating and circulating [15]. This clear

Table 2. Fundamental frequencies for the motion of the perihelion in $''$ per year after the Lagrange-Laplace theory (LLT), the theory of Bretagnon (B84) [1], the theory of Laskar (NGT) [14] and a new determination (NEW)

	LLT	B84	NGT	NEW
g_1	5.4633	5.6136	5.5689	5.6276
g_2	7.3477	7.4559	7.4555	7.4441
g_3	17.3283	17.2852	17.3769	17.5668
g_4	18.0023	17.9025	17.9217	17.9373
g_5	4.2959	4.3080	4.2489	4.2567
g_6	27.7741	28.1483	27.9606	28.2445
g_7	2.7193	3.1534	3.0695	3.0468
g_8	0.6333	0.6735	0.6669	0.6727

indication of chaos does not a priori mean that the planetary system will be unstable. Nevertheless these important results show that the orbital elements of the planetary system – especially the terrestrial planets – lie in a thin chaotic layer in phase space.

In a further step Laskar [14] numerically integrated the averaged equations of motion over even longer time scales (several 10^9 years with a time step of 250 years for this integration). The most interesting results are shown in Fig. 6 where the development of the eccentricities of the innermost three planets, respectively the maximum and minimum value over a time span of some billion years is plotted. The two lines on the bottom with quite a similar behaviour belong to Venus (respectively to the Earth) and show the dynamical coupling of these two planets. For Mercury we see a line of minimum values and a line of maximum

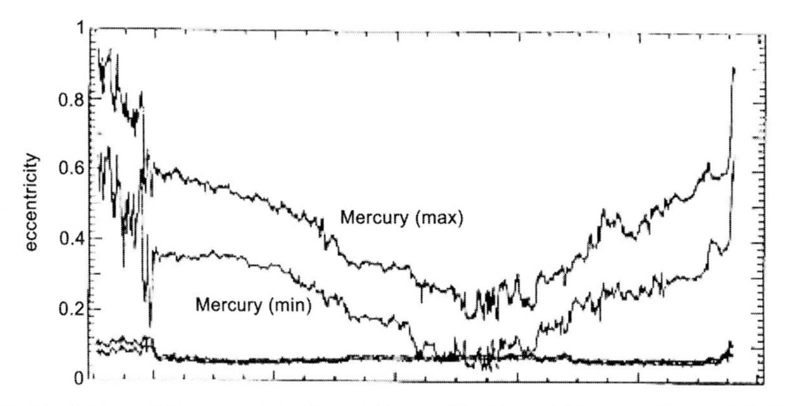

Fig. 6. Evolution of the eccentricities of Venus, Earth and Mercury from -6.6 Gyrs to 3.5 Gyrs (after [15])

values which are parallel (separated by $\Delta e \sim 0.15$) over the whole time span of 10^{10} years. The most interesting fact is that Mercury's orbit seems to be able to achieve values of e close to unity! Laskar concludes that this could lead (in the future, because the system is symmetric with respect to the time) to a close encounter with Venus. The most important point of criticism is that the development of the perturbation function is not convergent for large inclinations and eccentricities! Thus the solution which he computes does not describe the dynamical development of the planetary system. Nevertheless Laskar argues that the secular system, which keeps the semimajor axes of the planets constant, is even less chaotic than the real one and additional degrees of freedom" will probably lead to even stronger chaotic behaviour, as in general, addition of degrees of freedom increase the stochasticity of the motion" ([15]).

Because of the shortcomings in the equations of motions these computations cannot give a conclusive answer to the question of the stability of the planetary system, but there is no doubt that the motion of the planets is NOT regular (and not quasiperiodic). This fact is mainly due to Mercury and Mars in the inner solar system; Earth and Venus are strongly coupled in their dynamical behaviour and the outer planets show regular behaviour in their motions (with the exception of Pluto, which is a Kuiper belt object moving in a 2:3 resonance with Neptune).

The results discussed in this chapter raise several questions:

• is the sudden increase of Mercury's orbital eccentricity (in short time scales of $10^5 - 10^6$ years, compare Fig. 6) due to the shortcomings of the equations of motions, or is it a real phenomenon?

• what is Mercury's action on the other planets when its orbit is highly eccentric?

• what is the possible outcome of a close encounter between Mercury and Venus?

To clarify this question we have undertaken numerical integrations of the full equations of motion in different models: only the inner planets, the inner planets plus Jupiter, the inner planets plus Jupiter and Saturn and the complete planetary system.

4.3 Mercury as Perturber of the Inner System

Figure 7 shows the evolution of the eccentricity of Mercury for 1 million years, starting with a large eccentricity $e_{\rm ini} = 0.95$ in 2 different models, namely with and without Jupiter: in the first model after only 10^5 years the eccentricity can drop from 0.95 to 0.6 and then rise again (upper panel)! Thus the result of Laskar is confirmed that a "sudden" increase of Mercury's eccentricity is a possible scenario. On the lower panel of Fig. 7 we depict the behaviour in a model with Jupiter, and here one can see that, although a close encounter occurred, the eccentricity of Mercury did not drop below $e = 0.75$. Other test calculations confirm the result, that the outer planetary system stabilizes the inner one (a longer paper with computations for different dynamical models is in preparation). In all our computations the consequences of close encounters

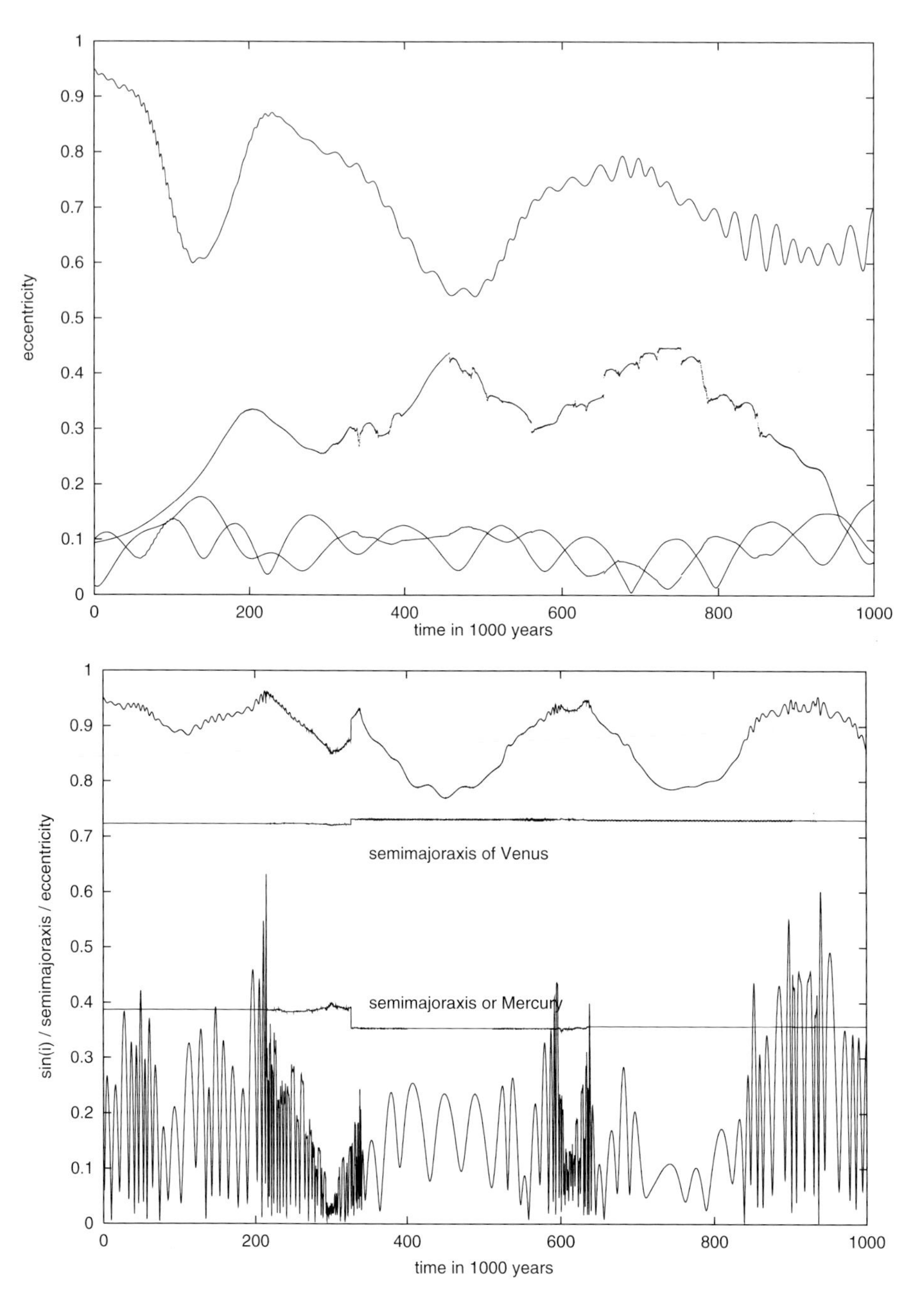

Fig. 7. Eccentricities of the inner planets for 1 million years in a model without the outer planets; the two bottom lines are for Venus and Earth, the upper one for Mercury and the one in the middle for Mars (upper panel). Eccentricity, semimajor axis, and $\sin i$ of Mercury and semimajor axis of Venus in a model with Jupiter for 1 million years (lower panel)

between Venus and Mercury never resulted in an escape, but only in a small change in the semimajor axis. We therefore conclude that escape can only happen after several cascade–like encounters, and then it may be that Jupiter could throw Mercury far out into the Kuiper belt! Regarding the action of the unstable orbit of Mercury on the other planets, it turned out that the influence is small, owing to the small mass of this planet. Only an unlikely close encounter of Mercury (or quasi–collision, which in fact we observed in one of our experiments) could lead to larger changes of the orbital elements.

4.4 The Obliquity of Planets

The results of an investigation of the obliquities [16] of the planets were quite surprising: with one exception all inner planets are – in what concerns the obliquity of their rotation axes – in a mode of chaos. That means that within relatively short time scales of millions of years their axes can vary substantially in space. For Mars the amplitude of this chaotic development of the obliquity is 15° within some hundred thousand years and in 50 million years it may gradually evolve from almost 60° to 15°. At the same time the precession rates also vary by a factor of two. Although Venus now possesses an orientation of the rotational axis very close to that of our Earth, the obliquity could reach a value close to 90° and therefore it would almost "roll" in its orbital plane (like Uranus is actually doing it). For the Earth the situation is quite different: in an 18 Myrs integrations of the precession equations it was found that the precession rate can change to a negative precession of $p = -39''/\mathrm{yr}$. The actual values for the Earth of $\epsilon_o = 23^o$ and $p = 50''/yr$ lie well inside a quiet region with only very small changes for the maximum and minimum values. For $55^o < \epsilon < 90^o$ there exist a large chaotic region allowing variations in the order of 35°. The same integrations were undertaken without the presence of the Moon and the obliquity turned out to lie in a region of large chaos (such as Mars is actually in). Thus it seems that only the presence of the Moon stabilizes the obliquity so that only small changes in ϵ occur and occurred in the last millions of years!

5 Conclusions

Let us summarize the most interesting manifestation of chaos in our planetary system:

• all small bodies in our solar system – asteroids and comets – suffer from chaotic motion either directly due to close encounters or after long term evolutions

• there is a thin layer of chaos in phase space, where the inner planets move

• on a long time scale diffusion may bring the planets into a state of larger chaos which may allow for large eccentricities of Mercury. The scenario which follows then is still not known and seems to be more speculation than outcome of scientific investigations; at any rate an immediate escape of a planet does not seem possible. A snowball like effect is more probable, where Mercury changes

Venus' orbit, Venus comes close to the Earth and changes its orbit, the Earth itself is then perturbing Mars, which now may achieve larger eccentricities ...

• Jupiter and Saturn keep the inner planets in stable orbits for quite long times

• large changes in the orientation of the obliquity of the planets are possible due to chaotic motion of the inner planets. On the contrary the Earth's spin axis is stabilized by the Moon which turns out to be in a nonchaotic region of configuration space.

6 Epilog

The knowledge of more than 100 planets orbiting other stars has opened new interest for the dynamics of our own planetary system, where it is difficult to detect the weak chaoticity the planetary orbits are in. This is different for extrasolar planets: in a recent work it was found that the planets in HD12661 are apparently in a chaotic region of phase space [11]. In Fig. 8a we show the time evolution of the semimajor axis of the inner planet, where it jumps from one high order mean motion resonance to another close by. In numerical experiments of the dynamical evolution of our planetary system with fictitious larger masses of the terrestrial planets a similar behaviour was found (Fig. 8b): taking for Mars the mass of the Earth leads to a chaotic behaviour of the semimajor axis of the

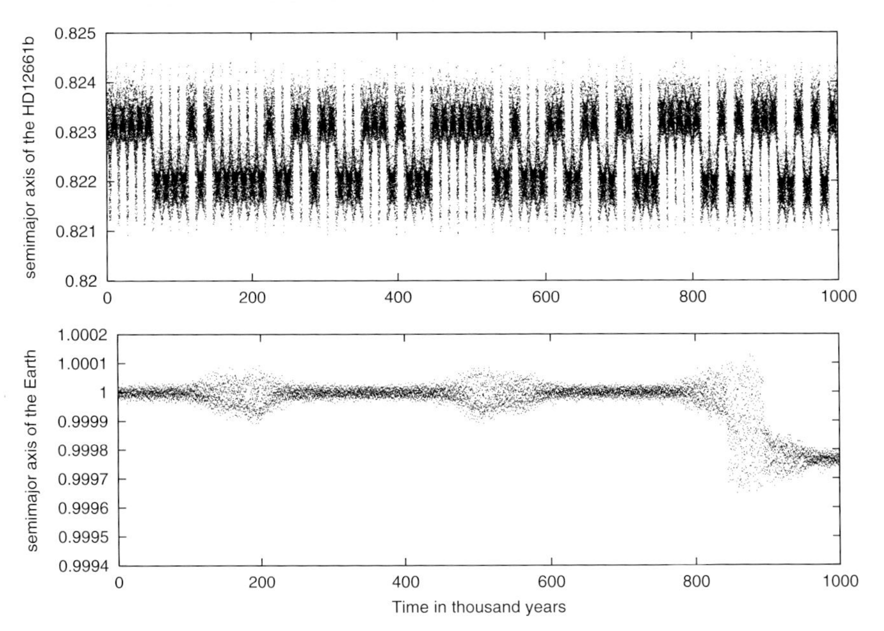

Fig. 8. Time evolution for 1 million years of the semimajor axis of the inner planet of HD12661 (upper panel) and of the Earth (in a model with a fictitious mars with 10 time its actual mass) (lower panel)

Earth (the masses of the other planets were unchanged). Because we can expect many more discoveries of planets even with masses comparable to our terrestrial planets in the future, the knowledge of our own system is a basis for the future research in the dynamics of other planetary systems.

Acknowledgements

For improving the paper I have to thank primarily J. Howard (Boulder). F. Freistetter and the graduate students of the Astro Dynamics Group in Vienna helped to finalize the manuscript. This work was partly carried out within the framework of the FWF project P14375-TPH.

References

1. P. Bretagnon, P.: Cel. Mec. Dyn. Astron. **34**, 193 (1982)
2. J. Chapront: Astron. Astrophys. **109**, 181 (1995)
3. R.V. Chirikov, V.V. Vecheslavov: Astron. Astrophys. **221**, 146 (1989)
4. G. Contopoulos: *Order and Chaos in Dynamical Astronomy* (Springer–Verlag, Berlin Heidelberg New York 2002)
5. R. Dvorak, J. Kribbel: Astron. Astrophys. **227**, 264 (1990)
6. S. Ferraz-Mello, D. Nesvorny: Astron. Astrophys. **320**, 672 (1997)
7. C. Gamsjäger: *Eine Neubestimmung der Basisfrequenzen in den Planetenbewegungen*, Master Thesis, University of Graz (2002)
8. A. Hanslmeier, R. Dvorak: Astron. Astrophys. **132**, 203 (1984)
9. M. Hénon, C. Heiles: AJ **69**, 73 (1964)
10. T. Ito, K. Tanikawa: MNRAS **336**, 483 (2002)
11. L. Kiseleva-Eggleton, E.Bois, N. Rambaux, R. Dvorak: ApJ **578L** 145 (2002)
12. Y. Kozai: AJ **67** 591 (1962)
13. J. Laskar: Icarus **88** 266 (1990)
14. J. Laskar: Astron. Astrophys. **287** L9 (1994)
15. J. Laskar: Cel. Mec. Dyn. Astron. **64**, 115 (1996)
16. J. Laskar, P. Robutel: Nature **361**, 608 (1993)
17. M. Lecar, F.A. Franklin, M.J. Holman, N.J. Murray: Ann.Rev. of Astron. and Astrophys. **39**, 581 (2001)
18. H. Lichtenegger: Cel. Mech. **34**, 357 (1984)
19. E. Lohinger, R. Dvorak, C. Froeschlé: Earth, Moon and Planets **71** 225 (1995)
20. A. Morbidelli: *Modern Celestial Mechanics : Aspects of Solar System Dynamics* (Taylor & Francis, London 2002)
21. P. Müller, R. Dvorak: Astron.Astrophys. **300** 289 (1995)
22. D. Nesvorný, A. Mordbidelli: AJ **116** 3029 (1998)
23. H. Poincaré: *Les Méthodes Nouvelles de la Mécanique Céleste*, tome I–III, (Gauthier–Villard et fils, Paris 1892)
24. P. Robutel, J. Laskar: Icarus **152** 4 (2001)
25. J. Wisdom: Icarus **56**, 51 (1983)

Dynamics of Extrasolar Planetary Systems: 2/1 Resonant Motion

John D. Hadjidemetriou and Dionyssia Psychoyos

University of Thessaloniki, Department of Physics, 541 24 Thessaloniki, Greece

Abstract. A systematic study of the dynamics of resonant planetary systems is made, based on the existence and stability character of families of periodic orbits of the planetary type. In the present study we consider planetary systems with two planets, moving in the same plane. We explore the whole phase space close to the 2/1 resonance, for the masses of the observed planetary system HD82943. We find four basic resonant families of periodic orbits at the 2/1 resonance, and show that large regions on the families correspond to stable motion, even for large values of the eccentricities of the two planets and for intersecting planetary orbits. The initial phase of the two planets plays a crucial role on the stability of the system. It is close to a periodic orbit that stable motion of a planetary system can exist. So, the study of the families of periodic orbits provides a systematic way to find all the regions of phase space where a resonant planetary system could exist in nature. Planetary systems with large eccentricities can exist in nature only if they are close to a resonance. Indeed, we show that the real planetary system HD82943 is close to a stable periodic orbit. The alignment of the line of apsides of the planetary orbits plays also a stabilizing role.

1 Introduction

The study of extrasolar planetary systems is an important new field of research in dynamical astronomy, following the discovery during the last decade of planetary systems around distant stars. A complete catalogue of extrasolar planetary systems can be found in the web site *http://www.obspm.fr/encycl/catalog.html*, maintained by Jean Schneider. There are 91 confirmed extrasolar planetary systems, with ten of them having two planets and two having three planets. In some cases, the eccentricities of the two planets are quite large, and their masses are comparable to the mass of Jupiter. In all these cases, the two planetary orbits are in a mean motion resonance.

There are several problems associated with the extrasolar planetary systems. First are the cosmogonic problems: how were these systems formed? Another problem is the stability of such systems. What makes them stay there, for millions or billions of years? These systems are very different from our own solar system, because the planetary eccentricities are large, but they do exist in nature. What mechanism keeps them stable? A third question refers to the possibility of existence of life is an extrasolar planetary system. This implies the existence of Earth-like planets, i.e. planets of the size of the Earth, in nearly circular orbits at a distance of about 1AU from the Sun. The observational techniques at the moment are not so accurate and consequently such planets cannot be observed.

Only larger planets are detected at present. But knowing the position of the larger planets, we can study the stability of a fictitious small planet at about 1AU, and thus find if such a planet could possibly exist. A study on this problem is in Celletti et al. (2002), Jones and Sleep (2002).

In this paper we address the problem of the stability of an extrasolar planetary system. There are several studies on the stability of the observed extrasolar planetary systems, based on numerical integrations, starting with initial conditions close to those corresponding to the observed values: (Beauge et al. 2003, Callegari et al. 2002, Ferraz-Mello 2002, Ford et al. 2001, Gozdziewski et. al. 2002, Kinoshita and Nakai 2001a,b, Kiseleva et.al. 2002, Laughlin and Chambers 2001, Lissauer and Rivera 2001, Malhotra 2002a,b, Murray et al, 2001, Peale and Lee 2002, Rivera and Lissauer 2001). In the present study we propose a new approach to detect stable motion. A systematic method is presented, based on periodic orbits, to find all the regions of the phase space where stable motion exists and consequently a real planetary system could be found in nature. A study on the dynamics, based on families of periodic orbits, is made in Hadjidemetriou (2002), for the 2/1 and 3/2 resonant systems.

We consider an extrasolar planetary system with two planets, moving in the same plane, taking into account the gravitational attraction between the two planets. This is a special case of the general planar three body problem. The study is based on the existence of periodic orbits in a rotating, synodic, frame, because it is the periodic orbits and their stability character, that determine critically the structure of the phase space. So, the knowledge of all the basic families of periodic orbits in a dynamical system will provide useful information on the evolution of the system. In particular, the position of the stable periodic orbits in phase space gives a systematic way to find those regions where stable motion can exist. This is so, because it is only close to a stable periodic motion where a real planetary system can exist. In this way, we are able to find all the resonant planetary systems that could exist in nature.

As we shall explain in the next section, there are two types of periodic orbits of the planetary type (in the rotating frame): Periodic orbits where the two planets move in circular orbits and periodic orbits where the two planets move in elliptic orbits (slightly distorted due to the gravitational interaction between the two planets). These latter periodic orbits are necessarily resonant, i.e. the ratio of the periods of revolution of the two planets is a rational number. The eccentricities of the two planets vary along such a resonant family of periodic orbits, starting from zero values and reaching high values, while the semimajor axes remain almost constant.

An extrasolar planetary system, even with large masses and eccentricities, can be stable, provided that it is close to a stable resonant periodic orbit. Indeed, all the extrasolar planetary systems with large eccentricities that have been observed, are close to a resonance. In particular, we study the 2/1 resonant case, and we find that stable families of resonant periodic orbits exist, with large eccentricities of the planets. As a model system we consider the observed extrasolar planetary system HD82943. In this system two planets have been observed (Israelinian et al. 2001): The masses of the two planets, expressed in Jupiter masses, are $m_1 \sin i = 0.88J$, $m_2 \sin i = 1.63J$, the semimajor axes are $a_1 = 0.73$AU, $a_2 = 1.16$AU, their periods are $T_1 = 221.6$d, $T_2 = 444.6$d, their eccentricities are $e_1 = 0.54$, $e_2 = 0.41$ and the mass of the sun is $m_{sun} = 1.05$ solar masses. We clearly see that this is a resonant planetary system, with 2/1 mean motion resonance of the planets, $T_2/T_1 = 2.006$.

We computed four basic families of 2/1 resonant periodic orbits, for the masses of the above system. Along each of these families the eccentricities of the two planets increase, starting with almost zero values, while the semimajor axes remain almost constant, corresponding to the 2/1 resonance. Large stable regions exist, even for large values of the two eccentricities. It is close to these stable periodic orbits that we should look for real planetary systems, because these are the only regions of phase space where bounded motion could exist. The exploration of the phase space close to a periodic orbit is made by studying perturbed orbits, by the method of Poincaré surface of section. It is shown that the phase (relative position of the two planets in their orbits) plays a crucial role on the stability of the system, and that for the same resonance and the same eccentricities chaotic motion appears, if the initial phase is changed, resulting to a quick disruption of the planetary system. On the other hand, a phase protection mechanism appears for suitable initial phase of the two planets, close to a stable periodic orbit, resulting to stable, ordered, motion.

2 Families of Periodic Orbits

2.1 Periodic Orbits in the Rotating Frame

As we mentioned above, the motion of an extrasolar planetary system is a special case of the general three body problem. We consider here the planar case. The best way to study this problem is to consider a rotating frame of reference (Hadjidemetriou 1975). This rotating frame is defined as follows:

Let us consider three bodies (point masses), S, P_1 and P_2, with masses m_0, m_1 and m_2, respectively, moving in the same plane under their mutual gravitational attraction. We define a rotating frame of reference, xOy, whose x axis is the line $S\,P_1$, and the origin O is the center of mass of these two bodies. The y axis is perpendicular to the x axis, in the plane of motion. The body P_1 moves on the x axis and the body P_2 moves in the xOy plane. This is a non uniformly rotating frame and the position of the three bodies are defined by the coordinate x_1 of P_1 on the x axis, the coordinates x_2, y_2 of P_2 on the xOy plane and the angle θ of the rotating x axis with a fixed direction in the inertial frame (where the center of mass of the system of the three bodies is at rest). We have four degrees of freedom, but it turns out (Hadjidemetriou 1975) that the angle θ is ignorable, so we are left with three degrees of freedom in the rotating frame.

The initial conditions defining the motion of the three bodies in the rotating frame are: x_{10}, x_{20}, y_{20}, $\dot{x}_{10}$, $\dot{x}_{20}$, $\dot{y}_{20}$. Two more initial conditions are needed to completely define the motion in space: θ_0 and $\dot{\theta}_0$. The value of $\dot{\theta}_0$ determines the angular momentum L, which appears as a fixed parameter in the differential equations of motion in the xOy frame. The value of θ_0 defines the initial orientation of the rotating frame and does not affect the motion. So, in order to study the motion in the rotating frame, we need the initial conditions x_{10}, x_{20}, y_{20}, $\dot{x}_{10}$, $\dot{x}_{20}$, $\dot{y}_{20}$ and the angular momentum L.

It can be proved that *periodic orbits* of the three body system exist in the rotating frame defined above (Hadjidemetriou 1975). Note that periodicity on the rotating frame means that the *relative* configuration of the three bodies is repeated after one period. The system *is not*, in general, periodic in the inertial frame.

It is proved (Hadjidemetriou 1976) that the periodic orbits in the rotating frame are not isolated, but belong to *monoparametric families of periodic orbits*, along which all the masses are fixed.

Of particular interest are the *symmetric periodic orbits*. The numerical integrations that follow showed that all the basic families of periodic orbits turned out to be symmetric with respect to the x axis of the rotating frame. A periodic orbit is symmetric with respect to the x axis of the rotating frame xOy if at $t = 0$ the body P_1, which moves always on the x axis, has zero velocity ($\dot{x}_{10} = 0$) and the body P_2 is on the x axis and its velocity is perpendicular to this axis ($y_{20} = 0$, $\dot{x}_{20} = 0$). Consequently, a symmetric periodic orbit is determined from three nonzero initial conditions only, namely x_{10}, x_{20} and $\dot{y}_{20}$. So, such a family can be represented by a continuous curve in the space $x_1, x_2, \dot{y}_2$.

2.2 Periodic Orbits of the Planetary Type

In section 2.1 we mentioned that the general planar three body problem can be reduced to a system of three degrees of freedom, in a rotating frame xOy. We shall now restrict ourselves to a planetary system with the sun S, and two planets P_1 and P_2, with the mass m_0 of the sun much larger than the masses m_1 and m_2 of the planets. This is a special case of the general three body problem, but now, since the masses m_1 and m_2 are small, we can consider this system as a *perturbed* system of the *integrable system* consisting of two uncoupled two body systems, $S - P_1$ and $S - P_2$.

Let us assume at first that $m_1 = 0$ and $m_2 = 0$ and that the planets P_1 and P_2 move in *circular* orbits, with radii a_1 and a_2, respectively, in the same plane and the same direction. It is clear that this motion is a symmetric periodic motion, with respect to the x axis, in the rotating frame xOy defined above, for any value of a_1 and a_2, i.e. for any value of the ratio of the periods of revolution of the two planets around the sun S. So, we have a monoparametric family of nonresonant, in general, periodic orbits in the unperturbed problem (zero planetary masses).

Next, we consider the case where the masses of the planets are zero and they move around the sun in *elliptic* orbits, but their semimajor axes are such that the ratio T_1/T_2 of their periods is a rational number. This means that we are in a *mean motion resonance*. We keep now the semimajor axes of the planets fixed and vary the eccentricities and the orientation of the two uncoupled planetary orbits. The resulting motion is periodic (not symmetric, in general) in the rotating xOy frame. So, we have families of unperturbed resonant periodic orbits in the rotating frame.

We give now to the planets nonzero masses. What happens to the above mentioned unperturbed families? It can be proved (Hadjidemetriou 1976) that all the members of the unperturbed family of circular orbits can be continued to the nonzero case, as *symmetric circular* periodic orbits. This is true for all orbits, except for those corresponding to the resonances of the form $(n+1)/n$, $n = 1, 2, 3, ...$, i.e. to the resonances 2/1, 3/2, .. At these points gaps appear and the single unperturbed family of circular periodic orbits breaks into an infinite number of families of symmetric nearly circular periodic orbits, which are separated at the resonances 2/1, 3/2, ...by gaps. At these gaps, a bifurcation of a family of *resonant elliptic* symmetric periodic orbits appears and the circular family continues as a resonant elliptic family of symmetric periodic orbits, along which the resonance remains constant, equal to the corresponding resonance at the bifurcation point, and the eccentricities of the two planets increase, starting from almost zero values, as we go outwards. We remark at this point that these families of elliptic periodic orbits are the continuation (from zero to non zero planetary masses) of the unperturbed families of elliptic periodic orbits that we mentioned above.

So, in the space x_{10}, x_{20}, $\dot{y}_{20}$ of nonzero initial conditions, the perturbed families of symmetric periodic orbits are represented by a set of continuous curves, having a circular, nonresonant in general, part and a resonant part, with a fixed resonance. This is shown in Fig. 1, for the region near the 2/1 and 3/2

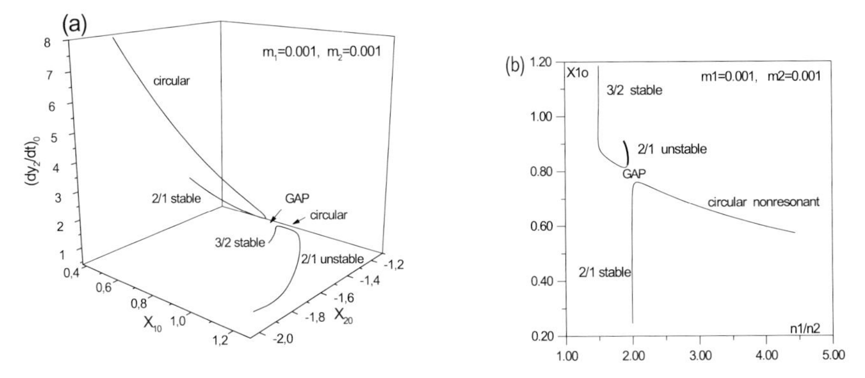

Fig. 1. Families of circular and elliptic periodic orbits of the general three body problem, of the planetary type, for $m_1 = 0.001$, $m_2 = 0.001$ and $m_{sun} = 0.998$. A discontinuity exists at the 2/1 resonance and branches of 2/1 resonant elliptic periodic orbits appear. Also, a stable 3/2 resonant branch of periodic orbits appears: (a) The families in the x_{10}, x_{20}, $\dot{y}_{20}$ space. (b) The families in the *resonance* - x_0 space. One 2/1 resonant branch (marked by a thick line) is unstable. There is also a small region on the circular branch close to the 3/1 resonance (not shown) which is also unstable.

resonances, for the masses $m_0 = 0.998$, $m_1 = 0.001$ and $m_2 = 0.001$. Note the gap at the 2/1 resonance.

The orbits of the two planets of a periodic system in the *inertial* frame are nearly Keplerian ellipses, due to their week gravitational interaction. These ellipses precess with a small angular velocity, which depends on the particular periodic orbit, if the motion is ordered (see Fig. 3b, as an example), but the precession may be chaotic in some cases.

So, summarizing all the above, we can say that there are two types of periodic orbits of the planetary type, *circular* and *resonant elliptic*:

- *Circular*: The orbits of the two planets are almost circular, and
- *resonant elliptic*: The orbits of the two planets have finite eccentricities, but the two semimajor axes are such that the ratio of the periods of the two planetary orbits is rational.

The basic periodic orbits are *symmetric* with respect to the rotating x-axis. This means that at $t = 0$ the initial conditions are x_{10}, $\dot{x}_{10} = 0$, x_{20}, $y_{20} = 0$, $\dot{x}_{20} = 0$, $\dot{y}_{20}$. In the study of the motion in the rotating frame, the angular momentum is kept as a fixed parameter and can be used to find $\dot{\theta}$.

2.3 Families of Periodic Orbits at the 2/1 Resonance

We make now a complete study of the *periodic orbits*, close to the 2/1 resonance, for a planetary system having the masses of the extrasolar planetary system HD82943, namely

$$m_1 = 0.88J,\ \ m_2 \sin i = 1.63J,\ \ m_{sun} = 1.05M_{\odot}, \tag{1}$$

(for $\sin(i) = 1$). The mass of the sun is in solar masses and the masses of the two planets are in Jupiter masses.

In order to avoid duplication in the numerical study, we must fix the units of mass, length and time. In the present study this normalization is made by taking the total mass of the system equal to one, $m_0 + m_1 + m_2 = 1$, the gravitational constant equal to one and we keep the angular momentum L fixed, equal to $L = 0.002$. This means that the unit of mass is slightly larger than one solar mass. Concerning the units of length and time, we note that they depend on the particular value of the angular momentum L that we are using in the numerical computations (we remind that L appears as a fixed parameter in the equations of motion in the rotating frame). So, a particular periodic motion of the planetary system may correspond to an infinite set of periodic planetary systems with *similar* planetary orbits. There are however some elements of the orbit whose values are independent of the units, namely the eccentricities of the planets or the ratio of the planetary semimajor axes (or planetary periods). Several plots that follow are in these elements of the orbit.

In normalized units, the masses of the planetary system that we will study are

$$m_1 = 0.0008,\ \ m_2 = 0.0014,\ \ m_0 = 0.9978. \tag{2}$$

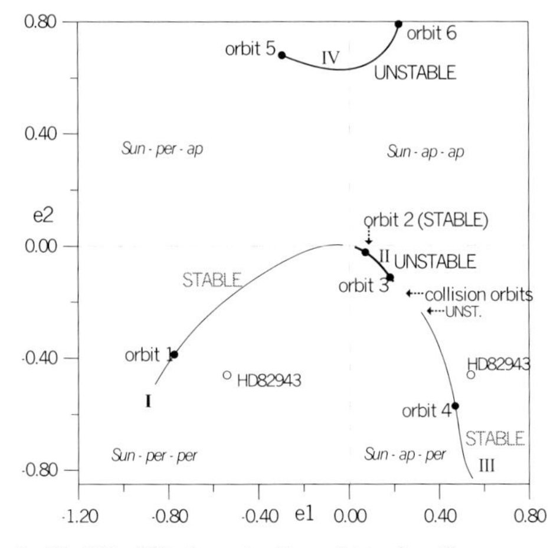

Fig. 2. The families *I, II, III, IV* of periodic orbits for the masses of HD82943, in the eccentricity space. The indication *per, ap* refers to the initial configuration at $t = 0$. (*per* denotes position at periastron and *ap* position at apoastron.)

We computed four different families of symmetric periodic orbits, all at the 2/1 resonance, which differ in the initial phase of the two planets at an initial epoch $t = 0$. These families are shown in Fig. 2. Along each family the angular momentum has a fixed value, $L = 0.002$, the same for all members of the family. To make the presentation clearer, we present these families in the space of the initial eccentricities of the two planets. These are in fact the osculating eccentricities at $t = 0$, but we note that they do not vary much along the periodic orbit, unless we are close to a collision orbit. At $t = 0$ the two planets and the sun are in the same line, the planets being either at periastron or apoastron. This is a consequence of the fact that the periodic orbits are symmetric with respect to the x axis of the rotating frame. Throughout this study we will assume that the planet P_1 is the inner planet and the planet P_2 is the outer planet.

In Fig. 2 we made the convention to use a positive value of the eccentricity if the corresponding planet is at apoastron at $t = 0$ and a negative eccentricity if it is at periastron. Note that due to the fact that we are in a 2/1 resonance, a simple geometric consideration shows that the outer planet P_2 after a half period $t = T/2$ changes position from periastron to apoastron, or vice versa, while the inner planet P_1 returns to the same position as at $t = 0$.

We remark that the families designated as *II* and *III* in Fig. 2 are in fact a single family, and the gap that appears between them is simply due to the existence of a collision orbit between the two planets, so that the computation presented numerical difficulties.

The whole family *I* is stable. Family *II* is mainly unstable, with the exception of a small region which is stable. Family *III* is mainly stable, and there is an unstable region close to the collision orbit. Finally, the family *IV* is unstable.

In each periodic planetary system we have two nearly Keplerian planetary orbits, which are not fixed in space, due to the gravitational interaction between the two planets. The line of apsides of both planets rotates slowly.

3 The Different Types of 2/1 Resonant Periodic Orbits

Some typical orbits of each of the above four families are presented in the Figs. 3–7. We find it more illustrating to present the orbits in the inertial frame, although, as we mentioned, there is a slow precession of the line of apsides, because these orbits are periodic in a rotating frame only, in general. The orbits thus shown are for a small time period, close to the period of the periodic orbit in the rotating frame. In addition, we also present some periodic orbits in the rotating frame, to obtain an idea how the motion looks in the synodic frame.

In all the periodic orbits, the line of apsides of the two planetary orbits coincide. The pericenters of the two orbits may be in the same direction, with respect to the sun, or in opposite directions. Concerning the phase, i.e. the position of the two planets in their orbits at $t = 0$, all possible combinations can appear. Due however to the 2/1 resonance between the two planets, if we start with a certain phase at $t = 0$, after half a period, $t = T/2$, the position of P_2 shifts from pericenter to apocenter, or vice versa. So we have, for each periodic motion, two equivalent configurations, at $t = 0$ and at $t = T/2$, respectively. To the above two different phases of the same orbit, we have two perpendicular crossings of the planet P_2 from the rotating x axis, and consequently we could use either perpendicular crossing to represent a periodic orbit in Fig. 2.

Table 1. All possible phases at $t = 0$ and $t = T/2$

	$t = 0$		$t = T/2$
Type 1:	Sun - P_1(per) - P_2(per)	$\rightarrow$	P_2(ap) - Sun - P_1(per)
Type 2:	Sun - P_1(ap) - P_2(ap)	$\rightarrow$	P_2(per) - Sun - P_1(ap)
Type 3:	Sun - P_1(per) - P_2(ap)	$\rightarrow$	P_2(per) - Sun - P_1(per)
Type 4:	Sun - P_1(ap) - P_2(per)	$\rightarrow$	P_2(ap) - Sun - P_1(ap)

All the possible initial phases of a periodic orbit, and the equivalent configuration at $t = T/2$, are summarized in Table 1. As we will see in the next subsections, the 2/1 resonant orbits with a phase of type 1 are stable, because a phase protection mechanism appears. These are the orbits of the family *I*. The orbits of the families *II* and *III* are of the type 4. These orbits are also stable, provided that we are not close to a collision orbit, and the eccentricities of the two planets are large. In contrast, the orbits of the type 4 are unstable when the eccentricities of the two planets are small, as is the case with the family *II* (with some exceptions), and when we are close to a collision orbit. The resonant orbits of the type 2 and 3 belong to the family *IV* and are all unstable.

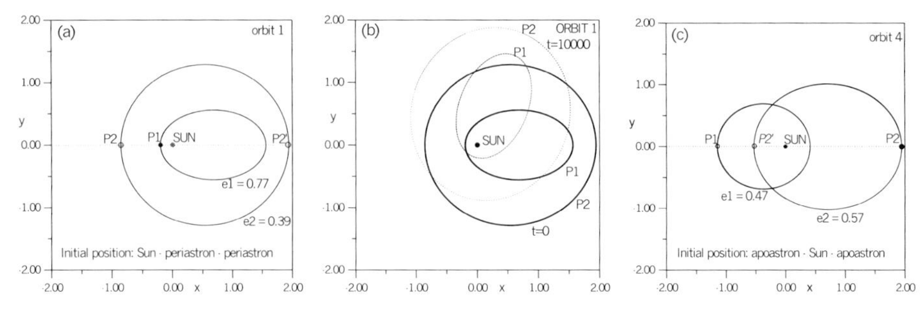

Fig. 3. (a) The *orbit 1* of Fig. 2 for a short time. (b) The same orbit at two different time intervals. The precession of the line of apsides is clear. (c) The *orbit 4* of Fig. 2. Note that both orbits 1 and 4 have large planetary eccentricities, but the system is stable, even for the *orbit 4*, where the orbits of the planets intersect.

3.1 Stable Orbits of the Families *I* and *IV*

A typical orbit of the family I is shown in Fig. 3a. The exact position of this orbit on the family I of Fig. 2 is indicated as *orbit 1*. The phase of this orbit is of type 1 of the Table 1. This motion is stable, although the eccentricities are large. The two planetary orbits do not intersect in space.

A typical orbit of the family III is shown in Fig. 3c. The exact position of this orbit on the family III of Fig. 2 is indicated as *orbit 4*. This orbit is also stable, with large values of the eccentricities, but now, in contrast to the orbits of the family I, the planetary orbits intersect. The initial phase is of type 4 in Table 1, and due to the fact that the two planets are locked to the 2/1 resonance, a phase protection mechanism exists, which prevents the two planets from close encounters, although their orbits intersect in space. A 2/1 resonant stable orbit close to the system HD82943, of the type of the Fig. 3c, where the two planetary orbits intersect, was presented last September 2002 by Ji *et.al* at the IAU Colloquium 189.

3.2 A Stable and an Unstable Orbit of the Family *II*

We present in Fig. 4 two orbits of the family II, one stable (Fig. 4a) and one unstable (Fig. 4b). The exact position of these orbits on the family II of Fig. 2 is indicated as *orbit 2* and *orbit 3*, respectively. Both orbits are of type 4 of the Table 1, as was the case with the orbit 4 of the Fig. 3b. In fact, the families II and III belong to the same family, as already mentioned before, but they appear as separated in Fig. 2 because there exists a collision periodic orbit between the two planets, and the gap which appears close to this collision orbit is due to numerical difficulties. All the orbits close to the collision orbit are strongly unstable.

We note that the eccentricities of the two planets are rather small, but now the *orbit 3* (Fig. 4b) is unstable. The *orbit 2* (Fig. 4a) has smaller eccentricities and is stable.

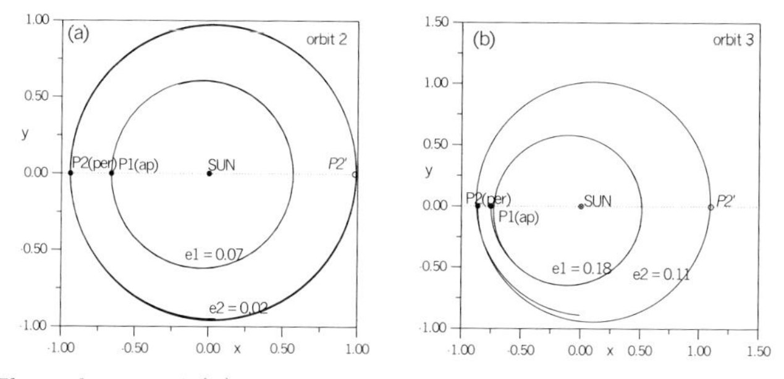

Fig. 4. (a) The *orbit 2* and (b) the *orbit 3* of Fig. 2. Note that in both cases the orbits of the planets have small eccentricities and the phase is the same, but the *orbit 2* is stable while the *orbit 3* is unstable.

It is interesting to compare the *orbit 3* (Fig. 4) with the *orbit 4* (Fig. 3). They have the same phase, but the *orbit 3* is unstable, while the *orbit 4* is stable. The instability of the *orbit 3* is due to the close approach between the two planets, as is clearly seen from the plot of Fig. 4. In the *orbit 4* the closest approach between the two planets is much larger, due to the large value of the planetary eccentricities. Thus, the increase of the eccentricities plays a stabilizing role, because, in relation also to a phase protection mechanism due to the fact that the two planets are locked to the 2/1 resonance, its effect is to increase the closest approach between the planets. This may explain why there are several actual extrasolar planetary systems with large eccentricities.

3.3 Two Orbits of the Unstable Family *IV*

We present in Fig. 5 two typical periodic orbits of the family *IV*, indicated as *orbit 5* (Fig. 5a) and *orbit 6* (Fig. 5b) of Fig. 2, at the two ends of the curve representing this family. The orbit 5 has a phase of type 2 in Table 2, while the orbit 6 has a phase of type 3. Note that the phase changes along the family from type 2 to type 3. This is due to the fact that the eccentricity of the inner planet starts with a positive value (*orbit 5*, position at apoastron) and as we proceed

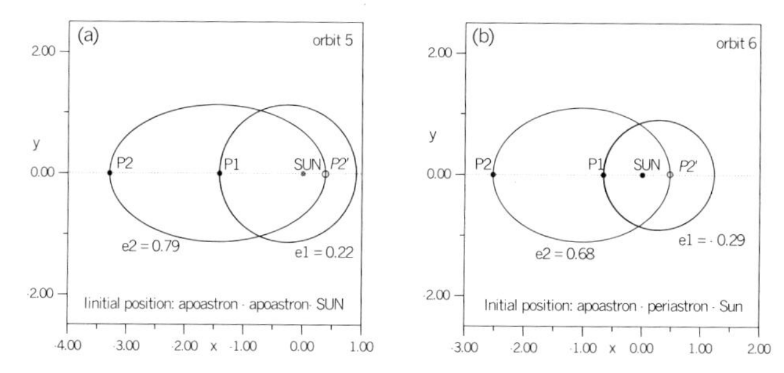

Fig. 5. (a) The *orbit 5* and (b) the *orbit 6* of Fig. 2. Note that in both cases the orbits of the planets have large eccentricities and, contrary to the *orbit 4*, both systems are unstable.

along the family it changes to a negative value (*orbit 6*, position at periastron), passing through a circular orbit, as is clearly seen in Fig. 2. Along the whole family *IV* the eccentricity of the outer planet is large. The orbits of the planets intersect in space, as in the family *III* (*orbit 4*), but now the phase is such that the phase protection mechanism is not operating and both orbits are unstable.

3.4 The Periodic Orbits in the Synodic Frame

In Fig. 6 we present the periodic orbits 1 and 4 of the Figs. 3a and 3c, respectively, in the rotating xOy frame, in which they are exactly periodic.

In Fig. 7 we present the periodic orbits 5 and 6 of the Figs. 5a and 5b, respectively, in the rotating xOy frame. Note that in the rotating frame the planet P_1 moves on the x axis only and its orbit is presented as a straight line. The length of this line depends on the eccentricity.

4 Perturbed Orbits Close to Stable and Unstable Periodic Orbits

In the previous sections we computed four basic families of resonant periodic orbits, at the 2/1 resonance. We shall explore now the phase space close to these periodic orbits, in order to detect the regions where stable motion exists.

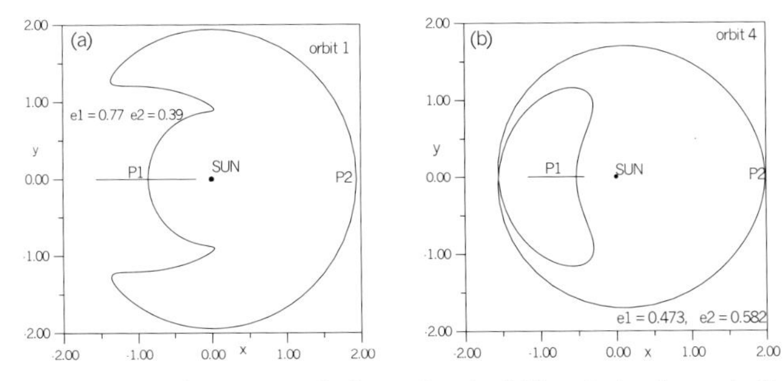

Fig. 6. The *orbit 1* of Fig. 2 and the *orbit 4* of Fig. 2, in the rotating frame.

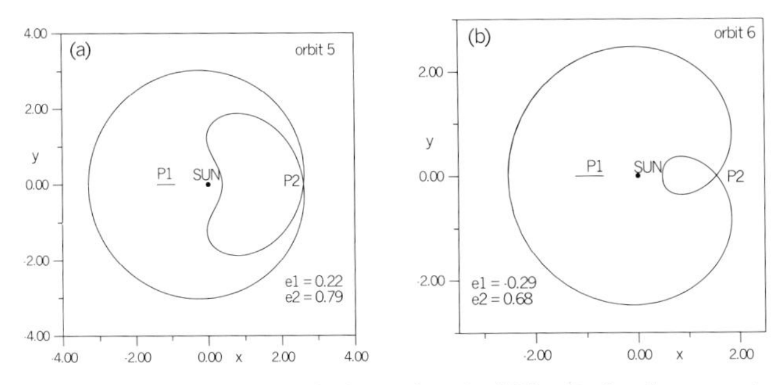

Fig. 7. The *orbit 5* of Fig. 2 and the *orbit 6* of Fig. 2, in the rotating frame.

We shall also study the generation of chaotic motion and the disruption of the planetary system.

The dynamical system we study has three degrees of freedom, in the rotating frame, so its phase space is six-dimensional. This can be reduced to a four-dimensional phase space, if we take the Poincaré map which is on a four-dimensional surface of section. A complete search of the whole phase space is very tedious, even if we restrict ourselves to the vicinity of a periodic orbit (a fixed point on the Poincaré map). In the following we study the effect of a phase shift of the planets, *in their orbits*, on the stability of the system. We start with a periodic orbit belonging to one of the four families presented in Fig. 2, and we change the initial position of the planet P_2, *on its orbit*. In his way we obtain a new planetary system, which is not periodic, but the two planetary orbits have the same elements. The study of the long term evolution of the perturbed system is made by computing the Poincaré map. We have taken as surface of section the surface $y_2 = 0$ and *Energy*=constant. So the phase space of the mapping is the four-dimensional space x_1, $\dot{x}_1$, x_2, $\dot{x}_2$ (the coordinates are in the rotating frame xOy).

In the following we show, for each orbit we studied, the projection of the mapping on one of the coordinate planes and also the time evolution of the eccentricities and the semimajor axes.

4.1 Perturbed Orbits Close to the *Orbit 1* and the *Orbit 4*

We start with two stable periodic orbits, *orbit 1* and *orbit 4*, belonging to the families *I* and *III*, respectively. Our aim is to see how the motion is affected if the planet P_2 is shifted along its orbit, thus changing the phase of the system. In Figs. 8a and 8b we present two shifted positions of P_2 along its orbit, indicated as P_{21} and P_{22}, for the periodic orbit 1 (Fig. 3a) and the periodic orbit 4 (Fig. 3b), respectively. The results are given in Figs. 9 and 10 for the perturbed orbits to the orbit 1 and in Figs. 11, 12 for the perturbed orbits to the orbit 4.

Note that for a small shift of P_2 on its orbit, to the position P_{21}, the perturbed motion is on a torus, as is clearly seen from the Poincaré maps in Figs. 9a and 10a. We remark that in these Figures we presented the projection of the

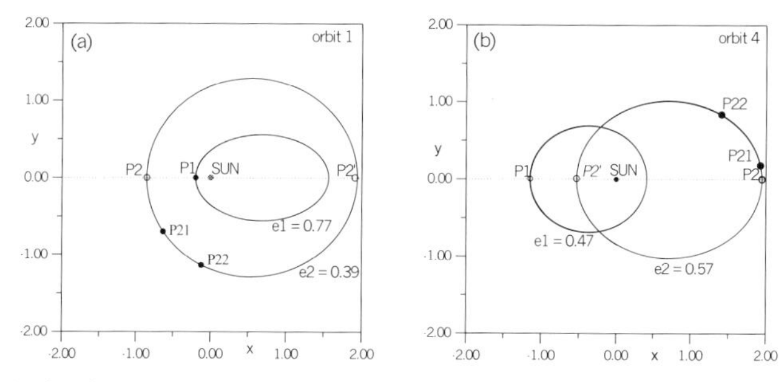

Fig. 8. The shift of the planet P_2 along its orbit, to the positions P_{21} and P_{22}, on the *orbit 1* and on the *orbit 4*.

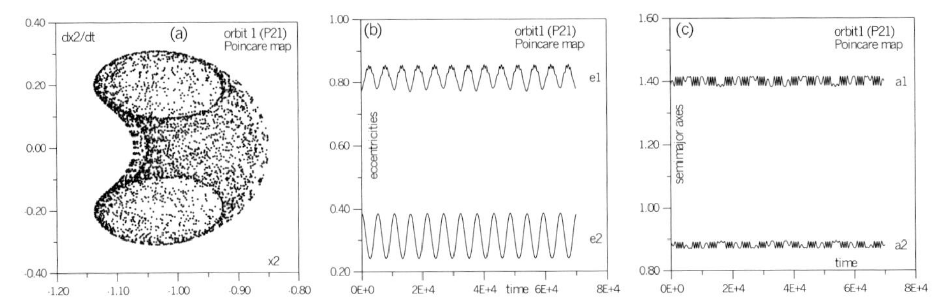

Fig. 9. A perturbed orbit of the *orbit 1*, due to a shift of P_2 along its orbit to the position P_{21} of Fig. 8a: (a) Projection of the four dimensional Poincaré map in the space $x_2\dot{x}_2$, (b) the evolution of the eccentricities in time, (c) the evolution of the semimajor axes in time. The perturbed motion is on a torus.

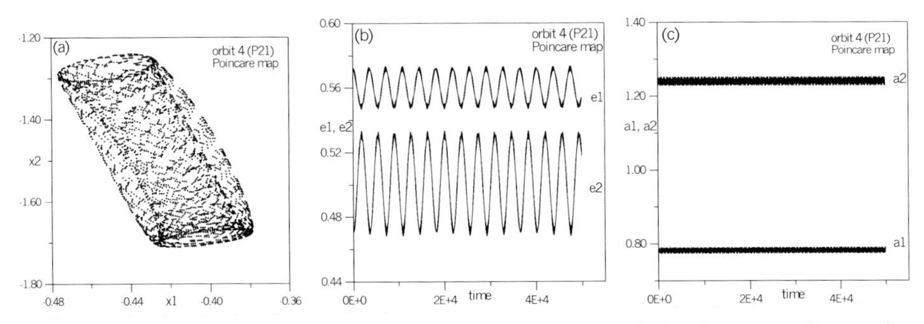

Fig. 10. A perturbed orbit of the *orbit 4*, due to a shift of P_2 along its orbit to the position P_{21} of Fig. 8b: (a) Projection of the four dimensional Poincaré map in the space x_1x_2, (b) the evolution of the eccentricities in time, (c) the evolution of the semimajor axes in time. The perturbed motion is on a torus.

four dimensional Poincaré map on a two dimensional coordinate plane. The projection of the Poinaré map in all the other coordinate planes (not shown here) is similar. So, indeed the perturbed motion on the surface of section is on a four dimensional torus. The evolution of the semimajor axes and the eccentricities of both planets undergo oscillations with a fixed amplitude. This is an indication that in both cases the motion is ordered and the two planets move on bounded orbits. This means that a small change of the phase of a stable periodic orbit results to ordered, bounded, motion on a torus in the four dimensional phase space.

If the shift of P_2 along its orbit is larger, to the position P_{22}, resulting to a larger change of the phase of the system, the stable periodic orbits 1 and 4 become now unstable and the system quickly disrupts into a close binary, consisting of the sun and one planet, while the other planet escapes to infinity. This is clearly seen in Figs. 11 and 12, for the orbits 1 and 4, respectively.

From the above we see that a small shift of the phase close to a periodic motion results to ordered motion on a torus, while a larger shift of the phase results to instability and escape of one planet. The distinction between ordered

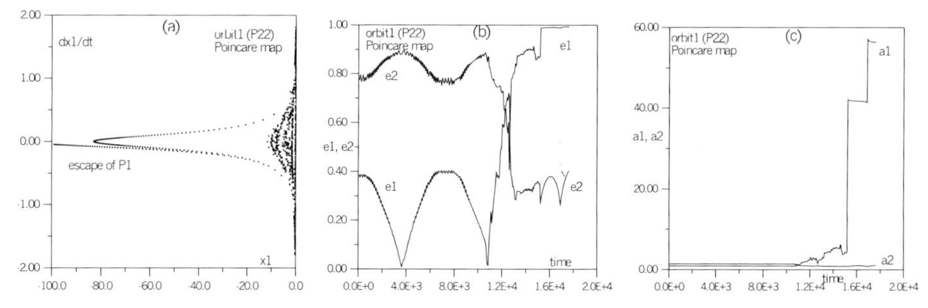

Fig. 11. A perturbed orbit of the orbit 1, due to a shift of P_2 along its orbit to the position P_{22} of Fig. 8a: (a) Projection of the four dimensional Poincaré map in the space $x_1\dot{x}_1$, (b) the evolution of the eccentricities in time, (c) the evolution of the semimajor axes in time. The perturbed motion starts on a perturbed torus, but soon the planet P_1 escapes.

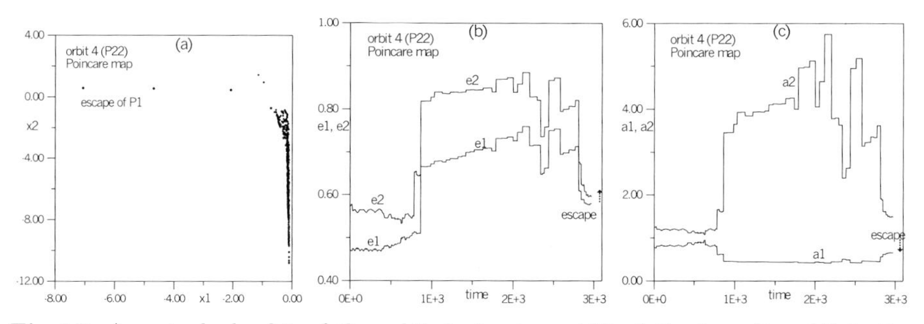

Fig. 12. A perturbed orbit of the orbit 4, due to a shift of P_2 along its orbit to the position P_{22} of Fig. 8b: (a) Projection of the four dimensional Poincaré map in the space x_1x_2, (b) the evolution of the eccentricities in time, (c) the evolution of the semimajor axes in time. The perturbed motion starts on a perturbed torus, but soon the planet P_1 escapes.

and chaotic motion was based on the geometric properties of the Poincaré map, as obtained from the numerical computations. A more rigorous analysis, based on spectral analysis of the time series of several elements of the orbit (eccentricity, semi major axis) verified the distinction between order and chaos, as obtained from the graphs. An typical example of ordered and of chaotic motion is presented in the next section.

4.2 Perturbed Orbits Close to the *Orbit 2* and the *Orbit 3*

As in the previous subsection, we study the effect of a phase shift of the planet P_2 on its orbit. We consider two orbits of the family II, *orbit 2* and *orbit 3*. The first is stable and the second is unstable. The shifted positions are shown in Fig. 13. The results of the numerical computations are shown in Figs. 14 and 15 for the *orbit 2* and in Figs. 15–18 for the *orbit 3*.

The numerical results show that a shift of P_2 along the stable periodic *orbit 2* results to ordered, bounded, motion on a torus, even for a large change of

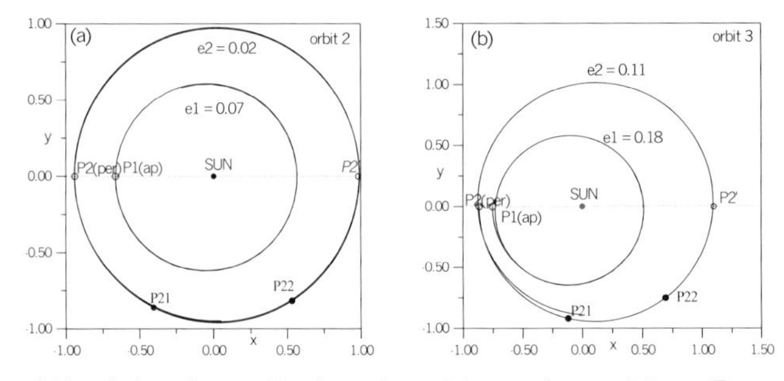

Fig. 13. The shift of the planet P_2 along its orbit, to the positions P_{21} and P_{22} on the *orbit 2* and on the *orbit 3*.

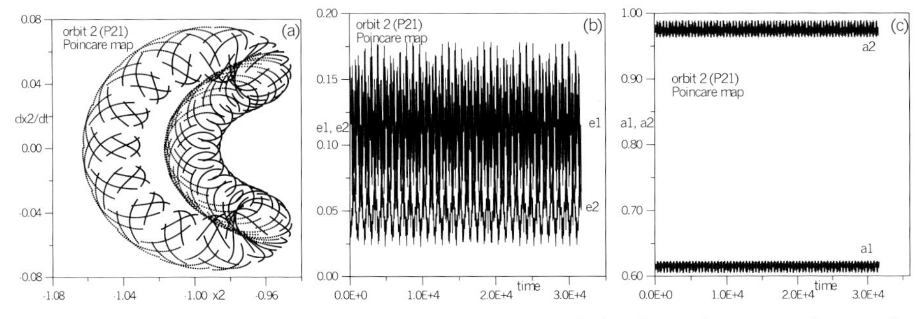

Fig. 14. A perturbed orbit of the *orbit 2*, due to a shift of P_2 along its orbit to the position P_{21} of Fig. 13a: (a) Projection of the four dimensional Poincaré map in the space $x_2\dot{x}_2$, (b) the evolution of the eccentricities in time, (c) the evolution of the semimajor axes in time. The perturbed motion is on a torus.

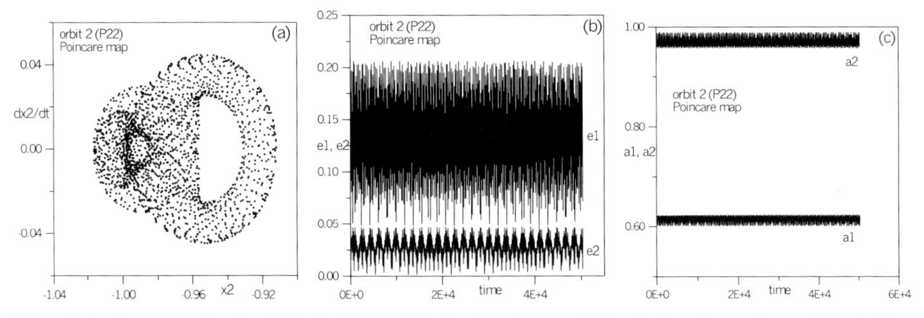

Fig. 15. A perturbed orbit of the *orbit 2*, due to a shift of P_2 along its orbit to the position P_{22} of Fig. 13a: (a) Projection of the four dimensional Poincaré map in the space $x_2\dot{x}_2$, (b) the evolution of the eccentricities in time, (c) the evolution of the semimajor axes in time. The perturbed motion is on a torus.

the phase (positions P_{21} and P_{22} of Fig. 13a), as shown clearly in the Figs. 14 and 15. (We also mention here that the projection of the Poincaré map to all other coordinate planes was similar to that shown in Figs. 15 and 15).

In contrast, a shift of P_2 along the unstable periodic orbit 3 (Fig. 13b) results to chaotic motion, although no escape was detected. Note that when the phase

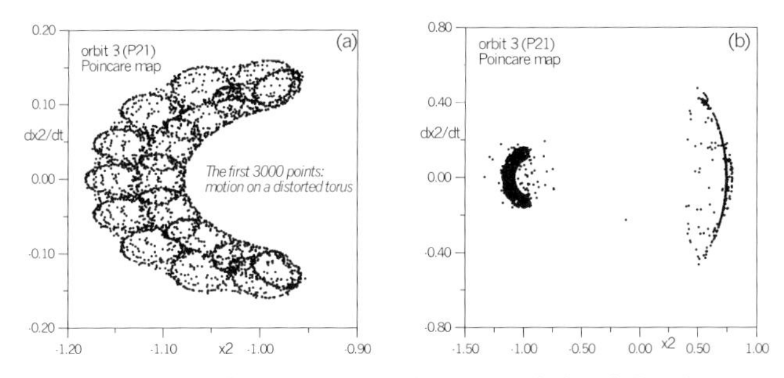

Fig. 16. A perturbed orbit of the *orbit 3*, due to a shift of P_2 along its orbit to the position P_{21} of Fig. 13b: Projection of the four dimensional Poincaré map in the space $x_2\dot{x}_2$ (a) for the first 3000 iterations, and (b) for long time. The motion starts on a torus, but later it becomes chaotic.

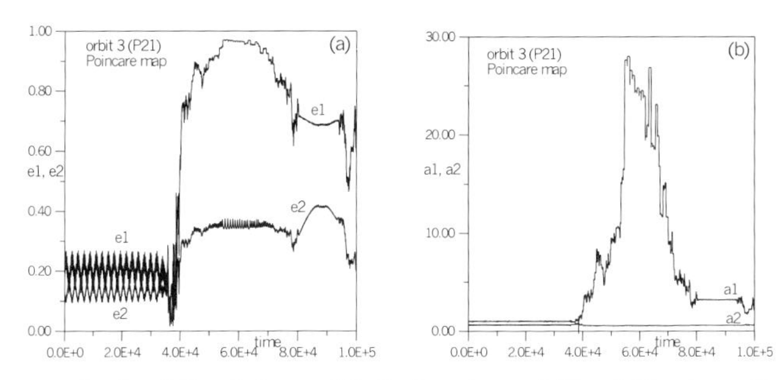

Fig. 17. A perturbed orbit of the *orbit 3*, due to a shift of P_2 along its orbit to the position P_{21} of Fig. 13b: (a) Poincaré map for the evolution of the eccentricities in time and (b) the evolution of the semimajor axes in time.

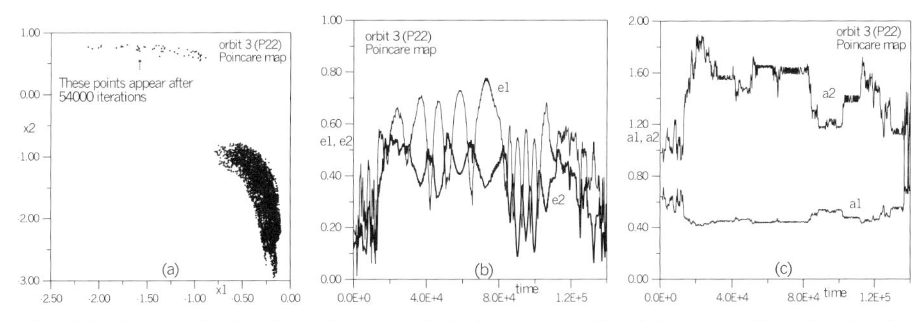

Fig. 18. A perturbed orbit of the *orbit 3*, due to a shift of P_2 along its orbit to the position P_{22} of Fig. 13b: (a) Projection of the four dimensional Poincaré map in the space x_1x_2, (b) the evolution of the eccentricities in time, (c) the evolution of the semimajor axes in time. The perturbed motion is chaotic.

shift is small (position P_{21}), the perturbed motion starts on a distorted torus (Fig. 16a), but later on the motion becomes chaotic (Figs. 16b, 17). Note that both the eccentricities and the semimajor axes are bounded in an oscillatory motion for a long time initially, corresponding to the motion on the distorted torus in Fig. 16a, but after a certain time chaotic motion starts. This phenomenon is in fact a manifestation of the *stickiness effect*, studied in Efthimiopoulos et. al. (1997) for two degrees of freedom.

For a larger phase shift of P_2, to the position P_{22} in Fig. 13b, the perturbed motion is chaotic from the start (Fig. 18).

4.3 Perturbed Orbits Close to the *Orbit 5*

In Fig. 19 we show the shift of the planet P_2 to the positions P_{21} and P_{22}, along the unstable periodic *orbit 5* of the family *IV*. The resulting motion is unstable for all positions of P_2. An example is given in Fig. 20, for a shift to the position P_{21}, where, after a strong chaotic motion of the system, the planet P_1 escapes. The same evolution appears for the shift to the position P_{22}.

5 Discussion

We have presented a systematic way to find all the regions of the phase space of a planetary system with two planets, moving in the same plane, where stable

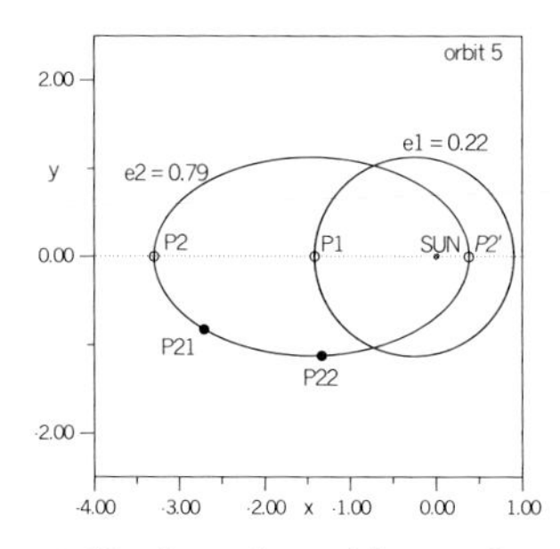

Fig. 19. The shift of the planet P_2 along its orbit, to the positions P_{21} and P_{22}, on the *orbit 5*.

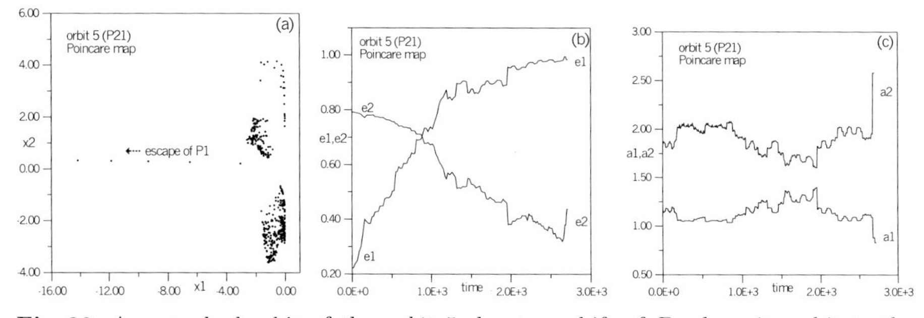

Fig. 20. A perturbed orbit of the *orbit 5*, due to a shift of P_2 along its orbit to the position P_{21} of Fig. 19: (a) Projection of the four dimensional Poincaré map in the space x_1x_2, (b) the evolution of the eccentricities in time, (c) the evolution of the semimajor axes in time. The planet P_1 escapes. In (b) and (c) the last points, leading to escape, are not shown.

motion could appear. This was achieved by studying the families of periodic orbits and their stability properties. The study was restricted to the region close to the 2/1 mean motion resonance between the planetary orbits, but a similar study can be carried out for all other important resonances. Each resonance however has its own characteristics, and the behaviour is different from one type of resonance to the other. The present study is restricted to planetary systems having the same masses as the observed planetary system HD82943. But we note at this point that a change of the planetary masses, for example keeping the total mass fixed and varying the ratio m_1/m_2, may result to a change of the stability (Hadjidemetriou, 2002).

The phase of the planets plays a crucial role on the stability of the system. This is so because, for some phases a *phase protection mechanism* is operating, protecting the two planets from close encounters. It was found that the system is stable, even for large eccentricities, provided that both planets start from their perihelia, situated in the same direction (*type 1* of Table 1). In this case the orbits of the planets do not intersect. This is the phase of all the orbits of the family I. Another stable phase, where the planetary orbits do intersect in space, is when the inner planet starts at aphelion and the outer planet at perihelion, both situated in the same direction (*type 4* of Table 1). It is interesting to note that for this phase, the increase of the eccentricities plays a stabilizing role. This is clear in Fig. 2, where the families II and III have the same phase (*type 4*), but almost all orbits of the family II, corresponding to small eccentricities, are unstable, while all orbits of the family III (with the exception of those close to the collision orbit) are stable. A typical unstable orbit is shown in Fig. 4b and a typical stable orbit is shown in Fig. 3a (both for the same phase of *type 4*).

Contrary to these stable phases, there are two more phases that are in all cases unstable (*type 2* and *type 3* of Table 1). These are the phases of the orbits of the family IV. We did not find any stable periodic motion corresponding to these phases. For the orbits of the family IV we found that a small deviation from the exact periodic motion results to a distorted motion and sooner or later a close encounter between the two planets appears, resulting to chaotic motion and in some cases disruption of the system (Fig. 20).

For the stable orbits, we studied the long term behaviour of the perturbed motion, obtained by shifting the planet P_2 along its orbit, thus changing the initial phase, but leaving all planetary elements unchanged. We found that for a small shift stable, quasi periodic, motion appears as is clear from the corresponding Poincaré maps that we have computed. Thus we found that there do exist zones of stability, where stable planetary systems could appear in nature. Some typical examples are in Figs. 9a, 10a, 14a and 15a. From the above it is clear that planetary systems with large eccentricities could be common on the sky. However, such systems with large eccentricities should necessarily be close to a resonance, with the proper phase, because it is the resonance that generates the phase protection mechanism which stabilizes the system. Far from the resonance close encounters are inevitable, resulting to chaotic motion.

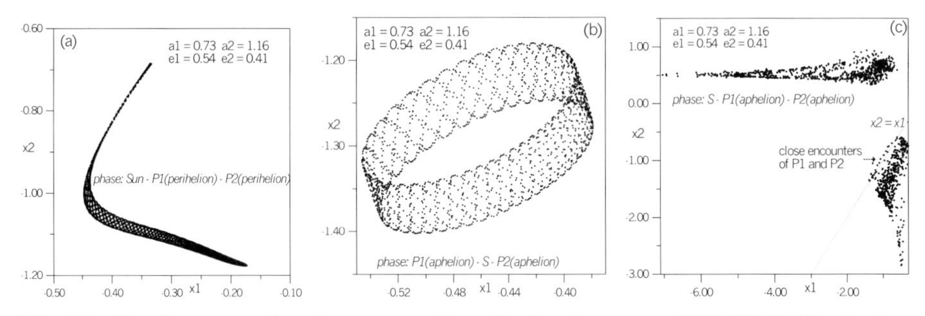

Fig. 21. The long run of a planetary orbit with the elements of HD82943 (Projection of the four dimensional Poincaré map in the space x_1x_2): (a) for the configuration Sun-P_1(perihelion)-P_2(perihelion), (b) for the configuration P_1(aphelion)-Sun-P_2(aphelion) and (c) for the configuration Sun-P_1(aphelion)-P_2(aphelion). The repeated close encounters in (c) are clearly seen, close to the line $x_2 = x_1$ (note that $y_1 = 0$, $y_2 = 0$).

The observed planetary system HD82943 is close to the stable configurations we found in our study, as shown in Fig. 2. In Fig. 21 we present the Poincaré map using the orbital elements of the system HD82943. Since the phase is not known, we used three different phases, one of the stable *type 1* of Table 1, corresponding to the family I (Fig. 21a), one of the stable *type 4*, corresponding to the family III (Fig. 21b) and one of the unstable *type 2*, corresponding to the family IV (Fig. 21c). We note that in the two stable phases the system is bounded, moving on a torus, while the motion for the unstable phase is chaotic, resulting to large variations of the planetary orbits and eventually to escape of one planet. Note that in this latter case close encounters between the two planets appear, as indicated in Fig. 21c, resulting to strong chaotic motion and large changes of the orbital elements.

The ordered or the chaotic nature of the motion in Fig. 21, as is clearly seen on the projections of the Poincaré map, is verified by a power spectrum analysis of the time series of the elements of the orbit, along the perturbed motion. In Fig. 22 we present two typical power spectra, one for the ordered motion of the Fig. 21b and one for the chaotic motion of the Fig. 21c, for the time series for the

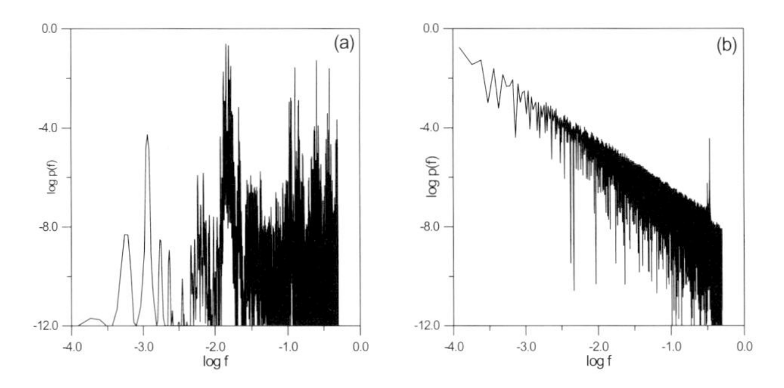

Fig. 22. Power spectra of the time series of the evolution of the semimajor axis: (a) for the ordered orbit of the Fig. 21b and (b) for the chaotic orbit of the Fig. 21c.

semimajor axis a_1. The distinction between order and chaos is clear. In Fig. 22a the power spectrum density has the typical features of ordered motion, (a few peaks of significant amplitude). On the other hand, the power spectrum density in Fig. 22b is typical of a chaotic motion, because it shows $1/f$-divergence, which indicates strong chaos.

A similar power spectrum density analysis was made for all the orbits that we studied in the previous sections and the geometric picture of order or chaos that we observed on the Poincaré map was confirmed.

Acknowledgements

We thank Dr. G. Voyatzis for his help in the the power spectrum analysis. We also thank the referee for his useful comments.

References

1. Beauge C., Ferraz-Mello S. & Michtchenko T. (2002): *Extrasolar Planets in Mean-Motion Resonance: Apses Alignment and Asymmetric Stationary Solutions*, preprint.
2. Callegari N., Michtchenko T. & Ferraz-Mello S. (2002): *Dynamics of two planets in the 2:1 and 3:2 mean motion resonances*, in 34th DPS Meeting, BAAS, 34, 30.01
3. Celletti A., Chessa A., Hadjidemetriou J.D. and Valsecchi J.B. (2002), *A systematic study of the stability of symmetric periodic orbits in the planar, circular, restricted three-body problem*, Cel. Mech. & Dyn. Astron. **83**, 239-255.
4. Efthimiopoulos C., Contopoulos G., Voglis N., and Dvorak R. (1997): *Stickiness and cantori*, Journal of Physics A - Mathematical and General **30**, 8167-8186.
5. Ferraz-Mello S. (2002): *Tidal Acceleration, Rotation and Apses Allignement in Resonant Extra-Solar Planetary Systems*, in 34th DPS Meeting, BAAS, 34, 30.08
6. Ford E.B, Havlikova M. and Rasio F.A. (2001): Icarus **150**, 303.
7. Hadjidemetriou J.D. (1975): *The continuation of periodic orbits from the restricted to the general three-body problem*, Cel. Mech. **12**, 155-174.
8. Hadjidemetriou J.D. (1976): *Families of Periodic Planetary Type Orbits in the Three-Body Problem and their Stability*, Astrophys. Space Science **40**, 201.-224.
9. Hadjidemetriou J.D. (2002): *Resonant periodic motion and the stability of extrasolar planetary systems*, Cel. Mech. & Dyn. Astron. **83**, 141-154.
10. Israelinian G., Santos N, Mayor M. and Rebolo R. (2001): *Evidence for planet engulfment by the star HD82943*, Nature **411**, 163.
11. Ji, J., Kinoshita, H., Liu, L., Li, G. and Nakai, H. (2003): *The apsidal antialignment of the HD 82943 system*, Proceedings of IAU Col. No 189, (accepted to Cel. Mech. & Dyn. Astron).
12. Jones B. and Sleep P. (2002a): *The stability of the orbits of Earth-mass planets in the habitable zone of 47 Ursae Majoris*, Astron. Astrophys. **393**, 1015.
13. Kinoshita H. and Nakai H (2001a): *Stability of the GJ 876 Planetary System*, PASJ,**53**,L.25-L.26.
14. Kinoshita H. and Nakai H.(2001b): *Stability Mechanism of Planetary System of υ' Andromedae*, Proceedings of the IAU Symposium 202, Manchester 2000.

15. Kiseleva-Eggleton L., Bois E., Rambaux N. and Dvorak R. (2002): *Global dynamics and stability limits for planetary systems around HD 12661, HD 38529, HD 37124 and HD 160691*, ApJ. Letters, 578, L145.
16. Krzysztof Gozdziewski, Eric Bois and Andrej J. Maciejewski (2002): *Glaobal dynamics of the Fgliese 876 planetary system*, Mon. Not. R. Astron. Soc.
17. Laughlin G. and Chambers J. (2001): *Short-term dynamical interactions among extrasolar planets*, Ap.J. **551**, L109-113.
18. Lissauer J.J. and Rivera E.J.(2001): *Stability analysis of the Planetary System orbiting Andromedae*, Ap.J. **554**, 1141-1150.
19. Malhotra R. (2002a): *A dynamical mechanism for establishing apsidal resonance*, ApJ. Letters bf 575.
20. Malhotra R. (2002b): *Eccentricity excitation and apsidal alignment in exoplanetary systems*, in 34th DPS Meeting, BAAS, 34, 42.05
21. Murray N., Paskowitz M. and Holman, M. (2001):*Eccentricity evolution of resonant migrating planets*, Ap.J. (preprint).
22. Peale S. and Lee M. (2002): *Extrasolar Planets and the 2:1 Orbital Resonances*, in DDA 33rd Meeting, BAAS, 34, #1.02
23. Rivera E.J. and Lissauer J.J. (2001): *Dynamical models of the resonant pair of planets orbiting the star GJ 876*, Ap.J. **558**, 392- 402.
24. Schneider Jean, (2002): *http://www.obspm.fr/encycl/catalog.html.*

The "Third" Integral in the Restricted Three-Body Problem Revisited

Harry Varvoglis, Kleomenis Tsiganis, and John D. Hadjidemetriou

Department of Physics, University of Thessaloniki, GR-541 24 Thessaloniki, Greece

Abstract. In 1964 M. Hénon and, independently, V. Szebehely with G. Bozis presented the first numerical results, indicating the existence of a "new" local integral of motion in the circular restricted three-body problem. The first terms of an asymptotic expansion of this integral were later calculated by Contopoulos [1]. Several years later, the Celestial Mechanics astronomical community started to develop a very successful theory on local integrals of motion in the restricted three-body problem, which however in the jargon of this field are called *proper elements* and are related to known analytical approximate solutions. The calculation of proper elements is based on the implicit assumption that the orbit traced by a planet (major or minor) is nearly-regular. Here we show that this method is also applicable, albeit partly, in a special case of chaotic motion in the Solar System, known as "stable chaos". Thus, the existence of an additional local integral of motion in the elliptic restricted three-body problem is responsible for the phenomenon of stable chaos.

1 Introduction

In 1964 the Laboratory of Astronomy of the University of Thessaloniki hosted IAU Symposium 25. This meeting was devoted to the interaction between astronomers working on two widely different fields of Dynamical Astronomy, namely Galactic Dynamics and Celestial Mechanics, in the hope that the methods used traditionally in one of the fields could prove useful in the other. Indeed, several papers presented in this meeting followed the above line. In two of them Hénon [2], on the one hand and, independently, Szebehely and Bozis [3] on the other, reported that they had found indications for the existence of a further integral of motion in the planar circular restricted three-body problem (a two-degrees of freedom dynamical system), besides the well known Jacobi integral.

Subsequently Contopoulos [1] showed how this integral could be constructed in a series form through an algorithm similar to the one he had proposed already [4] for the "third" integral in the case of a galactic type potential, in which (series) the zeroth order term is the angular momentum. At the same time Bozis [5] [6] studied extensively the properties of this new integral, as well as the computation, through its use, of "generalized" elements of motion (e.g. eccentricity, see next paragraphs).

Since Poincaré had shown that the three-body problem is non-integrable, it is obvious that this integral can only be a "local" (non-isolating) one. Therefore one should inquire in which regions of phase space this integral may be applied, as it was initiated by Bozis [6]. These regions should be called "regular", since the corresponding dynamical system has two degrees of freedom and, therefore, in the regions where there exist two integrals of motion, it behaves like an integrable one.

A three-dimensional elliptic orbit of the two-body problem is uniquely defined by three quantities, the three *elements* of the orbit a, e and I, where I is the *inclination* of the plane of the orbit with respect to a "reference" plane, a is the semi-major axis of the ellipse and e the *eccentricity*. In what follows we consider the motion of massless test-particles (i.e. asteroids) relative to a massive central body (i.e. the Sun) of mass M. The orbital elements of the minor planet are related to the energy, E and the angular momentum, h, of its orbit, through the relations

$$a = -\frac{GM}{2E} \tag{1}$$

$$e = \sqrt{1 + \frac{2Eh^2}{G^2M^2}} \tag{2}$$

For elliptic motion, the orbital energy, E, has to be negative.

It is worth to note that the two-body problem is an intrinsically degenerate dynamical system [9], a property that becomes obvious if we write the corresponding Hamiltonian in action-angle variables. One possible set of action angle variables in this case are the well known *modified Delaunay variables*, defined through the relations

$$\Lambda = \sqrt{G M a} \quad \lambda = \varpi + l \tag{3}$$

$$\Gamma = \Lambda(1 - \sqrt{1 - e^2}) \quad \gamma = -\varpi \tag{4}$$

$$Z = \Gamma(1 - \cos i) \quad \zeta = -\Omega \tag{5}$$

where the angles Ω, ϖ and l are the three Euler angles: the first two define the orientation of the ellipse in space and the third one the position of the the asteroid on the ellipse. In Celestial Mechanics the various angles have their own names: Ω is the *longitude of the ascending node* of the orbit, $\varpi = \Omega + \omega$ is the *longitude of the pericenter* and $\lambda = \varpi + l$ is the *mean longitude*. The *mean anomaly*, l, is related to time through the relation $l = nt$, where n is the *mean motion of the planet*, i.e. its mean angular frequency around the massive central body. The Hamiltonian of the two-body problem, written in the above variables, becomes simply

$$H = -\frac{G^2 M^2}{2 \Lambda^2} \tag{6}$$

i.e. it depends only on the action corresponding to the energy, which, according to (1), depends only on the semi-major axis.

The two-body problem is only a simple approximation of a planet's motion around the Sun. A better approximation is the restricted three-body problem. In this model a massless particle is moving in the gravitational field of two bodies, a central massive primary of mass M (the Sun) and a perturbing planet of mass m (say Jupiter). Moreover, the motion of the perturbing planet around the Sun is a Keplerian closed orbit (i.e. either a circle or an ellipse). The trajectory of the massless body is not anymore an ellipse, due to the perturbations induced by the planet. However, due to the small mass of the perturber relative to the Sun and for relatively large separation between the asteroid and the perturber, the trajectory can be described by means of the *osculating elements*, i.e. instantaneous values of the variables $a(t)$, $e(t)$ and $I(t)$, defined as the elements of an ellipse that is tangent to the real orbit at time t. The process is very easily implemented, since it reduces to the calculation of the elements of the orbit from the instantaneous values of the energy and the angular momentum (which, of course, are not anymore constants in the case of the three-body problem).

2 The Way Things Might Have Happened

2.1 Ordered Trajectories

From the form of the Hamiltonian alone and some educated guesses, one could relatively easily arrive at the form of the third integral, for the existence of which Hénon, Szebehely and Bozis had found numerical evidence, as follows. In the restricted three-body problem the Hamiltonian can be "split" into two parts, one of order zero with respect to the *mass ratio*, $\mu = \frac{m}{M+m}$, and one of order unity. In modified Delaunay variables the zeroth-order term depends only on Λ, while the other two actions appear only in the first order term, which therefore may be considered as a "perturbation". Thus, we have again a case of degeneracy, similar to the one appearing in the two-body problem. Due to this degeneracy, the Fourier expansion of the perturbation contains terms that do not depend on the angle λ. Therefore, if one ignores the terms involving λ and λ' [1], which become important only when they are almost resonant, the osculating semi-major axis, a, is constant, a famous result known as the *Laplace-Lagrange linear theory of secular motion*. Then E is constant to a linear approximation as well, since it depends only on the osculating semi-major axis through (1). As a consequence and, in view of (2), the osculating eccentricity, e, is, to a linear approximation, a function of h only, i.e. e depends, essentially, only on the angular momentum. Therefore it is natural to expect that, if one would attempt to calculate a "third" integral *for the full, non-linearized problem* as a series, using as a small parameter the mass ratio, μ, the zero-order term should be the angular momentum of the massless body on its (unperturbed) orbit around the central body. This is exactly the method used by Contopoulos [1]. In the same linear approximation as for a, the osculating eccentricity of the asteroid is given by

$$e^2 = e_f^2 + e_P^2 + 2\, e_f\, e_P \cos(g_P\, t + \beta_P), \tag{7}$$

where e_f, e_P, g_P and β_P (the phase at $t = 0$) are constants. In particular e_f (usually called *forced eccentricity*) and g_P (*proper frequency*) depend only on a and μ, while e_P is the constant amplitude of variation of the osculating

[1] Note that by a prime we denote the angles of the perturbing planet

eccentricity. In the full, non-linearized problem, e_P can be calculated through an algorithm similar to the one used by Contopoulos [1], and is called the *proper eccentricity*.

Since the circular restricted three-body problem is a two-degrees of freedom autonomous dynamical system, the existence of a second integral of motion would imply integrability. In this case all trajectories would be ordered and the secular solution would always remain $\mathcal{O}(\mu)$ close to the real solution. Note that (7) is the simplest secular theory of Celestial Mechanics (e.g. see Yuasa [7] or Milani and Knežević [8]). This result can be generalized for forms of the restricted three-body problem with more than two degrees of freedom, such as the elliptic (where the orbit of the perturber is an ellipse) or the three-dimensional (where the massless body moves outside the plane of the orbit of the perturber). In these cases one would need to calculate further integrals of motion, in the same spirit. As far as the total number of integrals is equal to the number of degrees of freedom of the corresponding (autonomous) dynamical system, all trajectories would be ordered. In this way we see that the three proper elements of the trajectory (or the associated modified Delaunay variables) constitute a set of action variables (and hence integrals of motion) of the *secular three-body problem*.

2.2 Chaotic Trajectories

The proper elements of ordered trajectories of asteroids are calculated through the secular theory at any desirable level of accuracy. However we know, from the work of Poincaré, that the restricted three-body problem does not admit any further integrals of motion, *analytic* in any variables. Therefore the corresponding dynamical system is non-integrable and the integrals in series form calculated through the method of Contopoulos (or some secular theory) can only be non-isolating, local ones. Hence in the vicinity of orbital resonances between the test-particle and the perturber (i.e. resonances between the angles λ and λ') the secular theory should fail, as a result of the small divisors problem and the appearance of chaotic motion. This means that all specific models of the restricted three-body problem (e.g. circular, elliptic or three-dimensional) should possess chaotic phase-space regions, besides the ordered ones. What can we say on the properties of chaotic trajectories? This problem was attacked by many authors through extensive numerical calculations, according to the available, at any period, computing power. The first model studied was the simplest one, namely the planar circular restricted three-body problem.

Soon it was realized, however, that this model does not represent the generic case, since it corresponds to an autonomous dynamical system with two degrees of freedom. But in this class of dynamical systems Arnold's diffusion (see e.g. [9]), which might play an important role in solar system dynamics, cannot be taken into account. Therefore, if we would like to consider a "generic" model for three-body dynamics, we should have at least three degrees of freedom! Consequently one should use as a "generic model" either the elliptic planar restricted or the circular three-dimensional restricted problem and not the planar circular. This was done by Contopoulos, who calculated the form of the "third" integral in the case of the three-dimensional restricted three-body problem [10] and the planar elliptical three-body problem [11].

The difference between the circular restricted three-body problem, on one hand, and the three-dimensional or elliptic restricted problem, on the other, is qualitative[2]. In both cases there exists a global (isolating) integral, which is the Jacobi integral in the first and the Hamiltonian of the extended phase-space in the second. But in the first case the situation is clear-cut: a specific trajectory is either ordered (if an additional local integral exists) or chaotic (if no local integrals exist). In the second case, however, there may exist from none to two local integrals of motion [12]. Two local integrals imply regular behavior and ordered trajectories, for which the secular solution would be an accurate approximation. The other two sub-cases correspond to chaotic motion, but with significant differences. If no local integrals exist, the chaotic trajectory covers densely a sub-manifold of the phase-space, defined by the constant "energy" surface. If one local integral exists, then the trajectory lies on a manifold which is the cartesian product of a two-dimensional torus with an annulus [18] (see Fig. 1). The motion on the two-torus corresponds to the ordered part of the trajectory, originating from the existence of the two integrals, while the motion on the annulus corresponds to the chaotic part.

In the case where no local integrals exist, the motion is "fully" chaotic, i.e. macroscopically it is equivalent to a random walk. Therefore, one might use methods of statistical mechanics (e.g. a Fokker-Planck-type equation) in order

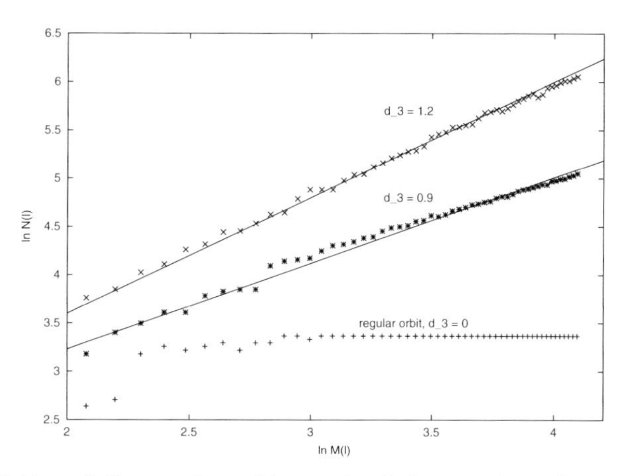

Fig. 1. Calculation of the number of integrals of three trajectories, one ordered and two stable chaotic, in the region of the 12:7 orbital resonance (from [15]). According to the theory, if we partition a 3-D space in M^3 bins of side l, N of which are occupied by a trajectory, then we have that $\log N(l) \sim d\_3 \log M(l)$, where $d\_3 = 3 - d$, and d is the number of integrals. The regular orbit yields $d\_3 = 0$, i.e. $d = 3$, while stable-chaotic orbits have $d\_3 \approx 1$, i.e. $d \approx 2$

[2] The 2-D elliptic and the 3-D circular problem are also by no means equivalent to each other.

to describe the evolution of a set of initial conditions as a diffusion process in the elements space. Since, according to what has been already said, the semi-major axis is constant to a linear approximation, we can select as a dependent variable either the eccentricity or the inclination. The eccentricity is our first choice, since it is intimately related to the escape of asteroids from the main belt.

It is easy to see that e increases on the average, since if we consider the chaotic motion as a random walk in eccentricity space, there is a reflecting wall at $e = 0$! Moreover, as e increases the resonances begin to overlap and chaotic motion becomes dominant. Therefore asteroids in fully chaotic trajectories follow more and more elongated orbits, until they hit a planet and are removed from the distribution. An analytic theory for the diffusion of asteroids was developed by Murray and Holman [13] and was recently applied, with considerable success, for the estimation of the age of the Veritas family of asteroids [14].

In the case where one local integral exists, the motion is "partially" chaotic, which means that some degrees of freedom are evidently chaotic and some appear as being ordered. From extensive numerical experiments it is relatively straightforward to show that the evolution of a is chaotic, while e and I change almost quasi-periodically with time, their proper values being almost constant [16] [17] [18] (Fig. 2). But, according to the secular theory, a only undergoes bounded erratic oscillations and does not change secularly, unless of course the trajectory escapes from the (non-isolated) region of the elements' space, where it is restricted by the level surfaces of the local integral. Since the usual way for the classification of trajectories is through the calculation of the Maximal Lyapunov Number, which in this case is positive, "partially chaotic" trajectories could be named, as well, "stable chaotic". Since for a stable chaotic trajectory e_P does not increase on the average, there are no collisions with other planets and, therefore, no escapes.

Extensive numerical work has shown that another important property of a phase-space region, besides the existence of local integrals of motion, is the existence or not of simple-periodic resonant trajectories. Although in the restricted circular three-body problem all orbital resonances with Jupiter correspond to periodic trajectories, this is not true for the elliptic problem. In general, orbital resonances do not correspond to periodic trajectories, unless their period is an exact multiple of Jupiter's revolution period [16]. Thus, the chaotic regions of phase space (i.e. the resonances' zones), in the planar elliptic (or the three-dimensional circular) restricted three-body problem, can be classified into three classes as follows, according to the type of trajectories they contain and the existence or not of periodic trajectories [16] [17] [18].

Stable chaotic regions constitute the first class. In such a region the evolution of trajectories is not diffusive. Chaotic trajectories are semi-confined by the level surfaces of the local integrals. Since, however, these surfaces are non-isolating, the trajectory eventually escapes from such a region through the "holes" of the "invariant" manifold. After such an escape, the eccentricity increases steeply. Numerical experiments have shown that the typical time-scale, T, for escape

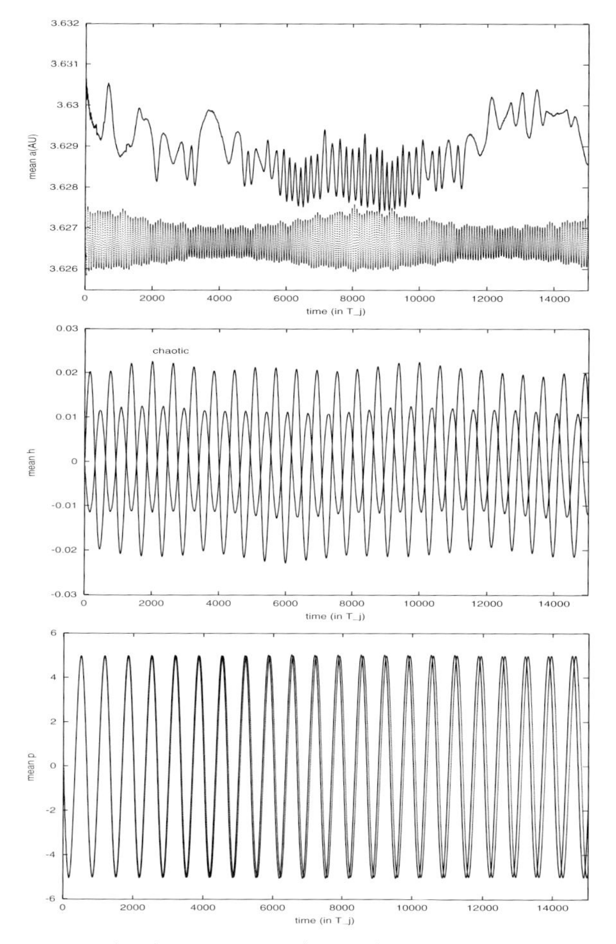

Fig. 2. The elements a (top), $h = e \sin \varpi$ (middle) and $p = I \sin \Omega$ (bottom) are given, as functions of time, for one regular and one stable-chaotic orbit of the elliptic three-body problem in the vicinity of the 12:7 orbital resonance (from [15]). The unit of time is the revolution period of Jupiter, $T_J \approx 11.86$ yr. It is easy to realize the different character of the motion between these two orbits, by monitoring the behavior of a. On the other hand, one cannot decide whether an orbit is regular or chaotic by just observing the graphs of h or p

through this process is $T \sim 1$ Gyr and can even exceed the age of the solar system (5 Gyrs), depending on the specific resonance.

Fully chaotic regions are divided into two classes, according to whether they support simple periodic orbits or not. If there are no periodic orbits, the evolution is diffusive, i.e. a trajectory undergoes many small “jumps” in eccentricity. This case is the one that can be described successfully through a diffusion equation and its typical time-scale, as can be calculated by the values of the diffusion coefficient, is of the order of 100 Myrs $< T <$ 1,000 Myrs (again, depending on the specific orbital resonance).

If there exist periodic orbits, then the evolution of chaotic trajectories is "fast" and intermittent, as the trajectory from time to time follows the unstable periodic orbit. This is the kind of motion found by Wisdom [19] and Hadjidemetriou [20]. The typical time-scale for the "jumps" is of the order of $5 \cdot 10^5$ yrs, while the escape time is of order $10^5 < T < 10^6$ yrs. There are only 5 such resonances in the phase-space region that corresponds to the main asteroid belt, in both the elliptic and the three-dimensional restricted three-body problems. These are the 2:1, 3:1, 4:1, 5:2 and 7:3 orbital resonances with Jupiter. Since the more well-known Kirkwood gaps lie exactly at these resonances, one arrives easily at the conclusion that the existence of a periodic trajectory is the common factor that differentiates between orbital resonances, associated with a Kirkwood gap, and those that are not.

Summarizing, we can say that stable chaos is the observational manifestation of the existence of a local integral of motion, while the Kirkwood gaps appear at resonances where periodic orbits exist, in the elliptic or the three-dimensional restricted three-body problem.

3 The Way Things Really Happened

Unfortunately, the evolution of ideas in science does not always follow the "obvious" path. The applicability of local integrals of motion presents another case of misunderstanding between theorists and applied-oriented astronomers. The scientific community of Celestial Mechanics did not capitalize on the work of Bozis and Contopoulos, related to the existence of local integrals of motion and the calculation of "primitive proper elements". Instead, for quite some time, the calculation of proper elements was only used for objects that move far away from the main resonances, where secular theory could apply.

Things started to change in the 1980's, when algorithms for the calculation of the maximal LCN were made available and Wisdom [19] found the "intermittent" behavior of the osculating eccentricity in the vicinity of the 3:1 resonance, which is characteristic of the existence of an unstable periodic trajectory. However, since as a rule only the maximal LCN was calculated, there was no way to differentiate between regions where none or one local integral exists. That is why the chaotic motion in the regions where local integrals exists was considered "peculiar" and termed *stable chaos*.

The first to point out that stable chaotic motion is not "fully chaotic" were Varvoglis and Anastasiadis [21]. This idea was subsequently elaborated in a series of papers by Tsiganis, Varvoglis and Hadjidemetriou [16] [17] [18]. In these papers it is shown, through the computation of autocorrelation functions, that stable-chaotic trajectories have almost constant proper elements, i.e. they possess local integrals of motion (see Fig. 1), and lie at the border between fully chaotic and regular phase-space regions. Consequently, stable-chaotic orbits represent cases of *sticky motion* in G and H (i.e. essentially eccentricity and inclination) and chaotic motion in L (i.e. semi-major axis), a type of motion for which no analogue exists in two-dimensional dynamical systems. The subsequent numerical calculation of the number of integrals, preserved by a large number of trajectories of the elliptic restricted three-body problem [15], confirmed this picture. In this way today we arrived finally, after thirty-six years, in the "re-discovery" of the work of Contopoulos-Bozis and its connection to proper elements, by understanding the phenomenon of stable chaos and its relation to local integrals of motion.

References

1. G. Contopoulos: Astrophys. J. **142**, 802 (1965)
2. M. Hénon: "Numerical exploration of the restricted three-body problem". In: *The theory of orbits in the Solar System and in stellar systems, IAU Symposium No. 25, Thesaloniki, Greece, 17–22 August 1964*, ed. by G. Contopoulos (Academic Press, London 1966) pp. 157–163
3. V. Szebehely: Discussion of the paper "Numerical exploration of the restricted three-body problem". In: *The theory of orbits in the Solar System and in stellar systems, IAU Symposium No. 25, Thesaloniki, Greece, 17–22 August 1964*, ed. by G. Contopoulos (Academic Press, London 1966) pp. 163–169
4. G. Contopoulos: Z. Astrophys. **49**, 273 (1960)
5. G. Bozis: Astron. J. **71**, 404 (1966)
6. G. Bozis: Astron. J. **72**, 380 (1967)
7. M. Yasa: Publ. Astron. Soc. Japan **25**, 399 (1973)
8. A. Milani, Z. Knežević: Cel. Mech. **49**, 247 (1990)
9. A.J. Lichtenberg, M.A. Lieberman: *Regular and Chaotic Dynamics*, 2nd edn. (Springer, New York 1992)
10. G. Contopoulos: Astron. J. **72**, 191 (1967)
11. G. Contopoulos: Astron. J. **72**, 669 (1967)
12. G. Contopoulos, L. Galgani, A. Giorgilli: Phys. Rev. A **18**, 1183 (1978)
13. N. Murray, M. Holman: Astron. J. **114**, 1246 (1997)
14. Z. Knežević, K. Tsiganis, H. Varvoglis: "The dynamical portrait of the Veritas family region". In: *Proceedings of the Conference Asteroids, Comets, Meteors - ACM2002, Technical University Berlin, 29 July-2 August 2002*, ed. by B. Warmbein (ESA SP-500, Noordwijk, 2002) pp. 335–338
15. H. Varvoglis, K. Tsiganis, G. Hadjivantsides: "Stable chaos and local integrals of motion". In: *Proceedings of the Conference Asteroids, Comets, Meteors - ACM2002, Technical University Berlin, 29 July-2 August 2002*, ed. by B. Warmbein (ESA SP-500, Noordwijk, 2002) pp. 355–357
16. K. Tsiganis, H. Varvoglis, J. Hadjidemetriou: Icarus **146**, 240 (2000)
17. K. Tsiganis, H. Varvoglis, J. Hadjidemetriou: Icarus **155**, 454 (2002)
18. K. Tsiganis, H. Varvoglis, J. Hadjidemetriou: Icarus **159**, 284 (2002)
19. J. Wisdom: Astron. J. **87**, 577 (1982)
20. J. Hadjidemetriou: Celest. Mech. Dyn. Astron. **56**, 563 (1993)
21. H. Varvoglis, A. Anastasiadis: Astron. J. **71**, 404 (1996)

Lecture Notes in Physics

For information about Vols. 1–582
please contact your bookseller or Springer-Verlag
LNP Online archive: http://www.springerlink.com/series/lnp/

Vol.583: W. Plessas, L. Mathelitsch (Eds.), Lectures on Quark Matter.

Vol.584: W. Köhler, S. Wiegand (Eds.), Thermal Nonequilibrium Phenomena in Fluid Mixtures.

Vol.585: M. Lässig, A. Valleriani (Eds.), Biological Evolution and Statistical Physics.

Vol.586: Y. Auregan, A. Maurel, V. Pagneux, J.-F. Pinton (Eds.), Sound–Flow Interactions.

Vol.587: D. Heiss (Ed.), Fundamentals of Quantum Information. Quantum Computation, Communication, Decoherence and All That.

Vol.588: Y. Watanabe, S. Heun, G. Salviati, N. Yamamoto (Eds.), Nanoscale Spectroscopy and Its Applications to Semiconductor Research.

Vol.589: A. W. Guthmann, M. Georganopoulos, A. Marcowith, K. Manolakou (Eds.), Relativistic Flows in Astrophysics.

Vol.590: D. Benest, C. Froeschlé (Eds.), Singularities in Gravitational Systems. Applications to Chaotic Transport in the Solar System.

Vol.591: M. Beyer (Ed.), CP Violation in Particle, Nuclear and Astrophysics.

Vol.592: S. Cotsakis, L. Papantonopoulos (Eds.), Cosmological Crossroads. An Advanced Course in Mathematical, Physical and String Cosmology.

Vol.593: D. Shi, B. Aktaş, L. Pust, F. Mikhailov (Eds.), Nanostructured Magnetic Materials and Their Applications.

Vol.594: S. Odenbach (Ed.),Ferrofluids. Magnetical Controllable Fluids and Their Applications.

Vol.595: C. Berthier, L. P. Lévy, G. Martinez (Eds.), High Magnetic Fields. Applications in Condensed Matter Physics and Spectroscopy.

Vol.596: F. Scheck, H. Upmeier, W. Werner (Eds.), Noncommutative Geometry and the Standard Model of Elememtary Particle Physics.

Vol.597: P. Garbaczewski, R. Olkiewicz (Eds.), Dynamics of Dissipation.

Vol.598: K. Weiler (Ed.), Supernovae and Gamma-Ray Bursters.

Vol.599: J.P. Rozelot (Ed.), The Sun's Surface and Subsurface. Investigating Shape and Irradiance.

Vol.600: K. Mecke, D. Stoyan (Eds.), Morphology of Condensed Matter. Physcis and Geometry of Spatial Complex Systems.

Vol.601: F. Mezei, C. Pappas, T. Gutberlet (Eds.), Neutron Spin Echo Spectroscopy. Basics, Trends and Applications.

Vol.602: T. Dauxois, S. Ruffo, E. Arimondo (Eds.), Dynamics and Thermodynamics of Systems with Long Range Interactions.

Vol.603: C. Noce, A. Vecchione, M. Cuoco, A. Romano (Eds.), Ruthenate and Rutheno-Cuprate Materials. Superconductivity, Magnetism and Quantum Phase.

Vol.604: J. Frauendiener, H. Friedrich (Eds.), The Conformal Structure of Space-Time: Geometry, Analysis, Numerics.

Vol.605: G. Ciccotti, M. Mareschal, P. Nielaba (Eds.), Bridging Time Scales: Molecular Simulations for the Next Decade.

Vol.606: J.-U. Sommer, G. Reiter (Eds.), Polymer Crystallization. Obervations, Concepts and Interpretations.

Vol.607: R. Guzzi (Ed.), Exploring the Atmosphere by Remote Sensing Techniques.

Vol.608: F. Courbin, D. Minniti (Eds.), Gravitational Lensing:An Astrophysical Tool.

Vol.609: T. Henning (Ed.), Astromineralogy.

Vol.610: M. Ristig, K. Gernoth (Eds.), Particle Scattering, X-Ray Diffraction, and Microstructure of Solids and Liquids.

Vol.611: A. Buchleitner, K. Hornberger (Eds.), Coherent Evolution in Noisy Environments.

Vol.612 L. Klein, (Ed.), Energy Conversion and Particle Acceleration in the Solar Corona.

Vol.613 K. Porsezian, V.C. Kuriakose, (Eds.), Optical Solitons. Theoretical and Experimental Challenges.

Vol.614 E. Falgarone, T. Passot (Eds.), Turbulence and Magnetic Fields in Astrophysics.

Vol.615 J. Büchner, C.T. Dum, M. Scholer (Eds.), Space Plasma Simulation.

Vol.616 J. Trampetic, J. Wess (Eds.), Particle Physics in the New Millenium.

Vol.617 L. Fernández-Jambrina, L. M. González-Romero (Eds.), Current Trends in Relativistic Astrophysics, Theoretical, Numerical, Observational

Vol.618 M.D. Esposti, S. Graffi (Eds.), The Mathematical Aspects of Quantum Maps

Vol.619 H.M. Antia, A. Bhatnagar, P. Ulmschneider (Eds.), Lectures on Solar Physics

Vol.620 C. Fiolhais, F. Nogueira, M. Marques (Eds.), A Primer in Density Functional Theory

Vol.621 G. Rangarajan, M. Ding (Eds.), Processes with Long-Range Correlations

Vol.622 F. Benatti, R. Floreanini (Eds.), Irreversible Quantum Dynamics

Vol.623 M. Falcke, D. Malchow (Eds.), Understanding Calcium Dynamics, Experiments and Theory

Vol.624 T. Pöschel (Ed.), Granular Gases

Vol.625 R. Pastor-Satorras, M. Rubi, A. Diaz-Guilera (Eds.), Statistical Mechanics of Complex Networks

Druck: Strauss Offsetdruck, Mörlenbach
Verarbeitung: Schäffer, Grünstadt

The Daily Trading Coach

Founded in 1807, John Wiley & Sons is the oldest independent publishing company in the United States. With offices in North America, Europe, Australia and Asia, Wiley is globally committed to developing and marketing print and electronic products and services for our customers' professional and personal knowledge and understanding.

The Wiley Trading series features books by traders who have survived the market's ever changing temperament and have prospered—some by reinventing systems, others by getting back to basics. Whether a novice trader, professional or somewhere in-between, these books will provide the advice and strategies needed to prosper today and well into the future.

For a list of available titles, visit our Web site at www.WileyFinance.com.

The Daily Trading Coach

101 Lessons for Becoming Your Own Trading Psychologist

BRETT N. STEENBARGER

John Wiley & Sons, Inc.

Published by John Wiley & Sons, Inc., Hoboken, New Jersey.
Published simultaneously in Canada.

For general information on our other products and services or for technical support, please contact our Customer Care Department within the United States at (800) 762-2974, outside the United States at (317) 572-3993 or fax (317) 572-4002.

Wiley also publishes its books in a variety of electronic formats. Some content that appears in print may not be available in electronic books. For more information about Wiley products, visit our web site at www.wiley.com.

Library of Congress Cataloging-in-Publication Data:

Steenbarger, Brett N.
The daily trading coach : 101 lessons for becoming your own trading psychologist / Brett N. Steenbarger.
p. cm. – (Wiley trading series)
Includes index.
ISBN 978-0-470-39856-2 (cloth)
1. Stocks–Psychological aspects. 2. Speculation–Psychological aspects. 3. Investments–Psychological aspects. 4. Self-help techniques. 5. Personal coaching. I. Title. II. Title: Becoming your own trading psychologist.
HG6041.S757 2009
332.6′4019–dc22

2008041524

Printed in the United States of America.

10 9 8 7 6 5 4 3 2 1

What? A great man? I always see merely
the play-actor of his own ideal.

—Friedrich Nietzsche

Contents

Preface xiii

Acknowledgments xvii

Introduction 1

CHAPTER 1 Change: The Process and the Practice 3

Lesson 1: Draw on Emotion to Become a Change Agent 4
Lesson 2: Psychological Visibility and Your Relationship with Your Trading Coach 7
Lesson 3: Make Friends with Your Weakness 9
Lesson 4: Change Your Environment, Change Yourself 11
Lesson 5: Transform Emotion by Trace-Formation 14
Lesson 6: Find the Right Mirrors 17
Lesson 7: Change Our Focus 20
Lesson 8: Create Scripts for Life Change 23
Lesson 9: How to Build Your Self-Confidence 25
Lesson 10: Five Best Practices for Effecting and Sustaining Change 29
Resources 32

CHAPTER 2 Stress and Distress: Creative Coping for Traders 33

Lesson 11: Understanding Stress 33
Lesson 12: Antidotes for Toxic Trading Assumptions 37

Lesson 13: What Causes the Distress That Interferes with Trading Decisions? 40
Lesson 14: Keep a Psychological Journal 43
Lesson 15: Pressing: When You Try Too Hard to Make Money 45
Lesson 16: When You're Ready to Hang It Up 48
Lesson 17: What to Do When Fear Takes Over 51
Lesson 18: Performance Anxiety: The Most Common Trading Problem 54
Lesson 19: Square Pegs and Round Holes 58
Lesson 20: Volatility of Markets and Volatility of Mood 61
Resources 64

CHAPTER 3 Psychological Well-Being: Enhancing Trading Experience 67

Lesson 21: The Importance of Feeling Good 67
Lesson 22: Build Your Happiness 71
Lesson 23: Get into the Zone 73
Lesson 24: Trade with Energy 77
Lesson 25: Intention and Greatness: Exercise the Brain through Play 79
Lesson 26: Cultivate the Quiet Mind 83
Lesson 27: Build Emotional Resilience 86
Lesson 28: Integrity and Doing the Right Thing 89
Lesson 29: Maximize Confidence and Stay with Your Trades 91
Lesson 30: Coping—Turn Stress into Well-Being 95
Resources 97

CHAPTER 4 Steps toward Self-Improvement: The Coaching Process 99

Lesson 31: Self-Monitor by Keeping a Trading Journal 99
Lesson 32: Recognize Your Patterns 103
Lesson 33: Establish Costs and Benefits to Patterns 106
Lesson 34: Set Effective Goals 109

Lesson 35: Build on Your Best: Maintain a Solution Focus 111
Lesson 36: Disrupt Old Problem Patterns 114
Lesson 37: Build Your Consistency by Becoming Rule-Governed 118
Lesson 38: Relapse and Repetition 121
Lesson 39: Create a Safe Environment for Change 123
Lesson 40: Use Imagery to Advance the Change Process 126
Resources 130

CHAPTER 5 Breaking Old Patterns: Psychodynamic Frameworks for Self-Coaching 131

Lesson 41: Psychodynamics: Escape the Gravity of Past Relationships 132
Lesson 42: Crystallize Our Repetitive Patterns 135
Lesson 43: Challenge Our Defenses 138
Lesson 44: Once Again, with Feeling: Get Distance from Your Problem Patterns 141
Lesson 45: Make the Most Out of Your Coaching Relationship 144
Lesson 46: Find Positive Trading Relationships 147
Lesson 47: Tolerate Discomfort 150
Lesson 48: Master Transference 153
Lesson 49: The Power of Discrepancy 156
Lesson 50: Working Through 158
Resources 161

CHAPTER 6 Remapping the Mind: Cognitive Approaches to Self-Coaching 163

Lesson 51: Schemas of the Mind 164
Lesson 52: Use Feeling to Understand Your Thinking 167
Lesson 53: Learn from Your Worst Trades 170
Lesson 54: Use a Journal to Restructure Our Thinking 172
Lesson 55: Disrupt Negative Thought Patterns 176
Lesson 56: Reframe Negative Thought Patterns 179

Lesson 57: Use Intensive Guided Imagery to Change Thought Patterns 182
Lesson 58: Challenge Negative Thought Patterns with the Cognitive Journal 185
Lesson 59: Conduct Cognitive Experiments to Create Change 188
Lesson 60: Build Positive Thinking 190
Resources 193

CHAPTER 7 Learning New Action Patterns: Behavioral Approaches to Self-Coaching 195

Lesson 61: Understand Your Contingencies 196
Lesson 62: Identify Subtle Contingencies 199
Lesson 63: Harness the Power of Social Learning 201
Lesson 64: Shape Your Trading Behaviors 204
Lesson 65: The Conditioning of Markets 207
Lesson 66: The Power of Incompatibility 211
Lesson 67: Build on Positive Associations 214
Lesson 68: Exposure: A Powerful and Flexible Behavioral Method 217
Lesson 69: Extend Exposure Work to Build Skills 220
Lesson 70: A Behavioral Framework for Dealing with Worry 223
Resources 226

CHAPTER 8 Coaching Your Trading Business 227

Lesson 71: The Importance of Startup Capital 227
Lesson 72: Plan Your Trading Business 231
Lesson 73: Diversify Your Trading Business 233
Lesson 74: Track Your Trading Results 236
Lesson 75: Advanced Scorekeeping for Your Trading Business 240

Lesson 76: Track the Correlations of Your Returns 244
Lesson 77: Calibrate Your Risk and Reward 248
Lesson 78: The Importance of Execution in Trading 250
Lesson 79: Think in Themes—Generating Good Trading Ideas 254
Lesson 80: Manage the Trade 257
Resources 259

CHAPTER 9 Lessons from Trading Professionals: Resources and Perspectives on Self-Coaching 261

Lesson 81: Leverage Core Competencies and Cultivate Creativity 261
Lesson 82: I Alone Am Responsible 264
Lesson 83: Cultivate Self-Awareness 271
Lesson 84: Mentor Yourself for Success 275
Lesson 85: Keep Detailed Records 279
Lesson 86: Learn to Be Fallible 283
Lesson 87: The Power of Research 286
Lesson 88: Attitudes and Goals, the Building Blocks of Success 290
Lesson 89: A View from the Trading Firms 295
Lesson 90: Use Data to Improve Trading Performance 300
Resources 305

CHAPTER 10 Looking for the Edge: Finding Historical Patterns in Markets 307

Lesson 91: Use Historical Patterns in Trading 308
Lesson 92: Frame Good Hypotheses with the Right Data 310
Lesson 93: Excel Basics 313
Lesson 94: Visualize Your Data 317
Lesson 95: Create Your Independent and Dependent Variables 320
Lesson 96: Conduct Your Historical Investigations 324

Lesson 97: Code the Data 327
Lesson 98: Examine Context 329
Lesson 99: Filter Data 332
Lesson 100: Make Use of Your Findings 334
Resources 336

CONCLUSION 339

Lesson 101: Find Your Path 339
For More on Self-Coaching 341

About the Author 343

Index 345

Preface

The goal of *The Daily Trading Coach* is to teach you as much as possible about coaching, so that you can mentor yourself to success in the financial markets. The key word in the title is "Daily." This book is designed to be a resource that you can use every day to build upon strengths and overcome weaknesses.

After writing two books—*The Psychology of Trading* and *Enhancing Trader Performance*—and penning more than 1,800 posts for the *TraderFeed* blog (www.traderfeed.blogspot.com/), I thought I had pretty well covered the terrain of trading psychology. Now, just three years after the publication of the performance book, I've once again taken electronic pen to paper, completing a trading psychology trilogy by focusing on the *process* of coaching.

Two realities led to *The Daily Trading Coach*. First, a review of the traffic patterns on the *TraderFeed* blog revealed that a large number of readers—about a third—were accessing the site during the hour or so immediately prior to the market open. I found this interesting, as most of the posts do not offer specific trading advice. Rather, posts deal with topics of psychology and performance—ones that should be relevant at any hour of the day.

When I asked a group of trusted readers about this pattern, they responded that they were using the blog as a kind of surrogate trading coach. Reviewing the posts was their way of reminding themselves of their plans and intentions before going entering the financial battlefield. This was confirmed when I gathered statistics about the most popular (and commented upon) posts on the blog. The majority were practical posts dealing with trading psychology. Most were uplifting in content, even as they challenged the assumptions of readers. It seemed as though traders were looking for coaching and finding some measure of it in the blog.

The second reality shaping this book involves digital publication and the rapid changes sweeping the publishing world. To this point, relatively few electronic books (e-books) have been offered to traders. When those books are available, they are little more than screen versions of the print text. Despite the allure and convenience of electronic publishing, few traders I consulted actually sought out or used e-books. The most common complaint among traders was that they did not want to spend hours devouring information in front of a screen after a full day of trading. I quickly realized that participants in the financial markets don't use the electronic medium in the same way that they engage print text. That led me to think about writing a different kind of book, one better suited to publishing's electronic frontier, but also useable in print.

When you overlay these two observations, you can appreciate the vision that led to this text: *a "trading coach in a book" that can be as easily read on the screen as on paper. The goal was to integrate blog and book content by creating practical "lessons" that help traders become their own trading coaches.* There are 101 lessons in *The Daily Trading Coach*, averaging several pages in length. Each lesson follows a general format, identifying a trading challenge, an approach to meeting that challenge, and a specific suggestion or assignment for working on the issue. The chapters are independent of one another: you can read them in order, or you can use the table of contents or index to read, each day, the lesson that most applies to your current trading. *Unlike a traditional book, the idea is not to read it through from front to back in a few sittings.* Rather, you take one lesson at a time and apply it to guide your development as a trader. Like the blog, it's an on-screen reminder of what to do when you're at your best, but—more than the blog—it's also a roadmap (and practical set of insights and tools) for discovering and implementing the best within you.

My ambition has been to pack into these 101 lessons more useable information and practical methods than might be found in any number of expensive seminars and coaching sessions, at far less expense. Too often, the goal of the seminar providers and coaches is to convert you into ongoing clients. The intent of this book is just the opposite: *to give you the tools to become your own coach*, so that you can guide your own professional and personal growth. In other words, this is a manual of *psychoeducation*: a how-to guide for improving yourself and your performance.

One thing I particularly like about the electronic format is that it enables a writer to link the book content to a vast array of material on the Web. I will be adding material to *The Daily Trading Coach* via a dedicated blog called *Become Your Own Trading Coach* (www.becomeyourowntradingcoach.blogspot.com), so that this book will grow over time. You will need only to click the e-book links to access free updated information and methods on the *Become Your Own Trading Coach* site. There is one master page on the blog for each chapter of this book containing the links relevant to that chapter's material. At the end of each chapter, there is also a resource page that alerts readers to further links and readings. I will be adding audio and video content to the new blog over time, which should be particularly helpful for those who learn best by seeing and hearing ideas. Once publishing becomes electronic, there's no reason that every text can't be a multimedia learning experience.

You'll notice from the table of contents that each of the 10 chapters contains 10 lessons. Those chapters cover a range of topics relevant to trading psychology and trading performance, including specific lessons for utilizing psychodynamic, cognitive, and behavioral brief therapy methods to change problematic behavior patterns and instill new, positive ones. The final two chapters are especially unique: Chapter 9 consists of self-coaching perspectives from 18 successful trading professionals who share their work online. Chapter 10 fulfills a long-standing promise to *Trader-Feed* readers, walking traders through the basics of identifying historical patterns using Excel. Each lesson is accompanied by homework activities and suggestions ("Coaching Cues") to help with application of the ideas. Major ideas are set apart within the text for quick review and scanning. At the end of each chapter is a list of resources to guide your further inquiry into the book's topics and ideas.

Yes, the aim of the book is to help you become your own trading coach, but a glance at the chapter and lesson titles reveals that the broader purpose is to help you coach yourself through life. The challenges and uncertainties we face in trading—the pursuit of rewards in the face of risks—are just as present in careers and relationships as in markets. Techniques that help you master yourself as a trader will serve you well in any field of endeavor. *In that sense, the goal is not just to make money in the markets, it is to prosper in all of life's undertakings.* I will be gratified and honored if this book is a resource toward your own prosperity, in and out of financial markets.

BRETT STEENBARGER

Acknowledgments

If, as the saying goes, it takes a village to raise a child, it takes a small army to write a book. The last lesson of the book is dedicated to my mother, Constance Steenbarger, who passed away last year. My deepest hope is that this book carries forward the nurturing spirit that she brought to her family and students.

If my mother represented nurturance in my life, my father, Jack Steenbarger, has embodied the virtues of hard work, achievement, and love of family. From the earliest days of my training as a psychologist, I have been fascinated by the psychology of exemplary achievement: what makes highly successful people tick. There's no question where that passionate interest originated, and it gives me the greatest of pleasure to acknowledge my father for that inspiration.

None of this would be possible, however, without the understanding, love, and support of my wife Margie. In 1984, I traded bachelorhood for a life with Margie and her family; to this day, it remains my one superlative trade. Twenty-five years later, I'm pleased to report we're still riding that trend, having taken no heat whatsoever!

I'm saddened, but happy at the same time, to be able to dedicate this book to the memory of my uncle, Arnold Rustin, MD, who also passed away during the year. A consummate teacher, Arnold represented everything I've admired and enjoyed in the world of academic medicine. It's the support of Arnold and his wife Rose, even amid their own challenges, which made the greatest impression on me, however. I hope their inspiration finds expression in this book.

Thanks, too, to Debi, Steve, Lea, Laura, Ed, Devon, and Macrae, the kids who aren't kids any more, but who have been remarkably understanding of my hours on the road meeting with traders and my even greater hours online, keeping up with a blog and dozens of e-mail and phone calls daily. I would not be so grounded without family, including my brother Marc and sister-in-law Lisa and our three feline friends: Gina, Ginger, and Mali.

To the traders and authors who contributed to Chapter 9, my deepest thanks and appreciation for your great work. You provide unparalleled resources for developing traders. Acknowledgments are also due to those whose work has inspired my own: philosophers Ayn Rand, Brand Blanshard, Colin Wilson, and G. I. Gurdjieff; the many psychologists and researchers who have contributed to the brief therapy and positive psychology literatures; and the traders who were formative in my development: Victor Niederhoffer, Linda Raschke, Chuck McElveen, and the many hedge fund traders I've been privileged to work with in the past few years. My colleagues at Upstate Medical University have been inspirational and supportive throughout my second career; special thanks to Mantosh Dewan, MD; Roger Greenberg, PhD; and John Manring, MD.

This is also my opportunity for a shout-out to those who write and play the music that kept me company through the writing of this book: Edenbridge, Armin van Buuren, Ferry Corsten, Cruxshadows, Assemblage 23, VNV Nation, and many others that you may discover on the *Become Your Own Trading Coach* blog.

Deepest thanks, as well, to the Wiley production staff and my fantastic and supportive editors, Pamela van Giessen, Kate Wood, and Emilie Herman. They've been tremendously helpful in bringing this book to life. My appreciation also goes out to the many readers of the blog, particularly those who have actively participated with their comments and insights. I hope this book contributes to your continued happiness and trading success.

Introduction

Too few of us are play-actors of our own ideals. We have strengths and talents, dreams and aspirations. But when we look hour-by-hour, day-by-day, not many of these ideals are concretely expressed. The days become months, then years, and—at some sad juncture—we look back on life and wonder where it went.

That could be you: the middle-age person looking back on how "I could've been a contender." Or, you could live a different life script. You could become the actor of your ideals and live their realization.

If you're thinking this is a strange introduction to a trading text, you're right. This book doesn't start with supply and demand, trading patterns, or money management. It begins with you and what you want from your life. Trading, in this context, is more than buying, selling, and hedging: it is a vehicle for self-mastery and development.

Every trader, whether he consciously identifies it or not, is an entrepreneur. Traders open their business and compete in a marketplace. They identify and pursue opportunity, even as they preserve their capital. Traders refine and expand their craft; they take calculated risks. As entrepreneurs, traders start with the premise that they bring value to the marketplace. Amid the inevitable disappointments and setbacks, the long hours and the limited resources, the risk and uncertainty, it can be difficult to sustain that optimism. It is so much easier than to keep one's visions on a shelf and forego the daily efforts of enacting ideals.

Some traders, however, cannot shelve their aspirations. Like the moth, they'll pursue distant lights even if it means an occasional singe. To those noble souls, I dedicate this book.

When I work with traders and portfolio managers at hedge funds, proprietary trading firms, and investment banks, I don't tell them how to trade. Most of them trade strategies different from my own and know far more about their markets than I ever will. Rather, I figure out their strengths. I learn what these traders and managers do well and how they do it, and I help them build a career out of what they're already good at. Just as fish cannot comprehend water, being immersed in it from birth, we typically lack an appreciation of our personal assets. Each of us is a curious mixture of skills, talents, strengths, conflicts, and weaknesses. But just as a new business must capitalize on the strengths of its founders, a career in the markets crucially hinges on the assets—personal and monetary—of the trader. As a coach, my role is to take traders out of their psychological water and help them see what has been around them all along: the assets that can provide a lifetime of dividends.

Never has self-coaching been more important for traders. As I write this, we have witnessed levels of market volatility unseen in the post-World War II period. Price volatility brings potential opportunity, but also risk. Traders who could not step back, recognize unfolding developments, and make adjustments have lost significant money. Those who have used the crisis to step out of the trading water, limit risk, and find fresh opportunity are the ones who are poised to reap those career dividends.

The book you are reading is intended to be your companion in this trading journey. It is organized in 101 lessons. Each lesson outlines a challenge and proposes a specific exercise for moving yourself forward with respect to that challenge. The lessons are intended as meditations to begin your trading day—coaching communications to help you enact the best within you. Eventually, as you read and live these lessons, the coaching communications become your own self-talk. You begin by play-acting the book's ideals and end up living them and shaping them into your own. *You become your own trading psychologist.*

If reading a short passage each day and planting the right ideas into your forebrain helps you prioritize your life and trading goals—and if that in turn helps you make one less bad trade per week and take the one good one you would have otherwise missed—think of how you will personally and financially profit. But just as pills can't work when they stay in a bottle, no one learns from an unopened book. The first step in becoming your own trading psychologist is to set time aside for self-mentorship—every day, every week—because that's how behavior patterns turn into habits. The great individual is simply one who has made a habit of self-development.

So there they are, staring at you from the shelf across the room: Your ideals, all those things you've wanted to do in life. You look longingly toward the shelf, but you can't reach it from your comfortable chair. Yet you hold a book in your hands. Perhaps that book can make that chair just a little less comfortable, place the shelf just a bit closer.

You turn the page.

The next step is ours.

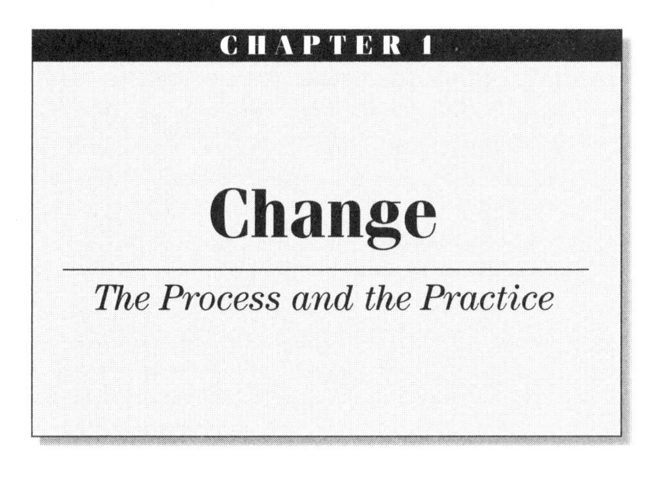

CHAPTER 1

Change

The Process and the Practice

The mind has exactly the same power as the hands;
not merely to grasp the world, but to change it.
—Colin Wilson

You are reading this book because you want to coach yourself to greater success in the financial markets. But what is coaching? At the root of all coaching efforts is change. When you are your own trading coach, you are trying to effect changes in your thoughts, your feelings, and your behavior. Most of all, you are trying to change how you trade: how you identify and act upon patterns of risk and reward, supply and demand.

There is a rich literature regarding change, grounded in extensive psychological research and practice. If you understand how change occurs, you are better positioned to act as your own change agent. In this chapter, we will explore the research and practice of change and how you can best make use of its sometimes-surprising conclusions. *Coaching is about making change happen, not just letting it happen.* It's about making the commitment to being a change agent in your own life, your own trading.

First, however, let's learn about the process and practice of change.

LESSON 1: DRAW ON EMOTION TO BECOME A CHANGE AGENT

For some of us, the status quo is not enough. We experience glimpses into the person we're capable of being; we yearn to be more than we are in life's mundane moments.

That yearning starts with the notion of *change*. We desire changes in our lives. We adapt—we grow—by making the right kinds of changes. All too often, however, we feel stuck. We're doing the same things, making the same mistakes again and again. Do we wait for life to change us, or do we become agents of our own life changes?

The easy part is initiating a change process. The real challenge is *sustaining* change. How many times does an alcoholic take the initial steps toward sobriety, only to relapse? How often do we start diets and exercise programs, only to return to our slothful ways? If we focus on starting a change process, we leave ourselves unprepared for the next crucial steps: keeping the flame of change burning bright.

The flaw with most popular writings and practices in psychology and coaching is that they are designed to initiate change. These writings and practices leave people feeling good—until it becomes apparent that different efforts are needed to sustain change. Successful coaching doesn't just catalyze change: it turns change efforts into habit patterns that become second nature. The key to successful coaching is turning change into routine; making new behaviors become second nature.

That's where emotion comes in.

For years I had attempted—unsuccessfully—to sustain a weight loss program. Then, in the year 2000, I was diagnosed with Type II diabetes. My diet *had* to change; I needed to lose weight. If I didn't, I realized with crystal clarity, I could lose my health and let my wife and children down. Literally that same day I began a dietary regimen that continues to this day. My weight dropped 40 pounds (I shed the pounds so quickly that friends were concerned that I had a wasting illness) and I regained control of my blood sugar.

What was the catalyst for the change? Years of telling myself to eat differently, exercise more, and lose weight produced absolutely no results. A single emotional experience of the necessity for change, however, made all the difference. I didn't just *think* I needed to change: I *knew* it with every fiber of my being. I *felt* it.

So it is with traders.

Perhaps you've told yourself that you need to follow your rules, that you need to trade smaller, or that you should avoid trading during certain market conditions or times of day. Still you make the same mistakes, lose money, and build frustration. Like my initial efforts at weight loss, your attempts at change fail *because they lack emotional force.*

Research into the process of successful versus unsuccessful therapy finds that emotional experience—not talk—powers change. No one ever felt valuable and lovable by standing in front of a mirror and reciting self-enhancing statements. The experience of a meaningful romantic relationship, however, yields the deepest of affirmations. Yes, you can tell yourself you're competent, but experiencing success in the face of challenge provides a lasting sense of efficacy. Pleasure, pain: nature hardwires us to internalize emotional experience so that we can pursue what enhances life and avoid what harms us. That ability to internalize our most powerful emotional experiences helps us to sustain the changes we initiate.

The enemy of change is relapse: falling back into old, unproductive ways of thinking and behaving. Without the momentum of emotion, relapse is the norm.

Are you going to work on yourself as a trader today? Are you going to use today as an opportunity to learn and develop yourself, regardless of the day's profitability? If so, you'll need a goal for the day. What are you going to work on: Building a strength? Correcting a weakness? Repeating something you did well yesterday? Avoiding one of yesterday's mistakes?

An important first step is to set the goal. We cannot succeed as change agents if we don't perceive a clear path from the person we are to the person we wish to become. A valuable second step is to write down the goal or talk out loud into a recorder. This step helps cement desired changes in your mind. But will the pursuit of your goal truly possess emotional force? Will it transform you from one who thinks about change to one who truly becomes a change agent?

The secret to goal setting is providing your goals with emotional force. If your goal is a want, you'll pursue it until the feeling of desire subsides. If your goal is a must-have—a burning need, like my dietary change—it becomes an organizing principle, a life focus. You won't become a better trader because you want to be. You will only coach yourself to success when self-improvement becomes your organizing principle: a must-have need.

Try this exercise. Before you start trading, seat yourself comfortably and enter into a nice slow rhythm of deep breathing. Imagine yourself—as vividly as you can—starting your trading day. Watch the market move on the screen; watch yourself tracking the market, your day's trading ideas at your side. Then turn your goal for the day into part of your visualization: imagine yourself performing the actions that concretely put that goal into practice. If your goal is to control your position sizing, vividly imagine yourself entering orders at the proper size; if your goal is to enter long positions only after a pullback, imagine yourself patiently waiting for the pullback and then executing the trade. As you visualize yourself realizing your goal, recall the feeling of pride that comes from realizing one of your objectives. Bask in the glow of living up to one of your ideals. Let yourself feel proud of what you've accomplished.

It's important not just to have goals, but also to directly experience yourself as capable of reaching those goals. Psychologists call that self-efficacy. You are most likely to experience yourself as a success if you *see* yourself as successful and *feel* the joys of success. You don't need to imagine yourself making oodles of money; that's not realistic as a daily goal. But you can immerse yourself in images of reaching the goals of good trading and experience the feelings of self-control, mastery, and pride that come from enacting the best within you.

> We are most likely to make and sustain changes when we perceive ourselves as efficacious: capable of making those changes.

Many traders only get to the point of self-coaching after they have experienced harrowing losses. The reason is similar to my experience with my diagnosis: it was the vivid fear of consequences—the intense *feeling* of not wanting to ruin my life—that drove my dietary change. Similarly, after traders lost a good deal of their capital, they never want to experience that again. They trade well, not because they talk themselves into discipline, but because they feel the emotional force of discipline's absence.

Contrary to the teachings of proponents of positive thinking, fear has its uses. Many an alcoholic maintains sobriety because of the fear of returning to the pain of drinking's consequences. Emotion sustains the change.

With guided imagery that you feel as well as see, you can create powerful emotional experiences—and catalyze change—every single day. That's when you become a change agent: one who sustains a process of transformation. The key is adding emotional force to your goals. Your assignment is to take those lifeless goals off the piece of paper in your journal and turn them into vivid, powerful movies that fill your mind. Try it with one goal, one movie in your head, before you start trading. It is not enough to set goals; *you must feel them to live them.*

COACHING CUE

To each of your goals, add an *or else* scenario. Vividly imagine the consequences of *not* sustaining your change. Relive in detail specific failure experiences that resulted from the faulty behavior you want to change. When you add an *or else* condition to your goal setting, you turn fear into motivation. The brain is wired to respond first and foremost to danger; you will not gravitate toward the wrong behaviors if you're emotionally connected to their danger. To this day, my diet is firmly in place. Fear has become my friend.

LESSON 2: PSYCHOLOGICAL VISIBILITY AND YOUR RELATIONSHIP WITH YOUR TRADING COACH

If you are to be your own trading coach and guide your trading development, we have to make you the best coach you can possibly be. That means understanding what makes coaching work—and what will make it work for you.

Research informs us that the most important ingredient in psychological change is the quality of the relationship between the helper and the person receiving help. Techniques are important, but ultimately those techniques are channeled through a human relationship. Studies find that in successful counseling, helpers are experienced as warm, caring, and supportive. When helpers are seen as hostile or disinterested, change processes go nowhere. There's a good reason for this: relationships possess magic.

The magic of relationships is that they provide us with our most immediate experiences of visibility. I recently took a phone call from a reader of the *TraderFeed* blog. Many readers have provided valuable feedback about the blog, but this caller went far beyond that. He read every single post and then explained to me why he was drawn to the site. He put into words the very values that have led me to publish some 1,800 posts in the space of less than three years: *the vision that, in cultivating our trading, we develop ourselves in ways that ripple throughout our lives.*

At the end of that conversation, I felt understood: I was visible to another human being. When my mother died, I kept my composure until I approached her gravesite; then I lost it. My two children instinctively reached out to comfort me. It's something I would have done for another person in that situation. At that moment, I saw a bit of myself in my children. Once again, I was visible.

An unfulfilling relationship is one in which we feel invisible. We can feel invisible because we're misunderstood or mistreated. We feel invisible when the things that matter most to us find no recognition among others. I recall one particularly unfulfilling relationship with a woman. We were on the dance floor at a club and I suddenly stopped dancing altogether. She didn't notice at all. She was in her own world. It was a perfect metaphor for everything I was experiencing at the time: I was there as a kind of prop, a rationale for being on the dance floor. No one was really dancing with me. The profound, wrenching emptiness that I felt at that time was a turning point; never again would I settle for invisibility.

In Iggy Pop's classic song, invisibility is a kind of "Isolation." But if there's anything worse than being isolated—crying for love—when you're with someone, it's being isolated from yourself. We are truly lost when we're invisible to ourselves.

> Many traders don't really know what they do best; they're invisible to themselves.

All of us have values, dreams, and ideals. How often, however, are these explicitly on our minds? To live mired in routine, day in and day out, estranged from the things that matter most to us: *that's a form of invisibility.* To compromise the things you love in the name of practicality, to settle for second best out of fear or convenience: those, too, leave us in isolation—from ourselves. Strange as it may seem, we spend much of our time invisible to ourselves. The day-to-day part of us dances away, oblivious to the other self, the one that thrives on purpose and meaning.

It's a real dilemma: How can we possibly coach ourselves to success if the very strengths that would bring us success are invisible to us? After all, the single best predictor of change is the quality of the helping relationship. What, then, is our relationship to ourselves? If we are to be our own trading coaches, the success of our efforts rests on our ability to sustain visibility and draw on the magic of a fulfilling relationship with ourselves.

To coach ourselves successfully, we must be visible to ourselves and sustain the vision of who we are and what we value. But how can we do that? There's a simple strategy that can build a positive, visible relationship with your inner trading coach: identify a single trading strength to express as a goal for the coming day's trading.

One way I do that when I coach others (and when I work on my own trading) is to ask traders to identify what they did best in yesterday's trading that they want to continue today. *Set a positive goal, based on strengths, to keep you in touch with the best within you.* It affirms your competencies and keeps these visible, even during challenging market times. Too many of our goals are negative: we declare that we won't do

X or that we'll do less of Y. Instead, frame a goal for today that says: "Here is what I'm good at, here's what I did best yesterday, and here's how I'm going to make use of that strength today."

> Trading goals should reflect trading strengths.

In the relationship between you the trader and you the coach, the quality of the relationship will play an important role in your development. *The best relationship is achieved when goals are linked to values and express distinctive strengths.* Relentlessly identify, repeat, and expand what you do best—even (and especially) after the worst of trading days. Only through repetition can we turn positive behaviors into habit patterns. When you are in the habit of identifying and building strengths, you will then be truly visible to you. The magic of that relationship—and the confidence it brings—will sustain you through the most challenging times.

COACHING CUE

Review the last week's entries in your trading journal. Count the number of positive, encouraging phrases in your writings and the number of negative, critical ones. If the ratio of positive to negative messages is less than one, you know you aren't sustaining a healthy relationship with your inner coach. And if you're not keeping a journal, your coach is silent. What sort of relationship is that?

LESSON 3: MAKE FRIENDS WITH YOUR WEAKNESS

The notion of change is a challenge and a trap. It challenges us to aspire to more than who we are, but it can also trap us in self-division. When we entertain the notion of change, we divide ourselves into qualities we like and those we don't. We parcel ourselves into strengths and weaknesses, good and bad, acceptable and unacceptable.

Once we make such a division, it is only natural to embrace the good and avoid the bad. We dismiss our shortcomings as mistakes, bad luck, or exceptions. That helps us identify with a partial image of ourselves and keep our frailties from our conscious awareness. Thus banished from the front of our minds, those frailties cannot guide our learning. *We do not sustain the motivation to grow, because we only contact the parts of ourselves that are relatively whole.*

Suppose I manage a position poorly because of frustration and I exceed my loss limit on the trade, leaving me in the red for the day. I finish flat for the week, however, and instead focus on that fact. The loss is soon forgotten. It doesn't bother me, but I also don't learn from it. The next time frustration hits, I repeat my earlier behavior and lose even more money. Disgusted, I decide to take a break from the markets and come back with a positive mindset. In reality, however, I merely return in denial, once again banishing the losses from my mind. Eventually those trading shortcomings catch up to us, forcing us to face them squarely.

Such self-division is often maintained with the fiction of positive thinking. By focusing on positive thoughts, we don't have to think about what we've done wrong; we don't have to achieve contact with the parts of ourselves we don't like. We become like rooms where the clutter is increasingly swept under the rug. Eventually our rooms bulge with mental clutter, making them uninhabitable.

The motivation for much positive thinking is a denial of weakness.

Our daughter Devon was born with a "strawberry" beside her nose: a hemangioma that was a bright red bump on her skin. We were told that it would eventually recede on its own, that no surgery was needed. During her early years, however, baby Devon had a large red mark on her face. We could have put a patch over the mark or insisted on surgery, but we didn't. It was *her* mark, and it was part of what made her who she was. When you love someone, even her personal blemishes become endearing. Before I was a parent, I used to wonder how I'd tolerate changing dirty diapers. When the time came, I actually enjoyed it. It was something I was doing with and for my child. The changing of the diaper became an opportunity for bonding.

So it should be when we deal with our own dirty diapers. Your weaknesses are part of you; someone who loves you will love the whole package, frailties and all. *And if you love yourself, you can reach that point of acceptance in which you are fully aware of your shortcomings and appreciate your very humanness.* Indeed, as with the diapers, those shortcomings become opportunities—to reach out to yourself and guide your own development. For the longest time, I was unsure of myself in social situations and avoided most of them. Then, in a college dorm party I pushed myself to organize, I noticed a few people standing around not talking with others. In a flash, I saw myself in them. I made a beeline for the stragglers, included them in the gathering, and introduced them around. Ever since, I've been able to reach out to that reticence in myself and use it as a prod to engage others. My development occurred not through positive thinking, but through an embrace of my vulnerability.

Have you lost money recently? Have your trading weaknesses cost you money and opportunity? Consider embracing your flaws: every losing trade is there to teach you something. At the close of today's trading, create a chart with three columns. The first column is a description of the losing trade you made; the second column will be what you can learn from the losing trade; and the third column will be how you will improve your trading the next day based on what you learned. What you learned from the losing trade might be an insight into the market—perhaps it was range-bound when you assumed it was trending. That insight could help you frame subsequent trades. Alternatively, what you learn from the losing trade might be something about yourself; perhaps an insight into how you can manage risk more effectively. *Either way, your losing trade is never a total loss as long as you embrace it and learn from it.*

> Much of self-coaching success is finding the opportunity in adversity.

When you create a trading diary, you bring yourself face to face with your worst trading and turn it into opportunity. It doesn't matter if blemishes mar your account statement. It's your account, red marks and all. You make yourself stronger when you reach out to your flaws. Embrace who you are and you take the first step to becoming the person you are capable of being.

COACHING CUE

As we'll see in the next chapter, the research of James Pennebaker suggests that giving voice to stressful events—in a journal or out loud—for at least a half hour a day is instrumental in our putting those events into perspective and moving beyond them. When you experience a horrific trading day, give it voice. Talk it through and sear its lessons in your mind. If you're in touch with how badly your trading makes you feel, you're least likely to repeat your errors. There can be gain in embracing pain.

LESSON 4: CHANGE YOUR ENVIRONMENT, CHANGE YOURSELF

Human beings adapt to their environments. We draw on a range of skills and personality traits to fit into various settings. That is why we can behave one way in a social setting and then seem like a totally different human being at work. One of the enduring attractions of travel is that it takes us out of our native environments and forces us to adapt to new people, new cultures, and new ways. *When we make those adaptations, we discover new facets of ourselves*. As we'll see shortly, discrepancy is the mother of all change: when we are in the same environments, we tend to draw upon the same, routine modes of thought and behavior.

A few months ago I had an attack of acute appendicitis while staying in a LaGuardia airport hotel awaiting a return flight to Chicago. When I went to the nearest emergency room at Elmhurst Hospital outside Jackson Heights, Queens, I found that I was seemingly the only native English speaker in a sea of people awaiting medical care. After some difficulty attracting attention, I was admitted to the hospital and spent the next several days of recuperation navigating my way through patients and staff of every conceivable nationality. By the end of the experience, I felt at home there. I've since stayed at the same airport hotel and routinely make visits into the surrounding neighborhoods—areas I would have never in my wildest dreams ventured into previously. In adapting to that environment, I discovered hidden strengths. I also overcame more than a few hidden prejudices and fears.

The greatest enemy of change is routine. When we lapse into routine and operate on autopilot, we are no longer fully and actively conscious of what we're doing and why. That is why some of the most fertile situations for personal growth—those that occur within new environments—are those that force us to exit our routines and actively master unfamiliar challenges.

> In familiar environments and routines, we operate on autopilot. Nothing changes.

When you act as your own trading coach, your challenge is to stay fully conscious, alert to risk and opportunity. One of your greatest threats will be the autopilot mode in which you act without thinking, without full awareness of your situation. If you shift your trading environment, you push yourself to adapt to new situations: you break routines. If your environment is always the same, you will find yourself gravitating to the same thoughts, feelings, and behaviors. We are mired in repetitive patterns of thought and behavior because we are mired in routines: the same emotional and physical environments. Indeed, we repeat the same patterns—for better or for worse—precisely because those patterns are adaptations to our current settings.

So how can we change our trading environments? The key is recognizing that our physical settings are only a part of our surroundings. Here are a few routine-busting activities that can alert us to risks and possibilities:

1. **Seek Out Divergent Views.** Conversations with traders who trade differently from you—different time frames, markets, or styles—can often help cement your views or question them. Similarly, reading

materials from fresh perspectives puts your ideas in a different light and pushes you to question your assumptions. I remained relatively bullish on the stock market's longer-term picture into the final quarter of 2007. Only when I pushed myself to read informed views that clashed with my own—and consulting data that did not fit my framework—did I modify my perspectives and avoid significant losses.

2. **Examine the Big Picture.** It's easy to get lost in the market's short-term picture; how it is trading that minute, that day. I find it important to periodically zoom out to longer-term charts and place the current action into context. Indeed, some of the best trading ideas start with a big picture view and then proceed to shorter-term execution. I especially find this to be the case when looking at longer-term support/resistance, trading ranges, and Market Profile value areas. Often, shifting my field of vision will help me avoid an ill-informed, reactive trade based on the market's last few ticks. If something seems obvious in the market, switch time frames and generate an entirely new perspective. What looks obvious from one view may well be obviously wrong from another.
3. **Examine Related Views.** Sometimes the action of a single stock or sector will illuminate what's happening in the broader market; one currency cross will break out ahead of others. Are we seeing a broad fixed income rally, or is the yield curve steepening or flattening? Looking across instruments and asset classes keeps us from getting locked into ways of thinking. I find myself tracking sector ETFs during the trading day to see if stocks are moving in a single direction (trending) or are taking different paths within a range. If I see bond traders fleeing to safety or assuming risk, I can anticipate selling or buying stocks. Seeing the entire financial playing field helps keep us from becoming wedded to preconceived ideas.
4. **Take the Break.** Just as we take vacations to return to work refreshed, a break from the screen can help us generate fresh market views. It is easy to become focused on what is most dramatic and salient in markets. Pull back and clear out the head to help you see what's not obvious and then profit by the time it's recognized by others. I find breaks especially helpful following losing trades, enabling us to reflect on the losses and what can be learned from them.

If your environment is comfortable, it probably isn't conducive to change.

In short, it's the mental routines—the mental environment—that we most need to change to break unwanted and unprofitable patterns of thought and behavior. When you're your own trading coach, you learn to think, but also to think about your thinking. Incorporate a fresh look at self and markets each day to inspire new ideas, challenge stale ones, and tap sources of energy and inspiration that otherwise remain hidden in routine. As with my adventure in Queens, you may find that the most exotic changes bring out your finest adaptations.

COACHING CUE

Many times it's the market views we most scorn that we need to take most seriously, because at some level we're finding them threatening. Seek out commentary from those you most disagree with and ask yourself what you would be seeing in the markets if that commentary proves to be correct. If you're quick to dismiss a market view, give it a second look. You wouldn't need to be so defensive if you didn't sense something plausible—and dangerous—in the views you're dismissing.

LESSON 5: TRANSFORM EMOTION BY TRACE-FORMATION

When traders seek coaching, they are usually troubled by a particular emotional state that affects their decision-making: anger, frustration, anxiety, or doubt. Their goal is to change how they feel, but they don't know how to accomplish that. Sometimes traders even view their emotions as fixed and unchanging aspects of personality: "It's just the way I am."

It is true that our traits and temperaments affect how we experience the world. They also play an important role in defining the range of our emotions. Some people feel things—good and bad—very strongly; others are quite even-keeled. Neuroticism, the tendency to experience negative emotions, is one of the big five personality traits identified by researchers. Like all such traits, it has a strong hereditary component. Though we like to think of ourselves as masters of our fates, the sobering reality is that much of our emotional experience is hardwired.

Does that mean we can't change how we feel in particular situations? Not at all. If psychological methods can help people overcome post-traumatic stresses and anxiety disorders, they certainly can help us master our feelings in normal life situations. For the most part, we cannot change personality, but *we can change how our personalities are expressed.*

The trap many traders fall into is trying to control feelings with thoughts. We attempt to talk ourselves into feeling better or differently. Rarely does that work. When people are grieving over losses, telling them they'll be okay doesn't really touch what they're experiencing. The feelings express a psychological reality; asserting a logical reality ignores the personal meaning and significance of the situation. Feelings are surprisingly refractory to willpower: if wanting to feel different—and talking ourselves into feeling different—were possible, there would be many fewer psychologists in the world.

If you serve as your own trading coach, *a great place to start is with the perspective that feelings contain information*. Research in cognitive neuroscience finds that emotion is an essential component of rational decision-making. When the brain is damaged and becomes unable to engage in emotional processing, the result is profoundly distorted behavior. Your coaching goal is not to banish the feelings associated with difficult trading—a strategy that only prevents resolution—or to blindly act upon them. Rather, the most constructive step you can take to change a feeling is to give it full acknowledgment and extract its vital information.

> Feelings inform us about our appraisals of self, others, and world.

The research of James Pennebaker, a professor at the University of Texas, is quite relevant here. He and his colleagues found that writing in a journal or talking aloud for a half hour a day had a powerful effect on enabling people to cope effectively with challenging emotional circumstances, including traumas and crises. When we make implicit feelings explicit, we view them from different angles and place them into a different context. For example, someone who has been angry and frustrated with himself over poor trading performance might journal about these thoughts and feelings at length. As he is writing—and reading over his writings—he suddenly realizes, "Whoa; I'm being awfully hard on myself. I'm not that bad!" With that, he is able to throttle back his negative self-talk and turn his attention back to markets.

When we fail to acknowledge emotions, we lose their information and thus the opportunity to shift perspectives. The frustrated, angry trader who brushes aside his tensions and forges blindly ahead finds them easily triggered the next day. This is particularly the case when the frustrations are triggered initially by trading mistakes. I recently met with a trader who fought a market trend all morning, built frustration through the day, and then blew up in the late afternoon. Had the trader used the frustration to examine his trading, he could have ridden the trend and made significant money. Brushing emotions aside doesn't change them. Ironically, acknowledging and accepting them, giving them free expression, sets the stage for transformation.

Does that mean that we should give full vent to whatever we're experiencing? No, psychological research also suggests that unbridled expression of emotion interferes with concentration and performance. Simply yelling when we're angry or pouting when we're discouraged does nothing to alter the feelings—and certainly does not place us closer to resolving the situations responsible for the upset in the first place. The trader from my example, for instance, spent much of his afternoon fuming, but never resolving his anger. Reflexive acting on such emotions only reinforces them; you can't overcome frustration by behaving in frustrated ways.

> Blindly venting or acting on emotion is as unproductive as blinding ourselves to emotion; both prevent learning from the information in our feelings.

The idea, then, is to transform feeling, not ignore it and not revel in it. One way to do this is to replace one emotional state with another: to substitute feeling for feeling, not thought for feeling.

In my *Psychology of Trading* book, I explained how I used the early music of Philip Glass to enter a meditative state and trance-form experience. Actually any stimulus that evokes calm, focused attention can be effective as a tool for shifting emotions. The key is to evoke and sustain the Yoda state—the calm focus—during periods of high frustration or discouragement. Biofeedback can be especially useful in this regard, as computer-based applications provide real-time feedback about your success in sustaining the altered state. It is virtually impossible to sustain a worked-up state—anger, anxiety, and stress—when keeping yourself calm and focused. Even better, in the relaxed state, you'll arrive at perspectives and insights that remain unavailable while you're immersed in the flight-or-fight mode.

One exercise I recommend to traders is to draw two thermometers side by side on a sheet of paper and then run off a number of copies of the paper. One thermometer records your emotional temperature with respect to frustration; the other records your temperature with respect to confidence. The sheet stays by your trade station; all you need to do is make a mark on each thermometer to indicate how frustrated and confident you're feeling at the time.

When we're most frustrated, but also most overconfident, we're likely to make our worst decisions and violate our trading principles. If you require yourself to "take your emotional temperature" during each trading session, you create a mechanism for catching your state of mind before it can disrupt trading performance.

Once you identify an elevated frustration temperature, a valuable, automatic rule is to take a few minutes away from the screen and enter into a trance-formation. This can be done by regulating your breathing—making it particularly deep and slow—and fixing your attention on something that captures your attention: music, imagery, or a picture in front of you. If you slow your body and take your attention away from the situations that may be elevating your emotional temperature, you shift your state and make it easier to act calmly, in a planned fashion. With practice, this can be accomplished in a matter of minutes, short-circuiting many disruptive patterns before they lead to poor trading decisions.

The key is to keep yourself aware of your emotional state throughout the day. The thermometers are an easy, visually arresting way of becoming your own observer—and coach.

COACHING CUE

Check out the insights about breathing in Chapter 9. Mike Bellafiore of SMB Capital explains how he and partner Steve Spencer teach the traders at their prop firm how to breathe as part of training them to trade. As practitioners of meditative disciplines understand, emotional self-control begins with physical control.

LESSON 6: FIND THE RIGHT MIRRORS

A mirror is an object that shows us our own image. Thanks to mirrors, we know what we look like. Far more goes into our self-image, however, than our physical reflection. *That is because virtually all of our experience serves as a psychological mirror.* We see ourselves reflected in the impacts we have upon the world around us. As a result, much of self-esteem—our sense of worth and competence—follows from finding the right mirrors in life.

Let's start with romantic relationships. When we select the right partner, we choose someone who knows and values the person we are. That love and support is ongoing; consistently reflected to us, it is a deep affirmation of self. In a similar fashion, parents constantly mirror a child's identity: "You're such a good boy!" and "What a smart girl!" *Our self-talk is born of just such early life conversations: we internalize the voices from significant relationships.*

This is why abusive relationships are so damaging. To share life with a spouse who attacks or demeans us—or who just doesn't care—or to endure parents who are neglectful is to continually face a distorting mirror. Over time, children absorb the distorted images and no longer feel lovable, secure, and important. Out of such twisted self-images, they select future partners that validate their identities, sadly finding others who repeat the messages and experiences of the past. That is how abused children find themselves in abusive relationships; how insecure people land in insecure marriages.

While relationships may be our most powerful psychological mirrors, given their emotional intensity and ongoing influence, they are far from the only determinants of self-image. The Devon Principle that I wrote about in the *TraderFeed* blog captures the understanding that *everything* we do is a psychological mirror. When my daughter Devon tackled work that she didn't like, she found the work frustrating and felt inadequate as a result. When she pursued work that she loved so much that it didn't feel like work, she felt fulfilled and gained confidence. The best work speaks to our interests and values, matching our abilities with challenges. Day after day, performing efficaciously at work that is important to us generates mirror-experiences of competence and worth. Conversely, when we're performing meaningless work that doesn't challenge our skills, it is difficult to feel anything other than boredom and meaninglessness. *A large portion of career success consists of finding the right mirrors*; it's much easier to get to the top when you're climbing the right ladders.

For more on the Devon Principle, check out my blog: http://traderfeed.blogspot.com/2006/12/devon-principle.html

For my work as a psychologist, nothing is a more powerful mirror than having a meaningful, positive impact upon people's lives, particularly when I get to know those people well and care about them. I enjoy giving a talk for a large audience or writing an article that's widely read, but the real joy is hearing back from someone who thought the ideas were of genuine value. And, to be honest, I find far more reward in helping a single person in counseling or coaching than in giving a keynote address for a large conference. When a person transforms her life via coaching or counseling, a mirror is created that validates and enhances both participants in the helping relationship. I have been most successful when I've immersed myself in these positive mirroring experiences, least successful when I have pursued activities that, ultimately, offer a limited sense of self.

When you serve as your own coach, your challenge is to structure your learning and development so that trading itself becomes an experience that mirrors your growing confidence and competence. Many traders limit their self-coaching to keeping a journal, and then limit their journaling to recounting all the things they've done wrong. As a result, self-coaching becomes little more than self-criticism. What is mirrored to a trader when the journal focus is so negative? What would be mirrored if we hired a teacher or coach who only offered criticism? Over time, such coaching would fail, reinforcing a sense of incompetence and failure.

One of the best means for creating positive mirrors is the structured pursuit of goals. When we create challenging, meaningful, and doable goals, we generate potential experiences of mastery and success. *When we make goal setting an ongoing feature of our self-coaching means, we continually construct opportunities for powerful, self-affirming emotional experiences.* We know from psychological research that such emotional episodes are processed more deeply and enduringly than normal, daily experience. A good therapist creates vivid experiences that challenge clients' old patterns; similarly, a good coach generates emotionally powerful and positive mirroring experiences for traders.

> Your goals should set yourself up for success and a building of confidence.

So here is your assignment: Each day this week your trading journal should include a specific goal for work that particular trading session, concrete actions that you will take to achieve that goal, and a self-evaluation at the end of trading to gauge your success in reaching that goal. *The goal should be a trading process that you wish to improve (i.e., something you have control over), not a profit target (which you ultimately don't control).* For example, your goal might be to increase your trading size incrementally, to implement a strategy for exiting trades in stages, or to limit trades to setups that align with the larger market trend. At the end of the day, you will give yourself a report card based solely on how well you achieved the goals you set for the day. These report cards can be displayed beside your monitor to reinforce your performance and progress. If you fail to achieve a good grade, improvement on that activity becomes your goal for the next day. If you receive a fine report card, you generate fresh goals for the next session. The idea is to never trade without consciously working on some aspect of your trading.

> It is not enough to set goals; you need ways of tracking your progress toward those goals and feeding that information into future goals.

Many traders only engage in such goal setting when they're trading poorly or losing money. The idea, however, is to make self-coaching and self-improvement an ongoing part of your trading career. Why? Because it's not just about making money, it's about creating the experiences that will sustain your sense of competence and confidence. Think of a young child: you don't offer positive feedback only when the child is hurting. Rather, your support and love are continuous, enabling the child to sustain a consistent self-image. As a developing trader, you are like that young child. Your ability to create powerful mirroring experiences will make a difference in your ability to sustain the optimism and courage to weather drawdowns and aggressively pursue opportunity.

Please take note of the following principle: If you limit your losses, pursue your strengths, and take concrete steps toward mastery, every single trading day can be a positive experience, even when you're not making money. You cannot eliminate losing days, but there should never be days that leave you feeling like a loser.

COACHING CUE

When you construct your report cards, grade yourself based on your improvement, not based on an abstract (or perfectionist) standard of success. If you manage your trades better today than yesterday, that merits a good grade. Your goal is to improve; by focusing on improvement, you create powerful mirrors of self-development. Relative, not absolute, goals will get you to your desired endpoint, and they will ensure an enjoyable and empowering journey.

LESSON 7: CHANGE OUR FOCUS

A valuable psychological rule is that if you wish to change the doing, you must change the viewing. How we see the world colors how we respond to life events. We don't just react to markets, but also to how we process those markets. Our thoughts are the filters between trading and trader.

Many times we respond in exaggerated ways to markets, not because there's anything unusual going on in the instruments we're trading, but because a set of negative thoughts have intruded into our performance. Let's say, for instance, that I notice that the ES futures are unable to surmount their overnight highs during the opening minutes of trade despite a few flurries of buying. I then observe that large traders are coming into the market hitting bids. I hypothesize that we cannot sustain strength and that the overnight highs will not be breached. I further infer that we will trade back into yesterday's price range and hit the average trading price from that session. I wait for a bounce higher in the NYSE TICK that cannot make a fresh price high and use the occasion to sell the futures. As the position moves my way, however, the thought enters my mind that I should take quick profits because I've had a losing week thus far. A buy program then hits the market and my position ticks higher, eroding some paper profit. Now especially concerned, I take a small profit—only to see the market weaken notably and eventually hit my initial price target.

What has happened in this scenario? Anxiety has interfered with my performance, turning a good trading plan into bad trading. But the anxiety has nothing to do with the behavior of the market: the market did absolutely nothing to disconfirm my idea. Indeed, when the buy program lifted the index futures briefly, the market was giving me a perfect opportunity to add a second unit to my trade! Not only did I miss an opportunity to profit from a good idea, I also missed an opportunity to hit a home run. Often, keeping losing trades small and hitting those few home runs generates profitability over the long run.

Many traders' problems show up in how they handle opportunity, not loss.

Sometimes anxiety is a legitimate and appropriate response to a market that behaves in violent and unexpected ways. After all, as I note in the blog, anxiety is our body's adaptive response to perceived danger. But danger can be a function of perception, not objective reality. I begin my trade immersed in market activity, framing hypotheses and executing an idea well. At some point, however, my thoughts veer from the present market and instead focus on how much money I've lost during the week thus far. That focus on loss creates a sense of danger and threat. Instead of responding to the market, I'm now reacting to my own concerns regarding profitability. My thought process has taken me out of the market immersion—and ultimately takes me out of my game.

In a cool, calm moment, I can see clearly that the validity of my trade idea/plan has absolutely nothing to do with how I've traded the past several days. If I introduce worries over profitability into my trading, however, I've now allowed the viewing to affect the doing. I'm no longer absorbed in the market; my focus is gone. I'm responding to my own uncertainties and insecurities.

How can we change our focus and stay grounded in our plans and in objective market activity? The first step is to recognize our triggers. These are the performance thoughts and worries that typically intrude during trading. Concerns over profitability are one trigger; excitement over anticipated profits could be another. Anything that gets you thinking about how well or poorly you're doing while you're doing it is a trigger that can nudge you out of your zone. When you know your triggers, you're in a much better position to intercept them when they occur and treat them the same way you would treat any ordinary distraction, such as road noise outside your window.

In other words, *it's not the thoughts of performance that take you away from your focus, but your identification with those thoughts.* This is an important distinction. Everyone experiences distracting thoughts at times. When we identify with those thoughts, however, they—not our markets, not our plans—become our focus.

> Negative thoughts are inevitable; the question is whether you buy into them.

Meditation can be a very helpful exercise. One purpose of meditation is to help people create a quiet mind by sustaining a single point of concentration and brushing aside all distracting internal dialogue and impulse. A simple adaptation of meditation for you to try is to take 15 minutes before the start of trading and seat yourself in a comfortable position, breathing slowly and rhythmically from the diaphragm. While in that position, focus your attention on quiet instrumental music played through headphones. You want to be as absorbed as possible in the music: as soon as your mind wanders, bring it back to the music and the sounds of the different instruments.

Once you've been able to sustain that focus for a few minutes, you then purposely bring to mind your greatest performance concerns—*while you stay seated, focused, and breathing rhythmically and deeply*. You evoke one concern at a time (for example, thoughts or fears about your recent profitability) and then dismiss the thoughts and bring yourself back to the music. Instead of having the thoughts intrude on you unexpectedly, you intentionally call them to mind and practice and then put them away, as you stay calm and focused. You might even guide yourself through imagery as you're breathing deeply and slowly, imagining your negative thoughts as trash that you decide to put in a garbage pail. Instead of avoiding your negative thoughts, bring them to mind as your own, inner trash-talking—and then visualize yourself taking out the garbage.

Do this every day for a few minutes and you can train yourself to gain control over negative thought patterns. Most importantly, you develop the capacity to become an observer to those thoughts, rather than to identify

COACHING CUE

Whenever you catch yourself thinking about your P/L during trading—how much you're making or losing—call a brief time-out; take a few deep, rhythmical breaths and talk out what you're seeing in the markets at the time. Your goal is to be market-focused, not self-focused. By repeatedly pairing a calm, relaxed state with an intense market focus, you can develop a positive habit pattern and ensure that the body keeps the mind in check.

with them. If you can observe something about yourself, you immediately introduce an element of psychological distance. Even the most negative thoughts and feelings cannot trigger unwanted behaviors if you don't identify with them. Daily meditation is a powerful strategy for building your own Internal Observer and sustaining a change process.

LESSON 8: CREATE SCRIPTS FOR LIFE CHANGE

There is a bit of a chicken-and-egg challenge associated with making changes in our lives. To change a behavior pattern, you have to be able to exit that pattern. If, however, you had the ability to avoid enacting those patterns, you wouldn't need to change in the first place.

This dilemma is a common barrier for traders who would like to be their own trading coaches, but don't know how to stand apart from the problem patterns that they repeat week after week, month after month.

To appreciate how we can shift ourselves out of old, problem patterns and into new, positive ones, we need to understand something about drama. Specifically, it's helpful to *start thinking about life in terms of the different roles that we enact during our life's performances.* "All the world's a stage," Shakespeare observed, and we are the sum of the roles that we play on that stage.

Some of our life roles have an automatic, scripted quality to them. Typically we learned these roles early in life and, for years, they may have worked well for us. As a result, these roles have become *overlearned.* For instance, we may have learned to gain attention from parents by complaining or by acting up and breaking rules. Over time, those behaviors can crystallize into fixed roles: we automatically find ourselves whining or acting out of frustration during times of personal conflict. What worked in childhood by bringing us attention now works against us, interfering with careers and romantic relationships.

Many trading problems have just such a scripted quality: We enact the same patterns repeatedly. We start by trading carefully and conscientiously. Then we lose money and become frustrated. Out of frustration we break trading rules, ignore stop-loss points, and undergo serious losses. Then we feel tremendous relief at exiting the losing positions and redouble our determination to trade carefully and conscientiously—until the next frustration comes around. Is this really so different from couples that are determined to get along with each other, then encounter frustrations, argue and fight to the point of being ready to break up, only to experience relief as they make up and vow ever stronger to stop hurting each other? Or the person who swears that he will stop gambling, only to make a few exceptions, lose money, and then out of relief step away from the casino once again, insisting that he won't go back?

> Trading requires a mind free to process data and select appropriate action. But we no longer have a free will if we are mechanically reliving scripts from the past.

A dramaturgic perspective suggests that these repetitive patterns are enactments—cyclical reenactments—of *roles* that we have learned in the course of life: the role of the down-and-out person who presses for success, the role of the aggrieved spouse, the role of the independent person who refuses to be bound by rules, and so forth. One trader I worked with grew up in an overprotective and controlling home. He rebelled as a teenager and subsequently found himself chafing at any constraints on his behavior. His violation of rules in relationships (monogamy) and trading (the risk-management rules of the firm) led to one failure after another. He was living out a script that could only provide unhappy endings.

But if we can acquire scripts through our relationship experiences, then surely we can cultivate new ones by placing ourselves in different roles. One trader I worked with experienced himself as sloppy and undisciplined. It showed not only in his trading, but also in his physical condition and the state of his apartment. His breakthrough occurred when he joined a fitness club and engaged a personal trainer. The regular series of classes and exercise sessions got him into shape and imposed a structure on his efforts at self-improvement. As he experienced more energy—and felt better about himself for getting in shape—he spontaneously took the initiative in cleaning his apartment and honing his trading rules. The sessions with a trainer provided him with a new script and positive experiences that mirrored a fresh identity. *By enacting a new role, he experienced himself in a new manner—and this infiltrated a variety of areas of his life.*

Here's another example: For years I tended to be impatient—with myself, with others, with trading, and even with the pace of change among people I met in therapy. When Margie and I had two children, however, I found myself in a new role that did not allow me to be impatient if I were going to be a good parent. Because it was clear that both our children had personalities very different from mine, I had to figure out ways to communicate with them on their terms. The new role as a patient parent provided me with a discrepant set of experiences; I subsequently found myself more patient in a variety of situations, whether it was behind the wheel of a car or during counseling work with a client who felt stuck. The new role generated novel, positive scripts. With the favorable mirroring over years of parenting, I've actually changed how I see myself. I now experience myself as a relatively calm, patient person: it has become an integral part of my identity.

For more on how new experiences generate new roles and scripts, check out my blog: http://traderfeed.blogspot.com/2006/11/cross-cultural-journey.html.

So here's your challenge and your assignment: Identify the person you would like to be and then throw yourself into a structured social activity—a role—that requires you to enact those ideals. If you want to be more disciplined, take on a discipline: martial arts, work with a personal trainer, etc. If you want to be more patient and focused, undergo meditation training or work with young children that you care about; if you want to become more socially confident, immerse yourself in public speaking; if you want to trade more aggressively, join a trading room that mirrors the style you want to adopt and actively participate in its discussions. *Create the roles that mirror your desired identity; live scripts of your choosing.* If you can place yourself in situations where you routinely practice being the person you want to be, you'll rapidly make that person your own. Change begins with novel experience, but is sustained through repetition.

COACHING CUE

To be the trader you want to be, consider taking on a student/trainee. When someone is observing you and learning from you, you'll be on your best behavior. With the teaching script, you'll access behavior patterns that you would never enact in isolation. Alternatively, take on a peer mentorship role. The social motivation to live up to your best for your trading buddies will enable you to access your best behavior patterns.

LESSON 9: HOW TO BUILD YOUR SELF-CONFIDENCE

Trading is one of the most challenging occupations, because traders routinely face working conditions that they cannot control. Psychological research suggests that one important basis for self-confidence is self-efficacy: the perception that we can control outcomes that are important to us. But how can we sustain self-confidence as traders if we cannot control whether we make money from day to day?

A trader recently called and expressed frustration with his performance. Markets were moving well and, for the most part, he was catching the direction correctly. He entered positions aggressively, but then was stopped out at the worst times when the market made sharp, short countertrend moves. When we reviewed his trading statistics, we found that his average win size exceeded his average loser, but that he had many more losing trades than winners. The steady drumbeat of losing trades was eroding his self-confidence.

So what was the problem?

Our trader was waiting for markets to begin moving in the anticipated direction and then entering with full size. By the time he lifted an offer or hit a bid, the market had already made a short-term move and was ripe for profit taking by scalpers. His full size made these countertrend moves intolerable, and his risk management rules ensured that he had no staying power with his ideas. The market was controlling him; he was not in control of his trading. The loss of self-confidence was inevitable.

The key to regaining self-confidence in such a situation is to turn the focus from making (or losing) money to the actual process *of trading.* We initiated a simple unit-sizing rule in which the trader could only enter positions with one unit (his maximum size was three units). If the position went his way, but then experienced a normal retracement, he added a second unit. If the position did not go his way, he maintained a defined stop-loss point and ensured a minimal loss, given that his size was one-third his maximum. The trader could not control the market's movement, but he *could* control his position sizing. This process focus promoted a sense of self-efficacy, which was essential to recapturing his confidence.

> You control how you trade; the market controls how and when you'll get paid.

This is one reason that trading with rules is so important. You can't control your P/L statement, but you *can* control whether you adhere to trading rules. Your focus becomes one of trading well, not one of making money. Every rule followed—every market traded well—is a success experience in the process sense. Over time, profits result (as long as the rules are sound!), but the confidence comes from self-mastery.

Another powerful source of self-efficacy is preparation. When you prepare your trading ideas for the day or week, you generate a sense of mental mastery. This is particularly the case when your preparation includes what-if scenarios that guide your decision-making under a variety of market possibilities. Successful experience is a powerful source of mastery, and the mental rehearsal of trading plans under various contingencies generates a form of experience. As a psychologist, I am impressed by the degree to which traders who prepare rigorously feel as though they *deserve* to win. That same sense is missing among those who casually scan newspapers, charts, or web sites and then plunk themselves down at the trade station to place their orders.

Many traders confuse self-confidence and positive thinking. *Self-confidence is not expecting the best; it's knowing, deep inside, that you can handle the worst.* The self-confident trader can look a stop-loss level in the eye and know that he will be okay if it is triggered. The self-confident trader knows that loss is part of the game—and that some of our best market information comes from good trade ideas that don't work out. Self-confidence is not cockiness, nor is it viewing the world through rose-colored glasses. It's the quiet sense of, "I've been here; I can handle this."

> Confidence doesn't come from being right all the time; it comes from surviving the many occasions of being wrong.

Nothing is so important in building self-confidence as successful experience in the face of adversity. When you serve as your own trading coach, a major task you face is generating your own positive trading experience. Just this morning I read an e-mail from a trader who had experienced harrowing losses in the markets over the past two years. Now he was having difficulty sustaining the optimism needed to weather normal losses. His failure was not as a trader, but as his own trading coach. When we traumatize ourselves, we generate negative experience. We create a sense of helplessness, rather than mastery. We create deep emotional connections between trading and loss, rather than between trading and self-efficacy. We undercut self-confidence.

A great way to build self-confidence is to focus on how you trade when you're in the hole. If confidence comes from successfully navigating adversity, you can build your confidence by working on how you trade when trades go against you. The idea is to focus on trading well by giving yourself a chance to dig out of the hole by not exiting a losing trade prematurely, but also to not allow the losing trade to move so far against you that it creates trauma. Every planned loss that you take provides you with an experience of control; every drawdown that you battle back from is an experience of mastery. When you come back from losses, you reinforce your emotional resilience. You can't control whether you win or lose on a particular trade, but you can control how much you lose and how you lose it.

Every trader needs a plan for losing. *Your stop-loss is your plan for a losing trade.* Cut your size after a series of losing days and focus your efforts on your highest probability trades as your plan for a drawdown period. In Ranger school, the Army exposes recruits to the most harrowing physical conditions possible. Once recruits have completed their training, they have the deep conviction that they can handle any and all battle conditions. You want to view your losing trades and your losing periods in markets as your Ranger School, your trial by fire.

> Your losing trades and losing periods are your trials by fire that build resilience and confidence.

How will you handle a significant losing trade? How will you handle a significant losing day? Week? Month? How will you ensure that you can draw upon your strengths and come back from these losses and build your resilience? Your assignment is to develop your plans for losing, to always know—and mentally rehearse—what will get you out of trades and out of markets, so that you can retool your efforts.

In psychology, crisis does offer opportunity: it shakes up our assumptions and forces us to make changes in how we think and act. Your challenge as your own coach is to find the opportunity in your crises by generating and rehearsing plans for anything and everything that can go wrong. To prepare for hurricanes and tornadoes, communities not only draw up disaster plans, but also conduct drills to put these into practice. Change is a function of preparation and training: drilling the right responses, so that they are second nature when market disaster looms.

COACHING CUE

Just before I wrote this section of the book, a savvy trader contacted me and explained that he broke some of his rules and lost the profits from the prior week. He was very upset and wrote a memo to himself to ensure that he learned from the experience. He sent the memo to me and insisted that we talk in a few days to ensure that he followed up on the memo. It's a great example of how a trader takes a losing situation and turns it into an opportunity for self-improvement. He won't let the issue go until he's rectified his errors.

That's how traders turn losing experiences into confidence builders. Next time you blow up in your trading, write yourself a detailed memo that explains what went wrong, why it went wrong, and what you will do to avoid the problem going forward. Then send the memo to a valued trading buddy for follow up to hold yourself accountable. That way, every big mistake becomes a catalyst for meaningful change.

LESSON 10: FIVE BEST PRACTICES FOR EFFECTING AND SUSTAINING CHANGE

The first two years of my career as a psychologist, I worked in a community mental health center, helping individuals, couples, and families with the full array of emotional disorders, from depression to drug abuse. The following year I shifted my practice to student counseling at Cornell University, which afforded my first opportunity to work with a relatively healthy population dealing with normal, developmental issues. I then took the community and student experiences to Upstate Medical University in Syracuse, New York, where I coordinated the counseling and therapy for medical, nursing, and other health sciences students and professionals for 19 years. It was in this latter setting that I learned to apply brief therapy methods to the challenges of young people who were in high-stress, high-achievement occupations. That experience would prove invaluable to my work with traders in the financial markets.

During my time in Syracuse, I met on average about 150 students a year for about eight sessions each. Multiply that times 19 and you have a sense for the changes I've seen happen and not happen. The shining successes, the disappointing failures: all stand out in my mind as if they occurred last week.

When you've worked with that many people over the course of a career, you develop a good sense for change processes and what makes them click or stall. It doesn't matter if you're working with a victim of abuse in the community, a student with test anxiety at an Ivy League school, or a medical student dealing with the first loss of a patient. Change has a particular structure and sequence; there are factors that speed it up and those that impede it. Below I share five of the most important change elements that affect my work as a trading coach. When you harness those elements, you will be well positioned to succeed in your self-coaching:

1. **Timing and Readiness.** Timing is everything, in psychology as in trading. The research of Prochaska and DiClemente suggests that people are most likely to make changes when they're *ready* to make changes. Many times we're conflicted about change; we're not really sure that we want to abandon old ways. I talked with a trader recently who lost much more money than he should have (and than his plan called for) because he simultaneously traded three positions with full size when those positions were highly correlated. He was wrong and he blew out. But when we reviewed the unit sizing by trade idea rather than simply by position, it was clear that he wasn't sure he wanted to make smaller bets. He was upset because he was wrong on the

idea, not because he had violated risk-management discipline. My job was then to help him become more ready for the change he needed to make, just as an alcohol counselor must help an alcoholic become more ready to commit to sobriety. *You will change when you're ready to change, and you'll be ready to change when you recognize that you* need *to change.* As we saw earlier, by becoming more emotionally connected to the consequences of our behavior, we cultivate that need to change.

2. **Ready, Steady, Go.** One of the traps that eager traders fall into when they start coaching is that they want to make many changes all at once. As a result, traders overload themselves with too many goals, water down their focus, and never adequately follow through with any of them. If you have a list of five changes to make, select the one that you're most ready to make (as noted above): the one that you're most committed to taking action on. *Work on that one goal intensively and daily until you make and sustain significant progress; then move to the next change.* The momentum from your success with the first effort will carry over and help your work with the others. If you begin with a goal that you should work on, but aren't fully committed to, you'll stall out on your entire coaching effort. Keep your work doable, but keep it steady. Build momentum and success, and that will help with your later change efforts

> When you coach yourself, focus your efforts and let one success fuel others.

3. **Double Down.** When you first make a desired change, don't let up. Rather, focus on what you did to make that change and redouble your efforts. Make your goal to sustain that change. All too often, traders let up once they make an initial improvement. That's like hurting your opponent in a boxing ring and then not moving in to finish him off. You want progress to double your motivation, not let your bad habits off the hook. W*e've seen that the enemy of change is relapse*: all of us too easily fall back into old patterns if we're not making conscious and sustained efforts to build new ones. The key to change is relapse prevention: repeating new patterns so often that they become natural to us. Any change worthy of pursuit is worth repeating 30 times in 30 days. In Alcoholics Anonymous, a committed new member will attend 90 meetings in 90 days; "bring the body in and the mind will follow" is the slogan. Make the change consistently enough and it will be your change.

> Successful coaching means working as hard at maintaining changes as initiating them.

4. **Stay Active.** The research literature in psychology finds that change is most likely to occur when we are active in its pursuit. That is, we change by enacting new patterns—by *doing* new things—not simply by talking about change or thinking about it. I often joke that traders approach coaching like many people treat church: they go once a week to feel virtuous and forget about it for the next six days. A truly religious person wants to live their beliefs every day; if you're going to get the religion of virtuous trading, it's no different. That is why each goal should be accompanied by specific daily activities that help you make progress. If your goal is better risk management, then you want to work on managing the risk of every trade. If your goal is a better mindset, then you want to perform specific exercises each day to keep calm and focused. You won't change your behavior by changing your mind; you'll start thinking differently once you enact new behavior patterns.
5. **Stay Positive.** "If it ain't broke, don't fix it," is the philosophy of those who fail to work on trading until they're broke. *If you're trading well, that's one of the best times to coach yourself*. Your goal isn't to change what's working; it's to become even more consistent in your efforts. Doing more of what works is a valuable goal that helps you press your advantage when you're doing well. The alternative, sitting back on your laurels when you're making money, will bring comfort, but not elite levels of success. I recently met with a prop trader who was trading very well on relatively small size. A quick look at his Sharpe Ratio and trading results suggested that he could be making much more money simply by running more risk in his portfolio and sticking to his bread-and-butter setups and markets. We developed a plan for doing that and he turned good success into outstanding success. By formulating positive goals—focusing on changes that involve doing more of the right things—he made the most of his strengths. It's for that reason that I often tell people that the best time for coaching is when you're doing really well and really poorly.

The best traders, I find, are those who have made self-improvement a way of life. Such traders are driven in their work lives, their physical fitness, and their recreations. They derive great meaning and satisfaction from being the best they can be. The same is true of great athletes: they love working out; they constantly challenge themselves. It's when change is a lifestyle that we see exemplary performance. At that point, self-coaching becomes a life philosophy—an organizing principle—not just a discrete activity among many during the day or week.

What is the one change you most want to make *outside of trading*? Develop a daily plan for action on the goal that will help your trading efforts. It's all about strengthening the coach within you, whether you're working on your finances, your relationships, your physical conditioning, or your chess game. The goal is to become a change agent, a master of change across all spheres of life. Working on your nontrading life is a way of building your self-coaching as a trader.

RESOURCES

The *Become Your Own Trading Coach* blog is the primary supplemental resource for this book. You can find links and additional posts on the topic of change at the home page on the blog for Chapter 1:
http://becomeyourowntradingcoach.blogspot.com/2008/08/daily-trading-coach-chapter-one-links.html

Some changes—particularly of problems that have been longstanding and that interfere with relationships and/or work in a significant way—may require more than self-coaching. Here is a referral list for cognitive therapists that I've found helpful:
http://www.academyofct.org/Library/CertifiedMembers/Index.asp?FolderID=1137

For a detailed, research-based summary of the literature on change, check out *Bergin and Garfield's Handbook of Psychotherapy and Behavior Change*, published by Wiley and in its fifth edition (2003). Of particular relevance are the chapters in Section 2, dealing with "Evaluating the Ingredients of Therapeutic Efficacy."

For an overview of short-term approaches to change, check out my chapter on "Brief Therapy" in the *Handbook of Clinical Psychology*, Volume 1 edited by Michel Hersen and Alan M. Gross, also published by Wiley (2008).

A worthwhile collection of creative approaches to change can be found in *Clinical Strategies for Becoming a Master Psychotherapist* edited by William O'Donohue, Nicholas A. Cummings, and Janet L. Cummings (Academic Press, 2006), including my chapter on "The Importance of Novelty in Psychotherapy."

If you plan on being anything less than you are capable of being, you will probably be unhappy all the days of your life.

—Abraham Maslow

When traders seek coaching, one of their most frequent requests is help with reducing stress. The assumption is that less stress is better and, if they could eliminate stress, trading would go well. But is that true?

In this chapter we'll take a look at stress and distress, as well as the difference between the two. We'll also explore coping and what makes for effective and ineffective responses to stressful situations.

One of the great challenges of coaching in a high-intensity, competitive field is making sure that normal, expectable stress doesn't turn into performance-robbing distress. In practice this means distinguishing between the stresses that are part and parcel of the trading profession and those that we unwittingly place on ourselves.

Let's take a look at how you can make stress work for you in your self-coaching efforts.

LESSON 11: UNDERSTANDING STRESS

It's common to hear suggestions that traders eliminate or greatly minimize psychological stress. This, of course, is impossible. The very act of trading requires daily encounters with risk and uncertainty. Psychological stress is assured when we operate in such environments. If minimal stress is your objective, trading should not be your vocation or avocation.

Many traders—and trading coaches—confuse stress and distress. *Not all psychological stress brings distress, and not all psychological stress is bad.* If you are going to coach yourself for trading success, it's important that you understand stress: how it helps performance, and how it can become distress and interfere with decision-making.

So let's start with an everyday example. You're driving on a highway on a long trip, and you're feeling a bit bored. Suddenly the wind picks up and snowfall becomes heavy. Your visibility is greatly reduced, and you can feel the road becoming slippery. Before you know it, you're hunched over the steering wheel, staring intently through the windscreen and reducing your speed. Your boredom has quickly turned into alertness. You're no longer operating on auto-pilot.

This is psychological stress: a heightened physical and cognitive state that prepares us for dealing with challenges. It's been called the flight or fight response, because it mobilizes mind and body to avoid challenging situations or to face them head on. Muscle tension, alertness, and the flow of adrenaline: these are but a few cues that we've entered a state of stress.

Note that this is an adaptive state in the example of driving in the snowstorm. Had you remained on autopilot, you might not have slowed your speed and taken measures to avoid an accident. The state of stress has mobilized your energy—shifted you from your boredom—to cope with the immediate situation. You can appreciate how silly it is to talk of minimizing stress in such a situation: when you're in a blinding snowstorm in a moving vehicle, your mind and body *should* mobilize!

> Stress is a mobilization of mind and body; it can facilitate performance.

But let's take our example one step farther. I lived in the Syracuse, New York, area for more than 20 years, so the snowstorm is challenging, but not unfamiliar. I've experienced similar conditions before, and I know what to do when they arise. My stress never becomes distress, because I never perceive the storm as an acute threat.

Suppose, however, I am from Florida and have never experienced such a storm. I've heard of cars getting into pileup accidents under those conditions, and I'm worried that my tires will lose their grip. The storm is highly threatening for me; I don't feel capable of mastering it. My stress quickly becomes distress, as alertness turns into anxiety.

The simple example of the car illustrates that perception and experience make the difference between stress and distress. When I was an undergraduate at Duke University, David Aderman and I performed an experiment. We had two groups of subjects deliver a speech. Both groups received negative feedback about their performance. The first group was told that speaking ability was linked to personality and could not be changed. The second group was told that they could improve their speaking skills relatively easily. At the end of the session, the first group reported significantly more distress than the second group. *It wasn't getting the negative feedback that generated distress; it was the perception that they were not competent to change the negative situation.*

> Our interpretations of situations turn normal stress into distress.

So how does this relate to trading? When you put your capital at risk, you're like the driver on the snowy road. You will be alert, processing information in real time to make quick mid-course corrections if needed. If you view losing as a natural part of trading, have experienced and bounced back from losses before, and have mechanisms in place to limit your losses, stress is unlikely to become distress. The losing trade is a mere annoyance, like having your trip delayed by a snowstorm.

If, however, you don't accept losing as normal and natural—and especially if you don't have position size limits and stop-loss levels in place to control your losses—the stress of a trade going against you is more apt to become distress. Your ability to focus and make those rapid mid-course corrections will become impaired. You'll be like the Florida driver in the Northeast snowstorm.

Position size limits, trading plans, and stop-loss levels are like snow tires on your vehicle: they may not seem to do a lot for you when things are going well, but they certainly help you deal with adverse conditions. A panicky driver in the snow doesn't feel that he can control his car; an experienced winter driver knows that he can. Similarly, the panicked trader feels unable to control losses; the experienced trader knows that losses can always be limited.

Your challenge as your own trading coach is to embrace stress—and always ensure that it does not become distress. A fantastic goal to work on for your trading is to start the day with position sizing guidelines, per-trade loss limits, per-trade price targets, and daily loss limits that you can readily live with. The amount of risk you're willing to take on each trade should be meaningfully smaller than the reward potential built into your profit targets. The amount of money you're willing to lose each day should be a fraction of the money you make on your best days. If you're a frequent trader, no single loss on a trade should prevent you from making money on the day; no single daily loss should be so large that you can't be profitable on the week. *Preparation and familiarity keep stress from becoming distress, because they enhance your sense of control.* If you plan your loss levels and review those plans thoroughly, they become familiar and you become prepared. (See the brief questionnaire in Figure 2.1 to assess whether your stress is turning into distress.)

> If you are prepared for adversity, you will respond with normal stress, not distress.

Trading will always be stressful, but self-coaching ensures that it won't be filled with distress. Remember, your job is to maintain a mindset that keeps your confidence and motivation high: under those conditions, you'll

Please respond to the following questions with the scale below:

1 = rarely or never
2 = occasionally
3 = sometimes
4 = fairly often
5 = most of the time

How often do the following interfere with your work and/or your relationships?

1) Nervousness or anxiety? —
2) A downbeat or depressed mood? —
3) Frustration or anger? —
4) Guilt or self-blame? —
5) Alcohol or other substances? —
6) Arguments or fights? —
7) Fatigue or exhaustion? —
8) Sleep problems? —
9) Headaches, stomachaches, muscle tension? —
10) Worry, negative thinking? —

Note: Even a single score of 4 or greater merits attention and possible consultation with a professional (see Resource page at the end of this chapter), particularly if the problem is long-standing (greater than one year in duration). Several scores of 3 or greater are also worthy of attention. Chapters 4 through 6 of this book will outline specific brief therapy techniques that can be useful in addressing these problems, particularly when these are situational, not chronic.

FIGURE 2.1 Brief Distress Questionnaire

work harder, learn more, and grow your trading account over time. *That doesn't mean repealing stress; it means creating active firewalls between stress and distress.* Risk management is one of the best psychological firewalls of all.

COACHING CUE

Plan for every possible glitch in your trading: if your broker can't be reached; if you lose your online connection; if your equipment fails; if your data vendor goes down. My own trading station is small, but there is redundancy in each of these areas. I have multiple brokers, multiple online connections, multiple computers, and multiple data streams. There's always a rehearsed Plan B if something fails when I have a position on. The glitches still cause stress, but not distress.

LESSON 12: ANTIDOTES FOR TOXIC TRADING ASSUMPTIONS

What we expect from life shapes our emotional experience. If we expect good things, we tend to be optimistic and energetic. If we expect negative outcomes, we tend to feel anxiety. If we expect that success will elude us, we'll feel discouraged and depressed. If we expect perfection, we'll be continually disappointed with reality.

To no small degree, our emotions are barometers of the degree to which we are meeting or falling short of our expectations. That is an important principle, because psychological research suggests that, if our expectations are biased, we're likely to experience skewed emotions.

The relationship between emotion and expectation is particularly important for the developing trader. If you are your own trading coach, one of your overriding priorities is to foster the kind of positive experience that will sustain your motivation and learning efforts. Discouraged, defeated, and fearful learners are not effective learners. If you are going to maintain that zone of focus and concentration that maximizes learning efforts, you have to be *absorbed* in markets. No one can be absorbed if they're also battling emotional distress.

In the spirit of Ayn Rand, who encouraged people to "check their premises" when they arrived at contradictory conclusions ("I should be happy" but "I shouldn't be selfish"), I'll now ask you to check your expectations—particularly if you're finding that distress is interfering with your learning and development. The following expectations are among the most trader-toxic:

1. **A Good Day Is a Winning Day.** Here's an assumption and expectation that ensures that our emotional experience will rise and fall with our P/L. We generally expect to have a good day; by equating a good day with a winning day, we set ourselves up for disappointment when the normal uncertainty of markets leaves us in the red. *A good day is one in which we follow sound trading practices, from skilled execution to prudent risk management.* Some good days will bring profits, others will not. We can trade poorly and stumble into a profit; we can place a trade with a two-thirds probability of success and lose money as often as an all-star baseball hitter gets a hit. We *should* expect to have good days, if those are defined by sound trading practice. If sound practices don't generate profits over time, we may need to tweak those practices. But going into trading expecting profits each day is a formula for emotional letdown.

Never set a goal if you're not in full control of its attainment.

2. **Working Harder at Trading Means Trading More Often.** This assumption is that, if you trade more, you'll learn more and build skills more quickly. The result is overtrading and a likely forfeiture of profits over time to market makers and brokers. *Every trade starts as a loser.* You're losing the bid-offer spread if you buy and sell at the market, and you certainly lose a transaction fee. A loss of a single tick per trade due to execution, on top of a substantial retail commission, easily ensures losses of thousands of dollars to day traders who otherwise break even on their trades. As we trade more often without a distinct edge to each trade, our broker becomes richer and we become broker. Pointing and clicking to execute trades is a small part of the process by which traders develop expertise. The lion's share of development occurs by tracking patterns in simulation and real time, practicing executing and stopping out trades on paper (and in simulation mode) prior to risking capital, and researching trading ideas. By expecting trading itself to generate learning, we ensure that our motivation for learning will lead to overtrading—and a loss of both capital and motivation.
3. **Success Means Making a Living from Trading.** Here is another expectation guaranteed to generate frustration and discouragement. No developing professional makes their living from their performances during the early years of expertise building. A golfer or tennis player may spend years on a college team and as an amateur before even starting the pro circuit. Star actors or actresses typically spend years

in lessons and regional theaters before they see their names under the Broadway lights. Before surgeons make a living from their trade, they spend four years as a medical student, four more years as a surgical resident, and even more years in subspecialty training. Expecting to make a sustainable living from trading within the first years of exposure to markets is wholly unrealistic. More realistic expectations would be to keep losses to a reasonable level, cover one's costs with regularity, and improve trading processes. *There is no path to expertise that doesn't first require time to develop mere competence.* If you expect to make a living from trading much more quickly than people in other fields are able to sustain a livelihood, you are setting yourself up for frustration and failure.

An excellent antidote to these toxic assumptions is to write out your expectations as part of keeping a trading journal. This includes your expectations for each day of trading—your goals for trading well—*but also your expectations for your development over time.* My goal for my own trading performance is relatively modest: I want to earn more than the riskless rate of return after trading costs, and I want to do that by risking less than that proportion of my capital. In other words, I'm targeting positive risk-adjusted returns; that integrates my performance and process goals. I will be very happy if I average one percent returns on my portfolio per month by risking significantly less than that.

> By focusing on risk-adjusted returns, not just absolute profitability, we blend process and outcome goals.

This may not be a realistic target for you, depending on your level of development and your risk appetite as a trader. What is important is not my target numbers, but rather the fact that I have developed realistic, attainable objectives for my trading. If I achieve my expectations, I feel a sense of pride and accomplishment. If I fall short, I can quickly identify that fact, pull back my risk exposure, and make necessary corrections.

A formula for positive trading development is: *Always expect success, and always define success so that it is challenging, but attainable.* Writing out your expectations for the day, week, month, and year—and ensuring that they're doable—is a powerful lever over your emotional experience as a trader.

COACHING CUE

A great entry in a daily trading journal: "What would make my trading day a success today, even if I don't make money?" That simple question leads directly to process goals—the things you can best control.

LESSON 13: WHAT CAUSES THE DISTRESS THAT INTERFERES WITH TRADING DECISIONS?

We become anxious and exit a good trade before it has an opportunity to reach its price target. We become frustrated and take a trade that completely contradicts our research and planning. We're afraid of missing a trade and enter at the worst possible time. We're reluctant to take a loss at our designated stop point and wind up with a much larger loss.

All of these are examples of trading behaviors for which all of us as traders can truly say, "been there, done that." In trading, as in much of life, we learn by making the mistakes our parents and mentors try to protect us from. *The key to longevity is making those mistakes early in your development, before you have too much on the line.* Mistakes in dating can lead to a great marriage. Mistakes on a simulator can lead to solid real-time performance. Making your errors when your risk is lowest is a large part of success.

But what causes these mistakes, in which the arousal of distress interferes with prior planning and consistent, sound decision-making? Perhaps if we can figure that out, we can shorten the painful learning curve just a bit.

> The fields of behavioral finance and neuroeconomics have illuminated how emotions affect financial decisions; check out references at the end of this chapter.

To hear traders talk, distress in trading is caused by markets. A market turns slow and range-bound: that is the supposed cause of a trader's boredom and overtrading. A market reverses direction: that allegedly generates frustration and impulsive trades for the trader. Because their emotions are triggered by a market event, traders assume that the market must be responsible for their feelings. A moment's thought, of course, dispels this notion. After all, if the market had the power to compel emotional reactions, we would see *all* traders respond identically in a similar situation. That, however, is not what happens. Not all traders respond to a slow market with boredom and overtrading; not all traders become frustrated and make impulsive decisions when a position reverses against them. There is more to the cause of our emotions than external events.

In order for a market event to generate a negative emotional response, we have to view it as a threat. Let's say that I'm a trader watching my market become slow and range-bound at midday. If I view that as an opportunity to update some of my ideas and prepare for the afternoon, the slow market will not trouble me in the least. Similarly, I may view the choppy, thin action as an opportunity to get away from the screen, clear out my head, and start the afternoon fresh. Again, the slow action affects neither my feelings nor my trading behavior.

If, however, I start my day telling myself that I must make at least several thousand dollars each day, then I'll now perceive the slow market as a threat to my trading goal. The narrow, whippy action translates into lack of opportunity, which translates into lack of profit, which translates into lack of success. It's easy to see, with that mental framework, how I could wind up viewing slow markets as threats to my career. It's no wonder that the slow market would trigger my distress and overtrading.

But, of course, it's not merely the slow market that is generating my negative mood and behavior; it's my *perception* of that market. Perception is the filter that we place between events and our responses to events. If we place a distorting lens over our eyes, we will see the world in distorted ways. If we adopt distorted perceptions of markets and ourselves, we'll experience trading in distorted ways.

Can we alter our perceptions? Chapter 6 deals with cognitive approaches to change that restructure thought processes.

So how do we change the filters that turn normal trading experience into abnormal events?

The rule is simple: if you don't know your filters, you cannot change them. Becoming aware of the expectations and beliefs that shape your perception is essential to the process of shifting your perception.

Here's a simple exercise that can aid you in becoming more aware of any distorted perceptions you might hold:

Every time you experience a distinctly negative emotional reaction to a market event, consciously ask yourself, "How am I perceiving the current market as a threat?" This turns your attention to your perceptual processes, giving you a chance to separate perceived threat from real threat.

A simple example comes from my recent trading. I really wanted to finish the week on a high-water note in my equity curve and found myself with a nice profit on a Friday morning trade to the long side. As the trade moved my way, I moved my stop to breakeven. The trade continued to go my way a bit before making a small reversal. It then chopped around for a few minutes. I found myself becoming nervous with the trade, as if it were on the verge of plunging below my stop point.

I quickly asked myself, "Why am I nervous with this trade? Why am I perceiving the trade with so much uneasiness?" A moment's reflection and review of the market told me that the trade was perfectly fine and progressing according to plan. It was my desire to end the week profitably (after an extended flat period of performance, I should add) that turned the potential reversal of a winning trade into a threat. If all the market can do after the run-up is chop around in a flat way, perhaps this is an opportunity to *add* to the position, I reasoned. I calibrated my risk/reward on the added piece to the position (and for the position overall, given my new average purchase price) and added a small portion to the trade. The added increment wasn't large enough to dramatically affect the profitability or risk of the trade, but it was an important psychological step: I turned a perceived threat into opportunity.

The key here is to distinguish between *actual* threats—markets that truly are not behaving according to your expectations—from *perceived* threat. That requires reflection about markets and about personal assumptions. Once I saw that the trade was proceeding normally, I was free to challenge the filters that were leading me to become nervous with a good trade.

> When you think about your thinking by adopting the perspective of a self-observer, you no longer buy into negative thought patterns.

After you identify a perception that turns a normal event into a threat, the next challenge is to find opportunity in that normal event. I might feel threatened by a difference of opinion with my wife, but that threat can be turned into an opportunity for fruitful communication and problem solving. We might feel threatened by a trade that starts modestly profitable but then stops us out, but that threat can be turned into an opportunity to flip our position or reassess our views of that market.

Identify the perceived threat; turn the perceived threat into an opportunity: that is a two-step process that addresses the true cause of emotional reactions that distort trading decisions. By keeping a journal specifically devoted to your thinking and perceiving, you can structure this two-step process and turn it into a habit pattern that you activate in real time.

COACHING CUE

When you talk or IM about the day's trading, pay attention to how you describe the markets: good, bad, quiet, active. Listen especially for the tone of your descriptions. Many times, your tone and language will give away whether you're in tune with markets or fighting them. If you become caught up in what you think the market *should* be doing, you're most likely to fight it when it does something else.

LESSON 14: KEEP A PSYCHOLOGICAL JOURNAL

When I first began trading, I kept a journal in the form of multiple annotated charts. I looked for every major turning point in the stock market and then investigated the patterns of indicators and price/volume patterns that could have alerted me to the changes in trend. After a while, I found that certain patterns recurred. It was out of those early observations that I learned to rely on patterns of confirmation and disconfirmation among such measures as the number of stocks making new highs versus lows, the NYSE TICK, and the various stock sectors. Later, as I gained new tools, such as Market Delta (www.marketdelta.com), I added to those patterns. For me, the journal was a tool for pattern recognition. Only after an extended time of recognizing patterns on charts, could I begin to see them unfold in real time. It was also only after an extended period of real-time observation that I felt sufficiently confident to actually place trades based on those patterns.

When we keep a psychological journal, the learning principles are not so different. *At first, the journal is simply a tool for recognizing our own patterns as traders.* These include:

- **Behavioral patterns**—Tendencies to act in particular ways in given situations.
- **Emotional patterns**—Tendencies to enter particular moods or states in reaction to particular events.
- **Cognitive patterns**—Tendencies to enter into specific thinking patterns or frames of mind in the face of personal or market-related situations.

Many of our trading patterns are amalgamations of the three patterns: in response to our immediate environment, we tend to think, feel, and act a certain way. Sometimes these characteristic patterns work against our best interests. They lead us to make rash decisions and/or interfere with our best market analysis and planning. It is in such situations that we look to a journal (and other psychological exercises) to help us change our patterns of distress.

For more on keeping trading journals: http://traderfeed.blogspot.com/2008/03/formatting-your-trading-journal-for.html

But why do such patterns exist? Why would a person repeat an unfulfilling pattern of thought and behavior again and again, even when she is aware of the consequences? Sometimes traders are so frustrated with their repeated, negative patterns that they swear that they are sabotaging their own success. This pejorative labeling of the problem, however, doesn't help the situation. It only serves to blame the frustrated trader, magnifying frustrations.

As I stressed in my Psychology of Trading *book, maladaptive patterns generally begin as adaptations to challenging situations.* We learned particular ways of coping with difficult events and those, at first, may work for us. As a result, these patterns become overlearned: they are internalized as habit patterns.

A good example is the tendency to blame oneself when there are conflicts with others. A child in a home fraught with arguing and fighting might adapt to the situation best by blaming himself for problems rather than risk conflict by blaming others. Later in life, with that pattern ingrained, even normal conflicts may trigger self-blame and depressed mood. Such a person, for example, might spend more time and energy beating up on himself after a losing day than learning from his losses.

When we repeat patterns in trading that consistently lose us money or opportunity, the odds are good that we are replaying coping strategies from an earlier phase of life: one that helped us in a prior situation but which we've long since outgrown. The task, then, is to unlearn these patterns—and that is where the psychological journal becomes useful.

Just as I used the trading journal to become keenly aware of market patterns, our psychological journal can alert us to the repetitive patterns of thinking, feeling, and acting that interfere with sound decision-making. Such a journal, like the annotated charts that I mentioned, begins with observation: we want to review our trading day and notice all of the patterns that affected our trading. The initial goal is not to change those patterns. Rather, we simply want to become better at recognizing the patterns, so that we'll eventually learn to identify their appearance in real time.

> The psychological journal is a tool for developing your internal observer: learning to recognize what you're doing, when you're doing it.

A favorite journal format that I use divides a normal piece of paper into three columns. The first column describes the specific situation in the markets. The second column summarizes the thoughts, feelings, and/or actions taken in response to the situation. The third column highlights the consequences of the particular cognitive, emotional, or action patterns.

The first two columns help us recognize the situational triggers for our patterns. This makes us more sensitive to their appearance over time. The third column emphasizes in our mind the negative consequences to our patterns. Those negative consequences could include emotional distress, losing money on a trade, or failure to take advantage of an opportunity. *When we clearly link maladaptive patterns to negative consequences, we develop and sustain the motivation to change those patterns.* That third column should spell out in detail the costs of the recurring pattern: how, specifically, the pattern interferes with your happiness and trading success. The clearer you are about the pattern and its occurrence and the more strongly you feel about the costs it imposes on you, the more likely you'll be to catch the pattern in real time and be motivated to interrupt and change it.

For now, however, your goal should be to identify your repetitive patterns and their consequences—not to try to change those patterns all at once. *You cannot change something if you're not aware of it.* The psychological journal is a powerful tool for building that awareness and understanding what is generating your distress. Keep the journal for 30 consecutive days to help you see just about every variation of your most common patterns. It will also begin the process of turning self-observation into a habit pattern—a positive pattern that can aid you in your personal life and in your trading.

COACHING CUE

Begin your psychological journal by tracking your individual trades and focusing on those situations in which your mindset took you out of proper execution or management of those trade ideas. In other words, these will be instances in which you failed to follow your trading rules, not ones in which you followed the rules and just happened to be wrong on your ideas. Replay these trades in your mind—or, better yet, consider videotaping your trading and observing those trades directly—and then jot down what set you off (Column A); what was going through your mind (Column B); and how it affected your trading (Column C). Zero in on how much money that trigger situation cost you. With practice, you'll build your internal observer and start noticing these situations as they are occurring. That will give you an opportunity to create a different ending to the script.

LESSON 15: PRESSING: WHEN YOU TRY TOO HARD TO MAKE MONEY

Traders call it pressing: forcing trades in an attempt to make money. Sometimes it takes the form of trading too large; other times traders press by trading too often. The hallmark of pressing is trying to *make things happen.* This is 180 degrees from a mindset in which you trade selectively,

when odds are with you. In the latter frame of mind, you let the market come to you and wait for your opportunity. In the mindset of pressing, you want things to happen and you want it now.

The irony of pressing is that it is often the most successful traders—those who are competitive and driven to succeed—that fall victim. They so hate losing that they'll do anything to win—including trading poorly!

Trading is a bit like flying a fighter plane or playing chess: it requires highly controlled aggression. In trading, the control element comes from knowing when markets present opportunity and when they don't. One of the best ways of instilling this control is to trade with rules. These may be rules related to position sizing, stop-loss levels, when to enter markets and when to stay out, trading with trends, etc. *When rules are repeated and followed over time, they are internalized and become mechanisms of self-control.* We can observe this process among children. They so often hear rules about respect for elders or cleanliness that (eventually!) those behaviors become automatic.

> The right trading behaviors start as rules and evolve into habits.

These automatic behaviors are important because they don't require effort and a dedication of mental resources. If we have to *make* ourselves follow rules each time we confront situations, we will be taxed—and our full attention will not be on those situations. One of the great strengths of the human mind is its ability to automatize rules, so that mental and physical resources can be fully devoted to challenges at hand. This enables us to face those challenges under self-control (i.e., under rule governance).

So how do we make our trading rules automatic? The answer is to turn them into habit patterns. At one time in our lives, "brush your teeth in the morning" might have been a rule that our parents had to impress on us. With repetition, it became habit; most of us need no reminder of the rule or special motivation to follow the rule. This is the kind of automaticity we aim for in trading: where our rules become so much a part of us that they require no special attention or effort.

When we're pressing to make money, the need to put on trades overwhelms our rule governance. Pressing normally occurs in situations in which we're frustrated with our performance. Perhaps we've lost money, missed out on opportunities, or are just going through a period of flat equity curve. The frustration leads us to try to *create* opportunities rather than respond to those presented by markets.

In our dance with markets, we want the market to lead. When we attempt to lead the market—when we try to *anticipate* what may happen instead of *identify* what is happening—we're most likely to be out of step with the next price movements. When we are pressing, we are trying to lead the markets, and that has the potential to turn normal losses and flat periods into veritable slumps.

So how do we make trading rules automatic? As with the tooth brushing, it is through repetition. By repeating your rules many times, in many ways, you gradually internalize them and turn them into habits. You will still experience the normal stresses of markets—no one can repeal risk and uncertainty—but you will be so grounded in your decision-making that you cannot fall prey to distress.

When you coach yourself, you can create opportunities for repetition before and during the trading day. This is a several step process:

1. **Make a list of your most important trading rules.** These rules should include, at minimum, your rules for risk management; taking breaks after large or multiple trading losses; entering at defined signal points; and preparing for the market day. *You can't expect to internalize trading rules if you haven't first made them explicit.*
2. **Create a routine before trading begins to review the rules.** Mental rehearsals are powerful vehicles for creating repetition. Every one of your trading rules can be captured as a visualized scenario that you walk yourself through mentally while you keep yourself calm and focused. You actually visualize yourself in different trading situations reminding yourself of rules and following those rules. The more extended and detailed the visualizations, the more likely it is that you'll internalize them as realistic experience.

> The more you think about rules and rehearse them, the more they become part of you. Repetition creates internalization.

3. **Create a break in your trading day to review your rule-following.** Midday break, when markets tend to slow down, is a perfect time to clear out your head, assess your trading to that point, and remind yourself of what you need to do in the afternoon. By turning your list of rules into a checklist, you can simply check yes or no for each rule depending on whether you followed it during the morning. If you did not follow a particular rule, you jot down that rule on a separate piece of paper, tape it to the monitor, and make it an explicit focus for the afternoon trade.
4. **Use the rules at the end of the day as a report card.** An end-of-day review will tell you how well you performed in your preparation for trading, your entries, your risk management, and your exiting of

positions. Each rule should receive an A, B, C, D, or F grade. Anything less than a B is a candidate to become an explicit goal for the next day's trading. In this way, the rules you most need to work on are assured of getting the most attention.

This approach undercuts the tendency to press during periods of frustration by helping you catch yourself in the act of deviating from rules. As a result, frustration is unlikely to escalate into ever-greater violations of sound trading principles. When you cement your rules through repetition, however, you also serve as your own trading coach by preventing frustration from affecting trading in the first place. After all, we can start our day on a frustrating note (perhaps we oversleep), but that won't lead us to shatter our rule-habits of morning personal hygiene. Behavior patterns, once overlearned, stick with us regardless of our emotional state. That is true self-control.

Good self-coaching is the ability to correct trading problems. Great self-coaching is to develop routines to prevent problems from occurring in the first place. You'll see the results in your mood—and in the dramatic reduction of large losing trades, days, and weeks.

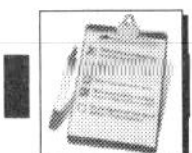

COACHING CUE

Don't work at internalizing too many rules at one time. Start with the most important rules that will keep you in the game: entry rules (getting good prices); position-sizing rules (limiting risk per position); and exit rules (setting clear profit targets and stop-loss levels). These three, along with the basic rationale for your trade, can be written down or talked aloud as a trade plan that becomes your guide for trading under control instead of pressing.

LESSON 16: WHEN YOU'RE READY TO HANG IT UP

One of the most difficult manifestations of distress that traders face is despair. I've seen it happen to the best of traders: you work hard, you feel as though you're on the brink of a positive breakthrough, and then you take several steps backward. It feels as though you're getting nowhere. You're tired of being wrong, tired of losing money. That excitement that used to greet the start of the market day is replaced with dread. It's difficult to sustain the research and the morning routines of preparation. If your body could talk, its posture would say, "What's the use?" You're ready to hang it up.

Let's face it: for many, there *is* a time to give up trading. I know quite a few traders who have been at it for years and have never developed the skills (and perhaps who never had the talent) to simply reach a point of competence where they cover their costs. *If you are meant to do something—something that speaks to your talents, skills, and interests—you will display a significant learning curve in the first year or two of effort.* If such a learning curve is not apparent, it's probably not your calling. Hang it up and pursue something that genuinely captures your distinctive abilities. It's not quitting, it's not being cowardly. It's cutting a losing position and getting into something better: a course of action that is as sound in life planning as in trading.

If, however, you've progressed steadily and have displayed genuine ability over time, discouragement and depression are your emotional challenges during difficult periods. Becoming your own trading coach requires shepherding yourself through the dark times.

One of my professors in graduate school, Jack Brehm, theorized that depression is a form of motivational suppression. When we perceive that meaningful goals are within our reach, we naturally experience a surge of optimism and energy. This surge helps us make those extra efforts to achieve our goals. Conversely, when we see that valued goals are beyond our reach, nature has provided us with the means to suppress that motivation. After all, it would make no sense to redouble our efforts in the face of unachievable ends. That motivational suppression, taking the form of discouragement and even mild depression, is unpleasant, but it is adaptive in its own way. It turns us away from ends we should not be pursuing, which frees us up for more energizing efforts.

In that sense, the feeling of wanting to give up contains useful information. It is not just a negative emotion to be overcome or minimized. *Discouragement tells us that, at that moment, we perceive an unbridgeable gap between our real selves (who we are) and our ideal selves (who we wish to be).* We no longer perceive that we have control over our future: our ability to attain goals that are important to us. If we are going to be effective coaches of our trading, we need to address this perception.

> Our real selves are always distant from our ideals: the question is whether we perceive ourselves to be competent to bridge the gaps.

The first way to address the real-ideal gap is to consider that, in its context, it may point to something based in reality. Perhaps an edge that we had counted on in trading is no longer present. Perhaps market patterns have changed, such that what was once working for us no longer has the same potential. In that event, our hang-it-up feeling is alerting us that maybe, temporarily, we *should* hang it up and instead focus our efforts in figuring out what is working in our trading and what isn't. The key word in that last sentence is *temporarily*. Just because we're discouraged doesn't mean we should feign optimism: perhaps there's good reason for our motivational suppression. By stepping back, we can investigate possible market-based reasons for our feelings.

Other times, our discouragement may be providing us with information that our expectations are too unrealistic. If, in the back of our minds, we're hoping or expecting to make money each trading day, we're setting ourselves up for considerable disappointment when we undergo a streak of losing trades or days that is entirely expectable by chance. In such cases, our motivational suppression provides a clue that we need to investigate, not just markets, but ourselves. When we expect the best, we leave ourselves poorly prepared for the worst.

There's a third source of reduced drive and motivation, and that's burnout. Psychological burnout occurs when we feel overwhelmed by the demands that we face. Very often, among traders, burnout signifies a lack of life balance: becoming so immersed in the stresses of trading that recreational, social, creative, and spiritual outlets are lost. While such immersion is possible—and sometimes necessary—for short stretches of time, the immersion leaves a trader impaired over the long run. This is not so much motivational suppression as motivational exhaustion. It is difficult to sustain energy and enthusiasm when we're operating on overload.

Burnout occurs when we feel that the demands on us exceed our resources for dealing with them.

In each scenario, the trader who serves as his own coach treats the lost drive as *information*. Maybe it's a reflection of changes in markets; maybe it's a sign of unrealistic self-demands or a signal that life is out of balance. If you feel discouraged about your recent trading, your first priority is to identify what that feeling is telling you, so that you can take appropriate action.

If market trends, themes, or volatility have shifted, altering the profitability of your trading setups and ideas, then your action should be a reduction in your risk-taking while you see which patterns, markets, and ideas are working, so that you can focus efforts on those. You also want to review your most recent trading performance to see if you can identify markets and patterns that have continued to work for you, even as others have shifted. *Reduce your risk, reassess your trading, and you preserve your capital and turn discouragement into opportunity.*

If the feeling like giving up is more a function of your own self-demands, then your challenge is to redouble your efforts at goal setting, making sure that each day and week starts with realistic, achievable goals. When basketball players get into slumps, their coaches will set up plays for high-percentage shots to get the players back on track mentally. Similarly, you want to set yourself up for psychological success by setting achievable goals that move you and your trading forward.

Finally, if burnout is contributing to the lack of optimism, then the challenge is to consciously structure your life outside of trading by ensuring proper time for physical exercise, social activity, and overall time away from markets. *An excellent strategy for achieving psychological diversification is to have significant life goals apart from trading.* If all the psychological eggs are in the trading basket, it will be difficult to sustain energy and enthusiasm when profits are scarce.

Being your own trading coach doesn't mean talking yourself into feeling good. Sometimes there are good reasons for lacking positive emotion. The superior coach will listen for those reasons and turn them into prods for constructive change.

COACHING CUE

How psychologically diversified are you? How much stress and distress are you experiencing in your social life, your family life, and in your general emotional state? How much satisfaction are you experiencing in each of these areas? What sustains you when trading goes poorly? What problems from your personal life creep into your trading day? How is your physical fitness? Your quality of sleep and concentration? Your energy level? It's worth evaluating the nontrading aspects of life as well as your market results with monthly reviews. If the other parts of your life are generating distress, it's only a matter of time before that compromises your focus, decision-making, and performance.

LESSON 17: WHAT TO DO WHEN FEAR TAKES OVER

Fear is a normal emotional response in the face of danger. Under conditions of fear, we are primed for flight or fight: running from the source of danger or confronting it. As we've seen, sometimes the dangers we respond to are not objective sources of threat, but ones that we interpret as threats. When we are prone to perceiving normal situations as threats, fear becomes anxiety. We experience the full flight or fight response, but there is nothing to run from or fight. The danger is in our perception, not in the environment.

When we feel nervous in a trade or feel nervous about putting on a trade, it's important to know whether our response is one of fear or one of anxiety. Is there a genuine danger in the market environment, or is the danger in our head?

Let's say that I am short the S&P 500 index (SPY) and we get to a prior level of support. There is a bout of selling as the NYSE TICK moves sharply to −500. SPY hits a marginal new low for the move, but I notice that other indexes—the NASDAQ and Russell—are not making fresh lows. The −500 TICK is well above the prior lows in the TICK for that move. I see signs that the selling is drying up. Nervously, I wait to see if any fresh selling will come into the market. My order to cover the position is ready to go, and my finger is poised over the mouse to execute the order. A few seconds go by and the market moves down a tick, up a tick, down a tick. Volume declines, my nervousness with the trade increases. In a flash, I act on my emotion and cover the position. The odds of the support holding, I decide, are too great to risk losing my profit in the trade.

In this scenario—based on a trade I made just a day ago—fear was adaptive. There was a real danger out there. (The market subsequently moved significantly higher over the next half hour). I knew to trust my feelings and act on my fear, because I could point to specific sources of danger: the established support level, the drying up of volume and TICK, and the nonconfirmations from the Russell 2000 and NASDAQ 100 indexes. Years of experience with intraday trading told me that moves are unlikely to extend in the short run if they cannot carry the broad market with them on increasing downside participation (volume, TICK).

> Fear is the friend of trading when it points to genuine sources of danger: a felt discomfort with a trade will often precede conscious recognition of a change in market conditions.

Note that had I identified fear as a negative emotion and tried to push past or ignore it, an important discretionary trading cue would have been lost. *When you are your own trading coach, your goal is not to eliminate or even minimize emotion.* Rather, your challenge is to extract the information that may underlie those emotions. This means being open to your emotional experience and, at times, trusting in that experience. Blind action based on emotion is a formula for disaster, particularly when what we're feeling is anxiety and not reality-based fear. But to ignore emotional experience is equally fraught with peril. When you ignore feelings, you cannot have a feel for markets.

So what do you do when nervousness enters your decision-making process?

In *The Psychology of Trading* book, I liken such feelings to warning lights on the dashboard of a car. The nervous feeling is a warning, a sign that something isn't right. When you see the light go on in your car, you don't ignore it or cover it up with masking tape. Rather, you use the warning to figure out: What is wrong? What should I do about it? Depending on the specific dashboard light, you might want to stop driving altogether and take the car into a repair shop. That's like exiting the market: the risk is just too great to proceed.

When nervousness hits, the first thing you want to do is simply acknowledge that fact. Saying to yourself, "I am not comfortable with the trade right now," cues you to extract the information from your experience. The next step is to ask, "Why am I not comfortable? Has something important changed in the trade?"

This latter question is crucial because it helps you distinguish realistic fear from normal anxiety you might feel in an unfamiliar or uncertain situation. I recently bumped up my average trading size per position and, at first, felt some nervousness in my trades. When I asked myself, "Why am I nervous?" I couldn't find anything wrong with the trades: they were performing as expected. This led me to acknowledge that I was just feeling some anxiety about the increased risk per trade. I reassured myself of my stop levels and overall trading plan, so I was able to weather the anxiety and benefit from the increased risk. I needed to carry out that internal dialogue, however, to see if my body's warning light—the nervousness—was based on a market problem or a problem of perception.

> Fear is a warning light; not an automatic guide to action. It is our mind and body's way of saying, "Something doesn't look right."

Similarly, if you're experiencing fear about entering a position, you want to ask, "*Why* am I uncomfortable with this idea? Do I really have an edge in this trade?" This questioning prods you to review the rationale for the trade idea: Is it going with the market trend? Can it be executed with a favorable reward-to-risk balance? Is it a pattern that I have traded successfully in the past? Is it occurring in a market environment with sufficient volume and volatility?

When I work with a trader in live mode (i.e., coaching them while they are trading), I have them talk aloud so that I can hear their thought process regarding a particular trade idea. I mirror back to traders the quality of their thinking about their decisions, so I can help them figure out when nervousness is warranted (the trade doesn't really have favorable expectancies) and when it might not be (it's a good idea, but you're feeling uncertain because you're in a slump).

When you are your own trading coach, you can place yourself in talk-aloud mode as well or—if you're trading in a room with other traders, as at many prop firms—you can use a brief checklist to review the status of your idea or trade. The checklist would simply be a short listing of the factors that should either get you into a trade or keep you in a trade. It's your way of using the nervousness to differentiate discomfort with a good idea and discomfort with an idea because it isn't so good. Many times, fear is simply fear of the unknown, the byproduct of making changes. As Mike Bellafiore notes in Chapter 9, that doesn't mean you shouldn't take the trade.

I have been trading for long enough that my checklist is well established in my mind. There was a time, however, when I needed to have it written out and in front of me. Is volume expanding in the direction of my trade? Is more volume being transacted at the market bid or at the offer? Is a growing number of stocks trading at their bid or at their offer? Even a very short-term trader can review such criteria quickly, just as a fighter pilot would check his gauges and radar screen in a dangerous combat situation. A fear response cues you to check your own gauges, to make sure you're making decisions for the right reason.

> You can use your fear as a cue to examine your trade more deeply and adjust your confidence in the idea, up or down.

If you can use fear in this way, even negative emotions can become trading tools and even friends. Your homework assignment is to construct a quick checklist that you can talk aloud or check off during a trade (or prior to placing a trade) when you are feeling particularly uncertain. This checklist should prod you to review why you're in the trade and whether it still makes sense to be in it. In this way, you coach yourself to make your best decisions even when you're at your most nervous. Confidence doesn't come from an absence of fear; it comes from knowing you can perform your best in the face of stress and uncertainty.

COACHING CUE

Find a positive change in your trading that makes you nervous and then pursue it as a trading goal. As I describe in *The Psychology of Trading*, anxiety often points the way to our greatest growth, because we feel anxiety when we depart from the known and familiar. Trading a new market or setup, raising your trading size, holding your trades until they reach a target—these are nerve-wracking situations that can represent great areas of growth and development. In that sense, fear can become a marker for opportunity.

LESSON 18: PERFORMANCE ANXIETY: THE MOST COMMON TRADING PROBLEM

Imagine you're about to give a presentation to a group of people as part of a job interview process. You very much want the job, and you've prepared well for the presentation. You're nervous going into the session, but you remind yourself that you know your stuff and have done this before.

As you launch into your talk, you notice that the audience is not especially attentive. One person takes out his phone and starts texting while you're talking. Another person seems to be nodding off. The thought enters your mind that you're not being sufficiently engaging. You're losing their interest, and you fear that you might also be losing the job. You decide to improvise something original and attention grabbing, but your nervousness gets in the way. Losing your train of thought, you stumble and awkwardly return to your prepared script. Performance anxiety has taken you out of your game, and your presentation suffers as a result.

Performance anxiety occurs any time our thinking about a performance interferes with the act of performing. If we worry too much while taking a test, we can go blank and forget the material we've studied. If we try too hard to make a foul shot at the end of a basketball game, we can toss a brick and lose the game. The attention that we devote to the outcome of the performance takes away from our focus on the process of performing.

This is a common problem among traders, probably the most common one that I encounter in my work at proprietary trading firms and hedge funds. Sometimes the performance anxiety occurs when a trader is doing well and now tries to take more risk by trading larger positions. Other times, traders enter a slump and become so concerned about losing that they fail to take good trades. A trader may feel so much pressure to make a profit that she may cut winning trades short, never letting ideas reach their full potential. As with the public speaker, the performance anxiety takes traders out of their game, leading them to second-guess their research and planning.

More on performance anxiety and how to handle it: http://traderfeed.blogspot.com/2007/04/my-favorite-techniques-for-overcoming.html

As we've seen in this chapter, our distressful emotions don't just come from situations: they are also a function of our *perception* of those situations. If I'm convinced that I'm a hot job candidate and believe that I have many job options, I won't feel unduly pressured in an interview or presentation. When I came out of graduate school, I went to a job interview in upstate New York and was asked by the clinic director to identify my favorite approach to doing therapy. I smiled and told him that I preferred primal scream therapy. That broke the ice, we had a good laugh, and the interview went well from there. I knew that, if this interview didn't work out, other opportunities would arise. That freed me up to be myself.

Had I told myself that this was the only job for me and that I *needed* the position badly, the pressure would have been intense. I would have been far too nervous to joke in the interview and probably would have come across as wooden and not very personable. If I had viewed the possibility of losing the job as a catastrophe, I would have ensured that I could not have interviewed well.

Traders engage in their own catastrophizing. Instead of viewing loss as a normal, expectable part of performing under conditions of uncertainty, traders regard losses as a threat to their self-perceptions or livelihood. When traders make money, they feel bright about the future and good about themselves. When they hit a string of losing days, they become consumed with the loss. Instead of trading for profits, they trade to not lose money. Like the anxious job interviewer, traders can no longer perform their skills naturally and automatically.

A common mistake that traders make is to try to replace catastrophic, negative thoughts with positive ones. They try repeating affirmations that they will make money, and they keep talking themselves into positive expectations. What happens, however, is that they are still allowing a focus on the outcome of performance to interfere with performing itself. *The expert performer does not think positively or negatively about a performance as it's occurring. Rather, he is wholly absorbed in the act of performing.* Does a skilled stage actress focus on the reaction of the audience or the next day comments of reviewers? Does an expert surgeon become absorbed in thoughts of the success or failure of the procedure? No, what makes them elite performers is that their full concentration is devoted to the execution of their skills.

> Thinking positively *or* negatively about performance outcomes will interfere with the process of performing. When you focus on the doing, the outcomes take care of themselves.

What gives these expert performers the confidence to stay absorbed is not positive thinking. Rather, they know that they are capable of handling setbacks when those occur. If a given night's performance doesn't go quite right, the actress knows that she can make improvements in rehearsal. If a surgery develops complications, the surgeon knows that he can identify those rapidly and take care of them. By taking the catastrophe out of negative outcomes, these experts are able to avoid crippling performance anxiety.

One of the most powerful tools I've found for overcoming performance anxiety in trading is to keep careful track of my worst trading days and make conscious efforts to turn those into learning experiences. This turns losing into an opportunity for self-coaching, not just a failure.

Let's say that you have a very reliable setup that tells you that a stock should be heading higher. You buy the stock and it promptly moves your way. Just as suddenly, however, it reverses and moves below your entry point. You note that the reversal occurred on significant volume, so you take the loss. In one frame of mind, you could lament your bad luck, curse the market, and pressure yourself to make up for the loss on the next trade. All of those negative actions will contribute to performance pressure; none of them will constructively aid your trading.

Alternatively, you could use the loss to trigger a market review. Are other stocks in the sector selling off? Is the broad market dropping? Has news come out that has affected the stock, sector, or market? Did your buy setup occur within a larger downtrend that you missed? Did you execute the setup too late, chasing strength? All of these questions offer the possibility of learning from the losing trade and quite possibly setting up subsequent successful trades. For instance, if you notice that a surprise negative earnings announcement within the sector is dragging everything down, you might be able to revise your view for the day on the stock and benefit from the weakness. When you are your own trading coach, you want to get to the point where you actually *value* good trading ideas that don't work. If a market is not behaving the way it normally does in a given situation, it's sending you a loud message. If you're not executing your ideas the way you usually do, you're getting a clear indication to target that area for improvement. A simple assignment that can instill this mindset is to identify—during the trading day (or during the week, if you're typically holding positions overnight)—at least one very solid trading idea/setup that did not make you money. That good losing trade is either telling you something about the market, something about your trading, or both. Your task is then to take a short break, figure out the message of the market, and make an adjustment in your subsequent trading.

By acting on the idea that losses present opportunity, you take a good part of the threat out of losing. That keeps you learning and

COACHING CUE

If you track your trading results closely over time, you'll know your typical slumps and drawdowns: how long they last and how deep they become. Know what a slump looks like and accept that they will arise every so often to help take away their threat value. Many times you can recognize a slump as it's unfolding and quickly cut back your trading and increase your preparation, thus minimizing drawdowns. Most importantly, if you accept the slump as a normal part of trading, it cannot generate performance anxiety. Indeed, it is often the slumps that push us to find new opportunity in markets and adapt to shifting market conditions. Much of success consists of finding opportunity in adversity.

developing, and it keeps you in the positive mindset that best sustains your development. Every setback has a purpose, and that's to help you learn: to make you stronger. Performance anxiety melts away as soon as it's okay to mess up.

LESSON 19: SQUARE PEGS AND ROUND HOLES

One of the central concepts of *Enhancing Trader Performance* is that each trader can maximize his development and profitability by discovering a niche and operating primarily within it. A trading niche has several components:

- **Specific Market and/or Asset Classes.** Markets behave differently and are structured differently. Some markets are more volatile; some are more mature and offer more market depth; some offer more information than others. *The personality of the market must fit with the personality of the trader.* Someone like myself, who thrives on data collection and the analysis of historical patterns of volume and sentiment, can do quite well in the information-rich stock market; not so well in cash currencies, where volume and moment-to-moment sentiment shifts are more opaque.
- **A Core Strategy.** The trader's core strategy or strategies capture her ways of making sense of supply and demand. Some traders gravitate toward trend trading; others are contrarian and more countertrend in orientation. Some traders rely mostly on directional trade; others on relational trades that express relative value, such as spreads and pairs trades. Some traders are highly visual and make use of charts and technical patterns; others are more statistically oriented and model-driven.
- **A Time Frame.** The scalper, who processes information rapidly and holds positions for a few minutes, is different from the intraday position trader and even more different from the swing trader who holds positions overnight. A portfolio manager who trades multiple ideas and markets simultaneously engages in different thought processes from the market maker in a single instrument. Your time frame determines what you look at: market makers will pay great attention to order flow; portfolio managers may focus on macroeconomic fundamentals. Time frame also determines the speed of decision-making and the relative balance between time spent managing trades and time spent researching them. My personality tends to be risk-averse: I trade selectively over a short time frame. I know many other, more aggressive traders who trade frequently and others who hold for longer periods and larger

price swings. Time frame affects risk, and it determines the nature of the trader's interaction with markets.

- **A Framework for Decision-Making.** Some traders are purely discretionary and intuitive in their decision-making, processing market information as it unfolds. Other traders rely on considerable prior analysis before making decisions. There are traders who are structured in their trading, relying on explicit models—sometimes purely mechanical systems. Other traders may follow general rules, but do not formulate these as hard-and-fast guidelines. My own trading is a combination of head and gut: I research and plan my ideas, but execute and manage them in a discretionary fashion. Each trader blends the analytical and the intuitive differently.

> What you trade and how you trade should be an expression of your distinctive cognitive style and strengths.

My experience working with traders—and my own experience in the school of hard trading knocks—is that much of the distress they experience occurs when they are operating outside their niche. Ted Williams, the Hall of Fame baseball slugger, offers a worthwhile metaphor. He divided the plate into a large number of zones and calculated his batting average for each zone. He found, for instance, that pitches low and away provided him with his lowest batting average. Other pitches, those high and directly over the plate, provided sweet spots where his average was quite high. With certain pitches, Williams was a mediocre hitter. With others, he was a superstar. *The source of his greatness, by his own account, was that he learned to see the plate well and wait for his pitches.*

A trader's niche defines his sweet spots. Certain markets I trade well, others I don't. Certain times of day I trade well; others fall short of breakeven. If I extend or reduce my typical time frame, my performance suffers. If I trade patterns outside my research, I suffer. Like Williams, I trade well when I wait for my pitches. If I swing at the low and away balls, I strike out.

> One theme that emerges from the experienced traders in Chapter 9 is that they know which pitches they hit, and they've learned the value of waiting for those pitches.

The implication is clear: Our emotional experience reflects the degree to which we're consistently operating within our niche. That is true in careers, relationships, and in trading. When there is an excellent fit between our needs, interests, and values and the environment that we're operating in, that harmony manifests itself in positive emotional experience. When our environments frustrate our needs, interests, and values, the result is distress. Negative emotions, in this context, are very useful: they alert us to potential mismatches between who we are and what we're doing.

When you are your own trading coach, your job is to keep yourself within your niche, swinging at pitches that fall within your sweet spots and laying off those that yield marginal results. This means knowing when to *not* place trades, *not* to participate in markets. Equally important, it means knowing when your advantage is present and making the most of opportunities. A common pattern among active traders is that they will trade too much outside their frameworks, lose money, and then lack the boldness to press their advantage when they find genuine sweet spots. It's easy to see careers lost by blowing up; less visible are the failures that result from the inability to capitalize on real opportunity.

I recently talked with a day trader who was convinced that he would make significant money if he just held positions for several days at a time. Though it looked easy to find spots on charts where such holding periods would have worked, in real time that trade was much more difficult. It was not in the trader's wheelhouse; it was outside his niche. Calibrated to measure opportunity and risk on a day time frame, he found himself shaken out by countertrend moves when he tried holding positions longer. Worse still, he mixed his time frames and tried to convert some losing day trades into longer-term holds. Outside his niche, he began trading like a rookie—with rookie results.

What is your wheelhouse? What do you do best in markets? If you could trade just one strategy, one instrument, one time frame, what would these be? Do you really *know* the answers to these questions: have you truly taken an inventory of your past trades to see which work and which have been low and outside?

YES 3:6

There is nothing wrong with expanding your niche in a careful and thoughtful way, much as a company might test market new products in new categories. But just as management books tell us that great companies stick to their knitting and exploit core competencies, we need to capitalize on our strengths in our trading businesses. As we will see in Chapter 8, you are not just your own trading coach: *you are the manager of your trading business*. That means reviewing performance, allocating resources wisely, and adapting to shifting market conditions.

> The greatest problem with overtrading is that it takes us outside our niches—and therefore out of our performance zones.

Here's a simple exercise that can move you forward as the manager of your trading business. At the time you take each trade, simply label it as A,

B, or C. A trades are clearly within your sweet spot; they're your bread-and-butter, best trades. B trades are your good trades: not necessarily gimmes or home runs, but consistent winners. C trades are more marginal and speculative: they're the ones that feel right, but are clearly outside your wheelhouse.

Over time, you can track the profitability of the A, B, and C trades and verify that you really know your niche. You can also track the relative sizing of your positions, to ensure that you're pursuing the greatest reward when you take trades in your sweet spots and assuming the least risk when you're going after that low, outside pitch.

The more clearly you identify your niche, the less likely you are to get away from it. That clarity can only benefit your profitability and emotional experience over time.

COACHING CUE

If you categorize your trades/time frames/setups/markets as A, B, or C as outlined above, you have the start for good risk management when you go into slumps. When markets do not behave as you expect and you lose your edge, cut your trading back to A trades only. Many times, slumps start with overconfidence and getting outside our niches. If it's the A trades that aren't performing, that's when you know you have to cut your risk (size) and reassess markets and trends.

LESSON 20: VOLATILITY OF MARKETS AND VOLATILITY OF MOOD

I recently posted something important to the *TraderFeed* blog. I took two months in the S&P emini futures (ES) market—January and May, 2008—and compared the median volatility of half-hour periods within the months. During that period, overall market volatility, as gauged by the VIX, had declined significantly. The question is whether this day-to-day volatility also translated to lower volatility for very short time frame traders.

The results were eye-opening: In January, when the VIX was high, the median 30-minute high-low price range was 0.60 percent. In May, with a lower VIX, the range dwindled to 0.28 percent. In other words, markets were moving half as far for the active day trader in May than January.

Let's think about how that affects traders emotionally. The trader whose perceptions are anchored in January and who anticipates much greater movement in May will place profit targets relatively far away. In the lower volatility environment, the market will not reach those targets in the time frame that is traded. Instead, trades that initially move in the trader's favor will reverse and fall well short of expectations. Repeat that experience day after day, week after week, and you can see how frustration would build. Out of that frustration, traders may double up on positions, even as opportunity is drying up. I've seen traders lose significant sums solely because of this dynamic.

Alternatively, the trader who is calibrated to lower volatility environments will place stops relatively close to entries to manage risk. As markets gain volatility, they will blow through those stops—even as the trade eventually turns out to be right. Once again, the likely emotional result is frustration and potential disruption of trading discipline.

Both of these are excellent and all-too-common examples of how poor trading can be the *cause* of trading distress. It may look as though frustration is causing the loss of discipline—and to a degree that is true—but an equally important part of the picture is that the failure to adjust to market volatility creates the initial frustration. Any invariant set of rules for stops, targets, and position sizing—in other words, rules that don't take market volatility into account and adjust accordingly—will produce wildly different results as market volatility shifts. For that reason, the market's changes in volatility can create emotional volatility. We become reactive to markets, because we don't adjust to what those markets are doing.

Poor trading practice—poor execution, risk management, and trade management—is responsible for much emotional distress. Trading affects our psychology as much as psychology affects our trading.

Personality research suggests that each of us, based on our traits, possess different levels of financial risk tolerance. Our risk appetites are expressed in how we size positions, but also in the markets we trade. When markets move from high to low volatility, they can frustrate the aggressive trader. When they shift from low to high volatility, they become threatening for risk-averse traders. *The volatility of markets contributes to volatility of mood because the potential risks and rewards of any given trade change meaningfully.* In the example from my blog post, that shift occurred within the span of just a few months.

Note that traders can experience the same problem when they shift from trading one market to another—such as moving from trading the S&P 500 market to the oil market—or when they shift from trading one stock to another. Day traders of individual equities will often track stocks on a watch list and move quickly from sector to sector, trading shares with different volatility patterns. Unless they adjust their stops, targets, and position sizes accordingly, they can easily frustrate themselves as trades get stopped out too quickly, fail to hit targets, or produce outsized gains or losses.

Many traders crow about taking a huge profit on a particular trade. All too often, that profit is the result of sizing a volatile position too aggressively. While it's nice that the trade resulted in a profit, the reality is that the trade probably represents poorly managed risk. Trading 1,000 shares of a small cap tech firm can be quite different from trading 1,000 shares of a Dow stock, even though their prices might be identical. The higher beta associated with the tech trade will ensure that its profits and losses dwarf those of the large cap trade. That makes for volatile trading results and potential emotional volatility.

Risk and reward are proportional: pursuing large gains inevitably brings large drawdowns. The key to success is trading within your risk tolerance so that swings don't change how you view markets and make decisions.

Do you know the volatility of the market you're trading right now? Have you adjusted your trading to take smaller profits and losses in low volatility ones and larger profits and losses when volatility expands? If you're trading different markets or instruments, do you adjust your expectations for the volatility of these? *You wouldn't drive the same on a busy freeway as on one that is wide open; similarly, you don't want to be trading fast markets identically to slow ones.*

One strategy that has worked well for me in this regard is to examine the past 20 days of trading and calculate the median high-low price range over different holding periods: 30 minutes, 60 minutes, etc. I also take note of the variability around that median: the range of slowest and busiest markets. With this information, I can accomplish several things:

- As the day unfolds, I can gauge whether today's ranges are varying from the 20-day average. That gives me a sense for the emerging volatility of the day that I'm trading. This helps me adjust expectations as I'm trading. For instance, the S&P e-mini market recently made a 12-point move during the morning. My research told me that this was at the very upper end of recent expectations, a conclusion that kept me from chasing the move and helped me take profits on a short position.
- When I see that volatility over the past 20 days has been quite modest, I can focus on good execution, place stops closer to entry points, and keep profit targets tight. That has me taking profits more aggressively and opportunistically in low volatility environments, reducing my frustration when moves reverse.

- When I see that volatility over the past 20 days is expanding, I can widen my stops, raise my profit targets, adjust my size, and let trades breathe a little more. Not infrequently, the higher volatility environment will be one in which I can set multiple price targets, taking partial profits when the first objective is hit and letting the rest of the position ride for the wider, second target.

Note that what's happening in the above situations is that I am taking control over my trading based on market volatility. *Instead of letting market movement (or lack of movement) control me, I am actively adjusting my trading to the day's environment.* That taking of control is a powerful antidote to trading distress, turning volatility shifts into potential opportunity.

> The Excel skills outlined in Chapter 10 will be helpful in your tracking average volume and volatility over past market periods.

As your own trading coach, you want to monitor your mood over time. When you see your mood turn dark and frustrated, you want to examine whether there have been changes in the markets that might account for your emotional shifts. Many times, these will be changes in the volatility of the markets and instruments you're trading. Establish rules to adapt to different volatility environments as a best practice that aids both trading and mood.

COACHING CUE

If you know the average trading volume for your stock or futures contract at each point of the trading day, you can quickly gauge if days are unfolding as slow, low-volatility days or as busy, higher-volatility days. If you know how current volumes compare to their average levels you can identify when markets are truly breaking out of a range, with large participants jumping aboard the repricing of value. If you can identify markets slowing down as that process unfolds, you can be prepared and pull your trading back accordingly.

RESOURCES

The *Become Your Own Trading Coach* blog is the primary supplemental resource for this book. You can find links and additional posts on the topic of stress and distress at the home page on the blog for Chapter 2:

http://becomeyourowntradingcoach.blogspot.com/2008/08/daily-trading-coach-chapter-two-links.html

Make sure that you structure your learning process as a trader in a way that will build success and confidence. This process will help greatly with the management of stress and distress. That topic—and especially the topic of how to find your trading niche—is covered in depth in the *Enhancing Trader Performance* book, including a section on psychological trauma. A detailed discussion of how emotional states affect trading can be found in *The Psychology of Trading*. Links to both books can be found on the Trading Coach blog (www.becomeyourowntradingcoach.blogspot.com).

If you're looking for a book specific to stress and trading, you might look into *Mastering Trading Stress* by Ari Kiev, MD (www.arikiev.com), which is written by an experienced trading coach.

How do emotions affect financial decisions? Two good books are Richard L. Peterson's *Inside the Investor's Brain*, published by Wiley (2007) and *Your Money and Your Brain*, written by Jason Zwieg (Simon & Schuster, 2007).

Information on the research of James Pennebaker regarding writing as a way of coping with stress and distress can be found on his page: http://pennebaker.socialpsychology.org/

Happiness is the meaning and the purpose of life, the whole aim and end of human existence.

—Aristotle

Recent research into what has been called *positive psychology* has yielded insights into the importance of well-being: positive emotional experience. It is not enough to cope with stress and minimize distress. If we're going to maximize performance, it means that we need to make the most of our concentration, motivation, and energy. How many times have I seen traders miss trades or fail to prepare adequately for the day, simply because they were worn down, not operating at their peak? An Olympic athlete would never think of coming to his event in less than the best shape. Too often, however, traders will risk their capital after being sleep-deprived, burned out, or distracted. How you manage your trading will depend, in part, on how you manage the rest of your life.

But what is well-being and how can we maximize it? A wealth of research, much of it conducted in the past decade and unknown to traders and even trading coaches, helps us answer these questions. Let's take a look at how you can coach yourself toward more positive experience—and performance.

LESSON 21: THE IMPORTANCE OF FEELING GOOD

One of the most overrated variables in trading psychology is passion for trading or passion for the markets. Self-reported passion means a great deal of things to different people; my experience is that it is only weakly correlated with how hard traders actually work at their craft. Traders who are desperate for profits, traders who approach markets addictively like gamblers, traders who live and breathe markets 24/7: all may claim a special passion for what they do. These passions may or may not support good trading and a positive learning curve. Desire and motivation are necessary, but not sufficient, for trading success.

Rather than focus on passion, traders would do well to reflect upon the overall emotional tone of their market experience. In researching and trading markets, as well as working on trading skills, do you experience meaningful happiness, contentment, and motivation? Are you truly enjoying what you're doing?

This positive side of emotional experience is what psychologists refer to as psychological well-being. A person with high well-being experiences the following on a frequent basis:

- A positive mood (happiness).
- Favorable expectations (optimism).
- A positive physical state (energy).
- A positive appraisal of self and life (fulfillment).
- Favorable relations with others (affection).

None of us feel all of these things all of the time; indeed, many of these five factors may wax and wane depending upon life circumstances. Still, psychological research suggests that some of us experience significantly more well-being than others. A portion of this variation can be attributed to inborn personality traits that are present across a range of life circumstances. Other variations can be attributed to our environment, especially the degree to which our settings satisfy or frustrate our needs and affirm or contradict our values.

> Much of psychological well-being is a function of the fit between a person and her social and work environments.

When people are chronically unable to experience positive emotions, we might suspect the presence of an emotional disorder, such as the form of depression known as dysthymia. Other emotional problems, including anxiety disorders, can be sufficiently pervasive as to prevent people from experiencing joy and life satisfaction. In such circumstances, it is important for people to seek the assistance of an experienced, licensed mental health professional. Many times, very hands-on approaches to therapy—sometimes in concert with medication—can make a huge difference when chronic problems interfere with positive life experience. You don't need to be actively depressed—crying, unable to get out of bed, suicidal—to benefit from psychological help when positive emotions are chronically absent.

The Resources for Chapter 1 include a source for brief therapy referrals through the Beck Institute for Cognitive Therapy and Research.

The majority of people, however, experience varying balances of positive and negative emotions as a function of life circumstances. When their lives affirm their values and needs, they enjoy the emotions listed above. When life fails to meet their needs, they respond with unhappiness, frustration, and diminished energy. *In this sense, positive emotions serve as life barometers, informing us of the degree to which we're doing the right things for us.*

You can see, then, why positive emotions are so important to trading. If you're trading well, learning and developing, and succeeding in your efforts, your positive emotions should outweigh the negative ones over time. The dominant emotional experience of your work should be the kind of pride, satisfaction, and sense of accomplishment that gives you energy and optimism. If you're not trading well, if you're not growing and developing as a trader, and if your efforts are not yielding success, your experience of trading is apt to be more negative. You'll spend less time feeling satisfaction, energy, and optimism than frustration, overload, and discouragement.

This is important because the energy and optimism generated by happiness and personal fulfillment are what sustain the trader's learning curve. They fuel concentration, and they help traders make the extra efforts that result in superior internalization of patterns. The trader who sustains high well-being is more apt to have the confidence to aggressively pursue good trades and lay off marginal ones. The trader with a positive emotional experience is less apt to make impulsive trades out of frustration and more likely to have the resilience needed to weather losing periods. In short, feeling good is a huge part of performing well, because we function best cognitively under conditions of emotional wellness.

Emotional well-being fuels cognitive efficiency. We think best when we feel good.

As your own trading coach, a constructive step you can take is to track your emotional experience over time. A simple adjective checklist filled out at the end of each trading day can provide you with a sense for whether you are in the emotional zone or swimming against emotional currents. One such simple checklist is shown in Figure 3.1.

Check the Adjective That Best Corresponds to How You're Feeling Now

- Happy, Neutral, Unhappy
- Satisfied, Neutral, Dissatisfied
- Energetic, Neutral, Run Down
- Optimistic, Neutral, Pessimistic
- Focused, Neutral, Scattered
- Calm, Neutral, Stressed Out
- Competent, Neutral, Incompetent
- Growing, Neutral, Stagnant

Scoring: For each of the first adjectives in the series that you check, give yourself a +1 score. For each "neutral" answer, count this as 0. For each of the third adjectives that you check, give yourself a −1 score. Add the total. If your total score is +4 or higher, you want to identify what you're doing right to keep your mood positive—and maybe even do more of it. If your score is −4 or lower, you want to identify what you might be doing wrong to keep your mood negative. If your score is relatively neutral, you want to take both of these steps, identifying what you're doing that gives you more positive experience and what you're doing that takes away from positive experience. By tailoring your daily activities, you can tip the balance between positive and negative emotions in a favorable direction.

FIGURE 3.1 Brief Emotional Checklist

It is not necessary that you feel great after every single trading day. That is unrealistic. Rather, what's important is the balance, over time, of positive and negative emotional experience. If you're feeling badly at the end of trading days more often than you're feeling good, that's a clear indication that either you're not trading well—that markets have shifted in ways that you have not adapted to— and/or that you're not doing a sufficiently good job of advancing your learning curve with clear, achievable goals.

When you take your emotional temperature at the end of each trading day, you also can become a better observer of the specific trading behaviors that aid and damage your mood. I learned to limit my losses on trades when I realized fully—in my own experience—that outsized losses were not only reducing my annual returns, but also ruining my love for and interest in trading. I also learned to focus on trading strengths when I realized that these brought me a deep sense of accomplishment over time. The idea of trading without emotion is hogwash: as long as we *care* about our performance, our feelings *will* be engaged when we put our capital at risk and pursue our goals. That emotional engagement can work to our favor if, through our self-coaching, we sustain a positive emotional state, bringing out the best in us.

COACHING CUE

Be particularly careful to track your well-being around times of great life transition: giving birth to a child, buying a new home, moving, undergoing a separation in a relationship, experiencing a death in the family, going through a major illness, etc. Many, many times these times of transition are also times of stress and diminished well-being. Even seemingly happy events, such as giving birth, can lead to enhanced performance pressures and diminished sleep! Adverse changes in financial status are particularly challenging in this regard. Consider reducing trading risk until these situations are placed into perspective and adequately addressed.

LESSON 22: BUILD YOUR HAPPINESS

Two of the essential components of psychological well-being are joy and contentment. It is important to have positive feelings about what we're doing, but it's also important to enjoy a degree of contentment with our lives. Together, joy and contentment yield an ongoing sense of happiness.

As the Aristotle quote at the beginning of this chapter indicates, happiness lies at the center of life. When the Greek philosophers referred to happiness, they did not simply mean a positive, sunny mood. Instead, happiness was intimately tied to fulfillment: the sense of actualizing one's potentials and being the person you are capable of becoming. A happy person in this sense can go through periods of sadness and loss, anxiety and frustration. Indeed, it is difficult to imagine a goal-directed life that does not encounter such obstacles. What contributes to the happiness, however, is the deep sense of being on the right path in life: the sense that you are doing what you're meant to be doing.

In that context, the opposite of Aristotle's happiness is not sadness, but rather a certain kind of emotional dis-ease: a vague but pervasive existential guilt that you're letting life's opportunities slip by; that you're settling for less than you rightfully should. Much is written about the negative emotional experiences of anger, anxiety, and depression, but little about this kind of gnawing guilt. It is not so dramatic as a panic reaction or an anger outburst, but it can be equally damaging in its haunting corrosiveness. Day after day, week after week, month after month, to feel like you're selling yourself short in your career, your romantic relationships, and your personal development: it's difficult to imagine a durable confidence and self-esteem built on such a foundation.

Conversely, there is a special glow of satisfaction when you're immersed in a truly fulfilling activity. As a psychologist, those moments when everything comes together and I'm able to make a meaningful difference in someone's life—those are affirmations that carry me through many sessions of slow, gradual, difficult work. Similarly, to prepare a challenging trade idea, execute it with a plan, and then see it make money yields a kind of happiness that cannot be achieved by the same profits from a dumb luck trade. Pride, not as in overweening arrogance, but as an inner sense of conviction about the rightness of one's choices, is an important manifestation of Aristotle's happiness.

Sadly, participation in the markets does not bring this fulfillment to many traders. Yes, they feel pleasure when they make money and pain when they lose it. But they lack the deep, inner sense of satisfaction and joy possible only to those who are pursuing a calling. *The reason for this is that they are hoping that profits will bring happiness, when in fact the relationship works in exactly the reverse order.* We profit from our life's endeavors when we pursue our happiness. Just as sexual conquests cannot provide the happiness of a fulfilling emotional relationship and winning a lottery cannot yield the depth of experience possible to one who builds a business, profits on trades do not provide the primary emotional fuel for a trading career. They are the happy results, not the causes of happiness.

Unfortunately, many traders pursue markets in ways that *cannot* lead to fulfillment. They pursue profits like a pick-up artist goes after sex. Sometimes they score, sometimes they don't. There is nothing cumulative in their efforts, however: they no more build a career than the barfly builds meaningful relationships. Many times, traders recognize that gnawing existential guilt: they realize that hours upon hours in front of a screen are contributing little or no economic value, but also yielding no ongoing sense of meaning and satisfaction. It's not that they're necessarily anxious, depressed, or frustrated. *They are just empty.*

We can recognize the happy trader because he is immersed in the process of trading and finds fulfillment from this process even when markets are not open. I track the traffic to the *TraderFeed* blog daily and have long noted that the visitor count drops precipitously when markets close on a Friday and through the weekend. After all, what fun are markets if they're not open and active? This is the mindset of the trading barfly. The happy trader finds joy in researching and understanding markets, in preparing for the next day and week, in reviewing trading results and tweaking performance, and in generating new ideas and methods. The distinction between week and weekend is as immaterial for such a trader as it would be for a dedicated artist or laboratory scientist. Indeed, it's not uncommon for these traders to increase their reading during evenings and weekends: they are immersed in the entire process of trading, not just the process of making money from trades.

You will know you're pursuing your happiness when you are so involved in what you're doing that you don't want it to be limited to business hours. When you're in love with the right person, that love pervades all your activities, not just those in the nightclub or bedroom. When you find deep fulfillment in markets, you live and breathe markets, not just when they can pay you out. A useful self-assessment exercise to carry out during the coming week is to log your hours spent on trading outside of formal market hours and how you feel during those times. Do you get bursts of joy when you find new patterns and ideas? Do you generate a deep sense of mastery and pride when you work on yourself and improve your craft? Do your ongoing efforts at figuring out markets and enhancing your performance bring a sense of pride and satisfaction?

If you're not spending as much time on trading outside formal market hours as during them, trading is probably not your calling. Can you imagine a priest who spent time on his religion only from nine to five? An artist who only painted when art shows were active? When trading is truly a part of you, contributing to your happiness, you're most likely to be immersed in the activities that build skills and yield pattern recognition. So much of trading success is finding the niche that sustains such immersion. Trading can be a great job and a potentially lucrative career, but only if it's also a calling.

COACHING CUE

Many traders confuse contentment and fulfillment with laziness and smug self-satisfaction. Out of a fear of becoming complacent, they resist the experience of contentment. As a result, traders are often not content with their progress and fall prey to frustration—and the effects of frustration on trading. You can be content with your progress to date and still be motivated to move forward. The key is setting shorter *and* longer-term goals, so that you can bask in satisfaction when you reach an immediate objective, but still stay hungry for the larger objectives.

LESSON 23: GET INTO THE ZONE

Most experienced traders know the feeling of being *in the zone*: seeing markets so clearly that the right decisions seem to be effortless. The psychologist Abraham Maslow referred to these occasions as peak experiences; researcher Mihalyi Csikszenmihalyi refers to them as periods of flow. The hallmark of being in the zone is that the performer feels at one with the performance, executing skills in a highly competent manner, seemingly without conscious effort.

The zone is not an emotional state, though it brings feelings of emotional well-being and certainly can be disrupted by negative emotional states. Entering the zone requires *immersion*: a total focus on what one is doing. In that sense, the zone is a state of heightened attention: it is the result of being fully focused and involved in an activity.

Note that we can behave automatically—and even in skilled ways—without being immersed in our activity and without experiencing the zone. Repetitive, routine tasks, such as driving a car on an empty road or walking a city street, don't require particular attention and also don't usually bring any experience of well-being. To enter the zone, one must expend mental effort at a task that absorbs all of one's attention. Though performance in the zone may seem effortless, it is far from robotic.

Earlier—and in *Enhancing Trader Performance*—I emphasized the importance of *niche*: performing in an area that captures one's talents, skills, and interests. A good barometer of whether you are operating within your niche is the relative proportion of time you spend in the zone while engaged in performance. For me, those flow experiences are relatively common when I'm writing. Rarely do I operate from a detailed outline. Rather, I think about a topic and let the thoughts and words flow as I type. Similarly, when I'm working with a person in counseling, I'm completely focused on what they're saying, what it means, and how to use the information to be of assistance. It's not at all unusual for time to fly by quickly while I'm in a meeting with someone; I'm so immersed in the interaction that I lose track of the passage of time.

I most often find a zone state in trading when I'm actively figuring out markets—absorbing myself in research—and applying those insights to short-term trades. In an important sense, it's the puzzle-solving aspect of trading and not the placing of trades themselves that captures my attention. If I try to trade in a mechanical fashion without engaging in the problem solving, I find trading psychologically noxious. It's like talking with people in a superficial social context. There's no meat on the bones cognitively; my attention remains unengaged, and I stay out of the zone.

One of the most damaging psychological patterns I see among traders is that they attempt to create a counterfeit zone by trading with too much risk. In other words, traders are not intrinsically interested in markets and the process of trading, so they attempt to create interest and attention by making large wagers on trade ideas. This is problematic for obvious reasons: it exposes traders to outsized losses and potential risk of ruin. Psychologically, however, it is also ruinous. Once the trader habituates to one level of risk, a higher level is needed to grab attention and interest—much as addicts require greater doses of a drug to achieve a high. Eventually the trader who needs the excitement of risk to sustain interest in trading has to blow up. This, truly, is addiction, not passion for markets.

More on trading addictions can be found in *Enhancing Trader Performance* and here: http://traderfeed.blogspot.com/2006/11/dr-bretts-heartfelt-plea-when-trading.html

One of the most effective ways to exit a flow state—or prevent one from emerging in the first place—is to focus attention on oneself rather than on one's performing. You can't be immersed in a sexual encounter if you're worried about your sexual performance. You can't find the zone in an athletic performance if you're pressuring yourself to set a record. The trader who focuses on P/L during the trade is, to that degree, no longer market focused. The dynamics of performance anxiety—thinking about the performance while you are performing—is a recipe for disaster if your goal is to operate within the zone.

When I wrote my first trading book, I decided to complete the entire manuscript before I ever had a signed contract from my publisher. I certainly wanted to see the text published, but I wasn't writing it for royalties or recognition. The book was written for me, to clarify my thoughts and contribute to the body of knowledge within the field. When I wrote in this fashion I didn't have to worry about how the readership would react to the ideas, whether editors would like my work, etc. I could just focus on the writing. It is very, very difficult to *need* to perform well and to stay absorbed in the performance. The surest way I could have ruined my writing experience (and my books) would have been to split my attention between generating/writing ideas and speculating about how those ideas would be received. Once performance becomes an acute *need*, not just a genuine desire, it is nearly impossible to place the outcome of performance in the back of your mind and solely focus on performing.

So it is for the trader. If a trader *needs* to make money, it is difficult to weather market ups and downs and stay focused on the execution of trade ideas and plans. If today's trade is needed to provide tomorrow's food and shelter, there can be no zone: anxiety naturally takes over whenever profits are threatened. Similarly, if I become psychologically attached to profitability, basing my self-esteem and identity upon my trading results, I no longer control my trading experience: market movements are likely to control how I feel. *The experience of flow requires a basic level of control over what we are doing.*

Perhaps a different analogy will illuminate the issue. My wife Margie and I recently invested a good amount of money in tax-free bonds, taking advantage of a situation in which their yields had skyrocketed above the corresponding yields on (taxable) Treasury instruments and certificates of deposit. Our plan was to lock in these positions as investments, not as short-term trades. Over time, we felt, we would benefit from attractive yields and possible capital appreciation as tax-free yields fell into line with those of taxable instruments. We could make this investment comfortably because it represented less than 10 percent of our savings. Our remaining capital was diversified in other investments, working for us even as our bonds might move against us in the short run. We could stay focused on our overall investment plan because we were so diversified that we didn't *need* any one position to perform wonderfully.

> Diversification, in life and markets, reduces performance pressure and allows us to become immersed in what we are doing.

Similarly, a trader who risks a small portion of her account on a trade can stay focused on executing an overall trading plan, because there is no acute need for that position to work out—and no acute threat if it fails to work.

This is a psychological paradox: To best focus on any single performance, it helps to be diversified among performances. If I have a successful experience as a father, husband, and psychologist, I don't *need* my books to sell well or my trades to make money. It is precisely that emotional diversification that enables me to stay focused on my writing and trading and achieve satisfying returns from them.

As your own trading coach you don't need to spend more and more time, effort, and emotion on markets. Indeed, if you place all your psychological eggs in the trading basket, it is a sure way to burn yourself out and stay out of a performance zone. Rather, you can best coach yourself by ensuring that trading is one among many fulfilling activities within your life. Other eggs might go into the baskets of spiritual interests, artistic activities, athletic pursuits, social life, intellectual life, family, community, and hobbies. If your life is full in those ways, you are best able to weather ups and downs in trading performance. You no longer need trading to work at any particular point in time, so you become more able to focus on the process of trading and generate and execute good trades.

So this is your assignment: Give yourself a grade for how much interest and satisfaction you've been achieving from the areas of life mentioned above. How diversified are you in your sources of well-being? Then select one area for cultivation to improve your emotional diversification. Not only will you find a new source of enjoyment and accomplishment, you will also lay the psychological groundwork—the inner sense of security and fulfillment—to find and stay in your performance zone.

COACHING CUE

Take short breaks from the trading screen. These breaks can be a great way to renew concentration and step back from difficult markets. Some of the best breaks are ones that are wholly absorbing, that get you into a different zone than trading. Physical exercise is one example, talking with people you enjoy can be another. I find myself wholly absorbed when I play with my cat Gina or when I swim in our pool. The activity completely takes my mind away from what I had been doing and lets me return with a new perspective. Switch gears—absorb yourself physically and/or emotionally—after you've been cognitively immersed in markets. These breaks can provide some of the most effective trades.

LESSON 24: TRADE WITH ENERGY

One of the important dimensions of psychological well-being is energy. Happiness, enthusiasm, motivation, and general contentment are difficult to sustain when you feel mentally and physically run down. Fatigue is the enemy of concentration; physical vibrancy fuels a positive, energetic mood.

We are like laptop computers running on batteries: after sustained operation, we run down. Concentration and attention require effort; eventually we drain our mental reserves and lose focus. This leads to trading mistakes: missing opportunities, overlooking important pieces of data, forgetting key aspects of trading plans. When we are run down, we're also most likely to fall back into old—and often negative—habit patterns. When we're drained, we might find ourselves eating out of boredom, becoming unusually irritated when things don't go our way, or getting caught up in negative ways of thinking.

Think of it this way: it requires sustained focus to remain goal-oriented. To actively direct ourselves, we need an alert, active mind. When we become fatigued, we lose this active direction. We become passive, responding to events rather than making them happen.

This distinction between active and passive trading is all-important. The active trader is one who researches markets, identifies distinct areas of opportunity, and consciously executes and manages trades to maximize that opportunity. For the active trader, nothing is left to chance: where to pursue opportunity, where to sit back, where to take profits, where to limit losses—all are preplanned. This takes time, energy, and a sustained focus. Good trading, in this sense, is pure intentionality: it is a directed act of will.

When we are physically drained, we lose the ability to sustain this intentional focus. We neglect our research; we fail to calibrate risk and reward. We fall back on simple heuristics and enter trades based on simple reasoning—chart patterns or price levels—that may well lack any true risk/reward edge. Worse still, when we're run down, we become emotionally reactive and find ourselves chasing price highs or lows or robotically enacting rules (fade weak stocks in a strong market) without taking the time and effort to assess the broader context our decisions (an trending day to the upside).

Managing your energy during the trading day may take little more than ensuring that you:

- **Get proper sleep and proper quality of sleep**. Interrupted sleep can deprive you of important stages of sleep and leave you feeling unrested, even though you've spent a full number of hours in bed.
- **Eat properly**. Highs and lows in blood sugar can make it difficult to sustain concentration; an excess of caffeine and sugar may provide temporary jolts, but can also lead to distracting rebound effects.
- **Maintain your mind properly**. I've seen alcohol and drugs take a fearsome toll on traders over time, as partying the night before leads to diminished performance the next day. Conversely, those who are focused and intentional in their personal lives tend to see this carry over into their trading.
- **Maintain your body properly**. Physical exercise is one of the most neglected facets of a trading plan. Hours upon hours sitting in front of a screen do not promote aerobic fitness. Over time, we lose conditioning—and our energy batteries lose their charge.
- **Take the breaks**. Not many people can stare at a screen and follow market action continuously through the day without losing focus. Breaks during slower market action can replenish the energy and concentration needed when markets become busier.

> A trading career is a marathon, not a sprint: the winners pace themselves.

None of the above considerations is earth shattering, but it's amazing how poorly many traders score if they incorporate the five factors above into a daily checklist. *We prepare our trades, but we often fail to prepare ourselves for trading.* How can we stick to disciplined trading decisions if we're inconsistent in our personal discipline?

When you are your own trading coach, you cannot afford to run yourself into the ground by working so hard that you can no longer work. Nor can you so neglect your physical state to such a degree that, like that laptop battery, your memory effects lead you to lower and lower energy states with each recharging. Your assignment is to track your daily profits and losses simply as a function of two factors: your energy level (high or low) and your trading mode (active/planned or passive/unplanned). Add a simple checklist to your trading journal to help you see the correlations among your physical state, your concentration level, your intentionality, and your trading results.

> If you lack energy, you will lack focus; if you lack focus, you'll lack intentionality; if you lack intentionality, you'll lack the ability to follow trading plans.

Unless you calculate and appreciate these correlations for yourself, you're unlikely to sustain the motivation to address—with consistency—the five areas above. Once you *see* that your energy level is directly correlated with the quality of your trading (and with your trading results), you will prod yourself to build a daily routine that addresses sleep, eating, exercise, and a healthy lifestyle. You'll also be able to overcome guilt or fear over leaving the screen and realize that *opportunity is not just a function of moving markets: it's also a function of your ability to capitalize upon those markets.*

COACHING CUE

Many traders neglect their family lives (spouse, children) in their absorption into their work. The resulting guilt and distraction from those unmet needs wind up interfering more with performance than the time it would have taken to spend the quality hours together. The mental rejuvenation from vacations—even weekend holidays—can renew family relationships and energize work. If you're too worn down for your personal life, you're probably not operating with good efficiency in your trading. It's not necessary to have a totally balanced life—few of us do—but if your life *feels* unbalanced, that will undermine energy, concentration, optimism, and effort.

LESSON 25: INTENTION AND GREATNESS: EXERCISE THE BRAIN THROUGH PLAY

One of the core concepts underlying *The Psychology of Trading* book is intentionality. We can define intentionality as the ability to sustain purposeful activity over time. The ability to sustain attention and concentration, coordinate a sequence of activities toward a chosen end, and persistently try different approaches toward solving a problem until a solution is reached: all of these are manifestations of intentionality.

As noted earlier, there is an intriguing connection between intentionality and psychological well-being. In studies of flow, Mihalyi Csikszentmihalyi found that these moments of being "in the zone" result from a complete absorption in one's activities. It is when we are completely focused on what we're doing that we reach a state in which performance seems almost effortless and completely natural. This is a highly pleasurable state and, among creative individuals, becomes a psychological reward in its own right. *In a real sense, the creator's passion for her work represents a passion for the flow state.* Exemplary performance thus provides its own rewards: a psychological feedback system that lies at the heart of greatness.

This helps to explain researcher Dean Keith Simonton's findings that great individuals across a variety of disciplines are unusually productive. They have mastered the art of working within their performance zones, so that sustained effort becomes a desired end in itself. Their productivity is a byproduct of a kind of positive addiction: a pursuit of the high of the performance zone for its own sake.

As Elkonon Goldberg notes in his excellent text *The Executive Brain*, the various facets of intentionality—attention, planning, reasoning—are functions of the brain's frontal cortex. His research also suggests a surprising degree of plasticity to the brain: *utilize brain functions and you exercise those brain regions and strengthen their functions, much as going to a gym builds our muscles and endurance.* At any given point in time, we may possess a relatively fixed quantity of intentionality: we can only exercise the brain so much before we become fatigued with the effort and need to rest. Over time, however, we can build our brain's capacity for intention by exercising those frontal cortex functions. Just as lifting weights is the best way to build our strength, engaging in sustained, directed effort is the best way to cultivate our intentional capacities.

When we pursue goals in an effortful manner, we build intentionality and free will.

One would think that trading should be an excellent form of mind exercise for this very reason. That is not necessarily the case, however. We can click a mouse and place trades without engaging in effortful, directed thought. This is the passive trading described in the previous lesson. When we trade without focused intent, we fail to use our mental muscles. At a broader level, when we live our lives on autopilot, those muscles atrophy. All of us know individuals who seem to drift from activity to activity, seemingly heedless of the longer-term consequences of their actions. They spend money and become mired in debt; they jump into relationships and reenact past conflicts. Gurdjieff described this as a tendency to live mechanically, as if we are stimulus-response machines. We see this among retirees: after functioning passively over time, even small efforts become taxing. Life becomes mechanical; the capacity for intentionality has atrophied.

Just as exercising one muscle group will not develop other ones (or building aerobic capacity will not confer muscular strength), cultivating one form of intentionality does not necessarily raise our self-directed capacities in others. Good examples of this are high-frequency day traders who develop an amazing capacity to sustain attention in front of a screen, as they track bids and offers, upticks and downticks, through the day. These same traders are often unable to sustain the kind of mental effort needed to observe themselves over time or systematically review markets to identify trends and themes. Clearly, we're most able to sustain concentration and effort during activities that interest us and that fit with our skill sets. The creative talent who stays in the flow is partly able to achieve this state because he sticks to the performance niche we discussed in the last chapter: a sphere of talent, skill, and interest. Outside of those niches, the sheer effort needed to produce results, the frustrations of not getting those results, and the boredom from operating outside our values and interests all conspire to interrupt flow and disrupt intention.

This is the dynamic described in *Enhancing Trader Performance*: talents lead to interests lead to immersion in skill building leads to competence leads to further flow and the eventual development of elite performance. *It is the interplay between the flow state and the development of intentionality that creates accelerated learning curves: without flow, talents have no place to go; they never evolve into elite skills.*

Many traders fail to succeed because they are operating outside of the niches defined by the intersection of talents, interests, and skills. Because they attempt something that doesn't intrinsically interest them and that doesn't play to their distinctive abilities, they rarely encounter flow states: their trading brings little well-being. Without the flow, these traders lack the motivational impetus to sustain efforts, and that prevents them from cultivating intentionality. Then they wonder why they can't stick to trading plans or why they sabotage themselves with impulsive trades.

> The learning curves of elite performers cultivate intentionality as they build skills, which means that—over time—elite performers can *do* more with their skills than others.

What is the first step in performance development? The most common response is *practice*, and that is important, of course. *But before practice, there should be play.* Play tells us which activities are fun and which are not. When we play with something, we discover its joys and frustrations: its intrinsic interest for us. Most traders who have not found their niche have never *played* with markets. They haven't tried to trade different styles, different instruments, and different time frames. They don't know what it is like to hold positions for weeks—or for only a few minutes. These traders can't appreciate the difference between trades made from rapid pattern recognition and those made from rigorous analysis. They try to imitate other traders, or they take the path of least resistance and trade from superficial chart or indicator patterns. Elite skills can never develop in such a learning environment; intentionality is stunted.

Elite performers never stop playing. Artists sketch; athletes play in scrimmages; actors improvise. Play is a means of self-discovery, and sometimes we discover passions and talents we didn't know we had. Your assignment for this lesson is to pick a market, trading style, or time frame different from your usual one and conduct paper trading in parallel to your usual trading. Your paper trading should document real trade ideas and real-time tracking of P/L. The simulated trades should be managed as real ones, with profit targets, stop-losses, and decisions about adding to or reducing positions.

For example, I maintain a separate, small trading account where I play with longer-term trading ideas. It's a way to test out my research and discover possible edges with very small amounts of money at risk. A majority of ideas in this sketchbook account may fail to bear fruit, but it only takes one promising effort to open new doors to opportunity. This keeps my mind and trading fresh; it also helps me stay in touch with the market's larger picture when placing bread-and-butter shorter-term trades. Most of all, it tells me which trading ideas and strategies truly capture my interest and imagination: which may form the promising basis of a new niche. When you play with trading, you avoid stagnation; you also discover niches that will sustain intentionality and performance. That's how you build a trading career—and that's how you build the mental muscles that propel performance to ever-higher levels.

COACHING CUE

If you structure your trading preparation like you would structure physical workout routines, then every day you are adding a bit to your capacity to sustain intention. I recently observed a trader enter a trade with a strong idea. He was stopped out, but reentered the same trade on a fresh signal. That was stopped out also. He entered a third time and then rode a trend for a very large winner. His resilience was a function of his persistence: his ability to sustain purpose over time, even through fatigue and discouragement. His diligent preparation each day conditioned him to make extra efforts when it counted, at a time when most others would have given up on the idea. *When you put effort into trading development, you not only prepare the mind, you condition the will.*

LESSON 26: CULTIVATE THE QUIET MIND

When we think of psychological well-being, we naturally think of joy, pleasure, and vigor. A different facet of well-being is serenity: a mind free of distracting thoughts and feelings. In many ways, serenity is vital to elite performance: *a mind at peace is one that can be fully focused on market patterns.*

Most of us spend too much of our time assaulted by stimuli to achieve a high degree of serenity. Social interaction, television, radio, music players, cell phones, billboards, and computers: much of our day is spent in a mélange of sights and sounds. Each calls to our attention, entertaining us from without, but leaving us ever more challenged to stimulate ourselves from within.

In the absence of the ability to generate our own stimulation, many of us equate the absence of stimulation with boredom. Boredom is an empty state, a frustrated state in which there is no-thing of interest. Upon reflection, however, we can see that boredom betrays a kind of inner emptiness, an inability to find objects of interest in our inner and outer worlds.

The aversion to boredom is the source of many trading problems. To erase boredom, traders will manufacture trades, overtrading—and sustaining losses—in the process. Traders will take unusual risks and size positions too daringly to sustain their excitement and interest. It is ironic that many traders consider emotion to be their enemy, when in fact it is the boredom of quiet markets that they particularly dread.

But the aversion to boredom damages trading in a much more subtle way. As I stressed in the *Enhancing Trader Performance* book, trading expertise hinges on the ability to detect and act on patterns that occur within noisy data. Experiments with implicit learning suggest that we can detect complex patterns in situations without being able to verbalize the specific nature of those patterns. This occurs routinely when we sense a market behaving differently from usual, or when we get an uneasy feeling about a conversation. Little children assemble grammatical phrases without knowing the rules of grammar: they've encountered so many examples of proper speech that they know what sounds right—and what sounds wrong. They, like traders and conversationalists, develop a feel for patterns and deviations from those.

This gut feeling, the basis of all valid intuition, is not mere hunch. It's the result of countless repetitions of complex patterns. When I first drove a car, I could barely stay in my lane. With experience, I now anticipate potential accidents several cars ahead of me. Many times I tap my brake or raise my alertness before I'm consciously aware of the troubling situation. If we needed to rely on explicit reasoning for all life's activities, we would never be able to respond quickly to danger. Evolutionarily, it makes sense for us to be able to develop a feel for reality, as well as a conceptual grasp.

> Access to intuition requires a still mind; highly intuitive people are not bored by stillness and, indeed, thrive on it.

When our attention is divided and we are distracted, we lose our feel. This is because the implicit pattern recognition manifests itself as a felt sense, a subtle kind of awareness. If I am not attending to those subtle cues of mind and body, I will miss signals altogether. In such a state, we cannot pick up on nuances of conversations or small, but significant shifts in traffic patterns. We lose valuable information, and we lose much of our ability to react quickly based upon internalized patterns.

Worse still, in a chronically distracted state, we never sustain the attention in the first place to internalize complex market patterns.

This is why serenity—the quiet mind—is so important. With a quiet mind, we can attend to the subtle cues of pattern recognition. Undistracted, our antennae are extended, able to pick up signals of situations that feel right and those that don't. The experienced trader has seen so many markets and perceived so many relationships among market variables that she learns to trust these gut signals. It is neither mystical nor irrational. Just as a horse whisperer can become one with the horse, understanding the most subtle communications, an experienced trader can hear the whispers of markets.

But if the mind is noisy, the whispers are drowned out.

> Those who fear boredom never achieve the still mind.

The two essential steps in achieving a quiet mind are a still body and focused thought. This is where biofeedback can be extremely helpful for the trader: it provides a structured method for learning to quiet the mind. The biofeedback that I use currently is the *emWave* unit from Heart Math (www.heartmath.com). It provides measures of both heart rate and heart rate variability (HRV). The user's finger goes into a small sensor, which is connected to a computer with the biofeedback software. The HRV readings are displayed in a chart; as readings rise, the chart readings become like sine waves. Lower readings create jagged, nonrhythmical patterns. The goal is to keep the patterns as sine wave-like as possible. There is also a feature that displays the proportion of high, medium, and low HRV readings over time, accompanied by audio beeps. You can thus close your eyes (for instance, while engaging in guided imagery) and still track your HRV. Even children can use the unit by clicking on video game features that play the game by keeping HRV readings high.

After a while using the biofeedback, you learn that keeping yourself still, focusing your attention, and keeping your breathing deep and rhythmical is the best way of generating high HRV scores. (Users can set the software for various levels of difficulty to build skills.) The emWave is thus a training tool—teaching users to control mind and body—and a way of tracking focused attention over time. There are other, similar units available (for example, *Journey to the Wild Divine*); ease of use and the appeal of the graphical interface will dictate most traders' preferences. If I had to invest in a single psychological tool to aid trading, this kind of biofeedback unit would be my choice. It is highly portable and can even be used in real time during trading, with the feedback screen minimized but sound enabled.

> Biofeedback is a tool for training yourself to control the arousal level of mind and body.

When you are your own trading coach, it's important to keep your mind in shape much as an athlete stays in proper conditioning. I find that 5 to 10 minutes each morning prior to the start of trading is useful in bringing a quiet mind to trading. During that time, you stay completely still in a comfortable seated position and breathe deeply, slowly, and very rhythmically. Your eyes can be closed throughout and you can focus your attention on your breathing, on soothing imagery, or on quiet music through headphones. The key is staying in that Yoda state described in *The Psychology of Trading*: very relaxed, yet very alert and focused. As you practice this each day, you build skills, so that you can eventually quiet your mind on demand, with only a few deep, rhythmical breaths. This is enormously helpful during hectic times during the trading day, keeping you out of situations in which you become impulsive and reactive in the face of moving markets.

Just as you prepare for the day's trading by studying recent market action, reviewing charts, and identifying areas of opportunity, it makes sense to engage in mental preparation to build the mind-set needed to capitalize on your ideas. Your assignment is to devote a portion of each morning to mental preparation and the generation of a quiet mind. If you have difficulty sustaining the effort or reaching that Yoda state, consider incorporating biofeedback into your morning routine, much as athletes work out daily on treadmills and weight machines. Mastering your mind state is a key component of mastering performance: if you can sustain serenity during the most boring market occasions, you'll be well prepared to catch moves when trading picks up.

COACHING CUE

If placing trades is your major source of stimulation in financial markets, you're bound to overtrade. By cultivating collaborative relationships with peer traders and developing routines for generating trade ideas and themes, you need not face boredom during slow markets. Other markets, other time frames: for the dedicated trader, there is always something of interest.

LESSON 27: BUILD EMOTIONAL RESILIENCE

Three traders place the exact same trades; all of them lose money. The first trader becomes discouraged, curses the market, and gives up for the day. The second trader reacts with frustration, vows to get his money back, trades more aggressively, and loses a bundle on the day. The third trader pulls back, reassesses her strategy, waits for a clear area of opportunity, and places a good trade that brings her even on the day.

What is the difference among these traders? The research literature in psychology refers to it as *resilience*: the ability to maintain high levels of functioning even in the face of significant stresses. A resilient person, for example, can lose his job, but still function well at home and implement an effective strategy for finding new work. The individual who lacks resilience is thrown for a loop by the lost job. This interferes with other areas of life and makes it difficult to find new opportunity.

A key reason why many people lack resilience is that they take negative events personally. Some portion of their self-worth is connected to their individual life outcomes. When events go well, they feel good. When they encounter roadblocks, they become discouraged, doubtful, and frustrated. Instead of dealing directly and constructively with the blocks, they react to the emotions triggered by their personalizing of events. An inspiring example of resilience is author Viktor Frankl's survival in a Nazi concentration camp. He set about writing a book (first on scraps of paper, then in his mind) during his internment, giving him a purpose: a reason to keep going. Others who experienced the same horrific conditions lacked such purpose and ultimately perished. The larger part of persistence is nurturing a reason to persist, a greater purpose and vision.

> The survivors are those who have a vision and purpose greater than themselves.

I recently researched a new pattern that I wanted to trade and saw an opportunity in early-morning trading. I vacillated between placing a small-sized trade and one more normal in size. I thought about the trade going against me and realized that I didn't really want to lose money on a relatively untested idea. With the smaller trade, I didn't care about the implications of the profitability of the trade for my portfolio. A larger trade could dent my week's performance, and that would have been frustrating to me. So I placed the small trade, observed the pattern in real time, made a small profit, and started the process of integrating the pattern in my usual trading.

In selecting trade size, I was letting my psychological resilience dictate my risk taking. When I traded to my resilience level, I kept myself in a favorable state regardless of the trade's outcome. "How will I feel if I'm stopped out?" dictated my trading size. To be sure, this can be taken to an unhealthy, risk-averse extreme. We can take so little risk on trades that we severely diminish potential returns. The key is to know yourself and especially the limits of your resilience. Occasionally I'll fantasize about placing an über trade on a promising idea and taking a mammoth profit. I realize, however, that such a trade can overwhelm my resilience. As soon as the position went against me—even in a normal, expectable adverse excursion—I would be stressing about the dollars lost. Undoubtedly this would prevent me from managing the trade effectively.

Successful traders learn to build their resilience over time and adapt to stresses that at one time might have been overwhelming. That small trade I recently placed would have qualified as a large trade back in the late 1970s when I placed my first trades. Now it is emotionally inconsequential. Experience builds adaptation: we can generally handle familiar situations with a high degree of resilience.

When we master one level of challenge, we build resilience for the next level.

The most effective way of to build emotional resilience is to undergo repeated, normal drawdowns and see—in your own experience—that you can overcome those. Our losses provide us with the deep emotional conviction that we can weather losses and ultimately prosper. Someone who has undergone many life setbacks and bounced back acquires the confidence that he can land on his feet in almost any situation. The trader who experiences repeated drawdown, only to later hit fresh equity highs, knows that she has nothing to fear during normal performance pullbacks.

When you are your own trading coach, your challenge is not only to sustain a high level of resilience, but also to build that resilience over time.

A worthwhile exercise is to expand on the routine from my recent trade and vividly visualize the worst-case scenarios for trades that you place. In other words, once you set your stop level, visualize how you would feel and how you would respond in that worst-case scenario. Most importantly, figure out what your next course of action might be, including your possible next trade. In other words, mentally rehearse the resilient behavior that you want to cultivate. You can think of this as play-acting the role of a highly resilient person. As you rehearse resilience and act on the rehearsal, that role becomes more a part of you. To paraphrase Nietzsche, you're finding your greatness by play-acting your ideal.

> In trading, we develop ourselves. Every gain is an opportunity to overcome greed and overconfidence. Every loss is an opportunity to build resilience.

Beware: resiliency does not mean that you jump into subsequent trades after you sustain losing ones. Rather, the resilient trader is one who can sustain well-being even after normal, expectable losing trades. When you lack resilience, you become backward-looking and respond to the last trade rather than the next market development. The resilient trader remains proactive, even in the face of loss. A resilient trader might thus stop trading or resume trading following a loss; it's the following of basic, time-tested plans and strategies—and not impulsively running toward or away from risk—that defines authentic resilience.

COACHING CUE

If I ask a trader how well he is doing and I receive a dollar figure as a reply, I usually know there's a problem afoot. Experienced traders think of their returns in percentage terms, not absolute dollars. Thus, for example, they might think of cutting their risk if they're down 5 percent on the year or limit their risk on a trade to 25 basis points (0.25 percent of their portfolio value). If you calibrate yourself in dollar terms, you will find it difficult to increase your trading size or to get larger as you grow your portfolio. Standardize your view in percentage terms and you make yourself more resilient; a $20,000 loss on a $2,000,000 portfolio won't feel significantly different from a $500 loss on a $50,000 portfolio. Similarly, when you cut your trading size, you'll standardize your risk management if you're calibrated by percentages, rather than let losses run because they seem small in absolute dollar terms.

LESSON 28: INTEGRITY AND DOING THE RIGHT THING

Many of the lessons in this book begin with a discussion of a trading issue and then proceed to suggestions about what you can do about the issue. This lesson will actually start with the recommendation and then work backward from there. Your assignment is to read Ayn Rand's novel *The Fountainhead*. If you've read it previously, the assignment is to reread and review it.

For those not familiar with the book, *The Fountainhead* is the story of architect Howard Roark, who is an unorthodox creative genius. He faces stiff opposition to his ideas, including the ambivalence of the woman he loves. Throughout, he must decide whether to abandon or compromise his ideals, especially as he sees lesser talents succeed commercially by pandering to public fashion. In many ways, *The Fountainhead* is a study in integrity and the difficulty and importance of doing the right thing.

There can be no self-esteem without a self: a well-defined sense of who one is and what one stands for. There are many false substitutes for self-esteem, including the approval of others and the size of one's trading account. Ultimately, however, self-esteem is a function of knowing yourself and remaining true to your values: possessing a vision of what can be and remaining faithful to that vision.

Many traders have no more vision than a desire to make money. There's nothing wrong with making money, of course, and for those who do so through the independent efforts of mind, such earnings are a rightful source of pride. Traders who attempt to latch onto holy grails instead of independently relying on their planning and judgment, however, substitute the desire for a quick, easy score for the more difficult challenge of developing competency in reading and acting on market patterns. Someone who consummates a long-term courtship and someone who hooks up for a one-night stand engage in the same physical act, but the meaning is completely different. One is an expression of esteem; the other is often a flight from self.

> In so many fields, we never see the fruits of our labors; we're part of a larger team and process. Trading is unique in that we alone are responsible for what we earn, and we see each day the outcomes of our efforts.

When you read *The Fountainhead*, it's instructive to reflect on how Howard Roark would approach the field of trading. Would he join a proprietary trading firm that frantically searches for stocks in play, robotically fading moves or chasing strength or weakness? The mere thought is ludicrous. Would he attend a few seminars or read a couple of books and trade the same untested chart patterns as other beginners? It's unthinkable.

No, Howard Roark the trader would be a keen student of markets, just as Roark the architect was a devoted student of building materials and methods. He wouldn't trade a single method in all markets, just as he didn't repeat the same design to fit all housing sites. Rather, he would carefully consider each unique situation and tailor the strategy to fit the present context. Roark the trader would work from carefully considered plans, just as he worked from blueprints that he had developed from scratch. In short, Roark would approach trading the same way he approached architecture: as an expression of his creative vision and the sheer joy of giving birth to something new and valuable.

Most of all, Roark the trader, like the architect Roark, would stand for something. He would have a view of markets and how and why markets move, just as he had a view of design and building. It would be *his* view, not something borrowed slavishly from tradition or current fads. The odds are good that this view would be unconventional and meet with more than a little skepticism by the self-appointed gurus of trading. That wouldn't matter. Roark the trader would remain faithful to his framework. When confronted with the choice of following the crowd versus act on his convictions, he wouldn't hesitate to do the right thing.

> Every great trader I have known has an outlook and a set of methods that are distinctively his own.

For that reason, economic success for trader Roark would be a tangible indicator of his efficacy and the rightness of his efforts. It is effect, not cause. He doesn't trade to simply make money any more than he builds to sell homes and office buildings. Roark the architect built because that was what he was meant to do. Even when he didn't have clients, he was designing buildings in his mind and in his sketches. Similarly, trader Roark would be tracking and investigating markets even if he wasn't placing orders. His work is an extension of who he is; his profits are the result of years of effort and integrity.

One's work could entail raising a child, building a business, designing a high-rise structure, or developing a unique framework for analyzing and trading financial markets. Each, to be accomplished well, requires sustained, dedicated effort; a vision of what can be; and a willingness to pursue that vision even when it's more comfortable to slide by. *This is the true source of emotional resilience: a pride and esteem so deep that one is unwilling to compromise oneself in the face of setbacks and disappointments.*

What about your trading is uniquely yours? What have you developed that most distinctly distinguishes you as a trader? What is the vision behind your trading? That is your core, your essence as a trader. When you're trading well, you're remaining true to that essence, and that will serve you well during the most challenging times. If you can't provide detailed answers to these questions, are you truly ready to be risking your capital? Will you really have the confidence to weather adversity, with only borrowed ideas and methods to draw upon? Read Chapter 9 of this book carefully; you'll see that experienced traders build a career from their work from figuring markets out for themselves and then remaining true to their ideas and to the evidence of their senses.

This lesson shows why it is so important to follow one's trading plans. The plans may not be perfect, and they may not work well at times. If, however, you are to build confidence in your judgment and train yourself to act with integrity, there's no alternative to following the ideas you believe to be correct. You cannot build confidence by abandoning your convictions and contradicting your perceptions. The clearer you are about your market views, mapping out your actions under various scenarios and your rationales for trades, the easier it will be to act on your judgment and see, in your own experience, your own progress and growth.

LESSON 29: MAXIMIZE CONFIDENCE AND STAY WITH YOUR TRADES

A great deal has been written about risk management and the importance of stop-losses. A stop-loss, ideally, is that point that tells you that your initial trade idea is wrong. Traders establish firm stops that are closer to the point of entry than the price targets and help ensure a favorable risk/reward profile to each trade. You can generally tell a professional trader by the way she closes out a losing trade. The exit is automatic, not a cause for consternation. Loss is an accepted part of the game. The good traders learn from those losses and use them to revise market views. A losing trade, as a result, can set up the next winning trade.

Much harder for many traders and far less remarked upon is something we might call stop-profits. Traders who religiously adhere to stop-losses can find it difficult to let profits run on winning trades. They stop those profits out prematurely, reducing the reward portion of the risk/reward profile. Over time, these traders have trouble succeeding, because their winning trades end up being not much larger than their losers—and sometimes smaller.

There are a few reasons that traders tend to cut profits short. One reason is that they fail to identify profit targets as clearly as stop-loss points. Such targets may be based on a number of factors, such as the market's overall volatility, the presence of distinct support and resistance levels, and the time frame of the pattern being traded. Many of my trades, for example, are based on historical analyses of the probability of hitting particular price levels (previous day's high or low; pivot point levels based on the prior day's high-low-close); those levels then serve as targets for setups. *It is much easier to stick with a trade when there is a firm target in mind, just as it's easier to get work done when you have a clear goal in mind.* Without a predefined target, it's easy to get caught up in the tick-by-tick ups and downs of the market, acting on fear and greed unrelated to the initial trade idea.

Another culprit in those stop-profit scenarios is a lack of confidence in one's trade ideas. One of the important advantages of testing one's trading setups is that you can estimate the historical odds of a market acting in your favor. That knowledge can provide the security necessary to see the trade through to its ultimate target. When trade setups and patterns are borrowed from others without prior testing (either through one's own paper trading or through historical analysis), it is difficult to have a deep, inner sense of confidence in the ideas. As markets experience normal retracements on the way to a profit target, those adverse excursions become difficult to weather. Instead of seeing them as potential opportunities to add to the trade at good prices, it's easy to perceive them as threats to paper profits.

Finally, a trader's risk aversion may play a role in prematurely stopping out profits. Suppose you have a choice between taking a sure $1,000 profit versus a 75 percent chance at $1,500 and a 25 percent chance at $500. Over time, taking the 75 percent chance will make you more money. Nonetheless, at any given point in time, a person may feel that it's foolish to walk away from a sure $1,000. In such a situation, the decision is made as much for the trader's peace of mind as for overall profitability. Similarly, traders may set stop-profit levels to achieve a sense of certainty, not to maximize returns.

Seeing a trade through to its target requires an unusual degree of security and ability to tolerate uncertainty. As the trade moves further in your favor, you have more money in paper profits that you're exposing to future risk, even if the risk/reward picture remains favorable. This ability to sit through a trade's uncertainty as profits accumulate requires particular confidence in the initial trade plan. Ironically, it takes more confidence to stay in the trade as it goes in your favor than if it remains in a narrow range, simply because more paper profits are at stake.

> It usually takes more confidence to sit in a winning trade than to enter it.

So how does one achieve the level of confidence needed to sit through good trades? Often it's not the loss of the paper profits per se that are the real threat for traders. After all, if a trade moves your way and you prudently raise your stop-loss level to breakeven, you'll never get hurt by a sudden, unusual adverse excursion. As disappointing as it may be to lose a paper profit, it's hardly, in itself, a threat to one's account.

Rather, the threat to traders lies in how they would *process* such a retracement. In many cases, their attitude would become quite negative in the face of lost profits. They might criticize themselves for the missed opportunity or lapse into an uncomfortable state of frustration. Instead of viewing the reversal of a gain as nothing lost—simply a scratched trade—they treat it as a situation calling for blame. It's the self-blame and the discomfort of second-guessing that traders are avoiding, not the (paper) dollar loss itself. "You're never wrong taking a profit," is an attitude that speaks more to this psychological reality than to the logical necessity of taking larger winners than losers.

> Traders often think they're managing a trade when they exit prematurely, when in fact they're managing their thoughts and feelings about that trade.

A large part of confidence is trust. You have confidence in your marriage because you trust your spouse. You have confidence in your driving because you trust your ability to maneuver the car under changing road conditions. If you don't act on your trade ideas—that is, by not seeing them through to their planned conclusion—you actually undercut your confidence by never allowing yourself to develop trust in those ideas. Just as mistrust of a spouse cannot lead to security in a marriage, a failure to trust your time-tested ideas cannot bring confidence to your trading. You can only endure the uncertainty of the trade that moves in your favor by seeing—in your own experience—that the discomfort is indeed endurable, and that you gain far more than you lose by sticking with your planned trades.

As your own trading coach, it's important that you instill both trust and confidence in your trading. This can be accomplished in two ways:

- **Instill the confident mindset.** Before trading starts, you want to mentally rehearse how you would talk to yourself in the event that you have to scratch a trade after having a paper profit. Specifically, you would rehearse a mindset of "nothing ventured, nothing gained"—it's okay to scratch a smaller percentage of trades if that allows you to let a larger percentage run—rather than a self-blaming, frustrated mindset. Prepare yourself in advance for adverse excursions so you remove much of their threat value.
- **Build on small change.** A useful brief therapy principle is to start making large changes by just starting with small changes. If you do just a little of the right thing, you will provide the feedback and encouragement necessary to expand those efforts. In the case of trading, this is easy: even if you take much of your position off ahead of a planned target, leave a small piece of the position on to either hit the target or scratch out. This preserves profits and assuages risk-aversion while it enables you to have the firsthand experience of seeing your ideas through to their conclusion. Over time, you can leave on larger pieces and build performance that way.

Confidence is not just a function of how you think, but also how you act. If you act in a way to trust your judgment, you'll have the opportunity to see your judgment work out—and that will build confidence. The stop-profit scenario, unfortunately, is a stop-confidence one as well. If you act with confidence—even in small measure—you coach yourself to self-trust and a deeper internalization of that confidence.

COACHING CUE

The flip side of the impulsive trader is the perfectionist. I've seen many traders come up with great trade ideas, only to never participate in them because the market never came to their desired entry levels. Coming up with a big, winning idea and then seeing it work out without you on board can be supremely frustrating. Don't let the perfect become the enemy of the good. If you have a fantastic idea—for instance, you see a market break out and enter a trending mode—get on board with at least a small piece of your maximum position size. If it's a good trend, you can always add to the position later on countertrend moves; if it's not a good trend, you can exit with a modest loss. *But always try to let your trading positions express your convictions*: you always benefit psychologically when you act on your confidence.

LESSON 30: COPING—TURN STRESS INTO WELL-BEING

We have seen that stress does not need to become distress if it is balanced with generous amounts of well-being. People can endure high levels of challenge, pressure, and uncertainty if their work is meaningful to them and they experience rewards tied to their efforts.

We can think of coping as a set of strategies for handling stressful situations so that they don't become distressful. By coping effectively with the risks and uncertainties of markets and the demands of the learning curve, traders can go a long way toward maintaining a favorable emotional balance.

Psychological research tells us that there is no single most effective coping strategy. Rather, people with different personalities and needs employ different coping patterns to best handle situations. When you are your own trading coach, it is important to know how you cope best with trading stresses, so that you can activate these strategies on demand.

This knowledge is especially crucial because, at times of greatest stress, we often lapse into old, well-worn coping patterns that may have worked at one time, but may not be appropriate to the current situation. An avoidant coping pattern may have worked in past work situations involving interpersonal conflict, but would be disastrous if employed in the middle of a losing trade in a fast-moving market. Doing what comes naturally is not necessarily the best strategy for handling stress. As we will see in Chapter 5, those past, overlearned modes of coping are often what keep us locked into cyclical problem patterns.

One example that I commonly encounter involves traders who utilize highly confrontive coping strategies. Many traders have aggressive personalities and succeed by facing challenges head on. This can work quite well in situations where one must negotiate a business deal or handle a piece of bad news. In the markets, however, the aggressive response is not always the best one. When facing a series of losing trades—something that happens to all of us—traders can become more aggressive and confront the situation by trading more and larger. This way of handling frustration leads a trader to take maximum risk when he is seeing the market least clearly—a virtual formula for catastrophic drawdowns.

> You can often recognize failed coping strategies when you look back on your actions and wonder what could possibly have led you to behave that way.

So how can you know which coping strategies work best for you in particular situations? Below is a checklist that will help you sort out your different ways of handling trading problems. For this lesson's exercise, I'd like you to think back to several situations in which you've handled trading problems effectively and several situations in which you've handled them poorly. Next to each coping strategy, place a checkmark if it's a mode of coping that you utilized when trading well. Then, next to each item on the list, place a circle if it's a mode of coping that you utilized when trading poorly. Here we go:

1. I reached out to others for ideas or feedback _____.
2. I took steps to make sure I didn't overreact _____.
3. I stepped back from the situation and figured out what to do next _____.
4. I tried to not make a big deal out of the problem _____.
5. I looked for what I could take away from the situation that would help me in the future _____.
6. I made a concerted effort to tackle the problem there and then _____.
7. I recognized my mistake and took action _____.
8. I decided to stop trading for a while and regain perspective _____.

Once again, the key is not to figure out the right and wrong coping strategies, but rather the ones that have worked best for you—and the ones that have been associated with problem patterns in your trading.

One important dimension of coping is action/reflection. Some people benefit most by taking prompt action to own and address challenges; others step back, get themselves under control, put things in perspective, think through plans, and/or consult with others. Another key dimension is problem-focused versus emotion-focused coping. Some traders respond best to situations by first venting and getting things off their chest, reaching out to others for input and support, and working actively to dampen negative emotions. Other traders fare best by putting feelings aside, analyzing situations, and engaging in active problem solving to address problems.

Often traders run into problems when they fail to enact their best coping strategies. The analytical trader can get hurt when he finds himself venting emotion and confronting problems without prior reflection and planning. The trader who thrives on social support and feedback from others is unlikely to cope effectively if she becomes discouraged and isolates herself from valued peers.

> If you contrast your best and worst coping—the times when you've handled trading problems most and least effectively—you identify what you need to do to sustain a favorable balance between well-being and distress.

When you track how you cope when you are trading well, you create a mental model of your best ways of handling trading challenges. This model can then become a script that you can draw on during times of difficulty. Make a coping checklist a part of your daily journal; it will alert you to behavior patterns that you can build on for the next market challenge.

COACHING CUE

Think of your best and worst coping patterns as being *sequences of actions*, not just isolated strategies. Thus, for instance, when I'm coping well, I first take steps to calm myself and get focused; then I engage in concrete problem solving. I best cope with losses by analyzing them to death—figuring out what went wrong—and then drawing positive learning lessons from those. When I'm coping poorly, I don't calm myself, and blame myself instead, adding a second bad trade to the first as a way of making up for the loss. In my poor coping mode, I don't analyze my losers, instead turning my attention to more promising markets, instruments, or setups. That ensures that I'll learn nothing from my loss—and that my error will repeat itself at some juncture. Think of coping as sequences of behaviors, so we can develop mental blueprints for the actions we need to take in the most challenging market conditions. This helps ensure that trading stress does not generate performance-robbing distress.

RESOURCES

The *Become Your Own Trading Coach* blog is the primary supplemental resource for this book. You can find links and additional posts on the topic of stress and distress at the home page on the blog for Chapter 3: http://becomeyourowntradingcoach.blogspot.com/2008/08/daily-trading-coach-chapter-three-links.html

One of the early texts summarizing research into positive psychology is *Well-Being: The Foundations of Hedonic Psychology*, edited by Daniel Kahneman, Ed Diener, and Norbert Schwarz and published by the Russell Sage Foundation (1999). Another worthwhile reference work is the *Handbook of Positive Psychology*, edited by C.R. Snyder and Shane J. Lopez and published by Oxford University Press (2003) and, by those same authors, *Positive Psychology: The Scientific and Practical Explorations of Human Strengths* (Sage Publications, 2006).

How our emotions affect our health and well-being is the topic of James W. Pennebaker's edited text *Emotions, Disclosure, and Health*, published by the American Psychological Association (1995).

A number of free articles covering topics of stress, coping, and emotions in trading can be found in the section "Articles on Trading Psychology" at www.brettsteenbarger.com/articles.htm.

CHAPTER 4

Steps toward Self-Improvement

The Coaching Process

Success does not consist in never making mistakes but in never making the same one a second time.
—George Bernard Shaw

What are the core processes of self-coaching? What concrete steps can we take to make changes in our trading to improve performance? These are some of the topics we'll tackle in this chapter.

Much of this chapter comes from research over the past several decades that has illuminated common effective ingredients across all counseling and therapy approaches. An interesting finding from that research is that all of the major approaches to counseling appear to be more effective than no counseling at all, but no single approach consistently shows better results across a range of people and problems. Not only do the major modes of helping seem to work equivalently, they also seem to work for many of the same reasons. Those reasons capture the essence of what creates change—and what can fuel our efforts to become our own trading coaches.

LESSON 31: SELF-MONITOR BY KEEPING A TRADING JOURNAL

Self-monitoring refers to methods that you use to track your own patterns of thought, feeling, and behavior over time. Self-monitoring is the foundation for many of the other self-coaching techniques described in this chapter, because it tells us what we need to change. We can't alter a pattern if we're not aware of its existence. Very often in brief therapy, the first homework exercises involve self-monitoring. Just as observing market patterns precede our ability to trade them, becoming aware of our own patterns is the first step in changing them.

One of the most common ways to monitor oneself is through the use of trading journals. Active intraday traders might make entries during a midday break and at the end of the trading day; others might simply write in the journal at the end of each day in which they are making trading decisions. The key is to catch your patterns as soon after they occur as possible, rather than rely upon fallible memory.

Note that self-monitoring is not a change technique in itself, but it often leads to changes. Once you see your patterns with crystal clarity—including their costly consequences—it becomes much easier to interrupt them and prevent their future occurrence. At other times, self-monitoring may alert you to patterns that you didn't know were present. This is exceedingly valuable, as it lays the groundwork for change that otherwise would have been impossible.

Any time you systematically review your performance over time—and the factors associated with successful and unsuccessful performance—you're engaging in self-monitoring. For example, I reviewed my recent trading results trade by trade and found that I was taking larger point losers on small trades than on my larger ones. This review led to the realization that, when trades were quite small, I was not as vigilant in setting and sticking to stop-loss levels. Although the total dollar loss for each small trade was not huge, over time the small losses added up. This led me to establish a new routine for setting and sticking to stop-losses with small trades by explicitly writing out my risk/reward ratio for each trade before I placed the order. The self-monitoring made me more conscious of what I was doing, which in turn kept me trading well.

> Self-monitoring is the foundation on which all coaching efforts are built.

My experience is that the best predictor of failure in the trading profession is the inability to sustain self-monitoring. This inability leaves traders unable to clearly identify their problem patterns, and it prevents them from reflecting and learning from their efforts at change. Goals without self-monitoring are but mere good intentions; they never translate into concrete actions to initiate and sustain change.

Why would a trader, seemingly desirous of success, not sustain efforts to monitor her own thoughts, emotions, and/or trading performance?

I believe it's because many traders are motivated by trading and making money, not by a desire to understand themselves and markets. This is an important distinction. To paraphrase coach Bob Knight, they are motivated to win, but not motivated to do the work it takes to become a winner. *In the best traders, self-mastery is a core motivation*. It's why they continue trading long after they could have comfortably retired.

The most common format for beginning a regimen of self-monitoring is keeping a journal. The basic components of a self-monitoring journal might look as follows:

Divide your journal page into three columns. The first column describes the trade that was placed, including the trade size and the time of day. If you scale into a single position, you would treat that as a single entry in the journal. Similarly, if you enter several positions to capitalize on a single trade idea (e.g., you want to be long precious metals, so you buy three different mining stocks), those would also be incorporated within a single journal entry. The first column might thus summarize what you did for each trade idea, how much you risked on the idea, when you placed the trades, the prices that you paid, and how you placed the trades (e.g., all at once or by scaling in; executed at the market or with limit orders). If you are a high-frequency trader, consider the possibility of automating your trade monitoring with tools such as StockTickr (www.stocktickr.com) or Trader DNA (www.traderdna.com).

The second column would summarize the outcomes of the trades, including the prices and times of your exits, your P/L for that trade (or trade idea), and how you exited the trades (e.g., all at once or scaling out, at the market or by working orders).

The third column would include all of your behavioral observations for that particular trade or trade idea: what you were thinking, how you were feeling, your preparation for that trade, your degree of confidence in the trade, etc. In other words, the third column takes a look at you and your state of mind, thought patterns, and physical state during the trades. The third column could also include observations about how well you entered, managed, and exited the trades. Whatever stands out for you—good or bad—about the trade would be included in that third column.

Keep your trading journal doable; many efforts at self-monitoring fail because they become onerous.

For a trader such as myself who only place, at most, a few trades per day, such a journal is relatively easy to keep. Prop traders who make dozens of trades per day or more, however, are likely to find such a journal to be onerous. One of the best ways to sabotage self-monitoring efforts is to make them so burdensome that you won't sustain the effort. If you're an active trader and cannot automate your monitoring of trades, you can streamline the journal in one of several ways:

- You can create a single entry for the morning trading and a second entry for the afternoon's trading, with the columns simply summarizing your positions, your P/L for the A.M. or P.M., and your associated observations of your trading at those times.
- You can create entries for selected positions only that stand out in your mind either because they were quite successful or quite unsuccessful. If you sample from your trades in this manner, make sure that you include best and worst trades, so that you can observe positive and negative patterns. These are the trades from which we learn the most.
- If your trading is complex, with many positions, hedges, and a flowing in and out of risk exposure over time (like a market maker or a very active portfolio manager), you can simply summarize your day with a single journal entry. The first column would review your major trading ideas, the second column would note P/L, and the third column would include your self-observations.

There is no single self-monitoring format perfect for all traders; the key is to adapt the format to your needs and trading style. The real work comes when you've accumulated enough entries to notice patterns in your trading: the factors that distinguish your best trades and days and those that accompany your worst trading. How to analyze your self-monitoring journal will be the focus of the next lesson. For now, however, your task is simply to sustain self-awareness: to be an active observer of your own trading process.

When you are your own trading coach, there is always a part of you that stands apart from your decision-making and execution, observing yourself and exercising control over what you do and how you do it. The real value of the trading journal is that it structures the process of self-awareness and helps make it more regular and automatic. If you were walking on a familiar street, you would hardly think about how you walk, everything would be on autopilot. If, however, you were taking the same walk in a minefield, you would be exquisitely self-aware, conscious of every step that you took. Trading is neither a walk in the park nor a minefield . . . perhaps it's more like a walk in a beautiful, but somewhat dangerous park. You want to be absorbed in the walk, but alert and aware at the same time. That is the function of the trading journal: it enables you to monitor yourself, even as you are immersed in what you're doing.

COACHING CUE

A great insight into journal keeping is offered by Charles Kirk (www.thekirkreport.com) in Chapter 9, who enters his observations into a database program so that he can readily retrieve journal entries on various topics. This is an effective way of monitoring specific trading challenges over time.

LESSON 32: RECOGNIZE YOUR PATTERNS

One of the keys to brief therapy is the creation of a concrete focus for change. One of the reasons that older forms of therapy took so long to implement—including years of psychoanalysis—was that they attempted broad personality changes. Our understanding of personality traits and their biological, hereditary components helps us to be a bit more modest in our aims. No form of coaching or counseling will restructure a person's personality; nor should it. *The goal of coaching is to help people work around their weaknesses and build their strengths, so that they can express their basic personalities and skills as constructively and successfully as possible.*

> You cannot change your personality, but you can change how it is expressed.

Many self-described coaches lack formal training in psychology and especially lack the experience and grounding of licensed helping professionals. They acquire a cluster of self-help methods and try to fit all problems to those. The result is a canned set of solutions for any given problem. This can be disastrous. The patterns that interfere with trading often lie well outside canned, self-help nostrums. One successful trader I recently met at a conference presentation was having a poor trading year and was even considering retiring. She complained of a loss of enthusiasm and excitement about her trading, as well as weight gain and more negative feelings about herself. She had seen three prior coaches and therapists, all to no avail. After a short discussion, I obtained enough information to suggest that she obtain a blood workup from her primary care physician. She did so, and the results suggested a low level of thyroid activity. Once she received proper hormonal supplementation, her mood and energy level returned, her concentration improved, and she resumed her successful career.

How many traders lose their careers because they never understand the patterns that underlie their problems?

You might ask the question, "How can I, as a trader relatively uneducated in applied psychology, ever hope to identify obscure patterns such as low hormone levels? If experienced coaches and counselors miss the pattern, how can I detect it?"

Ironically, I think that you're in better shape than most commercial coaches to identify and act upon unusual patterns that interfere with good trading. I could readily make the recommendation for the trader because I was not seeking her business. She was not paying me for my services and I did not stand to gain by making one recommendation versus another. (Note: my coaching is limited to a limited group of proprietary trading firms, hedge funds, and investment banks; I don't take on individual traders as clients.) Most coaches who focus on retail traders, on the other hand, need to constantly drum up business. It is not in their self-interest to raise a possible course of action (such as blood tests and thyroid medication) that doesn't lead to further services and additional fees. As a result, they focus on solutions that they can provide (i.e., that will bring them additional business). When all you have is a hammer, Maslow once remarked, you tend to treat everything as nails.

As your own trading coach, you have no such conflicts of interest. You can learn pattern recognition for yourself and diagnose your own concerns. If the problem still eludes you, even after you've reviewed your journal extensively, send an e-mail to the special address reserved for this book (coachingself@aol.com; see the Conclusion) and I will do my best to point you in a promising direction. But I think you'll be pleasantly surprised to find how readily you can tackle your own challenges once you learn a few basic techniques.

Once you learn to coach yourself, you have the skills to guide your development across a lifetime.

So let's see how you can become expert in recognizing your own patterns, building on the trading journal described in the previous lesson. Reviewing your journal entries, you want to divide them into two clusters: those describing your most successful trading and those that capture when you were trading at your worst. The first cluster will reveal what we call *solution patterns*; the second cluster alerts you to *problem patterns*. Many times, the difference between the solution and problem patterns will themselves point you to practical actions you can take to improve your performance. For instance, you might notice that during your successful trading you're more patient entering trades and take fewer trades, each with smaller size. When you're less successful, you trade more often and with maximum size.

The comparison between best and worst trading will also alert you to differences in your coping with market challenges. You may find, for example, that when you're trading well, you tend to be very problem-focused. When you are less successful, you trade while you are confused or frustrated, without waiting for clearly defined opportunity.

The key is to look for patterns, not just isolated instances of good or poor trading. For each good day of trading, you might jot down several things you did right and then see which entries appear day after day. Similarly, in reviewing the poor trading days, you would write down the key mistakes you made and observe which ones appear over time.

Your trading strengths can be found in the patterns that repeat across successful trades.

If you cannot find patterns that stand out, you may need to monitor your trading over a longer period, so that you have a rich sample of good and poor trading days. You're looking for common elements that jump out at you; don't be too quick to read subtleties into the patterns. The best things to work on are the ones that are most salient—that hit you between the eyes. For instance, when I have done the pattern-recognition work and compared my best and worst trading, I found stark differences in the sizing of my trades (initial positions neither too large nor too small performed best); the timing of the trades (positions too early in the morning or later in the afternoon underperformed those made after the market sorted itself out in the first minutes of trade); and the duration of my trades (better performance when held shorter, with clear targets and stops). I also found that I traded best when I had a clear longer-term picture of the market to guide shorter-term entries and trades. My absolute worst trading occurred when I held a strong view at the start of the market day and did not modify the view as the day progressed, continuing to trade against a market trend.

Notice how each of these patterns focuses a trader on *what* to change, which is the first step in deciding *how* to make changes. Many times, traders fail to make changes because they aren't clear on what to change. They rely on vague generalizations ("I need to be more disciplined") rather than identify specific behaviors to work on. By conducting detailed comparisons between your best and worst trading, you can find a focus for your self-coaching efforts and channel your energies in the most constructive directions. If you know your patterns—those that bring you success and failure—you're generally halfway home in making lasting changes. Below are some patterns to be especially aware of as you review your journal:

- **Emotional Patterns**—Distinct differences in how you feel when you're trading well and when you're trading poorly, particularly before and during trades.
- **Behavioral Patterns**—Notable differences in how you prepare for trades and manage them during your best and worst trading episodes.

- **Cognitive Patterns**—Meaningful differences in your thought process and concentration level during and after your best and worst trades.
- **Physical Patterns**—Differences in how you are feeling—your energy level, physical tension and relaxation, and posture—when you're trading at your best and worst.
- **Trading Patterns**—Differences in the sizing of trades, times of day when you're trading, mode of entering and exiting trades, and instruments being traded as a function of good versus poor trading.

COACHING CUE

Many times you'll observe more than one kind of pattern, and many times those patterns will be linked. For example, your cognitive patterns may lead to particular emotional patterns, which are then linked to specific trading patterns. Think of patterns in sequence—as a kind of positive or negative cascade—not as either/or phenomena. Excellent coaches see not only patterns, but also patterns of patterns. Here are some of the most common patterns among traders to watch out for:

- Placing impulsive, frustrated trades after losing ones.
- Becoming risk-averse and failing to take good trades after a losing period.
- Becoming overconfident during a winning period and taking more marginal and/or unplanned trades.
- Becoming anxious about performance and cutting winning trades short.
- Oversizing trades to make up for prior losses.
- Ignoring stop-loss levels to avoid taking losses.
- Working on your trading when you're losing money, but not when you're making money.
- Becoming caught up in the market action from moment to moment rather than actively managing a trade, preparing for a next trade, or managing a portfolio.
- Beating yourself up after losing trades and losing your motivation for trading.
- Trading for excitement and activity rather than to make money.
- Taking trades because you're afraid of missing a market move, rather than because of a favorable risk/reward profile for your idea.

LESSON 33: ESTABLISH COSTS AND BENEFITS TO PATTERNS

A huge step forward for traders who seek to mentor themselves is to know the patterns of behavior, thought, and emotion associated with successful and unsuccessful trading. Still, it is necessary, but not sufficient to produce lasting change. *This is because knowing a pattern is different from having—and sustaining—the motivation to alter that pattern.* It is in the motivational arena that many of our change efforts, personal and professional, fall short. We need only to look at the dismal record of people's attempts to diet, eat in healthy ways, or exercise regularly to see that knowing what you need to do and actually doing it are two different things.

When you are your own trading coach, your challenge is to *motivate* change, just as a sports coach motivates a team to keep them working hard and sustaining practice efforts. If keeping a journal and tracking your patterns of good and bad trading is nothing more than a routine exercise—another item on a to-do list—it will not inspire motivation, and you will not sustain the efforts. Stoking the desire to make changes is difficult, especially when trading is going reasonably well. "If it ain't broke, don't fix it," is a formula for eventually lacking fixes when things do break.

But there's a more important reason for sustaining self-coaching efforts. *When you're trading well is precisely the time you want to be the most self-aware of your strengths, so that you can maximize your earnings at those times.* The true competitors and most successful participants in sport, warfare, and games of skill such as chess are the ones who possess the killer instinct: they have a sense for when they have the advantage, and they press that advantage to the hilt. Making modest money when you are trading well is a great way to ensure poor returns when you're not seeing markets as well. For many traders, it's the relatively few, large gains from the best trading periods that contribute most to overall profitability.

> The measure of self-coaching is how hard you work on trading when you're making money.

It's as important to work on yourself when you're trading well as when you're trading poorly. It's not that you want to fix what isn't broken; rather, you want to crystallize what you're doing right so that you can do more of it and capitalize fully on it. Conversely, when you're trading poorly, you don't want to lapse into discouragement and defeat. Maintain the journal and the pattern recognition efforts to keep you in a constructive mode, even when all you may be able to do for the moment is to cut your risk when your negative patterns surface. That is still progress.

So how do excellent traders (and competitors in any field) sustain the motivation to operate at elite levels of performance? An important driver of that motivation is an intense competitive drive and a ferocious desire to win. The traders I've worked with personally who have been consistent, high earners have traded quite differently and viewed markets in radically different ways. Some traders have been loud and outgoing; others have been cerebral. Some have been uncannily intuitive; others have been analytically insightful. The common feature among all of them, however, has been their intense competitive nature. They compete against peers, they compete against markets: most of all they compete against themselves. They derive pride and validation, not just from making money, but also from getting better, which is what keeps successful traders in the game long after they could have retired comfortably.

The lesser traders? They trade to not lose; they trade to keep their jobs. They don't hunger to become more than they are; they want to do well, not to be their best.

The drive for self-improvement is different from the desire to make money and is far more rare.

These successful traders can sustain their drive by staying mindful of the costs of their negative patterns and the benefits of their positive ones. When a successful trader decides to avoid a particular trade or market, it's often because of a specific recollection that this idea has caused past losses. By staying emotionally connected to the pain created by their worst trading, traders stoke their motivation to avoid trading mistakes. Similarly, knowing their strengths is not just an abstract awareness for successful traders, but an emotional connection to the pride and sense of accomplishment over doing well.

A best practice in self-coaching—and a great assignment for this lesson—is to not only summarize the patterns of your best and worst trading, but to actually write down and visualize the costs associated with the most negative trading patterns and the benefits that accompany the best patterns. In other words, you don't finish your journaling until you achieve a state in which you are emotionally connected to the things you are writing about. You will want to change your negative patterns when you get to the point of hating those patterns and becoming disgusted with the ways in which they've set you back. You'll want to build on your positive patterns when you see and feel their benefits. When you're coaching yourself well, journaling is an emotional exercise, not merely a cognitive one.

There is little to be gained from abstract positive thinking. Reciting such affirmations as, "I will be a successful trader" is empty at best, self-delusional at worst. The reason such positive thinking doesn't work is that it is not connected to the day-to-day conduct of your trading. It's not enough to simply make yourself feel good, and, indeed, there can be real value in feeling so bad that you'll never repeat a mistake again. What is helpful is to associate the best emotional experiences of trading—your greatest moments of joy and achievement—with the specific practices that brought you such happiness. It is also tremendously helpful to re-create the pain of your worst trading with concrete vows to never go there again.

Think of football, basketball, and tennis coaches. Every practice session teaches skills, offers feedback, and supplies motivation. It's not a bad formula for self-coaching as well. You are most likely to change a negative thought or behavior pattern when you associate it with concrete costs and consequences; you're most likely to motivate a positive behavior by attaching to it specific, felt benefits.

COACHING CUE

Efforts at self-change break down when people start to make exceptions and allow themselves to revert to old ways. To accept exceptions, you have to accept the old, negative patterns. It's when our old patterns become thoroughly unacceptable that we are most likely to sustain change. When you keep a journal, you want to cultivate an attitude, not just jot down bloodless summaries of what you do. If you don't see plenty of emotion words in your journal—constructively expressed—the odds are that your journal will summarize your changes but not motivate them.

LESSON 34: SET EFFECTIVE GOALS

Successful coaching requires a focus for change. *Many self-help efforts among traders fail because they are unfocused.* One day the trader writes in a journal about position sizing and works on that; the next day he emphasizes emotional control; and the following day he stresses taking losses quicker. By jumping from one focus to another, no single direction is sustained.

This is problematic because most learning does not occur all at once. We know from research that learning typically requires many trials and considerable feedback. If we think about important skills that we've learned—from using a computer to driving a car—we can see that trial and error has been the norm. I can study a map of a new city for a long time and learn quite a bit, but I ultimately only learn to find my way around by driving on various roads, following signs, getting lost, and recognizing landmarks. If we pursue this learning intentionally—organizing our trials over a concentrated period of time with immediate feedback throughout—we can greatly shorten our learning curves. It's easier to learn to play a piano by practicing every day and taking lessons every week than by engaging in occasional efforts spread over years.

> The training of an Olympic athlete is a study in proper skill development: intensive work on specific aspects of performance,

accompanied by plenty of coaching feedback and corrective efforts.

When we lack a specific focus for change and jump from one trading goal to another from day to day, we don't really enter a learning curve: there is nothing cumulative in our efforts. As your own coach, you need to establish—and also sustain—a specific direction to guide your growth and development. This is the key to effective goal setting. If motivation provides the energy for self-improvement, goals supply the aim, the channeling of that energy.

When you recognize a problem pattern, it is not the same as establishing a goal for self-work, though the former usually will guide the latter. A problem pattern alerts you to what you're doing wrong. Goal setting requires an awareness of new patterns of thought, feeling, and/or action that will replace the problem pattern. A goal states what you are going to do or not do in specific situations. If you have constructed your trading journal and pattern-identification exercises in the ways suggested in recent lessons, many of your goals will naturally follow from the patterns associated with your best trading. Your overarching goal will be to trade in the way that you typically trade when you're trading well.

Notice that I am defining goals in a process sense, rather than as absolute outcomes. This is very important. Many traders think of such goals as making a million dollars or being able to trade for a living. These outcome goals may be motivating, and definitely have their place, but they do not focus a trader on what she needs to do today and tomorrow to become better. Traders on a short time frame do not have full control over their outcomes: a trader can make good decisions, placing all odds in their favor, only to lose money when markets behave in an anomalous fashion. What traders *can* control is the process of trading: how they make and implement decisions. The most effective daily goals emphasize trading well, not making oodles of money.

The same logic guides athletic coaches. A coach may stress the goal of victory over the next opponent, but the day-to-day practices will emphasize such fundamentals as swinging at good pitches (baseball), making the extra pass (basketball), and blocking effectively on running plays (football). These process goals provide the ongoing focus for practice over time and ensure that coaching leads to effective learning. Every coach is, at root, a teacher. When you're your own coach, you guide your own learning efforts.

Effective goals target effective trading practices that break trading down to component skills and then set targets for these, one at a time.

As you examine your journal entries for the patterns that distinguish your best and worst trading, the question you want to ask is, "What is the difference that will make the greatest difference in my trading?" Stated otherwise, your question is, "What are the one or two ways I can trade more like my best trading and less like my worst trading?" The answer to these questions will form the basis for your best process goals—and will guide your self-development efforts from day to day and week to week. Goals should not be so specific that they only apply to a limited set of circumstances ("I want to be a market buyer when put/call ratios hit 100-day highs"), and they should not be so broad or vague that they don't guide concrete actions ("I want to trade less often"). *The best goals are ones you can work on every day for a number of weeks.* If you do not work on a goal day after day for at least several weeks, it's unlikely that you will turn your new patterns into positive habits.

As mentioned earlier, you don't want to focus on too many goals at once. I've generally found three to be as many as I can work on with consistent intensity—and many times I will focus on fewer objectives. This means that a good self-coach will prioritize needed changes, emphasizing those differences that will most make a difference. A great exercise to try is to close your eyes and imagine yourself as a great trader. Visualize yourself as the best trader you can possibly be—or maybe visualize an absolutely perfect trading day. What are you doing differently from your usual trading when you're the great trader? How is your trading different on the perfect day than on poor one? In your visualizations, try seeing yourself doing all the right things when you're trading well. What are you doing? How are you doing it? Those visions, made highly specific, will form the backbones of your goals.

COACHING CUE

Make it a point to get to know successful traders who have had plenty of market experience. Many times you can form effective goals by simply trying to emulate their best trading qualities.

LESSON 35: BUILD ON YOUR BEST: MAINTAIN A SOLUTION FOCUS

To this point, the lessons in this chapter have emphasized the importance of tracking the patterns associated with both your trading shortcomings and your trading successes. It is common to focus on the former only. When we use coaching to build strengths rather than simply shoring up weaknesses, that solution focus produces surprising—and surprisingly rapid—results.

For more on the solution-focused approach to change, check out *The Psychology of Trading*

There are several reasons why a solution focus is helpful for traders who seek to coach themselves:

- **Motivation**—It is easier to stay motivated and optimistic when we emphasize what we're doing right, not just where we fall short. Imagine a sports coach who only harped on players' weaknesses. Over time, that would be demoralizing. When the focus is strengths, coaching can be empowering and inspiring without ignoring changes that need to be made or the urgency of making those changes.
- **Goal Setting**—Knowing what you've done wrong, in and of itself, does not tell you what you need to do right. When you focus on your best trading, you can identify specific patterns that are associated with your success and turn these into concrete goals for future work.
- **Bang for the Buck**—Working solely on improving weak areas is unlikely to create strengths; at best, you'll take a deficient area and bring it to average. It's making the most of strengths and learning how to work around shortcomings that produces optimal performance results.

One of the reasons I emphasize the importance of paper (simulation) trading and playing with different trading styles and markets is that this experience enables you to observe your own strengths firsthand. Many, many times, traders stumble across their performance niches when they discover something they're good at. If you don't know your strengths, it's unlikely that you'll be able to systematically build on them and turn them into drivers of elite performance.

But you might wonder, "How do I stay solution focused if, day after day, I'm in a slump and losing money?" It is difficult to stay connected to our strengths when our failings are written all over our P/L summaries!

We've seen that this question is a particular challenge when we identify winning days as good trading and losing days as bad trading. This thinking makes it difficult to observe and appreciate good trading when we're not making money. But traders can trade well—they can take setups with demonstrated edges, size positions well, and manage the risk of their trades—even if trades happen to go against them. After all, even a 60 to 40 edge per trade ensures that a trader, over time, will have sequences of consecutive losing trades and/or days. If you define good and bad trading in terms of process, not just outcome, you can observe strengths during performance slumps and also you can detect flaws even when you're making money.

To keep yourself solution-focused, you want to ask yourself, "What did I do well today? What about this trade did I do right?" You'll find that, over time, your performance is varied. Not all trades are poorly conceived, poorly executed losers. If you lost less today than the past several days, what did you do better? If you had a number of winning trades during several losing days, what distinguished those winners? Focus on the improvements in your trading and then isolate the specific actions you took to generate those improvements. These actions can be meaningful additions to a daily to-do list.

"What did I do better this week than last week?" is a great starting point for guiding next week's efforts. Do more of what works—it's the essence of the solution focus.

Another technique for sustaining the solution focus referenced at the end of the last lesson is to identify a mentor or trader you respect and ask yourself how he would be trading a particular idea. Sometimes it is very helpful to try out solution patterns that you borrow from others. Over time, you adapt these patterns to your own ways of thinking and trading, so that they become distinctly yours. For example, I've worked with quite a few hedge fund portfolio managers and have learned from them the importance of thinking thematically about markets: observing various sectors and asset classes and creating narratives that guide a longer-term perspective. The specific themes I track and the time frames I monitor are completely different from theirs, but there is a similarity of process. When I'm not trading well, I can model their processes and place myself more in line with market's trends.

Still another way to keep the solution focus is to make special note of mistakes that you *don't* make in your trading. These notes represent exceptions to problem patterns. If there's a mistake you've made lately, it helps to hone in on occasions when you haven't made the mistake. What are you doing differently at those times to avoid the mistake? Perhaps you're anticipating the problem and consciously doing something different. Maybe you're avoiding the mistake by following a particular rule or practice. Whatever helps you do less of the wrong things can also form the basis for solutions.

Look to situations in which you don't make your worst mistakes. Many times those situations hold the key to avoiding problem patterns more consistently.

The real power of the solution focus is that, when you discover what you do during your best trading, those positive patterns are uniquely yours. Instead of ceding the role of expert and guru to others, you're turning yourself into your own guru by finding the best practices that are unique to you. *You become your own role model.* This is one of the most promising facets of self-coaching: by discovering who you are at your best, you can create goals that are specific, unique, and relevant to you. The learning, as a result, will be more relevant and empowering, reinforcing strengths as you build them.

A great assignment is to review your trading journal and assess the ratio of problem entries (writings about your bad trading) to solution entries (writings about what you're doing best). If the ratio is lopsided in favor of the problem focus, consider structuring your writing to force yourself to address a few basic questions:

- What, specifically, am I doing best in my recent trading?
- How, specifically, am I avoiding old trading mistakes when I'm trading well?
- How, specifically, am I trading like my idea of the ideal trader during my best trading?

A good sports coach never loses sight of a player's potential, even when correcting weaknesses. The challenge of self-coaching is never losing sight of your strengths, and then working daily on how you can maximize them.

COACHING CUE

We have solution markets as well as solution patterns of behavior: specific markets and market conditions that facilitate our most successful trading. Knowing these intimately is very helpful in allocating risk to your trade ideas. It can also be helpful in staying out of certain kinds of trades and emphasizing others.

LESSON 36: DISRUPT OLD PROBLEM PATTERNS

As you continue your journaling over a period of weeks or longer, you will become attuned to your problem patterns. Usually, a trader does not have 10 different problems. Rather, he may have one or two problems that manifest themselves in 10 different ways. For instance, a trader may grapple with missing good trades, occasionally ignoring stop-loss levels, sizing positions too conservatively, and cutting winning trades too quickly. These examples may seem like different problems, each requiring a different coaching plan and process. As the trader examines his or her journals, however, they're likely to find that a single problem pattern—anxiety related to negative self-talk—is responsible for all of these. *It's not that the trader has many problems (though it may certainly feel that way); it's that there is a single, core problem that affects many aspects of the trading process.*

As you can see from the above example, self-coaching requires that you not only detect patterns but patterns among the patterns. It's these patterns of patterns that usually form the core focus of coaching efforts. This means that if you can accurately identify the core pattern, many different trading difficulties can fall into place in a reasonable period of time. Once the trader in our example learns to master anxiety and not channel it through self-defeating self-talk, he will miss far fewer opportunities, become more consistent in sticking to stop-loss levels, take appropriate risk, and let trades progress toward their designated targets.

> Asking yourself, "What is the common denominator behind my different trading mistakes?" begins the process of finding patterns of patterns.

HERE

Often the core pattern will involve a feeling state that recurs for the trader and that disrupts good decision-making. For example, the trader may lapse into periods of anxiety, frustration, or self-defeat. How this feeling state impacts trading may vary from day to day, which is what produces the multiple manifestations that lead traders to think that they have dozens of problems. By tracing each trading problem back to a particular cognitive (thinking) or emotional (feeling) state, we can then identify the events that typically trigger that state and design effective coaching interventions to tackle those situations and triggers.

Often, a trader will know what he is doing wrong, but won't know the right thing to do. This issue occurs when traders have not been sufficiently solution-focused. They know, for instance, that they should not double down on losing trades to make money back, but they don't know how to reenter a trade and exploit good research after having been stopped out of an initial position. This situation requires two important coaching steps: 1) interrupt the problem pattern so that it does not disrupt trading; and 2) develop rules and procedures for a possible solution pattern (which will form the theme of the next lesson).

Because traders don't always have solutions readily at hand and need to stop bleeding losses, interrupting problem patterns is often a first coaching objective. "Above all else, do no harm"—the Socratic oath in medicine—is relevant here. The ability to stop doing wrong things won't, by itself, generate good things, but it can keep you solvent long enough to find solutions!

Change starts when you stop yourself from doing what doesn't work.

A major point from *The Psychology of Trading* is that *the key to disrupting problem patterns is to alter the state that you're in when those problems first appear*. This means that it's important to become vigilant to the emergence of patterns, recognizing the characteristic ways that they appear. For example, some of my worst trading occurs when I focus on my P/L as the trade is unfolding. This focus may lead me to tolerate larger than normal losses in a small position, because it's not hurting me, or it may lead me to take profits too quickly on a large position to book a sure gain. I've learned that if I start counting profits during the trade, I need to refocus my attention. I accomplish this by turning briefly from the screen, fixing my gaze on something nearby, and taking a few deep breaths. Once I'm in a new state—more calm and focused—I find it easier to be detached from the P/L and let the trade unfold in its planned manner.

Another quick way of shifting state is simply to walk away from the screen temporarily and engage in a quick activity unrelated to trading. Some activities might be a few stretches or exercises; talking with a fel low trader; or getting something to eat or drink. Often, *doing* something different enables you to approach situations differently: the new activity helps you shift your frame of thinking and feeling. I find this activity particularly useful after taking losses: a quick walk outdoors, getting away from markets, allows me to return to the screen with a fresh perspective.

When you change your physical state, you alter how you experience the world and process information.

Still another mode of state shifting is to write in a journal or talk aloud during a particular situation. The latter is especially useful if you're trading alone and won't be a distraction to others if you process information aloud. If you write or talk about what is happening and give voice to what you think and feel at the time you're thinking and feeling it, you shift from being a person immersed in experience to being a person observing his own experience. If you're an active trader, in and out of markets quickly, you may not have time to write out journal entries and observe yourself in that manner. Talking aloud, however, can be accomplished while still watching the screen; in fact, that's how many traders in Chicago work with me during the trading day: they talk aloud about what is happening while they are engaged in trading.

To use my example from above, if I talk aloud my thoughts about my P/L while I'm in a trade, that alerts me to the fact that I'm no longer focused on the trade itself. If I hear myself talk about something other than the management of the present trade, it kicks me into a different mode and pushes me to make an effort to get back to the market itself. This shift becomes easier and easier as traders learn to make self-observation a habit.

> When you talk aloud your thoughts and feelings, you no longer identify with them; you listen to them as an observer.

A good self-coaching question to ask yourself is: How different would your P/L be if you could eliminate the 5 percent of your largest losing trades? Often, this percent by itself would make a huge difference to a trader's profitability. By interrupting the patterns that accompany those large losers that result from bad trading (not just being wrong on an idea), you can "above all else, do no harm." *It's usually a particular emotional state or pathway of thinking that triggers the bad trading.* If you recognize the state and thoughts as they're occurring, you can stop yourself and, at the very least, avoid disaster.

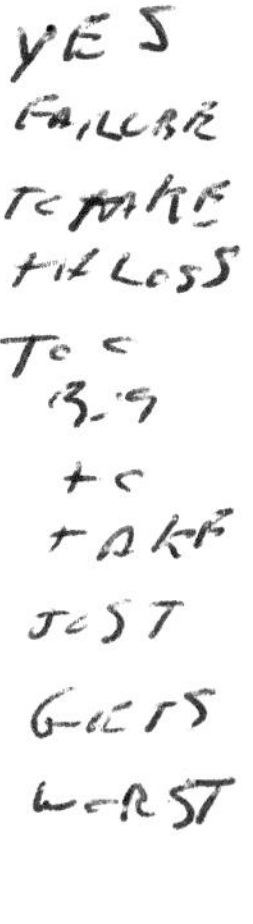

This recognition would make a wonderful goal to work on in your trading. Choose just one negative pattern that has accounted for many of your largest losing trades and then identify the common triggers for that pattern. Then select one method for interrupting the pattern when you notice one of the triggers occurring—even if that method is doing nothing more than placing no more trades until you regain emotional equilibrium. A good coach knows when to take his player out of the game for rest and a lesson. When you are your own trading coach, sometimes you need to do something similar. Remember: the goal is not to trade; the goal is to make money. Sometimes the best way to make money over the long haul is to ensure that you keep your money when all the wrong patterns are firing.

COACHING CUE

If I say something in a frustrated tone or make a frustrated gesture while I'm trying to get into a trade, while I'm managing a trade that's already on, or while I'm trying to exit, that is my signal that I need to interrupt an emerging pattern. I typically will slow my breathing considerably and focus on my breathing as I'm continuing with my business. As soon as is practical thereafter, I take a short break from the screen and don't enter any new positions until I have figured out why I'm frustrated, what that tells me (about me and/or the market), and how I want to factor that into my trading. If you use frustration as a cue to interrupt patterns you are prevented from acting mindlessly on the frustration, but also you are set up to become mindful of the reasons for the frustration.

LESSON 37: BUILD YOUR CONSISTENCY BY BECOMING RULE-GOVERNED

One of the major goals of coaching yourself is to turn positive trading behaviors into habit patterns. This is crucially important. You don't want to have to think and make yourself do the right things each time opportunity occurs. Rather, you want to do the right things automatically. Effort and energy you spend thinking about what to do—and trying to make yourself do it—is taken away from markets themselves. When you can do the right things automatically, your concentration can be wholly focused on what you are doing. That is essential if you're going to be sensitive to subtle market shifts in supply and demand.

Rules are the bridge between new behavior patterns and acquired habits. Children are not born with a developed sense of ethics and responsibility. They are taught rules by parents and teachers that are eventually internalized. Some of that internalization occurs by observing role models over time; much of it results from turning the desired behaviors into explicit rules that can be rehearsed. Such mental rehearsal allows people to keep old behavior patterns in check and make conscious efforts to engage in new ones.

We see such dynamics at work when traders are learning to control losses. Instead of exiting trades when the pain of loss is too great—a pattern that comes all too naturally—a trader will create a rule-based stop-loss level. The rule may be accompanied by other thoughts that emphasize the importance of the rule, the losses that will follow from not following the rule, and the benefits of adhering to the rule. In such an instance, traders choose to refuse to do what they feel like doing at the moment. Rather, they seek to be rule-governed. That is what keeps us driving on the proper side of the road, even when we're in a rush. Rules are checks on our impulses; they keep us doing the right things even when we're not inclined to act in our own best interests or the interests of others.

> We follow social rules without even thinking about the rules of proper social behavior because we've repeated the right behaviors so often and have internalized the rules so thoroughly over time. That's the goal with trading rules.

Trading is especially challenging, because the normal human response is rarely the one that makes money. One exercise from the *TraderFeed* blog examined periods of time that were up on a one-month, one-week, and one-day basis and compared those with periods that were down over those three time frames. In the first situation, almost anyone would identify the trend as rising; in the second situation, it's a clear downtrend. Had you bought the market after the up periods, however, you would have severely underperformed the market averages. Had you sold after the down periods, you would have lost considerable money. The obvious strategy fails precisely because it is so obvious. By the time a trend is readily apparent; all the momentum and trend followers are aboard. They're the ones scrambling to get out when the tide turns, leaving traders who acted on the obvious with losses.

When we create rules, we put a brake on those normal human tendencies. A rule might be as simple as "only buy if the market is down over X period." Surprisingly, in the broad stock market, such a simple rule works pretty well on average. Another rule might be, "Never enter a trade unless you first measure risk (stop-loss level) and reward (profit target) and have a reward-to-risk ratio of 2:1 or better." Such a rule would restrain a trader who is tempted to jump aboard late in a market move.

What traders call *setup* criteria often are simply rules for getting them into trades. When the criteria are not established as firm rules—and mentally rehearsed as such—there is a tendency to violate the setups. This violation often occurs because of the fear of missing a profit opportunity or because of risk aversion after a prior loss or series of losses. When the setups are structured as rules, trading may not be mechanical, but it can be much more consistent. Most frequently, the inconsistent trader is the one with the loosest rules.

Rules aid trading consistency.

When you are your own trading coach, you not only formulate your rules, but also must do so in a way as to maximize the odds that you will actually follow them. The key to successful rule-creation is the recognition that rules are more than thoughts that go into your head. *A good rule also comes with feelings attached*: an awareness of both the consequences of violating the rule and the benefits of following it. What keeps a diabetic person faithful to a diet or an eager child patiently waiting her turn to answer a question in class? It's not just the thought of the rule, but also the immediate sense of what would happen if the rule were violated. When people think, "I can get away with it," the rule loses its force: it's merely a set of empty words and good intentions.

This, then, is the secret to formulating trading rules, whether they relate to entries, exit, position sizing, stop-losses, diversification, or idea generation: *whenever you write down the rule or mentally rehearse it, make sure that you are emotionally connected to that rule.* Make yourself relive situations in which you've violated the rule. Focus your attention on successful episodes of trading in which you followed the rule. Make the rule more than a guideline; it should represent a belief and conviction. The best rules feel like *must*, not just *should*.

Your assignment is to take a thorough trading inventory of your trading rules. How many do you really have, and how explicit are they? Do you just remind yourself of them passively, or do you rehearse them with belief and conviction? If you're like most traders, you'll find that you have many loose guidelines, but few firm rules. That means that you haven't really drilled down to identify the patterns behind your best and worst trading, which really form the backbone of all good rules.

> Rules should reflect best practices in trading.

Remember: *you can't follow a discipline that you never formulated in the first place.* The clearer the rules and the more you feel them, the stronger they will serve as brakes to your impulses and guides to your best behavior. Rules are not straightjackets; they free you up to be your best. Think of professions in which consistency is a virtue: airline pilots landing an aircraft, surgeons making incisions, racecar drivers maneuvering in a pack. The best performers are rule governed: they are keenly aware of the dangers of ignoring the rules of their profession. It is in the internalization of their rules that they achieve flawless execution. That is your goal in coaching yourself: to make rules so routine that they make extraordinary performance the norm.

COACHING CUE

Rule-following is a great basis for self-evaluation. Creating checklist report cards to track your rule governance helps ground you day to day in best practices. Among the rules you should consider formulating and tracking for self-assessment are:

- Rules for position sizing.
- Rules for limiting losses, per trade, per day, per week, etc.
- Rules for adding to existing positions.
- Rules for when you stop trading or limit your size/risk.
- Rules for increasing your size/risk, per trade, per day, etc.
- Rules for entering and exiting positions.
- Rules for preparing for the trading day/week.
- Rules for diversification among positions.

Not all these rules will apply to all traders; the key is to focus on the rules that capture your best trading and turn those into report cards for daily/weekly self-assessment and goal setting.

LESSON 38: RELAPSE AND REPETITION

The greatest enemy of coaching is relapse. It is relatively easy to initiate change, but quite difficult to sustain it. Our old patterns are what come naturally: they're what we've been doing day after day, year after year. Those patterns are overlearned; they have been repeated so often that they have become automatic. If change efforts are not sustained, the automatic patterns naturally fill the void.

What this means in practice is that there is a certain series of stages to any change process:

Phase One—We repeat old patterns automatically, experience consequences, and try to avoid the consequences as much as possible.

Phase Two—Consequences of old patterns accumulate and we develop an awareness of the need to change, though we may not know how to change and may have ambivalent feelings about change.

Phase Three—We can no longer accept the negative consequences of our old ways and commit ourselves to making changes by trying to think and act differently.

Phase Four—We slide back into old patterns periodically when our change efforts lose momentum, creating oscillating periods of change and relapse.

Phase Five—We engage in new patterns sufficiently often that they become automatic, greatly reducing the relapse into old ways.

So let's take a practical example: In phase one, we are in an unfulfilling romantic relationship, but minimize the problems and try to go on from day to day. In phase two, we recognize that the problems are there, but wrestle with the question of whether we really want to rock the boat and raise concerns with our partner. Phase three brings clear awareness of the need for change and discussions at home to work out problems. Phase four sees periods of good times, interrupted by resumptions of the problematic interactions, perhaps aided by couples counseling. In phase five, we keep working on the counseling exercises, changing patterns of communication, until there are new and more constructive ways of engaging each other that have become routine.

This step-wise scheme suggests that relapse is not merely a problem, it is a step on the road to change. Few people change patterns all at once and for all time. More often, there is a tug-of-war between the old, overlearned patterns and the new, constructive ones we're working on. This tug-of-war occurs precisely because the new patterns have not yet been overlearned: it takes conscious effort to enact them. Early in the change process, we don't think about change; in the middle phases, we have to think about change in a very conscious manner. Only late in the process do the new behaviors come more naturally and automatically.

In every change process, there is an intermediate phase in which old problem patterns coexist with new, positive ones. Relapse at this stage is the norm—not necessarily a sign of failure.

So what enables us to make the transition from effortful change to an internalization of new patterns so thorough that those patterns become second nature? If you think back to when you learned to drive a car, there was a period when you had to consciously focus on every aspect of driving, from signaling and making turns to changing lanes. Only with repeated experience did these activities become automatic, freeing you up to focus on road conditions when necessary, converse with passengers, and locate unfamiliar destinations. Similarly, repeated experience with new cognitive, emotional, and behavioral patterns in trading is what cements these patterns and frees you up to focus on markets. Relapse is overcome through repetition.

Only when new behaviors have been repeated many times, in many contexts, do they begin to become automatic, overcoming the tendency to relapse.

We saw in the lesson dealing with goal setting that, as your own trading coach, you don't want to set a goal one day, another goal the next day, and still a different one later that week. This is a common flaw with many trading journals. Traders take steps to initiate change—phases three and four above—but fail to cement the changes through repeated experience. It is far better to focus on one or two changes and institute those with regularity over a period of weeks and months than to try to make many changes in a short period of time.

Your assignment, following the discussion of rules from the previous chapter, is to review your current trading goals and assess *how well you sustain work on them day after day*. Ideally your goals—and the changes you attempt to make—should be expressed in such a manner that you will necessarily have to work on them each and every trading day. One way to foster this consistency is to generate a daily report card, in which you grade yourself on your enactment of the behaviors you try to cultivate. The goal is to achieve good grades each day, not just hit your targets on a particular occasion.

Similarly, if you are writing about your goal performance each day, the mere act of thinking about your new behaviors and evaluating them will serve as a kind of repetition. You're much more likely to stick with new behaviors if they command top-of-the mind awareness. Talk about the changes you're making, write about them, grade yourself daily on them and—most of all—enact them during each day's trade. As with the driving, before too long you'll find yourself doing the right things automatically. At that point, you don't need motivation; you've turned goals into habits.

COACHING CUE

Engage in an important goal-oriented pattern as your first activity of the day to build momentum for a purposeful day. I've worked with traders who stuck with their trading goals much better after they began programs of physical fitness. Their fitness work forced them to be goal-oriented to start their day, which carried over into their trading. You're not just training yourself to trade better; you're training yourself to sustain change efforts across all facets of life.

LESSON 39: CREATE A SAFE ENVIRONMENT FOR CHANGE

In the last lesson, we took a look at the importance of repetition in cementing new patterns of thinking, feeling, and behaving. The single most common reason why skilled traders fail to coach themselves to higher levels of performance is that they *initiate* changes, but fail to *sustain* them. As soon as they make improvements, they relax their efforts and fall back into old ways. A successful coach knows when the opponent is on the ropes and doesn't let up. When you have a positive experience, you want it to be a motivation for further positive experiences, not a cue for complacence. The best coaching efforts develop a kind of momentum in that way, adding success to success and sustaining a sense of mastery and accomplishment.

The problem with experience is that it takes time. Particularly if you're a longer-time-frame trader, many weeks or months of trading may pass before you have the opportunity to build a large base of new experience. If only you could multiply your experience, you could accelerate your learning curve. Changes that would otherwise take months could be accomplished in a few weeks.

The way that coaches in sports and the performing arts multiply experience is through repeated practice. A team might only play opponents on the weekend, but will practice every day to prepare for the games. Similarly, actors and actresses will rehearse their lines every day before opening the curtains for the actual production. During those practice sessions, performers condense the coaching process: they learn what they're doing right and wrong, make conscious efforts to repeat their positive performances and correct their faulty ones, and eventually reach the point where their efforts become natural and automatic. Practice is valuable, because it creates a safe environment for making mistakes. The game won't be on the line or the play won't be ruined if a performer tries something new and it falls flat in practice.

Rehearsal speeds the learning curve.

Practice can be very helpful to your efforts to coach yourself. If you identify a specific change to make, the place to begin is in simulated trading where no capital is at risk. This can occur in several different ways:

- **Simple Chart Review**—Sometimes the changes you make to your trading involve decisions regarding how you would enter, exit, or manage risk. When those are your goals, you can review charts and simply talk aloud the decisions you'd be making at each juncture. This lacks the realism of real-time trading (and cannot substitute for real-life experience), but it does slow the decision-making process down to the point where you can try new things in a very conscious, reflective manner. My favorite way of engaging in the chart review is to advance the chart on my screen one bar at a time and then talk aloud my perceptions and decisions. This is like first learning to drive a car by driving very slowly in a large, empty parking lot. It gives the learner plenty of time to crawl before walking and running.
- **Simulated Trading**—My charting software comes with a simulation feature in which I can place orders and track my profits and losses over time. This is helpful because you're making decisions with real market data in real time, but placing no capital at risk. By trading in simulation mode, you can gain many days' worth of experience in a single day. You can also focus your attention on the most problematic and challenging market occasions, concentrating your skill rehearsal in contexts that most call for your new patterns.
- **Trading with Reduced Size**—Not all of the changes I seek in my own trading are revolutionary. Some are evolutionary tweaks. Recently I altered the criteria by which I set profit targets, allowing me to hold certain trades for a bit longer. I reduced my trading size in half when I traded with the new criteria, knowing that the extended holding times would, by themselves, be uncomfortable for me. Once I developed a comfort level with the lowered size—and made my mistakes with the

smaller risk exposure—I then gradually returned to my prior level of risk.

Learning is best started in very safe environments and only later tackled in riskier situations. If you violate safety and security, you create distractions that interfere with learning.

Note that if I were to make radical changes in my trading—say, switch from trading equity indexes to trading agricultural commodities—I would need an extensive period of time with chart review and simulation prior to putting any capital at risk. Those more substantial changes take longer to internalize; there is a more extensive learning curve. On average, we'll make more mistakes when we attempt large changes rather than small tweaks. When you are your own coach, you provide more security and safety for big change efforts; the smaller pattern shifts can proceed with live trading and reduced risk exposure.

One of the greatest mistakes traders make is to make a change once or twice and then jump immediately into larger risk-taking, giddy with the prospects of new returns from new habits. It is not unusual in my coaching and trading experience for trading results to get worse before they get better when tackling meaningful changes in trading practice. Just as you wouldn't learn how to use the car's brakes and gearshift in a couple lessons and then jump into highway driving, you don't want to greatly alter your decision-making process while running full risk. As the *Enhancing Trader Performance* book stresses, the worst psychological mistake you can make is to traumatize yourself. If you create large drawdowns in your account because you weren't prepared for your changes, the result will be damaging to both your trading performance and to your self-coaching. You want to structure the change process as much as possible to provide frequent successes and no emotionally damaging losses. This is how you sustain confidence and self-efficacy, even as you make your mistakes.

Many traders are too eager to trade. They crave excitement and profits and find it difficult to trade in observation, simulation, and reduced risk modes. This short-circuits the process of generating repetitions that cement new patterns. Once traders undergo losses while making changes, they become self-doubting and pull back from their change efforts. Instead of generating success and confidence, traders learn to fear change. An important key to coaching yourself is to turn yourself into a generator of concentrated experience by making maximum use of practice and feedback. We often seek change after periods of loss; it's human nature to want to jump back into markets and regain the lost capital. But the goal is to instill the right trading behaviors, not to make money back all at once. If you internalize the right patterns, the results will naturally follow.

Your assignment is to embed concentrated learning into your schedule by allocating time for daily rehearsal of new skills and patterns. *An excellent goal is to generate two day's worth of learning experience into every day by rehearsing new patterns outside of trading hours as well as during them.* Replay market days, either in video mode or through a simulation platform that has a replay feature, to accomplish this goal.

I know from the traffic statistics on my blog and others that traders spend less time gathering market information after the close of trading and especially during weekends and holidays. Traders use the hours outside of trading to get away from markets. No one argues with the need for and desirability of life balance, but a nine-to-five approach will work no better in trading than it would in running a business or building a career as an artist, scientist, or athlete. When you read about elite performers in any field, one fact stands out: they are not the clock punchers. They are absorbed in their interests and, as a result, learn far more than others. They develop new skills and competencies far more readily than their peers simply because they multiply experience.

> Many traders back away from the screen when they have trading problems, thereby reducing their experience. During the worst drawdowns, you want to minimize your trading and risk exposure, but maximize your work on markets.

When you create safe environments for changing yourself and your trading, you mimic the learning behaviors of the greats, from concert pianists to chess champions to Olympic athletes. When you start with practice that encourages errors and learning from them, development becomes a joy, not a burden. This is self-coaching at its finest.

COACHING CUE

The video recording of markets for later review is an excellent way to multiply experience. Replaying the market day allows you to watch patterns unfold again and again under different market conditions. Reviewing market moves that you missed sensitizes you to future occasions of opportunity.

LESSON 40: USE IMAGERY TO ADVANCE THE CHANGE PROCESS

The previous two lessons have emphasized the importance of repetition in cementing new ways of behaving and unlearning old patterns. By creating new opportunities to rehearse fresh skills, insights, and behavior patterns we accelerate their internalization, freeing our minds for the basic tasks of trading.

A huge advantage of the human brain in this context is the ability to generate experience virtually, through the use of imagination. If we vividly imagine a specific trading situation and visualize ourselves, step-by-step, enacting a new way of handling that situation, the mental rehearsal approaches the power of actual experience. You do not have to trade to rehearse many trading-related behaviors. By creating realistic situations in our minds and using imagination to summon our desired patterns, we can also cement these patterns.

A technique that makes use of this kind of visualization is the stress inoculation approach first described by Donald Meichenbaum. *By mentally summoning stressful market scenarios and imagining in detail how we want to respond to these, we inoculate ourselves against those stresses by priming our coping mechanisms.* This mental preparation can be applied to a range of situations, from ones that are psychologically challenging to those that require new trading methods.

Our coping is mobilized when we imagine anticipated stressful situations, preparing us for when those situations actually occur in trading.

The key to making effective use of imagery is ensuring that the imagery is vivid and evokes real feeling. Unlike a simple verbal repetition of a trading goal, imagery has the power to evoke the emotions associated with situations. This imagery turns the verbal recitation into a much closer approximation of trading experience. When we vividly imagine a trade going through our mental stop-loss level before we can execute an exit, we summon some of the fearful or frustrated feelings normally associated with unexpected loss. While we experience a mild version of the trading emotions (imagery can rarely fully duplicate actual experience), we can rehearse our best practices, keeping ourselves planful and disciplined in the face of stress. This mild exposure to the trading stress is like the body's exposure to a weak form of a virus: it inoculates because it is strong enough to arouse adaptive responses, but not so strong as to pose a major threat.

Few psychological techniques are as recognized and recommended as imagery, and few are executed as poorly. There are several facets of effective imagery exercises:

- **Specificity**—It's not good enough to imagine a stressful situation, such as losing money or missing an opportunity, in the abstract. The imagery should be very specific and guided, visualizing a specific market and market situation, specific price levels, and specific market action. It is the realism of the imagery that enables the exercises to serve as substitutes for actual experience.

- **Dynamism**—The imagery should be more like a detailed, realistic movie rather than a broad, static snapshot. If you read a newspaper description of a situation, your reaction isn't as strong as it would be if you were to see the same situation dramatized in a movie. The dynamic nature of imagery is essential to its realism, which in turn is essential to the inoculation process. Flat, unconvincing images won't arouse our coping, and they surely won't be effective approximations of real trading experience.
- **Elaboration**—If I had to identify the most common shortcoming in people's use of imagery as a change technique, it would be their tendency to cut the imagery work short. Longer, more elaborated exposures to imagined challenges are more effective than very brief exposures. Indeed, if the exposures are too brief, you may unwittingly reinforce the pattern of fleeing from stresses! The best practice is to imagine a situation from beginning to end in elaborate detail—*and then repeat the scenario until it no longer evokes emotion.* This practice not only reinforces coping, but mastery and success.

An interesting technique from the behavioral literature is *flooding*: prolonged exposure to imagined situations that are highly stressful. Traders learn to stay in control even during a flood of stressful imagery, so they prepare themselves for most anything the markets throw at them.

- **Variation**—Traders commonly imagine a challenge scenario, evoke the imagery, and then quickly move on to something else. As we've already seen, repetition cements learning. By failing to use the principle of repetition in the area of imagery, traders open themselves to the risks of relapse. Once you evoke a detailed, realistic trading scenario and how you would handle it and once you repeat the scene to the point of mastery, you then want to create variations on the scenario. For instance, you might begin by imagining a frustration associated with a fast-moving market. If your goal were to learn new coping patterns during periods of frustration, you would master this first scenario and then create variations, such as frustrations associated with slow markets or not getting filled on orders. Vary the scenarios and you can generalize your learning and make it increasingly applicable to actual trading conditions.
- **Consistency**—Any single imagery session will not affect behavior over days and weeks. *It is the daily repetition of the sessions that*

yield enduring results. Many traders derive some benefit from initial imagery work and then promptly return to business as usual. Consistency in the use of the exercises ensures that the new patterns you're rehearsing will remain top of the mind over time. Frequent use of imagery work is a great way to sustain mindfulness about change and the need for change.

When you are your own trading coach, you want to *see* and *feel* yourself to be successful, not just engage in occasional thoughts of success. In your internal world, you can practice skills, engage in new thought patterns, and achieve goals with consistency long before you actually accomplish all of those in live trading. Your assignment is to generate powerful, elaborate imagery scenarios of stressful, challenging market situations; the thoughts, feelings, and behavior patterns associated with those; and the specific steps you want to take to master those situations. When you construct your guided imagery, you want to feel the emotions of fear, greed, frustration, and boredom and imagine yourself tempted to engage in your usual, negative patterns in response to those states. In your imagined scenario, you will vividly envision yourself keeping those negative patterns in check and purposefully enacting your best practices. Thus, for instance, you might imagine yourself tempted to add to a position when it goes through your stop-loss level, but checking that temptation and instead acting on the stop.

Imagination for the trader is the equivalent of a practice field for an athlete: a place to prepare for performance by creating simulated performance situations.

You will be the trader you are capable of being in your imagery work and practice long before you consistently enact those ideals in your daily trading. There is no reason to take months or years to change behavior patterns when you can make every day an experience of concentrated learning. Much of successful self-coaching is the result of a creative and tenacious use of imagery and practice.

COACHING CUE

Use your imagery to imagine yourself as the kind of trader you aspire to be: the risk taker, the disciplined decision maker, the patient executioner of ideas, the canny trader who learns from losing trades. If you create a role and an image of yourself in that role, you enact scenarios that, over time, become part of you.

RESOURCES

The *Become Your Own Trading Coach* blog is the primary supplemental resource for this book. You can find links and additional posts on the topic of coaching processes at the home page on the blog for Chapter 4: http://becomeyourowntradingcoach.blogspot.com/2008/08/daily-trading-coach-chapter-four-links.html

Much of the framework discussed in this chapter comes from research into helping processes in brief therapy. Standard reference works in this area include the chapter on "Brief Therapy" written by Dewan, Steenbarger, and Greenberg for the volume *Textbook of Psychiatry* (*Fifth Edition, Volume 1*), edited by Robert E. Hales, Stuart C. Yudofsky, and Glen O. Gabbard (American Psychiatric Publishing, 2008) and the chapter on "Brief Psychotherapies" by the same authors for the reference work *Psychiatry* (*Third Edition*), edited by Allan Tasman, Jerald Kay, Jeffrey A. Lieberman, Michael B. First, and Mario Maj (Wiley, 2008).

An excellent framework for thinking about making the most of strengths is the Gallup research described in *Now, Discover Your Strengths*, written by Marcus Buckingham and Donald O. Clifton (The Free Press, 2001). See also the popular management text *Good to Great*, written by Jim Collins (Harper Business, 2001).

A full description of solution-focused work can be found in my chapter "Solution-Focused Brief Therapy: Doing What Works" in *The Art and Science of Brief Psychotherapies*, edited by Mantosh J. Dewan, Brett N. Steenbarger, and Roger P. Greenberg (American Psychiatric Publishing, 2004).

CHAPTER 5

Breaking Old Patterns

Psychodynamic Frameworks for Self-Coaching

The strangest and most fantastic fact about negative emotions is that people actually worship them.

—P.D. Ouspensky

To this point, we've been exploring processes that are common to all change efforts and all performance coaching. Now we begin a look at individual frameworks for coaching, beginning with psychodynamic modalities. These frameworks are approaches that emphasize the continuity of past and present—how old patterns replay themselves in the present—and the ways in which current relationships can remake patterns born of prior, negative relationship experiences.

The psychodynamic framework is more commonly connected to long-term psychoanalysis than to coaching, but I've found tremendous value in utilizing the medium of relationship experiences to help traders with their performance. This chapter will explain how psychodynamics are related to trading and how the dynamic framework can inform your own efforts at self-coaching.

I personally draw most upon dynamic modalities when I'm working on trading problems that are part of larger life challenges—particularly when those challenges have been part of our life histories prior to trading. Many issues that impact trading—concerns over success/failure, self-confidence/self-doubt, security/insecurity—predate our trading efforts, but play themselves out in how we make financial decisions.

The best way to describe psychodynamics is as a way of thinking, as well as a set of tools, to help ensure that people's pasts don't become their futures. And, in the context of this book, it all starts with a single insight:

we have relationships with our markets, and those relationships repeat patterns that we've enacted in other relationships. Let's take a look at how we break free of those patterns . . .

LESSON 41: PSYCHODYNAMICS: ESCAPE THE GRAVITY OF PAST RELATIONSHIPS

The psychodynamic approaches to change have evolved significantly from the psychoanalytic work of Sigmund Freud and the subsequent contributions of "neo-Freudians." In the process of this evolution, psychodynamic thought has turned away from viewing instincts and sexuality as the primary basis for behavior and toward an interpersonal understanding of recurring conflicts and patterns. At the same time, psychodynamic psychology has also departed from the analytic couch and the presumption of therapy as a long-term process, emphasizing more active, shorter-term methods of change. Surprisingly, this evolution has largely taken place outside the public eye, contributing to a perception that analytic methods are outmoded and outdated. Nothing could be further from the truth.

The core perspective of the psychodynamic model is that current problems are reenactments of conflicts and patterns from past relationships. The self is viewed as the result of years of internalization of relationship experiences. When we enjoy positive, affirming relationships, we internalize a positive self-image and concept. Negative, conflicted relationships are internalized as a negative and conflicted sense of self. This perspective touches on one of the themes from earlier in this book: relationships serve as mirrors by which we experience ourselves. We are the sum of our significant relationship experiences.

Over the lifespan, we learn coping strategies (defenses) for dealing with the anxieties brought by relationship conflicts and their consequences. These defenses may be successful in warding off discomfort, but they tend to become outmoded in future relationships and life situations. When current situations trigger the feelings evoked by past relationship problems, our outmoded defenses lead us to behave in ways that bring unwanted outcomes. These patterns of feelings evoked from the past along with the defensive reactions and negative consequences constitute what Lester Luborsky calls core conflictual relationship themes. These themes repeat themselves in various ways, across various situations, causing us to behave in unwanted ways. The purpose of counseling and therapy, from this vantage point, is to free us from these cyclical maladaptive patterns.

When we "overreact" to situations, the odds are good that we're reacting to themes from our past as well as the current situations that trigger those themes.

To use an example from *The Psychology of Trading*, I grew up in a very close family. My parents were both estranged from their parents and vowed to create a very different family environment for their children. They were so successful in that regard that I sometimes felt that the environment was *too* close. I sought privacy, taking long bicycle rides, walks, and even showers. Later, when I met my wife Margie and became part of a household with her three children, I found myself taking long showers and withdrawing from family life. This disrupted morning routines and led to a bit of tension in the household.

For years I had lived alone and had never experienced the feeling of being too close to others. When I entered a new family situation, the old feelings from my formative experiences came back to me and I coped rigidly in my old way: by withdrawing and seeking privacy. What worked in childhood, however, failed dismally in the new family. By reenacting old defenses in new situations, I created a fresh set of problems.

All of these defenses tend to happen automatically, outside of conscious awareness. Without some way of making these repetitive patterns conscious, we cannot change them. One goal of psychodynamic work is to make the unconscious conscious: to make us self-aware, so that we can find new endings to old conflicts. I was able to change my patterns within the family once I recognized that I no longer needed to cope in ways from my past: my present family wasn't my former one, and I was no longer a child.

The first goal of psychodynamic work is insight: recognition and understanding of one's patterns and awareness of their limitations. The key insight is simply: You had a reason for doing this in the past; you don't have to do it any more.

So how does this relate to trading? *The risk and uncertainty of trading—the gains and losses, victories and defeats—have an uncanny way of triggering feelings from our past.* A classic example is the trader who felt as though he never measured up to his parents' standards and expectations. He brings his sense of inferiority to the markets, taking imprudent risks to prove that he truly is a success. Of course, no amount of winning in markets can fill the emotional void, leading him to take ever-greater risks, until he finally blows up—and confirms his worst fears. In such a situation, no tweaking of trading methods will solve the trader's problem. Until he resolves the conflict at the heart of his dilemma, he will continually find himself acting out his personal dramas in his trading.

Many trading problems are the result of acting out personal dramas in markets.

The best way to identify situations in which past conflicts intrude into current trading is to consult the feelings evoked by your trading problems. If they are similar to feelings you've experienced in past career and relationship situations, there's a good chance that they represent the leading edge of cyclical maladaptive patterns. For instance, if your frustration gets you into trouble in your trading and has also gotten you into trouble with friends and romantic partners, there's a clear pattern that transcends markets. If trading leaves you with the same conflicted, hurt feelings that you've experienced in your past, that's a sign that the past is refusing to stay in the past.

As your own trading coach, you sometimes need to dig beneath the surface of problems to discover their origins. This discovery must be accomplished by reviewing your personal history and mapping that history against recent experience. *In other words, you want to look not only at your current patterns, but past ones as well.* It's in the overlap that we can make greatest use of psychodynamic change methods.

Your assignment is to take a sheet of paper and draw two sets of sine waves, each with at least four peaks and four troughs. For the first set of sine waves, you'll mark the peaks with the "peak experiences" from your life: your most positive and fulfilling experiences. These experiences can be taken from any life sphere, from relationships to career. You'll label the troughs with the most negative experiences of your life, again from any sphere. These experiences will be those filled with the most emotional pain and distress. By the time you're finished, the first sine wave chart should be filled with the highlights and lowlights of your life.

When you're finished with the first set of sine waves, you'll fill out the second set similarly, *only you'll limit your entries to trading-related incidents.* Thus, you'll mark the peaks of the sine waves with your most positive and fulfilling trading experiences and the valleys of the sine waves with your most painful and upsetting market-related experiences. Make sure that all of your entries from both sets of sine waves are described in sufficient detail that you can readily appreciate *why* each incident was a peak or valley experience.

The real work comes in when you compare the peaks across the two sets of waves and the valleys. *You're looking for common themes that link your life experience with your trading experience.* Many of these commonalities will be at an emotional level. Here are some common ones to look out for:

- Themes of adequacy and inadequacy.
- Themes of rebellion against rules and discipline.
- Themes of boredom and risk taking.

- Themes of achievement/hopefulness and failure/discouragement.
- Themes of recognition and rejection.
- Themes of contentment/acceptance and anger/frustration.
- Themes of safety and danger.

If the markets are making you feel the way you've felt during some of your life's valley experiences, consider the possibility that you've been caught in a web of repetitive conflict and coping. Recognizing that web and keeping it firmly in mind is half the battle of changing it. As your own trading coach, you want to be mindful in your trading, not unconsciously repeating your past. It is much more difficult to fall into old, destructive patterns when you're clear on what those patterns are.

COACHING CUE

So often, we're fighting the last psychological battle. Identify the most recent conflicted relationship experience in your life and the thoughts, feelings, and behaviors evoked by that relationship. Then take a look at your most recent trading difficulties to see if similar thoughts, feelings, and behaviors are involved. Many times, it's not just early childhood relationships that color our current behavior, but the most recent sets of conflicts. Identifying how you resolved those problems in the relationship (or are taking steps to resolve them if they are current) can be very helpful in minimizing their spillover into trading.

LESSON 42: CRYSTALLIZE OUR REPETITIVE PATTERNS

In the previous lesson, we used sine wave charts to identify peak and valley experiences from our life histories and from our trading. Once we have the sense that there are, indeed, commonalities, the next step is to crystallize these patterns so that we clearly understand why they exist and how they repeat themselves.

First, however, let's address the question of what it means if you cannot find common themes across the peaks and valleys of your charted experience. One possibility is that you need to draw more peaks and valleys on each chart before the themes will become salient. Another is that you are looking at the events logically, not psychologically. Examine the peaks and valleys for similarities of motivation and emotion, not surface detail. Many times the common elements will jump out when we think of the patterns as *emotional recurrences.*

Sometimes, though, there truly *aren't* links between our trading problems and our prior life's challenges. The trading problems may simply reflect shifting markets, a lack of skills and experience, or situational factors related to present-day distractions. If these are the most important sources of current trading difficulties, the psychodynamic approach will be less relevant to your change efforts. Rather, you may need to structure your learning process, as described in the *Enhancing Trader Performance* book, or you might need to adopt a framework that is more present-centered, such as the self-coaching approaches described in chapters 6 and 7.

> Not all trading problems have an emotional genesis. Sometimes we just need to hone our trading skills.

In my experience, the most fertile area to seek for similarities between trading problems and prior life conflicts is in significant relationships: those with parents and romantic partners. *Conflicts with parents and lovers are usually among the most emotionally powerful, and are the ones we are most likely to defend against—and thus reenact—in our trading.* Once we crystallize those patterns, we can become quite adept at recognizing their reappearance. That is the first step toward short-circuiting them and providing them with a new ending.

One trader I worked with had numerous relationships with girlfriends over the years, but he never committed to any of them. Indeed, he often kept one relationship on the side in case his primary one ever fell through. He had lost a sibling when he was young; his parents, devastated by the loss, tried to forge on by not dwelling on the tragedy. Our trader learned to defend against loss by never getting too involved with another person. This defense kept him from reexperiencing the pain of the childhood event, but it also kept him from having a fulfilling emotional life.

So what do you think his trading problem was? Although he described himself as being passionately interested in trading, he in fact spent surprisingly little time at it. He undertraded markets—in other words ignored trading signals, traded with almost ludicrously small size—and he was easily distracted by communications in chat rooms and reading web sites. While he justified the latter as "preparation" for his trading, he never actually got around to trading seriously. Just as he avoided commitment and loss in relationships, he dabbled at the edges of markets, never achieving anything close to his potential.

To crystallize this pattern, a good place to begin is with *our underlying need.* The trader in the example above has an overwhelming need for safety and consequently takes the path in relationships and trading that seems safest. The feeling that he is guarding against is the pain of loss: the vulnerability of investing oneself and losing that emotional investment. The trader was close to his sibling; he never truly recovered from the fallout of the loss. The way that he defends against that feeling is by keeping commitments at bay. He avoids becoming too involved with his relationships and trading so that he can never experience a loss as painful as that from his childhood. This leaves him with a host of negative consequences. Most of all, he feels empty, not truly fulfilled in love or career.

Most repetitive patterns can be broken down into this schematic of:

- **Need**—What we are missing, what we crave.
- **Feeling State**—Distress associated with not having that need met.
- **Defense**—What we do to cope and avoid the painful feeling state.
- **Repetitions**—How we replay defenses in current situations.
- **Consequences**—The negative outcomes from our current defensive efforts.

One trader I met with had an intense need for emotional support. His parents had divorced at an early age and his mother quickly remarried, forcing him to cope with joining a blended family. He frequently felt abandoned by his mother, and he felt not good enough to merit his (distant) father's involvement. His defense against these feelings was to take the role of overachiever, pushing himself to get the best grades at school, excel at athletics, and star in extracurricular activities. By doing more, he hoped to feel more worthy. As a result, however, he frequently drove himself to exhaustion and periods of hopelessness, as no amount of achievement could substitute for the parental love he missed. Unlike the trader from the prior example, he worked himself to the bone in trading, at times breaking down when his overachieving tendencies failed to bring results.

So this, as your own trading coach, is your psychodynamic challenge: *figure out your core need.* Most often, this will be something that you wanted in past relationships and could not consistently obtain. It might be autonomy, love, respect, or support. Once you identify that need, you want to review your valley experiences from your sine wave charts and clearly identify the feeling states that have been associated with the frustration of your core need. Maybe that feeling is depression and sadness, maybe it's anxiety, and maybe it's anger and frustration. This feeling, quite likely, will be the one that is repeated in your most difficult trading experiences.

> The needs that are most unmet in our personal lives are the ones most likely to sabotage our trading.

Now reflect on how you try to make that feeling go away. That is your defensive pattern, your way of coping with the pain of needs going unmet. Most likely, this coping is what most immediately brings you difficulties in your trading, causing you to overtrade, stand aside during times of opportunity, etc. *At these times, you're making trading decisions, not to manage your capital, but to manage the distress from those core conflicts.* Our worst trading occurs when we're managing our feelings rather than our positions.

Your assignment is to draw upon your sine wave charts from the previous lesson to map your trading problems as sequences of needs, feelings, coping/defenses, and consequences. You should be able to draw an accurate flow chart that shows how feelings associated with frustrated needs lead you to take actions that bring present-day problems. This chart will capture the focus of your psychodynamic self-coaching efforts. By breaking this pattern and inserting new elements, we can become more intentional in our trading, more self-determined in our results.

COACHING CUE

Most traders, I find, don't require long-term therapy. If traders are significantly troubled, they cannot sustain a trading career. Rather, they cope and function well, but periodically lapse into old patterns that interfere with their effectiveness. Many times, traders can identify their patterns by identifying their most frequent and costly departures from their trading plans and then noting the feelings and situations that accompanied these departures. By noting feelings from recent trading problems and then observing when you've felt those feelings in other areas of life, you can crystallize patterns that are most likely to impact future trading.

LESSON 43: CHALLENGE OUR DEFENSES

A cardinal idea within psychodynamic work is that the problems that motivate people to make change efforts are rarely the core conflicts that they are repeating across situations. Rather, it's their *defending* against the pain of those conflicts that brings the unfortunate consequences and the recognition of the need for change. Our problems, from this vantage point, are the result of our rigid, outmoded coping. What worked for us at one time of life now works against us.

A good example from the markets is "revenge trading." This occurs when we trade more aggressively following losses in an attempt to get our money back all at once. Pain and frustration from a loss lead to an angry defensive response, an effort to get rid of the hurt. At that point, the trading is not about opportunity; it's a defense to keep the feelings of disappointment and loss at bay. Of course, this often leads to further losses and a fresh set of consequences.

The defense is there because, deep down, the trader doesn't believe that she can tolerate the hurt or sadness of loss. Perhaps at one time of life such pain was indeed intolerable. Now, as a mature adult, the trader can handle normal losses, whether in business, markets, or love. It doesn't *feel* that way, however, and the trader continues to handle losses in the old, childhood ways. By crystallizing the pattern, the trader builds a self-observing capacity and a clearer awareness of the consequences of perpetuating old cycles. It is this heightened awareness that eventually enables the trader to challenge defenses and respond to old threats in new ways.

> Change in the psychodynamic mode means doing what doesn't come naturally: refrain from the old ways of coping that keep unpleasant feelings at bay.

In your role as your own trading coach, you interrupt cyclical problems by becoming their observer, not the one caught inside the patterns. *By observing, you stand outside the cycles; they no longer consume you.* The sine wave and flow charts described earlier are useful tools in sustaining this self-awareness. Your goal is to recognize repetitive patterns *before* they have an opportunity to play themselves out to their unhappy conclusions.

One way of accomplishing this will form the basis for your next coaching task. You'll typically observe a pattern beginning to emerge when an associated feeling state interrupts your trading. In the revenge-trading example above, that characteristic feeling might be frustration and tension. In other situations, the feeling may be one of loss or emptiness. As soon as you notice the characteristic feeling, you want to acknowledge it out loud, almost as if you are a play-by-play sports announcer. For instance, you might say aloud, "I just took a loss and now I'm feeling really frustrated. I'm feeling mad, and I want to get my money back. I want to find another trade, but that's what's hurt me in the past. I jump back into the market and it makes things worse."

If you cannot invoke this self-observation aloud—perhaps because you're trading in a room with others—a psychological journal can accomplish the same function. The key is to describe in detail what you are thinking and feeling, what you are tempted to do to make the thoughts and feelings go away, and how this course of action has hurt you in the past. "I'm disgusted with myself and I want to just quit trading for the day," would be one self-observation from a trader who rides a cycle of aggressive trading, losing money, guilt, withdrawal from markets, and renewed attempts at aggressive trading to make up for lost time and opportunity. You're taking the role of psychologist, identifying what is going on instead of identifying with it.

> When you describe a pattern of behavior, you're no longer identified with it.

At this point, we're not concerned with changing the pattern. Rather, we're becoming more alert to its appearance and more aware of its manifestations and consequences. This can take days and even weeks of consistent effort as you sustain the stance of the self-observer. With this effort, it's inevitable that, at times, you'll interrupt the pattern entirely: you'll avoid the revenge trade or the overaggressive trade. Before you do the right things, you'll stop doing the wrong ones. This builds a sense of mastery and self-control: *you're starting to control the patterns instead of having them control you.*

Talking aloud and writing in journals, by themselves, will not rid you of overlearned behavior patterns, but they do provide you with options. Just by doing something different—even in small degree, such as placing much smaller revenge trades—you achieve a measure of control.

One type of talking aloud to gain self-awareness is to actively remind yourself that what is going on in the present is not really the problem. The problem is what happened and hurt you in the past, not what you are reacting (overreacting) to in the present. For instance, you might be losing money on a position prior to hitting your stop-out level. You catch yourself feeling fear and you're tempted to bail out of the trade prematurely. In addition to talking this fear and temptation aloud, you would remind yourself, "The loss on this particular position is not the real issue. I'm reacting to my losses from last year (or the loss of my relationship). Abandoning my trade idea isn't going to take away those old losses."

(Notice, of course, how this example presumes that you've allocated a responsible size to your position and a reasonable stop-loss level. If not, it could be your poor trading and not any problems from your past that would be triggering—quite reasonably!—your fear and desire to exit.)

What you're really emphasizing in this talking aloud is the message of, "The problem isn't my trading; it's something else." That frees you up to deal with the "something else" and not act out the past in the present. *It also frees you to simply deal with your trade as a trade and not as something more emotionally laden.* You'll know that you've made significant progress when you consistently catch your patterns as they unfold, maintaining the role of observer rather than passive participant.

COACHING CUE

I've mentioned in this book and in *Enhancing Trader Performance* how videotaping markets can sensitize traders to patterns of supply and demand, aiding them in trading decisions. I've had traders turn the tables and turn the videorecorder on themselves, so that they're recording themselves as they trade. It's a great tool for self-observation and recognizing your emotional and behavioral patterns. After you've seen a few of your recurring cycles on tape, you become more sensitive to their appearance during real-time trading.

LESSON 44: ONCE AGAIN, WITH FEELING: GET DISTANCE FROM YOUR PROBLEM PATTERNS

Two crucial transitions that occur in psychodynamic work are viewing your past patterns as alien to you and experiencing them as your own, personal obstacles. *Once you've become skilled at recognizing your old ways and their disruption of your trading, you next step is to distance yourself from them.* You will not gravitate toward ways of thinking, feeling, and acting if these patterns feel alien to you and if you view them as sources of pain.

A term commonly used in psychodynamic writing is *ego alien.* What that means is that some way of viewing the world, experiencing it, or acting within it has become foreign to one's sense of self. When a person starts to repeat a pattern that has become ego alien, the resulting thoughts are:

"This is not the real me."

"This is what I used to do; not what I want to be doing."

"I'm reacting to the past, not to what's happening now."

Of course, no one consciously owns destructive behavior patterns. It is easy, however, to repeat them unthinkingly. When we make those patterns ego-alien, we're not just thinking about them, *we're actively rejecting them.*

It's helpful in this regard to start thinking about the old me and the new me. The old me was afraid of failure and equated falling short with losing the love and approval of others. The new me realizes that the outcome of any single trade will not make me a better or worse human being. The old me became angry and frustrated when I couldn't get my way, because I hated feeling out of control. The new me controls my trading through my planning and lets the market do what it will do. Notice how keeping a clear distinction between the old and new helps traders identify with positive, constructive patterns and maintain distance from old, automatic ones.

"Am I reacting to markets or to my feelings from the past?"—This is the question that follows from self-observation.

Another way to frame this distinction is to focus on "here's what I do to make money; here's what I do to lose money." Whenever you entertain a certain thought or course of action, ask yourself, "Is this how I think/act when I make money, or is it how I think/act when I lose money?" Once again, by focusing on the likely outcomes of what you're doing, you become the observer of your patterns and interrupt the automatic replaying of those patterns.

As we saw earlier in discussing the costs of patterns, focusing on the consequences of destructive patterns from the past not only keeps those patterns alien, but also heightens your motivation to not repeat them. If you recall a recent incident of losing money or opportunity as the result of falling into old habits, it is a great way to avoid making the same mistake again, particularly if that recollection highlights the monetary cost and emotional pain of the incident. In *The Psychology of Trading*, I emphasize the importance of treating old, negative patterns as an enemy. If you view something as an enemy, it's difficult to embrace it and it becomes easier to sustain efforts at changing it.

Most people are uncomfortable with the emotion of hate. We've been taught that it's not right to hate; that we should be nice to others. Hate, however, has its uses. Hate implies total rejection—a complete pushing away from the self. When we hate our past patterns, we are so attuned to their destructive consequences that we will move heaven and earth to not repeat them. The addict who has seen the destructive consequences of substance abuse learns to hate drugs; the person who has gone through a painful marriage and divorce to a narcissistic partner is so disgusted with the experience that she'll never make that same mistake again. There is no question in my mind that I would not have found my life partner had I not had prior, unsatisfying relationships. I so hated the way I felt in bad relationships that I was absolutely determined to find something better.

When we hate how our old patterns have hurt us, we find a source of positive motivation: the drive for a better life.

There's an old saw in Alcoholics Anonymous that alcohol abusers have to hit bottom before they sustain efforts at sobriety. Before serious consequences have accumulated, it's all too easy to deny and minimize problems, putting them out of mind. When you hit bottom the consequences have gotten to the point where they cannot be ignored; they have inflicted too much pain and damage. At that point, people give up entirely, or they reach the point of hating what their habit has done to their lives. "I can't go back there again," is the feeling of many people who have reached that point of hate and disgust. At that point, the old ways are held at arm's length; there is no identification with them.

A great application of this idea (and worthwhile homework exercise) is to focus on one enemy that you want to conquer over the next month of trading. Your choice should be a pattern of thinking, feeling, and acting that has noticeably hurt your trading in the past several months, and it should be a pattern that you've observed in others areas of your life. This is especially powerful if, with your sine wave charts, you've observed damaging consequences of the pattern not only in trading but also in those other life areas. You can then call to mind all the pain that this pattern has caused you over the years and the prices you've paid for repeating the pattern. You declare that pattern your enemy and then you purposefully look for opportunities to confront this enemy in each day's trading.

Notice that this is different from simply observing a pattern as it's happening. Instead, you're actively *anticipating* the pattern and even looking forward to its appearance so that you have the opportunity to reject it. The trader's frame of mind becomes, "I'm not the problem; it's this old pattern that's the problem." Once you frame outmoded ways of thinking and acting as leftovers from the past that no longer work for you, you've taken a giant step toward freeing yourself from those patterns.

It will take sustained effort to tackle your enemy, but it can be tremendously fun and empowering as well. Each time you notice your old ways of losing money and refuse to engage in them, you've won a victory against your enemy—and struck a blow for your own sense of confidence and

COACHING CUE

One of the first enemies to tackle in self-coaching is procrastination. We procrastinate when we know we need to change, but cannot summon and sustain a sense of urgency. In that situation, procrastination itself becomes the pattern we must battle, as it robs us of the power to change our lives. Often, procrastination is itself a defense—a way of avoiding anxieties associated with anticipated changes. By starting with small, nonthreatening changes that we undertake every single day, we do battle with procrastination and build a sense of control and mastery over change processes. Big goals and radical changes often meet with procrastination. If you focus on intermediate goals that can be pursued regularly, you rob change of its threat value.

mastery. *If you are highly competitive and achievement-oriented, ask yourself if you want to win your freedom of will or lose it to the repetition of your past.* The competitive soul does not want to lose and will fight an enemy to the death. A powerful aid to change is to use your competitive drive to defeat the enemies of your happiness: your internal demons.

LESSON 45: MAKE THE MOST OUT OF YOUR COACHING RELATIONSHIP

One of the cardinal principles of psychodynamic work is that change often occurs in the context of relationships. When you are your own trading coach, one of your challenges is to surround yourself with the right kind of relationships: ones that support your goals and mirror to you the person and trader you're capable of being.

Consider two coaching scenarios in which I'm working with a trader at a trading firm. In both situations, the trader has lost more money than planned by exceeding position and loss limits. An entire week's profit is wiped out by the single day's loss. In the first scenario, I chide the trader:

> *"How could you be so careless? You just ruined your week. Do you realize what will happen if you continue to do this? This is not a good time to be on the street looking for a firm hiring new traders."*

In the second scenario, I adopt a different tone:

> *"C'mon, you're a better trader than that! Remember last month when you put together a string of winning days? You never needed to put on big size to do that. Let's see if we can get back to that great trading."*

The first interaction mirrors a sense of failure; "I'm so disappointed in you!" is the tone of the message. The approach also focuses on the negative, emphasizing the worst-case scenario.

The second scenario doesn't avoid the problem, but starts from the premise of good trading. It mirrors encouragement and a reminder of the trader's strengths.

Imagine that these interactions occur day after day, week after week. It's not difficult to see how the first approach would undercut a trader's security and confidence, leading to further bad trading. The second interaction would likely help a trader get back to trading well. It would coach success, not just the avoidance of failure.

All traders inevitably coach themselves: they always talk to themselves about performance, take action to improve performance, and keep track of results. The only question is the degree to which this self-coaching is purposeful, guided, and constructive. From a psychodynamic perspective, your self-coaching is no different from a relationship with a professional coach: the dynamics of that relationship will serve as a mirror, and you'll tend to internalize what is reflected.

> Our self-talk *is* our self-coaching.

What this means is that *how* you focus on your problem patterns from the past is crucial to the success of your self-coaching. You *are* going to fall back into old patterns sometimes; you *are* going to miss opportunities to enact new, positive patterns. There will be occasions in which you work hard to avoid one pattern, only to fall into another one. All of these situations can be discouraging and frustrating. Yet, as your own coach, you are tasked with the responsibility of maintaining a constructive relationship with you, your student.

"Comfort the afflicted and afflict the comfortable," is the way I described my approach to working with people in *The Psychology of Trading*. It's not a bad formulation for coaching oneself. When you're afflicted—suffering, hurting, losing—you want to be your own best support. When you're winning, you want to be afflicting yourself by doubling down on your discipline, alert to any overconfidence that might allow old habits to reenter your trade.

A great exercise is to cull through your trading journal and examine the emotional tone of your writings. Do they sound like positive coaching communications, or are they negative, frustrated, and blaming? Do they place equal emphasis on your progress and achievements, or do they harp on what you didn't do right?

The worst thing you can do as a trading coach is reenact old personal patterns by adopting the voice of a past figure who was part of the destructive cyclical conflicts that you're trying to move beyond. If you had a parent who was hostile and critical, one who couldn't be pleased; if you had a spouse who could not acknowledge your achievements or a resentful sibling, you don't want to replay their voices in your own self-coaching. How you treat yourself may well be part of the very pattern you're trying to change: working on your coaching voice is a great way to move that work forward.

> Perfectionism is often a hostile, rejecting set of self-communications that masquerades as a drive for achievement.

One trader I work with actually talks to himself in the third person when he is in his self-coaching mode, reviewing his goals and performance in tape recordings that he then replays during breaks in the day. He might say, "Bill, today you need to be alert for your temptation to overtrade this market. That got you into trouble last week. We're coming up to a Fed announcement and it's unlikely that the market is going to move much until that's out of the way. Let's make sure you do it right this week!"

Day after day, listening to messages such as this, turns intentional coaching talk into internalized self-talk. After a long period of saying the right things, you'll start to feel them and repeat them to yourself automatically. The coaching role, at that point, has truly become part of you. Conversely, if your self-coaching voice is one of frustration, you are cultivating a relationship with yourself that can only rob you of motivation and confidence over time. Many traders think they are pushing themselves to succeed when, in fact, all they are doing is replaying critical, hostile voices from past relationships.

One trader I worked with would make significant money with many more winning trades than losing ones, but you would never know it from the way he talked. He always focused on the losing trades, the trades that he could have left on longer for greater profits, and the trades he could have exited sooner. The basic message of his talk was that anything he did was not good enough. When it came time for him to increase his risk and pursue larger profits, he hit the wall and could not make the change. Days, weeks, and months of telling himself that his trading was not good enough undercut his ability to be confident trading large size. He thought he was coaching himself for success by refusing to be content with his gains, but in reality he was wounding his self-esteem.

A far different approach to self-coaching was exemplified by the trader who set challenging, but reachable, performance goals and then promised himself (and his wife) a long-awaited vacation abroad if he reached those goals. When he recruited his spouse into his efforts to improve his trading and chose an incentive that meant a great deal to both of them, he was able to sustain a positive motivation. This made their trip especially rewarding, as it was a tangible reminder of his success. In this situation, he cultivated a relationship with himself that fed self-esteem and mastery.

The central message of the psychodynamic approach is that we are the sum of our significant relationships. None of our relationships is quite so central as the one that we have with ourselves. *In coaching ourselves, we take control over our relationship to our self*; the voice that we use as coach will be the voice that enters our head when we tackle the risk and uncertainty of markets.

COACHING CUE

Coaches typically address their teams before a game to emphasize important lessons and build motivation. Consider addressing yourself before the start of market days, stressing your plans and goals for the session. Tape record your address and then review it midday. Pay particular attention to how you speak to yourself: it's much harder to lapse into negativity when you take the time to make your self-talk explicit and then approach that self-talk from the perspective of a listener.

LESSON 46: FIND POSITIVE TRADING RELATIONSHIPS

Why do people hire a personal trainer to help them get into shape when they already know which exercises they should be doing? Why does a hedge fund hire a coach to work with experienced portfolio managers, when those managers know far more about the business than the coach? Why do elite athletes who have more skills than anyone they could possibly hire still rely on performance coaches?

An understanding of the psychodynamics of personal change makes the answers to these questions perfectly clear. In each case, hiring a coach or trainer takes an individual development process and turns it into an interpersonal process. This is one of the most powerful steps anyone can take toward accelerating their learning curves, but it is poorly understood.

When you pursue a goal with other people, you add a new source of motivation to your efforts. We know from the research in psychology that the most important ingredient in counseling and therapy outcomes is the quality of the relationship between the client and the helper. This makes sense: when the helper is valued, the client wants to not only make changes for himself, but also for the counselor. You don't want to let down someone you value and who's working for you. If you decide to go to the gym every other day to work out and get into shape, it's easy to skip a day here or there. But if you make that commitment with a close friend, you won't want to disappoint that person. You're more likely to stick to your plans.

> Making a commitment to change to others adds a layer of motivation and helps the other person motivate you.

Bring another person into change efforts and you will introduce a new source of mirroring. A good athletic trainer will provide you with feedback about how you're doing and will keep your motivation up, even as you seem to plateau in your efforts. One of the helpers at a diet center will help you track your weight loss, providing you with positive feedback when you're working the plan. Day after day, exposure to this encouragement makes it easier for you to generate your own encouraging self-talk. You become your own coach, in part, by internalizing the role of an actual helper.

To achieve this benefit, it is not necessary that your mentor be professionally trained or that your relationship with them be a commercial one. Alcoholics Anonymous is a great example of a peer organization in which experienced members aid newcomers. The group meetings provide a supportive environment for change; the slogans and readings provide a shared set of beliefs and commitments; and the relationship with a sponsor provides the motivation of working with someone who cares about your life. *The net effect is to mirror a new identity for the participant*: an identity as a recovering person, not simply that of a failed addict.

As a trading coach, I work hard to take money out of the equation with the traders and portfolio managers I work with. I routinely receive phone calls and e-mail from traders who update me on their progress. I wouldn't think of billing for that. That's not because I'm an altruist; it's because I want to emphasize that this isn't simply about the money. I want no impediments to a trader calling me, and I want the trader calling me because I care about what happens to them, not because I want to generate a billing. For me, it's about the relationship and doing everything I can to aid the trader's happiness and success, and my hope is that it becomes that way for the traders as well. Frequently, my motivation to see the trader succeed carries him through the rocky periods of self-doubt. It's easier for me to see his strengths than it is for him at such times.

A good coach is one who never loses sight of the best within you.

As your own trading coach, you don't need to hire someone like me in order to make meaningful change or to extend your change efforts to an interpersonal context. Rather, you can maximize your efforts at self-development *by creating your own performance team*—a group of like-minded, mutually concerned peers who help each other. If I were going into full-time trading, one of my first steps would be to scour blog comments, forum postings, conference attendees, and similar gatherings of traders to find people who trade my markets and take trading seriously. I wouldn't need clones, just traders who are compatible with me in their trading instruments and time frames. I would then reach out to form what I call *virtual trading groups*: a group of peers who trade their own capital, but freely share ideas and help one another. The group would have to be chosen carefully, and all participants would have to share their trade ideas and trading results completely freely. In such an environment, the group members could cross-fertilize each other's views, support one another during difficult periods, and learn from one another. A particularly valuable function within such a group would be peer mentoring, similar to the mutual assistance within Alcoholics Anonymous.

> StockTickr (www.stocktickr.com) has been active in facilitating trading groups and communities, with an eye toward improving performance.

Even if you just find one or two supportive peers to share ideas and results with, you've taken an important step to create a fresh, interpersonal context for the changes you're trying to make. If you choose your peers wisely, they will challenge you, support you, learn from you, and teach you. Because you value them and don't want to let them down, you'll be more likely to stick to your preparation, discipline, and goals.

A little-appreciated piece of psychological wisdom is to find social contexts to be the person (trader) you want to be. Over time, the feedback and responses from others will mirror the best of you to you and that ideal self will become an increasing part of you. When you're your own trading coach, you don't need to do everything yourself. A cardinal principle underlying the psychodynamic framework is that the best changes are the result of powerful, emotional relationship experiences.

Your assignment is to find just one person to be part of your team: someone whose developmental efforts you can support, and someone who will support your own. Out of that relationship may spring many more—a network of dedicated professionals mentoring and motivating each other. When you turn trading into a relationship experience, you gain role models, become a role model, learn from others, and benefit from teaching others. You add fresh ways to experience your strengths, even as you build on them.

COACHING CUE

Online trading rooms are excellent venues for meeting like-minded traders, and they can be powerful learning tools. Several long-standing ones are Linda Bradford Raschke's trading room, which emphasizes technical trading across multiple markets (www.lbrgroup.com); the Market Profile–oriented Institute of Auction Market Theory room run by Bill Duryea (www.instituteofauctionmarket theory.com); and the trading room run by John Carter and Hubert Senters (www. tradethemarkets.com). Also take a look at the Market Profile–related educational programs run by Jim Dalton and Terry Liberman (www.marketsinrprofile.com) as possible venues for connecting with like-minded traders. Two well-known trading forums are Elite Trader (www.elitetrader.com) and Trade2Win (www.trade2win.com). Both can be ways of sharing information with other traders and connecting with peers. For those developing trading systems, the community that has developed around the TradeStation platform (www.tradestation.com) is worth checking out. Indeed, if you have a favorite trading platform or application, connecting with others who are using the same tools can be quite valuable. Market Delta (www.marketdelta.com) runs educational programs for users and maintains a Web presence for users, as does Trade Ideas (www.trade-ideas.com); these are two trading applications I've found useful.

LESSON 47: TOLERATE DISCOMFORT

An important insight from psychodynamic psychology is that our defenses—the ways we cope with the pain from past patterns of conflict—can take a physical manifestation. Consider a trader who is caught in a losing position. He watches, tick after tick, as the trade grinds against him. Gradually, he becomes tenser: he hunches over the screen, tightens his neck and forehead muscles, and grips his mouse tightly. *This physical tension can be seen as a defensive strategy.* This strategy cuts off other, threatening physical and emotional experiences. Perhaps the trader would love to yell and curse, but is afraid to lose control. Perhaps the trader just wants to cry, saddened by a series of needless losses. Not wanting to seem weak, he holds back the tears with his tension.

There are many other physical manifestations of defense. Consider the trader who is conflicted about acting on a well-researched trading signal. As his anxiety mounts, he tells himself that the market is too uncertain and he walks away from the screen, only to find that his signal was valid after all. His avoidant defense—leaving the situation—temporarily defuses his nervousness, but it also keeps him from figuring markets out and acting on opportunity.

Still another physical defense occurs when traders act out of frustration, pounding a table, throwing their mouse, or cursing loudly and blaming unseen others for their losses. By venting their feelings, they avoid introspection and self-responsibility. Their defense is against the guilt and the awareness that they have been hurting their portfolios.

> Often we use our bodies to keep our feelings out of sight, out of mind.

One of my subtle defenses is that, when I sense a position isn't going my way, I'll begin a frantic scan of information to validate my idea. Of course, I'm defending against the feeling of being wrong and I'm looking desperately for reasons to stay in the trade and undo the loss. This reaction generally makes the situation much worse. I've learned that if I'm behaving frantically in a trade, there's usually good reason for my feelings and I need to listen to them.

Recall that in psychodynamic theory defenses are coping strategies that protect us from the emotional pain of past conflicts. One of the most basic defenses is repression: keeping thoughts, feelings, and memories out of conscious awareness so that they cannot trouble us. *The problem with repression, of course, is that a conflict repressed is a conflict that remains unresolved.* We can't overcome something if we remain unaware of its presence. Many traders use their bodies to repress their minds: their physical tension binds them, restricting the physical and emotional expression of feelings. I've met traders who were quite tight physically and yet who had no insight into the degree and nature of their emotional stresses. In an odd way, getting tense was their way of coping: they were always mobilized for danger, tightly keeping themselves in control. *It is difficult to stay in touch with the subtle cues of trading hunches—the implicit knowledge we derive from years of pattern recognition—when our bodies are screaming with tension and even pain.*

In your self-coaching, it takes more than a willingness to interrupt these defensive patterns to make the most of them. What is also needed is the ability to focus on the feelings being defended against. The questions you want to ask yourself are: "What feelings am I holding off when I'm tensing my muscles?" and "What am I trying to avoid by blaming others or by walking away from the screen?" The idea is to hold off on that defense—purposely relax the muscles, turn the focus inward, stay in front of the screen—and *simply experience the feelings that are threatening.*

In psychodynamic work this is known as facing or confronting one's defenses. In counseling, for example, a client might begin talking about her painful relationship experiences and then suddenly change the topic and begin talking about her children and how they are doing in school. I might gently point out the change of topic to the client, explaining that it's perhaps easier for her to talk about her children than about herself. She then resumes discussing her relationship, and new information—and a flood of feelings—comes forth, followed by memories of her bad relationship with her father. *Breaking through the defenses leads to an emotional breakthrough*: she becomes aware of suppressed and repressed feelings and their depth.

Psychodynamic therapists are quite familiar with this phenomenon: when you get in touch with repressed thoughts, feelings, and impulses, the result is a fresh emotional awareness of your situation. *Your perspective changes when your emotional state and awareness change.*

This shift often leads to new insights and new inspirations for dealing with difficult conflicts.

> When you feel in new ways, you often see in new ways as well.

This emotional work can be conducted effectively through guided imagery. If you vividly imagine a market situation that leads you to tense up or lash out in frustration, you can reenter your frame of mind at that time and see what it feels like to not engage in those defenses. Very often a different set of feelings will emerge in the situation: ones that you hadn't been aware of. For example, when you refuse to shout and blame others, you may find that you feel saddened for yourself, pained at your losses. This frees you to address the pain and support yourself, rather than bury the feelings beneath a show of anger.

Enhanced emotional awareness can lead to a feeling of empowerment, not greater distress. A good psychodynamic counselor or therapist will *challenge* our defenses, not letting us get away with the various strategies we use to keep difficult feelings at bay. *What results is awareness that the feelings we've been avoiding are not so devastating after all.* Perhaps at one time in life, when we were young and more vulnerable, we couldn't cope with those feelings and had to do our best to erase them. Now, as mature adults, we don't have to run. Feeling our most threatening emotions and seeing, at the end of it all, that we had nothing so terrible to fear after all is a tremendously powerful and empowering experience.

So what are you running from? *Think of your worst trading patterns as defensive maneuvers: actions you're taking to ward off emotional pain.* Then, when you refuse to act on those patterns, just sit with the experience and see what you feel. See if you can find a different way to handle that feeling. Very, very often, beneath our impulsive trading, our anxious avoidance of risk, our outbursts, and our mismanagement of risk are efforts to protect us from a painful emotional experience. Once you find that experience and contact those feelings, you find there's nothing to run from. You can handle loss, fears of failure, and disappointment. As your coach, you only need to prove that to you.

COACHING CUE

Massage can be an excellent tool for reducing physical tension, but also for learning about your body's pattern of tension. When you become more aware of your body, you can catch yourself tensing muscles and restricting your breathing and then make conscious efforts to relax. When you relax in this manner by loosening muscles and deepening breathing, you are opening yourself to emotional experience—and new ways of handling your feelings.

LESSON 48: MASTER TRANSFERENCE

When I was in graduate school, I entered psychoanalytic therapy during my internship year in New York. The therapist's office was in his home in a typical Manhattan high rise. Many of the issues I wanted to work on pertained to rebellion against authority, a pattern that appeared in various work and school settings. I generally tried to please those who supervised me. If I couldn't win their approval, I became quite rebellious. I particularly did not like the feeling of being controlled by others. If there was a hint of coercion in a working relationship, I pushed back—sometimes quite hard.

When I arrived at my therapist's office, I noticed that the keys had been left in the door. I pocketed the keys, knocked on the door, met my therapist, and began the session. A few minutes into the session, the therapist's wife—mortified to interrupt our session—entered the office to ask her husband if he had the house keys. I reached into my pocket, smiled at her, and handed them over with a wink. Not a bad start for a first session.

Well, you don't have to be a Freudian to have a field day with my therapy session, even apart from the sexual symbolism of the keys. I had the keys and I had control. When the wife needed something, I was the one to give it to her.

To my therapist's everlasting credit, he responded very well to the situation, didn't get defensive, but also didn't ignore the matter. He encouraged me to join him in reflecting on what had happened and why it happened. That's the psychodynamic way: you use your relationship as a medium for dealing with repeated conflicts from the past.

What the therapist recognized, of course, was *transference*. Transference, in psychodynamic jargon, refers to the transferring of conflicts from the past to the present day. In other words, I was reacting to my therapist the way I might have reacted to my father or boss.

Had the therapist become hostile and defensive (reactions he would have been entitled to, given that I had pocketed his keys!), I would have stayed in my rebellious stance. Indeed, such a reaction by the therapist would likely have fallen into the trap of confirming my worst fears about authority figures. By refusing to budge from his professionalism, he disconfirmed my expectations and took himself out of the authority role. That gave me the space to take a look at what I was doing in this and other relationships and why I was doing it. Eventually I came to recognize that my need to take control in relationships stemmed from weakness and fear, not from strength. I learned that I could achieve far better control by using my relationship skills to engage people constructively.

> The curative force in psychodynamic work is the use of the relationship to create new, positive, powerful emotional experiences.

It is not unusual for traders to personalize markets. Sometimes we view markets as dangerous, as out to get us, as rigged games, as treasure chests, as playgrounds, or as complex puzzles. When we attribute human characteristics to markets, we have to ask ourselves why we choose some qualities over others. *Just as I projected authority onto my therapist, we project the qualities we most struggle with onto markets.* This is a kind of transference. If we've lived for years with the sense that no one listened to us, we now feel that markets are irrational and capricious. If we've felt that others have taken advantage of us, we may just focus our frustration on the market makers who manipulate markets to our detriment.

The idea of transference suggests that what most frustrates us about markets is most likely to be something that has frustrated us in our past—and probably in relationships. In a very important sense, we have relationships to the markets we trade, and it's not unusual for us to imbue those relationships with the same qualities that bedevil our personal relationships. When we act out past patterns with current markets, we're no longer responding to objective demand and supply; lost in our own patterns, we become blind to those of our markets.

We can see examples of transference in our dealings with trading partners as well. Very often, how a trader interacts with me reflects how she is dealing with markets. Some traders will avoid interaction with me out of embarrassment due to recent losses. These same traders enact avoidant patterns in their trading, neglecting to limit losses, neglecting their preparation. Other traders take a helpless stance with me during coaching, almost begging to be spoon-fed answers rather than trying methods out for themselves. These same traders become passive in difficult markets, giving up readily, displaying little resilience after normal losses. One advantage of sharing your trading with a peer mentor or a group of like-minded traders is that you can monitor how you engage them and how they deal with you. Not infrequently, the patterns that show up in your dealings with traders will reflect the patterns you need to work on in your trading. Patterns of overconfidence, avoidance, rationalization—all come out in our social interactions.

> Your greatest shortcomings in dealing with relationships will find expression in markets.

How do you view the markets you trade? What do you say about markets when you're most upset about trading? If you were to draw a picture of your markets or describe them to a nontrader, how would you depict them? A worthwhile assignment is to review your journal and track your self-talk for anything you might say to personalize markets and trading.

As your own trading coach, you have the ability to create your own trading experiences. By controlling your risk exposure, executing trades only when there is a favorable reward-to-risk ratio, and limiting your setups to clear, tested patterns, you have the opportunity to create trading experiences that don't follow the transference script—much as my therapist refused to accept the role I had cast for him. If your problem is handling frustration, your challenge is to create manageable frustrations in how you approach markets. If your pattern is escapism, your task is to find safe ways of staying in trades, in accordance with plans.

> Create trading experiences in which you can safely face your fears and constructively give voice to frustrations. That experience provides new endings to old scripts.

When we project hated qualities onto markets, we divide ourselves. Part of us fights the trading process and part of us is tries to stay absorbed in it. Thus distracted and divided, we are less able to pick up on market patterns and shifts among those.

There's a saying traders use: "Let the market come to you." What that implies is that you should approach markets with an open mind, processing patterns as they unfold. This means being free of projections, free of conflicts that we transfer to our trading. By tracking how you talk to markets—and about them—you can step back from those repetitive themes and truly let markets come to you.

COACHING CUE

You can see from the foregoing discussion why it is so important psychologically to reduce your trading size/risk when you are experiencing trading difficulties. This provides a safe context for trying out new ideas and tweaking your methods, so that you can face market problems directly and constructively rather than respond with defensiveness. When we make our trading more planned and rule-governed, we create experiences of control. When we reduce risk, we create experiences of safety. The essence of self-coaching in the psychodynamic mode is to generate new and powerful experiences that change how we deal with self, others, and world. Fashioning new trading experiences enables you to experience your trading and your markets in fresh ways that open the door to opportunity.

LESSON 49: THE POWER OF DISCREPANCY

A cornerstone of the psychodynamic framework is that talk alone does not generate lasting change. Rather, as we saw in the last lesson, we change through new, powerful emotional experiences. These experiences are powerful precisely because they undercut our worst fears and expectations and show us that we can, indeed, master the past conflicts and feelings that were once overwhelming.

We can think of this change process as generating new endings to old stories. Perhaps my old story is that I am fearful of loss, having been through traumatizing losses in my past, either as a trader or predating my trading career. Out of this fear of loss, I find it impossible to hold trades until their logical stop-loss or profit targets have been hit. Invariably I become so concerned with protecting a gain or minimizing a loss that I front-run my trading plan and exit early.

Note that this pattern is a defensive one: I am trying to ward off the discomfort of loss by getting out of the market prematurely. The cost of that pattern is that I never fully participate in the upside of my ideas, leaving me to chop around in my P/L. As long as I continue to act on that pattern, I never stay in touch with that fear of loss and thus can never master that fear. Repeating the pattern simply reinforces it.

So what we need in psychodynamic work is a new ending to the scenario. We need to refuse to indulge in the defense and purposefully sit in the trade, while allowing the feeling of fear to remain. The idea is that, at one point of time in our lives, this fear was too threatening to sit with. Now, however, at a new life period when you have more resources, you can handle the fear. You don't *need* to continue to defend against it. Getting in touch with your basic conflict and its emotions—the things you have been most defending against—is crucial to the psychodynamic change process.

The psychodynamic change process can be schematized as a sequence:

- Identify your recurring problem patterns.
- Connect yourself to the costs and consequences of those patterns.
- Identify what those patterns are defending against (what you are avoiding).
- Create experiences, particularly in relationships, for facing what you've been avoiding.
- Repeat these experiences across different relationships to internalize new, constructive ways of coping.

Your problems, this view emphasizes, are simply ways of protecting you from fearful memories, feelings, or desires. Once you experience those fears and acknowledge them, the challenge is to channel the fear in a way that does not derail your trading plan. You could talk with a trading colleague to gain some perspective, perform some exercises to calm yourself down and reassure yourself, or use the opportunity to write in your journal and remain an observer to the anxiety rather than the one immersed in it.

By refusing to act on the old defensive pattern, you guarantee yourself a psychologically helpful outcome. This is very important. As long as your position is sized properly with appropriate risk/reward in the placement of stops and targets, one of two things will happen: you'll either hit your target and make your money or you'll get stopped out at your predetermined level.

While the latter scenario is less desirable than the first, neither scenario is catastrophic. By acting in a manner that is discrepant from your old pattern, you have created a win-win: either you make money, or you find out—in your own experience—that the loss was not so bad after all and nothing to fear. In some ways, it's this latter experience that is the most helpful. By facing your worst-case scenario and seeing that it's something you can, indeed, cope with, you generate a tremendous sense of confidence and mastery.

> Trading well is a powerful source of new, positive emotional experiences.

There are many ways of constructing discrepant experiences. One is to surround yourself with people who respond differently to you than those from your past and let them be a part of the changes you're working on. They can then support you in not repeating old patterns, but also in providing positive feedback when you enact new, positive ones. Another way to generate discrepancy is to face uncomfortable situations directly, refusing to engage in old defensive maneuvers. Once you get in touch with the feelings you've been holding at bay, you'll be surprised at how readily you can arrive at ways of coping with them that provide you with the new, constructive endings.

A term used by psychodynamic therapists Alexander and French captures the essence of this approach: *corrective emotional experience.* What enables people to overcome their problem patterns is a set of emotional experiences that correct the learning that occurred during times of past conflict. *It's not enough to perceive a problem and think about it; as your own trading coach, you need to create experiences that enable you to move past that problem.* Invariably this means refusing to keep difficult feelings buried and, instead, experience them fully and face them directly. When you see that you can live through your emotional worst-case scenarios and emerge with no lasting damage, *that* is the corrective emotional experience.

Your assignment, then, is to conduct a personal experiment and seek just one corrective emotional experience during the trading day. Identify the repetitive trading behaviors that most disrupt your trading and then figure out what you would be thinking and feeling if you *didn't* engage in those behaviors. When a situation arises in which you would normally repeat your pattern, make the effort to hold off and experience those feelings you identified. See what those feelings are like, see how you cope with them, and see how the new coping affects your trading. You may be surprised to find that facing your fears is the best way of moving past them.

COACHING CUE

If you have a trading mentor, someone you respect and admire, try trading like that person just for one day. Do everything as you think they would do it. See how you feel with the discrepant experience, as the enactment forces you to forego your own negative patterns. In enacting an ideal pattern, you create new experiences with markets that undercut old, repetitive patterns.

LESSON 50: WORKING THROUGH

The term "working through" has a specific meaning in psychodynamic work. Once you have made initial changes—breaking old patterns of defense, facing challenging emotional conflicts, and finding new ways of dealing with those—the working-through process involves extending these gains by repeating the process across a variety of situations. As we've seen, *repetition combats relapse*: working a problem through a number of relationship situations cements new, mature ways of handling core conflicts.

A classic example is the trader who avoids closeness in relationships out of fear of rejection. Sure enough, in his relationship with his counselor, he transfers his past fears to the present relationship and avoids intimate topics. While this provides a superficial sense of safety, it prevents the trader from talking about what is truly important—and moving beyond it. Once the trader makes a conscious effort to open up in the sessions, the counselor provides the discrepant response by not being rejecting. This makes it easier over time to break the pattern with the counselor in future sessions and remain open.

In the working-through process, the trader in our example would take the progress with the counselor and now apply it to other relationships, as appropriate: friendships, close work relationships, and romantic relationships. By working through the conflict in multiple situations, new, positive patterns are cemented. The positive mirroring of many relationships enables the trader to internalize a sense of safety and security, which can be carried forward to a variety of life situations. In other words, it's not a single corrective emotional experience but multiple such experiences that enable us to internalize a new sense of self.

> Successful self-coaching builds multiple corrective emotional experiences, so that new, constructive patterns can be internalized.

After all, this really is the essence of psychodynamic work: redefining the self by creating and absorbing so many impactful, constructive experiences that it becomes impossible to remain stuck in the past. I internalized the identity of an author not just by writing articles and books, but also by interacting with editors and readers over time. There was a time when I sat in front of a blank screen in writer's block mode, concerned that what I wrote would not find a receptive audience. Following multiple positive experiences with writings and readers, that is no longer a concern. The writing flows as naturally as conversation.

Similarly, there may have been a time when you thought of yourself as a small, beginning trader. Over months and years of trading experience, making money and building your account, you no longer see yourself as a newbie. Through the positive experiences, you absorb the identity of an experienced, skilled trader. Think back to the process of expertise development from *Enhancing Trader Performance*: the steps that build skills are also the steps that construct identity. Working the learning curve, moving from novice trader to competent trader to expert trader is more than building knowledge and skills. It is a transformation of self as the result of repeated, positive experience.

> Your training as a trader should provide ongoing corrective emotional experiences: training itself becomes a means of working through our shortcomings.

When you're at the point of working through, you want to be an active experience generator and tackle your patterns in as many situations as possible, giving yourself the opportunity to enact new ones. As prior lessons have emphasized, and as you'll detect from the advice of experienced traders in Chapter 9, this is particularly powerful if you're working problems through with the support of trading peers. Their mirroring of your success, like the feedback that solidified my identity as an author, will enable you to literally take your changes to heart.

Your efforts at self-coaching in the psychodynamic mode will find their greatest success if you can disrupt old patterns and enact new ones *on a daily basis*, with the active feedback of those you're working with. Many traders I've known have sought to keep their specific trading performance secret, obviously embarrassed that they're not making more money. They freely talk about winning days, but remain strangely vague or silent following bad ones. This is exactly the opposite approach to the one that will work for you. *You want to be visible, warts and all, because that will help you—emotionally—put those warts into perspective.* If your flaws (or your concerns about others' reactions to those flaws) are so threatening that you must hide them, then your defenses control you. When you can make yourself completely visible to others, you have nothing to hide. Their acceptance of you is complete and genuine, not a false reflection of a false self.

A while ago, when I was posting my trades live to the Web via the *TraderFeed* blog, I invited readers to join me and post their trades as well. The daily count of unique visitors at that time was around 2,000; I figured that, even if just one-half of 1 percent took me up on the offer, we could get 10 different models of trading to learn from. Well, out of 2,000 people, only one showed tentative interest. No one was willing to go public with his trading.

That, dear reader, is how losers react. If readers were taking a psychodynamic approach to change, they would freely share their trades in real time and make no effort whatsoever to maintain false selves. Over time, their progress would be evident and the massive positive feedback they would generate would cement a new identity, a deep sense of security, and an emotional fearlessness.

> Accountability provides powerful opportunities to work through our greatest insecurities.

Your challenge for this lesson is to open the kimono and conduct your working-through socially, with the feedback of people you respect. This would be part of a daily trading plan, ensuring that you're generating corrective emotional experiences every single day. Several traders I know have taken precisely that approach by starting their own blogs, posting their trades, and developing relationships with the traders who responded constructively to their ideas. *Relationships are a powerful medium for change, perhaps the most powerful.* If you harness the right relationships you will give your self-coaching a reality that transcends simple entries in a trading journal.

COACHING CUE

Find at least one person to whom you are accountable for your development as a trader. This should be someone you can trust in sharing your P/L, your trading journals, and your tracking ofpersonal goals. A major advantage enjoyed by traders at professional trading firms is that they are automatically accountable for performance and can thus openly discuss success and failure with mentors and risk managers. Accountability leaves no place to hide; it's an excellent strategy for combating defensiveness and removing the threat behind setbacks.

RESOURCES

The *Become Your Own Trading Coach* blog is the primary supplemental resource for this book. You can find links and additional posts on the topic of coaching processes at the home page on the blog for Chapter 5: http://becomeyourowntradingcoach.blogspot.com/2008/08/daily-trading-coach-chapter-five-links.html

An excellent overview of psychodynamic approaches to brief therapy can be found in Hanna Levenson's chapter "Time-Limited Dynamic Psychotherapy: Formulation and Intervention" in *The Art and Science of Brief Psychotherapies*, edited by Mantosh J. Dewan, Brett N. Steenbarger, and Roger P. Greenberg (American Psychiatric Publishing, 2004).

A worthwhile resource for building emotional self-awareness is Leslie S. Greenberg's *Emotion-Focused Therapy: Coaching Clients to Work Through Their Feelings* (American Psychological Association, 2002). See also the classic text by Leslie S. Greenberg, Laura N. Rice, and Robert Elliott, *Facilitating Emotional Change: The Moment-by-Moment Process* (Guilford, 1993).

Articles relevant to psychodynamics and self-coaching can be found in the "Articles on Trading Psychology" section of my personal site: www.brettsteenbarger.com/articles.htm. These articles include "Behavioral Patterns That Sabotage Traders" and "Brief Therapy for Traders."

It is rare to find trading coaches who know anything about psychodynamics; a notable exception is Denise Shull, who wrote the chapter, "What Would Freud Say: Stroll Down Freud's Mental Path to Profits" in the book edited by Laura Sether, *The Psychology of Trading* (W&A Publishing, 2007).

The greater danger for most of us lies not in setting our aim too high and falling short, but in setting our aim too low, and achieving our mark.

—Michaelangelo

In Chapter 5, we explored psychodynamic frameworks for self-coaching. These frameworks are especially relevant when we repeat unproductive patterns across a variety of situations over time. Fundamentally, the psychodynamic view is a historical one: it emphasizes linkages between how we coped in the past and how we now find ourselves responding to situations.

The cognitive framework, like the behavioral one that we'll visit in Chapter 7, is less historical: it emphasizes how we process the world in the here and now. Change the viewing and you change the doing is the essential message of cognitive approaches. While the past is not irrelevant to this task, cognitive self-coaching stresses what we can do in the here and now to alter how we process the world around us.

Cognitive coaching is most relevant if you find yourself battling negative thought patterns that interfere with your motivation, concentration, and decision-making. Some of the most common cognitive patterns that traders target for change include:

- Perfectionism
- Beating Up on Oneself After Losses
- Worry
- Taking Adverse Market Events Personally
- Overconfidence

Cognitive methods help us think about our thinking and restructure our perceptions of self and world . Let's take a look how. . .

LESSON 51: SCHEMAS OF THE MIND

Chapter 5 outlined psychodynamic approaches to the change process. That framework makes use of powerful emotional relationship experiences to break patterns of behavior left over from prior life conflicts. When applied to self-coaching, the psychodynamic perspective requires a dual look at past and present, with an eye toward recognizing occasions when we repeat the past in our current responses to trading challenges. The cognitive framework, on the other hand, is more present-oriented. Its focus is on how we think and the relationship between our thinking and the ways in which we feel and behave.

The cognitive approach to change, like the behavioral methods described in Chapter 7, is grounded in learning theory. Instead of emphasizing the creation of relationship experiences, the focus is on *skills building*. For that reason, homework exercises play a prominent role in cognitive work, which makes the cognitive modality particularly useful for self-coaching. In cognitive coaching, you learn skills for processing information more constructively.

Many cognitive psychologists draw on the analogy of the scientist when describing our thought processes. Scientists observe nature and look for patterns and regularities. Once scientists observed these relationships, they develop theories to explain their observations. Experiments test these theories and provide new observations that enable scientists to modify their theories. Over time, science arrives at ever more refined understandings of the world through the process of testing, observing, modifying, and testing further.

Cognitive researchers call the theories in our head *schemas*. These schemas are like mental maps, orienting us to the world around us. We interpret events and interactions with others through these schemas, assimilating new events to them when possible and accommodating our understandings to fit new events when needed. As developmental psychologist Jean Piaget explained, this process of assimilation and accommodation provides us with deeper and richer understandings of our world. We are always elaborating our maps of reality.

> We never experience the world directly; all perception is filtered through our mental maps. If our maps distort the world, our perceptions will be distorted.

Schemas are not just collections of thoughts, but are complexes of thoughts, feelings, and action tendencies. Let's say, for example, that I was beaten severely as a child and now perceive the world as a dangerous place. One of my schemas might be that, "You can't trust people; they'll hurt you." When others try to get to know me, that schema becomes a lens through which I view their behavior. Instead of responding with friendliness, I raise my guard and distance myself. Because of the schema, I've interpreted their behavior as dangerous.

Sometimes schemas, as the lenses through which we view events, are distorted. They lead us to view and respond to events in exaggerated ways, as in the example above. Take the example of the trader who views his worth through his profit/loss statements. He becomes overconfident and expansive when he's making money, and he turns risk-averse and self-doubting when he's in a slump. As long as his trading results are filtered through this schema, he's likely to think about and respond to his profitability in distorted ways.

Problem patterns develop when our distorted responses to the world become self-reinforcing. In the example above, because others hurt me, I now perceive people as dangerous and untrustworthy—even when they approach me in a friendly way. My guardedness makes me seem hostile or suspicious to others, and they naturally stop their friendly overtures. That, in turn, convinces me that my views of them were right all along, reinforcing my distorted schema. *When we are caught in such self-reinforcing patterns, we stop revising our mental maps*: we become locked into negative ways of perceiving—and responding to—the world.

Automatic thoughts are the habitual ways of thinking that result from our schemas. Once a schema is triggered, it usually sets off a series of thoughts and feelings that guide our action. A schema of vulnerable self-worth might, for instance, lead us to respond to a market loss with dejection and depression and a host of thoughts amounting to, "I'll never succeed." These thoughts and feelings are not objective assessments of the markets or our trading. Rather, they are automatic, learned reactions that have become habit patterns.

> We never directly observe our schemas; rather, we experience their manifestations through our automatic thoughts.

The goal of cognitive work is to unlearn these negative thought patterns and replace them with more realistic ways of viewing the world. This restructuring of our thinking means that we, like scientists, must revise our theories. The cognitive approach provides methods for accomplishing this revision.

There are many automatic thoughts that affect traders as they struggle with risk and uncertainty. Some of these thoughts are:

- "I need to make more money."
- "I'm so stupid; how could I have done that?"
- "I've got this market licked."
- "I can't afford to lose money."
- "The market is out to get me."
- "I've got to get my money back."
- "Nothing I do is right."

The first step toward becoming your own trading coach in a cognitive vein is to identify the thoughts that automatically appear during your trading. Several traders I've worked with have taken the unusual step of audio recording or videotaping themselves throughout the trading day and then reviewing the recording after the close of trading. It's a great way to identify the recurring thoughts and feelings associated with trading challenges. Many times, there are just one or two core automatic thoughts that dominate our experience. These are the thoughts that will form the initial focus of your coaching efforts.

Your assignment is to observe yourself in trading with either video or audio recording, making notes of recurring thoughts and feelings. At first, don't worry about changing these thoughts: simply observe how your mind is occasionally hijacked when events trigger particular schemas. It is crucial that you understand, from your own first-person experience, that you do not have complete freedom of will or mind. At times, all of us can be quite robotic, replaying thoughts that have become mere habits. By observing these habitual thought patterns, you begin the process of separating yourself from them.

COACHING CUE

Our most problematic automatic thoughts often come out when we're fatigued and/or overwhelmed. Think back to times when you've felt overloaded with work, responsibilities, and market challenges. What are the thoughts that go through your mind? How do those thoughts affect your feelings and behavior? Observing yourself when you're most psychologically vulnerable is a great way of clearly seeing the negative thought patterns and schemas that affect us.

LESSON 52: USE FEELING TO UNDERSTAND YOUR THINKING

One of the best ways to identify the automatic thinking that could most jeopardize your trading is to track your strongest feelings. In the cognitive framework, how we feel is a function of our perception: how we see things and how we interpret what we see shape our emotional responses. When our interpretations of events are extreme, we're most likely to respond with extreme feelings. Those occasions in which we look back at our behavior with embarrassment, wondering how we could have blown things so out of proportion, are most likely reflections of times in which we were controlled by the automatic thoughts from distorted mental maps.

If, say, I think about the times in which I most completely lose my temper, they would be occasions in which I want to accomplish something but find my path blocked for no apparent good reason. Perhaps I'm trying to get to an appointment on time and find myself behind a slow driver who is absorbed in a cell phone call. Or it could be a situation in which I'm trying to accomplish something with a trader at a firm, but find myself stymied by a bureaucratic response. The thought behind my temper outburst is, "I have to get this done, *now*!"

Often these are situations that aren't life or death: they don't truly *need* to be accomplished there and then. My schema says, however, that if something doesn't get done now, that would be *awful*; it would be a *catastrophe*. I am responding to my own internal should and must, not to the objective demands of the situation. The exaggerated emotional response is a tip-off to an entrenched thought pattern that distorts my perception.

> When we turn a desire into a demand, we mobilize the body and respond with stress.

In psychodynamic work, the focus would be historical: figuring out past relationship patterns that might have initiated my particular way of thinking and feeling. The cognitive framework, however, is less concerned with the origins of the thinking patterns than on what we do in the present to recognize and modify those. *By tracking our extreme emotional responses as they are occurring, we can learn to recognize the thought patterns that affect us in the present and eventually challenge these.* In the cognitive approach, this is accomplished literally by teaching yourself to think differently and filter the world through a different set of lenses.

For example, when I pressure myself about time and tasks that I want to accomplish, I recognize the mounting frustration and tell myself that this is going to get me nowhere. A different perspective on the situation is, "What's the worst that could happen? Will this really be a catastrophe?" By pushing myself to entertain the worst-case scenario, I see how foolish it is to get worked up. Rarely is the likely consequence commensurate with the extent of the pressure I'm placing on myself. That, after all, is what makes the schema distorted!

> When we change the lenses through which we view events, we change our responses to those events.

What are your most exaggerated emotional responses to markets? Do you feel angry when ideas don't work out; devastated after losses; stricken by fear during volatile periods? Or perhaps you swing from overconfident, cocky feelings to feelings of despair and worthlessness? Your strongest feelings are reflections of your most entrenched thought patterns, and those feelings reflect your core schemas:

- **Schemas of justice**—"I put in my work; I *should* make money."
- **Schemas of catastrophe**—"It would be terrible if my trade didn't work out."
- **Schemas of safety**—"I can't act; the market is too dangerous."
- **Schemas of self-worth**—"I'm a total failure; I can't make money."
- **Schemas of rejection**—"I'll look like such a fool if I can't succeed at this."

It's easy to see how these schemas naturally lead to exaggerated emotions of anger, frustration, fear, and depression. As your own trading coach, you want to use your most extreme feelings to figure out your most distorted ways of viewing yourself and your trading. *If you're managing risk properly, there should be nothing overly threatening about any single trade or any single day's trading.* If you find yourself responding to markets with a high degree of threat, then you know that the problem is not the markets themselves or even your trading, but the interpretations you've placed on your trading results.

Please read those last two sentences again, slowly. If you are trading well—with plans built on demonstrated edges, with proper risk control—trading will have its stresses, but should not be filled with distress. Markets cannot *make* us feel anxious, depressed, or angry; the threat lies in how we view our market outcomes.

One exercise I like to conduct when I find myself responding to trading with strong emotion is to simply ask, *Am I reacting to the situation as it really is, or am I reacting to what I'm telling myself about the situation?* That question forces me to confront my thinking and ask whether the magnitude of my emotional reaction is truly warranted. If it is not the objective situation that creates your feelings, then your emotion has to be internally generated, a function of how you are processing events. If the emotion is out of proportion to the situation, your thinking about the situation must be distorted.

> The greater the distortion in our thinking, the greater the distortion in our emotions.

Make yourself write down what you would have to say to another person—another trader—to make them react the way you've just reacted. What could you say to them that would lead them to respond so extremely? The odds are good that what you would tell another person to generate the emotion is what you are telling yourself:

- "You're no good!"
- "It's all your fault!"
- "You're going to lose your money!"
- "You can't win!"

If you write down these messages every time you catch yourself in the throes of an extreme emotional response, you'll come close to duplicating the output from your cognitive schemas. It's much easier to redraw mental maps when they're lying open in front of you.

COACHING CUE

A common thought pattern that distorts traders' reactions to markets is what we might call a "justice schema": the idea that markets *should* be fair, *should* offer opportunity, or *should* behave as they've behaved in the past. Once we lock ourselves into notions of how markets should behave, we open ourselves to frustration and disappointment when they take their own course. Many times, I've seen traders grow restive, fuming at markets that just aren't moving. Traders become impatient and jump all over any move to new highs or lows, hoping that this will be the breakout move—only to find the market return to its slow range. By challenging yourself when you catch yourself thinking or talking about how the market *should* behave (but isn't behaving), you can use the frustration to channel your energies elsewhere: toward longer time frames in the same market, toward fresh research, or toward other instruments or markets. When we react to our own sense of justice and injustice, we no longer objectively process actual market activity.

LESSON 53: LEARN FROM YOUR WORST TRADES

Alcoholics Anonymous teaches people to become aware of their stinkin' thinkin'. But we don't have to be alcoholics to process the world in distorted ways. We develop repetitive patterns of behavior, and we follow daily routines. Most of us are creatures of habit: we tend to go through consistent morning routines, eat at the same times of day, and go to sleep around the same hour. We take the same routes to and from work, and we listen to the same music, watch the same television shows. There's not much in our lives that isn't patterned.

So it is with our thinking. *We learn ways of processing information, and these become part of our routines.* We blame ourselves to help avoid conflict with others; we anticipate negative outcomes to help us not become surprised when things go wrong. In individual situations, such modes of thinking may suit us well. As engrained habit patterns, however, they impose distortions upon the world. After all, not everything really is our fault. Not every event does go poorly.

> Our negative thought patterns are learned habits; the key to cognitive work is unlearning them and replacing them with more constructive ways of processing events.

Once these modes of thinking become automatic, their accompanying feelings follow along. When we blame ourselves, we feel discouraged, diminished, and depressed. When we anticipate the worst, we feel anxious and uncertain. To the extent that we bring these schemas to trading, we no longer respond to markets objectively. We are like robots, responding with automatic thoughts and unwanted feelings.

As Gurdjieff noted, it is important to become emotionally aware of this reality: At some point you're deeply absorbed in trading, observing market patterns, and acting upon those. Then a shift occurs, and you are no longer in control of your thinking. It has become hijacked. An activated schema now sets off an avalanche of thoughts and feelings that may very well have nothing to do with the situation at hand. Suppose someone hijacked your computer as you were trading and suddenly switched the screen from your markets to some other, random ones? Suppose your mouse was taken out of your control and clicked on trades that you didn't want?

I guarantee, if that happened to you, you'd become very upset. You would not tolerate someone controlling your computer or your mouse. You would do everything in your power to regain control of your equipment. That has to become your attitude toward the hijacking of your mind. It's not enough to simply observe automatic thoughts taking control; you need to feel the horror of literally losing control of your mind and behavior. Much of the motivation to change faulty schemas will come from the awareness of the pain they inflict in all aspects of your life.

> Automatic thoughts don't just enter our mind; they take over. We change when we sustain the motivation to stay in control of our minds.

We've seen that reviewing your self-talk via audio or video recording and tracking your most extreme emotions during trading can alert you to your stinkin' thinkin'. Another powerful tool to help identify problematic schemas and thought patterns is to review your absolute worst trading decisions. Your worst trading decisions may or may not be your largest losing trades; they could be occasions in which you simply missed a golden opportunity. You'll know your worst trading decisions by your reaction to them, "How could I have done that?" That reaction is a fantastic tell, indicating that you truly were not in your proper mindset when you made the poor decision. At some level, when you're mystified how you could have been so mistaken or boneheaded, you are recognizing that your mind had been hijacked.

Once you identify these worst trades—and this will require a review of your journal, as well as a look back on your recent trading experience—you then want to re-create the thoughts and feelings that led to the faulty decision-making. Normally we like to put such episodes behind us, with a simple reassurance that next time we'll trade with better discipline and preparation. But in this exercise, you want to perform a psychological autopsy and exhume your faulty decision-making process in all its gory detail. What were you thinking at the time? What were you feeling? What were you trying to avoid or accomplish with your trading decision?

The common thoughts and feelings during these poor trading episodes will be your clue as to the schemas that were being activated at the time. Perhaps it was a safety schema: you were telling yourself that you could not afford to lose paper profits or to take a particular risk. Alternatively, it could have been a self-worth schema, as you told yourself how great it would be if this trade hit a home run. Your feelings during these trades—the fear, the overconfidence—will provide valuable clues as to the automatic thoughts that were generated.

> Our worst trades come from reacting to our automatic thoughts instead of markets themselves.

In my own trading, a common schema that is activated is a variation of the safety theme: avoid danger. To be sure, this can be a useful mode at certain market junctures, helping a trader size positions appropriately and limit losses on trades. Where the schema introduces distorted perception, however, is in defining danger as any drop from an equity peak, not as an outright loss of capital. This makes it particularly difficult to stay in winning trades, because relatively small retracements will stimulate desires for profit taking. A more realistic perspective would define danger not only in terms of lost paper profits, but also in terms of lost opportunity. Some of my worst trades have been ones in which I acted on a short-term perspective and subsequently missed the longer-term market move. The need to avoid danger exposed me to the equal danger of cutting profits short.

Notice that these worst trading episodes cut across patterns of thinking, feeling, and behaving. *Once we start from the (faulty) premise, "You must avoid risk," and once we define risk as any market movement against our positions, we shape our feelings and actions accordingly.* Reviewing your worst trades may be painful, but it is also liberating. It tells you where your mind has gone astray, and that can lead you to corrective action.

COACHING CUE

Many of our worst trades come from the demands we place on ourselves. Keep tabs of the times you tell yourself that you need to, must, and have to participate in market moves or make money. When these demands become rigid absolutes, we end up chasing market moves, refusing to take small losses, and otherwise violating principles of good trading. When we are more focused on those internal demands than on our trading rules, that's when we're most likely to lose money. You'll be able to identify those demands by the internal feeling of pressure that they generate. There's a different feeling when you trade from opportunity versus trade from pressure. Track your worst trades and the feelings associated with them to alert you to the ways in which your automatic thoughts can sabotage your best trading.

LESSON 54: USE A JOURNAL TO RESTRUCTURE OUR THINKING

Review your past emotional episodes and trading mistakes as a helpful way to identify your mental maps and the ways in which they can distort your perception. The goal of cognitive work, however, is to be able to catch your automatic thoughts—those bouts of stinkin' thinkin'—as they are occurring so that they cannot hijack your mind and your trading.

In *Enhancing Trader Performance*, I outlined how a cognitive journal can be used to help traders restructure their thought processes. The journal format I suggested took the form of a single page for each trading day or week (depending on the frequency of your trading), with each page taking the form of a table. The left-hand column of the table describes the events that occurred at the time you experienced a trading problem. This column would include what was happening in the market, what you were planning, and how you entered—or didn't enter—the market.

The second column is an account of how you are talking to yourself about the problem. In the book, I took the traditional approach of using the second column to describe your beliefs about the events. *What may be most helpful to your self-coaching, however, is to actually transcribe your thoughts about the events and capture what you're thinking and feeling.* This column should capture the ideas going through your head at the time as faithfully as possible, as in, "Why didn't I take the trade when I had it? I should have been up money today and instead I've lost more than I should. I am so disgusted with myself. I don't know if I even want to keep trading."

Many times, key phrases from your transcribed self-talk will alert you to the nature of the schemas being activated. For instance, in the example above, the word *should* is often a good sign that a perfectionist self-worth schema is playing itself out, leading to angry self-talk and a discouraged frame of mind. Once the musts and shoulds are triggered, they turn the trader's attention away from markets and toward the issue of self-worth. Note that this is not happening in a constructive context; rather, the self-talk is critical and punitive. It is difficult to see how such thinking could move a trader forward.

I like to think of these automatic thoughts from the second column as a kind of tape recorder in the brain that clicks on during particular situations (first column). Many times, the very same phrases and messages recur from situation to situation. This process becomes easy to observe when reviewing your cognitive journal: you see not only how negative the self-talk can be, but also how automatic and robotic it is.

> Pay particular attention to emotional words and phrases that recur in your self-talk: These words and phrases are shaped by our core schemas.

The third column describes what happens as a result of the self-talk: the feelings you have and the actions you take. For instance, in the example above, those angry, perfectionist thoughts might lead you to quit for the day and sulk, missing opportunity and a chance to learn about current markets. Alternatively, the angry self-talk might lead to subsequent revenge trades that lose even more money. The third column chronicles *all* the consequences of the automatic thinking, both personal and monetary.

Over time, a review of this third column will cement for you the absolute toll taken by the distortions in your thinking. When you are your own trading coach, it's necessary to sustain your motivation for change. Seeing that your thinking is both tape recorder-like in its mechanicalness and sabotaging in its consequences will sear into your mind that *change is not optional.* Reading entries from day after day after day that highlight the same thoughts, the same behaviors, and the same losses and lost opportunities focuses not only the mind, but also the motivation for change.

The most common mistake traders make in keeping such a journal is that they are not sufficiently specific in their entries and thus miss crucial details and understanding. Below is a sample of a journal that lacks detail and fails to help the trader understand the specific thought patterns and consequences that appear across various trading situations:

Situations	Self-Talk	Consequences
Didn't go over my charts in the morning	I need to get my rest	Missed a good move
Sized my trade too large	This could be a great trade	Took a large loss
Market moved through my stop	I'm losing too much money	Took a break
Didn't take profit on a trade	I think we're going lower	Market reversed
News came out	Stock is really moving	Stock failed to rally; I quit for the day

Now let's take a look at the same cognitive journal, but with detailed entries.

Situations	Self-Talk	Consequences
I was tired because I had been partying late at night and didn't go over my charts before the open.	"I have to get my rest, but I'll be able to figure things out after the open. I've got a good feel for the market"	I couldn't see the market well. I missed an opportunity that I had researched last week, but forgot about once the market opened. I could have made several thousand dollars on that trade.

Situations	Self-Talk	Consequences
I thought I had a good idea after I missed the morning trade. I doubled the position size to make up for the lost trade.	"This could be a great trade. If it goes for me, it would make my day. Maybe it was a good thing I missed that trade."	An economic report came out and the market moved against me. I didn't remember that the report was coming out. I panicked and sold out for a several thousand-dollar loss. This put me in the red after a good start to the month.
Right after the big loss, I saw that gold was moving. I went with it, but it reversed and blew through my stop.	"I'm losing my shirt today. If this keeps up, I'll be so far down for the month that I won't be able to come back. There's no way I can explain this to my wife."	I decided to take a break from trading, but I couldn't relax. I was worried about coming home and telling her about my day.
I had a winning trade in the solar stocks, but decided to hold the position past my profit target to make my day back.	"I can't afford another losing day. This trade can make me even on the day; then I'm going home. I have to stop doing this to myself."	The solar stocks reversed, and I only took a small profit on the position. I feel totally stupid for ignoring my exit.
One of the stocks on my list was favorably mentioned by an analyst and popped on the news. I decided to not take the trade, because I didn't want to lose more money.	"This is going to break out of its range; it could go much higher. I want to be on board, but I can't afford to lose any more."	The stock stalled near the top of its range and then broke through on high volume. I watched the stock move without being on board. I feel totally disgusted with myself, like I don't deserve to trade any more.

Note how the added detail makes it clear what is going on in the trader's mind. The elaboration of the trader's self-talk also clarifies the links among the events, as one trading mistake led to another, with one schema (self-worth) first triggering overconfident thoughts and feelings, then frustrated ones, then ones of defeat and failure. We can also see how the trader's home life is connected to the thoughts and feelings affecting trading decisions, as the trader is feeling a need to prove himself to his spouse as well as to himself.

When you are your own trading coach, you want to look *between* the journal entries as well as within each of them. That will often illustrate the links among your thoughts, as events trigger distortions in processing, which bring further events, and still additional distortions. Your assignment is to capture the flow of your thoughts and the connections of these to your feelings and actions. Only once you clearly see how the mental dominoes fall can you interrupt the process by turning your mind in a different direction.

COACHING CUE

What are the schemas and thoughts that accompany your *best* trades? Extending your journal to include how you think when you're trading very well helps you take a solution-focused approach to cognitive work. The last lesson in this chapter may provide some ideas along this line. Also keep an eye out for the hope schema, in which trades that are losing money trigger automatic thoughts of hope for a return to breakeven. Those thoughts often lead to violation of stop-loss rules and trigger subsequent schemas of regret and self-blame. There's a role for intuition in trading, but beware situations in which you are into wishin'. Those are usually excellent points to get flat and regain perspective. When you are your own observer, your negative thoughts can themselves become reliable trading indicators.

LESSON 55: DISRUPT NEGATIVE THOUGHT PATTERNS

How do you break a habit pattern? When we have a smoking habit, or when we find ourselves eating out of habit, one of the first steps toward change is simply catching ourselves in the act of repeating our unwanted actions. *By disrupting a habit pattern, we gradually make it less automatic, less capable of controlling us.*

So it is with our habitual thought patterns. When we interrupt and disrupt these patterns, they become less automatic. We gain a measure of control over them; they no longer take control from us.

The most basic technique to disrupt negative thought patterns is thought-stopping. Thought-stopping is exactly what it suggests: a conscious effort to stop a train of thinking while it is occurring. When you have used cognitive journals and reviews of your trading to clearly identify your pattern of automatic thoughts, you become increasingly sensitive to their recurrence. This enables you to recognize their appearance in real time. By giving yourself the command to Stop!, you disrupt the automatic nature of the thinking. This gives you time to calm down, change the focus of your attention, and engage in other useful cognitive exercises.

A good example from the previous lesson is situations in which you find yourself losing money on a trade and hoping or praying for a turnaround. I've even encountered traders who engage in a kind of bargaining (not unlike the dynamics of someone facing death in the Kubler-Ross work), promising to never violate their discipline again if they could just break even on this trade. The fact that hope is dominating the cognitive picture for the trader suggests that there is more than a little desperation. At some level, the trader is aware that this is not a good position to be holding. The underlying schema, however, says that it is not okay to lose money; that losing money equals failure. This makes even normal market losses unduly distressful, triggering maladaptive coping (holding positions beyond stop-loss points out of hope; doubling down on losing trades). If, however, the trader recognizes this pattern as it is occurring, he can use the appearance of hope to *stop* himself and disrupt the automatic thoughts and actions.

The more vigorous your efforts at *stopping*, the more successful you'll be in disrupting unwanted patterns of thought and behavior.

Thought-stopping is useful because it separates you from your ways of thinking. Instead of identifying with automatic thoughts and the feelings they engender, you separate yourself from them and remind yourself that this is what has gotten you into trouble in the past. In the beginning, as you coach yourself, you will find that you have to engage in thought-stopping numerous times during a trading day. As you become more expert at recognizing your negative thoughts and disrupting them, however, you find it easier and easier to stop yourself. A simple reminder of, "There I go again!" is sufficient to turn your mind to a different track. The interruption of habitual thoughts itself becomes a positive habit pattern.

I have found it helpful in my own trading to make the stop efforts particularly impactful, almost as if I'm shaking myself awake and mobilizing other ways of dealing with situations. One time I caught myself holding a winning trade beyond the point at which it had reversed and returned to breakeven. A cardinal rule I've learned to follow is to not allow trades that have moved a threshold amount in my favor to become losing trades. As I watched the winning trade hit breakeven and then turn red, I caught myself hoping that it would return to breakeven. The order flow was clearly suggesting, however, that large traders were hitting bids and driving the market lower. I gave myself a swift slap across the cheek and told myself to get out. That spontaneous act—admittedly not a coaching technique I use with other traders—woke me up and enabled me to take a small loss rather than a much larger one. However, I have remembered that slap over the years, and its impact has kept me out of trouble on multiple occasions.

(Now I take a break from the screen and douse my face in cold water when I need to shift how I'm thinking about markets. The physical jolt seems to facilitate a cognitive shift.)

When the thought-stopping is dramatic, the mind-shift can be equally radical.

Some traders I've worked with have found it useful to post signs on their computer monitors to remind themselves of the thoughts they most want to stop. *Stay Humble* is one sign a trader wrote after identifying a pattern of overconfident, arrogant thinking. Such signs help traders to think about their thinking, standing apart from the patterns that would normally trigger negative emotions and poor trading decisions. They also remind traders to periodically stop, interrupt automatic thought patterns, and reengage markets constructively. You can't be absorbed in a pattern of thinking if you're making yourself its observer.

Here is a simple thought-stopping exercise that I have found useful in my own trading. The idea is to be on the lookout for any thoughts during trading that include the words *I* or *me*. The goal is to interrupt and disrupt those thoughts. *The reason for this is that you don't want to be self-focused when you're concentrating on markets.* You neither want to be thinking overly positively about yourself and your performance or overly negatively. When your automatic thinking turns attention inward, that's the cue to immediately disrupt the pattern and become more market focused.

There are many examples of self-directed attention that divide your focus from markets. These include:

- "I'm doing great; this is the time to be aggressive."
- "I can't believe how badly I'm trading."
- "Why did I just do that?"
- "The market is killing me."
- "I'm going to make my money back."
- "Everything I do is wrong."
- "I hate this market."

Once the words *I* or *me* appear, you want to quickly close your eyes, take a deep breath, relax your muscles, and turn your attention to the markets. If you're already worked up to that point, a quick break from the screen—clearing your head and turning your attention elsewhere—can be useful.

I find it helpful, during trading, to periodically remind myself, "It's not about me."

With practice, you can become quite good at proactively steering your mind from self-focused thoughts. During one recent trading episode, I caught myself looking up my P/L in the middle of a trade, wondering how well I was doing and how much I was willing to give up of my week's gains. Of course, that had nothing whatsoever to do with the merits of the trade I was in. Because of prior practice, I was able to stop myself from clicking on the P/L summary before the numbers could get into my head. Reminding yourself that "it's not about me"—it's about the markets—is an excellent start to maintaining control over your decisions in the heat of market action. If you stop yourself from doing the wrong things, you clear the decks for implementing sound trading practices.

COACHING CUE

When I work with traders, one of my roles is to help them *stop* the flow of negative, automatic thoughts. Even when you're coaching yourself, however, you can derive the same benefit by keeping in touch with one or more trusted and valued trading peers during market hours. This can be in a single trading office or via Skype, Hotcomm, or even instant messaging. Many times your mates will pick up on your negative thinking before you're aware of their appearance. This can be very helpful in checking yourself and refocusing your attention.

LESSON 56: REFRAME NEGATIVE THOUGHT PATTERNS

Reframing is a psychological technique that takes a problem pattern and places it in a different context so that it can be viewed in fresh ways, opening the door to new responses. Suppose I'm meeting with a trader who is experiencing occasional bouts of performance anxiety that leave him unable to act on clear trading signals. He views himself as a weak person who can never succeed at trading. I take a different perspective, emphasizing his prudence about taking risk and his success at avoiding large losses. "Perhaps we can use that same good judgment to identify and act upon opportunity in a way that keeps you secure," I suggest. What the trader frames as weakness, I reframe as a potential strength. Instead of fighting against his own tendencies, the trader can use the reframing to help him figure out how to use those tendencies to produce acceptable risk-adjusted returns.

> Often we can find a strength underlying one of our weaknesses, enabling us to approach problem patterns in novel ways.

Reframing can often take a negative motivation and turn it positive. For a while, it seemed to me that my daughter was lazy in getting her schoolwork completed. I then hit on the idea that her primary motivations are social in nature: she's a real people-person. I proposed that we do homework together, and she readily rose to the occasion. This became a father-daughter tradition during the school year and a valued bonding experience. Similarly, when my son angrily confronted a teacher at school who "got in my face" about getting work done quickly, I started my conversation with him by congratulating him for using words only and not storming out of the class or laying hands on the teacher. Instead of framing the discipline problem as a failure experience, I reinforced the important lessons of self-control. He was much more able to hear my later advice, and he left the incident feeling better about himself, having learned from the confrontation. *The most negative thought patterns are there for a reason; identifying that positive reason and finding new ways to accomplish it makes self-coaching empowering.*

Consider, for example, the example of the trader who becomes lost in hope during a losing trade. Instead of flaying him for a lack of discipline, I will emphasize that he has found a valuable market indicator: the Hope-Meter. When hope enters the picture during a trade, it's a sign that deep down we know the trade is ill fated. By following the Hope-Meter, we can use the automatic thinking pattern to aid good trading, rather than interfere with it.

When you see problems in new ways, you gain the ability to respond in new ways. Novelty is a central element in all change efforts.

Many problems have a cyclical nature: I am afraid of losing money, so I set my stops too tight, lose more money on choppy action, and generate even more fear of losing money. When we reframe a problem, the new perspective can help us break the cycle. If I reframe the problem as one of position sizing, I can take the same monetary risk with wider stops, breaking the cycle of loss in choppy markets. *How we view our patterns very much determines how we respond to them.*

A useful exercise that I described in *The Psychology of Trading* is to reframe our inner dialogues by viewing them as actual dialogues. What we say to ourselves sounds very different when we imagine the same words spoken by someone else. This is particularly true when we are hard on ourselves after losses. What feels like worry when we are absorbed in our own thoughts sounds more like hostility when we reframe the same messages as part of an interpersonal dialogue.

Let's say a trader misses a good trade and then tells herself, "I can't believe I missed that trade. What is wrong with me? I wait all week for the right setup and get it handed to me on a silver platter and I don't take advantage of it. I'm never going to make it if I keep making mistakes like this."

Many traders actually consider such self-talk to be constructive. Traders think that being hard on themselves will help them avoid similar mistakes in the future. Suppose, however, we reframe the very same conversation as a dialogue from a friend to the trader:

"I can't believe you missed that trade. What is wrong with you? You wait all week for the right setup and get it handed to you on a silver platter and you don't take advantage of it. You're never going to make it if you keep making mistakes like this."

Clearly, when the dialogue is framed in such a manner we can appreciate that the tone is not at all constructive. Indeed, no true friend would ever talk to us in such a manner. The message is blaming and hostile, with no empathy or suggestions for improvement. Reframing the self-talk as an actual dialogue reveals the true emotion behind the automatic thoughts.

> Reframing thoughts as dialogues helps us view our thinking in a new light.

Such reframing is particularly effective if we imagine ourselves speaking to a good friend in such a manner. For instance, suppose a good friend of yours went through a series of losing trades and you were to say to that friend, "I can't believe you missed that trade. What's wrong with you?" We can be hard on ourselves and buy into all sorts of hostile ways of talking to ourselves, but if we imagine delivering those same messages to a friend, we don't buy into the scripts at all. When we reframe our thoughts as interactions with a friend, we draw upon personal strengths such as our ability to be supportive of others. Such strengths make it impossible to maintain a stance of angry blaming.

As your own trading coach, you want to maintain a consistent and constructive tone of voice with yourself, so that you don't damage your concentration or your motivation. An excellent exercise for working on this process is to close your eyes and imagine yourself as another trader that you are responsible for: perhaps an assistant or a student trader. Imagine that this valued assistant of yours has just made the same mistakes in the market that you've made. How would you talk with this person? What would you say? What would be your tone of voice? What emotions would you convey? Imagine these in as vivid detail as possible. Then note how your approach to this other trader differs from your own self-talk. If you wouldn't talk to a valued colleague in the way that you're addressing yourself, then you know that your automatic thinking is distorted in a negative way. Surely you deserve to talk to yourself the way you would talk with others in a similar situation!

> Many times it's the tone of our self-talk, not just its content, that disrupts our trading.

When you conduct this guided imagery exercise day after day, particularly after you've interrupted some of your negative thoughts, you gradually learn to talk with yourself in more positive ways. In one variation of the exercise, I have traders imagine that *they* are the other trader they are talking to, so that they literally are practicing talking to themselves in ways that they would support someone else they cared about. During these exercises, traders who have been the most volatile and angry can access a wealth of caring, supportive messages as they view themselves as people they truly care about.

You are your own trading coach. Do you want your coach to berate you, to focus on your every mistake, to threaten you with dire consequences? Or do you want your coach to bring out the best in you? When you cast your thoughts as dialogues, you have a chance to reframe the mental maps that guide your thinking, feeling, and acting. Inevitably, we do talk to ourselves. We *are* coaching ourselves. The only question is whether we do so consciously and constructively or automatically and destructively.

COACHING CUE

If you have a mentor or peer trader who you are working with, pay particular attention to how they talk to you when you are having problems. Many times we can internalize the voices of others in reframing our automatic thought patterns. "What would my mentor say to me?" or "What would a good coach say to me?" is an excellent start toward constructive reframing.

LESSON 57: USE INTENSIVE GUIDED IMAGERY TO CHANGE THOUGHT PATTERNS

The value of imagery is that it stands in for actual experience, with an unusual power to access emotional responses. Here is an effective cognitive exercise that makes active use of the emotional power of imagery.

The first step in the exercise, as emphasized in recent lessons, is to identify the repetitive, automatic thoughts that are disrupting your trading.

As noted earlier, these thoughts will generally form the self-talk that accompanies your most emotional trading episodes. The clearer you are in your capturing this self-talk, the more realistic and vivid your imagery work will be.

Let's take a specific example. A trader I recently worked with uncovered a pattern that was greatly holding him back in his success. He was a profitable trader over many years, but always had the nagging feeling that he was not fulfilling his potential. When we examined his trading and thinking patterns, it became clear that he became more risk-averse as he hit new peaks in his trading account and new P/L highs for the day. He was upset to finish off his highs (for the day, for the week), so he became unusually risk-averse as he hit these highs. He cut his size, traded less often, and behaved like a person who had just undergone a drastic drawdown.

His self-talk at these times was revealing. "You've had a good day," he'd tell himself. "Let's hang onto the gains. There will be more opportunity tomorrow." When it became clear that he missed opportunity because of this unusual caution, he felt vaguely guilty, but he reassured himself that his equity curve was positive and "you never go broke taking a profit."

This was a difficult pattern to break because, while the trader realized the pattern's limitations, he also bought into it at an emotional level. The trader who beats himself up truly suffers as a result of his self-talk and will want to change that, simply to feel better. The risk-averse thinking, however, kept our trader feeling safe. He *liked* the consequences of overcaution: it helped keep his emotions in check.

> We are less likely to change a pattern if we buy into it, if we're not in at least a moderate amount of distress over the consequences of that pattern.

A key step in changing his self-talk was to reframe the "let's not lose money" talk as "I don't think I can make money" talk. The trader thought of his self-talk as messages of safety; I reframed them as messages of low self-confidence. He doesn't want to finish off his highs, because at some level he's not sure that he can get past those peaks. He doubts his ability to bounce back from drawdowns, so he desperately tries to avoid drawing down. He settles for a steady, but modest equity curve because he doesn't have confidence that he can generate and sustain more robust returns. Perhaps he even feels, deep down, that he does not deserve large rewards.

Note how, in this situation, my reframing was not a way of helping a trader view a negative pattern more constructively; rather, I was framing the pattern in a way to accentuate the trader's discomfort. It's back to that notion that coaching is all about comforting the afflicted and afflicting the comfortable. Our trader was far too comfortable in his risk-averse ways;

my goal was to move him forward in his readiness for change by helping him highlight the costs of his ways of coping. In your own cognitive self-coaching, you are not simply reframing negative thought patterns positively; you also will frame them in ways that summon your motivation for change. Earlier, I mentioned the value of viewing negative patterns as personal enemies: this is an example of a reframing that afflicts our comfort.

When I reframed his self-talk as a lack of confidence, the trader's immediate response was to describe his mother. He loved her and felt close to her as a child, but said that she was overprotective. When other boys went out to play, she held him back, afraid he would into fights. She tried to dissuade him from dating later in life, because she thought that girls might "take advantage" of him. The trader explained, with some sadness, how he never had the opportunity to excel in sports despite early promise, because his mother worried about injuries.

This response provided me with the opening to use the guided imagery method. I asked the trader to close his eyes and vividly imagine a situation in which he has made money early in the day and now is considering packing it in for the day and sitting on his profits. I told him to imagine that the markets were moving and opportunity was present, *but to visualize his mother saying to him everything he has been saying to himself during his risk-averse self-talk.* He thus had to imagine his mother, with a worried, overprotective look on her face, warning him to not trade, to not lose the money he made, to not get hurt in the markets.

> Personalizing a pattern you want to change can heighten your motivation to change that pattern—and help you sustain that motivation.

Before we had finished the exercise, the trader opened his eyes and exclaimed, "Yuck!" The idea of being a little boy controlled by his mother disgusted him. "But isn't that your mother talking," I asked, "when you're telling yourself to not lose your money, to not get hurt in trades? Isn't that your Mom's voice within you?"

That reframing was what the trader needed to separate himself from his pattern. The last thing he wanted to do was repeat his overprotective childhood. Whenever he felt uneasy about participating in a market with opportunity, he simply closed his eyes and visualized what his mother would say in that situation. That visualization gave him the motivation to push the negative thinking aside and act on his trading instincts.

As I noted in *The Psychology of Trading*, most of us have someone in our past who we don't like or who we associate with negative thoughts and influences. If you imagine your least favorite person—someone who was mean to you, who hated you, who was abusive to you—saying to you what you have been saying to yourself in your worst self-talk, it will become much easier to push back. If that hated person were actually standing in front of you while you were trading, voicing negativity, you'd have no trouble telling him to shut up. When you use imagery to associate your worst thinking with the worst people from your past, you learn to shut yourself up. And that is a major triumph of self-coaching!

COACHING CUE

There are many other powerful applications of imagery in combating negative, automatic thought patterns. One trader I worked with took up martial arts and used the workouts to imagine striking out against the patterns that had held him back. After he rehearsed that mind frame in practice after practice, he only needed to adopt certain postures during the trading day to regain his fighting form. Another trader found that he was sharpest in his trading when he felt physically fit. He used his exercise routines to rehearse ways of thinking about himself and trading that reinforced his strengths. During the midday, he took a short exercise break, which helped clear his mind, but also placed him in greater touch with the mindset that he was cultivating. Not all effective imagery is visual; sometimes associating a way of thinking with a physical state can help us use body to affect mind.

LESSON 58: CHALLENGE NEGATIVE THOUGHT PATTERNS WITH THE COGNITIVE JOURNAL

One way to use a cognitive journal, we have seen, is to track automatic thought patterns, so that you can become highly aware of negative self-talk and how it is connected to your emotions and behavior. A simple extension of the journal, however, is useful in restructuring our mind maps.

In the journal described earlier, we divided pages into three columns, with the first column representing a description of situations in which automatic thoughts occur; the second column captures the self-talk; and the third column highlights the consequences (in mood, behavior, trading) of that self-talk. The added fourth column represents your systematic efforts to change the internal dialogue and replace the automatic, negative thoughts with more realistic, constructive alternatives.

Let's start with an example. Suppose a trader is dealing with a perfectionist pattern in which she frequently criticizes her performance as not good enough. Even when she makes money, she focuses on how much she left on the table by not catching highs and lows. The result leaves her feeling discouraged, and it also leads her to take bad trades in order to try to catch exact highs and lows.

In the fourth column of her journal she engages in a Socratic debate with those negative thoughts, challenging them and coming up with different ways of viewing the situation. This is truly self-coaching, because, just as the second column is the voice of the negative self-talk, the fourth column becomes the voice of the inner coach.

The cognitive journal can become a forum in which we vigorously and emotionally challenge our most negative thought patterns.

Here are sample entries and what they might look like:

Situation	Self-Talk	Consequences	Coaching Talk
I lost money when markets reversed on the surprise rate hike and my stop was hit.	"I don't know what I was thinking. I'm so stupid to be trading when there's a chance of news coming out and hurting me."	I felt down all morning and took too little risk in my next trades, even though they were good ideas that I had conviction in.	"This rate hike was not expected; c'mon, I can't anticipate everything. I had a high probability idea and kept the losses well within planned limits. That's a trade I should always take."
I got out of a winning trade too early, and the market ended up moving five points further in my direction.	"Why can't I stay in good trades? The only way I'll make money is by holding winners. I could have made so much more money."	I finished the day feeling rotten instead of feeling good about the money I made. I was in a bad mood all evening.	"This is just my frustration talking. The market had been choppy all day; it's part of my plan to take profits more quickly in choppy markets. Give me a break; I got a good price and made a nice profit. Beating myself up is just going to ruin my next trades."

Notice how the fourth column, the trading voice, is close to what you might say to someone else who might be going through your situation. It is an attempt to provide a perspective that is not so blaming and self-critical. As with other parts of the journal, it's important that this fourth column be detailed, so that you have an opportunity to really think about and absorb the alternative view. Writing, "I shouldn't be so hard on myself," is less effective than writing, "This is the same kind of talk I heard from the boss at my old job who couldn't stand me. I hated him, and I hate how he treated me. I don't deserve this and I'm not going to do it to myself." It is also effective to elaborate the negative consequences of the automatic thinking in that fourth column: "This kind of thinking has interfered with my trading all year long. I'm not going to let it cost me any more money!"

> You want to counter your automatic thoughts with emotional force; it is the emotional experience of challenging your ways of thinking that will cement the new patterns.

The reason the journal is effective is that it provides a regular, structured opportunity for you to take the self-coaching role: it's a great way to practice mentoring yourself. The use of the journal may feel artificial at first, but—with repeated entries—you'll begin to internalize that coach's voice and start challenging your negative thinking as soon as it pops up.

The cognitive journal also offers an excellent tool for reviewing your trading, particularly if you add a simple fifth column and track your profitability each day and/or grade the quality of your trading for that day. That column enables you to see how your progress in changing your self-talk is related to your trading progress, adding to motivation. Another alteration to the framework is to create an audio journal, so that all of your entries are spoken out loud in real time. This not only helps you restructure your thinking during breaks in the trading day, it also provides a useful day's end review and cements your lessons.

COACHING CUE

Where traders often fall short with the cognitive journal is in making it more of a logical exercise than a psychological one. Traders challenge their negative thoughts in a calm, rational manner, but that doesn't carry emotional force. The research literature in psychology suggests that we process emotional material more deeply than ordinary thoughts. You want to make your challenging of negative thought patterns into an emotional exercise where you vigorously reject the thinking that is holding you back. It helps to keep in mind that these are the thoughts and behaviors that have sabotaged your trading, cost you money, and threatened your success. If there was a person posing such a threat to you, you would surely confront him and reject his influence. When you personalize your automatic thoughts, you can create more powerful emotional experiences that aid the restructuring of your perception.

As your own trading coach, you want to use tools such as the journal in a way that helps your trading, not that becomes burdensome. It takes a bit of experimentation to see how the journal best fits into your workflow and routine. An excellent rule is that you won't make significant progress until the time you spend in the self-coaching mode exceeds the time you spend in the throes of negative, automatic thinking. The journal is a useful way to ensure that you get that coaching time.

LESSON 59: CONDUCT COGNITIVE EXPERIMENTS TO CREATE CHANGE

If people are like scientists, who construct their theories of the world based on their observations and experience, then it should be possible to treat their expectations as hypotheses that could either be confirmed or contradicted. *When you generate new observations and experiences that disconfirm negative thought patterns, you gradually modify those patterns and eliminate their distortions.*

Sometimes just a review of recent experience in a Socratic dialogue can be enough to challenge and modify negative views. "Whatever I do in the market is wrong!" might be one negative thought that automatically kicks in when the trader is losing money. A simple review of recent results, however, may bring the trader back to reality: "Wait a minute. I've had some excellent trades this week. I need to step back and figure out what's working for me."

When you're in the midst of negative thoughts, we've seen that it helps to take the role of the observer and ask, "Is this really true? Is this what I would be saying to someone else in my shoes? Is this what I would want someone else to be saying to me right now?" By disconfirming those negative thoughts, you make them less automatic—less able to take control of your decisions.

Sometimes, however, constructing specific experiments to challenge your negative thoughts and expectations can provide the right experience to jar and reshape your beliefs. One trader I worked with insisted that diversification didn't matter to him; he just wanted to be right on his trades. When he saw a good idea in a sector, he bought every name in the group, piling into the trade. Of course, the stocks moved in a correlated way; he probably would have been just as well off if he had bought the sector ETF and had saved some commissions. His thought pattern, "This is a great idea; I have to go all in," led him to risk a large amount of his capital on a single idea, even as he tried to convince himself that he was diversified because he held many names in his book.

For this trader, good enough wasn't good enough. He couldn't view his trade as a success unless it was a home run. For every home run he hit, however, he took a harrowing loss, leaving him discouraged and worried about his future. His pattern of needing to be "all in" to make his money back was taking an emotional as well as financial toll.

I suggested that we try an experiment. The gist of the experiment was that he had to divide his capital into four equal segments. No more than one segment, at his normal leverage, could go into any single trade idea. Thus, if he thought gold was going up, he could use up to a quarter of his normal buying power to buy the gold ETF and/or to buy gold miners. If he bought five names among the mining stocks, that quarter of his buying power would be divided among the five. To utilize the other quarters of his buying power, he needed to have different ideas. For instance, while he was long gold, he might have a short position on an individual stock or sector because of unfavorable news that had just been released.

An experiment, properly constructed, can provide a powerful, firsthand disconfirmation of our schemas.

What this meant, of course, was that our trader wouldn't be using all his buying power all the time, because he wouldn't always have four truly independent (noncorrelated) ideas. When he did deploy a good amount of his capital, it would be evenly distributed among setups and ideas. Some would be devoted to short-term scalps; other money would be used for longer-time-frame ideas. Some would be long; some might be short. This process would even out his returns, enabling him to benefit from the fact that he tended to have more winning trades than losers. By eliminating the large losers through diversification, the trader could actually take less risk (experience lower volatility of daily returns) and make more money.

The trader agreed to the experiment for a week. "What do I have to lose?" was his attitude. During the week, however, he actually saw that he made more money than he had during any week of the past several months. This result convinced him to continue the experiment. "I don't need to be banging my head against the wall," he explained after a few weeks. He was making more money—and he was happier doing it. Had he not actually conducted the experiment, however, he wouldn't have truly known—in his own experience—how wrong his thinking had been. Pointing out the destructiveness of a negative thought pattern (and the benefits of a more positive one) is one thing; actually seeing it for yourself and *experiencing the difference* is far more powerful.

The successful self-coach creates powerful and vivid experiences that undercut old habit patterns.

A common myth held by traders is that they need to be hard on themselves to maintain their motivation. This is another situation where a week's experiment can be helpful: make a conscious effort to stay constructive and positive every day for a week, and let's see how you feel and how you trade. When a trader sees that when he gives up the negative pattern he actually improves his concentration and the process of his trading, he gains considerable incentive to extend the experiment.

Your assignment, as your own trading coach, is to create a simple experiment—even if it's just for the span of a single day—in which you disrupt the negative thought patterns you've identified and just see what happens to your mood and your trading. If you don't like the results of the experiment, you can always go back to old ways and retool. If, however, you find that you can focus on trading better, that you stick to your plans better, and feel better about your work as a result, then you can decide to extend the experiment in time and perhaps also to other facets of your life. Our negative thought patterns have been the result of learning; surely we are capable of acquiring new ways of viewing our trading and ourselves. Well-constructed experiments provide us with the catalyst for changing that viewing—and that can change our doing.

COACHING CUE

Every trading rule can be turned into a cognitive experiment: See what happens when you follow the rule religiously—your trading results, your mood, and your decision-making. Many times, traders harbor fears in the backs of their minds as to negative consequences of sticking by their rules. By constructing experiments around the rules, we can see, firsthand, that these consequences are manageable and nothing to be feared.

LESSON 60: BUILD POSITIVE THINKING

The lessons for cognitive coaching thus far have emphasized ways to identify and restructure negative, automatic thoughts. What, however, of positive thought patterns? How can we become more intentional in building these? Fortunately, many of the cognitive techniques that work well to unlearn negative thought patterns can also be used effectively to cement positive ways to view self and world.

Note that the positive thinking we're looking to build is not necessarily positive in the superficial sense. Look into a mirror and tell yourself how you're the best trader, how you're going to make so much money, etc. This process is not positive thinking; it is delusional. It also reinforces unrealistic expectations, setting traders up for disappointment.

Rather, positive thinking is thinking that leads to constructive responses to challenging situations. For instance, a trader may make a rookie error and might chide himself for the mistake, using the incident to firm up his execution and attention to detail. This is very positive. A trader might also simply tell himself, "You're really not trading well; you can do better than this." That might be an accurate assessment and a prod toward greater motivation.

Positive thinking is not necessarily optimistic thinking; it is *constructive* thought.

How do you know the schemas and thinking patterns that are best for you and your trading? Fortunately, we can create a customized cognitive journal precisely for this purpose. Recall that in the traditional journal, the first column describes specific incidents of problematic trading; the second column summarizes the self-talk associated with the incidents; and the third column lays out the consequences of the self-talk. To create a format to track positive thinking, we use the journal to highlight episodes of positive trading. The first column describes what was happening in the markets at the time of the exemplary trading. The second column features the self-talk that occurred before and during these incidents; the third column identifies how the self-talk contributed to good trading practice. *In other words, you use the journal to highlight what you're doing when you're trading at your best.*

Observe that this doesn't mean that you only focus on your profitable trades, though many of your positive journal entries will be profitable occasions. Rather, you want to focus on *all* occasions when you traded well, even if you took normal losses. For instance, if you took a trade with very favorable risk/reward but were stopped out at your preset level and later reentered the idea for a gain, which would be a very positive episode of trading. The role of the journal is to isolate the thought processes that enabled you to keep your losses small and your trading flexible.

The cognitive journal can be used to identify the best practices in our thinking and trading.

One example of such a positive-oriented journal appears below. Once again, we are keeping the journal entries detailed so that we can crystallize in our minds the kinds of self-talk that are associated with our best trading. Some of the most useful entries will come from occasions when we *don't* make our usual mistakes and manage to break free of old, unhelpful patterns.

Situation	**Self-Talk**	**Trading Outcome**
The market gapped up at the open and continued to rise before I could get into the trade.	"I'm afraid I'm going to miss this move, but I've seen what happens when I chase a runaway market. I'm going to wait for the first pullback toward zero in the NYSE TICK and then see if that price level holds when I enter on the next bounce. It's more important to get a good price on my trades than to be involved in every move."	The market pulled back more sharply than I expected and made a feeble bounce. I saw that the buying was not continuing and actually sold the bounce for a nice scalp. Staying out of the market, being okay with the possibility of missing a trade, and sticking to my execution rules kept me flexible and made me some money.
I was stopped out of a trade within a few minutes of entry, as large selling pressure entered the market	"We just broke important levels in all the indexes. I just paid for useful information. The market is going to test its overnight lows if we see continued selling pressure."	I waited for the first bounce and sold the market, riding the trade down to the overnight lows. I looked at the losing trade as a useful piece of market data instead of as a failure. That led to a good trade.

Notice how the journal highlights the specific thoughts that led to the good trading decisions. By rehearsing this thinking, you can turn it into a positive set of habit patterns. Some of the best ways of thinking, I've found, have come from my interactions with successful traders. Talking with them has provided a model for how I can talk to myself during challenging trading occasions. For instance, one trader set his entry price at a level that would ensure a favorable risk/reward for the trade and said, "The market has to come to me." Instead of telling himself that he had to chase after opportunity, he insisted that he would only play when the market action fit his parameters. This kept him out of bad trades, but it also gave him an ongoing sense of control over his trading. It is difficult to feel stressed out by markets if you feel in control of your risk. I eventually adapted this way of thinking to my own entries, simply by never entering long trades on a high NYSE TICK reading and never selling the market on low readings. By making the market come to me, I found that I greatly reduced the heat I took on trades, maximizing profits. "The market has to come to me," became one of my cognitive best practices.

Your assignment is to identify the ways of thinking that put you into your best trades and that enable you to manage risk most effectively. Once you identify how you think at your best, you have a model that you can replicate day after day in your trading, turning virtues into positive habits. You don't have to be mired in cognitive distortions to benefit from a cognitive journal. Use the journal as a discovery tool for your best practices. It's an exercise even the most experienced, successful traders can benefit from.

COACHING CUE

A trader I worked with used the phrase *make them pay* (and other, choice colorful phrases) when he saw that the longs or shorts were overextended in a market. He would not exit until he saw evidence of high-volume puking from the traders running from cover. The idea of *make them pay* engaged his competitive instinct and kept him in winning trades. Frequently, he would add to his position on retracements, eager to make them pay even more. You may find that you use similar phrases during your best trades. Cement those phrases into cognitive patterns that you can rehearse. The phrases keep you grounded in best practices.

RESOURCES

The *Become Your Own Trading Coach* blog is the primary supplemental resource for this book. You can find links and additional posts on the topic of coaching processes at the home page on the blog for Chapter 6: http://becomeyourowntradingcoach.blogspot.com/2008/08/daily-trading-coach-chapter-six-links.html

More material on cognitive approaches to change can be found in my chapter on "Cognitive Techniques for Enhancing Performance" in *Enhancing Trader Performance*. See also the chapter on "Cognitive Therapy: Introduction to Theory and Practice" by Judith S. Beck and Peter J. Bieling in *The Art and Science of Brief Psychotherapies* (American Psychiatric Press, 2004). Of additional interest might be the article "Remapping the Mind" from the articles section of my personal site: www.brettsteenbarger.com/articles.htm

I like books that interview successful traders and portfolio managers; these books provide positive models for how to view markets and trading decisions. Among the most popular are the *Market Wizards* books by Jack Schwager; *Inside the House of Money* by Steven Drobny (Wiley, 2006), and *Hedge Hunters* by Katherine Burton (Bloomberg, 2007). Other models can be found in the writings of the contributors to Chapter 9: http://becomeyourowntradingcoach.blogspot.com/2008/08/contributors-to-daily-trading-coach.html. See also the Daily Speculations site (www.dailyspeculations.com) for interesting ways to think about markets and trading.

Without self-knowledge, without understanding the working and functions of his machine, man cannot be free, he cannot govern himself and he will always remain a slave.

—G.I. Gurdjieff

Behavioral methods in psychology are the outgrowth of early research into animal learning, emphasizing the roles of conditioning and reinforcement in the unlearning and learning of action tendencies. Modern cognitive-behavioral approaches to change treat thinking as a kind of behavior, making use of such methods as imagery and self-statements to modify our reactions to situations. Like the cognitive restructuring framework from Chapter 6, behavioral methods make extensive use of homework assignments in fostering change. The focus is on here-and-now skills-building, not explorations of past conflicts and their repetition in present-day relationships. Behavioral methods have been especially powerful in addressing anxiety problems, as well as issues of anger and frustration. In this chapter, we'll explore behavioral techniques that you can master as part of your own self-coaching. You'll find these techniques particularly relevant to help you deal with performance pressures and impulsive behavior.

Because the essence of the behavioral approach is skills-building, you will benefit from these methods to the degree that you are a diligent student. Frequent practice of the techniques and application of skills to new situations is crucial in making the behavioral efforts stick. Pay particular attention to the exposure methods discussed later in the chapter and also summarized in *Enhancing Trader Performance*. Pound for pound, so to speak, I find these the most useful methods in the coaching arsenal. Let's take a look at how you can master these methods for yourself . . .

LESSON 61: UNDERSTAND YOUR CONTINGENCIES

The essence of behavioral psychology is that we share many of the learning mechanisms found in the animal world. My cats Gina and Ginger, for instance, have learned that, when I get up at 5 A.M., I will give them some moist food for their breakfast. As soon as they hear me walking about, they come from wherever they've been sleeping and hustle into the kitchen, looking up at me with expectation. Because of repetition, they have learned to associate my walking around after a period of quiet with being fed. This is the essence of stimulus-response learning: animals learn to associate response patterns to stimulus situations. The contingencies between situation and response are reinforced over time, strengthening the learned pattern.

> Much of our behavior consists of simple responses to particular situations.

In traditional behaviorism, it is not necessary to explain these learned connections with reference to the mental states of the learners. The cats don't explicitly reason that it must be morning and I am rising for the day, so they should go to the kitchen. Nor does reason enter their decision to come to the portion of the house where we keep the moist food rather than the dry. Rather, the stimulus of hearing me awaken triggers their anticipation, much as hearing a particular old song may trigger associated memories. In the cognitive restructuring approach to coaching, we look to remap the mind and shift the explicit thinking of traders. In the behavioral mode, the goal is to unlearn associative connections that bring negative outcomes and acquire new connections that will be more adaptive.

In behavioral psychology, unlearning is the flip side of learning: if we don't reinforce a particular contingency over time, the associative links are weakened, and the response patterns eventually die out. If I were to ring a bell each evening and feed the cats but then not feed them in the morning, eventually they would stop coming to the kitchen in the morning. Instead, they would learn to come running at the sound of the bell. You build a behavior pattern by reinforcing it; you divest yourself of the pattern by removing reinforcement.

Many negative behavior patterns in trading, from this perspective, occur because they are either positively reinforced or negatively reinforced. This distinction is important and not well appreciated. Positive reinforcement is like the feeding of the cats: people come to associate something favorable with a particular stimulus situation. Thus, for example, I may associate my early-morning market preparation with a particular emotional state of readiness and mastery. That linkage has me looking forward to the preparation time and sticking with my routines. The contingencies between being prepared and feeling good (or being prepared and making good trades) are reinforced over time until they become ingrained habits.

Negative reinforcement is a bit subtler, and it lies at the heart of why traders seem to cling to patterns that bring them losses. In negative reinforcement, it is the removal of a negative set of consequences, not the appearance of something positive, that strengthens the bond between stimulus situation and response. Let's say I am in a trade that is going against me and I bail out of the trade at the worst possible time, when everyone else is selling. Intellectually I may know that, on average, this is an inopportune time to join the crowd, but the trade is so painful at that point that the exit feels, for the moment, like a relief. Drug addicts commonly begin their habits first by seeking a high (positive reinforcement), then by seeking to avoid withdrawal (negative reinforcement). The avoidance of pain is a powerful human reinforcement, and it shapes learned action patterns just as effectively as the introduction of pleasure.

> Many destructive trading behaviors are the result of pain-avoidance.

One of the most devastating examples of learned behavior patterns in trading is the association of thrills and excitement with the assumption of risk. When traders take too much risk, they experience profits and losses that are very large relative to their portfolio size. Some traders may find these swings stimulating, to the point where they become their own reinforcements. These traders find themselves trading, not for profits, but for thrills. Inevitably, the law of averages catches up to such traders. When these traders go through a series of losing trades, days, or weeks, their high leverage works against them and they blow up. This is not because thrill-seeking traders are inherently self-destructive. Rather, it is because they have learned, through repeated emotional experience, the linkage between risk-taking and excitement.

Research suggests that contingencies between situations and responses are more quickly and deeply learned if they are accompanied by strong emotion. This process is how people can become addicted to powerful drugs after only a few uses. It is also how we can become fearful and paralyzed by a single traumatic incident. An animal that would take weeks to learn a new trick can learn to avoid tainted food after getting sick on it just once. *Emotion accelerates behavioral learning.* This is the source of many trading problems, and it also opens the door to powerful behavioral coaching methods.

As your own trading coach, it is important that you understand your own contingencies: the linkages between your expectations and your behaviors. Instead of thinking of your trading problems as irrational, think of them as learned patterns that are supported by something positive that you gain or something negative that you avoid. Something is reinforcing your worst trading behaviors: once you understand that contingency, you are well-positioned to remove the reinforcement and introduce new reinforcers of desired trading patterns.

> Behavioral coaching is about reinforcing the right behaviors and removing reinforcement from the wrong ones.

To get started, think back to your most recent episode of truly bad trading. I can recall, for instance, a recent incident in which I so convinced myself of a turnaround in a falling market that I held a position well beyond the original stop-loss point. What is the reinforcement in that situation? In my case, I had been on a nice winning streak and I didn't want that to end. I associated getting out of the trade with breaking my streak; as long as I was in the trade, I could retain hope that my streak would be intact.

That reasoning makes no sense of course and, if pursued to its logical conclusion, I could have given back every ounce of profit I made during the streak just by holding the one bad trade. But the emotional connection was strong: I was attached to the winning—so much so that its pull was greater than the pull of simply trading well.

So take a look at your most recent episode of horrific trading. What gain were you associating with the bad trading? What negatives were you looking to avoid? What was the contingency at work? As previous lessons

COACHING CUE

Identify the emotions that are most painful for you and then track their occurrence during your trading. For some traders, this will be the pain of losing. Other traders respond negatively to boredom or to the helplessness of uncertainty. Many times your worst trading decisions will be the result of trying to rid yourself of those emotions. This negative reinforcement leads to hasty, unplanned trading behaviors that seem to make no sense in retrospect. If you can identify the negative reinforcement at work, you can more consciously and constructively deal with the difficult feelings.

have emphasized, the first step in the change process is to become our own observers and recognize the patterns that hold us back. Your behaviors, as irrational and destructive as they seem, are there for a reason. A careful behavioral analysis will reveal the reasons—and will position you well for changing those.

LESSON 62: IDENTIFY SUBTLE CONTINGENCIES

The linkages between situations and our behavioral responses to those linkages are sometimes quite clear. When traders experience fear in a volatile market and prematurely exit a position, we can readily appreciate that they are managing their emotions, not their capital. The relief at being out of a fast market outweighs objective considerations of risk and reward.

Other times, the contingencies that govern our behavior are far more subtle and difficult to identify. For that reason, such patterns can be extremely challenging to change. If we don't know what we're responding to, it's difficult to shape a different response pattern.

Subtle shifts of mood are one example of stimulus situations that could affect decision-making without our awareness. For instance, some individuals are emotionally reactive to the amount of sunlight they receive and can experience winter blues or even seasonal affective problems during periods of low sunlight. This disruption can affect a trader's concentration and motivation, interfering with his research and preparation. Similarly, family conflicts can affect mood, which in turn affects trading. One trader I worked with found himself less patient with his ideas, entering and exiting before his signals unfolded. When we looked into the problem, it was clearly episodic—not something that occurred every day. During periods of conflict at home, he was more irritated, and that manifested itself as impatience in his trading.

> The problem patterns in our trading are often triggered by subtle shifts in mood and energy level.

Many physical cues can also affect mood and cognitive functioning. These cues include fatigue, hunger, muscle tension, and fitness. I know that I process market data much more effectively and efficiently when I am alert. Anything that affects my energy level adversely will also impair my ability to synthesize large amounts of market information. This is not only because I am less cognitively efficient, but because lack of alertness also affects my mood. In a more fatigued state, I tend to feel less emotionally energetic and optimistic. I won't look for that creative market idea; I'll become more discouraged and risk-averse after losses. If I'm not clued into my physical state and its relationship to my mood, I'll simply think that these periods of lesser performance are random. In fact, most mind and body shifts are as stimulus-response bound as any animal behavior in a learning experiment.

As I emphasized in *The Psychology of Trading*, a great deal of our learning is state-based: *what we know in one state of mind and body can be quite different from what we process in another state.* When I am listening to favorite music, my mind is expansive, I can see broad market relationships, and formulate big picture ideas to guide the week's trade. In a state when I'm pressured for time or distracted by an irritating situation, I suffer from tunnel vision, losing the large perspective. On those occasions, I'm much more likely to make impulsive trades, responding to recent price action rather than broad market dynamics. Often those trades lack good risk/reward qualities; they are much more likely to be losers than winners.

I refer to these subtle environmental cues as triggers, because they can set off behaviors that are unplanned and unwanted. When I'm irritated, for instance, I've learned to rid myself of the feeling by simply pushing aside whatever is bothering me. This reaction is a classic example of negative reinforcement. If the thought of doing errands irritates me because I have other things I want to be doing, I quickly push the errands away and focus on what I want to do. The errands don't get done, of course, and loom as a chronic irritant. My pattern of procrastination is clearly negative reinforcement–based, but it is not helpful: it leaves me with a lingering negative mood and a backlog of unfinished business. Worse still, continually reinforced, the negative mood can become a pattern in my trading. It's not too great a leap from procrastinating over errands to procrastinating about acting on a losing position.

Many of the behavioral patterns that interfere with our day-to-day lives also find expression in our trading.

Sometimes, when coaching yourself, you won't know what is triggering your most troubling trading behaviors. They seem to come out of nowhere. That's when it's most important to use a trading journal to catalog all the possible factors—physical, situational, emotional, relationship-based, trading-based—that might be associated with your problematic trading. When you engage in this cataloging, you want as open a mind as possible; often, the patterns will be different from ones that you've been considering. One trader I worked with experienced trading problems for no apparent reason; only after considerable review did we figure out that these problems occurred when he experienced problems with the firm's management. The frustration led him to seek gratification from his trading, impelling him to overtrade. He didn't make a conscious connection between the two; rather, he was trading to manage an emotional state in a simple stimulus/response manner.

The cataloging you undertake in your journal may need to cover a considerable period of time before you notice patterns. What you're likely to find, however, is that how you trade is affected by your physical and emotional state—which is affected by situational factors at home and work. Understanding these contingencies enables you to build some firewalls into your trading practice, as we will see in the next several lessons. Without such understanding, however, you're likely to blindly repeat history, losing a measure of self-determination. It's when we create our own contingencies that we truly possess free will and the ability to pursue our chosen goals.

COACHING CUE

Track your daily physical well-being—your state of alertness, your energy level, your overall feeling of health—against your daily trading results. Many times fatigue, physical tension, and ill health contribute to lapses in concentration and a relapse into old, unhelpful behavior patterns. It is difficult to make and sustain mental efforts when you lack proper sleep or feel run down from a lack of exercise. Very often, our moods are influenced by our physical states, even by factors as subtle as what and how much we eat. When you keep a record of your daily performance as a function of your physical condition, you can see these relationships for yourself and begin preventive maintenance by keeping body—and thus mind—in peak operating condition.

LESSON 63: HARNESS THE POWER OF SOCIAL LEARNING

One of the greatest mistakes traders can make in coaching themselves is to work on their craft in isolation. It is easy to become isolated as a trader, particularly given that all it takes is a computer and Internet connection in one's home to access the most liquid markets. During my work with trading firms, however, I have consistently seen how access to other professionals aids the learning process. From peer professionals you obtain role modeling, encouragement, and valuable feedback on ideas. A social network of traders also offers powerful behavioral advantages that can aid your self-coaching. Thanks to Web 2.0 and the many online resources available, such social networking can occur virtually, not just within a trading firm.

Psychologist Albert Bandura was one of the first behaviorists to observe how reinforcement in a social context can aid the acquisition of new behaviors. *When we observe others rewarded for positive behaviors, the vicarious experience becomes part of our learning.* Similarly, when we see others making mistakes and paying the price for these, we learn to avoid a similar fate. In this way, your learning becomes a model for others and theirs provides models for you. Experience is multiplied many times over, accelerating the learning process.

> Social learning multiplies experience and shortens learning curves.

Since I first began full-time work as a coach for trading firms in 2004, my own trading has changed radically. I have learned to factor intermarket relationships into my trading, and I have learned to think in terms of risk-adjusted returns, with each trade carefully calibrated for both risk and reward. I am keenly aware of the effect of position sizing on my returns, and I carefully track my trading results to identify periods of shifting performance that might be attributable to market changes. These changes all resulted from observing successful professionals across a variety of trading settings, from proprietary trading shops to investment banks. Since instituting these changes, I've enjoyed greater profitability with smaller drawdowns. Seeing how the best traders managed their capital provided me with powerful lessons that I could apply to my own trading.

Perhaps my most effective learning, however, came from observing failed traders. I have seen many traders lose their jobs (and careers) as the result of faulty risk management and an inability to adapt to market shifts. Those failure experiences were painful for the traders, but also for me, as I developed close relationships with many of them. Their pain and the crushing of their dreams was powerful learning for me. I vowed to never make those mistakes myself.

> We learn most from emotional experience, including the experiences of others.

When you share ideas in a social network, including self-coaching efforts, you obtain many learning experiences that become your own. Vicarious learning is still learning, whether it's learning concrete trading skills or learning ways of handling performance pressures. One of the real values of published interviews with successful traders, such as found in the *Market Wizards* series, is that you can learn from the experience of others. When you actually observe this experience in real time, however, the contingencies are much more immediate and powerful. How a trader in the hole pulls himself out, or how a trader adapts to a changing market, or how a trader successfully prepares for a market day—all provide models for your own behavior. You learn, not just from their actions, but also from observing the results of their actions.

Once you enter a social network of capable and motivated peers, the praise and encouragement of the group become powerful reinforcers. Most of us want to be respected by our fellow professionals, and the support of valued peers can be a meaningful reward. This reinforcer occurs among children, who find that they are praised by teachers, parents, and peers for good behavior and not praised when they behave badly. Over time, this differential reinforcement creates associative links for the child, so that he will do the right things even if no immediate praise is available. Similarly, young, developing traders will absorb the praise of mentors like a sponge; this helps them associate the right trading behaviors with favorable outcomes. When you share successes with fellow professionals, you turn social interaction into social learning.

Find experienced traders who will not be shy in telling you when you are making mistakes. In their lessons, you will learn to teach yourself.

For this lesson, I encourage you to locate online networks of traders (or assemble one of your own) in which there is openness about trading successes and failures. Online forums are a possible venue; you can also connect with readers of trading-oriented blogs who participate in discussions. Or perhaps you will choose to write your own blog, openly sharing your trading experiences and attracting like-minded peers. When you network with traders who have similar levels of motivation, commitment, and ability (as well as compatible trading styles and markets), you can establish a framework in which learning follows from shared ideas and experiences. We've seen in Chapter 5 how relationships can be powerful agents of change. In the behavioral sense, you want to be part of the learning curves of other traders, so that you can absorb their lessons. A great start is to establish such a mutual learning framework with just one other compatible trader. Their emotional learning experiences become yours; yours become theirs. Their victories spur your ability to do the right things; your accomplishments show them the path to success. This effectively doubles your behavioral learning, supercharging your self-coaching efforts.

COACHING CUE

An increasing number of professional trading firms—particularly proprietary trading shops—are creating online access to their traders, trading, and resources. Several of these firms are mentioned in Chapter 9. Read the blogs from these firms and participate in their learning activities as an excellent way to connect with other traders and model their best practices.

LESSON 64: SHAPE YOUR TRADING BEHAVIORS

Two children, two different homes: both improve their test grades in math; both fail in English. In the first home, the parents praise the improvement in math and encourage similar progress in English. In the second home, the parents call attention to the English grade and demand to know why the child couldn't pick up that grade as well. Which child will be most likely to show further school progress?

Behavioral psychologists who utilize behavior modification as a means for altering action patterns would support the first set of parents. Positive reinforcement, as a whole, works better than punishment. If we reinforce the right behaviors, the child will learn to do the right thing. If we punish the wrong behaviors, the child will learn to fear us. Nothing positive is necessarily learned.

> Punishment fails because it does not model and reinforce the right behaviors.

Many traders seek to motivate themselves more through punishment than praise. These traders focus more on their losing trades than on their winners. They spend more time on weak areas of trading than building and extending their strengths. Such traders learn to associate unpleasant things with trading. These traders anticipate criticism and punishment and find it difficult to stay wholeheartedly engaged with the learning process.

We can see such dynamics at work in the journals many traders keep. One page after another details what the trader did wrong and what he needs to do to improve. Self-evaluations emphasize the bad trading, everything that could have been better. It's little wonder that these traders find it difficult to sustain the process of maintaining a journal. After all, who wants to face negativity and psychological punishment every working day?

Many traders fail to sustain work on their trading because they find little positive reinforcement in their work.

Trainers use frequent rewards to teach animals tricks. The trainers don't expect the dog to, say, jump through the hoops all at once. Rather, they will first give a reward each time the dog approaches the hoop. Then the trainer will wait for the dog to go through the hoop before they dole out the reward. Then they'll lift the hoop just a couple of inches and reward the dog when it jumps through the hoop. Then they'll add a second hoop and a third . . . they'll raise the hoops a little at a time . . . all the while requiring new behaviors that are closer to the desired endpoint before giving the reward.

This process is known as *shaping*. Trainers shape animal behaviors by rewarding successive approximations to desired ends. In a classroom, a teacher might first reward a disruptive student for five minutes of quiet attention. Later, it will take 10 minutes for the student to earn the reward; eventually the reward will require an entire class period of good behavior. Frequent-flyer programs at airlines aren't so different. At first, you earn bonuses for simply joining the programs. Only after you ride the airline regularly, however, do you earn later rewards. If you want the greatest perks, you have to shape your riding habits to fit the program.

Shaping is a testament to the power of positive reinforcement. Imagine punishing the dog for not going through the hoops. The chances are good that the dog would simply cower in the presence of the trainer; it certainly wouldn't figure out the right behaviors from the punishment of the wrong.

When you are your own trading coach, you are the trainer as well as trainee. You are teaching yourself to jump through the hoops of good trading. For this reason, you need an approach to coaching that is grounded in positive reinforcement. Your coaching must stay relentlessly positive, building desired trading behaviors—not punishing the wrong ones.

You can keep a positive tone to the learning process by shaping your trading behaviors: rewarding small, incremental progress toward the desired ends.

The first place to implement the shaping approach is in a journal. As an experiment and a worthwhile exercise, try keeping a *positive trading journal* for a few weeks. Divide your trading into several categories, such as:

- Research and preparation.
- Quality of trade ideas (ideas that carry conviction).

- Number of diversified (uncorrelated) trade ideas.
- Quality of entries (favorable risk/reward for trades; low amount of heat taken per trade).
- Sizing/management of trades (scaling in/out by planned criteria).
- Execution of exits (following profit targets/stop losses).

Each journal entry then focuses on *what you did right* in each of those categories each week. You write down, in specific detail, your best performance in each of these areas and then you review your entries before trading the next week, with the aim of continuing the positives.

As the previous lesson noted, this use of positive reinforcement and shaping is even more powerful if you conduct your assessment in a social framework, where you exchange your weekly positive report cards with one or more valued peers. This framework allows you to support the progress of others, even as they reward yours.

One of the better pieces of self-coaching from early in my trading career occurred when I set a goal of reaching a certain size in my trading account. I normally don't emphasize P/L goals, but in this case I wanted a tangible focus on steady profitability. Once I reached the goal, my commitment was to withdraw a portion of money from the trading account and use it for something enjoyable for the family. This emphasis rewarded my longer-term progress, but also brought my family into the positive reinforcement. When I finish this book, one of my personal goals will be to lose some weight—long hours in airplanes and hotels between working with traders have taken their toll. I've promised myself a new wardrobe from a Chicago tailor if I reach my weight goals. Each week I'll be weighing myself and tracking my progress. With every opportunity to snack, I'll be thinking about that new wardrobe and how I would feel if I didn't make weight that week. There's little doubt in my mind that I'll reach my goal.

> Tangible rewards for your success are among the strongest positive reinforcers.

The key to making a positive journal work is shaping. At first, you jot down entries for even very small things that you did right. Later, you only make notations of larger examples of virtuous trading. *If you conduct the shaping process properly, you'll always have good things to write about—even on losing days.* This process ensures that you're always learning, always building on strengths, always keeping your motivation up. The difficult part about self-coaching isn't just making progress, it's sustaining the progress. Progress is much easier to accomplish when your focus is building yourself up, not tearing yourself down.

COACHING CUE

What is meaningful for you as a tangible reward for your self-coaching progress? A vacation with loved ones? A new car? One trader I work with donates a portion of his profits to a charity he deeply believes in; helping them out inspires his own efforts. It helps to reinforce the small steps of progress via shaping, but it also helps to have a larger goal that you're working toward; a goal that is meaningful for you. Remind yourself periodically of the goal; track your progress toward the goal. The psychologist Abraham Maslow recognized clearly: we perform at our best when we are impelled toward positive goals, not driven by deficits and unmet needs.

LESSON 65: THE CONDITIONING OF MARKETS

A large part of money management follows from a deep appreciation of fat tails. Market returns are not normally distributed; they show a higher proportion of extreme occurrences than you would expect from a simple flipping of coins. This is true across all time frames. The odds of a multiple standard deviation move against you (or for you) are sufficiently high that, if you're in the market frequently over a long period of time, you will surely encounter those periods in which markets stay irrational longer than you can stay solvent.

The distribution of market returns is also leptokurtic: it is far more peaked around the median than a normal distribution. This implies that market moves revert to a mean more often than we would normally expect by chance. Just as a market seems to be moving in one direction—trending—it reverses course and finishes little changed.

It is difficult to imagine a situation better designed to create frustration. Markets produce large moves more often than would be expected if returns were distributed normally, which leads traders to seek large, trending moves. But markets also revert to mean returns more often than we would expect in normal distributions, creating many false trends. If you trade a countertrend strategy, you run the risk of being blown out by a multiple standard deviation move. If you try to jump aboard trends, you'll find yourself chopped to pieces during false breakouts.

> The very structure of market returns ensures a high degree of psychological challenge for traders.

The tendency of markets to make extreme moves amid frequent mean reversion creates interesting and important psychological challenges that affect self-coaching. To fully appreciate this, we need to understand the dynamics of behavioral conditioning.

Let's say that, each time I ring a bell, I hit you over the head. Soon, you'll learn to duck as soon as you hear the bell. That is a *conditioned response*. Days later, you might be in a different location and will still duck if the bell sounds. It's automatic; not a behavior guided by explicit reasoning. You've learned to associate bell and pain, just as Pavlov's dogs associated a ringing bell with the appearance of meat. Bell rings, dogs salivate. Bell rings, you protect yourself.

Now let's take our experiment a step further. I ring a similar but different bell and once again hit you over the head. Before long, you learn to duck whenever you hear any bell. This is called *generalization*. Your conditioned responses (the ducking) have now extended to a class of stimuli similar to the original one.

Much of what we call traumatic stress is the result of such conditioning. In the *Psychology of Trading* book, I mentioned my car accident in which I was thrown from a vehicle while riding as a passenger. Just as a result of that single, powerful event, I developed an anxiety response anytime I subsequently sat in the passenger seat of a car—even when the vehicle wasn't moving! I had learned an associative connection between being a passenger and extreme danger; the conditioning stuck with me even though I intellectually knew it made no sense.

Many of our extreme reactions to market events are the result of prior conditioning.

Powerful positive emotional events can yield the same kind of conditioning. The high obtained from certain drugs can be so strong that some people will develop addictive patterns after a single use. Underlying the addiction is the learned connection between the high and the use of the substance. That, too, overrides reason and reorganizes behavior.

One of my greatest failures as a trading coach occurred with a young trader who experienced early market success. He took the time to observe markets, learn short-term patterns, and track his own trading. He started trading small and learned the important lessons about waiting for good entry points, cutting losing trades, and letting his winning trades run to their target points. The trading firm was happy with his progress and gave him significantly greater size to trade. That was where I went wrong. I should have stepped in and demanded that the trader's increase in risk be more graduated. Instead, armed with his new size, the young trader decided he would try to compete with the more experienced traders at the firm. He traded full size in his positions and his profits and losses swung wildly. Unprepared emotionally for those swings, he became impulsive and, one day, abandoned all discipline, blowing himself up on a single trade he allowed to get away from him. He never recovered from that loss and eventually had to start over at another firm.

> It is impossible to remain emotionally stable if you greatly amplify your P/L swings.

When traders are undercapitalized and still hope to trade for a living, they too are impelled to take high levels of risk to achieve their desired returns. The result is that their portfolio swings wildly, with gains and losses that represent a large portion of total account value. These financial swings bring emotional swings, both positive and negative. The larger the financial swings, on average, the larger the emotional swings. The larger the emotional swings, the greater the potential for the development of learned, conditioned responses that disrupt future trading.

When a trader undergoes an emotionally harrowing loss, many of the situational factors associated with that trade may become associated with the emotional pain. Some of these situational factors, from the trader's physical state to the particular type of movement in the market, may be quite random. Nonetheless, they can trigger the emotional pain, much like sitting in a passenger seat triggered my anxiety following the automobile accident. A trader who consulted me about problems pulling the trigger on good trade setups experienced precisely that problem. He had lost significant money shorting the market during an uptrend, incurring several large losses. Subsequently, even when his trades were small in size, he felt fear whenever he tried to short the market. The feelings associated with his loss came back as a conditioned response, inhibiting his trading. This is the dynamic behind the flashbacks that occur during post-traumatic stress: stimuli associated with the initial trauma trigger memories and feelings from that painful incident.

The problem may have been just as severe had this trader made large money on the initial trade instead of losing. The emotional impact of a windfall profit, like the impact of a crack cocaine high, would bring its own conditioning, leading him to pursue similar gains (and highs) in future trades. It is poorly understood by traders that, psychologically, outsized gains are just as problematic as outsized losses. The fat tails of returns threaten fat tails of psychological response, interfering with sound perception and decision-making.

For this reason, when you're your own trading coach, you don't want patterns of extreme returns. Steady, consistent profits are far better for psychological performance than wild swings up and down, even though they may lead to the same ultimate returns. Stated otherwise, good risk-adjusted returns are better for the psyche than extreme patterns of returns. *It's not how much you make, but how much you make per unit of risk taken that will keep you in or out of the performance zone.*

Your assignment for this lesson is to track the variability of your returns as intensively as your overall profitability. By variability of returns, I mean the absolute value of daily/weekly changes in your portfolio value: how much your account swings up or down on average each day. As markets change in their volatility and as you shift in your level of conviction about trades, you'll see changes in this variability. This tracking will tell you when you run more and less risk. On the whole, you'll want greater variability when you trade well and have many solid ideas; you'll want to cut your risk (lower the variability of returns) when you don't see markets well and when good trading ideas and moves are scarce.

> Track the volatility of your returns, not just their direction. Volatility affects trading psychology every bit as much as winning and losing.

When you track the variability of returns, you'll also be able to see when your swings in profit/loss are outliers from your historical norms. This will be an excellent alert that your levels of risk may be sufficient to generate those large emotional swings that will produce unwanted conditioned responses. Traders tend to love volatility when they're making money and hate volatility when they're losing. Psychologically, it makes sense to keep the volatility of your returns within bounds: markets may possess fat tails, but with prudent position sizing, your returns can remain stable. You don't want markets conditioning your learning; you want to be your own coach, directing your own learning.

COACHING CUE

The psychological research on trauma suggests that processing a very stressful event verbally—out loud or in writing—can be extremely helpful in making sense of that event and divesting it of enduring emotional impact. When we repeat something again and again, it becomes familiar to us and no longer evokes powerful emotion. If you encounter outsized gains or losses in your portfolio, double down in your use of the trading journal or in your conversations with peer traders to thoroughly process what happened and why. As noted above, this process is just as important following large gains as following large losses. When highly emotional events bypass explicit processing, that is when we are most vulnerable to the effects of conditioning.

LESSON 66: THE POWER OF INCOMPATIBILITY

Earlier we saw how much of what we learn is state-dependent. We associate particular outcomes with specific physical and emotional states. These associative links trigger unwanted behavior patterns when we enter those states. The classical conditioning mentioned in the previous lesson is an excellent example: if we experience overwhelming anxiety due to large losses, exiting the market may provide immediate relief. Subsequent experiences of anxiety in the market may trigger the same exiting behavior even when it would be in our financial interest to hold the position. The association between the anxiety and perceptions of danger may be so strong that it overwhelms our prior planning.

Boredom, for many active traders, can be as noxious as strong anxiety. It may be associated with failure to make money, or it may have much earlier negative associations: being lonely or feeling abandoned as a child. If you get into a trade—particularly a risky one—you immediately relieve the boredom, but you create a new trading problem. In such cases, the trading behaviors triggered by the state are more psychological in their origins than logical.

> If trading is associated with an aversive state, we tend to do what is necessary to alleviate the state, even at the expense of our portfolios.

One of the simplest behavioral techniques for breaking these bonds of conditioning is to place yourself in a state that is incompatible with the one that triggers your problematic trading. Thus, for instance, if you find that anxiety triggers hasty and ill-timed market exits, you would work on placing yourself in a calm, relaxed physical condition that is incompatible with anxiety. If boredom were your nemesis, you would cultivate activities that hold your interest during slow markets. When I am fatigued, I find that a round of vigorous exercise not only makes me more alert, but also triggers positive action patterns, as I tackle work that had previously seemed overwhelming. *If you're not in a state that supports sound decision-making, your self-coaching focus turns from the markets to yourself and doing something different to shift your state.*

Two of the methods I have found particularly helpful in maintaining states incompatible with one's triggers are controlling breathing and muscle tension during trading. When I focus on the screen and breathe deeply and slowly while I follow the market, I minimize the physical manifestations of any form of overexcitement—from overconfidence to fear—and stay in a highly focused mode that I have learned to associate with good trading. When we slow ourselves down through deep, rhythmical breathing, it is difficult to be simultaneously speeded up and excited. The careful breathing thus acts as a dampener on extreme emotion. It reinforces self-control and discipline at the most elemental level.

In my own trading, I've found that problematic trading tends to occur when I am physically tense, especially when I tense the muscles of my forehead. I rarely knit my eyebrows and wrinkle my forehead when I am comfortable in a situation. Conversely, I am prone to headaches and associate forehead muscle tension with tension headaches, which can pose a considerable distraction. By purposely keeping my forehead relaxed—widening my eyes slightly and going into a temporary stare—as I maintain the slow, deep breathing, I can sustain a state incompatible with the ones that occur when I'm on edge. Instead of waiting to become tense or nervous and then performing exercises to reduce these feelings, I proactively pursue and maintain an incompatible state *before* problematic trading occurs.

Control the arousal level of the body as a powerful means of controlling the arousal level of the mind.

I can often recognize my physical level of tension by my seating position. When I am comfortable, confident, and relaxed, I sit in the chair firmly, with my lower back and behind flush with the seat back. When market events trigger a stress response, however, I find myself leaning forward, with my seat near the end of the chair. Over time, this position gives me a backache in my lower back. I know that I'm not comfortable with my trading or with the markets when I feel that pain. Often, I'll readjust my seating, reorient my breathing, and find it easier to view the markets from a different—and more promising—angle.

The principle of incompatibility can also extend to thinking behaviors. Cognitive-behavioral work treats thinking as a discrete behavior that can be conditioned and modified just like any muscular behavior. If we tend to engage in negative thinking during trading, we can enter a mode of thinking that is incompatible with negativity before trading problems occur. I frequently have one of my cats sitting beside me as I'm trading, usually Gina. It's nearly impossible for me to become consumed with negative or angry thoughts when I am petting Gina. She alternates between licking me and rubbing her face against mine, all the while purring loudly and making kneading movements with her front paws. Stroking the cat helps me stay in touch with loving, caring feelings that are incompatible with the nastier emotions that can emerge during frustrating market periods.

One of the states that is most disruptive to my trading is what I would call a *chaotic* state, in which I feel as though I'm a step behind markets, not really understanding what is going on. It's a confused state, but also a frustrated one, as I don't feel in control in the situation. I've learned that if I place myself in environments that are incompatible with chaos, I am in a much more balanced frame of mind. Such environments are ordered and well organized—my notes and materials are readily at hand—and they are designed to evoke positive feelings. Music is particularly effective for me in this regard. It is also harder for me to feel chaotic if I have gone through a routine of research and track markets prior to the New York stock market open. I organize my ideas in advance to help me feel more organized, settled, and in control.

You can structure your trading routines to make them incompatible with stress and distress.

When you are your own trading coach, you have wide latitude in modifying your environment—inner and outer—so that it does not trigger states that are associated with poor trading. One trader I worked with loved trading in a room with other traders (he joined a prop firm) because, in the social setting, he was too embarrassed to engage in behaviors he might lapse into on his own. He found that he was much more prudent about risk-taking and much less emotionally volatile when he was accountable to others. *The key is to find a state or situation that is incompatible with the triggers for your worst trading and then build that into your normal trading routine.*

A simple way to get started is to complete the following sentence:

I trade my worst when I ______________________________.

Once you write your answer, your assignment is to create the incompatible situation. For example, I would complete the sentence with "don't do my homework." I know that my day's preparation for the trade has a huge bearing on my odds for success that day. I also know that I'm least likely to do my homework diligently if I oversleep or am fatigued. When I build stretching and physical exercise into my early mornings, I enter an energized state that prepares me for the homework: I've learned to associate the vigorous, energetic state with being prepared and engaging in my preparation. After you observe the differences in your states when you trade your best and worst, you'll be able to construct similar activities that proactively keep you in an optimal trading mode.

COACHING CUE

I mentioned above how *chaotic* feelings are a trigger for my worst trading. If markets aren't making sense to me, my mind feels scrambled and trading seems rushed. I've learned through hard experience that a powerful way to create a state incompatible with that chaos is to temporarily lower my trading size until I regain a feel for markets. With much less at risk, I don't feel pressured and yet can stay actively engaged in markets. When we control our risk we can control our emotional reactions to markets: it's tough to panic when you have little on the line. Markets seem to move slower—and our feel for them returns—when we're not distracted by emotions triggered by risk and uncertainty.

LESSON 67: BUILD ON POSITIVE ASSOCIATIONS

In the cognitive-behavioral framework, we can utilize imagery as a stimulus to evoke desired responses, triggering our own positive, learned patterns. Making use of imagery in this fashion can help us create *positive* associative links, triggering our best trading behaviors.

Let's say we have a trader who anticipates an early-morning entry into the market based upon a researched setup. Before the market opens, she visualizes the setup and her execution, noting the feelings of satisfaction from making a good decision. This positive mental rehearsal acts as a preparation for the actual trade, as she follows the behavioral pathway she has laid down in advance. I call this *anticipatory reinforcement*: by imagining the positive benefits of doing the right things, we strengthen positive associative links and make it easier to act on our learning in real time.

Many traders conduct anticipatory reinforcement in reverse: they dwell on negative outcomes and feared scenarios, undercutting their own sense of efficacy. This, in essence, is anticipatory punishment, and it leads traders to miss opportunities or to not act on them. I've found over the years that much of what separates the excellent traders from the average ones is not so much their ideas, but what they do with those ideas. Two traders will have positions go their way and then pull back a bit. The first trader, anticipating punishment, fears losing his gain and takes a quick, small profit. The second trader, anticipating reward, adds to the position on the pull back and reaps large gains. Same idea, different outcomes, all as the result of conditioned patterns of thinking.

> Our ways of thinking can reflect conditioned responses; that's how markets can control our minds.

When we reinforce positive patterns, we not only strengthen these but also begin the process of extinguishing negative patterns. In behavioral theory, a stimulus-response connection is extinguished over time if it is not reinforced. The animal that was given food each time it performed a trick will eventually stop performing the trick if food is not forthcoming. *Behavioral patterns, in this way, not only have to be learned but also actively reinforced to find active expression in our trading.* We can unlearn negative behavior patterns simply by withdrawing their reinforcement and by introducing more powerful rewards elsewhere. This is a powerful principle.

One common learned pattern among traders is the connection between anger/frustration and aggression. When traders become frustrated by market conditions—say, a choppy, directionless trade—they react out of anger and lash out by placing trades to get even with the offending market. This pattern—relieving anger by lashing out—may make traders feel better for the moment (negative reinforcement), but it leads to poor decisions and losing trades.

How can we use positive associations to unlearn this pattern of revenge trading?

Suppose a trader engages in a thorough examination of his trading during the choppy markets of the past month. He investigates charts to identify the choppy periods and then reviews all his trades from these periods, pulling out the most successful ones. What he may find is that his successful trades in choppy conditions are more selective (fewer in number); that they are placed near the edges of trading ranges; and that they are held for shorter periods of time to capitalize either on breakouts/false breakouts or on moves back within the range. His losing trades, on the other hand, tend to be placed in the middle of the range and are held for longer periods, reversing before they can hit distant price targets.

Armed with this bit of self-coaching information, the trader now can view the choppy period as one of opportunity, not threat. When he notices a trading range going into the day's trade, he can use imagery to rehearse calm caution when the market is trading near the center of the range. He can also mentally rehearse entries near range extremes, including his placement of modest price targets. When he rehearses these trade ideas, it is with the feelings associated with his prior winning trades. Over time, with repetition, he learns a *positive* association with range-bound, choppy markets. His prior behavior pattern, built on frustration and its removal, is no longer reinforced. It faces gradual extinction, as he builds the more constructive associative patterns.

> Find the market conditions that are most challenging for you and then identify how you trade them best. This process turns threat into opportunity.

Since the late 1970s, I have traded actively and gained a pretty good feel for many short-term market patterns, including patterns of volume and intraday sentiment. Many times I first become aware of these patterns with a gut feeling: something seems right or not right about the market action. I have learned through hard-won experience that I suffer in my performance if I ignore these intuitions. They are not based on hopes or fears; they are the result of implicit learning over a period of years. In my mental rehearsals, I include scenarios in which I act upon this feel for the market, recalling specific, recent trades in which I saved myself considerable grief by not overriding my judgment. This rehearsal of positive associations has created a kind of intrinsic reinforcement: I actively look forward to the emergence of those gut cues and am mentally prepared to act on them when they arise.

As I mentioned in the previous section, I have many positive associations to music. Indeed, as I'm writing this, I'm listening to music from a group called Edenbridge, a kind of music that I find both energetic and uplifting. My writing today began at 6:30 A.M., and it is now two hours later and I'm going strong. The association of the music with the writing keeps me in a positive state of mind. It keeps me looking forward to the writing, even when the editing process can become tedious. With these positive connections activated regularly, my more negative patterns of procrastination are not reinforced and gradually lose their strength. It is not necessary for me to fight my tendency to procrastinate; such internal conflict would likely create writer's block. Rather, I create a positive source of motivation that outweighs the negative reinforcement value of avoidance.

A good example of the power of anticipatory reinforcement is occurring right now as I am writing this. I'm on a 15-hour flight to Hong Kong on my way to working with traders in Asia. The cabin is dark, and I'm feeling tired. I've promised myself, however, that I can take a long-awaited rest after I finish this chapter. I find myself more motivated as I get closer to my goal; by the time I get to rest, I will have earned it. Ultimately, the positive reinforcement of living up to my deal and earning the rest outweighs any negative reinforcement value of avoiding the writing out of tiredness.

Find your strongest motivations and link those to your best behaviors.

Your coaching assignment for this lesson is to create what-if scenarios for the day's trading, rehearsing the good, planned trades you would make in each scenario. These rehearsals should be detailed and vivid, accompanied by a visualization of the pride and satisfaction you experience when you trade well. For every single what-if outcome, you should envision a concrete response that embodies good trading. In this way, you plan your trading as well as your trades, but you also strengthen the bonds of positive learned patterns and extinguish the negative patterns. As your own trading coach, you have the power to be teacher as well as student: the shaper of behavior as well as the one whose behavior is shaped. If you take the active learning role that lies at the heart of the behavioral approach, you become the programmer of your own patterns.

COACHING CUE

I find it helpful to help traders identify the highlights of their trading from the past week: what they did especially well. From these highlights, we frame ideas about what the trader is really good at—what makes her successful. We then use this *what I'm good at* idea to frame positive goals for the coming week: how the trader is going to enact those strengths in the next few days. Because these are goals we track together, we create a situation of anticipatory reinforcement and a momentum for continuing best practices. This is a process you can carry forward with trading colleagues: share what you do best and how you put your best talents and skills into practice. Focus on your best trading and you begin the process of extinguishing your worst practices.

LESSON 68: EXPOSURE: A POWERFUL AND FLEXIBLE BEHAVIORAL METHOD

If I had to name a single behavioral method that is of greatest value to traders, it would be exposure. As I described in my chapter on behavioral methods in the *Enhancing Trader Performance* book, exposure is a technique that enables you to reprogram those stimulus-response triggers that set off faulty trading.

A major idea underlying exposure is that the avoidance of negative experience itself becomes a reinforcer, preventing people from overcoming learned fears. Let's say, for instance, that I have taken large losses on a short position and now experience fear whenever a buy program hits the tape and moves the index higher by several ticks. I can avoid that fear by simply exiting the position. While that avoidance is a relief, it never addresses the learned connection impelling my behavior. Indeed, it reinforces my fear by acting on it. It is impossible to overcome a fear when you give into it.

> We overcome fear by facing it successfully.

In exposure work, we intentionally expose ourselves to the situations that set us off. Generally, this process begins with imaginal exposure (facing situations in realistic imagery) and progresses to in vivo (real time) exposure. These exposures pair the trigger situation with learned skills that invoke a state incompatible with the bad trading. Thus, in the example above, we might rehearse a calm, focused state of mind while vividly imagining the market moving higher against us.

Think about what this accomplishes. On one hand, we immerse ourselves in thoughts and images of something we find threatening. We force ourselves to experience our worst fears. At the same time, however, we make special efforts to keep ourselves calm and controlled. We talk to ourselves in calming ways, slow our breathing, and keep our bodies relaxed. We do this again and again, repeating the imagined scenarios until we are able to stay completely calm and focused throughout. In that way, we extinguish the learned connection between the situation and the fear.

> Exposure methods are ways to reprogram our emotional responses to situations.

Two steps are important to make exposure effective:

1. Before you try to expose yourself to imagery of your trigger situations, make sure you've thoroughly learned the coping skill that you'll be using as part of the pairing. For example, you want to practice a deep breathing, muscle relaxation routine every day for at least a week to ensure that you can focus and relax yourself on demand. At first, practicing the technique may take 20 minutes or so to get quite relaxed; later it will take only 15, then 10. Eventually, with enough practice, you'll be able to relax and focus yourself quite effectively with just a few deep breaths. You want to get to that point *before* undertaking the imagery work. The idea is to internalize the coping skill before you try to pair it with threatening situations.
2. Repetition is the key to effective exposure work. You don't just imagine a stressful situation, keep yourself calm, and then go on with your day. Rather, you imagine the situation in great detail, with multiple variations. You won't imagine very stressful situations until you've been able to keep yourself thoroughly relaxed with less stressful imagined scenarios. If that means you repeat a single scenario five times until it no longer elicits anxiety, that is fine. The goal is to unlearn the connection between the situation and the unwanted response and train yourself to a new connection: between the trigger situation and staying in the zone.

As a beginning exercise, here is a very basic exposure routine that you can apply to almost any trading patterns that you wish to change. I have found that this works very well for reprogramming anxiety responses to market situations and frustration/anger responses. Any time a situation evokes an exaggerated emotional and/or behavioral response from you, exposure methods can be used to alter your reactions:

Step 1. Seat yourself comfortably, listening to relaxing music through headphones. While listening, close your eyes and breathe deeply and slowly. Keep yourself still physically and keep your mind focused on the music.

Step 2. Start with lower part of your body and gradually tense and relax the muscles, performing several repetitions with each muscle group before moving higher along your body. Thus you tense and relax your toes several times, then your flex your foot, then your lower leg, etc. All the while you are tensing and relaxing, you are breathing deeply and slowly and staying focused on the music.

Step 3. Once you reach the top of your body, tensing and relaxing the muscles of your face, you then take a few more deep breaths and notice your body's relaxation.

Step 4. With the music still playing, imagine in detail a trading situation that you anticipate. Visualize the position you're in and the market movement. Imagine the market behaving in a way that normally would trigger your fear, frustration, etc. All the while, you are breathing deeply and slowly, keeping your muscles relaxed, and playing the music in the background.

Step 5. When you feel yourself tense up or experience fear or frustration, stop the visualization (freeze the frame) and simply go back to breathing deeply and slowly and listening to the music. Once you're relaxed again, continue the scenario from where you left off. Make sure you freeze things and keep yourself calm and focused when trigger responses start to affect your visualization.

Step 6. If you had to interrupt the visualization to keep yourself calm, repeat the exact same scenario the same way until you can get all the way through without having to freeze the scene. At that point, you've extinguished the response to the situation.

Step 7. After you master one scenario, construct variations of the scenario, perhaps making each one a bit more stressful. Once again, don't move on to another scene until you've been able to keep yourself fully relaxed and focused during the scene you've been rehearsing. If you can stay calm and focused during a visualization of a moderate stress, make the next visualizations more threatening. Don't stop your exposure work until you have tackled your absolute worst fears.

This basic exercise enables you to extinguish emotional and behavioral responses to trading situations that can lose you money. If you rehearse staying calm and focused in stressful situations, you build a new learned connection and reprogram your behavioral responses. This process is effective for situations in which you've been through extreme losses, and it is also quite useful in reprogramming patterns of overtrading. When you coach yourself to face and overcome your worst fears, you build confidence, resilience, and a sense of efficacy, empowering yourself in situations where you had seemed powerless.

COACHING CUE

Imagery can be powerful in programming new responses, but consider extending your exposure work to live trading. Such in-vivo exposure, beginning with small trading size and gradually ramping up to full size with success, is the single most effective technique for reprogramming traumatic experiences in trading, such as large losses that overwhelm mood and confidence. By re-creating the market conditions that caused the trauma in imagery—and then facing those conditions in simulated and actual trading—all the while rehearsing self-control skills, we can regain a sense of mastery over trading. It takes repeated experiences of safety during exposure work to undo traumatic stresses. Eventually, the emotional learning that we can face our fears without terrible things happening sinks in and contributes to a newfound confidence.

LESSON 69: EXTEND EXPOSURE WORK TO BUILD SKILLS

In the previous lesson, we saw how exposure methods can be used to deprogram negative behavior patterns. Just a small adjustment in the technique is needed to create positive learning by rehearsing and reinforcing proper trading behaviors.

The fundamental difficulty of trading is that we know what to do (enter on pullbacks in a trend, size positions appropriately) when we are out of the heat of battle. When stressed, however, or when we face unusual opportunity, we find that other behavior patterns are triggered and it is much more difficult to do the right things. I work with a good number of experienced portfolio managers and proprietary traders, and even they make the occasional rookie errors, in which they are swayed by situational influences. Techniques that reinforce the right actions can be useful for the pros as well as beginners.

One trader I worked with was bedeviled with the problem of regret. He would enter a longer-term position and, while it was going his way, he was fine. As soon as the position retraced some of the gains, however, he began to regret that he hadn't lightened up at the more favorable price levels. This regret was a very tangible psychological influence for him. At times, it became outright guilt as he convinced himself that he had done the wrong thing.

What happened as a result of this pattern is that he would inevitably assuage this guilt by waiting for the profits on the trade to move back to their high water mark so that he had an exit approximating the one he had missed. The problem was that this cut his original trade idea short. Many times he would take his profit on the first rebound from the retracement, only to see the position move toward his initial target without him on board. Then the trader experienced massive regret and guilt. This led him to seek additional home run trades (to relieve his newest guilt), only to make the same mistakes on these trades as well. By the time I met with this trader, all he could talk about was how much he could have made if he had just traded the way he planned.

> Many traders are shaken out of good trades when they aim to not lose, rather than aim to maximize profits.

The exposure work for this trader was straightforward. As the previous lesson outlined, we first just worked on the skill of staying calm and focused. I used the heart-rate variability (HRV) biofeedback unit for this work (www.heartmath.com). He had to concentrate and breathe rhythmically and deeply while keeping his HRV readings high. The trader was able to use the biofeedback unit for practice at home and he could track his skill-building by keeping the majority of his readings in the highest bin for a continuous period of five minutes or more. He found that he could keep his readings high by focusing his attention (counting in his head), keeping physically still and relaxed, and breathing from his diaphragm in a smooth, gradual fashion.

Once he had mastered the skill of keeping himself in the HRV zone, he used visualization to walk himself through his trade setup, including his profit target and stop. He vividly imagined the market moving in his favor, but instead of imagining himself being pleased with this outcome (which was what happened in his usual trading), he mentally reviewed his original trade plan and told himself that nothing had changed to alter the plan: it was working as anticipated. I asked the trader to simply repeat this part of the visualization over and over until he no longer reacted to the initial gain with excitement (and with a mental accounting of his paper profits). Instead, he visualized staying calm by reaffirming his plan for the trade.

Getting excited by gains in a trade is the first step toward getting panicky when those gains are threatened.

Only after the trader had mastered this aspect of the trading situation did we proceed to imagining that the market retraced some of its initial move, eroding a portion of his gain. Again and again, he imagined this retracement while breathing deeply and slowly and staying focused on the computer screen (which displayed his biofeedback readings), until the imagery of the retracement no longer brought fear or concern. At that point we mentally rehearsed the pullbacks all over again, this time while not only staying calm and focused but also while mentally reviewing his trade idea and his exits. Our trader spontaneously began to focus his attention on how proud he would feel if he just stuck with his ideas and saw them through. This pride, for him, was the opposite of the guilt he had been feeling. When he invoked this sense of pride, he not only extinguished his old behavior, but also positively reinforced his discipline.

The key to making this work is mentally rehearsing the right trading behaviors while you're in the state that normally triggers the wrong ones. When you're your own trading coach, your challenge isn't simply to figure out the right things to do. Rather, your job is to be able to act in the right ways in situations that normally pull for all the wrong trading behaviors. If you practice good trading when you're not in realistic trading situations, it is much less powerful than overcoming learned connections as they're occurring.

Of course, we can extend the power of exposure by shifting from imagery-based work to actual trading. Typically I'll have a trader start trading small size at first while engaging in the deep breathing and concentration and implementing trading plans. While the trade is on, the trader keeps her biofeedback readings in the optimal range and rehearses the plan for that trade. During the troublesome retracements, the trader simply repeats what she had practiced in the imagery: staying focused on the trading plan and keeping physically calm through the regular, deep breathing. Once this process is successful with small trades, the trader can gradually increase size back to the normal level of risk, performing the biofeedback work at each new size level.

Utilize biofeedback during trading and you can often detect departures from the performance zone before you are consciously aware of them.

If there are challenging market situations, the best approach to master them is to face them directly while you stay grounded in your best trading practices. With the use of imagery, this conditioning work can be accomplished outside of market hours and without taking risks. With repetition, the mentally rehearsed patterns feel increasingly natural, as the old, learned connections fall away. It is not always comfortable doing exposure work—and the better you do it, the less comfortable it will be—but you cannot coach yourself through discomfort unless you're willing and able to tackle it directly. Rehearse your best trading practices while you're in your most stressful situations; it is one of the most effective training techniques you can employ.

COACHING CUE

It is helpful to formulate your best trading practices as specific, concrete rules so that these rules can be rehearsed in detail during the exposure work. Among the rules I've found most helpful for this work are:

- Generating trading ideas by identifying themes that cut across sectors and/or asset classes.
- Waiting for pullbacks in a trend before entering a position.
- Establishing my target price at the outset of the trade, so that I can enter the trade with a profit potential that exceeds the loss I'm willing to take.
- Sizing my trade so that I'm risking a fixed, small percentage of my portfolio value on the idea.
- Adding to longer-term trades on pullbacks after they have gone my way and remain profitable.
- Exiting trades on my planned stop-loss points or at my designated profit target.

Trading rules will differ for each trader depending on their markets and trading style. The important thing is to know what you do when you are most successful, so that you can cement these positive patterns, even as you expose yourself to challenging trading conditions.

LESSON 70: A BEHAVIORAL FRAMEWORK FOR DEALING WITH WORRY

We hear a great deal about fear and greed, and all of us have experienced bouts of overconfidence and frustration. On a day-in and day-out basis, however, few problems are as thorny for traders as worry.

Worry occurs when we anticipate an adverse outcome and its consequences. We can worry about missing an opportunity or about being wrong in a trade. We can worry about the future of our trading career or, sometimes, worries from personal life outside of trading can affect decision-making. It is common, for instance, for young traders to experience more stress after they have married, had children, or purchased a new home. With the added financial responsibilities come worries.

Worry is problematic for traders for several reasons:

- **It undercuts confidence.** It is difficult to maintain optimism and focus on progress while anticipating negative outcomes.
- **It interferes with concentration.** Thought and emotion directed toward worries are taken away from tracking market patterns.
- **It leads to impulsive decisions.** For most people, worry is so noxious that they will take action to reduce their concerns. Such action is not necessarily in the best interest of one's trading account.
- **It is not productive.** Rarely does worry lead to concrete, constructive problem solving. Worrying about negative outcomes does not generally help people achieve positive ones.

It is difficult to make sense of worry from a behavioral vantage point. No one truly enjoys worry, so it is unclear why the behavior persists. This is especially puzzling for chronic worriers. They do not enjoy focusing on negative things and typically are not happy people. So what keeps them worrying?

> Visualizing worst-case scenarios and how you would handle them is constructive; worry reinforces a sense of hopelessness and helplessness in the face of those scenarios.

To make sense of worry, let's review the difference between thinking about a negative event and actually experiencing that event. I can think about losing money in my trading and the thought does not bring particular anxiety or concern. If, however, I vividly imagine a particular trade that I am planning and visualize myself taking a loss on a large position, I can generate palpable experiences of nervousness. Abstract thought rarely generates strong emotion. Imagery, on the other hand, acts as a surrogate for reality. Think about sexuality and nothing happens; imagine an erotically charged scene and the body responds.

From a behavioral vantage point, worry is a form of thinking and, as such, it can function as a negative reinforcer. Let's say that I anticipate a stressful meeting with the risk manager at my trading firm. My underlying fear is that he will reduce my capital and express a loss of confidence in me. Rather than experience the hurt and resentment that such a meeting would engender, I worry about making the meeting on time, what I'll say in the meeting, what I might miss in the markets while the meeting is going on, etc. None of these worries has the power to evoke strong emotion. Rather, the worries serve as distractions from the difficult feelings I would experience if I actually visualized outcomes of the meeting. If I avoid experiencing these feelings, worry serves as a negative reinforcer. Strange as it might seem, worry is not so noxious when the alternative is facing scary outcomes.

Worry can possess reinforcement value in other ways, as well. If I were feeling out of control in my trading, that feeling would be unpleasant to dwell upon. If I worry about details in the work I'm going to have done on my house, I shift my focus to something more controllable. While it may seem that I worry about negative outcomes—and, in the example, I am—the psychological reality is that I substitute a lesser concern for a greater one when I worry. *What we worry about is usually not what is scariest to us.* Indeed, it is a diversion from the scariest scenario—and therein lies its reinforcement value.

> Worries about small things usually mask larger concerns.

Exposure work can be a great antidote to worry. When we expose ourselves to our greatest concerns—our worst-case scenarios—we can plan for these possibilities and mentally rehearse positive coping. If, for instance, I'm threatened by an upcoming meeting with the risk manager at my firm, I'll look at the worst case outcome—a large cut in my capital—and figure out a trading plan that will focus on my most successful trading and bring me back to my prior portfolio size. Once I anticipate the worst and figure out how I'd deal with it, I take the catastrophe out of the situation. That eliminates the need for worry-based diversions. Worry thinking can't be a negative reinforcer if it is more noxious than the alternative of facing possible outcomes constructively.

A great way to coach yourself past worry is to make note whenever you catch yourself worrying and ask, "What am I *really* fearful of? What's the real issue here?" What you generally find is that there's an unresolved situation looming in the background. Until the situation is faced squarely, it intrudes in your work and affects your mood. Suppose you find yourself worrying about whether a specific trade will work out. When you stop and reflect, you realize that you've sized the trade and placed your stop-loss point in such a way as to make such worry unnecessary. So what is the real concern? Perhaps the fear is of one's future as a trader. Perhaps it's a conflict at home. Whatever the real problem is, you want to visualize the situation vividly and walk yourself through your most constructive response. Then visualize the situation and solution again—and again. With repetition, the worst-case scenario will become routine. It will no longer evoke strong emotion. And that will leave you with little reason for worry.

Worry can be a great signal that we are harboring larger concerns about our basic trade ideas. When I find myself glued to the screen, following the market tick by tick during a longer-term trade, I know that something is wrong. Beneath the worries about the market's moment-to-moment action, I have deeper concerns—perhaps that my basic idea is wrong all along. This can be a useful signal: when we're comfortable with trades, we don't need to worry over every tick in the market. And when we are worrying about those ticks, it's a good sign that we're not comfortable with our position—and that can lead to constructive reevaluation and planning.

RESOURCES

The *Become Your Own Trading Coach* blog is the primary supplemental resource for this book. You can find links and additional posts on the topic of coaching processes at the home page on the blog for Chapter 7: http://becomeyourowntradingcoach.blogspot.com/2008/08/daily-trading-coach-chapter-seven-links.html

Chapter 9 of *Enhancing Trader Performance* details several strategies for changing behaviors that interfere with trading decisions, including a step-by-step description of exposure-based methods. See also Chapter 8 of that book for cognitive and cognitive-behavioral techniques.

A detailed account of behavioral approaches to change can be found in the chapter "Brief Behavior Therapy" by Hembree, Roth, Bux, and Foa in *The Art and Science of Brief Psychotherapies*, edited by Dewan, Steenbarger, and Greenberg (American Psychiatric Publishing, 2004).

Articles relevant to behavioral views of trading can be found among my collected articles, including the articles on "Behavioral Patterns That Sabotage Traders" and "Techniques for Overcoming Performance Anxiety in Trading": www.brettsteenbarger.com/articles.htm

Articles on emotional intelligence, staying in the zone, and balancing trading with the rest of life can be found in *Psychology of Trading*, edited by Laura Sether (W&A Publishing, 2007).

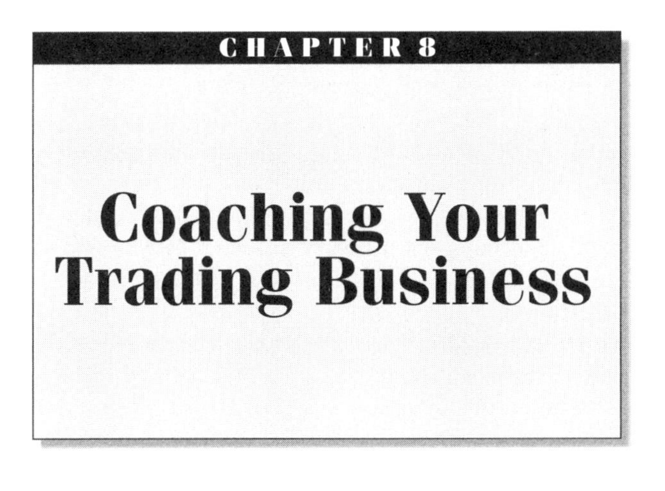

CHAPTER 8

Coaching Your Trading Business

He is not great who is not greatly good.
—William Shakespeare

In the previous chapters, we have explored ways of coaching yourself by becoming your own trading psychologist. Now we will turn to another facet of self-coaching: guiding your trading business. You, as a trader, are a business person no less than someone who offers goods and services to the public. You have overhead to cover, and you have returns you need to make to stay in business. Like any business owner, you risk your time, effort, and capital to earn returns higher than you could obtain from other activities. But are you getting the best return for your efforts? Are you taking the right amount of risk at the right times? Are you devoting the majority of your efforts to the activities that will provide the best returns? When you are your own business coach, you focus both on doing the right things and upon doing things right. There's much you can do as your own psychologist. Now let's see how you can thrive as your own business consultant . . .

LESSON 71: THE IMPORTANCE OF STARTUP CAPITAL

If you consistently break even in your trading, you will eventually lose all your capital. This is because there are costs embedded within trading, such as commissions and fees for data services, software, and computer support. It is no different in any business: an entrepreneur has to at least make enough to cover the overhead to stay afloat.

Many businesses fail because they lack adequate startup capital and cannot keep their overhead under tight control. They don't realize how long it will take to build a large and loyal customer base. As a result, they burn through their cash before they can sustain breakeven operations. To preserve capital, they cut back on essentials such as marketing and advertising. This creates a death spiral of fewer customers, lower income, and further belt-tightening.

Adequate startup capital enables the entrepreneur to make a beginner's mistakes and address the holes in his business plan before going out of business. Business plans are like battle plans in times of war: they are indispensable, but also subject to frequent change. Without sufficient resources, businesses cannot weather those changes.

Much of the stress that new traders experience is the result of an inadequate capital base: they are trying to do too much with too little.

So it is with traders. When they begin their business with modest capital, they cannot survive their learning curves when markets change and inevitable slumps take hold. Like failing businesses, they then begin to cut back on essential overhead, such as needed data and redundant systems. With little more to trade with than the same charts that everyone else looks at, the undercapitalized trader in overhead reduction mode virtually ensures that she will never maintain a distinctive edge.

So how much startup capital is sufficient for a trader? If you are just learning about markets, very little capital is needed to advance your learning curve. I began trading in late 1977 with a $2,500 stock market account at a regional brokerage in Kansas City. That enabled me to trade 100-share lots of individual stocks and test out my ideas without undue risk. Now, with the advent of simulation platforms, as discussed in the *Enhancing Trader Performance* book, it is possible to realistically test strategies and gain a feel for markets without placing money at risk.

Some commentators downplay the value of paper trading and simulation-based trading with live data because the psychological pressures of losing real money (and the overconfidence that comes with winning) are not present. This, however, is precisely why simulated trading is perfect for traders early in their development. Simulation enables the beginner to simply focus on the mechanics of trading and the recognition of trading patterns without having to worry about losing startup capital.

After all, if traders can't succeed in simulated trading, there's no way they'll succeed when those psychological pressures are added to the mix!

What makes sense, therefore, is to require yourself to earn consistent money in practice trading before you assume modest risk with 100 shares (or one lot of a futures contract). You thus need to sustain success with that small size before you trade larger. Just as a business should sustain success with one store before opening other outlets, a trader should have to earn his way toward trading size. If beginning traders stick to this one self-coaching rule, many could stay in the game long enough to become experienced traders.

A great business rule: Make yourself earn increases in trading size/risk by trading well and consistently with smaller size/risk.

When the aim is trading for a living, far more startup capital is required. At the money management firms I currently work with, for example, a portfolio manager is quite a star if she can sustain 30 percent returns year after year without taking undue risk. That portfolio manager will inevitably be given more capital to manage and, if success follows, may even strike out on her own with her own fund. In truth, a consistent 15 percent annual return, achieved with modest risk, will keep a portfolio manager well employed at most firms. True, any particular trader may achieve outsized returns in a given year, especially if taking large risk. The question, however, is: What kinds of average returns are sustainable over time?

A developing trader who expects to outperform seasoned money managers year after year substitutes fantasies for business plans. But if consistent 30 percent returns after expenses are stellar, how much trading capital would be required to sustain a living—and to keep the trading account growing at the same time? It's not difficult to see that an account well into the six figures would be a minimum startup for a trader that wanted a good living from his work.

Not many traders early in their careers have access to that kind of liquid capital. As a result, they start with much less capital and try to trade it aggressively to generate returns large enough to support a household. For a while, that might work out. Eventually, however, such traders sustain grievous losses that cannot be surmounted. After all, once you lose 50 percent of your capital, you have to double your remaining money just to get back to where you were. An undercapitalized trader, like an undercapitalized business, can't weather many adverse events—especially if taking large and frequent risks.

> Long before you seek to trade for a living, you should work at trading competence: just breaking even after costs.

When you are your own trading coach, you're also the manager and entrepreneur of your own trading business. That means that you have to start with a viable plan for success. Among other things, that plan should address:

1. How you'll learn markets and obtain trading competencies.
2. How you'll capitalize your business so that you can make a good income from realistic, solid risk-adjusted returns.

If you cannot raise the capital needed to make a living from realistically good returns, then your challenge is to make yourself attractive to trading firms that can front you sufficient trading capital. Take the steps needed to become attractive to such a firm and make that part of your business plan; this will form the basis for the next lesson. For now, your assignment is simply to learn markets before you put significant capital at risk, establish success over a range of market conditions and cycles, and ensure adequate access to capital before you give up your day job. Stress test the startup plan for your trading business: calculate how you would get by if returns were modest for the first couple of years. Run your plan by seasoned traders who make their living from markets; find the weak spots and address them. As the old saying goes, failing to plan is tantamount to planning to fail.

COACHING CUE

One of the smartest business decisions I made in my trading was to begin with an account small enough that it would not impact my family's lifestyle if I lost every penny. Early in my development, I had no illusions of trading for a living; my goal was to simply get better. A major milestone in my development came when I could consistently keep losing trades smaller than winners. Early in my trading efforts, it was a few large losers that impaired my overall performance. Had I been trading with money needed for my family's well-being, the stress of making rookie mistakes would have been overwhelming. Margie referred to my trading account as play money; she never counted on it or the income from it in our financial planning. Without the pressure of profitability demands early in my development, I was free to make mistakes and learn from them. A sure way to maximize stress and lower your odds of success is to put your capital at risk before you have cultivated your skills.

LESSON 72: PLAN YOUR TRADING BUSINESS

When you're your own trading coach, you are also the manager of your trading business. What is your business plan for success? How are you going to achieve your goals as a trader?

The first step toward good planning is to know *why* you are trading. That sounds silly: doesn't everyone trade to make money? Yes and no; I'm continually surprised at traders' fuzziness about their goals. If you're a beginner, your goal is simply to learn the ropes, internalize market patterns, and practice skills related to good execution and risk management. If so, as the previous lesson emphasized, you can accomplish those ends with little or no capital at risk. What you need is a learning plan and a platform from which you can observe markets and trade them in simulation mode. (The elements of learning plans are covered in *Enhancing Trader Performance* and will be addressed in later lessons in this chapter).

If you are like me and don't trade full time for a living, your goal is different. Your objective is to make more than the riskless rate of return (i.e., the amount you could earn, say, from a savings bond or bank certificate of deposit) after expenses. In that case, you allocate a portion of your savings to your trading account and use that portion of your money to improve returns from other investments and savings vehicles. This process means that you will be particularly sensitive to risk-adjusted returns, as you won't want to place your savings at undue risk. Trading, in that context, is part of diversification of your capital and is part of a larger financial plan.

> Your business plan will look different if you're trading as an avocation rather than as your vocation.

If you're trading for a living, then you're truly in the mode in which your trading is your business. A retail business needs to know how it will make money: what products they will sell, how they will sell them, how much it will cost them to sell them, and how much they can charge for them in order to make an acceptable return on investment. In your trading business, the questions become:

- What will you trade and how will you trade it? What simulated and live trading experience tells you that this will be successful?
- What will your overhead be? This includes software, hardware, commissions, and other expenses related to full-time trading, from the cost of data to your electronic connections and educational materials.

- How much can you expect to make per trade? Per month? Per year? What is the likely variability of your income? Will this be manageable?

These questions require hard data based on experience, not guesses or hopes.

Before you attempt to trade for a living, you should have a sufficient base of experience to tell you four things:

- What is the average size of my winning trade?
- What is the average ratio of my winning trades to losing trades?
- What is my average percentage of winning versus losing trades?
- What is my average variability (volatility) of returns per day, week, and month?

The answers to these questions will determine the likely path of your returns: the income generated from your trading business. These questions lead you to ask other questions:

- What kind of trader am I: do I tend to make money by being right more often than wrong, by having larger winning trades than losers, or a combination of the two?
- How much variation in my winning percentage and in the ratio of the size of winners versus losers is normal for me?
- How large would my trading need to be to generate acceptable returns and how much capital would I need to support that trading without undergoing drastic swings?

It is surprising—and dismaying—how few traders really look under the hood of their trading to understand how *they make money.* Because traders don't have a grasp on how they perform on average and how much variation from average can be expected, they are poorly equipped to distinguish normal drawdowns from troublesome slumps. They are also in a poor position to identify those occasions when the patterns of their returns shift due to changing markets.

> **If your trading experience does not extend to a variety of market cycles and conditions, your trading business will be ill prepared to weather shifts in volatility and trend.**

Your assignment for this lesson is to go to Henry Carstens' Vertical Solutions site (www.verticalsolutions.com/tools.html) and check out his two forecaster tools, using your own trading data as inputs. His first tool will show you how the path of your returns will vary as a function of changes in volatility. These volatility shifts could be attributable to market changes or to your taking more risk in each of your trades. You'll see clearly how much drawdown is associated with a given level of volatility, which will help you gauge your own tolerance.

The second forecaster tool asks you to input the average size of your winning trade, your average ratio of winning to losing trades, and the ratio of the size of your average winners to losers. Run the forecaster many times with your data and you'll see a variety of plausible sets of returns. This will give you a good sense for the expectable runs (to the upside and downside) in your trading, as well as the expectable returns over a 100-trade sequence.

Finally, tweak the parameters from the second forecaster to simulate the paths of possible returns if your average winning trade shrinks (maybe due to slow markets) or if your ratio of winning to losing trades declines (perhaps because of misreading markets). Tweak the ratio of the size of your average winners and losers to see what happens to your returns if you lose discipline and hold losers too long or cut losers short, creating poor risk/reward per trade.

All of these what-if scenarios will give you a good sense for what you can expect from your trading. It is much easier to deal with business adversity if you've planned for it in advance. When you're coaching yourself, the more you know about your trading business, the better you'll be able to make it grow.

COACHING CUE

An important element of success in building a trading career is being able to identify periods of underperformance as quickly as possible, before they create large drawdowns. The more you know about your trading—the average sizes and durations of drawdowns and the variability around those averages—the better prepared you'll be to identify departures from those norms. Keep statistics on your trading so you can also highlight periods in which you're trading particularly well and learn from these episodes. The single most important step you can take to further your trading performance is to keep detailed metrics on your trading. These steps will highlight what you're doing right and wrong, informing your self-coaching efforts.

LESSON 73: DIVERSIFY YOUR TRADING BUSINESS

Suppose you have a passion for coffee and decide to start your own coffeehouse as a business. You develop reliable sources for high quality beans, purchase a roaster, and rent space in a well-trafficked area. You furnish the café attractively and purchase all the cups, saucers, and utensils you will need. Altogether you sink $100,000 into your new enterprise, which is loaned from a bank with your home as collateral. Your average cost to serve a cup of coffee, just based on materials and labor expenses alone, is 50 cents. At $1.50 per cup, you're making a dollar for each cup you sell. At 300 customers per day, that's $300 per day or about $90,000 per year. That doesn't leave you with much to take as a salary once you pay off your overhead.

In this scenario, you can only make a go of the business by increasing the number of customers coming to the café, by increasing the average expenditure per customer, or both. So let's say you try to increase the average check size per customer by adding something additional to the menu. In addition to coffee, you now also serve tea.

Unfortunately, this doesn't help your business greatly. A few more customers enter the café who are tea drinkers, but few customers order both coffee and tea. As a result, you've increased your overhead (for tea equipment and supplies), but haven't greatly added to the bottom line. Tea overlaps coffee too much to add much to the menu; it doesn't really diversify the offerings of your café.

Suppose, however, you add pastries to the menu, sourcing them from a local bakery. Now you find that many people interested in your coffee also like a pastry to go with their drinks; this increases the size of their checks and enables you to make profits from two sources instead of one: the beverage and the pastry. You also now attract people who are interested in a snack or who just want a bite to eat after a concert or theater. The increased traffic also adds to the bottom line.

What has happened is that you've made your business more diversified. You have multiple profit centers, not just one. If you offered evening entertainment, sandwiches, and breakfast items, you would be even more diversified. Instead of attracting 300 patrons per day at one dollar each, you might attract 800 a day at $2.50 each. With $2,000 a day of gross income after labor and materials costs, you now have the basis for a thriving business. Moreover, should a café open elsewhere in the neighborhood, your business will be protected because of its other unique offerings.

> Diversification leverages talent.

The same business principles that impact the viability of the café apply to trading. When you trade different markets, time frames, and patterns, you generate multiple potential profit sources. This protects you when markets shift and place any single idea or pattern into drawdown mode. It also leverages your trading productivity, as you now can generate profits from many centers instead of a very few.

There are many ways of diversifying your trading business. If you are an intraday trader, you'll be diversified by trading long and short and by allocating your trades to different stock names and/or sectors. You may also hold some positions overnight, creating a degree of diversification by time frame as well. If you are trading over a longer time frame, you may trade different markets or strategies, each with different holding periods.

The key, for the trader as well as the café, is to make sure that diversification truly adds diversity. Adding tea to a coffee menu did not achieve adequate diversification for the café. Adding a Dow trade to an S&P 500 trade similarly fails to add unique value. Your diversification should provide a truly independent and reliable income stream. When the coffee business is slow, for example, customers may come to the café for a bite to eat. This keeps the flow of customers strong through the day. Similarly, when one of your trading strategies is drawing down, other ones that are not correlated can sustain the flow of profits to your account.

Of course, when you diversify, you need to make sure you stay within your range of expertise. Adding fresh entrées to a menu would make little sense for a café owner who lacked cooking skills. Similarly, it doesn't help your profitability as a trader to add strategies that are not well tested and known to be successful. Diversification only makes sense when it adds unique value to what you're already doing.

Many beginning traders think they'll find a way of trading that is profitable and then trade that for a career. Rarely are markets so accommodating. If a café brings in a huge number of customers, you can be sure competitors will soon follow. If a trading strategy is successful, it will find wide interest. Successful businesses must always innovate, staying ahead of the competitive curve. Adding new sources of revenue to exploit changing markets is essential to long-term survival.

I cannot emphasize this strongly enough: *markets change.* Edges in markets disappear. Trends change. The participants in markets change. The themes that drive markets change. The levels of volatility and risk in markets change. I have heard many promoters hype trading methods that they claim are successful in all markets, but I have yet to see documentation of such success. Every trader I have known who has sustained a long, successful career has evolved over time, just as successful businesses evolve with changing consumer tastes and economic conditions. Quite a few traders I've known who have been successful with a single method have failed to sustain that success when that strategy no longer fit market conditions (momentum trading of tech stocks in the late 1990s) or when it became so overcrowded that the edge disappeared (scalping ticks on the S&P 500 index by reading and gaming order flow). It's difficult to learn how to trade; even harder to unlearn old ways and cultivate new ones.

> The successful trading business, like elite technology, pharmaceutical, consumer, and manufacturing firms, devote significant resources to research and development: staying ahead of their markets.

When you are your own trading coach, it's not enough to learn markets. You're an entrepreneur; you're always developing new strategies, new ways of building upon your strengths. What products do you have in your pipeline? What markets, strategies, or time frames are you looking to expand to? You can adapt your current trading approaches to new markets or cultivate new strategies for familiar markets. *Your challenge is to develop a pipeline: to always be innovating, always searching for new sources of profit that capitalize on what you do best.*

Many traders sit down at their stations a little before markets open, trade through the day, and then go home, repeating the process day after day. That schedule is like coming to work at your café, putting in your hours, and then going home until the next workday. That is what you do if you're the employee of the business, not the owner. Your challenge, as your own coach, is to actively own and manage your trading business, not just put in hours in front of a screen. You need an edge to succeed at trading, but you need to develop fresh sources of edge to sustain your trading business.

COACHING CUE

I find there is value in learning trading skills at time frames different from your own. Short-term, intraday traders can benefit from looking at larger market themes that move the markets day to day, including intermarket relationships and correlations among stock sectors. If you identify those themes and relationships you can catch market trends as they emerge. Conversely, I find it helpful for longer-term traders, portfolio managers, and investors to learn the market timing perspectives of the short-term trader. This process aids execution, helping traders enter—and add to positions—at good prices. The views from different time frames can fertilize the search for new sources of edge: the perspectives of big-picture macro investors and laser-focused market makers can add value to one another.

LESSON 74: TRACK YOUR TRADING RESULTS

You cannot coach your trading to success if you do not keep score. Keeping score is more than tracking your profits and losses for the day, week, or year. It means knowing how you're performing and how this compares with your normal performance.

Score keeping makes sense if you once again think of your trading as a business. A sophisticated retail clothing firm tracks sales closely every week. Retailers know not only how much they've sold in total, but how much of each product. Perhaps the economy is slow, so women's accessories—which are lower-priced—are hot, but high-priced clothing is not. The company that tracks these trends regularly will be in the best position to shift their product mix and maximize profits. Similarly, if one store is dramatically underperforming its peers despite a favorable location, managers can use that information to see what might be going wrong at the store and make corrections.

Score keeping in the business world can be extremely detailed. There are good reasons for the investments in information systems that we observe among the world's most successful corporations. Firms may track sales by hour of the day to help them determine when to open and close. Purchasing patterns based on gender and age are factored into advertising messages and promotional campaigns. Score keeping provides the business with knowledge; in the business world, knowledge utilized properly is power.

> You can't properly manage your business if you don't understand what it is doing right and wrong.

Nowhere do we see this power more dramatically than in quality control. Firms such as Toyota collect reams of data on their manufacturing processes to help them identify lapses in quality, but also to make continuous improvements in manufacturing processes. If you don't collect the data, you can't establish the benchmarks that enable you to track progress. *It's not just about ensuring that you do well; the best businesses are driven to do better.*

When you keep score in your trading business, a few metrics are absolutely essential. These include:

- Your equity curve, tracking changes in portfolio value over time.
- Your number of winning versus losing trades.
- The average size of your winning trades and the average size of your losers.
- Your average win/loss per trade.
- The variability of your daily returns.

Let's take a look at each metric in a bit of detail.

Equity Curve

Here you're interested in the slope of your returns and changes in the slope. As we saw with Henry Carstens' tools that simulate trading returns, a great deal of directional change in your portfolio can be attributed to chance. For that reason, you don't want to overreact to every squiggle in your equity curve, abandoning hard-won experience. Too many traders jump from one promised Holy Grail to another, shifting whenever they draw down. *A far more promising framework for your self-coaching is to know the equity curve variation that is typical of your past trading, so that you can compare yourself against your own norms.* If you have learned trading properly, you will have a historical curve of your returns from simulation trading and small-size trading before you begin trading as an income-generating business. When your current equity curve varies meaningfully from your historical performance, that's when you know you may need to make adjustments. If the variation is in a positive, profitable direction, you'll want to isolate what is working for you so that you can take full advantage. If the variation is creating outsized losses, you may need to cut your risk (reduce the size of your trades) and diagnose the problems.

> Knowing your normal performance is invaluable in identifying those periods when returns are significantly subnormal.

Winning Versus Losing Trades

This is a basic metric of how well you're reading markets. Again, the emphasis is not on hitting a particular number, but on comparing your current performance to your historical norms. Let's say, for instance, you're a trend follower. You tend to make money on only 40 percent of your trades, but you ride those winners for relatively large gains compared to your losers. If your win percentage suddenly drops to 25 percent, you'll want to diagnose possible problems. Has your market turned choppy and directionless? Have you altered the way in which you're entering trades or managing them? The more the drop to 25 percent is atypical of your historical trading, the more you'll want to enter a diagnostic mode. If, however, you've had past periods of 25 percent winners just as a function of slow, directionless markets, you may choose to ride things out without making major changes in your trading simply by focusing on markets or times of day with greater opportunity.

Average Sizes of Winning and Losing Trades

It doesn't help to have 60 percent winning trades if the average size of your losers is twice that of your winners. Keeping score of the average sizes of winners and losers will tell you a great deal about your execution of trade ideas—whether you're entering at points that provide you with favorable returns relative to the heat that you're taking. The data will also tell you how well you're sticking to your risk management discipline, particularly stop-losses. If your average win size *and* loss size are expanding or contracting at the same time, you may simply be dealing with greater or lesser market volatility (or you may be sizing your positions larger or smaller). It is the relative shifts in size of average wins and losses that are most important for managing your business. If your winners are increasing in average size and your losers are decreasing, you're obviously trading quite well. It will be important to identify what you're doing right so that you can be consistent with it. Conversely, when losers are increasing in average size and winners are not, you want to figure out where the problem lies. Are you reading markets wrong, executing and managing trades poorly, or both?

Average Win/Loss Per Trade

Suppose you make a particular amount of money in January and the same amount in February. You might be tempted to conclude that you traded equally well in the two months. That would be a mistake, however. If you had placed 50 trades in the first month and 100 in the second month, then you can see that more trading did not produce more profit. Your average profit per trade actually declined. This suggests that at least some of your trading is not providing good returns, and that bears investigation. The situation is similar to that of a business that opens five new stores in a year, but reports the same sales volume year over year. The average sales per store have actually declined, an important factor masked by the increase in overall activity. Average win/loss per trade will vary with your position sizing and with overall market volatility. Be alert for occasions in which market volatility may increase, but your average win per trade goes down: you may not be trading as well in shifting market conditions. Many times, traders make as much or more money if they simply focus on their best ideas and reduce their total number of trades. This selectivity shows up as a soaring average win per trade. It's a great measure of the efficiency of your trading efforts.

> When you trade more often, make sure that the incremental trades are adding economic value.

Variability of Daily Returns

If you take your wins and losses for each day and convert those to absolute values, you'll then have a distribution of your returns. You'll see how much your equity curve moves per day on average. You'll also observe the variation around this average: the range of daily swings that is typical for your trading. The variation in your daily returns will ultimately shape the size of drawdowns you experience in your portfolio. Given that you're going to experience runs of losing days over your career, you'll have larger drawdowns when those runs are 2 percent each than if they're ½ percent. *Indeed, if you investigate the losing periods that are historically typical of your trading, you can use these to calibrate the daily variability you want to tolerate in your trading.* This is central to risk management. If you want to keep total drawdowns in your portfolio to less than 10 percent, for example, you cannot risk average daily swings of 2 percent. Of course, if you're keeping drawdowns to less than 10 percent, you also cannot expect to be making 50 percent or more per year: risk and reward will be proportional. By calibrating the average swing in your portfolio per day, you target both overall risk and reward. If you're trading very well (i.e., very profitably) with a relatively low variation in your portfolio size from day to day, you can probably afford to gradually pick up your risk (increase trade size to generate larger returns). If you're trading poorly and losing money beyond your norms, you may want to reduce your daily variation and cut your risk.

What all of this means is that, when you're your own trading coach, you are also your own scorekeeper. The metrics above are, in my estimation, an essential part of the journals of any serious trader. The more you know about how you're doing, the more prepared you'll be to expand on your strengths and address your vulnerabilities.

COACHING CUE

See David Adler's lesson in Chapter 9 for additional perspectives on trader metrics. A particular focus that is helpful is to examine what happens to your trades after your entry and what happens to them following your exit. Knowing the average heat that you take on winning trades helps you gauge your execution skill; knowing the average move in your favor following your exit enables you to track the value of your exit criteria. Sometimes the most important data don't show up on a P/L summary: how much money you left on the table by not patiently waiting for a good entry price or by exiting a move precipitously.

LESSON 75: ADVANCED SCOREKEEPING FOR YOUR TRADING BUSINESS

After the last lesson, you may be feeling overwhelmed by the data you need to keep to truly track and understand your trading business. I'm sorry to inform you that such items as equity curve, average numbers of winning/losing trades, and average size of winning/losing trades are just a start to serious scorekeeping. Professional traders at many firms obtain much more information than that about their performance. Access to dedicated risk-management resources is one of the great advantages of working in such firms. Although it is unlikely that you could duplicate the output of a dedicated risk manager, it is possible to drill down further into your trading to uncover patterns that will aid your self-coaching.

In this lesson, we'll focus on one specific advanced application of metrics that can greatly enhance your efforts at performance improvement: tracking results across your markets and/or types of trades. This tracking will tell you, not only how well you're doing, but also which trading is most contributing to and limiting your results.

> Of the different types of trades that you place, which most contribute to your profits? These are the drivers of your trading business success.

We've already seen how important it is for your trading to be diversified. Among the forms of diversification commonly found are:

- Trading different instruments, such as individual stocks versus index ETFs.
- Trading different markets, such as crude oil futures and stock indexes.
- Trading different setups, such as event trades and breakout trades.
- Trading different times of day and/or different time frames.

To truly understand your trading business, you want to tag each of your trades by the type of trade it represents. You will thus segregate trades based on the criteria above. All your overnight trades, for instance, will go in one bin; all your day trades in another. All our stock index trades will fall into one category; all our fixed income trades in another. If you're a day trader, you may want to segregate trades by time of day, by sector traded, or even by stock name.

Each of these categories is a product in your trading business. Each category is a potential profit center. When you think of your trading as a diversified business, you want to know how each of your products is contributing to your bottom line. Simply looking at your bottom line will mask important differences in results among your trades.

In practice this means that the metrics discussed in the last section—equity curve; proportion of winning/losing trades; average size of winning/losing trades; win/loss per trade; and variability of returns—*should be broken down for each portion of your trading business.* My own trading, for example, consists of three kinds of trades: intraday trades with planned holding times of less than an hour; intraday swing trades with planned holding times of greater than an hour; and overnight trades. The short intraday trades are based on short-term patterns of price, volume, and sentiment. The swing trades are trend following and are based on historical research that gauges the odds of taking out price levels derived from the prior day's pivot points. The overnight trades are also trend following and are based on longer-term historical research that shows a directional edge to markets. These trades comprise different kinds of trades, because they are based on different rationales and involve different positioning in the markets. For instance, I can take a short-term intraday trade to the long side, but short the market for an overnight trade that same day.

> You can identify the inventory of your trading business by segregating trades based on what you're trading and how you're trading it.

For other traders, the division of the trading business will be by asset class (rates trading versus currency trades versus equity trades) or by trading strategy/setup (trades based on news items; trades based on opening gaps; trades based on relative sector strength). Relatively early in my trading career, I broke my trades down by time of day and found very different metrics associated with morning, midday, and afternoon trades. That finding was instrumental in focusing my trading on the morning hours. In other words, the divisions of your trades should reflect anticipated differences in your income streams. If the returns from the trading strategies or approaches are likely to be independent of one another, they will deserve their own bin for analysis by metrics.

One of the breakdowns I've found most helpful in my trading is based on tagging the market conditions at the time of entry. To accomplish this, you would tag each trade based on whether it was placed in an upward trending market, a downward trending market, or a nontrending market. You will want to be consistent in how you identify market conditions; my own breakdowns (because I trade short-term) were simply based on how the current market session was trading relative to the previous one. It is common that traders will perform better in some market environments than others. For instance, based on a different breakdown, I learned that I was most profitable in moderate volatility markets, as tagged by the VIX index at the time of entry. When markets were relatively nonvolatile or highly volatile, my returns were significantly reduced. This was useful to me in knowing when and where to take risk.

> **Your trading business is most likely to succeed if you play to your strengths and work around your weaknesses.**

What you'll find by breaking your trading down in this fashion is areas of relative strength and weakness in your performance. You will have a historical database of the normal trading in each area of your trading business, and you will have a way to see how your current trading compares with those norms. Thus, for example, you might make money overall, but your event-based trades (those based on fast entries after news items and economic reports) perform much better than your trend-following trades. This may say something about the market you're in and how you want to be trading that market by focusing more on what's working than what is not. Similarly, you may find that your average numbers of winners versus losers is fine overall, but lagging in one particular market. This could lead you to fine tune what you're doing in that market.

The goal of keeping score is to identify your own patterns and use those patterns to your advantage.

When you are coaching yourself, I encourage you to think of yourself as a collection of different trading systems. Each system—each market or strategy that you trade—contributes to the overall performance of your portfolio. Your job is to track the results of each system, see when each is performing well, and determine when each is underperforming. Armed with that information, you can thus allocate your capital most effectively to the systems that are working rather than those that are not. Diversification can't work for you if you are not diversified in how you view and work on your trading performance. Just as a football coach breaks down his team's performance into segments—running, kicking, defense against the run, defense against the pass, throwing—and works on each, you want to analyze and refine your team of strategies. Many a drawdown can be avoided if you stay on top of each segment of your trading business.

COACHING CUE

When you add to positions or scale out of them, how much value does your management of the trade provide? What is the performance specific to the pieces that you add to positions? How much money do you save or leave on the table by removing pieces from your positions? If you hedge your positions, how much do you gain or save through those strategies? Are you better trading from the long or short side? Do you perform better when trades are entered at certain times of day or held for particular time frames? If you drill down with metrics you can become specific when you work on various facets of your trading performance.

LESSON 76: TRACK THE CORRELATIONS OF YOUR RETURNS

A department store is an example of a diversified business. The store sells a variety of goods, so it attracts a wide range of customers. If consumers aren't buying seasonal items, they may come in to shop for clothes or housewares. The departments that offer products for children, teens, men, and women ensure that there will be a mix among customers, evening out peaks and valleys in traffic patterns for any of these individual groups.

In your trading business, diversification provides you with multiple profit centers. You can make money from intraday stock index trades and longer-term moves in the bond market, for example. If you divide your capital among different ideas and strategies, you smooth out your equity curve, much as the presence of many departments keeps traffic flowing to the department store. When any one or two strategies fail to produce good returns, others contribute to the bottom line.

> Diversification in your trading enables you to stay afloat when any one of your strategies stops working for a while or becomes obsolete.

But how do you know your trading business is truly diversified? *Just because you are trading different setups or markets doesn't mean that you necessarily possess a diversified portfolio.* The only way you can ensure true diversification is by tracking the correlations among the returns of your different strategies.

Suppose you are a day trader who trades two basic patterns: moves on earnings news and breakouts from trading ranges. The idea is that your returns would be diversified because you would be long some names (earnings surprises and breakouts to the upside) and short others (earnings surprises and breakouts to the downside). You would also be diversified across market sectors and perhaps even by the time frame of your holdings. If you follow the logic of the previous lesson, you can track performance metrics for your earnings-related trades and your breakout trades. You can also track performance across your long trades and your shorts.

When you track the correlation of your returns, you take the analysis a step further. You calculate the daily P/L for each of your strategies over a period of time. You then evaluate the correlation between the two number series. If the strategies are truly independent, they should not be highly correlated. A slow market, for example, may yield little in the way of breakout trades, but you could still make money on selected stocks with earnings surprises. Similarly, you may get little earnings news on a particular day, but the market may provide a number of breakout moves due to economic reports.

Many traders think they are diversified, when in reality they are trading the same strategy or idea across multiple, related instruments. This means they're taking much more risk than they realize.

One stock market trader I worked with had a phenomenal track record and then stopped making money all of a sudden. On the surface, he was well diversified, trading many issues and utilizing options effectively for hedging and directional trade. When we reviewed his strategies in detail, however, it was clear that all of them relied on bull market trending to one degree or another. When the market first went into range-bound mode and then into a protracted bear move, he no longer made money. He was trading many things without much in the way of diversification. His trading business was like a car dealership that sells many kinds of trucks, vans, and SUVs. Once large vehicles go out of favor, perhaps due to high fuel costs, the business becomes vulnerable. It's not as diversified as it looks.

Suppose you track your performance historically and find that your strategies correlate at a level of 0.20. The outcome variance shared by the strategies would be the square of the correlation—0.04, or 4 percent. That means that results from one strategy only account for 4 percent of the variation in the other strategy—not a high level. Imagine, however, that during the last month, the correlation soars to 0.70. Now the overlap between strategies is close to 50 percent. They are hardly independent.

What could cause such a jump in correlation? In a strongly trending market, the breakouts might all be in one direction—the same direction as earnings surprises. Alternatively, you might get caught with an overall market opinion and select your trades to fit that opinion. In either case, your diversified business is no longer so diversified, just like the car dealership.

Correlations among your strategies and trades vary over time because of how markets trade and because of your own hidden biases.

That lack of diversification matters, because it means that your capital is concentrated on fewer ideas. Your risk is higher, because instead of dividing your capital among two or more independent strategies, it is now concentrated in what is effectively a single strategy. The same problem occurs when different markets or asset classes begin trading in a more correlated fashion, perhaps due to panicky conditions among investors. During those periods of high risk-aversion, traders will often shun stocks and seek the perceived safety of bonds and gold. The three asset classes (equities, fixed income, and gold) will thus trade in a highly correlated manner. Instead of being diversified across three markets, your capital is effectively concentrated in one.

When your trading results are becoming more correlated, your self-coaching will lead you to ask whether the correlations are due to biases in your trade selection or due to shifts among the markets. In the first instance, you can make special efforts to seek out uncorrelated ideas; in the second, you may want to reduce the size of each of your trades so that you have less concentrated risk. What you want to avoid are situations in which all your trading and investment eggs are in a single package. That can produce fine returns for a while, but leaves a trading business vulnerable when market conditions change.

My coaching experience with traders suggests that they most often achieve sound diversification in one or more ways:

- Blending intraday trading with swing trading or blending longer-term trading/investing with shorter-term swing trading.
- Blending directional trading (being long or short a particular instrument) with relative value trading (being long one instrument and short another, related one).
- Blending one strategy (such as trading around earnings events) with another (trading opening range breakouts).
- Blending the trading of one market or asset class (such as a currency pair) with another (U.S. small cap stocks).

For example, a trader might be long high-yielding stocks as one idea. A second idea would have the trader selling the front end of the yield curve and buying the long end, perhaps in anticipation of yield curve flattening ahead of Fed tightening. A third idea would have the trader short value stocks and long growth issues, anticipating that a historically wide spread in performance between the two will revert to its norm. A fourth idea might be a short-term long trade on small cap stocks. The good thing about spread/pairs trades, such as ideas two and three, is that they can work regardless of the direction of the underlying market, providing a measure of diversification. Trading horizons widen considerably when you don't just ask if a market is going up or down, but start to think about what will go up or down *versus other things*.

High frequency day traders will tend to achieve diversification and low correlation in other ways. They will demonstrate a relatively even mix of long and short trades over time. They may also manage positions over different time frames or place different trades in different stocks and sectors. The risk for the day trader is getting caught up in fixed opinions about the market, biasing trades in a single direction. If the day trader is diversified, the correlations among returns from his different setups and the serial correlations of returns among trades (and among returns across times of day) should be relatively modest over time. Each trade or type of trade for the day trader should, in a sense, be a separate product in the business mix.

You don't have to diversify by trading many markets. You can diversify by time frame (longer-term, shorter-term), directionality (long, short), and setup pattern (trending, reversal).

Is your trading business adequately diversified? Is its diversification expanding or narrowing? If you're like most traders, you don't know the answers to these questions. The data, however, can be at your fingertips. All you need is to divide your trades by strategy, track the results of each strategy daily, and enter the information into Excel. From there, it's simple to calculate correlations over varying time periods. And if you don't trade every day and hold most positions for days? No problem: simply calculate the returns of each strategy as if you had sold all positions at the end of each trading day. That will tell you if your trades are moving in unison or independently. And *that* will tell you if you have many profit centers supporting your business or only a very limited few.

COACHING CUE

It is not too difficult to turn good directional ideas into good pairs trades. Once you determine that an index, sector, commodity, or stock is going to go up or down, ask yourself what related indexes, sectors, commodities, or stocks are most likely to maximize this move. You would then buy the instrument most likely to maximize the move and sell the related one that is likely to lag. For instance, you might think the S&P 500 Index is headed higher. You note strength in the NYSE TICK ($TICK) relative to the Dow TICK ($TICKI) and so buy the Russell 2000 small caps and sell the Dow Jones Industrials in equal dollar amounts. This gives you an idea that can be profitable even if your original idea about directionality in the S&P Index doesn't work out. As long as there's more buying in the broad market than among the large caps in relative terms, you'll make money. Learn to think and trade in terms of relationships so you increase your arsenal of ideas.

LESSON 77: CALIBRATE YOUR RISK AND REWARD

I recently used Henry Carstens' P/L Forecaster (www.verticalsolutions.com/tools.html) to simulate possible equity curves under scenarios for two small traders:

1. Trader A has a small negative edge – The trader wins on 48 percent of trades and the ratio of the size of winners to losers is 0.90.
2. Trader B has a small positive edge – The trader wins on 52 percent of trades and the ratio of the size of winners to losers is 1.10.

I viewed the simulation as tracking returns over a 100-day period. The average size of winning days was $100 in both scenarios. That means that, if we assume that the traders began with a portfolio size of $20,000, that the daily variability of their returns was somewhere around 50 basis points ($^1/_2$ percent, or $100/$20,000). If the traders averaged just one trade per day, then it's plausible that they were risking roughly 2 S&P 500 emini points per trade and making about that much per trade.

By running the scenarios 10 times each, I was able to generate an array of returns for the two traders:

Forecast Number	Trader A	Trader B
1	–$904.2	$769.9
2	–727.4	667.5
3	–718.5	614.7
4	–763.5	783.8
5	–786.1	528.7
6	–551.0	830.4
7	–518.9	933.0
8	–610.5	500.7
9	–760.5	791.5
10	–812.6	884.2

What we see is that small edges over time add up. When the small edge is negative, as in the case of Trader A, the average portfolio loss over 100 days is around 3 percent. When the edge is positive, we see that Trader B averages a 100-day gain of about 3 percent.

Clearly, it doesn't take much to turn a modest positive edge into a modest negative one: the distance between 52 percent and 48 percent winners and the difference between a win size that is 10 percent smaller versus larger than the average loss size are not so great. Just a relatively small change in how markets move, how we execute our trades, or how well we concentrate and follow our ideas can turn a modest winning edge into a consistent loser.

> You don't need to have a large edge to run a successful trading business; you do need to have a consistent edge.

Had we tracked results every single day, we would have found that Trader A had some winning days and Trader B had some losers. Over a series of 100 days, however, the edge manifests itself boldly. There are no scenarios in which Trader A is profitable and none where Trader B loses money. Just as the edge per bet in a casino leads to reliable earnings for the house over time, the edge per trade can create a reliable profit stream when sustained over the long run. Once you have your edge, your greatest challenge as your own trading coach is to ensure your consistency in exploiting that edge, every day, every trade.

But let's take the analysis a step further and explore risk and reward. Our Traders A and B are not great risk takers: they are only trading one ES contract for their $20,000 account. This provides them with average daily volatility of returns approximating 50 basis points, which is not so unusual in the money management world. We can see, however, that the 3 percent return for Trader B over 100 days would amount to about 7.5 percent per year (250 trading days). While that's not a terrible return, it is before commissions and other expenses are deducted. The consistent small edge does not produce a large return when risk-taking is modest.

So how could our Trader B juice his returns? A simple way would be to trade 3 contracts instead of 1. Assuming that this does not change how Trader B trades, the average daily variability of returns would now be 1.5 percent, with an average win size of $300. The return of over 20 percent per year would now look superior. If you take more risk when you have a consistent edge, this certainly makes sense, just as betting more makes sense at a casino makes sense if the odds are in your favor. You simply need to have deep enough pockets to weather series of losses that are expectable even when you do have the edge. When I ran the scenarios for Trader B, peak to trough drawdowns of $700 to $800 were evident. Multiply that by the factor of 3 and now Trader B incurs drawdowns of 12 percent.

But what if Trader B took pedal to the metal and traded 10 contracts with a small edge? The potential annual return of 75 percent looks fantastic. The potential drawdowns of more than 40 percent now seem onerous. That small, consistent edge suddenly generates large losses when risk is ramped to an extreme.

> The variability of your returns will tend to be correlated with the variability of your emotions.

In that aggressive scenario, Trader B would ramp the variability of daily returns to 5 percent. Average daily swings of that magnitude, particularly during a slump, are bound to affect the trader's psyche. Once the trader becomes rattled, that small positive edge can turn into a small negative one. Amplified by leverage, the trader could easily blow up and lose everything, all the while possessing sound trading methods.

The size of your edge and the variability of your daily returns (which is a function of position sizing) will determine the path of your P/L curve. Management of that path is crucial to emotional self-management. Your assignment is to utilize Henry's forecaster in the manner illustrated above, using your historical information regarding your edge and average win size to generate likely paths of your returns. Then play with the average win size to find the level of risk, reward, and drawdown that makes trading worth your while financially, but that doesn't overwhelm you with swings in your portfolio.

Few traders truly understand the implications of their trading size, given their degree of edge. If you know what you're likely to make and lose in your trading business, you'll be best able to cope with the lean times and not become overconfident when things are good. Match your level of portfolio risk to your level of personal risk tolerance for a huge step toward trading success.

COACHING CUE

Doubling your position sizing will have the same effect on the path of your returns as keeping a constant trading size when market volatility doubles. This process is a dilemma for traders who hold positions over many days and weeks, but is also a challenge for day traders, who experience different patterns of volatility at different parts of the trading day. It is common for traders to identify volatility with opportunity and even raise their trading size/risk as markets become more volatile. This greatly amplifies the swings of a trading account, and it plays havoc with traders' emotions. Adjusting your risk for the volatility of the market is a good way to control your bet size so that a few losses won't wipe out the profits from many days and weeks.

LESSON 78: THE IMPORTANCE OF EXECUTION IN TRADING

You can have the greatest ideas in the business world, but if they're not executed properly, they won't be worth much. A great product marketed poorly won't sell. A phenomenal game plan by a top coach won't work on the basketball court if the players don't pass well and can't establish position underneath the basket for rebounds. The quarterback can call a great play, but if the line doesn't block, the pass will never get off.

So it is with trading: execution is a much larger part of success than most traders realize. The average trader spends a great deal of attention on getting into a market, but it's the management of that trade idea that often determines its fate. When you are the manager of your trading business, you want to focus on day-in and day-out execution, just as you would if you were running your own store.

A trade idea begins with the perception that an index, commodity, or stock is likely to be repriced. For example, we may perceive that a stock index is trading at one level of value, but is likely to be trading at a different level. Our rationale for believing this may be grounded in fundamentals: at our forecasted levels of interest rates and earnings growth, the index should be trading at X price rather than Y. Our rationale might be purely statistical: the spread between March and January options contracts on the index is historically high and we anticipate a return to normal levels. Too, we may use technical criteria for our inference: the market could not break above its long-term range, so we expect it to probe value levels at the lower end of the range. In each case, the trade idea takes the form: "We're trading here at this price, but I hypothesize we'll be trading there at that price."

What this suggests is that a fully formed trade idea includes not just an entry setup, but also a profit target. Too often that target is not made clear and explicit, but still it lies at the heart of any trade idea. A trade only makes sense if we expect prices to move in an anticipated way to an extent that is meaningful relative to the risk we are taking.

> Just as businesses set target returns on their investments, traders target returns on their trade ideas.

It is in this context that every trade is a hypothesis: our belief regarding the proper pricing of the asset represents our hypothesis, and our trade can be thought of as a test of that hypothesis. As markets move, they provide incremental support for or disconfirmation of the hypothesis. That means that, as trades unfold, we either gain or lose confidence in our hypothesis.

Any good scientist not only knows when a hypothesis is supported, but also when it is not finding support. A hypothesis is only meaningful if it can be objectively tested and falsified. The outcome that would falsify our trade hypothesis is what we set as a stop-loss level. It is the counterpart to the target; if the target defines the possible movement in our favor, the stop-loss point captures the amount of adverse movement we're willing to incur prior to exiting the trade and declaring our hypothesis wrong.

The trader who lacks clearly defined targets and stop-losses is like the scientist who lacks a clear hypothesis. You can trade to see what happens, and scientists can play around in the laboratory, but neither is science and neither is likely to prove profitable over the long run. A firm hypothesis and objective criteria for accepting or rejecting the hypothesis advances knowledge. Similarly, a clear trading idea and explicit criteria for validating or rejecting the idea can guide our market understanding. Frequently, if you get stopped out of a seemingly good trade idea you can reframe your understanding of what is going on in the market. After all, scientists learn from hypotheses that are not confirmed as well as those that are.

With the target and stop-loss firmly in mind, we now have the basis for executing our idea. Good execution mandates that we enter the trade at a price in which the amount of money we would lose if we were wrong (if we're stopped out) is less than the amount of money that we would make if we were right (if we reach our target). When traders talk about getting a good price, this is what they mean: they are entering an idea with relatively little risk and a good deal more potential reward. A good way to think about this is to think of each trade as a hand of poker: where we place our stop-loss level reflects how much we're willing to bet on a particular idea.

> Many traders make the mistake of placing stops at a particular dollar loss level. Rather, you want to place stops at levels that clearly tell you that your trade idea is wrong.

Let's say my research tells me that we have an excellent chance of breaking above the prior day's high price of $51 per share. We are currently trading a bit below $49 after two bouts of morning selling took the stock down to $48, which is above the prior day's low of $47.50. A news item favorable to the sector hits the tape and I immediately buy the stock at $49, with $51 as my immediate target. My hypothesis is that this news will be a catalyst for propelling the stock higher, given that earlier selling could not take out yesterday's low. I'm willing to lose a point on the trade (stop myself out at $48) to make two points on the idea (target of $51).

Suppose, however, that the stock was trading at $50, rather than $49 when the news came out. Now my risk/reward is not weighted in my favor. If I'm willing to accept a move back to $48 before concluding I'm wrong, I now have two points of potential loss for a single point of targeted profit. While the idea is the same, the execution is quite different. It is difficult to make money over the long haul if you're consistently risking two dollars to make one. If, however, you're risking a dollar to make two or more, you can be right less than half the time and still wind up in the plus column.

> Good execution means that you calibrate risk as a function of anticipated reward.

Execution provides proactive risk management. If you control how much you can win and lose based on your price of entry, you keep your risk known and lower than your potential reward. You can track the quality of your execution if you calculate the amount of heat you take on your average trades. Heat is the amount of adverse price movement that occurs while you're in the trade. If you're taking a great deal of heat to make a small amount of money, you're obviously courting disaster. When your execution is good, you should take relatively little heat compared to the size of your gains. Your assignment for this lesson is to calculate heat for each of your recent trades and track that over time. This assignment will tell you how successful you are at executing your ideas, and it will provide a sensitive measure of changing risk/reward in your trading.

Good execution, psychologically, is all about patience. To get a good price, you will have to lay off some trade ideas that end up being profitable. Like the poker player, you want to bet when the odds are clearly in your favor. That means mucking a lot of hands. Similarly, a business doesn't try to be all things to all people. A business owner passes up certain opportunities to sell products in order to focus on what she does best. When you're running your trading business well, you don't take every conceivable opportunity to make money; you wait for *your* highest probability opportunities. The clearer you are about risk and reward, the easier it will be to stick to trades that offer favorable expected returns.

COACHING CUE

A simple rule that has greatly aided my executions has been to wait for buyers to take their turn if I'm selling the market and wait for sellers to take their turn before I'm buying the market. Thus, I can only buy if the NYSE TICK has gone negative and if the last X price bars are down. Similarly, I can only sell if the NYSE TICK has gone positive and the most recent X price bars have been up. This reduces the heat I take on trades by entering after short squeezes and program trades juke the market up or down. Once in a while you'll miss a trade if you wait patiently for the other side to take their turn, but the extra ticks you make on the trades you do get into more than make up for that.

LESSON 79: THINK IN THEMES—GENERATING GOOD TRADING IDEAS

Notice how successful businesses are always coming up with new products and services to meet the changing needs and demands of consumers. A great way to become obsolescent in the business world is to remain static. If a current product is a hit, competitors and imitators are sure to follow. What was hot at one time—large vehicles during periods of cheap gasoline, compact disc players for music—can go cold quite suddenly when economic conditions or technologies change.

It is the same way in the trading world. For a while, buying technology stocks was a sure road to success. Thousands of day traders gave up their jobs to seek their fortunes. After 2000, that trade vanished with a severe bear market. Now, selected technology firms are doing relatively well, while others languish. Markets, like consumer tastes, are never static.

As we saw in the lesson on diversification, if you base your trading business on a single type of trading—a limited set of ideas or patterns—it is a vulnerable place to be. I recall when breakouts of opening ranges were profitable trades; in recent years, those breakouts have tended to reverse. We all hear that the trend is your friend, but in recent years buying after losing days, weeks, and months has on average—been more profitable than selling. Market patterns, as Niederhoffer emphasized, are ever-changing. That means that successful traders, like successful business people, must continually scout for fresh opportunity.

> The best system developers don't just develop and trade a single system. Rather, they continually test and trade new systems as part of a diversified mix.

One of the skills I see in the best traders is the ability to synthesize data across markets, asset classes, and time frames to generate trading themes. These themes are narratives that the trader constructs to make sense of what is going on in markets. These themes are the trader's *theories* of the financial marketplace. Their trades are tests of these theories. The successful trader is one who generates and acts on good theories: themes that truly capture what is happening and why.

This thematic thinking is common in the portfolio management world, but I see it also among successful short-term traders. The scope of the theories—and the data used to generate them—may differ, but the process is remarkably similar. A portfolio manager might note high gasoline prices and weakening housing values and conclude that a lack of discretionary income among consumers should hurt consumer discretionary stocks. That might lead to a trade in which the manager shorts the consumer discretionary sector and buys consumer staples stocks that are more recession-proof.

The short-term trader may look at a Market Profile graphic covering the past week and observe a broad trading range, with the majority of volume transacted in the middle of that range. As the market moves to the top of that range, the trader sees that many sectors are nowhere near breaking out. Moreover, overall market volume is light during the move, with as much volume transacted at the market bid as at the offer. From this configuration, the short-term trader theorizes that there is not sufficient demand to push the market's value area higher. She then places a trade to sell the market, in anticipation of a move back into the middle of the range.

Thematic thinking turns market data into market hypotheses.

The key to thematic thinking is the synthesizing of a range of data into a coherent picture. Many traders, particularly beginners, only look at their market in their time frame. Their view is myopic; all they see are shapes on a chart. The factors that actually catalyze the movement of capital—news, economic conditions, tests of value areas, intraday and longer-term sentiment—remain invisible to them. Without the ability to read the market's themes, they trade the same way under all market conditions. They'll trade breakout moves in trending markets and in slow, choppy ones. They'll fade gaps whether currencies and interest rates are impelling a repricing of markets or not. Little wonder that they are mystified when their trading suddenly turns from green to red: they don't understand why markets are behaving as they are.

There is much to be said for keeping trading logic simple. The synthesis of market data into coherent themes is a great way to distill a large amount of information into actionable patterns. In the quest for simplicity, however, traders can gravitate to the simplistic. A setup—a particular configuration of prices or oscillators—may aid execution of a trade idea, but it is not an *explanation* of why you think a market is going to do what you anticipate. The clearer you are about the logic of your trade, the clearer you'll be about when the trade is going in your favor and when it is going against you.

Intermarket relationships are particularly fertile ground for the development of themes. You'll notice in the *TraderFeed* blog that I regularly update how sectors of the stock market are trading relative to one another. This alerts us to themes of economic growth and weakness, as well as to lead-lag relationships in the market. These sector themes change periodically—financial issues that had been market leaders recently became severe laggards—but they tend to be durable over the intermediate-term, setting up worthwhile trade ideas for active traders.

Other intermarket themes capture the relative movements of asset classes. When the economy weakens and the Federal Reserve has to lower interest rates, this has implications not only for how bonds trade, but also the U.S. dollar. Lower rates and a weaker dollar might also support the prospects of companies that do business overseas, as their goods will be cheaper for consumers in other countries. That situation might set up some promising stock market ideas.

Track rises and falls in correlations among sectors and markets as a great way to detect emerging themes—and ones that are shifting.

Short-term traders can detect themes from how markets trade in Europe and Asia before the U.S. markets open. Are interest rates rising or falling? Commodities? The U.S. dollar? Are overnight traders behaving in a risk-averse way or are they buying riskier assets and selling safer ones? Are Asian or European markets breaking out to new value levels on their economic news or on decisions from their central banks? Often, these overnight events affect the morning trade in the United States and set up trade ideas about whether a market is likely to move to new highs or lows relative to the prior day.

The short-term trader can also develop themes from breaking news and the behavior of sectors that are leading or lagging the market. If you keep track of stocks and sectors making new highs and lows often, you will highlight particular themes that are active in the market. If oil prices are strong, you may notice that the shares of alternative energy companies are making new highs. That can be an excellent theme to track. Similarly, in the wake of a credit crunch, you might find that banking stocks are making new lows. *Catching these themes early is the very essence of riding trends; after all, a theme may persist even as the broad market remains range bound.*

As your own trading coach and the manager of your trading business, you need to keep abreast of the marketplace. That means that considerable reading and observation must accompany trading time in front of the screen. On my blog, I try to highlight sources of information that are particularly relevant to market themes. I update those themes daily with tweets from the Twitter messaging application (www.twitter.com/steenbab). Ultimately, however, you need to figure out the themes that most make sense to you and the sources of information for generating and tracking those themes. The more you understand about markets, the more you'll understand what is happening in *your* market. Like a quarterback, you need to see the entire playing field to make the right calls; it's too easy to be blindsided by a blitzing linebacker when your gaze is fixed!

COACHING CUE

It is particularly promising to find themes that result from a particular news catalyst. For instance, if a report on the economy is stronger than expected and you see sustained buying, track the sectors and stocks leading the move early on and consider trading them for a trending move, especially if they hold up well on market pullbacks. Many of those themes can run for several days, setting up great swing moves.

LESSON 80: MANAGE THE TRADE

Business success isn't just about the products and services a firm offers to the public. Much of success can be traced to the management of the business. If you don't hire the right people, supervise them properly, track inventories, and stick to a budget, you'll fail to make money with even the best products and services.

So it is with trading. The best traders I've known are quite skilled at *managing trades*. By trade management, I mean something different from generating the trade idea and executing it. Rather, I'm referring to what you do with the position *after* you've entered it and *before* you've exited.

Your first reaction might be: You don't do anything! It's certainly possible to enter a position and sit in it, waiting for it to hit your profit target or your stoploss level. That reaction, however, is inefficient. It's like selling the same mix of products at all stores even though some products sell better and some sell worse at particular locations.

To appreciate why this is the case, consider the moment you enter a trade. At that point you have a minimum of information regarding the soundness of your idea. As the market trades following your entry, you accumulate fresh data about your idea: the action is either supporting or not supporting your reasons for being in the market. For instance, if your idea is predicated on falling interest rates and you see bonds break out of a price range, that would be supportive of your trade. If you anticipate an upside breakout in stocks and you see volume expand and the NYSE TICK move to new highs for the day on an upward move, that is similarly supportive. *If you track market action and themes while the trade is on, you can update the odds of your trade being successful.*

> Trade management is the set of decisions you make based on the fresh information that accumulates during the trade.

One way that I see traders utilizing this information is in the way that they scale into trades. Their initial position size might be relatively small, but traders will add to the position as fresh information validates their idea. My trading capital per trade idea is divided into six units. I typically will enter a position with one or two units. Only if the idea is finding support will a third or fourth unit enter the picture. I have found that, if my trades are going to be wrong, they're generally wrong early in their lifespan. By entering with minimal size, I incur small drawdowns when I'm wrong. If I add to a position as it is finding support, I maximize the gains from good trades. *This trade management, I find, is just as important to many traders' performance as the quality of their initial ideas.* Indeed, I've seen traders throw lots of trade ideas at the wall and only add capital to the ones that stick: the management of their trades makes all the difference to their returns.

Such trade management means that you have to be actively engaged in processing information while the trade is on, not just passively watching your position. Good trade management is quite different from chasing markets that happen to be going in your favor. It is a separate execution process unto itself, in which you can wait for normal pullbacks against your position to add to the position at favorable levels. If you are long, for example, and the market is in an uptrend, the retracements should occur at successively higher price levels. By adding after the retracement, you gain the profit potential when the market returns to its prior peak, but you also ensure that risk/reward will be favorable for the piece of the trade that you're adding.

> Each piece added to a trade needs a separate assessment of risk and reward to guide its execution.

While I'm in a short-term trade, I'm closely watching intraday sentiment for clues as to whether buyers or sellers are more aggressive in the market. I will watch the NYSE TICK (the number of stocks trading on upticks minus those trading on downticks), and I will track the Market Delta (www.marketdelta.com; the volume of ES futures transacted at the market offer minus the volume transacted at the market bid). These trackings tell me if the balance of sentiment is in my direction. If the sentiment turns against me, I may decide to discretionarily exit the position prior to hitting my stop-loss level. That, too, is part of trade management. To be sure, it is risky to front-run stop-levels: it invites impulsive behavior whenever markets tick against you. I've found, however, that if I'm long the ES futures and we see a breakdown in the Russell 2000 (ER2) futures or a move to new lows in a couple of key market sectors, the odds are good that we will not trend higher. By proactively exiting the position, I save myself money and can position myself for the next trade.

What this suggests is that it is important to be right in the markets, but it is even more valuable to *know when you're right*. Very successful traders, I find, press their advantage when they know they're right, and they're good at knowing when they're right. This means that they are keen observers of markets in real time, able to assess when their trade ideas—their hypotheses—are working out and when they're not. They are good traders because they're good managers of their trades.

Your assignment for this lesson is to assess your trade management as a separate profit center. Do you scale into trade ideas? Do you act aggressively on your best ideas when you are right? If you're like many traders, this is an underdeveloped part of your trading business. It may take a return to simulation mode and practice with small additions to trades to cultivate your trade management skills. It may also mean that you structure your time while you're in trades, highlighting the information most relevant for the management of your particular idea. Most of all it means cultivating an aggressive mindset for those occasions when you *know* you have the market nailed.

As your own trading coach, you want to make the most of your assets. It's easy to identify traders who exit the business because they lose money. It's harder to appreciate the equally large number of traders who never meet their potential because they don't make the most of their winning ideas. A great exercise is to add to every position at least once on paper after you've made your real-money entry. Then track the execution of your added piece, its profitability, the heat you take on it, etc. In short, treat trade management the way you would treat trading a totally new market, with its own learning curve and need for practice and feedback. You don't have to be right all the time; the key is to know when you're right and make the most of those opportunities.

COACHING CUE

Track your trades in which you exit the market prior to your stops being hit. Does that discretionary trade management save you money or cost you money? It's important to understand your management practices and whether they add value to your business.

RESOURCES

The *Become Your Own Trading Coach* blog is the primary supplemental resource for this book. You can find links and additional posts on the topic of coaching processes at the home page on the blog for Chapter 8: http://becomeyourowntradingcoach.blogspot.com/2008/08/daily-trading-coach-chapter-eight-links.html

Jim Dalton's books are excellent resources when it comes to developing a conceptual framework for trading. I recommend *Mind Over Markets* as a first read, then *Markets in Profile*. Both books are available on the site where Jim and Terry Liberman offer training for developing traders: www.marketsinprofile.com

An excellent book on risk and risk management is Kenneth L. Grant's text *Trading Risk* (Wiley, 2004).

Exchange Traded Funds (ETFs) are an excellent tool for achieving diversification. Two good introductions to ETFs are David H. Fry's book *Create Your Own ETF Hedge Fund* (Wiley, 2008) and Richard A. Ferri's text *The ETF Book* (Wiley, 2008).

I receive a number of questions regarding seminars, courses, and other resources that are offered for sale under the general rubric of trading education. My impression, and the feedback I get from blog readers, is that these offerings are often expensive and of limited relevance to the particular strengths and interests of individual traders. Similarly, I receive many inquiries into proprietary trading firms, which allow traders to trade the firm's capital in exchange for a share of profits. Some of these firms are quite professional and ethical; others are not. I strongly encourage due diligence: talk to a number of people who have taken these courses or who are trading at these firms. If you cannot get direct feedback from actual users (not just one or two stooges), move on. Don't pay thousands of dollars for something unless you know *exactly* what you're getting.

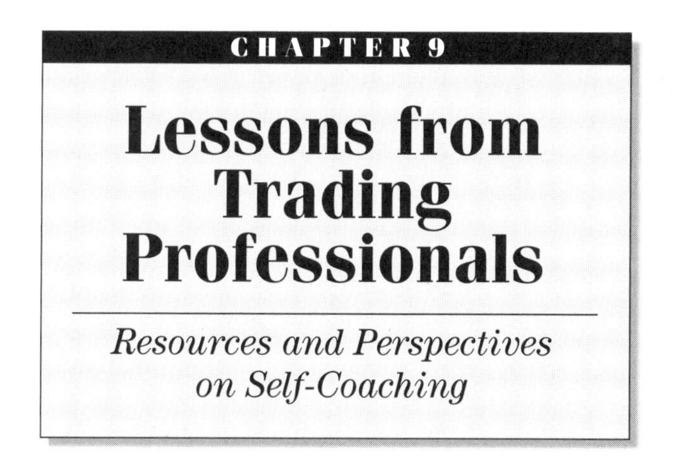

CHAPTER 9

Lessons from Trading Professionals

Resources and Perspectives on Self-Coaching

Anyone who fights for the future lives in it today.

—Ayn Rand

In this chapter, I sought the perspectives of experienced traders who share their views on the Web. The question I asked them was simple:

> "What are the three things you have found most helpful in mentoring/coaching yourself as a trader?"

I think you'll find their outlooks worthwhile, and I know you'll benefit from checking into the resources these traders provide. To coach yourself, you don't need to reinvent wheels. The guidance of those who have come before you can be invaluable. Think of this chapter as lessons in the best practices of self-coaching. Look for the themes emphasized by a majority of the contributors; these themes include the best of the best practices.

LESSON 81: LEVERAGE CORE COMPETENCIES AND CULTIVATE CREATIVITY

We encountered Henry Carstens when we took a look at profits and losses as a function of one's trading metrics. Henry's web site (www.vertical solutions.com) is filled with valuable information for discretionary traders, but his distinct specialty is the design of trading systems. Henry eats his own cooking: he builds and trades his own systems, manages money for clients with his systems, and develops trading systems for other traders and trading firms. When I posed the question to him of what he has found most helpful in coaching himself, he quickly offered three responses:

1. Leveraging core competencies and interests
2. Learning about learning
3. Collaboration and work ethic

Let's take a look at each.

From childhood, Henry learned about building and the use of machinery from his grandfather, Floyd Grahm. "I am, and have always been, a heavy equipment operator," Henry observes. Out of his fascination with machines and moving things around, Henry tried his hand at a different kind of building: *he built a machine that traded.* After building several such machines, he figured out a way to quantify rates of change in market behavior. When he tossed aside all but one of his machines, Henry developed others, so that his machines would never rust and become obsolete. His core competence and interest is building: markets just happen to be the earth he is currently moving. Henry leveraged what he knows and loves and found success and fulfillment.

> Success is found by leveraging distinctive interests, talents, and skills: doing what you love, and doing what you do well.

Note, however, that the flip side of building is destroying. Sometimes you have to take down a structure to erect a new one; you have to remove a mound of earth to lay a foundation for something new. It was Henry's willingness to put aside his old machines and develop new ones that provided him with his edge. It's not about finding a perfect system and taking cash from it like an ATM. It's about having the integrity to acknowledge when ideas are no longer working and having the desire to be the best machine operator possible.

With respect to learning about learning, Henry credits his chess lessons with grandmaster Nigel Davies for helping him become sensitive to patterns and their nuances. "What I learned from those lessons was the power of nuance—the cascading insights brought by experience and the continuous pursuit of a single objective," he explains. Henry has learned to scan markets like he scans a chessboard for moves. "I continually search for the next insight now, the nuance that means just a bit of an edge for just a little while longer."

I've observed this interdisciplinary cross-fertilization among many successful traders. Those traders with mathematical backgrounds learn to quantify value and divergences from value. Those with athletic training use their experience to guide their training and direct their competitive drive. My own work in psychology has sensitized me to the ways in which human behavior is patterned. This work has been a tremendous aid in recognizing similar patterns in markets. Coaching yourself begins with knowing yourself, and especially knowing your distinctive strengths: how you best process information, how you think, what you love. It is difficult to imagine being successful at trading if your market activity does not tap into these core competencies and values.

Finally, Henry identifies collaboration and a strong work ethic as key to his self-development. "A diverse network of people with broad and similar interests in trading both pushed and supported me," Henry explains. "The very strong work ethic I got from my grandfather always makes me do just a little bit more." The keys here are that the network is diverse—providing input from many angles to aid creativity—and the network is both challenging and supportive. I've often been asked why I don't charge money for my blog content. The answer is that, in developing a community of readers, I also cultivate the diverse network described by Henry. Readers push me to think more deeply about markets, and they support my efforts at learning more. All of us tend to fall into the trap of operating within our comfort zones. We benefit when those who care about us also push us to move beyond those self-imposed boundaries.

Like many successful creative individuals, Henry exemplifies what I call an *open source* approach to networking. When others share ideas and research with him, he freely passes his work on to them. This approach is very common among professional portfolio managers, particularly the successful ones. A surprising number of traders, however, are convinced that they must keep their work secret. The result is that they become isolated and stagnant. "When a colleague sends me something that I build upon or that sets me off on a new and interesting direction, I always try to reciprocate with a relevant finding or study as way of thanks," Henry points out. "What I give up in secrecy comes back many-fold in strong relationships and flow of ideas that are the lifeblood of the creator."

> The trader must be a creator if his business is to avoid stagnation.

I cannot emphasize this latter point strongly enough: as a trader, you are only as strong as your flow of ideas. Those ideas are your lifeblood: the novel mutations that will enable you to evolve and adapt to changing market conditions. Success in trading *is* a kind of evolution; without creativity and the ongoing flow of ideas, extinction is a real threat—as it is for any business that fails to keep up with the marketplace.

Your assignment for this lesson is to set a goal and plan for becoming just a bit more creative in your approach to markets. You don't have to be a system developer to learn about learning, build upon your distinctive strengths, and fertilize your creativity within a network of colleagues. For your assignment, I encourage you to add just one source of new ideas that will add to and challenge your own. This source can be a web site, a newsletter, or a peer trader—but it should be a source that you can consult as a regular part of your generation of trading ideas. Henry freely shares his experience on his site, including a framework for testing trading ideas; making use of the fruits of his self-coaching will be invaluable for your own.

COACHING CUE

When we're successful, our work expresses who we are. Henry emphasizes that he is a builder and creator. How does your trading express who you are: your greatest talents and interests? Identify the recent times in the markets when you have been your happiest and most fulfilled. What made those times special? How can you bring those special elements into your trading more regularly?

LESSON 82: I ALONE AM RESPONSIBLE

Chris Czirnich, author of the Globetrader blog (www.globetrader.blogspot.com), was one of the first contributors I considered for this chapter. Uniquely among market writers, Chris is sensitive to the psychology of trading. His blog consistently combines market insights and psychological ones; it's a useful resource.

I posed the three-pronged question to Chris, asking him what has been most helpful for his self-coaching as a trader. The floodgates opened and Chris came up with 10 ideas, not three. His response was so well considered that I decided to summarize all 10 of his insights and their implications for self-coaching:

1. Being able to observe myself.
2. Discipline.
3. Having an edge.

4. Accepting that I alone am responsible.
5. There is no holy grail.
6. Having trust in myself.
7. Being able to stand up again, after being beaten down.
8. Being lucky from time to time.
9. Keeping a trading journal and writing a blog.
10. Don't panic.

Building the External Observer

"The external observer is a concept I came across after reading your first book," Chris explains. "It allows me to see myself trading, to observe and interact in case I'm in a position from a teacher's perspective. It allows me to argue about my trade, to see the pros and cons, to notice price behavior I might miss otherwise. It's an invaluable resource in a trade because it is not affected by emotions and will guide me, even in a fast-moving market. I rely on it to always know what I should do. I might override it, but it is the clear voice in a turbulent market, where I can always turn to find the safe way out."

Maintaining Discipline

"Discipline is so elusive, so difficult to maintain, and yet it is something without which you won't succeed at trading," Chris explains. "Am I a disciplined trader? Alas no, unfortunately not, but after all these years I have learned that I have to follow certain rules or I will do a lot of damage to my account. Let's look at a concept that every trading book tells you is wrong and will lead to disaster: adding to a losing position. Admit it, you have done it at times, because you were certain that you were right and the market was wrong. My biggest loss came from adding to a losing position; still on a range day, adding to a losing position is the way to trade, because otherwise, you will die the death of a thousand stops. Yes I add to losing positions, but now I have the discipline to stick to the twice wrong and I'm out rule. Meaning, if the add-on does not work, I'm out. Of course you can be wrong and a trade signal might go against you, but if you have two signals in a row that go against you, then you have to take a step back and see why your signals aren't working properly."

Rules promote discipline.

Having an Edge

"Without an edge you are doomed," Chris asserts. "Very simple. Of course you might throw a coin and trade a 50:50 chance system. But only a professional will have the discipline to stick to the necessary money management rules to trade a 50:50 system successfully. Having an edge means you can statistically prove that your trading system works, that it would have paid your commissions and costs of doing business and made you profits in the long run. You might be a mechanical trader, you might trade fundamentals, you might trade price action or some arcane indicator or a combination of all of them. It all comes down to one thing only: You must be able to prove to yourself that your trading system works and will make you profits in the long run. If you can't prove that to yourself, if you don't understand the mechanics behind your trading system, then you won't trust your trading system and you will not be able to trade it. The best trading system will produce losses in the hands of a trader who does not believe that that trading system works."

I Alone Am Responsible

"I came to trading because I came into real financial difficulties after some clients of mine went bankrupt and did not pay their bills" Chris recounts. "Trading seemed the solution to me, because only trading gave me the promise of instant payment, of knowing that when I did it right I would get paid. But that promise came with a responsibility I did not fully understand: I alone am responsible for any action taken. There is no one but me to blame, if I have a red day, a red week, a red month or year. Each and every day I can look back and tell you where I was wrong, what trade I should have taken, where I missed the opportunity to make it back. It's only me who is responsible. And if I alone am responsible, then there can't be any guru to whom I can turn to tell me what to do. I trade my system. I can tell you what I do, but whether you will be able to use that knowledge depends on you alone."

> The need for a guru confesses an absence of self-guidance.

There Is No Holy Grail

"The charts of the best traders have price bars only or they trade without charts at all, like many forex traders," Chris explains. "They are not magicians, but they follow the price of the instruments they trade for such a long time, they no longer need indicators or charts. But if you ask them, they will tell you that there is divergence on price and that a bottom or top might be near, that right now will be a great buying or selling opportunity. All these traders have looked at charts, they have used all the common and not so common indicators or oscillators or volume analysis and after a while they removed them from their charts until they were back at the beginning looking at a bare chart, but now knowing that there is no holy grail among these indicators. Nothing will give you 100 percent winning trades, so it's futile to search for it. You need to focus your efforts somewhere else to succeed."

Having Trust in Yourself

"I'm a discretionary trader," Chris points out. "This means that I have certain trade rules, which provide me with a trade setup, but I decide on a, let's call it *gut feeling* whether I take the trade or not. I have tried a few times already to build a successful mechanical trading system, but I was never able to boil my trade rules down to a mechanical system that I could trust enough to trade. On the other hand I have accepted that my subconscious mind is a better computer than my mechanical skills will ever be. Maybe if I tried my hands at neural nets, I could come up with a working mechanical trade system, but then I would not understand the rules any longer and that means I would not trust it enough to trade it. The subconscious mind will not give clear instructions; it communicates through feelings. You need to learn to listen to them, if you want to use its power. But if you do, it can be a nearly unlimited resource you shouldn't ignore. To teach or program your subconscious mind to do its job, you need to invest a lot of screen time. You need to expose it to as many situations as possible . . . to develop that trust in yourself, the trust that you will always do what's right for you."

> If you don't trust yourself or your methods, you will not find the emotional resilience to weather periods of loss.

Standing Up after Being Beaten Down

"Being bankrupt, having a real bad day: yes it happens," Chris explains. "Many traders trading for their own account will have gone through such a slump not once but multiple times before they manage to develop their account. When I started trading, I had lost a huge amount of money in funds. I had trusted the fund managers to do a good job; instead they did a lousy job and I decided I could do better! It wasn't easy. Somewhere I read that if you can't trade 1 futures contract successfully, why do you think you can trade 10 contracts successfully? That was a statement I could accept and actually still follow to this day. So I gave myself a $3,000 account and started trading futures. (I never encountered the problems I had in my trading when trading in demo mode, so I traded real most of the time). I went bankrupt (actually below $2,000, which was the limit I had to maintain to continue trading) at least five times. I funded my account with about $20,000 over the last seven years and made it all back within three months, when I finally got it right. I'm not out of the woods today, but I have started taking out money for my living from my account. I still have days where I screw up big time and need to build myself up again; where I need to question my plan, myself, and my approach to the markets. But I know today that I can trade and that I have an edge. I trust myself to do what is necessary to do, even when I screw up. I know I will stand up again and make it back."

Mastering great challenges yields great confidence.

Luck

"There is no room for luck in trading? Don't believe that for one second," Chris asserts. "How many trades did you do and looking back you know you were just lucky to get out breakeven or make a huge windfall profit? I always think I'm entitled to two or three lucky trades per month. But make sure you know you got away lucky. Don't bask in the glory of that wonderful trade, when all you did was violate your rules, add to that lousy entry, and then have the luck to ride a spike against the prevailing trend right to the tip."

Keep a Detailed Trade Journal and Write a Blog

"You need to be totally honest with yourself," Chris advises. "There is no rock to hide under if you screw up. It shows in your account and you need to document it. Otherwise you will do the same mistakes over and over. Believe me, you will still do the same mistakes over and over again, even when you write a journal, but at least now you know you made the same mistake again. A trading journal can provide you with the statistics necessary to develop trust in yourself. It will tell you if you have an edge. It can tell you which approach to the markets works and which was a big failure. The trading journal I use today goes back more than four years now, and I made about 5,500 trades in that time. It is an invaluable source of information about myself and the ways I handle certain types of markets. If I encounter a rough patch in the markets I can look back and see if I had a similar experience in the past. I can see how I handled the situation then, whether my solution was successful, or whether I should better try a different approach today. Usually before I screw up big time, I have a few days with smaller and smaller profits. Looking back I see that I felt insecure in the markets: something was changing and I was not changing with the market. So I struggled to keep the green until something snapped and suddenly I was totally and absolutely wrong. The next day or two, I often make it back before I have a second deep red down day. After that I usually get back on track with smaller profits. The account starts to consolidate before I manage the next trend move.

"Writing the *Globetrader* blog I maintain to this day has made me accountable. I started the blog because I hoped that by sharing my approach to the markets, older, wiser traders would read it and question me or point me in a different direction by commenting on my ideas. Fortunately for me some of the comments I received proved invaluable and are now an integral part of my trading system. You don't need to write a public blog, but writing about your thoughts in a trade, how you see the markets, or what constitutes a trade setup structures your approach to the markets. Right now I'm at a point in my development as a trader where I try to dissect that gut feeling I wrote about earlier, so I can consciously see why my subconscious mind just gave me a clear *Go ahead and take that trade* signal. Or why it just questioned an otherwise wonderful looking signal and is proven right a minute later. By writing about these trade setups, I can relive the feelings I had when the trade opportunity presented itself in real time. Eventually I can see why the trade setup actually was not an opportunity. The blog is also the place to deal with all the demons and obstructions you will encounter in your trading. Writing about the problems is the first step to solving them. As long as you have no mechanical automated trade system, you have to accept that you are human and will make mistakes. You need to deal with them and you will have to find ways to avoid or integrate them or you will not make it in trading. But the first step is always to bring them in the open, so they can no longer hide."

Start a blog as a great way to journal your ideas and interact with others about them.

Don't Panic

"These are the famous words found on the cover of the *Hitchhikers Guide to the Galaxy* by Douglas Adams," Chris explains. "They are so true in trading. If you panic, your instincts take over, and these instincts will surely cause the maximum possible damage to your account. If the market suddenly starts to drop big time and you are long, don't be frozen; believe what you see and act. Or decide not to act and execute your contingency plan. You need to have a plan for every situation. Usually you will pay for every lesson the market gives you. How much you pay is totally up to you . . . So if I'm suddenly in an unwanted position, I look at the chart and see if I like the position or not. If not, I'm out. Simple as that. Otherwise I manage the trade. But never ever allow panic to take over."

Chris's lessons are the result of hard-won experience. His attitude of taking full responsibility for all aspects of trading lies at the heart of self-coaching: you are the author of the story of your trading career. Your actions will determine the plot and ending of that story. One of Chris's lessons that I like best is the notion of standing up after being beaten down. His success came as a result of resilience: he lost small amounts of money many times before he started to trade well and trade larger. Your assignment for this lesson is to create a disaster plan for your trading that explains how and when you will cut your trading size/risk when you are not trading well, but also how you will stand up and persevere with your best trading ideas to bring yourself out of drawdown. The best traders are quick to pull in their horns when they're not trading well, but they are not quick to give up on their trading. If you develop and follow your disaster plans, you take responsibility for your trading and place yourself in control of your market participation. As Chris notes, we cannot repeal uncertainty, but we can avoid the poor decisions that come from panic and lack of preparation.

COACHING CUE

Make sure your trading journal highlights important lessons learned, so that it becomes a constructive tool for review months and years later. The value of a journal is in its review, not just its initial writing. If you ensure that every journal entry has a lesson for the future, you also ensure that today's learning can enrich tomorrow.

LESSON 83: CULTIVATE SELF-AWARENESS

Trevor Harnett enjoys an interesting perspective on the trading world. He is a seasoned trader, and he is someone who runs a software firm that provides tools for traders. As a result, he has observed his learning curve, but also the curves of traders who utilize his Market Delta software (www.marketdelta.com). While Market Delta incorporates a number of charting features, the heart of the program is its ability to separate volume traded at the market's offer price and volume transacted at the bid. This distinction enables traders to obtain instantaneous readings of short-term sentiment. This is very valuable for intraday traders, and it provides a useful execution tool for longer timeframe traders.

Michael Seneadza is a full-time trader and author of the *Trader Mike* blog (www.tradermike.net). When I first entered the world of blogging, Michael's was one of the very first blogs I read regularly. It seemed to me that he had a fine grasp for short-term trading, including the news items that move markets. His blog posts the stocks he's following for the day, as well as market and news updates. Michael keeps it real—his site is devoid of hype and self-promotion—which helps account for its popularity. I grouped Trevor and Michael together for this lesson because they both touched on an important facet of self-coaching: being self-aware and acting as one's own psychologist.

When I asked Trevor for the three things that have most contributed to his self-coaching, he replied, "Upon looking back at my trading career, the three factors that have influenced my trading the most are 1) the environment I put myself into, 2) the discipline I exercised as a trader and else where in my life, and 3) self-awareness in terms of personality and how I viewed the markets." Let's take a look at those.

Environment

When Trevor started trading, he made sure that he was close to the action and rented an office in the Chicago Mercantile Exchange building. "I had plenty of desire but little knowledge and few friends or mentors to show me the way," Trevor explains. "Being in an environment with lots of seasoned traders was important to me." Trevor learned most from these traders' mistakes. "A majority of what I learned by being around other traders was what *not* to do," Trevor pointed out. "I learned valuable lessons from other traders, but the lessons that have kept me trading over the years are the lessons I learned from other traders on not what to do. For me, being in an environment that consisted of more than just my own experiences increased my rate of learning tremendously. I was able to share in others successes and defeats and learn from what they did right and wrong. To me this was invaluable because some of the experiences they had to endure were ones that I hoped to never find myself in. I could see what happened and try and learn from the situation."

I find this theme again and again with successful professionals: much of their success comes from the accelerated learning curve afforded by being in the right settings. Trevor's experience suggests that it's not necessary to be employed by a trading firm to find that environment. Just being around experienced peer professionals can multiply learning experience.

If you want to experience yourself as successful, place yourself in settings and situations where you can interact with successful people.

Self-Awareness

Trevor emphasizes the importance of knowing yourself as a trader. "For me," he explains, "when I entered into trading in 1998 after graduating from college everyone was telling me pit trading was what I needed to do. I had always been much more of an introvert and very proficient on the computer." He quickly found pit trading not to his liking and became involved in the emerging electronic Globex trading platform. He also recognized that his trading style was naturally risk-averse and stayed within his comfort zone. "My personality was much more oriented to taking frequent trades and keeping my losses under control," Trevor recalls. "This worked very well for me because very rarely would I have a day that would get away from me." We hear many generalizations about the best or right way to trade. *Trevor's insight was that his trading had to fit who he was.* His success came from sticking to his basic strengths and interests.

Discipline

"This makes or breaks traders from what I saw," Trevor points out. "When you are trading your own money with no risk manager breathing down your neck, you better have discipline. Without it, it will be just a matter of time before you blow up and are unable to recover." Indeed, this discipline is what makes the learning curve possible. "I was often early on many trades because I lacked the patience to let the trade play out," Trevor explains. "Fortunately I had the discipline to work my way out of the trade and try over." This is an excellent point: *discipline doesn't mean not making mistakes; it means making mistakes the right way.* Particularly when you're building your competence and confidence, it's important to learn how to "work your way out of the trade." Positive experience can build optimism, but it's the ability to work out of difficult situations that yields the confidence that you can handle most anything the market can throw your way.

First and foremost, good traders are good risk managers.

Michael described his three most valuable steps in self-coaching as: 1) keeping a detailed trading journal; 2) becoming an amateur trading psychologist; and 3) listening to trading affirmations. This reflects a balance that I see among many experienced, successful traders: always working on their trading, and always working on themselves.

Keeping a Journal

"The thing that's helped my development the most," Michael explains, "is keeping a proper, detailed trading journal. I've always prided myself on my ability to remember and learn things solely by memory. Writing things down just seemed like unnecessary work . . . That all changed a few years ago when I hit a rough spot and decided to reassess things. I went back to basics and did things according to the advice of what I'd read in all those books over the years—mainly create a detailed business plan and keep a journal. In just a few weeks of keeping a detailed journal, a few self-defeating behaviors jumped out at me. I was amazed at how those behaviors never registered with me previously. For example, I discovered that I had a bad habit of adjusting my initial stop-losses too soon. That often resulted in stopping myself out of winning trades at breakeven. Fixing that one thing has added considerably to my bottom line."

I have experienced the same thing: in keeping records of my trading, I learn things about my performance that I had never recognized earlier. Often, it's just a few tweaks to what you're doing that makes the difference between breakeven and profit.

The advice you see repeated in one trading book after another is often the best advice, because it's the result of years of experience.

Becoming an Amateur Trading Psychologist

"I had picked up bits and pieces about trading psychology over time," Michael notes, "but it wasn't until I read a book dedicated to the topic that things really jelled for me. The book I read was *Trading in the Zone* by Mark Douglas. The book crystallized all those bits and pieces I'd picked up, as well as forced me to take a look at my own beliefs and behaviors. It's one of those books in which I get something different each time I read it. Long before I read *Trading in the Zone*, I knew logically that trading was nothing but a game of probabilities. I knew all about expectancy and that I could still make money, even if I had more losing trades than winning trades. Yet there was a disconnect between my knowledge and my actions while actually trading. The book made it very clear to me that I needed to accept that I won't know, nor do I need to know, how any given trade is going to turn out. It made me realize that, as long as I stuck to my business/trading plan and kept taking good setups, I would make money over time."

> When we become our own psychologists, we bridge the gap between what we know and what we feel.

Listening to Trading Affirmations

"A few years ago, I purchased a CD called *Trader Affirmation* from the Day Trading Course site (www.DayTradingCourse.com/cd/)," Michael recalls. "The CD has about 30 minutes of someone reading a list of affirmations to help the trader's state of mind. I try to listen to the affirmations at least twice a week, usually while I'm showering in the morning. The affirmations help me to remember all the things I've read in the aforementioned book on trading psychology. Listening to them has been a great help in keeping my head on straight." To be honest, I've never been a big fan of the whole idea of positive thinking and affirmations (I haven't heard the CD that Michael uses), but I have to say that what Michael says makes a great deal of sense. It's not enough to read a book on trading psychology and file away the lessons. Rather, you need to repeat those lessons in order for them to sink in. That has become a part of Michael's weekly routine, helping him cement his efforts at becoming his own psychologist.

Your assignment for this lesson is to identify and implement one weekly routine to help you internalize sound trading practices. This routine could be listening to a CD (or even your own self-recorded messages), or it could be a structured review of your trading journal with a colleague.

The idea is to make right thinking and right action a regular part of your experience, so that you become your ideals.

COACHING CUE

Check out Michael's insights into trading journals at http://tradermike.net/2005/08/on_trading_journals/ and http://tradermike.net/2005/08/thoughts_on_day_trading/#moving_stops. See also Charles Kirk's insights in this chapter.

LESSON 84: MENTOR YOURSELF FOR SUCCESS

Brian Shannon is a trader and an educator of traders. His *AlphaTrends* blog (www.alphatrends.blogspot.com) utilizes video to illustrate trading patterns each day, a unique resource for developing traders. He has captured many of the principles from these videos in his book *Technical Analysis Using Multiple Timeframes*. Brian's work is a great illustration of what I call contextual thinking: placing observed patterns into larger contexts to gauge their meaning and significance. I find that many short-term traders run into problems when they become so focused on the patterns over the past few minutes that they miss the larger picture of what the market is doing from hour to hour, day to day. By gauging patterns within larger contexts, we stand a greater chance of aligning ourselves with longer timeframe trends.

Corey Rosenbloom is a full-time trader who chronicles his work in his *Afraid to Trade* blog (www.afraidtotrade.com). What I like most about Corey's work is that he blends an awareness of trading psychology with an understanding of the psychology of markets. His site provides a number of trading insights, as well as insights into the minds of traders. I grouped Brian and Corey for this lesson because both described the ways in which they mentor themselves—guide their own learning processes—as part of coaching themselves for success.

Brian's response to my query about the three things that have been most valuable to his self-coaching reflects his trading as well as his teaching. Let's take a look.

Tuning Out Opinions

Brian stresses that he doesn't completely ignore what he hears from others, but he's learned to emphasize his own views from what he's learned over the years. "The edge in trading is so small and quite often elusive that it is imperative to understand market dynamics/structure and where my personal edge lies," he points out. This is very important: As Brian's videos illustrate, he is extremely open to information from the markets, but he filters out opinions. He has learned to rely on his own judgment and experience to maintain his advantage in the marketplace. This reliance is essential in building and maintaining confidence. It's difficult to imagine sustaining the resilience to weather drawdowns if you don't have a basic trust in how you process information and make decisions. It is better to make a mistake with your own judgment—and learn from that—than to make a lucky trade based on the tips of others.

Review

When you view markets and review them day after day, week after week, you develop an intimacy with market relationships and trading patterns. An internalization of this intimacy is what traders refer to as a feel for markets. It is not mystical inspiration; it's the result of repeated exposure to information under proper learning conditions. Brian explains, "I review hundreds of stocks using multiple timeframes in an attempt to find what I believe to be the lowest risk/highest potential trades according to my entry and exit parameters." This review provides him with good trade ideas, but it also feeds a learning curve. After so much review across multiple timeframes, he has learned what a good stock looks like. This internalized expertise helps him deploy his capital in the most efficient manner possible.

> We learn our patterns—and the patterns of markets— through intensive review. It is the intensity of the review that enables us to internalize those patterns and become sensitive to their occurrence.

Mental Checklist

Here I'll let Brian speak for himself: "This one is somewhat new and came about as a result of letting my guard down on a few occasions earlier in the year, which resulted in losses which were larger than what I would normally take. Each day before the market opens I go through a mental checklist of: how do I feel (tired, anxious, excited, etc.) to identify any possible weakness before I commit money. I also try to visualize how I will react to what I view as either normal or abnormal trading conditions. I am trying to spend more time on the mental preparation than I have in the past and it seems to be working well for me." *Time and again, I find that this is what winners do: they learn from their losses and adapt.* Trading requires an active mindset; it's a bit like patrolling enemy territory, where you have to be on the alert for surprises at all times. If you're not prepared—and haven't rehearsed that preparation—you won't be able to act on instinct when those surprises hit. Brian let down his guard, and he was surprised. He created a mental checklist and incorporated visualizations of what-if scenarios and sharpened his active focus, anticipated what could go wrong, and enhanced his results.

Corey's three best practices for self-coaching were: 1) find a trading partner/group; 2) think in terms of concepts; and 3) keep an idealized trade notebook. All three practices reflect the progression of his learning as a trader.

Find a Trading Partner/Group

"The first thing I learned when I began trading full-time," Corey recounts, "was that trading could be an extremely lonely, isolating experience. It can be difficult to sustain motivation when you're the only one who knows what you're doing, and friends and family may not understand what trading is all about. Trading can be quite difficult, and it is immensely helpful to have at least a handful of solid friends or colleagues who understand your strengths and weaknesses while supporting one another for mutual benefit, such that the whole is greater than its parts. I began writing the blog initially as a way to reach out to others who had similar experiences... That has made an ultimate difference in my trading, mostly from the interactions and idea-sharing with others, which has broadened my awareness... I also have one experienced trader locally with whom I meet almost every evening to discuss the day's events and share ideas and study markets. This interaction has challenged us both, and we bring a combination of skills that benefit us academically (combined research), emotionally (motivation), and financially (improved trading tactics)."

Form a team to make trading personally rewarding and stimulate ongoing learning.

Think in Terms of Concepts

"I think the largest shift in my performance came when I began to view markets and price behavior conceptually, rather than being driven by indicators or news reports," Corey explains. "This was a process that took time and was difficult for me. Previously, I viewed multiple indicators and believed those were the secret to trading success. However, too much conflicting information was not only frustrating, but unprofitable. Even when I decreased the number of indicators, I still struggled to find profitability. My results were often no better than random entries, which was endlessly disappointing. The shift came when I was able to view markets and price behavior conceptually . . . The shift happened slowly and was attributable in part to studying Market Profile information, such as the concepts of trend day, bracketing markets, auction dynamics, timeframe participation, etc. Other concepts were based in the teachings of the early founders of technical analysis, including momentum, price range (expansion/contraction), broader trend structure, dynamics of price behavior, and price patterns (with their underlying reasons: accumulation or distribution, reversal or continuation). Essentially the shift was one towards greater understanding of price behavior and participation by all sorts of market participants . . . To further the conceptualization switch, I also began researching the broader concepts of intermarket analysis, which compares markets to each other, and sector rotation, which details performance of equity sectors and expectations . . . I began to see markets as a grand chess game, which opened up a new method of perception. Markets clearly do not trade in isolation."

When you think in concepts, you understand why markets move, and that helps you formulate promising trade ideas.

Keeping an Idealized Trade Notebook

Corey explains, "In addition to keeping a simple spreadsheet that tracks trading performance (which is essential in knowing when you're making mistakes and correcting them), I use a different kind of trading journal that I call my idealized trade notebook. In this notebook, I print off the intraday chart (I use the five-minute chart most frequently) of the stock or index I traded for the day. Also, if there are particular charts I find interesting, I annotate by hand what I deem to be ideal (or best) trades based on my understanding of price behavior and opportunity. Through looking at the charts at the end of the day without the pressure of real-time trading, I am able to see new patterns that I had missed . . . I then overlay my fills to see how close I came to achieving the total potential move . . . This serves a dual purpose of deeper visualization of my performance, but more importantly, helps clarify the distinct patterns and trade setups I use for trade entry, management, and exit."

By tracking ideal trades, we internalize best practices.

Following Brian and Corey, your task for this lesson is to structure your process of review. One task should include a review of the market day, comparing your trades to the actual moves in the market, so that you are learning both about you and about the patterns you want to be trading. Mentoring yourself is not an occasional activity to be performed during losing periods. Rather, among the best traders, it is a regular process embedded into each trading day. Compare what you did with what you could have done as a great way to track your progress and bring yourself closer to your ideals.

COACHING CUE

When you coordinate your learning with a trading partner, compare your ideas of the best setups for the markets you are trading. If you see markets through the eyes of others, you can enrich your own pattern recognition.

LESSON 85: KEEP DETAILED RECORDS

Two of the respondents to my question—two whose work I've followed for years now—independently arrived at similar answers. This is not because they trade similarly. Instead, it's a reflection of the wisdom they've accumulated over years of tackling markets and honing their own performance.

Charles Kirk is a trader, portfolio manager, and author of *The Kirk Report* blog site (www.thekirkreport.com). He also maintains a portion of his site for members, who are treated to his stock picking tools and selections. Much writing focuses on when to trade; Charles's forte is selecting *what* to trade. His blog is among the few on the Web that comprehensively links to articles on key themes that deal with markets and the economy. This is a particularly valuable resource for those who want to stay on top of the market's larger picture. If you want to see how institutional money might move markets, it makes sense to focus on the themes tracked by institutional money managers. Charles seems to have a knack for identifying those themes.

Jason Goepfert is the editor of the Sentimentrader site (www.sentimentrader.com), which—as its name suggests—focuses on measures of market sentiment. Jason freely offers his perspectives on markets and also shares the results of his tests of historical market patterns. He collects a large amount of data on markets and assembles the data in unique ways to uncover possible edges. These data provide information to guide traders' thinking, as well as food for specific trade ideas. A particularly interesting facet of his service is the tracking of relative smart and dumb money, including unique ways of reading options sentiment.

In response to my question of what has most helped his self-mentoring, Charles Kirk provided a single, detailed response: his BOO book. BOO stands for Book of Observations, and it is a collection of his trading experience. "In this book," he explains, "I keep a detailed track record for every trade I've made, along with observations about the market and things I've learned from others and from monitoring my own success and failures ... My BOO book contains specific and detailed information on every strategy and screen(s) I use, along with detailed performance information over different periods of time. In essence, everything I've learned up until now can be found in this book." Significantly, the contents of the book are organized in a database called do-Organizer (www.gemx.com), which enables him to readily access any idea that he's written about.

A database turns a trading journal into an active research tool.

Charles indicates that the database keeps his thoughts organized, as a scientist might systematically record data and observations from laboratory investigations. "Treating the market, and a strategy, from a scientific, evidence-based approach in this manner was helpful to me to keep me focused, disciplined, and on the right track," he explains. "This also helped me to test new strategies and to recognize early when certain strategies stopped working in specific market conditions, so I could adjust and transition my trading as needed." He also uses the BOO book to track new strategy ideas that he wants to integrate into his own trading. "I consider myself a perpetual student of the market and maintaining and using my BOO book has been incredibly helpful in this regard," he notes.

Indeed, Charles explains, "Looking back, the biggest mistake of my trading career was not starting my BOO book sooner. It took me several years to understand the importance of keeping notes in an organized manner while using a scientific, evidence-based approach to test and improve my skills and strategies."

The BOO book is a great example of the creative strategies that successful traders utilize to identify and hone their strengths. I believe

Charles's key insight is that the ideas in his book must be organized to be maximally useful. By placing his journal in a database format, he is able, with a few keystrokes, to access relevant experience from a broad time period. Journals can become unwieldy over time, and it is difficult to pull material from past entries. Increasing his access to his experience has enabled Charles to keep the past relevant to the present as a source of learning.

When I asked Jason Goepfert to share his three greatest sources of self-coaching, he started with an idea similar to Charles Kirk's.

Write Down Every Idea

"I've written close to 5,000 comments publicly over the past six years," Jason explains, "and also keep a personal journal that tracks more soft subjects such as how I'm feeling, anecdotal evidence, clips of headlines on news sites, etc. I review all of these periodically and find that they are exceptional tools in several respects. They keep me honest (not getting too ahead of myself when trading well and not too down when not), and they also serve as a check for when I'm anxious about a trade. I've looked at how I felt right before putting on past winning trades and saw that I was anxious then, too, so what I'm feeling now isn't necessarily some sixth sense subconsciously hinting that I not put on a trade."

Talking to More Accomplished and Experienced Traders

Jason notes that he meets many successful traders through his market service. "I am always struck that most of them suffer through the same travails as the rest of us," he points out. "They all get emotional at times, but they never let that seep into their risk control discipline. And that discipline is constant—there is no deviation from the strict principle that no one trade will sink them or their career. That is something I have written down in front of me. Risk control is paramount, and it is something I use as a mantra.

It's okay to be emotional; it's not okay to let emotions change your management of risk.

Always Learning New Things

"It's a cliché," Jason acknowledges, "but I've found that the more I learn, the more I discover how little I know. That helps tremendously in trading, as it has helped me to find new ways to approach old problems. Market dynamics are always changing, so we need to find ways to adjust as conditions change. Learning new trading strategies or new ways to test old ones can be very fruitful. It's a lot of work, but anyone afraid of hard work shouldn't be risking his or her capital. It isn't just trading-related stuff, either. I try to push myself into uncomfortable situations and experience new places, new people. That helps broaden my perspectives so I don't get closed-minded to new approaches."

What most struck me about Jason's insights was that *writing thoughts down led to self-discovery.* When he tracked his trades and emotions, he found that he was often nervous prior to winning trades. This tracking helped him not succumb to nerves when putting on a trade. It is this constant desire to learn new things—about self and market—that keeps trading challenging and interesting as a career. Tracking also helps the trader adapt to shifting market conditions. Jason Goepfert and Charles Kirk are not afraid of the hard work: they spend a great deal of time developing, reviewing, and testing their strategies. There is nothing *get rich quick* about their approaches to markets. Their record keeping is their way of sustaining a learning curve.

Your assignment for this lesson is to create an indexing system for your own trading journal, so that you can track themes associated with what you're doing and how well you're doing it. Tracking means categorizing your trades by strategy/setup, by markets, by results, and by your specific market and personal observations. Keep a journal in electronic form, such as through the StockTickr service (www.stocktickr.com), as one way of indexing your ideas. Another method is to turn your journal into a trading blog, with tags for various topics. Still another approach is to maintain your journal in a formal database, like Charles Kirk. *Your records need to be living, breathing entities that you can frequently review for insight and perspective.* Imagine your trades organized by market, market condition, trade setup, time of day, and size, so that you can pull up your results for any given market situation. Organized in this manner, your experience may just become your greatest trading coach—as it has been for Charles and Jason.

COACHING CUE

Consider a portion of your journal devoted solely to research: developing and tracking new trade ideas. Charles and Jason continuously search and research for trade ideas as market conditions change. What is working in the current market environment? Which stocks are moving? Which patterns are showing up? Journal about the markets as well as about your trading to help you anticipate opportunity.

LESSON 86: LEARN TO BE FALLIBLE

Dave Mabe is a trader, system developer, and founder of the StockTickr service and site (www.stocktickr.com). StockTickr is a unique resource because it enables traders to track their ideas and performance in an online format that can be shared with selected groups of traders. This Web 2.0 approach to developing ideas and tracking progress enables traders to build their own community of like-minded peers. The StockTickr site also includes an informative blog featuring interviews with traders who share their work online. I particularly like how StockTickr has created a true online trading journal, making journaling a social activity. It gives traders control over what they share and with whom. Indeed, there is huge potential simply in the idea of sharing a real-time journal with a trading coach.

Chris Perruna is a full-time trader and blogger whose work can be found on the site that bears his name (www.chrisperruna.com). His site is devoted to "successful investing through education" and covers topics ranging from screening for fundamentals among stocks to position sizing and charting. He shares his stock screens with readers, along with specific trade ideas. I like how the site enables traders to learn from his example.

Let's take a look at how these two pros responded to my question about what has been most helpful to their self-coaching, starting with Dave.

Trading Journal

Dave asserts, "A trading journal is by far more powerful than any indicator or platform. It provides the foundation for everything I do as a trader. Your mind can play tricks on you, but your execution data don't lie. Being able to reflect upon my trading results allows me to step back and view results in aggregate to see how I'm measuring up to my goals." I've seen this with many successful traders: the journal offers a layer of accountability and focus that would otherwise be missing. Dave also stresses the importance of flexibility in goal setting via journals. "Instead of setting a single goal (for example, a certain dollar amount over a time period), I find it much better to set a range of goals from conservative to radical. A lot of traders will set high goals, which set them up for devastation when they aren't achieved."

> We often focus on what we want to see. Statistics on our trading patterns don't lie; they focus us on what we need to see.

Learning to Be Wrong

"Most beginning traders have a tremendous need to be right and are resistant to admitting they might be wrong," Dave observes. "I learned quickly that being right (that is, having a high win rate) doesn't correlate well with making money. Win rate is overrated—in fact, one of the most profitable strategies I've traded had a winning percentage below 30 percent." Dave is right: most good traders I've worked with are not far from a win rate of 50 percent. Their success comes from knowing when they're right—and taking full advantage—and knowing when they're wrong—and minimizing losses. Overcoming the psychological need to be right is essential to success; without that, it's too easy to take profits early and remain stubborn in losing trades.

Automate

Dave recounts, "I've spent my trading career trying to remove as much of my discretion as possible from my trading. Many aspects of manual trading systems can be automated. The benefits of automation are numerous: more consistency, less time spent doing trading grunt work, and fewer mistakes. I've found that the more automation I have in my systems, the better my results. This includes my manual trading systems all the way to 100 percent completely automated trading systems that I trade." His point is well taken: even with discretionary trading, execution can be automated so that decisions can remain strictly rule-governed. The simple step of trading with limit orders rather than at the market can make a meaningful difference in performance over time, as traders enter and exit trades at favorable levels, rather than chase markets and get themselves in and out at the worst possible times.

Chris Perruna's responses will ring true to traders who have traversed their initial learning curves; he focuses on some of the universals of successful trading:

Understand Me

"The most powerful tool I have found in life and in this specific case, the market," Chris explains, "is what I, as a person, am capable of doing. I finally understand that personal characteristics that are ingrained in my DNA will only allow me to trade successfully under specific circumstances. For example, I am much more consistent and profitable as a medium-term and longer-term trend trader than as a day trader (even more so on the long side). I don't need to be everything all the time as long as I continue to focus on the areas that bring me the greatest success. Understanding *me* has been my holy grail of understanding how to trade the market with consistency and profitability." Chris's insight is critically important: *you don't make yourself fit a market or trading style; you find the markets and styles that best fit you.* Successful traders trade within themselves: they stick to what they do best and ignore the rest.

Find what you do best and fashion trading strategies around that.

Learning to Cut Losses

"It's almost cliché," Chris points out, "but not many people can do it in any aspect of life. I have learned to cut losses in my trading, my career, my hobby of competitive poker, and everywhere else in life where the rule applies. Without this rule, there wouldn't be a third rule." Chris makes a valuable point: *you can't live one way and trade another.* It's hard to imagine being totally disciplined in trading and lax in other areas of life. Good trading practice is a philosophy of living: pursuing opportunity, managing risk, limiting losses, and diversifying positions. Chris trades the way he lives.

Study and Work Hard

It is difficult to find a successful trader who does not place hard work at the center of what she does. "It is extremely important to my success for me to continuously study the markets on a fundamental and technical level and learn from my successes and mistakes," Chris points out. "Applying the knowledge gained from past experience allows me to properly analyze similar situations in the future with slightly greater odds of success. Never stop learning is a phrase I will never stop saying, as it proves to be truer as I get older."

Notice how both Dave and Chris emphasize the importance of *knowing when they're wrong.* This is an important difference between beginning traders and experienced ones. The beginners focus on being right, as Chris pointed out. Beginners try to avoid being wrong. The experienced traders know they'll be wrong on a significant proportion of their trades and fully accept that. *Their self-coaching is designed to help them anticipate and manage losses, not avoid them.* They have a plan for each trade to deal with loss, and they have an overall trading plan to deal with periods of drawdown. Your assignment for this lesson is to use your trading journal to flag your fallibility, identifying the five largest losing days in the past year. What did you do wrong on those days; what could you have done differently? What were the problems that occurred on more than one of these losing occasions? The idea is to embrace your fallibility by turning it into an engine of learning. If you clearly identify the mistakes from your worst trading days, you'll be better prepared to avoid them in the future. Your worst trades can be your best tool for self-understanding—and your best guide for self-coaching.

COACHING CUE

As you review your trading, make a special study of how you exit trades. Do you tend to exit too early, so that you leave potential profits on the table? Do you tend to overstay your welcome, so that potential profits are retraced? Take it a step further: what could you have looked at to stay in the trade longer or to exit sooner? How can you best adjust your exits to the market's level of volatility? If you refine your exits, you can break your trading down into components and turn observations into goals for improvement.

LESSON 67: THE POWER OF RESEARCH

Rob Hanna is a trader and the writer of the Quantifiable Edges blog (www.quantifiableedges.blogspot.com), a unique site that tracks historical patterns in the stock market. His electronic newsletter goes out daily, detailing the trades he places from his research. For traders, the blog and newsletter are unique tools that can extend their market edge. I particularly find historical patterns relevant to discretionary trading, as we'll see in Chapter 10. With a sound understanding of historical performance, we are in a great position to identify markets that are following their usual patterns and those that are not. Both scenarios can generate excellent trade ideas.

Jeff Miller is a money manager in Naperville, Illinois, who also shares his ideas about markets and trading through his blog, *A Dash of Insight* (http://oldprof.typepad.com). He frequently challenges accepted trading wisdom and offers perspectives on markets that reflect his disciplined analysis and understanding of economics. He has researched trading systems and uses those systems in his portfolio management. Like Rob, Jeff's edge is that he tests ideas before he trades them, giving him confidence in risking his capital.

Rob's answer to my question of what he's found most helpful to his self-mentoring as a trader was quite simple: "Research, research, and research." He explains, "I know that sounds like one answer repeated three times, but it's not. It's the top three answers . . . I've been through several stages in my career and traded using different methodologies. The one constant I've found with any method of trading I've employed is that it takes a substantial amount of research outside of market hours to successfully implement them."

Rob Hanna started his career day trading, focusing on short-term setups, such as those described in Jeff Cooper's *Hit and Run* books. What Rob found most useful was not the setup patterns, but the screening for volatile, trending stocks to implement the strategies. He began creating his own scans based on trading patterns, focusing on trades with the best risk/reward. "I wrote down each potential trade in a notebook along with the trigger price for the next trading day," Rob described. "When the trade was done, I would log the results in my accounting software. I kept a field in the accounting database called *reason*. It was there I entered the name of the setup I used to initiate the trade." This is a theme we see time and again with successful traders: they track their results meticulously to aid their learning curves.

> Keeping records of trades that work cements success patterns in your mind.

Rob Hanna points out that keeping his trades in a database accomplished two goals: it forced him to have a reason for every trade, and it enabled him to track his results as a function of the trade setups employed. "It made it extremely easy for me to determine which setups worked best and which ones struggled," he explains. "By doing this, I knew which setups I should continue to focus on and which ones I should scrap altogether." When Rob didn't have a solid reason for a trade, he simply entered the word *Hunch* as the reason for the trade. "It didn't take long for me to figure out how much the Hunch trades were costing," he recounts. He quickly scrapped trading from hunch alone.

The second phase of Rob's research occurred when he switched to longer-term trading that utilized patterns inspired by William O'Neil's CANSLIM approach. Rob began to utilize technical screens with the TC2000 software from Worden Bros., with his final lists filtered through fundamental criteria. "With TC2000," he describes, "I am able to easily place notes on the chart. This is incredibly useful. If a stock pops up with an interesting pattern, I may have already researched that stock in recent weeks. With the note feature, I can see if I already checked the fundamentals and rejected it for some reason. No need to waste time looking up the same symbol over and over." As with the day-trading patterns, the setups were less important than the work that went into implementing them. "Once again," he notes, "I found it was the research and not the trading that made me the money."

In Rob's third phase of research, he has formally back-tested his trading ideas, using Excel and TradeStation as primary tools. This back-testing has enabled him to generate actionable trade ideas. As he puts it, "I love taking these trades because I have a good idea of my success rate and profitability expectation going in." Reviewing a variety of price, breadth, volume, sentiment, and other indicator data helps him develop a view on markets, but it also "helps me to unveil what is truth and what is lore with regards to conventional market wisdom. There is so much information out there. It's difficult to know what's valuable and what information is simply hype... There is great value in being able to test ideas and understand what indicators and setups actually provide a quantifiable edge, and what ones don't."

> Much conventional trading wisdom does not stand the scrutiny of objective analysis.

For Rob, research has been his source of edge. "Whether my focus has been day trading, intermediate-term momentum trading, or quantitative swing trading," he explains, "I have consistently found that it's been the nightly research that has facilitated my growth as a trader more than anything... The ideas are all constructed at night. Market hours are simply used for executing those ideas. Research (stock screening and charting), research (quantitative analysis), and research (results analysis) are the three things I've found most helpful in coaching myself as a trader."

What I most like about Rob's perspective is that it highlights the relevance of research for every kind of trader, not just those that trade mechanical systems. Rob has tested and traded setups, *but he also has treated himself—and his trading—as a subject for study.* When Rob identified the ideas that work best for him, he has been able to maximize opportunity and eliminate the hunches and market lore that cost him money.

Jeff Miller approaches markets differently from Rob, but his perspectives on self-mentoring are surprisingly similar. "The most important thing—by far—in my trading is having a system and/or method," he stresses. "Without a system in which you have confidence, you are adrift. You second-guess yourself on every occasion. This leads to selling winners too soon, holding on to losers too long, and many other errors. You need to know that your basic method works. If you really understand and believe this, you can focus on making the correct decisions, which may not always be the winning decisions."

The key word Jeff uses is *know*. Many traders don't know their edge; they don't know how well their methods work in different market conditions. They have beliefs, but not deeply held convictions about the ways in which they make decisions. As a result, traders lose discipline. This loss is not because they cannot follow rules. It's because they don't deeply believe in the rules to begin with.

Many times, poor discipline is the result of shallow conviction. If we don't truly know our edge, how can we believe in it?

A second important element for Jeff Miller's self-coaching is analysis and review. "Having a system means testing it properly," he explains. "This is not back-fitting for a short time period. It is developing the method in one era and testing over out-of-sample data covering different markets. Only then can you be confident. Even with this method, you must do regular performance reviews to make sure that something in the world has not changed. There may be a tough decision about whether you are in a predictable slack period or circumstances that are really different." In other words, trading is fraught with uncertainty; even the best ideas have a limited shelf life. *The purpose of analysis and review is to generate an edge, but also to monitor changes in that edge over time.* It's not as simple as finding systems that make money for all time and all markets.

Jeff's third area of self-coaching is learning to recognize exceptions. "Understanding your method means knowing when something truly exceptional is happening," he points out. "We all know the danger in saying that, 'This time is different.' Keeping this in mind, there are exceptional trading opportunities lasting a day or a week, even for those of us with longer time horizons." This thought gets us back to the excellent point raised by Henry Carstens: knowing when to turn your trading off. When markets are behaving in historically abnormal ways, the usual methods may not produce their usual results. Exceptional markets yield exceptional risks as well as rewards.

The takeaway from Rob and Jeff is the value of self-knowledge. *The more you know about your trading methods, the more you can play to their strengths and avoid their weaknesses.* Note that for both Rob and Jeff, this has meant considerable time and effort outside of trading hours to hone their edges and stay on top of how they change. I consistently find that a major predictor of trading success is the amount of time devoted to markets outside of trading hours proper. The time spent in defining and refining trading methods is a major part of this commitment. When you are your own trading coach, you are no different from a basketball or football coach: much of your success will come from the hours you put into recruiting new talent, practicing, and planning.

Your assignment for this lesson is to treat yourself as a trading system, so that you can research, research, research your performance over various markets and market conditions. For this assignment, pay particular attention to the kinds of trades that make you most of your money. Do they occur at particular times of day, or in particular market conditions? Do they occur primarily in a few markets or stock names? Are they primarily short-term trades or longer-term ones? Are they mostly reversal trade, or are they trend following? Your goal is to clearly identify your bread and butter as a trader, so that you can allocate most of your risk to what you do best and reduce the risk associated with trades that are outside your wheelhouse. Understand what you do—especially what you do best—as it is the most effective means for developing and sustaining confidence in your work. Weekly and monthly reviews of each of your trades by categories are a great start in this direction.

COACHING CUE

Track the number of trades you place that break even or that make or lose only a small amount of money. Many times, this is a sign of good discipline in cutting losers (although it can also reveal problems with exiting winning trades far too early). Recognize quickly when a trade is wrong to help keep the average size of losing trades below that of winners, an essential ingredient of trading success.

LESSON 88: ATTITUDES AND GOALS, THE BUILDING BLOCKS OF SUCCESS

Ray Barros wears many hats as a money manager, trader, blog writer, book author, and trading coach. He is one of the very few coaches that I know who incorporates a keen awareness of psychology with a sound understanding of markets. His *Trading Success* blog (www.tradingsuccess.com/blog) is notable for trading and psychological insights, and his book *The Nature of Trends* is an excellent tool for mentorship. It explains his ways of analyzing markets, his risk/money management ideas, and his trading psychology tools.

John Forman similarly combines a wealth of roles. An athletic coach, he also mentors traders and offers his insights in *The Essentials of Trading* blog (www.theessentialsoftrading.com/Blog). His book by that same name is an excellent introduction and orientation to markets, with valuable views on analyzing markets, executing trades, and developing trading systems. He is keenly aware of the importance of skill development and psychology to the evolution of traders.

When I asked Ray for the three things that have most contributed to his self-coaching, his number one factor was attitudes. Among the attitudes he views as essential to trading success are:

Honesty

Ray defines this as "the value of never consciously faking reality. If there is one trait that has proven critical to my success and to the success of my students, it has been this one. Successful students are brutally honest with themselves. Failed students tend to provide excuses and rationalizations for their failures."

Responsibility

"I learned to take full responsibility for my successes and failures," Ray explains. "With successes, my question is: How can I repeat this? With failures, my question is: What can I learn from this?"

Tenacity

"I'll do whatever is necessary to achieve my goals," Ray emphasizes. "This includes constant learning by first learning the material, then adapting it to suit my needs. I notice that failed students tend to resist the material whenever that leads outside their comfort zones."

Learning requires a willingness to venture outside one's comfort zone to see and do things in new ways.

Discipline

Ray stresses, "I am disciplined enough to write out my trading rules and execute the rules consistently. I am disciplined enough to keep my psychological and equity journals so that I can learn from my trades. And I am disciplined enough to celebrate my successes and take time from the markets to recharge."

Citing Linda Bradford Raschke, who has mentored many traders over the years through her online trading room, seminars, and books, Ray Barros stresses that coaching is only valuable if its insights are implemented by traders in their *3-Rs*: routines, research, and reviews.

Among the routines that Ray finds essential to his lifestyle as a trader are:

- Updating data and journals.
- Reviewing trades.
- Preparing for the coming day, "including visualization of entries and exits."
- Balancing other duties and responsibilities with trading regimes.
- Staying on top of personal and business finances.
- Completing commitments, including writing articles, preparing for talks, and so on.

He notes that his self-mentoring blends review and research. "I may notice a pattern in my journals that needs attention," he explains. "It may be that I am suffering losses or experiencing profits beyond the norm. In the former case, I would need to research the context that is leading to the loss. I'd need to determine whether it is incompatibility between my plan and current market conditions, or whether I am breaching discipline. In either case, I have to decide what to do. If current market conditions do not suit my plan, I'll cut down size or take a break. If I am breaching discipline, I identify the context/contexts within which the breach/breaches occur and take remedial actions. I then review the actions to see if they have had the desired results. If not, I change the actions."

Ray also carefully reviews and researches periods of unusually positive trading performance. "If I am making above normal profits, I determine if current market conditions happen to suit my plan or if there has been a fundamental shift that is leading to greater profitability. In the former case, I increase my size and ensure that I maintain my discipline before I take a trade. I have learned that, in my case, I need to be more vigilant when I am having a great run than when I am suffering a drawdown. If there has been a fundamental shift, I seek to identify what I have done to cause the shift, and I seek consciously to continue the new behavior."

Vigilance during a run of profitability is an effective way to prevent overconfidence and lapses of discipline.

"I also constantly research new ideas," Ray Barros notes. He is an avid reader and seeks insights that will impact his life and trading. When he encounters a new trading idea, he outsources the testing of the idea to determine whether or not it truly possesses an edge. "My review provides a solid foundation for my activities," he explains. "I set goals, take action, and then see if the action is leading toward or away from the desired outcome."

One of Ray's best practices is the separation of his daily journals into trading and personal components. In his trading journal, he grades his entry and exit discipline, giving himself three points if he entered and exited according to plan; one point if either the entry or exit broke discipline; and zero points if he broke discipline on both. "I look to maintain a 90 percent threshold," he explains. "I must garner 90 percent of the total possible points. If I drop below 90 percent but above 85 percent, I start looking for causes, and I start remedial actions. If I drop below 85 percent, I take time off from trading." In the trading journal, he also tracks the excursion of each of his trades, expressing how much he took out of the trade as a proportion of what he possibly could have made. "I seek to capture around 65 percent of a possible move," he elaborates. "If I find that I am consistently capturing significantly less than 65 percent, I take this as a warning I am entering an ebb state."

> Like successful manufacturing businesses, traders can engage in continuous quality improvement by evaluating their processes and correcting shortcomings.

In the personal portion of the journal, Ray notes event, feelings, and behaviors that accompany each of his trades. "The aim here," he points out, "is to have enough details so that I can spot the patterns that warn of fundamental shifts, breaches of discipline, and ebb-and-flow conditions." In other words, he is tracking his performance much as he tracks a market, looking for signs of trends emerging from the data. When he is flowing, he wants to be more aggressive in his trading; when his execution is ebbing, he wants to cut his risk. Toward this end, he also tracks his trading metrics, including his average win and loss sizes; his win and loss rate; the standard deviations of profits and losses; consecutive wins and losses; average holding periods for winners and losers; his expectancy ratio; his drawdowns; and his recovery periods from drawdowns. The key to Ray's self-coaching is to study himself as intensively as he studies markets.

John Forman echoes Ray's point about making sure that one's trading life fits into her personal life. "The first thing a trader needs to do," he emphasizes, "is step back and take a big picture view of things. This is extremely important for new traders, as they need to figure out how trading is going to fit into their lives. Even folks who have been doing it for a while need to do this from time to time as well. Trading is part of one's life, not separate from it. What part it plays must necessarily define how it is approached, and that can change over time. Periodically taking the 30,000-foot view allows one to maintain perspective." I wholeheartedly agree with John's insight. Even successful professional traders can become overloaded by work responsibilities, tracking markets and themes day and night. If traders allow trading to consume them, they lose concentration and efficiency—and eventually that takes a toll on performance. *Successful trading means knowing when to not trade and when to conserve and renew personal energy.* Often, the best trading decision is the decision to take risk off and go on a holiday from markets. This reprieve can spark good thinking about markets and performance from the 30,000-foot view, aiding performance once trading commences.

"A second important thing," John Forman notes of his self-coaching, "is the commitment to performance improvement. That may seem to be an obvious thing, but it's something easy to stray from at times. It's often hard to not become complacent with one's trading, especially when a level of success has been achieved. In order for the self-coaching to have any value, though, the realization that one can keep getting better, and the desire to do so, must be at the fore all the time." I have noticed this time and again among the firms where I work. The best traders and portfolio managers seek out coaching when they're doing well, not just when they're losing. They have a continual drive for self-improvement; not just a temporary desire to remedy deficiencies.

> The measure of a trader is how hard he works on trading during winning periods.

"Finally," John concludes, "setting good goals and assessing how one is progressing toward them is critical. These are things coaches in other activities like athletics do as external observers. The advantage there, however, is that they don't have the direct link to the individual's psyche, which complicates self-assessment. The most challenging aspect of this process for the individual is not allowing it to adversely impact one's confidence level. That means the process needs to be as objective as possible, and the trader needs to be able to disconnect their ego from it." Forman raises an excellent point here: *goal setting and review must be pursued in a manner that does not damage confidence or motivation.* Vague or distant goals offer insufficient feedback and learning; difficult goals can yield frustration. Tracking goals with a negative mindset—emphasizing shortfalls—makes self-coaching a punitive activity. The good self-coach, like the good athletic coach, uses goals to facilitate learning and build confidence. No one will sustain a process if, over time, it leads them to feel worse about themselves.

Your assignment for this lesson is to conduct a self-assessment from 30,000 feet. We've talked about tracking your trading, *but now the goal is to track your self-coaching.* Is trading fitting into your life, or do you find yourself fitting your life into the markets? Is most your time consumed with trading, or are you spending at least equal time in performance improvement—the reviews and research of markets and trades—and the routines that help you develop new ideas and hone skills? How much of your efforts are goal-focused, and how much are you drifting from day to day? Do you get down to the hard business of grading your performance and tracking your ebbs and flows, and do you use this information to guide your risk-taking? In short, if you're going to be a good self-coach, you have to be as aware of your own coaching performance as your trading results. The value of such meta-coaching—training yourself to be a better mentor of yourself—is a key lesson we can take away from Ray and John.

COACHING CUE

Just as you can develop a report card on your trading to track your progress, you can grade your self-coaching efforts by assessing how much time you spend in self-coaching mode; how clearly you set goals for yourself; and how well you sustain work toward those goals. You can't develop as a trader without working on trading skills, and you can't develop as your own coach without working on your coaching skills.

LESSON 89: A VIEW FROM THE TRADING FIRMS

Mike Bellafiore is a partner at SMB Capital, a proprietary trading firm in New York City that specializes in the short-term trading of individual equities. He is also a successful trader and a mentor of traders within the firm. Most recently, SMB has extended its training to the trading public via a blog (www.smbtraining.com/blog) and a formal trading curriculum. I had the pleasure of visiting SMB Capital and was impressed by Mike, Steve Spencer, and the others in the firm. There was a good buzz on the trading floor throughout the day as traders shared ideas and breaking developments.

Larry Fisher is a co-owner of Trading RM, a proprietary trading firm in Chicago that specializes in trading individual stocks and options on those stocks (http://tradingrm.com). Larry and his partner Reid Valfer started the firm with the desire of providing a mentoring and teaching environment for traders. An unusual feature of the firm is that Larry and Reid call out all their trades, illustrating to their traders what they're doing throughout the day. Teaching and mentorship are thus woven into the fabric of daily trading. In visiting Trading RM, I was impressed by the learning environment. Larry and Reid have developed a web site and blog so that they can share their insights with the trading public (http://blog.tradingrm.com).

It is typical of Mike that, when I asked him for the three things that most help his self-coaching, he emailed me a 14-page document. He is attuned to the mentoring process and practices it in his own trading. Number one on his list is keeping trading statistics. "Statistics are very important for my trading," Mike explains. "I must know what trading plays are working best for me, what stocks I am trading profitably, my win rate, my liquidity stats, etc."

The head trader at SMB, Gilbert Mendez (GMan) created a tool for the desk called the SMB Chop Tracker. It summarizes trading statistics each day for each trader, so that they can see how well they're doing and where their profits and losses are coming from. "Most often I struggle with my trading because I am in the wrong stocks," Mike Bellafiore notes. He tells the story of how he traded one particular stock, MBI, quite well in the fall and then consistently lost money in it. "I felt like someone else was inhabiting my trading body," he jokes. "So I looked at my stats. They were screaming, 'Hey, Mike, maybe another stock for you?' I figured out some adjustments I could make, concluded there were better stocks for me to trade, and decided to move on. I went right back to making money."

> Statistics on our trading alerts us to hidden patterns, both problems and solutions.

Mike also tells about a particularly vicious loss he took in trading SNDK. "I will always remember 11/21/05," he recalls. "What a bloodbath. For weeks I walked around cussing SNDK underneath my breath and swore to never trade it again. But one day I checked my statistics and surprisingly learned that I actually traded SNDK well, save that one day. I was overvaluing that last rip. While trading, you develop a perception of how well you are trading a stock. That perception can be incorrect. When you study your trading statistics, you may discover that the stocks you thought you were killing, you weren't. And you may discover that the stocks you thought were a disaster weren't."

Mike Bellafiore's second coaching practice is something we all do, but not with intention: breathe. "While trading, it is essential to quiet your mind so that you accurately process the data that the market offers," he observes. "Some traders think they just need to focus better and shut out unneeded stimuli. These traders believe they can will themselves to focus better. But quieting your mind is an acquired skill. Mariano Rivera [a fastball pitcher] can't just start throwing a changeup because he really wants to. He would have to spend hundreds of hours working on his grip, motion, and control. It takes 15 to 30 minutes of deep breathing a day to develop and maintain this skill. My partner and co-founder of SMB Capital, Steve Spencer, taught me how to properly breathe. Steve teaches this to our new traders on our prop desk. I used to think I accurately processed the data that the market offered. But after I learned how to properly breathe, I recognized that this was not accurate . . . You must develop the skill of quieting your mind so that you accurately process your market data and, as a result, fulfill your trading potential."

Mike tells the story of a young trader next to him who cheered whenever his stocks moved in his favor. The veteran traders merely smirked; that trader soon blew up. "For old-school traders like us," he points out, "there is no celebrating intraday. You are now rooting for your stocks and not just interpreting the data that the market is offering." By controlling his breathing, he is better able to let the market data come to him, improving his decision-making.

> If you're celebrating or bemoaning a trade while you're in it, you're not focused on the market itself.

Mike Bellafiore's third self-coaching best practice is watching his trading tapes. "Watching my trading tapes has improved my trading more than any other self-improvement technique," he asserts. "Many great athletes such as Alex Rodriguez use video to improve their performance. I record all my trades and watch back the important plays. Doing so has helped me particularly with my two biggest weaknesses: closing out a winning position prematurely, and adding size." By watching tapes of his trading, Mike developed rules for recognizing when he should hold positions, when he should get back into a position he has exited, and when he should get out of positions. These rules were compiled into lists that became his system. "It gave me the confidence to add size when I see a great risk/reward opportunity that is on my list," he explains. "I learned from my trading tapes that adding size in certain spots offered favorable risk/reward trading opportunities, and that perhaps it was even irresponsible to not add size with certain trades. So when I spot a trade from my list of When to Add Size, I just execute."

Once again, self-coaching boils down to directed, hard work. "In my trading space," Mike insists, "if you are not willing to come into the office on the weekends and/or find some time after the close to watch your trading tapes, then you are not competing as a trader. Trading is a sport. It's a competition. And the results of your trading are often determined by the effort you put in before the open."

Larry Fisher's responses to the question of the three self-coaching practices that have most aided his trading reflect his teaching practices at his firm, which in turn reflect his years of trading experience. Here's what he has to say:

Writing a Trading Journal

"Over the years, I have used a journal as a medium to make sure that I am in tune with my emotions," Larry explains. "The journaling process has become a very important part of my trading routine. I have realized that writing in my journal pays huge dividends, especially when I am trading well and I am trading poorly. The process keeps me grounded, while often limiting the duration of trading slumps and extending periods of trading successes." Notice that Larry employs the journal effectively both when he's trading well and when he's not. This keeps him attuned to emotions in a positive way—it grounds him in confident trading when he's seeing markets well—and it enables him to take corrective action quickly when he's not in tune with the stocks he's trading. So often the difference between the successful trader and the unsuccessful one is how they handle being very right and very wrong. The journal, properly constructed, can be a tool for adjusting to these extremes, enabling you to add risk when you're trading well and pull back when you're not.

The trading journal is a means for sustaining self-observation.

Communicating with Peers

Larry Fisher notes, "I have a network of friends and colleagues with whom I make an effort to communicate on a regular basis. This allows me to learn from others while sharing real-time market experiences. These conversations aid me in dealing with the ebb and flow associated with being a professional trader." This theme arises again and again with the best traders: they have a rich network of contacts that help them personally and professionally. Larry's observation echoes what we heard from Ray Barros: there is always an ebb and flow to trading; profitable times and lean times. Being able to connect with traders who have been through the cycles and know how to move beyond them can be a tremendous support. We also underestimate the power of social interaction as a means of cognition: *some of us simply think more effectively when we think aloud.* Sounding boards for our ideas helps us hone our market views and make better decisions.

Trading in a Good Environment

"In order for me to be able to coach myself," Larry explains, "I need to trade in an environment that is conducive for success. We built our firm with that in mind. All the traders at my firm are on the same page. Willingness to be a part of a team combined with the desire to learn are characteristics each trader possesses." I have visited many firms in which traders operate in almost total isolation of one another. One person's learning experiences become just that: opportunities to learn for that individual alone. *When a firm is founded upon a team concept, everyone's learning becomes learning for the group.* This is Larry Fisher's central insight, and it is the greatest strength of his firm. When everyone calls out her trades, there's no place to hide. That is tremendously freeing. You can learn from the successes of your peers and also from their mistakes. Their ideas spark yours, and your heads-up on news or breakouts aids everyone else. In an environment in which all traders are their own coaches, all traders inevitably contribute to each other's coaching.

Learning cannot occur without accountability.

These are the real words of real traders who really trade for a living and really run successful trading firms. Their best practices can become your own, even if you don't work for SMB Capital or Trading RM. How do your trading practices compare with those at these firms? How does your trading atmosphere compare with theirs? When you're coaching yourself, you are—in a sense—creating your own trading firm. You are coach, risk manager, researcher, and trader rolled into one. How well you fulfill these roles depends on the time and effort you devote to each. A world-class basketball player works on offense and defense; on passing, dribbling, shooting, rebounding, and physical conditioning. There are many facets to one's game—in sports and in trading. The successful firms pay attention to all of them.

Mike and his partner Steve are correct to emphasize breathing in their training of traders. This exercise makes a worthwhile assignment for your development. *The first step toward controlling emotional and cognitive arousal is controlling the level of arousal in the body.* When we are filled with stresses and worries, we bring those to markets. When we sustain a quiet mind, we let markets come to us and free our minds to respond to the patterns we perceive. Write in journals, communicate with peers, and consult your trading statistics. These actions are all ways to make sense of your market experience so that you can then sit in front of the screen with a quiet, confident mind. It does take dedicated time each day to sustain the quiet mind, but it comes more easily with experience. Find a room with no distractions—no noise—and keep yourself totally still as you fix your attention on something in the room: an object on the wall, music in headphones, etc. Then breathe very deeply and slowly, keeping your attention as fixed as possible. You'll find yourself able to tune out fear and greed, anxiety and overconfidence as you sustain a high level of concentration and fix your attention on an emotionally neutral stimulus. The best trading practices and environments cannot benefit you if you are not in a state to make good use of them. Quite literally, with each breath, you can be coaching yourself.

COACHING CUE

I have found that if I start my day with physical exercise and biofeedback, I can sustain calm concentration as an effective strategy for maximizing my energy and focus. If you start your day run down and distracted, you're likely to become even more fatigued and scattered during the trading day. Part of preparation is to study the market; part is also to keep yourself in a physical and cognitive mode that maximizes performance.

LESSON 90: USE DATA TO IMPROVE TRADING PERFORMANCE

Rainsford "Rennie" Yang is the author of the Market Tells web site and newsletter (www.markettells.com), which generates trade ideas through historical analyses of stock market behavior. His service is unusually helpful in finding trading edges, particularly with respect to generating trend-catcher alerts during the day. The ideas can either be traded outright or can be used to inform discretionary decisions from favored setups. For traders who don't have the time, skills, or inclination to conduct their own historical research (see Chapter 10), a service such as Market Tells is invaluable.

David Adler is the Director of Trader DNA (www.traderdna.com), which markets a program for tracking trading performance over time. The software captures information about futures trades and generates a series of metrics that reveal areas of trading strength and weakness. This information is especially helpful for high-frequency traders, who would find it impossible to manually enter trades into a log for analysis. Results are charted as well as summarized in print, providing easy-to-understand reports.

When I asked Rennie to summarize his most useful self-coaching practices, he most generously shared some of the patterns from his historical research. He included daily/weekly analysis and intraday analysis in his response, which I quote extensively in the following pages.

Daily/Weekly Analysis

"When advancers recently outnumbered decliners by more than a 3:1 margin on the NYSE and the market continued to push higher over the next few sessions," Rennie recounts, "I brought up the Master Spreadsheet where I conduct all of my testing and research. It contains the daily data back to 1980 and weekly data back to 1950 on all of the major averages and all of the market internals (breadth, volume, new highs/lows, etc.) to make testing quick and easy. In this case, I searched for instances when the S&P was higher three days after a 3:1 positive breadth session and examined the market's performance over the next two weeks. Such lopsided breadth days can mark *buying climaxes*, in which buying power is exhausted and the market trades lower short-term. But when the market remains on firm ground in the days following such a lopsided positive breadth session, I would expect the S&P to continue moving higher over the next two weeks."

Rennie Yang explains how to conduct this analysis: "To keep things simple, let's assume I have a spreadsheet containing daily S&P 500 and NYSE advance/decline data. Column A has the date, while columns B and C contain the daily S&P 500 closing price and NYSE advance/decline ratio, respectively. Starting at the fifth row in column D, enter the following formula:

= if(and(c2 > 3, b5 > b2),(b15-b5)/b5,"")

"This states: *if the advance/decline ratio from three days ago was over 3.0 and the S&P closed above its three-day ago close, show the percentage gain for the S&P over the next ten trading days.* Fill this column down to the point where the data ends and quickly scan the results. You can immediately see that the hypothesis seems correct. Over the last 30 examples, the S&P has been higher 10 trading days later in 25 out of 30 cases, or 83 percent of the time."

> To get a sense for whether an edge is present, it is important to compare a historical pattern over X days with the market's general tendency over X days.

"That may look like a bullish edge," Rennie points out, "but first you need to check the S&Ps at-any-time odds of posting a higher close 10 trading days later. Here's a quick and easy method. Go back to the fifth row in Column D of the sample spreadsheet above and change the formula to read "=if(b15>b5,1,"")." This means that *if the S&Ps close two weeks later is greater than today's closing S&P, print a one, otherwise print nothing.* Then fill this column down to the point where the data ends. In most spreadsheet applications, such as Excel, you'll see in the lower right corner the summation of all those 1s. Dividing that result by the number of days in the sample reveals the at-any-time odds—57 percent. In other words, on any given day, the chances that the S&P will be higher two weeks later have been 57 percent. That is far less than the 83 percent odds when the S&P is higher three days after a 3:1 breadth session. This confirms the original hypothesis that the chances for a market rally over the intermediate-term are far better than average, meaning there's a clearly bullish edge.

"Instead of relying on traditional indicators, most of which merely manipulate and regurgitate price action, look beneath the surface of the major averages at the market internals such as breadth, up/down volume, new highs/lows, NYSE TICK action, etc," Rennie advises. "This is the area in which I've found the majority of trading setups that stand up to historical testing. Does a surge in new 52-week lows portend an intermediate-term bottom? Is a 90 percent up volume day bullish? How about a cluster of 80 percent up volume days in a short time frame? Just about any concept you can imagine can be quickly researched and tested with the proper preparation. By maintaining your own version of a master spreadsheet and conducting your own testing and research, you'll know when a concept truly provides a bullish or bearish edge. Consistently exploit that edge, and you'll have a leg up on the competition."

It is powerful when you find a pattern with an edge, but even more powerful when your edge is the ability to find and trade many such patterns.

Intraday Analysis

"The NYSE TICK is probably the single most helpful intraday indicator," Rennie Yang asserts. "It tells you, at a glance, how many issues last traded on an uptick versus a downtick. A reading of +500, for instance, means that, at that moment, 500 more issues last traded on an uptick. When you first view a chart of the NYSE TICK, it will look as if it's too noisy to be of any use... But change your viewpoint and you'll see an entirely different picture... You can actually hide the NYSE TICK itself and just plot the 20-period moving average of the TICK to gain considerable insight into the supply/demand equation. Is the average holding above zero, meaning generally more buying power, or is the average holding below zero, reflecting better selling pressure? That's something every day trader should know.

"Here's another technique for utilizing intraday TICK readings," Rennie offers. "Many data feeds such as e-Signal allow you to export data in real time to a spreadsheet. Through a technology known as DDE (dynamic data exchange), it's a relatively simple process to have one-minute NYSE TICK data updating constantly in your spreadsheet. Once this has been accomplished, you can easily create your own cumulative TICK. Here's how: Set up a spreadsheet with columns A, B, and C containing the date, time, and close of the NYSE TICK ($TICK in e-Signal). It should start at row 2 and contain the last 390 one-minute bars, the equivalent of one full trading day. In the first row of column D, enter a zero. In the first row of column E, enter a space, followed by today's date (the space is due to a quirk on e-Signal's part). Then jump down to the second row of column D and enter the following formula:

= if(a2 = E1,c2+d1,d1)

"and fill it down through all 390 rows. This states that, *if the date matches today, then take the most recent closing one-minute TICK and add it to the running total for the session.* As the data comes in, this will automatically build a cumulative TICK in column D, which can then be charted to provide a real-time intraday chart of the cumulative TICK. Draw a line at the zero mark and watch the cumulative TICK reveal the underlying buying and selling pressure that is hidden in the noise of the NYSE TICK."

> The cumulative TICK reveals the trend of daily sentiment.

David Adler approaches the use of data for self-coaching in a different manner, focusing on the assessment of trading performance itself. "The philosophy behind TraderDNA, which I firmly believe," he explains, "is the idea of being cognizant of what happened (in terms of your performance) within a given session, week, month, etc. of your trading, so that, going forward, the negative aspects can be minimized and the positive can be maximized. The fundamental idea is that, if the trader is able to look back on a certain time period of his trading and understand more about the overall result, then he can be proactive . . . in preventing the same mistakes going forward. Likewise, he can identify strengths and focus on situations that are likely to result in a profit based upon what his past trading has shown.

"Our users extract their order data from their front-end software," David Adler notes, "and import the data into TraderDNA. This affords them the opportunity to thoroughly analyze their data in order to understand more about the strengths and weaknesses of their trading: specifically, their performance trends, where their losses came from, characteristics of their trades, the differences in their winners and losers, amongst other things." Here are some of the analytics provided by the software, along with David's commentary:

1. **Hour of day analysis.** "Because markets trade differently throughout the day," David explains, "many of our users measure their performance (in terms of average P/L, risk taken, profit opportunity, number of winners/losers, size of winners/losers) by the time of day the trade occurred. This helps them to use their past performance to determine the most ideal times for them to trade a given market."
2. **Winning trades versus losing trades.** "In looking for differences in winning and losing trades, it's helpful—and necessary—to group all winners together and group all losers together and then apply metrics to each category," David points out. "The value in doing so is the opportunity for you to discover the differences in your winning trades and your losing trades." Among the metrics applied to both the winning and losing trades are the number of winners and losers; the average win and average loss size; the number of times a trader added to winning and losing positions; the average amount of heat taken in a trade before it was covered for a profit or loss; and the average time it has taken to hit the point of maximum heat. The latter is an especially interesting metric in that, by comparing winning and losing trades, it can help guide traders to formulate rules for the proper amount of time to be holding positions.

If you know how much heat you take on winners versus losers and how long it takes you to reach that point of maximum heat, you can set guidelines for when and where it might be prudent to cut your losers.

3. **Comparing results among market/product traded.** "Traders that trade more than one market/product sometimes have difficulty interpreting how their performance compares in their trading of each market," David Adler observes. "Oftentimes the trader will be very profitable in one market but have consistently less profit or even losses in other markets. If you trade more than one market, it's important to split up any analysis or performance reporting that you do by the markets you trade." Among the metrics he applies to different markets are: the total amount earned/lost per market; the average win and average loss for each market; the number of consecutive winning and losing trades for each market; the average risk incurred among trades for each market; the average lost profit opportunity for each market; the number of times you added to losing and winning positions per market; the average amount of time spent in losing trades per market; the maximum losing and winning trades per market; and breakdowns by hour of the day for each market. My experience with metrics is that these

breakdowns by market will often shift over time, as certain markets yield greater opportunities and others go dry. *Tracking results over time can be a great way of seeing, in real time, when and how markets are changing.*

"I've seen numerous traders increase their P/L by minimizing trading losses and increasing the frequency and/or size of their winners after applying one or more of the techniques above," David concludes. "From what I've seen, it's most effective to conduct your analysis or review of your trading no more than once a week, and ideally once every two weeks or even once a month."

Intuition—the result of implicit learning that occurs after long periods of observing market patterns—may play an important role in getting traders into and out of positions. Even the most intuitive and discretionary trading, however, can benefit from analytics: knowing which markets and time frames offer opportunity and measuring how well you're taking advantage of that opportunity. *Ultimately, you are your own trading system.* Your task, as your own performance coach, is to know how your system operates, avoid its shortcomings, and maximize its strengths. The insights and tools provided by Rennie and Dave are excellent guides in the quest to become more scientific in the management of our trading business.

COACHING CUE

The contributors to this chapter have provided a wealth of insights, derived from firsthand experience, as to the principles and practices that can improve your trading. A worthwhile exercise is to review each of the contributor's ideas and identify the overlap: the points emphasized by more than one contributor. These points of overlap represent important best practices that can guide your efforts going forward.

RESOURCES

The *Become Your Own Trading Coach* blog is the primary supplemental resource for this book. You can find links and additional posts on the topic of coaching processes at the home page on the blog for Chapter 9: http://becomeyourowntradingcoach.blogspot.com/2008/08/daily-trading-coach-chapter-nine-links.html

The contributors to this chapter maintain their own web sites, which offer a wealth of resources to developing traders. Here are links to the contributors to this chapter and their web sites:

http://becomeyourowntradingcoach.blogspot.com/2008/08/contributors-to-daily-trading-coach.html

For background on technical analysis, Brian Shannon's book is a useful resource for mentorship:
www.technicalanalysisbook.com/

Ray Barros's book *The Nature of Trends* details his approach to trading and trading psychology; see also the seminars he offers on these topics:
www.tradingsuccess.com/

John Forman's book *The Essentials of Trading* is an excellent introduction to the practice and business of trading: www.theessentialsoftrading.com/Blog/index.php/the-essentials-of-trading/

The NewsFlashr site is a great way of staying on top of many popular trading-related blogs, as well as news: www.newsflashr.com/feeds/business_blogs.html

Science is the great antidote to the poison of enthusiasm and superstition.

—Adam Smith

Traders commonly refer to having an *edge* in markets. What this means is that they have a positive expectancy regarding the returns from their trades. Card counting can provide an edge to a poker player, but how can traders count the cards of their markets and put probabilities on their side? One way of accomplishing this is historical investigation. While history may not repeat exactly in markets, we can identify patterns that have been associated with a directional edge in the past and hypothesize that these will yield similar tendencies in the immediate future. By knowing market history, we identify patterns to guide trade ideas.

So how can we investigate market history to uncover such patterns? This has been a recurring topic of reader interest on the *TraderFeed* blog. If you're going to mentor yourself as a trader, your efforts will be greatly aided by your ability to test the patterns you trade. After all, if you know the edge associated with what you're trading, you're most likely to sustain the confidence needed to see those trades through.

A thorough presentation of testing market ideas would take a book in itself, but this chapter should get you started. Armed with a historical database and Excel, you can greatly improve your ability to find worthy market hypotheses to guide your trading. Let's get started . . .

LESSON 91: USE HISTORICAL PATTERNS IN TRADING

A trading guru declares that he has turned bearish because the S&P 500 Index has fallen below its 200-day average. Is this a reasonable basis for setting your trading or investing strategy? Is there truly an edge to selling the market when it moves below its moving averages?

The only way we can determine the answer is through investigation. Otherwise, investing and trading become little more than exercises in faith and superstition. Because markets have behaved in a particular way in the past does not guarantee that they will act that way now. Still, history provides the best guide we have. Markets, like people, will never be perfectly predictable. But if we've observed people over time in different conditions, we can arrive at some generalizations about their tendencies. Similarly, a careful exploration of markets under different historical conditions can help us find regularities worth exploiting.

As it turns out, moving average strategies—so often touted in the popular trading press—are not so robust. As of my writing this, since 1980, the average 200-day gain in the S&P 500 Index following occasions when we've traded above the 200 day moving average has been 8.68 percent. When we've been below the 200-day moving average, the next 200 days in the S&P 500 Index have returned 7.32 percent. That's not a huge difference, and it's hardly grounds to turn bearish on a market. When David Aronson tested more than 6,000 technical indicators for his excellent book, *Evidence-Based Technical Analysis*, he found a similar lack of robustness: not a single indicator emerged as a significant predictor of future market returns.

> Investigate before you invest: Common trading wisdom is uncommonly wrong.

One way that historical patterns aid our self-coaching is by helping us distinguish myth from fact. "The trend is your friend" we commonly hear. My research on the blog, however, has consistently documented worse returns following winning days, weeks, and months than following losing ones. It is not enough to accept market wisdom at face value: just as you would research the reliability of a vehicle before making a purchase, it makes sense to research the reliability and validity of trading strategies.

There are traders who make the opposite mistake and trade mechanically from historical market patterns. I have seen an unusual proportion of these traders blow up. Market patterns are relative to the historical period that we study. If I examine the past few years of returns in a bull market, I will find significant patterns that will completely vanish in a bear market. If I include many bull and bear markets in my database, I will go back so far in time that I will be studying periods that are radically different from the current one in terms of who and what are moving markets. Automated, algorithmic strategies have completely reshaped market patterns, particularly over short time periods. If you study precomputer-era markets you would miss this influence altogether. Select a look-back period for historical analysis that is long enough to cover different markets but not so lengthy as to leave us with irrelevant data is as much art as science.

My approach to trading treats historical market patterns as qualitative research data. In a nutshell, qualitative research is *hypothesis-generating research*, not hypothesis-testing research. I view the patterns of markets as sources of trading hypotheses, not as fixed conclusions. The basic hypothesis is that the next trading period will not differ significantly from the recent past ones. If a pattern has existed over the past X periods, we can hypothesize that it will persist over the next period. Like any hypothesis, this is a testable proposition. It is an idea backed by support, not just faith or superstition, but it is not accepted as a fixed truth to be traded blindly.

Historical testing yields hypotheses for trading, not conclusions.

For this reason, I do not emphasize the use of inferential statistics in the investigation of historical patterns. I am looking for qualitative differences much as a psychologist might look for various behavior patterns in a person seeking therapy. In short, I'm looking to generate a hypothesis, not test one. The testing, in the trading context, is reflected in my trading results: if my returns significantly exceed those expected by chance, we can conclude that I am trading knowledge, not randomness.

When we adopt a qualitative perspective, the issue of look-back period becomes less thorny. As long as we consider the results of historical investigations to be nothing more than hypotheses, we can draw our ideas from the past few weeks, months, years, or decades of trading. The basic hypothesis remains the same: that the next time period will not differ significantly from the most recent ones. With that in mind, we can frame multiple hypotheses derived from different patterns over different time frames. One hypothesis, for example, might predicate buying the market based on strong action at the close of the prior day, with the anticipation of taking out the previous day's R1 pivot level. A second hypothesis might also entail buying the market based on a pattern of weakness during the previous week's trading. When multiple independent patterns point in the same direction, we still don't have a certain conclusion, but we do have a firm hypothesis.

> When independent patterns point to similar directional edges, we have especially promising hypotheses for trading.

Of course, if we generate enough hypotheses, some are going to look promising simply as a matter of chance. We could look at all combinations of Dow stocks, day of week, and week of year and the odds are good that we'd find some pattern for some stock that looks enticing, such as (to invent one possibility) IBM tends to rise on the first Wednesday of months during the summer season. Good hypotheses need to make sense; you should have some idea of *why* they might be valid. It makes sense, for instance, to buy after a period of weakness because you would benefit from short covering and an influx of money from the sidelines. It doesn't make sense to buy a stock on alternate Thursdays during months that begin with M—no matter what the historical data tell you.

When you're first learning to generate good hypotheses, your best bet is to keep it simple and get your feel for the kinds of patterns that are most promising. Many of your initial candidates will emerge from investigations of charts. Perhaps you'll notice that it has been worth selling a stock when it rises on unusually high volume, or that markets have tended to bounce following a down open that follows a down day. Such ideas are worth checking out historically. What patterns have you noticed in your trading and observation? Write down these patterns and keep them simple: these patterns will get you started in our qualitative research.

COACHING CUE

Several newsletters do an excellent job of testing historical patterns and can provide you with inspiration for ideas of your own. Check out the contributions of Jason Goepfert, Rob Hanna, and Rennie Yang in Chapter 9, along with their links. All three are experienced traders and investigators of market patterns.

LESSON 92: FRAME GOOD HYPOTHESES WITH THE RIGHT DATA

In the previous lesson, I encouraged you to keep hypotheses simple. This is not just for your learning; in general, we will generate the most robust hypotheses if we don't try to get too fancy and add many conditions to our ideas. Ask a question that is simple and straightforward, such as, "What typically happens the week following a very strong down week?" This question is better than asking, "What typically happens the week following a very strong down week during the month of March when gold has been up and bonds have been down?" The latter question will yield a small sample of matching occasions—perhaps only three over many years—so that it would be difficult to generalize from these. While I will occasionally look at patterns with a small N simply as a way to determine if the current market is behaving in historically unusual ways, it is the patterns that have at least 20 occurrences during a look-back period that will merit the greatest attention. The more conditions we add to a search, the more we limit the sample and make generalization difficult.

> The simplest patterns will tend to be the most robust.

Of course, the number of occurrences in a look-back period will partly depend on the frequency of data that you investigate. With 415 minutes in a trading day for stock index futures, you would have 8,300 observations of one-minute patterns in a 20-day period. If you were investigating daily data, the same number of observations would have to cover a period exceeding 30 years. Databases with high frequency data can become unwieldy in a hurry and require dedicated database applications. The simple historical investigations that I conduct utilize database functions in a flat Excel file. When I investigate a limited number of variables over a manageable time frame, I find this to be adequate to my needs. Clearly, a system developer who is going to test many variables over many time frames would need a relational database or a dedicated system-testing platform, such as TradeStation. The kind of hypothesis-generating activities covered in this chapter are most appropriate for discretionary traders who would like to be a bit more systematic and selective in their selection of market patterns to trade—not formal system developers.

Before you frame hypotheses worthy of historical exploration, you need to create your data set. This data set would include a range of variables over a defined time period. The variables that you select would reflect the markets and indicators that you typically consult when making discretionary trading decisions. For instance, if you trade off lead-lag relationships among stock market sectors, you'll need to include sector indexes/ETFs in your database. If you trade gap patterns in individual stocks, you'll need daily open-high-low-close prices for each issue that you trade at the very least. Some of the patterns I track in my own trading involve the number of stocks making new highs or lows; this is included in my database with separate columns on a sheet dedicated to each.

As you might suspect, a database can get large quickly. With a column in a spreadsheet for each of the following: date, open price, high price, low price, closing price, volume, rate of change, and several variables (indicators) that you track, you can have a large sheet for each stock or futures contract that you trade—particularly if you are archiving intraday data. *I strongly recommend that beginners at this kind of historical investigation get their feet wet with daily data.* This process will keep the data sets manageable and will be helpful in framing longer timeframe hypotheses that can supplement intraday observation and judgment. Many good swing patterns can be found with daily data and clean, affordable data are readily available.

Some of the most promising historical patterns occur over a period of several days to several weeks.

There are several possible sources for your historical database. Many real-time platforms archive considerable historical data on their servers. You can download these data from programs such as e-Signal and Real Tick (two vendors I've personally used) and update your databases manually at the end of the trading day. The advantage of this solution is that it keeps you from the expense of purchasing historical data from vendors. It also enables you to capture just the data you want in the way you want to store them. This is how I collect most of my intraday data for stock index futures and such variables as NYSE TICK. My spreadsheet is laid out in columns in ways that I find intuitive. The entire process of updating a sheet, including built in charts, takes a few minutes at most.

A second way you can go, which I also use, is to purchase historical data from a vendor. I obtain daily data from Pinnacle Data (www.pinnacledata.com), which includes an online program for updating that is idiot-proof. Many of their data fields go back far in market history, and many of them cover markets and indicators that I would not be able to easily archive on my own. The data are automatically saved in Excel sheets, with a separate sheet for each data element. That means that you have to enter the different sheets and pull out all the data relevant to a particular hypothesis and time frame. The various fields can be copied onto a single worksheet that you can use for your historical investigations (more on this later). Among the data that I find useful from Pinnacle Data are advance-decline information; new highs/lows; volume (including up/down volume); interest rates; commodity and currency prices; and weekly data. These data are general market data, not data for individual equities. When I collect individual equity data, I generally find the historical data from the real-time quotation platforms to be adequate to my needs.

For the collection of clean intraday data, I've found TickData (www.tickdata.com) to be a particularly valuable vendor. The data management software that accompanies the historical data enables you to place the data in any time frame and store them as files within Excel. This is a great way to build a historical database of intraday information quickly, including price data for stocks and futures and a surprising array of indicator data.

If you go with a historical data vendor, you'll have plenty of data for exploration and the updating process will be easy. Manual updating of data from charting platforms is more cumbersome and time-consuming, but obviously cheaper if you're already subscribing to the data service. *It is important thing that you obtain the data you most want from reliable sources in user-friendly ways.* If the process becomes too cumbersome, you'll quickly abandon it.

As your own trading coach, you want to make the learning process stimulating and enjoyable; that is how you'll sustain positive motivation. Focus on what you already look at in your trading and limit your initial data collection to those elements. Price, volume, and a few basic variables for each stock, sector, index, or futures contract that you typically look at will be plenty at first. Adding data is never a problem. The key is to organize the information in a way that will make it easy for you to pull out what you want, when you want it. As you become proficient at observing historical patterns, you'll be pleasantly surprised at how this process prepares you for recognizing the patterns as they emerge in real time.

COACHING CUE

Consider setting up separate data archives for daily and weekly data, so that you can investigate patterns covering periods from a single day to several weeks. You'd be surprised how many hypotheses can be generated from simple open-high-low-close price data alone. How do returns differ after an up day versus a down day? What happens after a down day in which the day's range is the highest of the past 20 days? What happens after three consecutive up or down days? How do the returns differ following a down day during a down week versus a down day during an up week? You can learn quite a bit simply by investigating price data.

LESSON 93: EXCEL BASICS

In this lesson, I'll go over just a few essentials of Excel that I employ in examining historical market data. If you do not already have a basic understanding of spreadsheets (how cells are named, how to copy information and paste it into cells, how to copy data from one cell to another, how to create a chart of the data in a sheet, how to write simple formulas into cells), you'll need a beginning text for Excel users. All of the things we'll be reviewing here are true basics; we won't be using workbooks linking multiple sheets, and we won't be writing complex macros. Everything you need to formulate straightforward hypotheses from market data can be accomplished with these basics.

So let's get started. Your first step in searching for market patterns and themes is to download your historical data into Excel. Your data vendors will have instructions for downloading data; generally this will involve copying the data from the charting application or from the data vendors' servers and pasting them into Excel. If, for instance, you were using e-Signal (www.esignal.com) as a real-time data/charting application, you would activate the chart of the data you're interested in by clicking on that chart. You then click on the menu item Tools and then click on the option for Data Export. A spreadsheet-like screen will pop up with the chart data included. Along the very top row, you can check the boxes for the data elements you want in your spreadsheet. If there are data in the chart that you don't need for your pattern search, you simply uncheck the boxes for those columns.

On that spreadsheet screen in e-Signal, if you click on the button for Copy to Clipboard, you will place all of the selected data on the Windows clipboard, where the data elements are stored as alphanumeric text. You then open a blank sheet in Excel, click on the Excel menu item for Edit, and select the option for Paste. That will place the selected data into your Excel spreadsheet.

If you had wanted more historical data than popped up in the e-Signal spreadsheet-like screen, you would have to click your chart and drag your mouse to the right, moving view of the data into the past. Move it back as far as you need and then go through the process of clicking on Tools, selecting Data Export, etc. If you need more historical data than e-Signal (or your current charting/data vendor) carries on their servers, that's when you'll need to subscribe to a dedicated historical data source such as Pinnacle Data (www.pinnacledata.com).

If you need data going back many years for multiple indicators or instruments, you'll want to download data from a historical data vendor who has checked the data for completeness and accuracy.

If you're using Pinnacle Data, you can automatically update your entire database daily with its Goweb application. The program places all the updated data into Excel sheets that are stored on the C drive in a folder labeled Data. The IDXDATA folder within Data contains spreadsheets with each instrument or piece of data (S&P 500 Index open-high-low-close; number of NYSE stocks making 52-week highs) in its own spreadsheet. Once you open these sheets, you can highlight the data from the historical period you're interested in, click on the Edit menu item in Excel, click on the Copy option, open a fresh, blank spreadsheet, click on Edit, and then click on the Paste option. By copying from the Pinnacle sheets and pasting into your own worksheets, you don't modify your historical data files when you manipulate the data for your analyses.

Personally, I would not subscribe to a data/charting service that did not facilitate an easy downloading of data into spreadsheets for analysis. It's also helpful to have data services that carry a large amount of intraday and daily data on their servers, so that you can easily retrieve all the data you need from a single source. In general, I've found e-Signal and Pinnacle to be reliable clean sources of data. There are others out there, however, and I encourage you to shop around.

When you download data for analysis, save your sheets in folders that will help you organize your findings and give the sheets names that you'll recognize. Over time, you'll perform many analyses; saving and organizing your work will prevent you from having to reinvent wheels later.

Once you have the data in your sheet, you'll need to use formulas in Excel to get the data into the form you need to examine patterns of interest. Formulas in Excel will begin with an = sign. If, for example, you wanted to calculate an average value for the first 10 periods of price data (where the earliest data are in row 2 and later data below), you might enter into the cell labeled D11: "=average(C2:C11)," without typing the quotation marks. That will give you the simple average (mean) of the price data in cells C1 through C10. If you want to create a moving average, you could simply click on the D10 cell, click the Excel menu item for copy, left-click your mouse and drag from cell D11 down, and release. Your column D cells will update the average for each new cell in column C, creating a 10-period moving average.

As a rule, each column in Excel (labeled with the letters) will represent a variable of interest. Usually, my column A is date, column B is time (if I'm exploring intraday data), column C is open price, column D is high price, column E is low price, and column F is closing price. Column G might be devoted to volume data for each of those periods (if that's part of what I'm investigating); columns H and above will be devoted to other variables of interest, such as the data series for another index or stock or the readings of a market indicator for that period. *Each row of data is a time period, such as a day.* Generally, my data are organized so that the earliest data are in row 2 and the later data fall underneath. I save row 1 for data labels, so that each column is labeled clearly: DATE, OPEN, HIGH, LOW, CLOSE, etc. You'll see why this labeling is helpful when we get to the process of sorting the data.

Here are some simple statistical functions that I use frequently to examine data in a qualitative way. Each example assumes that we're investigating the data in column C, from cells 1 through 10:

- =median(C2:C11) – The median value for the data in the formula.
- =max(C2:C11) – The largest value for the data in the formula.
- =min(C2:C11) – The smallest value for the data in the formula.
- =stdev(C2:C11) – The standard deviation for the data in the formula.
- =correl(C2:C11,D2:D11) – The correlation between the data in columns C and D, cells 1–10.

Much of the time, our analyses won't be of the raw data, but will be of the changes in the data from period to period. The formula =(C3-C2) gives the difference from cell C2 to cell C3. If we want to express this difference as a percentage (so that we're analyzing percent price changes from period to period), the formula would read = ((C3-C2)/C2)*100. This takes the difference of cells C3 and C2 as a proportion of the initial value (C2) multiplied by 100 to give a percentage.

When we want to update later cells with the percentage difference information, we don't need to rewrite the formulas. Instead, as noted above, we click on the cell with the formula, click on the Excel menu item Edit, click on copy, then left click the cell below the one with the formula and drag down as far as we want the data. The spreadsheet will calculate price changes for each of the time periods that you selected by dragging. This means that if you save your formulas into worksheets, updating your data is as simple as downloading the fresh data from your vendor, pasting into the appropriate cells in your sheets, and copying the data from formulas for the cells representing the new data period. Once you've organized your sheets in this manner, it thus only takes a few minutes a day to completely update.

Once you create a spreadsheet with the appropriate formulas, updating your analyses is mostly a matter of pasting and copying. As a result, you can update many analyses in just a few minutes.

Once again, the basic formulas, arrangement of rows and columns, and copying of data will take some practice before you move on to actual analyses. I strongly encourage you to become proficient with downloading your data from your vendor/application and manipulating the data in Excel with copying, pasting, and formula writing before moving on. Once you have these skills, you'll have them for life, and they will greatly aid your ability to generate promising trading hypotheses.

COACHING CUE

Trading platforms that support Dynamic Data Exchange (DDE) enable you to link spreadsheets to the platform's data servers, so that the spreadsheets will populate in real time. This is helpful for tracking indicators as you trade, and it can also be a time-efficient way to archive data of interest. See Rennie Yang's segment in Chapter 9 for an illustration of the use of DDE.

LESSON 94: VISUALIZE YOUR DATA

One of the best ways to explore data for possible relationships is to actually see the data for yourself. You can create simple charts in Excel that will enable you to see how two variables are related over time, identifying possible patterns that you might not have noticed from the spreadsheet rows and columns. For instance, when charting an indicator against a market average, you may notice divergence patterns that precede changes in market direction. Should you notice such patterns frequently, they might form the basis for worthwhile historical explorations.

Again, a basic introductory text for Excel users will cover the details of creating different kinds of charts, from column charts to line graphs to pie charts. You'll also learn about the nuances of changing the colors on a chart, altering the graphics, and labeling the various lines and axes. In this lesson, I'll walk you through a few basics that will get you started in your data exploration.

> Many times, you can identify potential trading hypotheses by *seeing* relationships among data elements.

A simple chart to begin with will have dates in column A, price data in column B, and a second set of price data in column C (see Table 10.1). This chart is helpful when you want to visualize how movements in the first trading instrument are related to movements in the second. For basic practice, here are some hypothetical data to type into Excel, with the data labels in the first row. Column A has the dates, column B contains closing prices for a market index, and column C has the closing prices for a mining stock.

To create the chart, highlight the data with your mouse, including the data labels, and click on the Excel menu item for Insert. You'll select

TABLE 10.1 Sample Market Data

Date	Market Index	Mining Stock
12/1/2008	1200.25	9.46
12/2/2008	1221.06	9.32
12/3/2008	1228.01	9.33
12/4/2008	1230.75	9.30
12/5/2008	1255.37	9.18
12/8/2008	1234.44	9.24
12/9/2008	1230.24	9.22
12/10/2008	1228.68	9.27
12/11/2008	1235.56	9.20
12/12/2008	1240.20	9.22
12/15/2008	1251.98	9.12
12/16/2008	1255.50	9.15
12/17/2008	1239.88	9.29
12/18/2008	1226.51	9.39

Chart and a menu of different kinds of charts will appear. You'll click on Line and select the chart option at the top left in the submenu. That is a simple line chart. Then click Next, and you will see a small picture of your chart, the range of your data, and whether the series are in rows or columns. Your selection should be columns, because that is how you have your variables separated. Click Next again and you will see Step 3 of 4 in the Chart Wizard, allowing you to type in a chart title and labels for the X and Y axes. Go ahead and type in Market Index and Mining Stock for the title, Date for the Category (X) axis label, and Price for the Value (Y) axis label. Then click Next.

The Step 4 of 4 screen will ask you if you want the chart as an object in your spreadsheet, or if you want the chart to be on a separate sheet. Go ahead and select the option for "As new sheet." Then click Finish.

What you'll see is that the Wizard has recognized the date information from column A and placed it on the X-axis. The Wizard has also given us a single Y-axis and scaled it according to the high and low values in the data. Unfortunately, this leaves us unable to see much of the ups and downs in the mining stock data, since the price of the stock is much smaller than the price of the index.

To correct this problem, point your cursor at the line on the chart for your Market Index and right click. A menu will pop up, and you will select the option for Format Data Series. Click the tab for Axes and then click on the button for "Plot Series on Secondary Axis." When you do that, the picture of the chart underneath the buttons will change, and you'll notice now that you have two Y-axes: one for the Market Index price data and one for the Mining Stock data. You'll be able to see their relative ups and downs much more clearly. Click on OK and you will see your new chart. If you'd like the Y-axes to have new labels, you can place your cursor on the center of the chart (away from the lines for the data) and right click. A menu will pop up, and you'll select Chart Options. That will give you a screen enabling you to type in new labels for the Value (Y) axis (at left) and the Second value (Y) axis (at right).

If you right click on either of the two lines in the chart and, from the pop-up menu select Format Data Series, you'll see a tab for Patterns. You can click the arrow beside the option for Weight and make the line thicker. You can click the arrow beside the option for Color and change the color of the line.

If you right-click on the X- or Y-axes, you'll get a pop-up menu; click on Format Axis. If you select the tab for Font, you can choose the typeface, font style, and size of the print for the axis labels. If you select the tab for Scale, you can change the range of values for the axis. With a little practice, you can customize the look of your charts.

So what does your chart tell you? You can see that the Mining Stock is not moving in unison with the Market Index. When the index shows large rises or declines, the stock is tending to move in the opposite direction. By itself, over such a short period, that won't tell you anything you'd want to hang your hat on, but it does raise interesting questions:

- Why is the mining stock moving opposite to the market index? Might the mining stock be moving in unison with the gold market instead?
- If the mining stock is moving with gold, is gold also moving opposite to the market index? If so, why might that be? Might there be a common influence on both of them: the strength of the U.S. dollar?
- Does this relationship occur over intraday time frames? Might we be able to identify some buy or sell signals in the mining stock when we see selling or buying in the broad market?

> Reviewing charts that you create helps you see intermarket and intramarket relationships.

Many times, investigating relationships through charts leads you to worthwhile questions, which may then lead you to interesting and profitable trading ideas. The key is asking "Why?" What might be responsible for the relationship I am observing? Remember, in your own self-coaching, you want to be generating hypotheses, and there is no better way than plain old brainstorming. When you can actually see how the data are related to each other in graphical form, it is easier to accomplish that brainstorming. You won't arrive at hard and fast conclusions, but you'll be on your way toward generating promising trading ideas.

COACHING CUE

Plot charts of the S&P 500 Index (SPY) against the major sector ETFs from the S&P 500 universe as a great way to observe leading and lagging sectors, as well as divergences at market highs and lows. The sectors I follow most closely are: XLB (Materials); XLI (Industrials); XLY (Consumer Discretionary); XLP (Consumer Staples); XLE (Energy); XLF (Financial); XLV (Health Care); and XLK (Technology). If you want to bypass such charting, you can view excellent sector-related indicators and charts at the Decision Point site (www.decisionpoint.com). Another excellent site for stock and sector charts is Barchart (www.barchart.com).

LESSON 95: CREATE YOUR INDEPENDENT AND DEPENDENT VARIABLES

When I organize my spreadsheets, I generally place my raw data furthest to the left (columns A, B, C, etc.); transformations of the raw data into independent variables in the middle; and dependent variables furthest to the right. Let's take a look at what this means.

Your independent variables are what we might call *candidate predictors*. They are variables that we think have an effect on the markets we're trading. For example, let's say that we're investigating the impact of price change over the previous day of trading (independent variable) on the next day's return for the S&P 500 Index (dependent variable). The raw data would consist of price data for the S&P 500 Index over the look-back period that we select. The independent variable would be a moving calculation of the prior day's return. The dependent variable would be a calculation of the return over the next day. The independent variable is what we think might give us a trading edge; the dependent variable is what we would be trading to exploit that edge.

> If I keep my raw data to the left in the spreadsheet, followed by transformations of the raw data to form the independent variable, and then followed by the dependent variable, I keep analyses clear from spreadsheet to spreadsheet.

Let's set that up as an exercise. We'll download data for the S&P 500 Index (cash close) for the past 1,000 trading days. That information will give us roughly four years of daily data. If I obtain the data from Pinnacle Data, I'll open a blank sheet in Excel; click on the Excel menu item for File; click Open; go to the Data folder in the C drive; double-click on the IDXDATA folder; select All Files as the Files of type; and double-click the S&P 500 file. I'll highlight the cells for the past 1,000 sessions; click on the Edit menu item in Excel; click on Copy; open a new, blank spreadsheet; then click the Edit menu item again; and click Paste, with the cursor highlighting the A2 cell. The data from the Pinnacle sheet will appear in my worksheet, leaving row A for data labels (Date, Open, High, Low, Close).

If you download your data from another source, your menu items to access the data will differ, but the result will be the same: you'll copy the data from your source and paste them into the blank spreadsheet at cell A2, then create your data labels. As a result, your raw data will occupy columns A–E. (Column A will be Date; column B will be Open; column C will be High; column D will be Low; and column E will be Close.) Now, for the data label for column F (cell F1), you can type (without quotation marks): "SP(1)." This is your independent variable, the current day's rate of change in the index. Your first entry will go into cell F3 and will be (again without quotations marks): "=((E3-E2)/E2)*100)." This is the percentage change in the index from the close of the session at A2 to the close of the session at A3.

Now let's create our dependent variable in cell G7, with column G labeled SP+1 at G1. Your formula for cell G3 will be "=((E4-E3)/E3)*100)." This represents the next day's percentage return for the index.

To complete your sheet, you would click and highlight the formula cells at F3 and G3; click on the Excel menu item for Edit; and select the option for Copy. You'll see the F3 and G3 cells specially highlighted. Then, with your cursor highlighting cells F4 and G4, drag your mouse down the full length of the data set and release, highlighting all those cells. Click again on the Excel menu item for Edit, then select Paste. Your spreadsheet will calculate the formulas for each of the cells and the data portion of your spreadsheet will be finished. The raw data will be in columns A–E. The independent variable (our candidate predictor) will be in column F; and our variable of trading interest—the dependent variable—will reside in column G. Save this spreadsheet as Practice Sheet in an Excel folder. We'll be using it for future lessons.

Note that we downloaded 1,000 days worth of data, but the actual number of data points in our sample is 998. We could not compute SP(1) from the first data point because we didn't have the prior day's close; hence we had to begin our formula in the third data row. We also could not compute SP+1 from the last data point because we don't know tomorrow's closing price. Thus our analyses can only use 998 of the data points of the 1,000 that we downloaded. If you want an even 1,000 data points, you'd have to download the last 1,002 values.

With a bit of practice, all of this will become second nature. It will take only a minute or two to open your data files, copy and paste the raw data, write your formulas, and copy the cells to complete your sheet. In this example, we are exploring how the prior day's return is related to the next day's return. We're setting the spreadsheet up to ask the question, "Does it make sense to buy after an up day/sell after a down day; does it make sense to sell after an up day/buy after a down day; or does it make no apparent difference?" I call the independent variable the *candidate* predictor, because we don't really know if it is related to our variable of interest. It's also only a candidate because we're not conducting the statistical significance tests that would tell us more conclusively that this is a significant predictor. Rather, we're using the analysis much as we used the charting in the prior lesson: as a way to generate hypotheses.

Remember, in the current examples, we're using historical relationship to describe patterns in markets, not to statistically analyze them. We're generating, not testing, hypotheses.

If I had been interested in examining the relationship between the prior week's price change with the next week's return, the spreadsheet would look very similar, except the raw data would consist of weekly index data, rather than daily. *In general, it's neatest for analysis if you are investigating the impact of the prior period's data on the next period.* This ensures that all observations are independent; there are no overlapping data.

To see what I mean, consider investigating the relationship of the prior week's (five-day) price change on the price change over the next five trading days utilizing daily market data. Your independent variable in column F would now look like "=((E7-E2)/E2)*100)"—price change over the past five days. The dependent variable in column G would be written as "=((E12-E7)/E7)*100": the next five-day's price change. *Note, however, that as you copy those cells down the spreadsheet per the above procedure, that each observation at cells F8, F9, F10, and so on and G8, G9, G10, and so on, is not completely independent.* The prior five-day return overlaps the values for F8, F9, and F10, and the prospective five-day return overlaps for cells G8, G9, and G10. This will always be the case when you're using a smaller time period for your raw data than the period that you're investigating for your independent and dependent variables.

Inferential statistical tests depend on each observation in the data set being independent, so it is not appropriate to include overlapping data when calculating statistical significance. For my purpose of hypothesis generation, I am willing to tolerate a degree of overlap, and so will use daily data to investigate relationships of up to 20 days in duration—particularly if the amount of overlap relative to the size of the entire data set is small. I would not, say, investigate the next 200 days' return using daily data for a sample of 1,000 trading days. I wouldn't have a particular problem using the daily data to investigate, for instance, the prior five-day price change on the next five-day return with a four-year look-back period.

> Your findings will be most robust if your look-back period (the period that you are drawing data from) includes a variety of market conditions: rising, falling, range bound, high volatility, low volatility, and so on.

In general, my dependent variable will consist of prospective price change, because that is what I'm interested in as a trader. The independent variable(s) will consist of whatever my observations tell me might be meaningfully related to prospective price change. Normally, I look at dependent variables with respect to the next day's return (to help with day trading ideas) and the next week's return (to help with formulating swing hypotheses). If I want a sense of the market's possible bigger picture, I'll investigate returns over the next 20 trading days. Traders with different time frames may use different periods, including intraday. Overall, I've found the 1 to 20 day framework to be most useful in my investigations.

Once again, practice makes perfect. I would encourage you to become proficient at downloading your data and assembling your spreadsheets into variables before you try your hand at the actual historical investigations. Your results, after all, will only be as valid as the data you enter and the transformations you impose upon the data.

COACHING CUE

Note how, with the Practice Sheet assembled as in the above example, you can easily look at the next day's average returns following opening gaps. Your independent variable would be the opening gap, which would be written as =((b3-e2)/e2)*100 (the difference between today's open and yesterday's close as a percentage). The day's price change would be =((e3-b3)/b3)*100 (the difference between today's close and today's open as a percentage). You would need to use stock index futures data or ETF data to get an accurate reflection of the market open; the cash index does not reflect accurate opening values, as not all stocks open for trading in the first minute of the session.

LESSON 96: CONDUCT YOUR HISTORICAL INVESTIGATIONS

Once you have your data downloaded and your independent and dependent variables calculated, you're ready to take a look at the relationship between your two sets of variables. In the last lesson, you saved your spreadsheet of the S&P 500 Index data with the prior day's price change in column F and the next day's change in column G. Open that sheet, and we will get started with our investigation.

Your first step will be to copy the data from the sheet to a fresh worksheet. We will first copy the data to the Windows clipboard, then paste into the new sheet. This eliminates all formulas from the sheet, because the clipboard saves only alphanumeric text data. This process is necessary for the data manipulations that will be required for our investigation.

So, highlight all the cells in your sheet with the exception of the last row (the most recent day's data). We don't include that row in our analysis because there won't be any data for the next day's return. With the cells highlighted, click the Excel menu item for Edit, then select Copy. You'll then exit out of the spreadsheet and instruct Windows to save the data to the clipboard. Open a fresh, blank sheet; click on cell A1; click the Excel menu item for Edit; and select Paste. Your data will be transferred to the new sheet, with no formulas included.

Once you've done this, you'll delete the first row of data below the data labels, because there will be no data for the change from the prior day. When you delete the row by highlighting the entire row and clicking Edit and selecting Delete, the rows below will move up, so that there are no empty rows between the data labels and the data themselves.

You'll now highlight all the data (including the first row of data labels), select the Excel menu item for Data; then select Sort; and select the option SP(1) in the Sort By drop down menu. You can click the button on the drop down menu for Descending: this will place your largest daily gain in the S&P 500 Index in the first row, the next largest in the second row, etc. The last row of data will be the day of the largest daily drop in the S&P 500 Index.

> The Sort function separates your independent variables into high and low values, so that you can see how the dependent variables are affected.

Now we're ready to explore the data. For the purpose of the illustration, I'll assume that your data labels are in row 1 and that you have 999 rows of data (998 days of S&P data plus the row of labels). Below your bottom row in column G (say cell G1002), type in "=average(g2:g500)" (without the quotation marks) and hit Enter. In the cell below that (G1003), type in "=average(g501:g999)" and hit Enter. This gives you a general sense for whether next day returns have been better or worse following the half of the days in the sample that were strongest versus the half of days that were weakest. Note that you could analyze the next day returns roughly by quartiles simply by entering "=average(g2:g250)"; "=average(g251:g500)"; "=average(g501:g750)"; and "=average(g751:g999)".

What your data will show is that next day returns tend to be most positive following weak days in the S&P 500 Index and most restrained following strong days in the Index. How much of a difference makes a difference for your trading? As I emphasized earlier in the chapter, I am not using this information to establish a statistically significant mechanical trading system. Rather, I'm looking qualitatively for differences that hit me between the eyes. These will be the most promising relationships for developing trading hypotheses. If the difference between average next day returns following an up day and a down day is the difference between a gain of 0.01 percent and 0.03 percent, I'm not going to get excited. If the average returns following the strong days are negative and those following the weak days are positive, that's more interesting.

> As you conduct many sortings, you'll gain a good feel for differences that may form the basis for worthwhile hypotheses.

So how might I use the information? Perhaps I'll drill down further, examine those quartiles, and find that returns are particularly muted following strong up days. If that's the case, I will entertain the hypothesis of range-bound trading the morning following a strong daily rise in the S&P 500 Index. If I see that particularly weak days in the S&P 500 Index tend to close higher the next day, I may entertain the notion of an intraday reversal the day following a large drop. *The data provide me with a heads up, a hypothesis—not a firm, fixed conclusion.*

The data might also help sharpen some of my trading practices. If I'm holding positions for intermediate-term swing positions, I might be more likely to add to a long position after a daily market dip than after a strong daily rise. I might be more likely to take partial profits on a short position following toward the end of a weak market day than toward the end of an up day.

And suppose we find no apparent differences whatsoever? This, too, is a finding. It would tell us that—at this time frame, for this time period—there is no evidence of trend or countertrend effects. This would help us temper our expectations following strong and weak market days.

We would not assume that trends are our friends; nor would we be tempted to fade moves automatically. We would also know to look for potential edges elsewhere.

> Keep a record of the relationships that you examine and what you find; this will guide future inquiries and prevent you from duplicating efforts later on.

If your analysis does not identify a promising relationship within the data, you're limited only by your own creativity in exploring alternate hypotheses. For instance, you might look at how prior returns affect next returns for weekly or monthly data, rather than daily data. You might explore next day returns for a different instrument or market. Perhaps you'll see greater evidence of trendiness in commodities or small stocks than in the S&P 500 Index.

Where your creativity can really kick in is in your selection of independent variables. The same basic spreadsheet format outlined above could be used to examine the relationship between the current day's put-call ratio and the next day's S&P 500 returns; the current day's volume and next day returns; the current day's financial sector performance and next day S&P 500 returns; the current day's bond yield performance and next day returns. Once you have the spreadsheet analysis process mastered, it's simply a matter of switching one set of variables for another. This way, you can investigate a host of candidate hypotheses in a relatively short period of time.

The key to making this work is the Sort command in Excel. This sorts your independent variable from high to low or low to high so that you can see what happens in your dependent variable as a result. Along with visualizing data in charts, sorting is a great way to get a feel for how variables may be related, highlighting important market themes. But save your original spreadsheet with the formulas—the one you had titled Practice Sheet in the previous lesson. We're not finished with Excel tricks!

COACHING CUE

Here is one fruitful line of investigation: Take a look at next day returns as a function of weak up days versus strong up days. You can define weak versus strong with indicators such as the daily advance/decline ratio or the ratio of up volume to down volume. Limit your sort to the rising days in the sample and sort those based on market strength. What you'll find for some markets is that very strong markets tend to continue their strength in the near term; weaker rising markets are more likely to reverse direction. Later, you can limit your sorting to the declining days and sort them by very weak and less weak markets. Many times, the patterns you see among the rising days are different from those that show up among the falling days.

LESSON 97: CODE THE DATA

Sometimes the independent variable you're interested in is a categorical variable, not a set of continuous values. If I wanted to investigate the relationship between a person's weight (independent variable) and their lung capacity (dependent variable), all of my data would be continuous. If, however, I wanted to investigate the relationship between gender (male/female) and lung capacity, I would now be looking at a categorical variable in relationship to a continuous one. Conversely, if I wanted to simply identify whether a person had normal versus subnormal lung capacity, I would wind up with a categorical breakdown for my dependent variable.

There are times in market analysis when we want to look at the data categorically, rather than in a continuous fashion. In my own investigations, I routinely combine categorical views with continuous ones. Here's why:

If you reopen the spreadsheet we created, Practice Sheet, that examined current day returns in the S&P 500 Index as a function of the previous day's performance, you'll see that we had Date data in column A; open-high-low-close data in columns B–E; the present day's price change in column F; and the next day's price change in column G. For the analysis in the last lesson, we sorted the data based upon the present day's price change and then examined the average price change for the next day as a function of strong versus weak days. Our dependent measure, next day's price change, was continuous, and we compared average values to get a sense for the relationship between the independent and dependent variables.

Averages, however, can be misleading: a few extreme values can skew the result. These outliers can make the differences between two sets of averages look much larger than they really are. We can eliminate this possible source of bias by changing our dependent variable. We'll keep the next day's price change in column G, but now will add a dummy-coded variable in column H. This code will simply tell us whether the price change in column G is up or down. Thus, in cell H2, I would type in (without quotation marks): "=if(G2>0,1,0)" and hit Enter. This instructs the cell at H2 to return a "1" if the price change in cell G2 is positive; anything else—a zero or negative return—will return a "0." I will then click on H2; click the Excel menu item for Edit; click Copy; click cell H3 and drag all the way down the length of the data; and then click Enter. The 0,1 dummy code will populate each of the column H cells.

We want to know whether the independent variable is associated with greater frequency of up/down days, as well as the magnitudes of change across those days.

Now we go through the same sorting procedure described in the previous lesson. We highlight all the cells in the worksheet—including the new column H—and click Edit and Copy. We exit the spreadsheet, instructing Excel to save changes and to save the highlighted data. We open a new sheet; click Edit; click Paste; and all the spreadsheet data—again minus the formulas—will appear on the sheet. Once again we sort the data by column F (current day's price change) in descending order, as described in the previous lesson. Again we divide the data in half and, below the last entry in column G, we type in "=average(g2:g500)" and, below that, "=average(g501:g999)."

This, as noted in the previous lesson, shows us the magnitude of the average differences in next day's returns when the current day is relatively strong (top half of price change) versus relatively weak (bottom half of the price change distribution).

In column H, next to the cells for the two averages in column G, we enter the formula "=sum(H2:H500)" and, below that, "=sum(H501:H999)." *This tells us how many up days occurred following relatively strong days in the market and how many up days occurred following relatively weak days.* Because we're splitting the data in half, we should see roughly equal numbers of up days in the two sums if the current day's performance is not strongly related to the next day's price change. On the other hand, if we see considerably fewer up days following the strong market days than following the weak ones, we might begin to entertain a hypothesis.

If the average next day changes in column G look quite discrepant, but the number of winning days in the two conditions in column G are similar, that means that the odds of a winning day may not be significantly affected by the prior day's return, but the *size* of that day might be affected. In general, I like to see clear differences in both criteria. Thus, if the average size of the next day's return are higher following a falling day than a rising one *and* the odds of a rising day are higher, I'll be most likely to use the observation to frame a possible market hypothesis.

Note that we can dummy code independent variables as well. If, for example, I wanted to see whether an up or down day (independent variable) tended to be followed by an up or down day (dependent variable), I could code column F (current day's price change) with a code as above in column H and also code column G (next day's price change) identically in column I. I would then copy the spreadsheet to a fresh sheet and sort the data based on column H, so that we'd separate the 1s from the 0s. We'd then examine the column I sum for the cells in column H that were 1s and compare with the column I sum for the cells in column H that were 0s.

Dummy coding is especially helpful if we want to examine the impact of events on prospective returns. For instance, we could code all Mondays with a 1; all Tuesdays with a 2, etc., and then sort the next day's return based on the codings to tell us whether returns were more or less favorable following particular days of the week. Coding is also useful when we want to set up complex conditions among two or more independent variables and examine their relationship to future returns. This kind of coding gets a bit more complex and will form the basis for the next lesson.

COACHING CUE

If you include volume in your spreadsheet, you can code days with rising volume with a 1 and days with falling volume with a 0. This would then allow you to compare next day returns as a function of whether today's rise or decline were on rising or falling volume. All you'd need to do is sort the data once based on the current day's price change and then a second time separately for the rising and falling occasions as a function of the rising and declining volume.

LESSON 98: EXAMINE CONTEXT

Philosopher Stephen Pepper coined the term *contextualism* to describe a worldview in which truth is a function of the context in which knowledge is embedded. A short-term price pattern might have one set of expectations in a larger bull market; quite another under bear conditions. A short-term reversal in the first hour of trading has different implications than one that occurs midday. To use an example from *The Psychology of Trading*, you understand *Bear right!* one way on the highway, quite another way in the Alaskan wilderness.

We can code market data for contexts and then investigate patterns specific to those contexts. What we're really asking is, "Under the set of conditions that we find at present, what is the distribution of future expectations?" We're not pretending that these will be universal expectations. Rather, they are contextual—applicable to our current situation.

> Many of the most fruitful trading hypotheses pertain to certain kinds of markets—not to all markets, all the time.

Let's retrieve the Practice Sheet historical daily data for the S&P 500 Index that we used in our previous lessons. To refresh memory: Column A in our spreadsheet consists of the Date; columns B through E are open-high-low-close data. Column F is the independent variable, the current day's price change; entered into cell F22, it would be: "=((F22-F21)/F21)*100." Column G will serve as our contextual variable. In G22, we enter the following:

= if(E22>average(E3:E21),1,0)

This will return to cell G23 a "1" if the current price is above the prior day's simple 20-day moving average for the S&P 500 Index; a "0" if it is not above the average. The data label for cell G1 might be MA. Our dependent measure will be the next day's price change. In cell H22, this would be "=((E23-E22)/E22)*100" and H1 would have the label SP+1.

To complete the sheet, we would highlight cell G22 and H22; click the Excel menu item for Edit; click Copy; highlight the cells below G23 for the full length of the data set; and hit Enter. We highlight and copy all the data in the sheet as before; save the sheet as Practice Sheet2; and instruct Windows to save the data to the clipboard for another application. We open a fresh spreadsheet; click on cell A1; click the Excel menu item for Edit; select Paste; and our sheet now fills with text data. Note that, in this case, we'll have to eliminate rows 2 to 21, since they don't have a value for the 20-day moving average. We'll also eliminate the last row of data, because there are no data for the next day. You eliminate a row simply by highlighting the letter(s) for the row(s) at left; clicking the Excel menu item for Edit; selecting Delete. The row will disappear and the remaining data below will move into place.

Note that using a moving average as a variable of interest reduces the size of your data set, since the initial values will not have a moving average calculated. You need to take this into account when determining your desired sample size.

We now double sort the data to perform the contextual investigation. Let's say that we're interested in the expectations following a rising day in a market that is trading above versus below its 20-day moving average. We sort the data based on column F "SP(1)" as we did in our previous lesson, performing the sort on a Descending basis, so that the largest and positive price changes appear at the top of the sheet. Now we select only the data for the cells that show positive price change and copy those to another sheet. Our second sort will be based on column G (MA), again on a Descending Basis. This will separate the up days in markets trading above their 20-day moving average (those coded "1") from all other days.

As before, we'll examine the average next day's return by calculating the average for the cells in column H that are coded in column G as "1" and comparing that to the average for the cells in column H that are coded in column G as "0."

Thus, let's say that there are 538 cells for up days; 383 of these are coded "1" in column G and 155 are coded "0." You would compare "=average(H2:H384)" and "=average(H385:H539)." You could also code the cells in column H as either "1" or "0" in column I based on whether they are up or down "=if(H2>0,1,0)" and then compare "=sum(I2:I384)" and "=sum(I385:I539)" to see if there are notable differences in the number of up days following up days in markets that are above and below their moving averages.

Just for your curiosity, using cash S&P 500 data as the raw data, I found that the average next day change following an up day when we're above the 20-day moving average to be −0.04 percent; the average next day change following an up day when we're below the 20-day moving average was −0.18 percent. This is a good example of a finding that doesn't knock my socks off, but is suggestive. I would want to conduct other investigations of what happens after rising days in falling markets before generating trading hypotheses that would have me shorting strength in a broader downtrend.

Many times you'll see differences in the sorted data that are strong enough to warrant further investigation, but not strong enough to justify a trading hypothesis by itself.

This combination of coding and sorting can create a variety of contextual views of markets. For example, if we type in, "=if(E21=max(E2:E21), 1,0)" we can examine the context in which the current day is the highest price in the past 20 and see how that influences returns. If we include a second independent variable, such as the number of stocks making new 52-week highs and lows, we can examine how markets behave when new highs exceed new lows versus when new lows exceed new highs. For instance, if new highs go into column F and new lows into column G, we can code for "=if(F21>G21,1,0" in column H, place our dependent measure (perhaps the next day's price change) in column I and sort based on the new high/low coding.

As mentioned earlier, it is wise to not create too many contextual conditions, because you will wind up with a very small sample of occasions that fit your query, and generalization will be difficult. If you obtain fewer than 20 occasions that meet your criteria, you may need to relax those criteria or include fewer of them.

As your own trading coach, you can utilize these contextual queries to see how markets behave under a variety of conditions. The movement of sectors, related asset classes—anything can be a context that affects recent market behavior. In exploring these patterns, you become more sensitive to them in real-time, aiding your selection and execution of trades.

COACHING CUE

If you're interested in longer-term trading or investing, you can create spreadsheets with weekly or monthly data and investigate independent variables such as monthly returns on the next month's returns; VIX levels on the next month's volatility; sentiment data on the next month's returns; price changes in oil on the next month's returns, etc. You can also code data for months of the year (or beginning/end of the month) to investigate calendar effects on returns.

LESSON 99: FILTER DATA

Let's say you want to analyze intraday information for the S&P 500 Index futures. Now your spreadsheet will look different as you download data from sources such as your real-time charting application. Your first column will be date, your second column will be time of day, and your next columns will be open, high, low, and closing prices. If you so select, the next column can be trading volume for that time period (one-minute, five-minute, hourly, and so on).

Suppose you want to see how the S&P 500 market has behaved at a certain time of day. What we will need to do is filter out that time of day from the mass of downloaded data and only examine that subset. Instead of sorting data, which has been a mainstay of our investigations to this point, we will use Excel's filter function.

To illustrate how we might do this, we'll start with a simple question. Suppose we want to know how trading volume for the current first half-hour of trading compares with the average trading volume for that corresponding half-hour over the prior 20 days of trading. This will give us a rough sense of market activity, which correlates positively with price volatility. The volume also gives a relative sense for the participation of large, institutional traders. If, say, we observe a break out of a range during the first 30 minutes of trading, it is helpful to know whether or not these large market-moving participants are on board.

Volume analyses can help you identify who is in the market.

For this investigation, we'll examine half-hourly data for the S&P 500 emini contract. I obtain my intraday data from my quote platforms; in the current example, I'll use e-Signal. To do this, we create a 30-minute chart of the ES futures contract; click on the chart and scroll to the right to move the chart backward in time. When we've covered the last 20 days or so, we click on the menu item Tools; select Data Export; then uncheck the boxes for the data that we won't need. In this case, all we'll need is Date, Time, and Volume. We click the button for Copy to Clipboard and open a fresh sheet in Excel. Once we click on the Excel menu item for Edit and select Paste, with the cursor at cell A2, we'll populate the sheet with the intraday data. We can then enter names for the columns in row 1: Date; Time; and Volume. (If you're downloading from e-Signal, those names will accompany the data and you can download the data with the cursor at A1).

Our next step is to highlight the entire data set that we want to cover. We click on the Excel menu item for Data; select Filter; and select AutoFilter. A set of small arrows will appear beside the column names. Click the arrow next to Time and, from the drop down menu, select the time that represents the start of the trading day. In my case, living in the Chicago area in Central Time, that would be 8:30 A.M. You'll then see all the volume figures for the half-hour 8:30 A.M. to 9:00 A.M. Click on Edit; select Copy; open a blank sheet; click on Edit; and select Paste. This will put the 8:30 A.M. data on a separate sheet. If you have 20 values (the past 20 days), you can enter the formula "=average(c2:c21)" and you'll see the average trading volume for the first half-hour of trading. Of course, you can filter for any time of day and see that half-hour's average volume as well.

> When you know the average trading volume for a particular time period, you can assess institutional participation in real time—particularly with respect to whether this volume picks up or slows down as a function of market direction.

The filter function is helpful when you want to pull out data selectively from a data set. Let's say, for instance, that you had a column in which you coded Mondays as 1; Tuesday's as 2; etc. You could then filter out the 1s in the historical data set and see how the market behaved specifically on Mondays. Similarly, you could code the first or last days of the month and filter the data to observe the returns associated with those.

In general, I find filtering most helpful for intraday analyses, when I want to see how markets behave at a particular time of day under particular conditions. *Frankly, however, this is not where I find the greatest edges typically, and it's not where I'd recommend that a beginner start with historical investigations.* Should you become serious about investigating such intraday patterns, I strongly recommend obtaining a clean database from a vendor such as Tick Data. You can use their data management software to create data points at any periodicity and download these easily to Excel. Serious, longer-term investigations of historical intraday data need tools far stronger than Excel. Limits to the size of spreadsheets and the ease of maneuvering them make it impossible to use Excel for long-term investigations of high frequency data.

Still, when you want to see how markets behave in the short run—say, in the first hour of trading after a large gap open—investigations with intraday data and filtering can be quite useful. You'll find interesting patterns of continuation and reversal to set up day-trading ideas or to help with the execution of longer timeframe trades.

COACHING CUE

Filtering can be useful for examining patterns of returns as a function of time of day. For instance, say the market is down over the past two hours: how do returns compare if those two hours are the first versus the last two hours of the day? How are returns over the next few hours impacted if the day prior to those two hours was down? Such analyses can be very helpful for intraday traders, particularly when you combine price change independent variables with such intraday predictors as NYSE TICK.

LESSON 100: MAKE USE OF YOUR FINDINGS

This chapter has provided only a sampling of the kinds of ways that you can use simple spreadsheets and formulas to investigate possible patterns in historical data. Remember: these are qualitative looks at the data; they are designed to generate hypotheses, not prove them. Manipulating data and looking at them from various angles is a skill just like executing trades. With practice and experience, you can get to the point where you investigate quite a few patterns all in the hour or two after market close or before they open.

The key is to identify what makes the current market unique or distinctive. Are we well below or above a moving average? Have there been many more new lows than highs or the reverse? Has one sector been unusually strong or weak? Have the previous days been strong or weak? It is often at the extremes—when indicators or patterns are at their most unusual—that we find the greatest potential edges. But sometimes those unique elements are hard to find. Very high or low volume; strong or weak put/call ratios; large opening gaps—all are good areas for investigations.

We find the greatest directional edges following extreme market events.

Once you have identified a pattern that stands out, this becomes a hypothesis that you entertain to start a trading day or week. If, say, I find that 40 of the last 50 occasions in which the market has been very weak with a high put/call ratio have shown higher prices 20 days later, this will have me looking for a near-term bottoming process. If, after that analysis, I notice that we're making lower price lows but with fewer stocks and sectors participating in the weakness, this may add a measure of weight to my hypothesis. Eventually, I might get to the point where I think we've put in a price bottom and I'll buy the market, giving myself a favorable risk/reward should the historical pattern play out.

But equally important, consider the scenario in which we see good historical odds of bouncing over a 20-day period, leading us to search for a near-term bottoming process. My fresh data, however, suggest that the market is weakening further: more stocks and sectors are making lows, not fewer. *The historical pattern does not appear to be playing itself out.* This, too, is very useful data. When markets buck their historical tendencies, something special may be at work. Some very good trades can proceed from the recognition that markets are not behaving normally.

This is the value of considering patterns as hypotheses and keeping your mind open to those hypotheses being supported or not. A historical pattern in markets is a kind of script for the market to follow; your job is to determine whether or not it's following that script.

Our analyses only inform us of historical tendencies. If a market is not behaving in a manner that is consistent with its history, this alerts us to unique, situational forces at work.

All of this suggests that historical investigations are useful logical aids, but my experience is that their greatest value may be psychological. Day after day, week after week, and year after year of investigating patterns and running market results through Excel have given me a unique feel for patterns. It also has given me a keen sense for when patterns are changing: when historical precedents may no longer hold.

One routine that has been very helpful has been to isolate the last five or so instances of a potential pattern. *If the market has behaved quite differently in the last several instances than it has historically, I entertain the possibility that we're seeing a shift in market patterns.* If I see the last few instances behaving abnormally across many different variables and time frames, those anomalies strengthen my sense of a market shift.

When I see how results have played out over the years, I become a less naïve trend follower. I don't automatically assume that rising markets will continue to skyrocket or that falling markets will continue to plunge. I've developed tools for determining when trending markets are gaining and losing steam; these have been helpful in anticipating reversals. Seeing how these indicators behave under various market conditions over time—and actually quantifying their track record—has provided me with a measure of confidence in the ideas that I would not have in the absence of intimacy with the data.

Much of the edge in trading comes from seeing markets in unique ways, catching moves before they occur or early in their appearance. It is easy to become fixed in our views, with vision narrowed by looking at too few markets and patterns. As your own trading coach, you need to keep your mind open and fresh. Read, talk with experienced traders, follow a range of markets closely, test patterns historically, and know what's happening globally: you'll see things that never register on the radar of the average trader. You'll be at your most creative when you have the broadest vision.

COACHING CUE

When you examine historical patterns, go into your data set and specifically examine the returns from the occasions that didn't fit into the pattern. This will give you an idea of the kind of drawdowns you could expect *if* you were to trade the pattern mechanically. Many times, the exceptions to patterns end up being large moves; for instance, most occasions may show a countertrend tendency with a relative handful of very large trending moves. If you know this, you can look for those possible exceptions, study them, and maybe even identify and profit from them.

RESOURCES

The *Become Your Own Trading Coach* blog is the primary supplemental resource for this book. You can find links and additional posts on the topic of coaching processes at the home page on the blog for Chapter 10: http://becomeyourowntradingcoach.blogspot.com/2008/08/daily-trading-coach-chapter-ten-links.html

My own interest in historical patterns owes a great deal to the work of Victor Niederhoffer. His Daily Speculations web site is a source of many testable ideas regarding market movements:
www.dailyspeculations.com

Henry Carstens' online resource, *An Introduction to Testing Trading Ideas*, is a worthwhile and popular resource:
www.verticalsolutions.com/books.html

Mike Bryant's trading systems work is quite good; here's a collection of free downloads from his site:
www.breakoutfutures.com/PreDownload.htm

Rob Hanna's blog tests a number of historical trading patterns and is a great stimulus for your own research:
www.quantifiableedges.blogspot.com

Two subscription services that do a fine job of testing trading ideas are the SentimenTrader site from Jason Goepfert (www.sentimentrader.com) and the Market Tells letter from Rennie Yang (www.markettells.com).

Henry, Rob, Jason, and Rennie all contributed segments to Chapter 9 of this book, offering insights into the relevance of testing ideas for self-coaching.

Conclusion

Science strives to achieve unity of fact. Art strives to achieve unity of feeling.

—Stephen Pepper

This is our final lesson; let's see if we can achieve a bit of unity of feeling as well as fact.

LESSON 101: FIND YOUR PATH

My mother, Constance Steenbarger, passed away last year. She was an artist and an art teacher. Her greatest work of art, however, was her family. She provided her children—and her husband—with the one, irreplaceable psychological gift: the knowledge and the feeling that they were special. It's amazing how much you can achieve when you know that you're not ordinary. Out of that awareness, you're unwilling to settle for average in your work, your relationships, or your returns from markets. When you create a work of art out of your family, you empower human beings to want to make works of art of their lives. What greater accomplishment could there be? If I can achieve, as a psychologist and parent, a bit of what Connie Steenbarger accomplished with her family, it will count more than any degree after my name, any great trade I might place.

But isn't that what becoming your own coach is all about? It doesn't matter if the focus is trading, sales, parenting, or athletics: *the goal is to make a work of art of your life by becoming the best you can possibly be.*

The great disease that afflicts most people is their inability to think greatly of themselves. It's not about narcissism (which reflects an absence of self, not authentic greatness), and it's not about new-age self-esteem palliatives. Rather, thinking greatly of oneself is charting a path in life that makes a difference. It's living a goal-oriented life, not a life of drifting from day to day. It's remaining true to values and purposes, so that life has worth and meaning. It's about making such a profound impact that someone, somewhere will want to conclude their book with a dedication to you.

> Your life is a partially finished work of art.

There's an old saw that we tend to marry people like our parents. In my younger days, I would have been horrified at the prospect. Looking back on my mother's impact on her family and my wife Margie's impact on hers, I know that the rule holds true for me. Margie's greatest talent is that she is secure enough within herself to help others feel special about themselves. When one of our children went through a difficult marriage, I never once worried. I knew that she would eventually find happiness, because she had the experience of being special to her mother. *When you have that deep feeling of not being ordinary, you ultimately gravitate toward the best within you, the best for you.*

If you are going to be successful as your own coach, you will need to be like Connie and Margie; you'll need to sustain a relationship with yourself in which you are always special, no matter how daunting immediate obstacles may seem. You'll need to focus on your successes every bit as much—if not more—than your failures. You'll need to structure specific goals and concrete activities designed to achieve those, so that every day is an affirmation of drive and competence. Self-coaching is not about keeping journals or tracking your profits and losses. It's about forging a relationship with yourself that is as empowering as a mother's with a family.

At the end of all of this, you may decide that trading is not your path in life. Have the courage to embrace that and find the work that truly captures who you are and what you do best. I love trading—the intellectual challenge, the endless opportunities for improvement, and the immediacy of the feedback. You know when you've done well; you know when you've let yourself down. While trading has made me money, it's not truly what I do best. I once tried to be a full-time trader and quickly felt a large hole in my life where psychology—and working with people—had vanished. So now I trade markets on the side, work as a coach to professional traders, apply my greatest interests and talents in the most challenging settings, and write books that maybe, just maybe, will help others find what is special within them.

> Let your strengths define your path.

Know what you do best. Build on strengths. Never stop working on yourself. Never stop improving. Every so often, upset the apple cart and pursue wholly new challenges. The enemy of greatness is not evil; it's mediocrity. Don't settle for the mediocre. You don't have to be an artist and art teacher to make a work of art out of your life. And if trading is your path, learn from those who have blazed the trail ahead of you. Your final assignment is to absorb the resources from the various chapters of this book and select the few that will best support your self-coaching. They will provide the brushes and paint with which you'll create your life's artwork.

FOR MORE ON SELF-COACHING

The contributors to Chapter 9 have assembled their own mentoring resources, all linked on the Trading Coach site: http://becomeyourowntradingcoach.blogspot.com/2008/08/contributors-to-daily-trading-coach.html

My latest project is a free electronic book on trading theory and technique entitled Introduction to Trading that I am writing one blog post at a time: http://becomeyourowntradingcoach.blogspot.com/2008/09/introduction-to-trading.html

The TraderFeed blog covers a range of topics, from the psychology of traders to the psychology of markets: www.traderfeed.blogspot.com

I'll be adding coaching resources to the Trading Coach site over time; if you have questions or particular interests, by all means feel free to leave a question or comment on one of the blog posts: http://becomeyourowntradingcoach.blogspot.com. Also feel free to contact me at the e-mail address specific to this book: coachingself@aol.com

For my books on trading psychology and trader performance, as well as related materials, check out the Amazon site: www.amazon.com/s/ref=nb_ss_b?url=search-alias%3Dstripbooks&field-keywords=Brett+Steenbarger&x=8&y=18

About the Author

Brett N. Steenbarger, PhD is clinical associate professor of psychiatry and behavioral sciences at State University of New York Upstate Medical University in Syracuse, NY. As a clinical psychologist, Dr. Steenbarger has co-authored a training text and written numerous book chapters and peer-reviewed journal articles on the topic of brief therapy. Dr. Steenbarger has traded equity markets since the late 1970s; he works as a trading coach for hedge funds, investment banks, and proprietary trading firms in the U.S., U.K., and Asia. He is also the author of two books on trading psychology (*The Psychology of Trading*, John Wiley, 2003; *Enhancing Trader Performance*, John Wiley, 2006), and writes a daily blog on trader and market psychology (www.traderfeed.blogspot.com).

Index

A Dash of Insight blog, 286
acceptance, 10–11
addiction, 75
Adler, David, 240, 300, 303–305
Afraid to Trade blog, 275
Alcoholics Anonymous, 30, 142, 148–149, 170
Alpha Trends blog, 275
anxiety, 211, 300
Aronson, David, 308
automatic thoughts, 165–166, 170–171, 173–174, 176, 182, 185–188

Bandura, Albert, 202
Barchart.com, 320
Barros, Ray, 290–293
behavioral coaching, 195–226
  conditioning and, 207–210
  contingencies and, 196–201. *See also* reinforcement
  exposure, 217–223
  incompatible states and, 211–214
  positive associations and, 214–217
  shaping and, 204–207
  social learning and, 201–204
  worry and, 223–226
Become Your Own Trading Coach blog, xiv, 32, 64–65, 97, 130, 161, 193, 226, 259, 305, 336, 341
Bellafiore, Mike, 295–297
biofeedback, 16, 84–85, 221, 300
boredom, 83, 211
breathing, 5–6, 17, 211–212, 222, 296–297
brief therapy, 94, 100, 103
burnout, 50, 76
business plans, 228–234, 273

Carstens, Henry, 232, 248, 261–264, 289
Carter, John, 149
catastrophizing, 56, 167–168, 225
change, 4–32
  emotion in, 5
  environment and, 12–14
  focused, 30
  readiness for, 29–30
  routine and, 12–14
chart review, 124–125
cognitive coaching techniques, 163–194. *See also* schemas.
  challenging thought patterns, 182–187
  cognitive journal, 172–176, 185–188, 191–193
  disrupting thought patterns, 176–179
  emotion in, 167–169
  experiments, 188–190
  imagery and, 182–185
  positive thought patterns and, 190–193
  reframing, 179–182
collaboration, 263
communication, 293, 300
concentration, 77, 84, 118, 199, 222, 224
conditioning, 207–211, 223
confidence, 54, 91–94, 125, 273, 290
contextualism, 329
contingencies, 196–201
Cooper, Jeff, 287

coping, 44, 95–97, 127, 133, 135, 138, 151, 156, 225
core competencies, 261–263
core needs, 136–137
corrective emotional experiences, 157–159
correlations of returns, 244–247, 256
creativity, 80, 261–264
Csikszentmihalyi, Mihalyi, 73, 80
Czirnich, Chris, 264–270

Dalton, Jim, 149
defenses, 133, 137–141, 150, 157–158, 160
despair, 48–51
Decision Point, 320
Devon Principle, 18
discipline, 62, 120, 180, 209, 265, 272–273, 281, 285, 289–293
discrepancy, 11, 156–158
diversification, 51, 76, 120, 233–236, 243–247, 254
  psychological, 51, 76
Douglas, Mark, 274
Dow TICK (TICKI), 247
drawdowns, 87, 125, 249–250, 336
Duryea, Bill, 149
Dynamic Data Exchange (DDE), 317

e-Signal, 312, 314–315
Edenbridge, 216
ego alien, 141
elitetrader.com, 150
emotion, 14–17, 20–22, 41, 52–54, 59–60, 68–71, 115, 119, 129, 135–137, 140, 151–152, 156, 167–169, 178, 181, 187–188, 197, 199, 201, 208–210, 214, 225, 281–282, 298–299. *See also* mood.
  changing, 20–22
  behavioral coaching and, 197, 199, 201, 208–210, 214, 225
  cognitive coaching and, 167–169, 178, 181, 187–188
  fear, 51–54
  imagery and, 129
  journaling and, 281–282, 298
  niche and, 59–60
  perception and, 41
  positive, 68–71
  psychodynamic coaching and, 151–152, 156,
  repetitive patterns of, 135–137, 140
  states, 115, 119
  transforming, 14–17
energy, 77–79, 199
Excel, 64, 247, 288, 307–336
  basics, 313–317
  coding data in, 327–328
  sorting data in, 324–327
  visualizing data in, 317–320
execution, 236, 250–253, 259, 284
expectations, 37–39
expertise development, 159
exposure, 217–223, 225
external observer, 265

fatigue, 166, 199–210, 300
fear, 51–54, 156–157, 171, 218, 225
Fisher, Larry, 295, 298–300
flow, 73, 80–81. *See also* zone
forecasting P/L, 232–233, 248–250
Forman, John, 290, 293–295
Frankl, Viktor, 86
frustration, 117, 128, 134, 150, 175, 207, 215, 219, 223

generalization, 208
Globetrader blog, 264
goals, 5–9, 19–20, 39, 50, 109–112, 117, 122–123, 147, 231, 294–295
  emotion in, 5–7
  process, 110–111
Goepfert, Jason, 280–282, 310
Goldberg, Elkonon, 80
greatness, 339–340
Gurdjieff, G. I., 80, 170, 195

habit, 176, 189
Hanna, Rob, 286–288, 310
happiness, 71–73
Harnett, Trevor, 271–273
hate, 142
honesty, 291

imagery, 126–129, 181–185, 195, 214, 218–220, 222–223. *See also* visualization
Institute of Auction Market Theory, 149
integrity, 89–91
intentionality, 77, 79–82
intuition, 83–84

Kirk, Charles, 275, 279–281

lbrgroup.com, 149
Liberman, Terry, 149
Luborsky, Lester, 132

Mabe, Dave, 283–284
Market Delta, 43, 150, 258, 271
Market Profile, 255, 278
Market Tells, 300
Marketsinprofile.com, 149
Maslow, Abraham, 73, 207
meditation, 22–23
Meichenbaum, Donald, 127
mental checklist, 276–277
Miller, Jeff, 286, 288–290
mirroring, 17–20, 24–25, 148, 159
mood, 61–64, 68, 199. *See also* emotion
motivation, 49–50, 107–109, 112, 123, 147, 171, 199–200, 216
  suppression of, 49–50

niche. *See* trading niche
Niederhoffer, Victor, 254
novelty, 180
NYSE TICK, 43, 52, 192, 247, 253, 257–258, 302–303, 312

O'Neil, William, 287
overconfidence, 163, 165, 171, 175, 223, 300

patterns of behavior, 104–118, 121, 133–138, 141–144, 154, 156–158, 170, 199, 201, 215–217, 220, 227–236, 276, 279, 302
  extinguishing, 215–217, 220
  problem, 104, 110, 114–117, 141–144, 156–158, 199
  repetitive, 133, 135–138, 170
  solution, 104, 111–114, 227–236
  thinking, 167–193. *See also* schemas
Pennebaker, James, 11, 15
Pepper, Stephen, 329, 339
perception, 20–22, 41, 51
  emotion and, 20–22
  fear and, 51
performance anxiety, 54–58, 75, 202
Perruna, Chris, 283–285
personality, 14
physical tension, 150–152
Piaget, Jean, 164
Pinnacle Data, 312, 314–315
play, 81–82
position size, 35, 53, 62, 120, 124, 229, 270, 283, 292
positive psychology, 67
positive thinking, 190–193
pressing, 45–48
price targets, 62, 92
procrastination, 143–144, 200, 216
proprietary trading, 204, 213, 221, 295–300
psychodynamic coaching, 131–161
  challenging defenses, 138–141
  coaching relationship and, 144–147
  discomfort and, 150–152
  discrepancy and, 156–158
  emotion and 141–144
  past relationships and 132–135
  positive relationships, 147–150
  repetitive patterns, 135–138
  transference and, 153–155
  working through, 158–161

qualitative data, 309
Quantifiable Edges blog, 286

Rand, Ayn, 37, 89
Raschke, Linda Bradford, 149, 291
Real Tick, 312
reframing, 179–184
reinforcement, 196–200, 202–207, 214–216, 220, 224
relapse, 5, 30, 121

relationships, 7–8, 132–136, 144–150, 340
with self, 340
repetition, 122–126, 128, 218, 226
research, 286–290, 292
resilience, 86–88, 90, 267, 270, 276
responsibility, 266, 291
review, 276
risk, 47, 50, 53, 61–63, 74, 91–92, 94, 100, 120, 125, 133, 152, 155, 157, 165, 168, 183, 200, 202, 210, 222, 229–231, 233, 240, 248–250, 252–253, 256, 258, 270, 272, 290
adjusted returns, 210
allocation, 290
aversion, 62, 92, 94, 165, 183, 200, 272
excessive, 74, 125, 133
increasing, 53
management, 62, 91–92, 94, 152, 168, 202, 240
measuring, 248–250
reducing, 50, 155
reward and, 63, 100, 157, 233, 248–250, 252–253, 258
rules and, 47, 120
tolerance, 62–63
roles, 23–25
rules, 46–48, 62, 118–120, 223

SMB Capital, 295
SMB Training blog, 295
schemas, 164–166, 170–177, 191
self awareness, 272
self confidence, 25–28
self efficacy, 6, 25–26, 125
self esteem, 89
self mastery, 101
self monitoring, 99–103, 139
self talk, 17, 115, 155, 171, 173, 175, 179–188
self understanding, 284–286
Seneadza, Michael, 271, 273–275
Senters, Hubert, 149
Sentimentrader.com, 280
serenity, 83–86
Shannon, Brian, 275–277
shaping, 204–207, 217
shoulds, 169, 173
Simonton, Dean Keith, 80
simulation trading, 112, 124–126, 228, 231, 259
slumps, 61, 113, 232, 267
social learning, 201–204
Spencer, Steve, 295
startup capital, 227–236
state, 116–117, 200, 211–214
incompatible, 211–214
stimulus-response, 196, 217
Stock Tickr, 101, 149, 282–283
stop loss, 27–28, 35, 48, 62, 93, 100, 114, 118, 129, 140, 157, 223, 225, 239, 251–252, 258. *See also* risk
strengths, 8–9, 31, 103, 105, 112, 340
stress, 33–65, 95, 127, 138, 213, 218, 220, 230. *See also* coping
distress and, 34–37, 40, 62, 138, 213, 220
inoculation, 127
perception and, 41–42

tenacity, 291
tension, 212
The Essentials of Trading blog, 290
The Kirk Report blog, 279
thought stopping, 176–179
Tick Data, 313
Trade Ideas, 150
trade2win.com, 150
trade management, 257–259
TradeStation, 150, 288, 311
tradethemarkets.com, 149
Trader DNA, 101, 300
TraderFeed blog, xiii, 341, 343
Trader Mike blog, 271, 275
trading,
affirmations, 274
automated, 284
business, 60, 227–260
concepts, 277–278
edge, 248–250, 266, 276, 288–289, 292, 301–302, 307–337
environment, 11–14, 123–126, 271–272, 299–300
historical patterns, 307–336

journal, 10–11, 19, 39, 43–45, 100–103, 116, 154, 160–161, 171, 200, 205–206, 268–270, 273, 278–279, 281–282, 286, 292–293, 298, 300
metrics, 237–243, 261, 304
niche, 58–61, 81
partner, 277
plans, 35, 53, 221
records, 279–282
size, *see* position size
statistics, 296, 300. *See also* trading metrics
target, 251–252
themes, 133–135, 239, 254–257
volume, 64, 216, 271
Trading RM, 295
Trading Success blog, 290
transference, 153–155
trauma, 125, 220
triggers, 21, 200, 211, 213, 218, 220, 222
trust, 93–94
Twitter, 256

Valfer, Reid, 295
variables, independent and dependent, 320–323, 326
variability of returns, 210, 239–240, 249–250
video recording, 126, 141, 297
virtual trading groups, 148
visualization, 5–6, 47, 88, 108, 111, 184, 216, 219–222, 224–226, 276–278, 292, 317–320
of data, 317–320
VIX, 242, 332
volatility, 61–64, 210, 232–233, 235, 239, 242, 250, 286

well being, 67–98. *See also* happiness
working through, 158
worry, 163, 223–226

Yang, Rennie, 300–303, 310, 317

Zone, 73–77, 80, 222

WHERE WE DIFFER.

1. MOVING STOPS TO BREAKEVEN

2. RISK/REWARD — BASED UPON STP - NEGLECTS PROBABALITY

3. TRAILING STOPS

4. SHOULD KNOW HOW MANY TIMES STOPS TOOK YOU OUT

5. COMPARING WITH OTHERS

6. BLDG ON YOUR STRENGTHS

7. SETUPS — TIES INTO COMMENTS ABOUT CHANGING CONTRACTS YOU WILL TRADE. HE SAID IT WILL TAKE A LONG TIME TO MAKE THE CHANGE — PRINCIPLES ARE THE SAME FOR ALL COMMODITIES

8. HE IS SYSTEMATIC WE ARE NOT

9. GOOD PT. BEFORE LOOKING FOR ANSY ANSWERS YOU HAVE TO UNDERSTAND MKTS AND HAVE ACCUMULATED EXPERIENCE